Flora of North America

Contributors to Volume 6

María Mercedes Arbo
Kerry A. Barringer
David M. Bates
Orland J. Blanchard Jr.
Luc Brouillet
Garrett E. Crow
Laurence J. Dorr
Aaron Floden
Paul A. Fryxell†
Susannah B. Fulton
John F. Gaskin
Aaron Goldberg†

Douglas H. Goldman
Margaret M. Hanes
Steven R. Hill
Walter C. Holmes
Brian R. Keener
John La Duke
David E. Lemke
R. John Little
John M. MacDougal
Landon E. McKinney†
T. Lawrence Mellichamp
Meghan G. Mendenhall

Guy L. Nesom
Lorin I. Nevling Jr.
C. Thomas Philbrick
Norman K. B. Robson
Janice G. Saunders
Bruce A. Sorrie
John L. Strother
John W. Thieret†
Gordon C. Tucker
Linda E. Watson
Molly A. Whalen
George Yatskievych

Editors for Volume 6

David E. Boufford
Taxon Editor for Apodanthaceae
and Thymelaeaceae

Kanchi Gandhi
Nomenclatural Editor

Ronald L. Hartman
Co-Taxon Editor for Violaceae

Robert W. Kiger
Bibliographic Editor, Lead
Editor, and Taxon Editor for
Calophyllaceae, Cistaceae,
Clusiaceae, Cucurbitaceae,
Hypericaceae, Malvaceae,
Muntingiaceae, Passifloraceae, and
Tamaricaceae

Nancy R. Morin
Taxon Editor for Droseraceae

Jackie M. Poole
Taxon Editor for Begoniaceae,
Cochlospermaceae, Datiscaceae,
Frankeniaceae, and Turneraceae

Richard K. Rabeler
Co-Taxon Editor for Violaceae

Heidi H. Schmidt
Managing Editor

Mary Ann Schmidt
Technical Editor

Leila M. Shultz
Taxon Editor for Podostemaceae

John L. Strother
Reviewing Editor

James L. Zarucchi
Editorial Director

Volume Composition

Kristin Pierce
Compositor and Editorial Assistant

Heidi H. Schmidt
Production Coordinator and
Managing Editor

Dionaea muscipula

Flora of North America

North of Mexico

Edited by FLORA OF NORTH AMERICA EDITORIAL COMMITTEE

VOLUME 6

Magnoliophyta: Cucurbitaceae to Droseraceae

NEW YORK OXFORD · OXFORD UNIVERSITY PRESS · 2015

Oxford University Press is a department of the University of Oxford.
It furthers the University's objective of excellence in research,
scholarship, and education by publishing worldwide.

Oxford New York

Auckland Cape Town Dar es Salaam Hong Kong Karachi
Kuala Lumpur Madrid Melbourne Mexico City Nairobi
New Delhi Shanghai Taipei Toronto

With offices in

Argentina Austria Brazil Chile Czech Republic France Greece
Guatemala Hungary Italy Japan Poland Portugal Singapore
South Korea Switzerland Thailand Turkey Ukraine Vietnam

Oxford is a registered trademark of Oxford University Press in the UK and certain other countries.

Published by Oxford University Press, Inc.
198 Madison Avenue, New York, New York 10016
www.oup.com

Library of Congress Cataloging-in-Publication Data
(Revised for Volume 6)
Flora of North America North of Mexico
edited by Flora of North America Editorial Committee.
Includes bibliographical references and indexes.
Contents: v. 1. Introduction—v. 2. Pteridophytes and gymnosperms—
v. 3. Magnoliophyta: Magnoliidae and Hamamelidae—
v. 22. Magnoliophyta: Alismatidae, Arecidae, Commelinidae (in part), and Zingiberidae—
v. 26. Magnoliophyta: Liliidae: Liliales and Orchidales—
v. 23. Magnoliophyta: Commelinidae (in part): Cyperaceae—
v. 25. Magnoliophyta: Commelinidae (in part): Poaceae, part 2—
v. 4. Magnoliophyta: Caryophyllidae (in part): part 1—
v. 5. Magnoliophyta: Caryophyllidae (in part): part 2—
v. 19, 20, 21. Magnoliophyta: Asteridae (in part): Asteraceae, parts 1–3—
v. 24. Magnoliophyta: Commelinidae (in part): Poaceae, part 1—
v. 27. Bryophyta, part 1—
v. 8. Magnoliophyta: Paeoniaceae to Ericaceae—
v. 7. Magnoliophyta: Salicaceae to Brassicaceae—
v. 28. Bryophyta, part 2—
v. 9. Magnoliophyta: Picramniaceae to Rosaceae—
v. 6. Magnoliophyta: Cucurbitaceae to Droseraceae

ISBN: 978-0-19-534027-3 (v. 6)
1. Botany—North America.
2. Botany—United States.
3. Botany—Canada.
I. Flora of North America Editorial Committee.
QK110.F55 2002 581.97 92-30459

1 2 3 4 5 6 7 8 9
Printed in the United States of America
on acid-free paper

Contents

This volume of the

Flora of North America North of Mexico

is dedicated to the memory of

Paul Arnold Fryxell

1927 – 2011

FOUNDING MEMBER INSTITUTIONS

Flora of North America Association

Agriculture and Agri-Food Canada
Ottawa, Ontario

Arnold Arboretum
Jamaica Plain, Massachusetts

Canadian Museum of Nature
Ottawa, Ontario

Carnegie Museum of
  Natural History
Pittsburgh, Pennsylvania

Field Museum of Natural History
Chicago, Illinois

Fish and Wildlife Service
United States Department of
  the Interior
Washington, D.C.

Harvard University Herbaria
Cambridge, Massachusetts

Hunt Institute for Botanical
  Documentation
Carnegie Mellon University
Pittsburgh, Pennsylvania

Jacksonville State University
Jacksonville, Alabama

Jardin Botanique de Montréal
Montréal, Québec

Kansas State University
Manhattan, Kansas

Missouri Botanical Garden
St. Louis, Missouri

New Mexico State University
Las Cruces, New Mexico

The New York Botanical Garden
Bronx, New York

New York State Museum
Albany, New York

Northern Kentucky University
Highland Heights, Kentucky

Université de Montréal
Montréal, Québec

University of Alaska
Fairbanks, Alaska

University of Alberta
Edmonton, Alberta

The University of British Columbia
Vancouver, British Columbia

University of California
Berkeley, California

University of California
Davis, California

University of Idaho
Moscow, Idaho

University of Illinois
Urbana-Champaign, Illinois

University of Iowa
Iowa City, Iowa

The University of Kansas
Lawrence, Kansas

University of Michigan
Ann Arbor, Michigan

University of Oklahoma
Norman, Oklahoma

University of Ottawa
Ottawa, Ontario

University of Southwestern
  Louisiana
Lafayette, Louisiana

The University of Texas
Austin, Texas

University of Western Ontario
London, Ontario

University of Wyoming
Laramie, Wyoming

Utah State University
Logan, Utah

For their support of the preparation of this volume,
we gratefully acknowledge and thank:

Franklinia Foundation

The Philecology Foundation

The Andrew W. Mellon Foundation

The David and Lucile Packard Foundation

an anonymous foundation

Chanticleer Foundation

The Stanley Smith Horticultural Trust

Chris Davidson and Sharon Christoph,
    Botanical Research Foundation of Idaho

For sponsorship of illustrations included in this volume,
we express sincere appreciation to:

Erick Adams, Olympia, Washington
   Drosera intermedia, Droseraceae

David Bates, Ithaca, New York
   Malacothamnus palmeri, Malvaceae

Jennifer Bryant, Hobart, Oklahoma
   Callirhoë involucrata, Malvaceae

Dorothy King Young Chapter, California Native Plant Society, Gualala, California
   Viola adunca, Violaceae

Friends of University of North Carolina, Charlotte, Botanical Gardens, Charlotte, North Carolina
   Volume 6 Frontispiece, *Dionaea muscipula*, Droseraceae – in honor of T. Lawrence Mellichamp

Arthur V. Gilman, Marshfield, Vermont
   Viola rostrata, Violaceae

Douglas Goldman, Greensboro, North Carolina
   Passiflora tenuiloba, Passifloraceae – in memory of Hobbes Goldman

Nancy Morin, Point Arena, California
   Hypericum ascyron subsp. *pyramidatum*, Hypericaceae – in honor of Norman K. B. Robson
   Fremontodendron californicum, Malvaceae – in appreciation of the California Native Plant Society
   Melochia corchorifolia, Malvaceae – in memory of Aaron Goldberg

The Polly Hill Arboretum, West Tisbury, Massachusetts
   Dirca palustris, Thymelaeaceae

Jackie Poole, Austin, Texas
   Fryxellia pygmaea, Malvaceae – in memory of Paul Fryxell

Barbara and David Rice, Gualala, California
   Viola glabella, Violaceae

Jennifer H. Richards, Miami, Florida
   Tilia americana, Malvaceae
   Passiflora incarnata, Passifloraceae
   Piriqueta cistoides subsp. *caroliniana*, Turneraceae

Linda Watson, Stillwater, Oklahoma
   Amoreuxia palmatifida, Cochlospermaceae

*Project Staff — past and present
involved with the preparation of Volume 6*

Barbara Alongi, *Illustrator*
Burgund Bassüner, *Information Coordinator*
Ariel S. Buback, *Editorial Assistant*
Trisha K. Distler, *GIS Analyst*
Pat Harris, *Editorial Assistant*
Linny Heagy, *Illustrator*
Ruth T. King, *Editorial Assistant*
John Myers, *Illustrator and Illustration Compositor*
Kristin Pierce, *Editorial Assistant and Compositor*
Mary Ann Schmidt, *Senior Technical Editor*
Hong Song, *Programmer*
Yevonn Wilson-Ramsey, *Illustrator*

Contributors to Volume 6

María Mercedes Arbo
Instituto de Botánica del Nordeste
Corrientes, Argentina

Kerry A. Barringer
Jersey City, New Jersey

David M. Bates
Cornell University
Ithaca, New York

Orland J. Blanchard Jr.
Florida Museum of Natural
*   History*
University of Florida
Gainesville, Florida

Luc Brouillet
Université de Montréal
Montréal, Québec

Garrett E. Crow
University of New Hampshire
Durham, New Hampshire and
Michigan State University
East Lansing, Michigan

Laurence J. Dorr
National Museum of Natural
*   History*
Smithsonian Institution
Washington, DC

Aaron Floden
University of Tennessee
Knoxville, Tennessee

Paul A. Fryxell†
Claremont, California

John F. Gaskin
Northern Plains Agricultural
*   Research Laboratory*
United States Department of
*   Agriculture*
Sidney, Montana

Aaron Goldberg†
National Museum of Natural
*   History*
Smithsonian Institution
Washington, DC

Douglas H. Goldman
U.S.D.A. Natural Resources
*   Conservation Service*
Greensboro, North Carolina

Margaret M. Hanes
University of Wisconsin
Madison, Wisconsin

Steven R. Hill
Illinois Natural History Survey
Champaign, Illinois

Walter C. Holmes
Baylor University
Waco, Texas

Susannah B. Johnson-Fulton
Shasta College
Redding, California

Brian R. Keener
The University of West Alabama
Livingston, Alabama

John La Duke
University of Nebraska at Kearney
Kearney, Nebraska

David E. Lemke
Southwest Texas State University
San Marcos, Texas

R. John Little
Sacramento, California

John M. MacDougal
Harris-Stowe State University
St. Louis, Missouri

Landon E. McKinney†
Environmental Quality
*   Management, Inc.*
Cincinnati, Ohio

T. Lawrence Mellichamp
University of North Carolina
Charlotte, North Carolina

Meghan G. Mendenhall
Texas Department of Public Safety
Austin, Texas

Guy L. Nesom
Fort Worth, Texas

Lorin I. Nevling Jr.
Illinois Natural History Survey
Champaign, Illinois

C. Thomas Philbrick
Western Connecticut State
  University
Danbury, Connecticut

Norman K. B. Robson
The Natural History Museum
London, England

Janice G. Saunders
Instituto de Botánica Darwinion
Buenos Aires, Argentina

Bruce A. Sorrie
Whispering Pines, North Carolina

John L. Strother
University of California
Berkeley, California

John W. Thieret†
Northern Kentucky University
Highland Heights, Kentucky

Gordon C. Tucker
Eastern Illinois University
Charleston, Illinois

Linda E. Watson
Oklahoma State University
Stillwater, Oklahoma

Molly A. Whalen
Flinders University
Adelaide, Australia

George Yatskievych
Missouri Botanical Garden
St. Louis, Missouri

Taxonomic Reviewers

Laurence J. Dorr
*National Museum of Natural
  History
Smithsonian Institution
Washington, DC*

Margaret M. Hanes
*University of Wisconsin
Madison, Wisconsin*

Steven R. Hill
*Illinois Natural History Survey
Champaign, Illinois*

Peter M. Jørgensen
*Missouri Botanical Garden
St. Louis, Missouri*

Peter H. Raven
*Missouri Botanical Garden
St. Louis, Missouri*

Zachary S. Rogers
*Missouri Botanical Garden
St. Louis, Missouri*

Regional Reviewers

ALASKA / YUKON

Bruce Bennett
*Yukon Department of
  Environment
Whitehorse, Yukon*

Robert Lipkin
*Alaska Natural Heritage Program
University of Alaska
Anchorage, Alaska*

David F. Murray
*University of Alaska
  Museum of the North
University of Alaska Fairbanks
Fairbanks, Alaska*

Carolyn Parker
*University of Alaska
  Museum of the North
University of Alaska Fairbanks
Fairbanks, Alaska*

Mary Stensvold
*U.S.D.A. Forest Service
Sitka, Alaska*

PACIFIC NORTHWEST

Edward R. Alverson
*The Nature Conservancy
Eugene, Oregon*

Adolf Ceska
*British Columbia Conservation
  Data Centre
Victoria, British Columbia*

Richard R. Halse
*Oregon State University
Corvallis, Oregon*

Eugene N. Kozloff
*Shannon Point Marine Center
Anacortes, Washington*

Jim Pojar
*British Columbia Forest Service
Smithers, British Columbia*

Cindy Roché
Talent, Oregon

Peter F. Zika
*University of Washington
Seattle, Washington*

SOUTHWESTERN UNITED STATES

H. David Hammond†
Northern Arizona University
Flagstaff, Arizona

G. Frederic Hrusa
Courtland, California

Charles T. Mason Jr.†
University of Arizona
Tucson, Arizona

James D. Morefield
Nevada Natural Heritage Program
Carson City, Nevada

Donald J. Pinkava
Arizona State University
Tempe, Arizona

Jon P. Rebman
San Diego Natural History
Museum
San Diego, California

Margriet Wetherwax
University of California
Berkeley, California

WESTERN CANADA

William J. Cody†
Agriculture and Agri-Food Canada
Ottawa, Ontario

Lynn Gillespie
Canadian Museum of Nature
Ottawa, Ontario

A. Joyce Gould
Alberta Tourism, Parks and
Recreation
Edmonton, Alberta

Vernon L. Harms
University of Saskatchewan
Saskatoon, Saskatchewan

Elizabeth Punter
University of Manitoba
Winnipeg, Manitoba

ROCKY MOUNTAINS

Curtis R. Björk
University of Idaho
Moscow, Idaho

Bonnie Heidel
University of Wyoming
Laramie, Wyoming

B. E. Nelson
University of Wyoming
Laramie, Wyoming

NORTH CENTRAL UNITED STATES

William T. Barker
North Dakota State University
Fargo, North Dakota

Anita F. Cholewa
University of Minnesota
St. Paul, Minnesota

Neil A. Harriman
University of Wisconsin Oshkosh
Oshkosh, Wisconsin

Bruce W. Hoagland
University of Oklahoma
Norman, Oklahoma

Robert B. Kaul
University of Nebraska
Lincoln, Nebraska

Deborah Q. Lewis
Iowa State University
Ames, Iowa

Ronald L. McGregor†
The University of Kansas
Lawrence, Kansas

Lawrence R. Stritch
U.S.D.A. Forest Service
Shepherdstown, West Virginia

George Yatskievych
Missouri Botanical Garden
St. Louis, Missouri

SOUTH CENTRAL UNITED STATES

David E. Lemke
Southwest Texas State University
San Marcos, Texas

EASTERN CANADA

Sean Blaney
Atlantic Canada Conservation
Data Centre
Sackville, New Brunswick

Jacques Cayouette
Agriculture and Agri-Food Canada
Ottawa, Ontario

Frédéric Coursol
Mirabel, Québec

William J. Crins
Ontario Ministry of Natural
Resources
Peterborough, Ontario

John K. Morton†
University of Waterloo
Waterloo, Ontario

Marian Munro
Nova Scotia Museum of Natural
History
Halifax, Nova Scotia

Michael J. Oldham
Natural Heritage Information
Centre
Peterborough, Ontario

GREENLAND

Geoffrey Halliday
University of Lancaster
Lancaster, England

NORTHEASTERN UNITED STATES

Ray Angelo
New England Botanical Club
Cambridge, Massachusetts

Preface for Volume 6

Since the publication of *Flora of North America* Volume 9 (the eighteenth volume in the *Flora* series) in late 2014, the membership of the Flora of North America Association [FNAA] Board of Directors has undergone changes. Jay A. Raveill and Elizabeth H. Zacharias have retired from the board. New board members include: Thomas G. Lammers (Taxon Editor) and Yury Roskov. As a result of a reorganization finalized in 2003, the FNAA Board of Directors succeeded the former Editorial Committee; for the sake of continuity of citation, authorship of *Flora* volumes is still to be cited as "Flora of North America Editorial Committee, eds."

Most of the editorial process for this volume was done at the Hunt Institute for Botanical Documentation at Carnegie Mellon University in Pittsburgh, Pennsylvania. Final processing and composition took place at the Missouri Botanical Garden in St. Louis; this included pre-press processing, typesetting and layout, plus coordination for all aspects of planning, executing, and scanning the illustrations. Other aspects of production, such as art panel composition plus labeling and occurrence map generation, were carried out in Gaston, Oregon, and Miami, Florida, respectively.

Line drawings published in this volume were executed by four very talented artists: Barbara Alongi prepared illustrations for *Corchorus*, *Tilia*, and *Triumfetta* (Malvaceae); Linny Heagy illustrated taxa of Begoniaceae, Cistaceae (except *Crocanthemum*), Cochlospermaceae, Datiscaceae, Droseraceae (including the frontispiece depicting *Dionaea muscipula*), Frankeniaceae, Malvaceae (except the three genera illustrated by B. Alongi and *Napaea* by Y. Wilson-Ramsey), Muntingiaceae, Passifloraceae, Thymelaeaceae, and Turneraceae; John Myers illustrated Apodanthaceae, *Crocanthemum* (Cistaceae), and Podostemaceae; and Yevonn Wilson-Ramsey prepared illustrations for Calophyllaceae, Clusiaceae, Cucurbitaceae, Hypericaceae, *Napaea* (Malvaceae), Tamaricaceae, and Violaceae. In addition to preparing various illustrations, John Myers composed and labeled all of the line drawings that appear in this volume.

Starting with Volume 8 published in 2009, the circumscription and ordering of some families within the *Flora* have been modified so they mostly reflect that of the Angiosperm Phylogeny Group [APG] rather than the previously followed Cronquist organizational structure. The groups of families found in this and future volumes in the series are mostly ordered following E. M. Haston et al. (2007); since APG views of relationships and circumscriptions have evolved, and will certainly change further through time, some discrepancies in organization will occur. Volume 30 of the *Flora of North America* will contain a comprehensive index to the published volumes.

Support from many institutions and by numerous individuals has enabled the *Flora* to be produced. Members of the Flora of North America Association remain deeply thankful to the many people who continue to help create, encourage, and sustain the *Flora*.

Introduction

Scope of the Work

Flora of North America North of Mexico is a synoptic account of the plants of North America north of Mexico: the continental United States of America (including the Florida Keys and Aleutian Islands), Canada, Greenland (Kalâtdlit-Nunât), and St. Pierre and Miquelon. The *Flora* is intended to serve both as a means of identifying plants within the region and as a systematic conspectus of the North American flora.

The *Flora* will be published in 30 volumes. Volume 1 contains background information that is useful for understanding patterns in the flora. Volume 2 contains treatments of ferns and gymnosperms. Families in volumes 3–26, the angiosperms, were first arranged according to the classification system of A. Cronquist (1981) with some modifications, and starting with Volume 8, the circumscriptions and ordering of families generally follow those of the Angiosperm Phylogeny Group [APG] (see E. Haston et al. 2007). Bryophytes are being covered in volumes 27–29. Volume 30 will contain the cumulative bibliography and index.

The first two volumes were published in 1993, Volume 3 in 1997, and Volumes 22, 23, and 26, the first three of five volumes covering the monocotyledons, appeared in 2000, 2002, and 2002, respectively. Volume 4, the first part of the Caryophyllales, was published in late 2003. Volume 25, the second part of the Poaceae, was published in mid 2003, and Volume 24, the first part, was published in January 2007. Volume 5, completing the Caryophyllales plus Polygonales and Plumbaginales, was published in early 2005. Volumes 19–21, treating Asteraceae, were published in early 2006. Volume 27, the first of two volumes treating mosses in North America, was published in late 2007. Volume 8, Paeoniaceae to Ericaceae, was published in September 2009, and Volume 7, Salicaceae to Brassicaceae appeared in 2010. In 2014, Volume 28 was published, completing the treatment of mosses for the flora area, and at the end of 2014, Volume 9, Picramniaceae to Rosaceae was published. The correct bibliographic citation for the *Flora* is: Flora of North America Editorial Committee, eds. 1993+. Flora of North America North of Mexico. 19+ vols. New York and Oxford.

Volume 6 treats 545 species in 104 genera contained in 19 families. For additional statistics please refer to Table 1 on p. xx.

Contents · General

The *Flora* includes accepted names, selected synonyms, literature citations, identification keys, descriptions, phenological information, summaries of habitats and geographic ranges, and other biological observations. Each volume contains a bibliography and an index to the taxa included in that volume. The treatments, written and reviewed by experts from throughout the systematic botanical community, are based on original observations of herbarium specimens and, whenever possible, on living plants. These observations are supplemented by critical reviews of the literature.

Table 1. *Statistics for Volume 6 of Flora of North America.*

Family	Total Genera	Endemic Genera	Introduced Genera	Total Species	Endemic Species	Introduced Species	Conservation Taxa
Cucurbitaceae	23	1	10	56	11	24	2
Datiscaceae	1	0	0	1	0	0	0
Begoniaceae	1	0	1	2	0	2	0
Calophyllaceae	2	0	2	2	0	2	0
Clusiaceae	1	0	0	1	0	0	0
Podostemaceae	1	0	0	1	0	0	0
Hypericaceae	2	0	0	58	47	3	4
Violaceae	2	0	0	78	55	9	4
Turneraceae	2	0	0	4	0	1	0
Passifloraceae	1	0	0	18	2	5	0
Apodanthaceae	1	0	0	1	0	0	0
Muntingiaceae	1	0	1	1	0	1	0
Malvaceae	52	2	10	250	77	52	45
Cochlospermaceae	1	0	0	3	0	0	1
Thymelaeaceae	4	0	3	7	3	4	0
Cistaceae	5	1	2	40	28	6	6
Frankeniaceae	1	0	0	5	1	1	1
Tamaricaceae	1	0	1	8	0	8	0
Droseraceae	2	1	0	9	4	0	0
Totals	**104**	**5**	**30**	**545**	**228**	**118**	**63**

Italic = introduced

Basic Concepts

Our goal is to make the *Flora* as clear, concise, and informative as practicable so that it can be an important resource for both botanists and nonbotanists. To this end, we are attempting to be consistent in style and content from the first volume to the last. Readers may assume that a term has the same meaning each time it appears and that, within groups, descriptions may be compared directly with one another. Any departures from consistent usage will be explicitly noted in the treatments (see References).

Treatments are intended to reflect current knowledge of taxa throughout their ranges worldwide, and classifications are therefore based on all available evidence. Where notable differences of opinion about the classification of a group occur, appropriate references are mentioned in the discussion of the group.

Documentation and arguments supporting significantly revised classifications are published separately in botanical journals before publication of the pertinent volume of the *Flora*. Similarly, all new names and new combinations are published elsewhere prior to their use in the *Flora*. No nomenclatural innovations will be published intentionally in the *Flora*.

Taxa treated in full include extant and recently extinct native species, hybrids that are well established (or frequent), and waifs or cultivated plants that are found frequently outside cultivation and give the appearance of being naturalized. Taxa mentioned only in discussions include waifs or naturalized plants now known only from isolated old records and some non-native, economically important or extensively cultivated plants, particularly when they are relatives of native species. Excluded names and taxa are listed at the ends of appropriate sections, for example, species at the end of genus, genera at the end of family.

Treatments are intended to be succinct and diagnostic but adequately descriptive. Characters and character states used in the keys are repeated in the descriptions. Descriptions of related taxa at the same rank are directly comparable.

With few exceptions, taxa are presented in taxonomic sequence. If an author is unable to produce a classification, the taxa are arranged alphabetically and the reasons are given in the discussion.

Treatments of hybrids follow that of one of the putative parents. Hybrid complexes are treated at the ends of their genera, after the descriptions of species.

We have attempted to keep terminology as simple as accuracy permits. Common English equivalents usually have been used in place of Latin or Latinized terms or other specialized terminology, whenever the correct meaning could be conveyed in approximately the same space, for example, "pitted" rather than "foveolate," but "striate" rather than "with fine longitudinal lines." See *Categorical Glossary for the Flora of North America Project* (R. W. Kiger and D. M. Porter 2001; also available online at http://huntbot.andrew.cmu.edu) for standard definitions of generally used terms. Very specialized terms are defined, and sometimes illustrated, in the relevant family or generic treatments.

References

Authoritative general reference works used for style are *The Chicago Manual of Style,* ed. 14 (University of Chicago Press 1993); *Webster's New Geographical Dictionary* (Merriam-Webster 1988); and *The Random House Dictionary of the English Language,* ed. 2, unabridged (S. B. Flexner and L. C. Hauck 1987). *B-P-H/S. Botanico-Periodicum-Huntianum/Supplementum* (G. D. R. Bridson and E. R. Smith 1991), *BPH-2: Periodicals with Botanical Content* (Bridson 2004), and *BPH Online* [http://fmhibd.library.cmu.edu/fmi/iwp/cgi?-db=BPH_Online&-loadframes] (Bridson and D. W. Brown) have been used for abbreviations of serial titles, and *Taxonomic Literature,* ed. 2 (F. A. Stafleu and R. S. Cowan 1976–1988) and its supplements by Stafleu et al. (1992–2009) have been used for abbreviations of book titles.

Graphic Elements

All genera and approximately 28 percent of the species in this volume are illustrated. The illustrations may show diagnostic traits or complex structures. Most illustrations have been drawn from herbarium specimens selected by the authors. Data on specimens that were used and parts that were illustrated have been recorded. This information, together with the archivally preserved original drawings, is deposited in the Missouri Botanical Garden Library and is available for scholarly study.

Specific Information in Treatments

Keys

Dichotomous keys are included for all ranks below family if two or more taxa are treated. More than one key may be given to facilitate identification of sterile material.

Nomenclatural Information

Basionyms of accepted names, with author and bibliographic citations, are listed first in synonymy, followed by any other synonyms in common recent use, listed in alphabetical order, without bibliographic citations.

The last names of authors of taxonomic names have been spelled out. The conventions of *Authors of Plant Names* (R. K. Brummitt and C. E. Powell 1992) have been used as a guide for including first initials to discriminate individuals who share surnames.

If only one infraspecific taxon within a species occurs in the flora area, nomenclatural information (literature citation, basionym with literature citation, relevant other synonyms) is given for the species, as is information on the number of infraspecific taxa in the species and their distribution worldwide, if known. A description and detailed distributional information are given only for the infraspecific taxon.

Descriptions

Character states common to all taxa are noted in the description of the taxon at the next higher rank. For example, if sexual condition is dioecious for all species treated within a genus, that character state is given in the generic description. Characters used in keys are repeated in the descriptions. Characteristics are given as they occur in plants from the flora area. Characteristics that occur only in plants from outside the flora area may be given within square brackets, or instead may be noted in the discussion following the description. In families with one genus and one or more species, the family description is given as usual, the genus description is condensed, and the species are described as usual. Any special terms that may be used when describing members of a genus are presented and explained in the genus description or discussion.

In reading descriptions, the reader may assume, unless otherwise noted, that: the plants are green, photosynthetic, and reproductively mature; woody plants are perennial; stems are erect; roots are fibrous; leaves are simple and petiolate; flowers are bisexual, radially symmetric, and pediceled; perianth parts are hypogynous, distinct, and free; and ovaries are superior. Because measurements and elevations are almost always approximate, modifiers such as "about," "circa," or "±" are usually omitted.

Unless otherwise noted, dimensions are length × width. If only one dimension is given, it is length or height. All measurements are given in metric units. Measurements usually are based on dried specimens but these should not differ significantly from the measurements actually found in fresh or living material.

Chromosome numbers generally are given only if published, documented counts are available from North American material or from an adjacent region. No new counts are published intentionally in the *Flora*. Chromosome counts from nonsporophyte tissue have been converted to the $2n$ form. The base number ($x =$) is given for each genus. This represents the lowest known haploid count for the genus unless evidence is available that the base number differs.

Flowering time and often fruiting time are given by season, sometimes qualified by early, mid, or late, or by months. Elevations over 200 m generally are rounded to the nearest 100 m; those 100 m and under are rounded to the nearest 10 m. Mean sea level is shown as 0 m, with the understanding that this is approximate. Elevation often is omitted from herbarium specimen labels, particularly for collections made where the topography is not remarkable, and therefore precise elevation is sometimes not known for a given taxon.

The term "introduced" is defined broadly to refer to plants that were released deliberately or accidentally into the flora and that now exist as wild plants in areas in which they were not recorded as native in the past. The distribution of non-native plants is often poorly documented and presence of the plants in the flora may be ephemeral.

If a taxon is globally rare or if its continued existence is threatened in some way, the words "of conservation concern" appear before the statements of elevation and geographic range.

Criteria for taxa of conservation concern are based on NatureServe's (formerly The Nature Conservancy)—see http://www.natureserve.org—designations of global rank (G-rank) G1 and G2:

G1 Critically imperiled globally because of extreme rarity (5 or fewer occurrences or fewer than 1000 individuals or acres) or because of some factor(s) making it especially vulnerable to extinction.

G2 Imperiled globally because of rarity (5–20 occurrences or fewer than 3000 individuals or acres) or because of some factor(s) making it very vulnerable to extinction throughout its range.

The occurrence of species and infraspecific taxa within political subunits of the *Flora* area is depicted by dots placed on the outline map to indicate occurrence in a state or province. The Nunavut boundary on the maps has been provided by the GeoAccess Division, Canada Centre for Remote Sensing, Earth Science. Authors are expected to have seen at least one specimen documenting each geographic unit record (except in rare cases when undoubted literature reports may be used) and have been urged to examine as many specimens as possible from throughout the range of each taxon. Additional information about taxon distribution may be presented in the discussion.

Distributions are stated in the following order: Greenland; St. Pierre and Miquelon; Canada (provinces and territories in alphabetic order); United States (states in alphabetic order); Mexico (11 northern states may be listed specifically, in alphabetic order); West Indies; Bermuda; Central America (Belize, Costa Rica, El Salvador, Guatemala, Honduras, Nicaragua, Panama); South America; Europe, or Eurasia; Asia (including Indonesia); Africa; Atlantic Islands; Indian Ocean Islands; Pacific Islands; Australia; Antarctica.

Discussion

The discussion section may include information on taxonomic problems, distributional and ecological details, interesting biological phenomena, and economic uses.

Selected References

Major references used in preparation of a treatment or containing critical information about a taxon are cited following the discussion. These, and other works that are referred to in discussion or elsewhere, are included in Literature Cited at the end of the volume.

CAUTION

The Flora of North America Editorial Committee **does not encourage, recommend, promote, or endorse** any of the folk remedies, culinary practices, or various utilizations of any plant described within this volume. Information about medicinal practices and/or ingestion of plants, or of any part or preparation thereof, has been included only for historical background and as a matter of interest. Under no circumstances should the information contained in these volumes be used in connection with medical treatment. Readers are strongly cautioned to remember that many plants in the flora are toxic or can cause unpleasant or adverse reactions if used or encountered carelessly.

Key to boxed codes following accepted names:

 C̄ of conservation concern
 Ē endemic to the flora area
 F̄ illustrated
 Ī introduced to the flora area
 W̄ weedy, based mostly on R. H. Callihan et al. (1995) and/or D. T. Patterson et al. (1989)

Flora of North America

CUCURBITACEAE Jussieu

• Cucumber Family

Guy L. Nesom

Plants usually vines, sometimes shrublike in *Cucurbita*, or perennial [annual] herbs (*Melothria*), usually monoecious or dioecious, rarely andromonoecious (*Cucumis*). **Stems** prostrate, procumbent, sprawling, trailing, or climbing; tendrils usually present, unbranched or branched. **Leaves** simple (also compound in *Cyclanthera, Momordica*), alternate, estipulate, petiolate (sessile or subsessile in *Sicyos*); blade unlobed or palmately, pedately, or pinnately lobed. **Inflorescences** paniculate, racemose, umbellate to subumbellate, fasciculate, corymbose, or solitary flowers. **Flowers** unisexual [bisexual]; sepals (4)5(6), sometimes vestigial (*Cyclanthera*), connate, calyx rotate, campanulate, saucer-shaped, or tubular, adnate to corolla, producing hypanthium; petals 5(6), distinct or connate, imbricate or induplicate-valvate, usually yellow, orange, or white, sometimes green, margins entire, rarely fimbriate, corolla rotate, cupulate, campanulate, salverform, or funnelform; stamens (2)3–5, with 4 mostly connate in pairs, appearing as only (1–)3 stamens; anthers connate or distinct, monothecous when all stamens distinct, or stamens 3, with 1 monothecous and 2 dithecous; thecae straight, curved, flexuous, replicate, or conduplicate, connective often present; staminodes often present in pistillate flowers; ovary inferior [semi-inferior], (2)3(–5)-carpellate, syncarpous, 1–3(–8)-locular; placentas parietal, fleshy, mostly meeting in middle; ovules 1–150, anatropous; styles 1(–3), distinct; stigmas 1–3(–5), each 2- or 3-lobed. **Fruits** usually pepos, rarely capsules, elongate to globose, exocarp usually hard, sometimes fleshy and berrylike, glabrous or hairy, smooth or bristly, echinate, aculeate, muricate, tuberculate, or furrowed, indehiscent or dehiscent. **Seeds** mostly compressed, sometimes winged, arillate in *Coccinia, Ibervillea, Momordica*, and *Tumamoca*, exalbuminous; embryos straight.

Genera ca. 120, species ca. 825 (23 genera, 56 species in the flora): nearly worldwide; mostly tropical.

Sechium edule (Jacquin) Swartz (chayote) has been attributed to Louisiana in the PLANTS database on the basis of a putative record in R. D. Thomas and C. M. Allen (1993–1998); there appears to be no such record therein. The only specimen from Louisiana in the NLU herbarium was collected from a cultivated plant in Ouachita Parish. *Sechium edule* is native to Central America and is grown nearly worldwide as a food crop.

Cucurbitaceae, together with Begoniaceae and other families, is currently placed in Cucurbitales (P. F. Stevens, www.mobot.org/MOBOT/research/APweb), in the same clade as Fabales and Rosales.

The familial classification of Cucurbitaceae by C. Jeffrey (1980b, 1990, 2005) is widely used. Recent phylogenetic studies partly support the groupings outlined by Jeffrey but indicate that the cladistic topology of the genera differs significantly from what might be expected from Jeffrey's classification (A. Kocyan et al. 2007; H. Schaefer et al. 2008); the studies suggest that tribes recognized by Jeffrey are largely monophyletic and his subtribes are not. The sequence of genera here generally corresponds with relationships suggested by the molecular data.

Initial classifications by C. Jeffrey (1980b, 1990) emphasized fusion of stigmas and tendril morphology; his latest revision (2005) incorporated information from studies of seed coat anatomy by D. Singh and A. S. R. Dathan (1998). In Cucurbitoideae Kosteletzky, the main sclerenchymatous layer of the seed coat is conspicuous and distinct; in Nhandiroboideae Kosteletzky, it is inconspicuous and not clearly distinct from adjacent sclerified hypodermal layers.

The largest genera in the family are *Trichosanthes* Linnaeus (ca. 100 species), *Momordica* (ca. 60 species), *Zehneria* Endlicher (20–50 species), *Cucumis* (ca. 55 species), *Sicyos* (ca. 50 species), *Cayaponia* (ca. 75 species), and *Gurania* (Schlechtendal) Cogniaux (ca. 40 species). About 40 of the genera are monospecific.

Cucurbitaceae generally is easy to recognize: the fruit is usually a pepo (a hard-shelled berry). Fruits are produced in a wide array of shapes and sizes, especially as the result of millennia of selection. With intense watering, custom fertilization, and selection for size increase, squashes have been grown to 900 pounds; pumpkins can reach 1800 pounds.

Cucurbitaceae were important in early agriculture in the Americas as one of the three main staple food crops—squash, corn, and beans. Squashes are all native to the New World and provide edible flesh and seeds rich in amino acids. Melons are native to Africa and Asia and are used primarily as dessert fruits.

Species domesticated for food include *Benincasa hispida* (Thunberg) Cogniaux (wax gourd), *Citrullus lanatus* (watermelon), *Coccinia grandis* (ivy gourd), *Cucumeropsis mannii* Naudin (white-seeded melon), *Cucumis* (three species—bur gherkin, melon, and cucumber), *Cucurbita* (five species of squash, pumpkin, and gourd), *Cyclanthera pedata* (slipper gourd), *Lagenaria siceraria* (bottle gourd), *Luffa* (two species of loofah), *Momordica charantia* (bitter melon), *Praecitrullus fistulosus* (Stocks) Pangalo (tinda), *Sechium edule* (Jacquin) Swartz (chayote), *Sicana odorifera* (Vellozo) Naudin (casabanana), *Telfairia* Hooker (two species of oyster nut), and *Trichosanthes* (two species of snake gourd).

Species planted as ornamentals include *Cucumis dipsaceus* Ehrenberg (teasel gourd, for its yellow, densely aculeate fruits), *C. metuliferus* E. Meyer ex Naudin (African horned cucumber, for its bright yellow, coarsely aculeate fruits), *Cucurbita pepo* (for its colorful and oddly shaped gourds), *Echinocystis lobata* (balsam-apple, for its massive displays of small, white flowers), *Lagenaria siceraria* (for its bottlelike gourds), and *Thladiantha dubia* (golden creeper, for its large, golden-yellow flowers).

Species employed for other economic uses include *Lagenaria siceraria* (bottle gourd, for containers, birdhouses, floats, and musical instruments), *Luffa cylindrica* M. Roemer (loofah, as a sponge, scrubber, and filter), *Sicana odorifera* (casabanana, as an air freshener, the ripe fruits producing a long-lasting, fruity fragrance), and *Siraitia grosvenorii* (Swingle) C. Jeffrey ex A. M. Lu & Zhi Y. Zhang (luo han guo, as a sweetener, the fruit flesh 300 times sweeter than sugar and low in calories).

SELECTED REFERENCES Bates, D. M., R. W. Robinson, and C. Jeffrey, eds. 1990. Biology and Utilization of the Cucurbitaceae. Ithaca, N.Y. Heiser, C. B. 1979. The Gourd Book. Norman. Jeffrey, C. 1962. Notes on Cucurbitaceae, including a proposed new classification of the family. Kew Bull. 15: 337–371. Jeffrey, C. 1971. Further notes on Cucurbitaceae: II. Kew Bull. 25: 191–236. Jeffrey, C. 1975. Further notes on Cucurbitaceae: IV. Some New World taxa. Kew Bull. 33: 347–380. Jeffrey, C. 1980b. A review of the Cucurbitaceae. Bot. J. Linn. Soc. 81: 233–247. Jeffrey, C. 1990. An outline classification of the Cucurbitaceae. In: D. M. Bates et al., eds. 1990. Biology and Utilization of the Cucurbitaceae. Ithaca, N.Y. Pp. 449–463. Jeffrey, C. 2001. Cucurbitaceae. In: P. Hanelt, ed. 2001. Mansfeld's Encyclopedia of Agricultural and Horticultural Crops.... 6 vols. Berlin and New York. Vol. 3, pp. 1510–1557. Jeffrey, C. 2005. A new system of Cucurbitaceae. Bot. Zhurn. (Moscow & Leningrad) 90: 332–335. Jeffrey, C. and W. J. J. O. De Wilde. 2006. A review of the subtribe Thladianthinae (Cucurbitaceae). Bot. Zhurn. (Moscow & Leningrad) 91: 766–776. Kocyan, A. et al. 2007. A multi-locus chloroplast phylogeny for the Cucurbitaceae and its implications for character evolution and classification. Molec. Phylogen. Evol. 44: 553–577. Lira, R., J. L. Villaseñor, and P. D. Davila. 1997. A cladistic analysis of the subtribe Sicyinae (Cucurbitaceae). Syst. Bot. 22: 415–425. Nayar, N. M. and T. A. More, eds. 1998. Cucurbits. Enfield, N.H. Nesom, G. L. 2011c. Toward consistency of taxonomic rank in wild/domesticated Cucurbitaceae. Phytoneuron 2011-13: 1-33. Nesom, G. L. 2012f. Terms for surface vestiture and relief of Cucurbitaceae fruits. Phytoneuron 2012-108: 1–4. Robinson, R. W. and D. S. Decker. 1997. Cucurbits. New York. Schaefer, H., C. Heibl, and S. S. Renner. 2008. Gourds afloat: A dated phylogeny reveals an Asian origin of the gourd family (Cucurbitaceae) and numerous oversea dispersal events. Proc. Roy. Soc. Biol. Sci. B, 276: 843–851. Schaefer, H. and S. S. Renner. 2011. Cucurbitaceae. In: K. Kubitzki et al., eds. 1990+. The Families and Genera of Vascular Plants. 10+ vols. Berlin etc. Vol. 10, pp. 112–174. Schaefer, H. and S. S. Renner. 2011b. Phylogenetic relationships in the order Cucurbitales and a new classification of the gourd family (Cucurbitaceae). Taxon 60: 122–138. Singh, D. and A. S. R. Dathan. 1998. Morphology and embryology. In: N. M. Nayar and T. A. More, eds. 1998. Cucurbits. Enfield, N.H. Pp. 67–84. Stocking, K. M. 1955. Some considerations of the genera *Echinocystis* and *Echinopepon* in the United States and northern Mexico. Madroño 13: 84–100. Teppner, H. 2004. Notes on *Lagenaria* and *Cucurbita* (Cucurbitaceae)—Review and new contributions. Phyton (Horn) 44: 245–308.

1. Inflorescences: pistillate flowers 1–16, (clustered at peduncle apex or in fascicles in distal axils or in racemoid to corymboid panicles).
    2. Inflorescences: pistillate flowers sessile to subsessile in umbelliform clusters at peduncle apex; seeds 1; plants annual . . . . .11. *Sicyos*, p. 25
    2. Inflorescences: pistillate flowers in axillary fascicles or in axillary racemoid to corymboid panicles; seeds 2–80; plants perennial.
        3. Leaf surfaces eglandular; fruiting peduncles 6–9 cm; fruits 6–10 cm. . . . . 13. *Apodanthera*, p. 28
        3. Leaf surfaces eglandular or glandular; fruiting peduncles 0–5 cm; fruits 0.6–2 cm.
            4. Pistillate flowers 2–6(–10) in corymboid to racemoid panicles; fruits 0.6–1 cm . . . . 4. *Bryonia*, p. 11
            4. Pistillate flowers 1–5, in axillary fascicles or on racemoid branches proximal to staminate flowers; fruits 1–2 cm. . . . . 22. *Cayaponia*, p. 46
1. Inflorescences: pistillate flowers 1(–3).
    5. Fruits echinate, muricate to muriculate, subaculeate, or tuberculate.
        6. Plants trailing, tendrils absent; leaf blade margins sinuate-undulate or sinuate-crisped and whitish because of light-colored abaxial surfaces; peduncles recurved-nodding in fruit. . . . . .3. *Ecballium*, p. 10
        6. Plants climbing or trailing, tendrils present; leaf blade margins not crisped-undulate and whitish; peduncles straight (not recurved-nodding) in fruit.
            7. Inflorescences: staminate flowers solitary.
                8. Seeds red-arillate; filaments inserted near hypanthium rim; anther thecae distinct or initially loosely connate and forming central body, sigmoid-triplicate; fruits tuberculate, muricate, muriculate, or irregularly smooth-ridged . . . . 1. *Momordica*, p. 7
                8. Seeds not arillate; filaments inserted near hypanthium base; anther thecae replicate (usually 2-folded), distinct; fruits densely echinate or aculeate to muricate. . . . . .16. *Cucumis* (in part), p. 32
            7. Inflorescences: staminate flowers (1)2–200 in racemes, racemoid panicles, panicles, or fascicles.
                9. Anther thecae distinct; petals yellow . . . . .16. *Cucumis* (in part), p. 32
                9. Anther thecae connate into head or fused into ring; petals white to cream-yellow, greenish yellow, greenish, or greenish white.
                    10. Leaves 3–7-foliolate; fruits fleshy-capsular; anther thecae connate, fused into ring. . . . . 6. *Cyclanthera*, p. 15
                    10. Leaves 3–7-lobed; fruits dry, thin-walled; anther thecae connate, forming a head but not fused into ring.

[5. Shifted to left margin.—Ed.]

1. MOMORDICA Linnaeus, Sp. Pl. 2: 1009. 1753; Gen. Pl. ed. 5, 440. 1754 • Balsam-apple, bitter melon, cundeamor [Latin *mordicus*, biting, alluding to sculptured seed surfaces and margins, appearing as though bitten] ⊺

Plants annual [perennial], monoecious [dioecious], climbing or trailing; stems glabrous or hairy; roots fibrous; tendrils unbranched [2-branched]. **Leaves** simple or compound; blade broadly ovate or reniform to orbiculate, palmately 3–7[–9]-lobed, usually pedate, lobes broadly ovate, rhombic-ovate, ovate-oblong, or ovate-elliptic, margins coarsely and widely dentate to crenate-dentate or sinuate-dentate [mucronulate-dentate or remotely denticulate], surfaces eglandular. **Inflorescences:** staminate flowers solitary [corymbose, racemose, or umbellate], axillary; pistillate flowers solitary, from different axils than staminate; peduncles erect at apex; bracts persistent, ovate-cordate, reniform, or orbiculate-cordate [ovate, rhombic-ovate]. **Flowers:** hypanthium obconic; sepals 5, linear to ovate-lanceolate or ovate-acuminate; petals 5, distinct, yellow to bright yellow [greenish yellow], obovate or oblong [suborbiculate], [5–]7–25[–32] mm, glabrous, corolla rotate to broadly and shallowly campanulate. **Staminate flowers:** stamens (2)3; filaments inserted near hypanthium rim, distinct; thecae distinct or connate, forming central body, sigmoid-triplicate, connective broad; pistillodes glandular or absent. **Pistillate flowers:** ovary 3-locular, ellipsoid to elongate-fusiform, short-beaked; ovules ca. 3–7 per locule; styles 1, columnar; stigmas 3, 2-lobed; staminodes absent or 3, glandular. **Fruits** berrylike or capsular, pendent, red or red-orange to yellow-orange, oblong-fusiform to ellipsoidal, cylindric or ovoid, beaked, fleshy, thick-walled, irregularly tuberculate to muricate, muriculate, or irregularly smooth-ridged, glabrous, apically dehiscent by 3 valves [indehiscent or irregularly dehiscent]. **Seeds** 10–50, oblong to ovoid-oblong beyond narrowed base, turgid or flattened, red-arillate, margins grooved, surfaces smooth or sculptured. $x = 11, 14$.

Species ca. 60 (2 in the flora): introduced; Asia, Africa, Australia; introduced also in Mexico, West Indies, Central America, South America, Europe, se Asia, Pacific Islands, Australia.

SELECTED REFERENCE Schaefer, H. and S. S. Renner. 2010. A three-genome phylogeny of *Momordica* (Cucurbitaceae) suggests seven returns from dioecy to monoecy and recent long-distance dispersal to Asia. Molec. Phylogen. Evol. 54: 553–560.

1. Bracts of staminate flowers near peduncle apex, margins dentate to denticulate; leaf lobes and teeth apiculate; fruits 2.5–4(–7) cm. . . . . 1. *Momordica balsamina*
1. Bracts of staminate flowers at or proximal to middle of peduncle, margins entire; leaf lobes and teeth sometimes submucronate but not apiculate; fruits 7–25 cm . . . . .2. *Momordica charantia*

1. Momordica balsamina Linnaeus, Sp. Pl. 2: 1009. 1753 [I]

Stems pubescent to glabrescent. **Leaves:** petiole 1–4(–6) cm; blade broadly ovate or reniform to orbiculate, palmately 3–5-lobed, 1–9(–12) cm, base cordate, lobes broadly ovate or rhombic-ovate, sinuses 80–90% to base, margins sinuate-dentate, leaf lobes and teeth apiculate, surfaces glabrous or sparsely hairy. **Inflorescences:** staminate peduncles bracteate near apex, bracts sessile, broadly ovate-cordate to reniform, margins dentate to denticulate; pistillate peduncles ebracteate or bracteate at base to submedially. **Petals** yellow, obovate, 8–15 mm. **Fruits** orange-red, broadly ovoid, 2.5–4(–7) cm, beak becoming less prominent at maturity, surface minutely tuberculate, muriculate in longitudinal rows. **Seeds** ovate-oblong, 9–12 mm. $2n = 22$.

Flowering May–Sep. Hammocks, disturbed areas, roadsides, fencerows; 10–200 m; introduced; Ala., Fla., La., N.Mex., Okla., Tex.; Asia; Africa; introduced also in Mexico, West Indies, Australia.

Reports of *Momordica balsamina* from Alabama and Texas are not documented. Naturalized occurrences of the species elsewhere in the flora area are scattered and uncommon.

2. Momordica charantia Linnaeus, Sp. Pl. 2: 1009. 1753 • Balsam pear [F] [I]

Subspecies 2 (1 in the flora): introduced; Africa; introduced widely.

SELECTED REFERENCE   Marr, K. L., X. Y. Mei, and N. Bhattarai. 2004. Allozyme, morphological and nutritional analysis bearing on the domestication of *Momordica charantia* L. (Cucurbitaceae). Econ. Bot. 58: 435–455.

2a. Momordica charantia Linnaeus subsp. charantia [F] [I]

Momordica charantia var. *abbreviata* Seringe

Stems glabrous or hairy. **Leaves:** petiole 1.5–3.5(–6) cm; blade suborbiculate to orbiculate, deeply palmately 5–7-lobed, 5–10(–12) cm, base cordate, lobes ovate-oblong or ovate-elliptic, sinuses 50–90% to base, margins coarsely and widely dentate to crenate-dentate, leaf lobes and teeth sometimes submucronate, not apiculate, surfaces glabrous or sparsely hairy. **Inflorescences:** staminate peduncles bracteate at or proximal to middle, bracts sessile, reniform or orbiculate-cordate, margins entire; pistillate peduncles ebracteate. **Petals** bright yellow, obovate to oblong-obovate, 7–25 mm. **Fruits** red or orange to yellow-orange, cylindric or ellipsoidal to oblong-fusiform or ovoid, 7–25 cm, beak becoming less prominent at maturity, surface usually minutely tuberculate to muricate and muriculate, sometimes irregularly smooth-ridged. **Seeds** oblong, 10–15 mm. $2n = 22$.

Flowering Apr–Nov. Disturbed areas, roadsides, fencerows, thickets, hammock edges, fields, ditches, canal banks, shell mounds, orange groves, sand pine scrub, flatwoods, floodplain woods; 10–200 m; introduced; Ala., Conn., Fla., Ga., La., Pa.; Africa; introduced also in Mexico, West Indies, Central America, South America, Europe, se Asia, Pacific Islands, Australia.

No documentation has been found for a report of *Momordica charantia* in Texas. The species is apparently becoming common in Florida, where it is known to be naturalized in many counties. It is grown as a crop in many countries for the young fruits, which are edible before an extreme bitterness develops in the mature fruits.

Two forms of *Momordica charantia* in the New World can be distinguished: the ubiquitous weed of the humid lowland tropics and subtropics, growing spontaneously and not cultivated or eaten except occasionally and opportunistically for the pulp; and the domesticated forms, which are seen in Asian markets. The cultivated forms do not escape (M. Nee, pers. comm.).

Momordica charantia subsp. *macroloba* Achigan-Dako & Blattner is endemic to the Dahomey Gap and Sudano-Guinean phytoregion of Benin and Togo in west-central Africa.

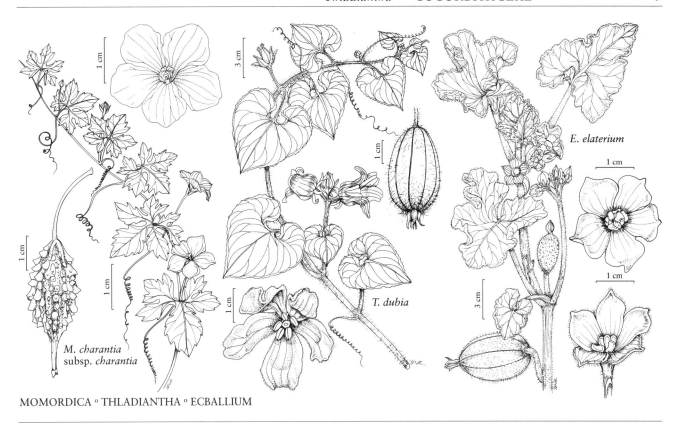

E. elaterium

T. dubia

M. charantia
subsp. charantia

MOMORDICA ∘ THLADIANTHA ∘ ECBALLIUM

2. THLADIANTHA Bunge, Enum. Pl. China Bor., 29. 1833 • Golden creeper [Greek *thladias*, eunuch, and *anthos*, flower, alluding to staminodes in neuter flowers] I

Plants perennial, dioecious, climbing; stems annual, hirsute; roots tuberous; tendrils unbranched or 2-branched. **Leaves:** blade broadly ovate to suborbiculate, not lobed or angulate, margins serrate to denticulate [serrulate], surfaces eglandular. **Inflorescences:** staminate flowers 5–15 in axillary racemes or umbelliform fascicles, less commonly solitary; pistillate flowers solitary [2 or 3], axillary; bracts absent [present]. **Flowers:** hypanthium shallowly campanulate to subrotate; sepals 5, recurving, linear-lanceolate to narrowly lanceolate; petals 5, distinct nearly to base, yellow, ovate-oblong, 15–25 mm, villous abaxially, papillose-glandular adaxially, corolla campanulate. **Staminate flowers:** stamens 5; filaments inserted near hypanthium base, distinct; thecae connate, forming central oblong body, sigmoid-flexuous, connective broad; pistillodes present. **Pistillate flowers:** ovary 3-locular, oblong; ovules 30–70 per locule; styles 1, columnar; stigmas 3, 2-lobed; staminodes 5 (4 in approximate pairs), linear. **Fruits** pepos, orange-red [red or green], oblong-ovoid [ellipsoid], longitudinally furrowed [smooth], densely hispid-hirsute to hirsute or hirsute-villous, irregularly dehiscent. **Seeds** 100–200, obovoid, flattened, not arillate, margins not differentiated, surface smooth. $x = 9$.

Species ca. 23 (1 in the flora): introduced; e Asia, Africa.

SELECTED REFERENCE Lu, A. M. and Zhang Zhi Y. 1982. A revision of genus *Thladiantha* Bunge (Cucurbitaceae). Bull. Bot. Res., Harbin 1: 61–96.

1. Thladiantha dubia Bunge, Enum. Pl. China Bor., 29. 1833 • Manchu tuber gourd, red hailstone, thladianthe douteuse F I

Leaf blades 5–10 × 4–9 cm, base cordate, apex short-acuminate, surfaces hirsutulous. **Inflorescences:** peduncle 5–15 mm. **Flowers:** hypanthium 2–4 mm; sepals recurving, 12–13 mm; petals ca. 25 mm, apex acute. **Pepos** 4–5 cm, surface with 10 obscure furrows. $2n = 18$.

Flowering Jun–Sep. Roadsides, thickets, pastures, waste places; 10–200 m; introduced; Man., Ont., Que.; Ill., Mass., N.H., N.Y., Wis.; Asia (China, Russia); introduced also in South America (Ecuador), Europe, elsewhere in e Asia (Japan), Pacific Islands (Galapagos Islands).

Thladiantha dubia is characterized by its ovate-cordate, unlobed leaves; relatively large, solitary flowers with narrow, recurving sepals, campanulate corollas, yellow petals, usually with recurving apices; and pendulous, hirsute-villous, orange-red fruits. It is grown as an ornamental, especially for its large, bell-shaped flowers and brightly colored fruits. Few populations are apparently outside of cultivation at present, but even staminate plants are potentially invasive because of the spread by tubers.

3. ECBALLIUM A. Richard in J. B. Bory de Saint-Vincent et al., Dict. Class. Hist. Nat. 6: 19. 1824, name conserved • Squirting or exploding cucumber [Greek *ekballein*, to throw or cast out, alluding to seed discharge] I

Plants perennial, monoecious, prostrate, trailing; stems annual, villous-hirsute; roots tuberous; tendrils absent. **Leaves:** blade triangular to hastate or ovate-triangular, unlobed, base cordate, margins crenate to crenate-dentate, usually prominently sinuate-undulate or sinuate-crisped, appearing whitish because of light-colored abaxial surfaces, surfaces eglandular. **Inflorescences:** staminate flowers 1 or 2–8 in axillary racemes; pistillate flowers solitary, from same or different axils as staminate, peduncles recurved-nodding in fruit; bracts absent. **Flowers:** hypanthium broadly campanulate; sepals 5, lanceolate; petals 5, distinct nearly to base, [cream] pale yellow or white with [darker and brighter] yellow [greenish] center, oblong-ovate, abruptly acuminate apically, 6–16 mm, puberulent, corolla campanulate to rotate-campanulate. **Staminate flowers:** stamens 5, 2 connate in pairs, 1 distinct, appearing as 3 stamens; filaments inserted near hypanthium rim, distinct; thecae distinct, sigmoid, connective broadened; pistillodes absent. **Pistillate flowers:** ovary 3-locular, narrowly ovoid-ellipsoid, hispid to hispid-hirsute; ovules ca. 50 per locule; styles 1, thick, barely evident at base of stigmas; stigmas 3, each with 2 long appendages; staminodes absent. **Fruits** berrylike, green to greenish glaucous, or slightly yellow, ellipsoid or oblong to ovoid, muricate-hispid to muricate-hispid-hirsute, dehiscing explosively, expelling seeds through ruptured pedicel attachment. **Seeds** 40–100, ovoid, compressed, not arillate, obscurely margined, surface smooth. $x = 9$.

Species 1: introduced; e, se, sw Europe, w Asia, n Africa; introduced also in w Europe, Australia.

Ecballium plants are prostrate and without tendrils, and the hastiform leaves have crisped-undulate margins that appear whitish because of the light-colored abaxial surfaces. The erect peduncles are sharply nodding near the base of the fruit such that the mature fruit is pendent.

The explosive dehiscence is stimulated by touching the peduncle near the fruit. The ripening fruit becomes turgid with mucilaginous fluid. Increasing pressure from the maturing and enlarging seeds forces the fruit to break away explosively from the plant and the seeds are ejected in a mucilaginous stream to several meters. The specific epithet (Greek, *elaterios*, driving) either reinforces the generic name or alludes to the violently purgative properties of the plant.

1. **Ecballium elaterium** (Linnaeus) A. Richard in J. B. Bory de Saint-Vincent et al., Dict. Class. Hist. Nat. 6: 19. 1824 [F] [I]

Momordica elaterium Linnaeus, Sp. Pl. 2: 1010. 1753

Stems 15–80 cm. **Leaves** long-petiolate; blade bicolor, 4–10 cm, fleshy, abaxially lighter because of densely hirsute to hirsutulous vestiture. **Staminate flowers** 1 or 2–8 in racemes 5–35 cm. **Pistillate flowers:** peduncles erect, sharply recurved-nodding at apex so that mature fruit is pendent, 1–8 cm. **Corollas** (14–)18–20 mm diam.; petals 8–16 mm (staminate) or 6–14 mm (pistillate). **Fruits** 4–5.5 cm, pendulous on peduncles 1–8 cm. **Seeds** 4–5 mm. $2n = 18, 24$.

Flowering Jun–Aug. Disturbed sites; 20–100 m; introduced; Ala., N.Y., Pa.; e, se, sw Europe; w Asia; n Africa; introduced also in w Europe (England), Australia.

Some, perhaps all, collections of *Ecballium* from New York and Pennsylvania may represent garden curiosities rather than escaped or established populations.

4. BRYONIA Linnaeus, Sp. Pl. 2: 1012. 1753; Gen. Pl. ed. 5, 442 [as Bronia]. 1754

• Bryony [Greek *bruein*, to burgeon or sprout, alluding to rapid growth of herbaceous stems produced annually from large perennial roots] [I]

Plants perennial, usually dioecious, rarely monoecious, climbing or sprawling; stems annual, hispidulous to glabrate; roots tuberous; tendrils unbranched. **Leaves:** blade orbiculate to ovate or hastate, 3–5-angled, or palmately 3–5-lobed, lobes ovate to deltate or triangular, margins coarsely toothed, surfaces eglandular. **Inflorescences:** staminate flowers 4–16 in axillary racemes or fascicles; pistillate flowers [1]2–6(–10) in axillary corymboid to racemoid panicles; bracts absent. **Flowers:** hypanthium campanulate to shallowly campanulate; sepals 5, lanceolate to deltate; petals 5, distinct, white to cream, greenish white, or yellowish green, or yellowish, usually green-throated and faintly green-lined, oblong-ovate to ovate-lanceolate, 3–7 mm, glabrate, corolla broadly funnelform-campanulate to nearly rotate. **Staminate flowers:** stamens 5, 2 connate in pairs, 1 distinct, appearing as 3 stamens; filaments inserted near hypanthium rim, distinct; thecae distinct, C-curved to S-curved, connective broadened and sheetlike; pistillodes absent. **Pistillate flowers:** ovary 1-locular, ovoid to broadly ellipsoid; ovules 3–10 per locule; styles 1, slender-columnar; stigmas 3, 2-lobed; staminodes 3–5 or absent or vestigial. **Fruits** berrylike, red to orange or black, globose, smooth, glabrous, indehiscent. **Seeds** 2–8, ovoid or oblong, compressed, not arillate, margins thickened-corrugated or not differentiated, surface roughened.

Species ca. 12 (2 in the flora): introduced; Europe, Asia, Africa, Pacific Islands.

SELECTED REFERENCES Jeffrey, C. 1969. A review of the genus *Bryonia* L. (Cucurbitaceae). Kew Bull. 23: 441–461. Volz, S. M. and S. S. Renner. 2009. Phylogeography of the ancient Eurasian medicinal plant genus *Bryonia* (Cucurbitaceae) inferred from nuclear and chloroplast sequences. Taxon 58: 550–560.

1. Plants monoecious; fruits black; stigmas glabrous . . . . 1. *Bryonia alba*
1. Plants dioecious; fruits red to orange; stigmas hairy. . . . . 2. *Bryonia dioica*

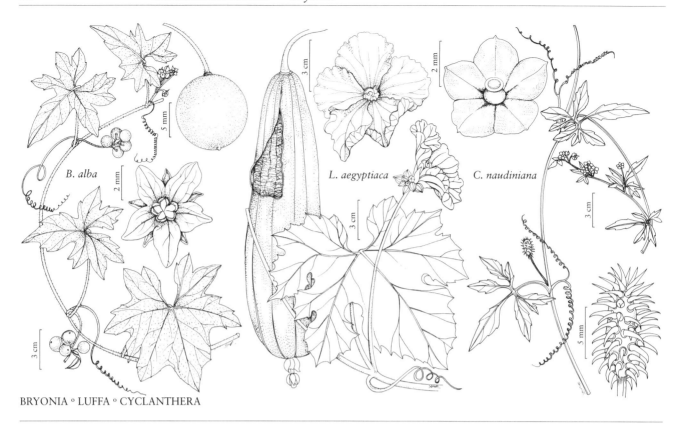

B. alba

L. aegyptiaca

C. naudiniana

BRYONIA ∘ LUFFA ∘ CYCLANTHERA

1. Bryonia alba Linnaeus, Sp. Pl. 2: 1012. 1753

• White bryony   F I

Vines monoecious, high-climbing or forming dense mats over vegetation, to 7 m. **Leaves:** petiole 1.5–3.5 cm; blade hastate to 5-angular, palmately 3–5-lobed, 3–8(–15) × 2–6(–8) cm, lobes ovate to deltate or triangular, central lobe largest, base deeply cordate, margins coarsely dentate to remotely dentate or serrate, surfaces hispidulous to scabrous or pustulate, not white-sericeous abaxially. **Flowers:** calyx campanulate; sepals recurving, whitish, green-veined, lanceolate, 3.5–5 mm; petals yellowish to greenish white, oblong-ovate, 3–7 mm; stigmas glabrous. **Fruits** 4–6(–10), black, 0.6–1 cm, peduncles 2–3[–14] cm. **Seeds** 2–6(–8), 4–5 × 2 mm. *2n* = 20.

Flowering Jun–Aug. Sagebrush, riparian zones, thickets, stream terraces, irrigation ditches, seeps, moist slopes, draws, lawns, roadsides, fencerows, picnic areas, disturbed sites; 100–1800 m; introduced; Idaho, Mont., N.J., Utah, Wash.; Europe; Asia; introduced also in n, e Europe (Czech Republic, Denmark, Finland, Lithuania, Poland), Pacific Islands (Hawaii).

Bryonia alba has been naturalized in the United States since the 1880s and may have been introduced through commercial seed trade. A tincture from the roots has been widely used as an effective diuretic. The seeds are dispersed by birds.

As noted by C. Jeffrey (1969), *Bryonia alba* is monoecious in northern Europe, as it seems consistently to be in the flora area; it is commonly dioecious in the southeastern part of its native range (Macedonia to Turkey and southwestern Russia).

Measurements from relatively few collections suggest that peduncles of *Bryonia alba* are longer in Europe (4–14 cm) than in North America (2–3 cm).

SELECTED REFERENCES Novak, S. J. and R. N. Mack. 1995. Allozyme diversity in the apomictic vine *Bryonia alba* (Cucurbitaceae): Potential consequences of multiple introductions. Amer. J. Bot. 82: 1153–1162. Novak, S. J. and R. N. Mack. 2000. Clonal diversity within and among introduced populations of the apomictic vine *Bryonia alba* (Cucurbitaceae). Canad. J. Bot. 78: 1469–1481.

2. Bryonia dioica Jacquin, Fl. Austriac. 2: 59, plate 199. 1774 • Cretan or red bryony, wild hop [1]

Bryonia cretica Linnaeus subsp. *dioica* (Jacquin) Tutin

Vines dioecious, high-climbing. **Leaves:** petiole 2–4.5 cm; blade orbiculate to ovate or 3–5-angular, palmately 5-lobed, 3–5 × 3–6 cm, lobes deltate to triangular, central lobe largest, base cordate, margins remotely dentate, surfaces hispidulous. **Flowers:** calyx campanulate; sepals green, deltate, 0.5–1 mm; petals white to cream or yellowish green, ovate-lanceolate, 3–5 mm; stigmas hairy. **Fruits** 1 or 3 or 4(5), red to orange, 0.6–0.8 cm, peduncles (0–)1–5 cm. **Seeds** 3–6, 3 × 2 mm. $2n = 20$.

Flowering Jun–Aug. Disturbed places; 0–100 m; introduced; Calif.; Europe; w Asia; n Africa; introduced also elsewhere in Europe (Czech Republic, England), Pacific Islands (New Zealand).

Naturalized plants of *Bryonia dioica* were first collected in Golden Gate Park in San Francisco in 1919 and 1920 and have been recollected as recently as 1965. C. Jeffrey (1969) noted that the glandular inflorescences of *B. dioica* distinguish it from *B. cretica*.

5. LUFFA Miller, Gard. Dict. Abr. ed. 4, vol. 2. 1754 • Vegetable sponge, loofah [Arabic *lufah*, name for *Luffa aegyptiaca*] [1]

Plants annual, monoecious, climbing or trailing; stems villous or hirsute to scabrous; roots fibrous; tendrils [2]3–6-branched. **Leaves:** blade suborbiculate to ovate-cordate [triangular], palmately (3–)5–7-lobed, [rarely subentire], lobes broadly ovate to triangular, suborbiculate, oblong-ovate, oblong-triangular, or depressed-ovate, margins entire or sinuate, sinuate-toothed, sublobulate-dentate, or coarsely dentate, surfaces eglandular or gland-dotted. **Inflorescences:** staminate flowers (5–)15–20 in axillary racemes, opening one at a time in sequence acropetally; pistillate flowers solitary, in same or different axils as staminate; bracts caducous, linear. **Flowers:** hypanthium campanulate; sepals 5, erect, straight, triangular-lanceolate; petals 5, distinct, yellow, obcordate to obovate, oblong-obovate, or obovate-cuneate, 20–45 mm, glabrous, corolla shallowly campanulate [rotate, ± flat]. **Staminate flowers:** stamens (3–)5, one 1-locular, others 2-locular or all monothecous; filaments inserted on or near hypanthium rim, distinct; thecae connate, sigmoid, contorted, connective slightly broadened; pistillodes absent. **Pistillate flowers:** ovary 3(–6)-locular, subglobose to cylindric; ovules 30–100 per locule; styles 1, columnar; stigmas 3, 2-lobed; staminodes 3(–5). **Fruits** pepos, usually greenish, sometimes white-striped, drying brownish, cylindric to narrowly clavate [subglobose], (6–)15–50 cm, smooth [ribbed, aculeate-scabrous, or densely echinate], glabrous, with rigid internal network of fibrovascular bundles, dehiscent by apical operculum. **Seeds** 100–300, broadly ellipsoid to oblong-ellipsoid or ellipsoid-ovoid, compressed, not arillate, sometimes with broad beak, margins winged or not, surface smooth or roughened. $x = 13$.

Species 7 (2 in the flora): introduced; Central America, South America, Asia.

Young and tender fruits of both species of *Luffa* in the flora area are used as vegetables. The mature fruit is inedible due to the hardened network of fibrovascular bundles that forms the loofah "sponge." The sponges are commonly used for scrubbing and cleaning.

Four *Luffa* species are native in the Old World (including the two treated here) and three in the New World. Both of the widely cultivated species, *L. acutangula* and *L. aegyptiaca*, apparently originated in the India-Pakistan region and are closely related to each other (B. Dutt and R. P. Roy 1990; C. B. Heiser and E. E. Schilling 1990). The Asian species are more distinctly differentiated among themselves than are the American ones.

SELECTED REFERENCES Ali, M. A., S. Karuppusamy, and F. M. A. Al-Hemaid. 2010. Molecular phylogenetic study of *Luffa tuberosa* Roxb. (Cucurbitaceae) based on internal transcribed spacer (ITS) sequences of nuclear ribosomal DNA and its systematic implication. Int. J. Bioinf. Res. 2: 42–60. Heiser, C. B. and E. E. Schilling. 1990. The genus *Luffa*: A problem in phytogeography. In: D. M. Bates et al., eds. 1990. Biology and Utilization of the Cucurbitaceae. Ithaca, N.Y. Pp. 120–133.

1. Pepos narrowly clavate, 10-angled; petals pale yellow; seeds beaked, beak margins obscurely winged . . . . 1. *Luffa acutangula*
1. Pepos cylindric, not angled; petals bright yellow; seeds not beaked, margins with wings ca. 1 mm wide . . . . .2. *Luffa aegyptiaca*

1. **Luffa acutangula** (Linnaeus) Roxburgh, Fl. Ind. ed. 1832, 3: 713. 1832 • Sponge or ridged gourd, Sinkwa towel sponge, Chinese squash I

Cucumis acutangulus Linnaeus, Sp. Pl. 2: 1011. 1753

Varieties 2 (1 in the flora): introduced; Asia (India, Pakistan); introduced also in West Indies (Puerto Rico).

Variety *amara* (Roxburgh) C. B. Clarke is native in India and Pakistan; compared to the typical variety, it has leaves small and softly villous, petal apices round or scarcely emarginate, and fruits smaller and obovoid-oblong. The range and magnitude of difference suggests that var. *amara* may be justifiably treated as a distinct species, depending on the degree of isolation.

1a. **Luffa acutangula** (Linnaeus) Roxburgh var. **acutangula** I

Vines climbing to 10 m; tendrils mostly 3-branched; stems 5-angled, scabrous along ribs. **Leaves:** petiole 8–12 cm; blade light green, suborbiculate, shallowly 5–7-lobed, 15–20 × 13–20 cm, lobes suborbiculate to broadly ovate, oblong-ovate, oblong-triangular, or depressed-ovate, margins irregularly and coarsely dentate, apex rounded, surfaces glabrous. **Staminate racemes** 10–20-flowered, 10–15 cm; peduncles and pedicels erect to spreading. **Flowers** 4–5 cm diam.; hypanthium 5–10 mm, sepals equaling to slightly longer than tube; petals pale yellow, 20–25 mm, apex truncate to shallowly emarginate; stamens 3; filaments 4 mm. **Pepos** narrowly clavate, 10-angled, (10–)15–30(–45) × 5–8 cm. **Seeds** black, beaked, 10–12 × 6–8 mm, surfaces rugose, glabrous, beak margins obscurely winged. 2*n* = 26.

Flowering Aug–Oct. Gardens, fields, trash heaps, cultivated in home gardens, abandoned plantings; 10–200 m; introduced; La., Va.; Asia (Pakistan); introduced also in West Indies, Central America, South America, elsewhere in Asia, Africa, Indian Ocean Islands, Pacific Islands, Australia.

2. **Luffa aegyptiaca** Miller, Gard. Dict. ed. 8, Luffa no. 1. 1768 • Sponge or dishcloth gourd, smooth luffa F I

Momordica luffa Linnaeus, Sp. Pl. 2: 1009. 1753

Vines climbing or trailing to 15 m; tendrils 3–6-branched; stems not angled, finely hairy or glabrous. **Leaves:** petiole 2–15 cm; blade dark green, ovate-cordate, (3–)5-lobed, 6–20(–30) × 6–25 cm, lobes triangular to ovate, margins entire or sinuate to sinuate-toothed or sublobulate-dentate, apex acute-apiculate, surfaces scabrous. **Staminate racemes** (5–)15–20-flowered, 12–35 cm; peduncles and pedicels erect to spreading. **Flowers** 5–10 cm diam.; hypanthium 4–8 mm, sepals longer than tube; petals bright yellow, 25–45 mm, apex shallowly obtuse to obtuse-apiculate; stamens (3–)5; filaments 6–8 mm. **Pepos** cylindric, not angled, (6–)20–50 × (2.5–)6–10 cm. **Seeds** dull black, not beaked, 10–12 × 6–8 mm, surface smooth, glabrous, margins with wings 1 mm wide. 2*n* = 26.

Flowering Aug–Oct. Gardens, fields, trash heaps, cultivated in home gardens, abandoned plantings; 10–100 m; introduced; Fla., Ga., Ill., La., N.C., S.C., Tex., Va.; Asia (India, Pakistan); introduced also in Mexico, West Indies, Central America, South America, elsewhere in Asia, Africa, Pacific Islands, Australia.

The later name *Luffa cylindrica*, incorrectly attributed to (Linnaeus) M. Roemer, has been widely and erroneously applied to *L. aegyptiaca*.

6. CYCLANTHERA Schrader, Index Seminum (Göttingen) 1831: 2. 1831 • Bur cucumber [Greek *kyklos*, circle, and *antheros*, blooming, alluding to ringlike anthers]

Cremastopus Paul G. Wilson

Plants annual, monoecious [dioecious], climbing; stems annual, glabrous or hairy; roots fibrous; tendrils unbranched or 2[3]-branched, branches unequal in length. **Leaves** simple or compound; blade lanceolate to orbiculate, [unlobed or shallowly 3–9-lobed to palmately 3–5-lobed] 3–7-foliolate (3–7-palmatisect) [divisions narrowly broadly lanceolate to elliptic-lanceolate or oblanceolate to obspatulate], base attenuate to cordate, margins coarsely serrate to shallowly or deeply lobed [to serrulate or denticulate], abaxial surface glabrous, adaxial surface scabrous with hair bases. **Inflorescences:** staminate flowers in axillary racemes or narrow panicles with or without well-differentiated main axis; pistillate flowers solitary, usually from same axils as staminate, peduncles erect at apex; floral bracts caducous, filiform. **Flowers:** hypanthium campanulate to shallowly cupulate; sepals vestigial [5, ovate-triangular to triangular]; petals 5, connate $\frac{1}{3}$[–$\frac{2}{3}$] length, white [greenish or yellowish], oblong-ovate to triangular, 1–1.5[–5] mm, glabrous, minutely papillate-glandular adaxially, corolla rotate [campanulate]. **Staminate flowers:** stamens 5; filaments inserted at receptacle base, connate into central column, 0.2–2 mm (anthers sessile or subsessile to short-columnar); thecae connate-fused into continuous, horizontal ring as a single structure (synanther), forming a head, dehiscing by continuous slit, connective broad; pistillodes absent. **Pistillate flowers:** ovary usually 1-locular, ovoid to elliptic-ovoid or subglobose; ovules (1)2–6 per locule; styles 1, short-columnar; stigmas 1, depressed-subglobose; staminodes absent. **Fruits** fleshy-capsular, green to light green, narrowly obliquely ovoid to obliquely globose or obovoid (or subglobose to broadly ellipsoid), gibbous, acute to obtuse at both ends, densely echinate [to smooth or very sparsely echinate], irregularly and explosively dehiscent. **Seeds** (1–)4–12, ovoid to oblong, compressed, not arillate, margins not winged but sculptured, surface tuberculate [smooth, roughened, muriculate]. *x* = 16.

Species 35 (4 in the flora): w, c United States, Mexico, Central America, South America.

Species of *Cyclanthera* are characterized by their annual duration, relatively small flowers, androecium of a single, horizontal, continuous, circular theca, and gibbous, echinate, relatively few-seeded fruits with explosive dehiscence. The seeds are thrown by a catapultlike placental arm. Nectaries are derived from secretory hairs on the hypanthium (S. Vogel 1981). The northern Mexican *Cremastopus* was distinguished by the production of a single ovule, but in all other diagnostic features it is similar to *Cyclanthera* (D. M. Kearns and E. C. Jones 1992).

Fruits of *Cyclanthera* are usually echinate; only *C. pedata* (Linnaeus) Schrader and two other species lack spinules. Wild relatives of *C. pedata* are spiny-fruited and the occurrence of spinules within the species is variable. *Cyclanthera pedata* is cultivated in Central America and South America for its large, virtually hollow, edible fruits, which are used like bell peppers, mostly to be filled and baked (B. E. Hammel 2009).

SELECTED REFERENCES Jones, C. E. 1969. A Revision of the Genus *Cyclanthera* (Cucurbitaceae). Ph.D. dissertation. Indiana University. Nesom, G. L. 2014. Taxonomy of *Cyclanthera* (Cucurbitaceae) in the USA. Phytoneuron 2014-11: 1–17.

1. Staminate inflorescences (1–)7–27-flowered, usually without lateral branches, floriferous portion 0.2–2.2 cm.
   2. Fruiting peduncles 10–30 mm; staminate corollas 4.2–6.3 mm diam.; anther heads 1.4–2.2(–2.8) mm diam., consistently ciliate . . . . 2. *Cyclanthera naudiniana*
   2. Fruiting peduncles 2–6 mm; staminate corollas 2.5–2.8 mm diam.; anther heads 0.6–0.8 mm diam., glabrous . . . . 3. *Cyclanthera gracillima*
1. Staminate inflorescences (12–)18–90(–130)-flowered, laterally branched, floriferous portion 3–7 or (0.5–)1.2–6 cm.

3. Staminate inflorescences (50–)65–90(–130)-flowered, lateral branches 3–25 mm; terminal leaflet margins coarsely serrate to lobed . . . . 1. *Cyclanthera dissecta*
3. Staminate inflorescences (12–)18–70-flowered, lateral branches 3–7 mm; terminal leaflet margins shallowly to coarsely serrate but not lobed . . . . 4. *Cyclanthera stenura*

1. Cyclanthera dissecta (Torrey & A. Gray) Arnott, J. Bot. (Hooker) 3: 280. 1841 • Cut-leaf or Texas coastal plain cyclanthera [E]

Discanthera dissecta Torrey & A. Gray, Fl. N. Amer. 1: 697. 1840

Stems glabrous except for minutely villosulous nodes; tendrils 2-branched, rarely (shortest tendrils) unbranched. **Leaves** 3-foliolate, lateral pair of leaflets deeply divided, petiolules 3–5 mm, terminal leaflet 3–4.5 cm, blade broadly lanceolate, petiolule 7–15 mm, linear, abruptly broadening into leaflet base, leaflet margins coarsely serrate to shallowly or deeply lobed; petioles 10–21 mm. **Staminate inflorescences** 5–11 cm, floriferous portion 3–7 cm, (50–)65–90(–130)-flowered, paniculate-racemoid, lateral branches 3–25 mm, longest proximally; flowers solitary, in fascicles of 2–4, or along short axes. **Staminate corollas** 3.5–4.9 mm diam. **Anther heads** 0.6–0.8 mm diam., subsessile, glabrous. **Fruiting peduncles** 4–6 mm. **Capsules** narrowly ovoid, slightly oblique, short-beaked, 15–25 mm, spinules 3–4 mm.

Flowering Sep–Nov. Riparian woods and thickets, bottomlands; 20–70 m; Tex.

Cyclanthera dissecta is a rarely collected endemic of the southeastern Texas coastal plain (G. L. Nesom 2014), known only from five counties (Austin, Brazoria, Brazos, Grimes, Victoria). It differs from the plants of central Texas (previously identified as *C. dissecta*, here as *C. naudiniana*) in its smaller staminate flowers and anthers, short fruiting peduncles, and especially in its paniculate-racemoid staminate inflorescences with an abounding number of flowers.

2. Cyclanthera naudiniana Cogniaux in A. L. P. P. de Candolle and C. de Candolle, Monogr. Phan. 3: 831. 1881 • Central Texas cyclanthera [E] [F]

Stems glabrous except for minutely villosulous nodes; tendrils unbranched, less commonly 2-branched. **Leaves** 3-foliolate, lateral pair of leaflets deeply to nearly completely divided (appearing 5-foliolate), petiolules 1–3(–5) mm, terminal leaflet 3–7 cm, blade lanceolate to narrowly or broadly lanceolate or elliptic-lanceolate, petiolule 5–15 mm, narrowly oblanceolate, gradually broadening into leaflet base, leaflet margins coarsely serrate to shallowly or deeply lobed; petioles (5–)15–40 mm. **Staminate inflorescences** 1.2–10.5 cm, floriferous portion 0.3–2 cm, (1–, on shortest axes)8–27-flowered, racemoid, without lateral branches or lateral branches rarely 2 mm; flowers solitary or in fascicles of 2 or 3. **Staminate corollas** 4.2–6.3 mm diam. **Anther heads** 1.4–2.2(–2.8) mm diam., subsessile, consistently ciliate with a ring of short, white hairs arising just inside thecal ring. **Fruiting peduncles** 10–30 mm. **Capsules** ovoid, barely oblique-gibbous or not, short-beaked, 15–25 mm, spinules 3–5 mm.

Flowering May–Oct. Canyons, rocky slopes, streamsides, riparian woods of elm-hackberry, sycamore, willow-cottonwood, juniper, oak-juniper, pinyon-oak-juniper, live oak, roadsides, open woods; 200–1800 m; Colo., Kans., Nebr., N.Mex., Okla., Tex.

Cyclanthera naudiniana is distinct in its short, few-flowered staminate inflorescence, relatively large corollas and anther heads, and relatively long fruiting peduncles. In Texas it is most abundant on the eastern edge of the Edwards Plateau.

3. Cyclanthera gracillima Cogniaux, Diagn. Cucurb. Nouv. 2: 71. 1877 • Slender cyclanthera

Cyclanthera naudiniana Cogniaux var. *tenuifolia* Cogniaux

Stems glabrous; tendrils unbranched, less commonly 2-branched. **Leaves** 3- or 5-foliolate, lateral pair of leaflets completely or nearly completely divided to base, petiolules 1–4 mm, terminal leaflet 3–5 cm, blade lanceolate to broadly lanceolate or elliptic-lanceolate, petiolule 5–10 mm, linear, abruptly broadening into leaflet base, leaflet margins coarsely serrate to shallowly lobed; petioles 5–10 mm. **Staminate inflorescences** 1.5–10.5 cm, floriferous portion 0.2–2.2 cm, 7–26-flowered, racemoid, without lateral branches; flowers solitary or in fascicles of 2–3. **Staminate corollas** 2.5–2.8 mm diam. **Anther heads** 0.6–0.8 mm diam., subsessile, glabrous. **Fruiting peduncles** 2–6 mm. **Capsules** narrowly ovoid, slightly oblique, short-beaked, 12–25 mm, spinules 2–4[–6] mm.

Flowering Aug–Oct. Canyons, arroyos, canyon walls, slopes, streamsides, riparian woods, sycamore, cottonwood-sycamore-willow, sycamore-Arizona cypress-oak, oak-maple-pine, pine-oak-Arizona cypress, pine-juniper-cottonwood, pine-juniper woodlands; 1400–2000(–2500) m; Ariz., N.Mex.; Mexico.

Cyclanthera gracillima is distinct in its short, few-flowered staminate inflorescence and short fruiting peduncles. The few occurrences in southern Arizona and New Mexico are at the northwesternmost tip of its geographic range. From there it continues southward into Chihuahua and Sonora and then apparently is disjunct to a much broader distribution in the states of central and southern Mexico, from Nayarit and Jalisco to the northeast and southeast.

4. **Cyclanthera stenura** G. L. Nesom, Phytoneuron 2014-11: 15, fig. 9. 2014 • Trans-Pecos cyclanthera E

Stems glabrous; tendrils unbranched. **Leaves** 3-foliolate, lateral pair of leaflets deeply lobed, petiolules 2–5 mm, terminal leaflet 3–5 cm, blade narrowly lanceolate, petiolule 3–12 mm, narrowly oblanceolate, gradually broadening into leaflet base, leaflet margins coarsely serrate; petioles 16–28 mm. **Staminate inflorescences** (0.5–)2–12 cm, floriferous portion (0.5–) 1.2–6 cm, (12–)18–70-flowered, narrowly racemoid, with racemose or fasciculate lateral branches 3–7 mm; flowers solitary or in fascicles of 2–3. **Staminate corollas** 3.8–6.3 mm diam. **Anther heads** 0.6–1 mm diam., sessile, glabrous. **Fruiting peduncles** 2–7 mm. **Capsules** ovoid, distinctly oblique-gibbous, short-beaked, (12–) 15–20 mm, spinules (2–)3–5 mm.

Flowering (May–)Aug–Oct. Canyons, rocky slopes, among boulders, igneous soil, roadsides, pinyon-oak-juniper woodlands; 1100–2500 m; Tex.

Cyclanthera stenura is distinct from *C. gracillima* in the more elongate and profusely flowered floriferous portion of its staminate inflorescence and its geography. It occurs in Brewster, Jeff Davis, and Presidio counties of the trans-Pecos region of Texas.

7. **BRANDEGEA** Cogniaux, Proc. Calif. Acad. Sci., ser. 2, 3: 58. 1890 • [For Townshend Stith Brandegee, 1843–1925, California botanist, explorer and collector, civil engineer, topographer]

Plants annual (sometimes short-lived perennial), monoecious, sprawling, trailing, or climbing; stems annual, glabrate; taprooted or roots slender-fibrous; tendrils unbranched. **Leaves**: blade hastate, 4- or 5-angular, or suborbiculate, shallowly to deeply palmately 3(–5)-lobed, lobes triangular or ovate to linear-oblong, central lobe usually longest, margins entire, surfaces eglandular. **Inflorescences**: staminate flowers (1)2(3) in axillary racemes or racemoid panicles; pistillate flowers solitary in same axils as staminate, irregularly produced; peduncles erect at apex; bracts absent. **Flowers**: hypanthium cupulate; sepals 5, barely differentiated as apiculae; petals 5, distinct or nearly so, white, triangular to ovate or narrowly oblong-triangular with acute apices, 1–1.5[–3] mm, glabrous, corolla rotate to shallowly cupulate. **Staminate flowers**: stamens 3–5 (appearing 1–3 from connation); filaments inserted at hypanthium base, connate; thecae connate, forming a head but not fused into ring, horseshoe-shaped, twisted-contorted, connective broad; pistillodes absent. **Pistillate flowers**: ovary 1-locular, broadly fusiform-rostrate; ovules 1(2) per locule; styles 1–3, columnar; stigmas 1, depressed-globose to hemispheric, sometimes 2-lobed; staminodes absent. **Fruits** capsular, light tan, obovoid to suborbicular, gibbous, slightly compressed, beaked, dry, thin-walled, sparsely short-echinate or subaculeate, spinules thick-based, antrorsely upturned, irregularly dehiscent. **Seeds** 1(2), subcylindric-clavate to obdeltoid, compressed, not arillate, margins not differentiated, surface muriculate to warty.

Species 1: sw United States, nw Mexico.

J. N. Rose (1897b) recognized five species of *Brandegea* (without a key to distinguish them). Of the five, *B. minima* (S. Watson) Rose was later transferred to *Cyclanthera* (D. M. Kearns and E. C. Jones 1992); the others have been relegated to synonymy of *B. bigelovii*.

1. Brandegea bigelovii (S. Watson) Cogniaux, Proc. Calif. Acad. Sci., ser. 2, 3: 58. 1890 • Desert starvine F

Elaterium bigelovii S. Watson, Proc. Amer. Acad. Arts 12: 252. 1877; *Brandegea monosperma* (Brandegee) Cogniaux; *B. palmeri* (S. Watson) Rose; *B. parviflora* (S. Watson) S. Watson ex Rose; *Vaseyanthus insularis* (S. Watson) Rose var. *palmeri* (S. Watson) Gentry

Vines sprawling, trailing, or climbing over associated plants, often completely covering them, stems to 2(–4) m. **Leaves:** petiole 8–25 mm; blade 1.5–5(–7) cm, base cordate to convex, lobes broader and shallower on juvenile growth, narrower (to linear-oblong) on distal stems, abaxial surface glabrous, adaxial surface and margins pustulate-scabrous to pustulate-hirsutulous. **Staminate inflorescences** 2–4 cm. **Capsules** 0.5–1.2 cm, beak (3–)4–7 mm. **Seeds** 4–5 mm.

Flowering (Feb–)Mar–Apr(–later sporadically). Desert wash bottoms and banks, cliffs, canyons, moist areas from tank seeps, roadsides, *Larrea* flats, Sonoran desert scrub, upland Sonoran desert scrub, desert riparian zones, commonly with *Acacia, Ambrosia, Atriplex, Cercidium, Larrea, Lycium, Olneya, Parkinsonia, Prosopis*; 20–600(–800) m; Ariz., Calif.; Mexico (Baja California, Baja California Sur, Sonora).

Plants of *Brandegea bigelovii* are cultivated for the beauty of their clusters of white staminate flowers. Garden plants in Tucson flower from February through the summer, except during extreme heat. According to R. S. Felger (2000), *B. bigelovii* is the most common and widespread vining plant in northwestern Sonora.

8. ECHINOCYSTIS Torrey & A. Gray, Fl. N. Amer. 1: 542. 1840, name conserved • Wild or wild mock cucumber, wild balsam apple [Greek *echinos*, hedgehog, and *kystis*, bladder, alluding to prickly, hollow fruits] E

Plants annual, monoecious, climbing or trailing; stems glabrate; taprooted with branching secondary roots or roots slender-fibrous; tendrils 3-branched. **Leaves:** blade depressed-orbiculate to suborbiculate or ovate, palmately (3–)5(–7)-lobed, lobes triangular to oblong-triangular, margins entire or serrulate-cuspidate, surfaces eglandular, glabrous or scabrous. **Inflorescences:** staminate flowers 50–200 in axillary racemes or racemoid panicles; pistillate flowers 1(–3) from same axils as staminate, peduncles erect at apex; bracts absent. **Flowers:** hypanthium shallowly cupulate; sepals (5)6, filiform, almost pricklelike; petals (5)6, connate ¼ length, white to greenish white, narrowly triangular to linear-lanceolate or oblong-lanceolate, 3–6 mm, minutely glandular-puberulent, corolla rotate. **Staminate flowers:** stamens 3; filaments inserted at corolla base, connate, column 0.4 mm; thecae connate, forming subsessile capitate androecium but not fused into single ring, sigmoid-flexuous, connective broad; pistillodes absent. **Pistillate flowers:** ovary 2-locular, subglobose; ovules 2 per locule; styles 1, nearly vestigial; stigmas 3, forming a subglobose head; staminodes absent. **Fruits** pepos, light green to blue-green, globose-ovoid to ovoid or ellipsoid, bladdery-inflated, symmetric, dry, thin-walled, surface moderately to densely echinate, spinules weak, flexible, whitish glaucous with green mottling at bases, dehiscence apically irregularly lacerate, sometimes explosively. **Seeds** 4, broadly oblong-ellipsoid, strongly flattened, not arillate, margins not differentiated, surface slightly pitted-roughened. $x = 32$.

Species 1: North America.

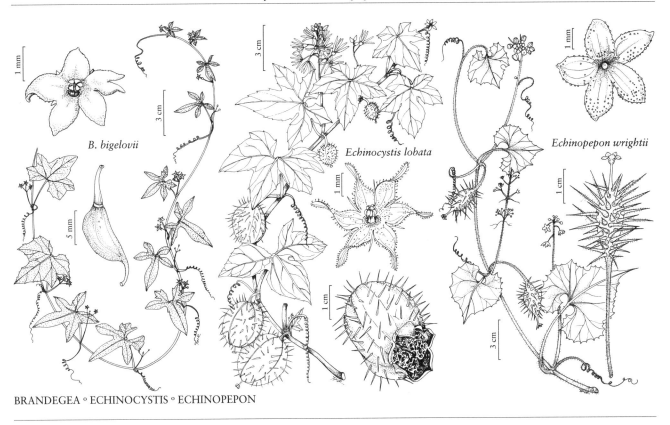

B. bigelovii

Echinocystis lobata

Echinopepon wrightii

BRANDEGEA ∘ ECHINOCYSTIS ∘ ECHINOPEPON

1. Echinocystis lobata (Michaux) Torrey & A. Gray, Fl. N. Amer. 1: 542. 1840 • Concombre grimpant E F

Sicyos lobatus Michaux, Fl. Bor.-Amer. 2: 217. 1803 (as lobata); *Micrampelis lobata* (Michaux) Greene

Leaves: petiole 1–4 cm; blade 2–8(–12) cm, lobe apex acute, sinuses rounded, surfaces glabrous or slightly scabrous, hair bases pustulate. **Inflorescences:** staminate racemes 8–14 cm; pistillate peduncles 2–5 cm. **Flowers** lightly fragrant; corolla 8–12(–16) mm diam. **Pepos** 3–5 cm, spinules 4–6 mm, glabrous or slightly scabrous. **Seeds** 12–20 mm. $2n = 32$.

Flowering May–Sep(–Oct). Bottomland forests and thickets, riparian woods, marshes and marsh edges, thickets in pastures, fencerows, ditches, lake shores, railroad banks, dunes; 0–2000 m; Alta., B.C., Man., N.B., N.S., Ont., P.E.I., Que., Sask.; Ariz., Colo., Conn., Del., D.C., Idaho, Ill., Ind., Iowa, Kans., Ky., Maine, Md., Mass., Mich., Minn., Mo., Mont., Nebr., N.H., N.J., N.Mex., N.Y., N.C., N.Dak., Ohio, Okla., Pa., R.I., S.Dak., Utah, Vt., Va., Wash., W.Va., Wis., Wyo.

Echinocystis lobata is sometimes cultivated in arbors for its showy white flowers (staminate) in long inflorescences. Its occurrence in the western United States is sporadic.

9. ECHINOPEPON Naudin, Ann. Sci. Nat., Bot., sér. 5, 6: 17. 1866 • Wild balsam apple [Greek *echinos*, hedgehog, and *pepon*, melon or pumpkin, alluding to prickly fruits]

Apatzingania Dieterle

Plants annual [perennial], monoecious, climbing; stems annual, hairy [glabrous]; taprooted with branching secondary roots or roots slender-fibrous; tendrils [2]3-branched. **Leaves:** blade reniform to orbiculate, deeply to shallowly palmately 3–5-lobed or dissected, rarely subentire, lobes deltate or ovate-triangular or rounded, margins entire or denticulate [dentate], surfaces hispid to hispidulous, eglandular.   **Inflorescences:** staminate flowers 50–100 in branched or unbranched axillary racemes; pistillate flowers solitary, from same or different axils as staminate; peduncles erect at apex; bracts absent.   **Flowers:** hypanthium campanulate; sepals 5(6), linear to subulate or narrowly triangular [shallowly deltate, shallowly triangular, or widely ovate-deltate]; petals 5[6], connate ¼–⅔ length, white, triangular-ovate or oblong to deltate, [oblong-ovate to narrowly oblong], 3–6[–9] mm, glabrous or sparsely pubescent and eglandular or punctate-glandular to stipitate-glandular, corolla rotate or campanulate.   **Staminate flowers:** stamens (4)5; filaments inserted at base of corolla, connate into column; thecae arcuate or recurved [sigmoid-replicate], connate, forming a head, not fused into ring, connective broadened; pistillodes absent.   **Pistillate flowers:** ovary 2-locular, ovoid; ovules 2–5 per locule; styles 1, inserted at base of hypanthium, columnar; stigmas 1, a subglobose head; staminodes absent.   **Fruits** capsular, tan to brown, obovoid [ovoid or cylindric], symmetric, beaked, dry, thin-walled, densely echinate, dehiscence through apical pores, beak sometimes caducous, calyptralike.   **Seeds** 4–10, ovoid to quadranguloid beyond narrowed base, flattened, not arillate, margins not differentiated, surface corrugate to rugose or tuberculate. $x = 12$.

Species 19 (2 in the flora): sw United States, Mexico, Central America, South America.

The monospecific Mexican genus *Apatzingania* was treated as distinct by R. McVaugh (2001b) and J. C. Rodríguez (1995) and included within *Echinopepon* as *E. arachoideus* (Dieterle) A. K. Monro & Stafford by A. K. Monro and P. J. Stafford (1998).   It is distinct from other species of *Echinopepon* in its geocarpy and unilocular, single-seeded, indehiscent fruit.   Monro and Stafford viewed these morphological features as adaptational consequences of geocarpy and noted that, on the basis of pollen morphology, seed ornamentation, and other features, *E. arachoideus* belongs in the species group that includes *E. coulteri*.

SELECTED REFERENCES   Monro, A. K. and P. J. Stafford. 1998. A synopsis of the genus *Echinopepon* (Cucurbitaceae: Sicyeae), including three new taxa. Ann. Missouri Bot. Gard. 85: 257–272. Rodríguez, J. C. 1995. Distribución geográfica del género *Echinopepon*. Anales Inst. Biol. Univ. Nac. Autón. México, Bot. 66: 171–181.

1. Leaf blades 4–5(–7) cm wide; corollas 8–12 mm diam., petal apices emarginate; capsule surfaces and prickles eglandular-pubescent, prickles mostly 3–5 mm; seeds with elliptical depression on each surface. . . . . .1. *Echinopepon coulteri*
1. Leaf blades 5–8(–15) cm wide; corollas 6–8 mm diam., petal apices acute to slightly obtuse; capsule surfaces and prickles hirsute, hairs stipitate-glandular, prickles 10–20 mm; seeds without depressions on surfaces. . . . . .2. *Echinopepon wrightii*

1. Echinopepon coulteri (A. Gray) Rose, Contr. U.S. Natl. Herb. 5: 116. 1897 • Coulter's wild balsam apple

Elaterium coulteri A. Gray, Smithsonian Contr. Knowl. 5(6): 61. 1853; *Echinopepon confusus* Rose

Stems 0.5–4 m, sparsely short-villous, hairs eglandular.   **Leaf blades** mostly orbiculate, deeply to shallowly lobed, 4–5(–7) cm wide, base broadly cordate, margins entire or slightly sinuate-denticulate, surfaces sparsely hispid abaxially, hispid especially on veins adaxially.   **Inflorescences:** staminate flowers in simple racemes 5–14(–20) cm.   **Flowers:** sepals linear to subulate; petals oblong to deltate, 4–6 mm, apex emarginate, surfaces glabrous or sparsely pubescent, eglandular, not punctate, corolla campanulate, 8–12 mm diam.; anther thecae recurved.   **Capsules** 2–3 × 1.2 cm, surface and prickles eglandular-pubescent, beak 5–8 mm, prickles mostly 3–5 mm.   **Seeds** with elliptical depression on each surface.

Flowering Aug–Sep.   Rocky slopes, washes; 1500–2200 m; N.Mex.; Mexico (Chihuahua, Durango, Zacatecas, s to Oaxaca).

Echinopepon coulteri is known in the flora area from Doña Ana, Grant, Hidalgo, Sierra, and Socorro counties (as cited by K. M. Stocking 1955, from vouchers at CAS, GH, POM, and US).

2. Echinopepon wrightii (A. Gray) S. Watson, Bull.
Torrey Bot. Club 14: 158. 1887 [F]

Elaterium wrightii A. Gray,
Smithsonian Contr. Knowl.
5(6): 61. 1853; *Echinocystis
wrightii* (A. Gray) Cogniaux

Stems 0.5–4 m, hirsute, hairs
stipitate-glandular. **Leaf blades**
reniform to orbiculate, angular
or undulate to shallowly lobed,
5–8(–15) cm wide, base broadly
cordate, margins entire or denticulate, surfaces hispid to
hispidulous. **Inflorescences:** staminate flowers in simple
or compound racemes 6–12 cm. **Flowers:** sepals narrowly
triangular; petals triangular-ovate, 3–4 mm, apex acute
to slightly obtuse, surfaces stipitate-glandular abaxially,
punctate-glandular adaxially, corolla rotate, 6–8 mm
diam.; anther thecae arcuate. **Capsules** tapering to base,
2.5–3 × 1.4–1.8 cm, surface and prickles hirsute, hairs
stipitate-glandular, beak 10 mm, prickles 10–20 mm.
Seeds without depression on surfaces. *2n* = 24.

Flowering Jun–Oct. Canyon bottoms, washes, cliff
bases, desert grasslands, oak woodlands, desert scrub,
riparian scrub and woodlands, roadsides, orchards;
500–1500 m; Ariz., N.Mex., Tex.; Mexico (Chihuahua,
Sonora, Zacatecas).

Records for *Echinopepon wrightii* in the flora area
are from four counties in New Mexico and from seven
counties in Arizona. The species is included here for
Texas on the basis of the statement by D. S. Correll and
M. C. Johnston (1970) that it occurs in extreme west
trans-Pecos Texas; a voucher has not been seen.

The distribution of *Echinopepon wrightii* was
described by A. K. Monro and P. J. Stafford (1998) as
only central Mexico; this obviously was in error. The
inclusion of Zacatecas as part of the range of *E. wrightii*
follows their citation of a collection from that state, the
only one cited by them apart from the type.

10. MARAH Kellogg, Proc. Calif. Acad. Sci. 1: 38. 1854 • Manroot [Hebrew *marah*,
bitter, alluding to taste of all parts]

Plants perennial, monoecious, sometimes temporarily dioecious, trailing or climbing; stems
annual, usually glabrous, sometimes scabrous; roots tuberous, globose to fusiform; tendrils
unbranched or 2- or 3-branched. **Leaves:** bracts absent; blade suborbiculate, shallowly to
deeply palmately 5(–7)-lobed, lobes deltate to oblong-ovate or ovate, margins entire or remotely
and coarsely dentate-lobulate, surfaces eglandular. **Inflorescences:** staminate flowers 5–100 in
axillary panicles or racemoid panicles; pistillate flowers solitary, in same axils as staminate;
peduncles erect at apex; bracts absent. **Flowers:** hypanthium campanulate to cupulate; sepals 5,
linear or filiform to subulate or deltate, sometimes vestigial; petals 5, connate ½ length, usually
white, cream-yellow, or greenish yellow, rarely greenish, triangular to ovate or oblong-ovate, (1–)
2.5–10(–12) mm, glandular-villous adaxially, corolla campanulate, cupulate, or rotate. **Staminate
flowers:** stamens 3(4); filaments inserted near hypanthium base, connate, sometimes vestigial;
thecae not fused into ring, forming a head, arched-flexuous, connective broad; pistillodes absent.
Pistillate flowers: ovary (2–)4(–8)-locular, ovoid to globose; ovules 1–4 per locule; styles 1,
short to nearly vestigial; stigmas 1, discoid to subglobose head; staminodes present or absent.
Fruits capsular, yellowish green to orange-yellow, often green-striped, globose to subglobose or
depressed-globose, short-ellipsoid, ovoid, or oblong, symmetric, often short-beaked, dry, thin-
walled, moderately to sparsely or densely echinate (smooth to sparsely echinate or muriculate in
M. watsonii), irregularly dehiscent by splitting or dropping of beak. **Seeds** 1–16(–24), orbicular
to suborbicular, ellipsoid, oblong, ovoid, obovoid, or oblanceoloid, slightly compressed, or
globose to subglobose, not compressed, not arillate, margins usually not differentiated, slightly
grooved in several species, surface smooth. *x* = 16.

Species 8 (7 in the flora): w North America, nw Mexico.

Marah micrantha Dunn, endemic to Cedros Island of Baja California, is the only species of the genus without at least part of its range in the United States. See comments below under 6. *M. macrocarpa*.

The only consistent distinctions of *Marah* from *Echinocystis* and *Echinopepon* noted by K. M. Stocking (1955) are the hypogeous (versus epigeous) germination of *Marah* seeds and its massively tuberous (versus mostly fibrous) roots. The common name manroot alludes to the large tubers, which in *M. macrocarpa* can weigh more than 160 kg.

R. A. Schlising (1993) noted without documentation that presumed hybrids occur where species overlap, but K. M. Stocking (1955b) found no evidence of hybridization between species of *Marah* in nature or in experimental gardens.

A remarkable adaptation in seedling establishment of *Marah* to the Mediterranean climate of California was described by R. A. Schlising (1969). In seed germination, in the cooler and wetter months of November and December, the radicle and epicotyl are carried 5–25 cm down into the soil by the elongating fused bases (the petioles) of the fleshy hypogeal cotyledons. As they elongate, the fused cotyledon bases form a tube that carries the embryonic axis at its tip. From that point, already deep in the soil, the radicle grows downward while the epicotyl grows upward through the petiole tube. The epicotyl reaches the soil surface by early spring, completes the first season's growth, and dries up by the beginning of the arid summer season. Concurrently with epicotyl growth, the hypocotyl begins to enlarge, forming a tuber. The subterranean cotyledons transfer nutrients to the tuber, which produces shoots in following seasons.

SELECTED REFERENCE Stocking, K. M. 1955b. Some taxonomic and ecological considerations of the genus *Marah*. Madroño 13: 113–137.

1. Capsules globose to subglobose or depressed-globose to short-ellipsoid.
   2. Corollas deeply cupulate to campanulate; leaf blades glaucous abaxially; capsules smooth to sparsely echinate or muriculate, spinules weak, 1–2 mm . . . . 3. *Marah watsonii*
   2. Corollas rotate; leaf blades not glaucous; capsules sparsely to densely echinate, spinules rigid or flexuous, 2–12 mm.
      3. Corollas yellowish green to cream-yellow or (especially inland) white; capsules sparsely to densely echinate, spinules 4–12 mm . . . . 1. *Marah fabacea*
      3. Corollas white; capsules densely echinate, spinules 2–3(–5) mm . . . . 2. *Marah gilensis*
1. Capsules usually oblong, ellipsoid, or ovoid.
   4. Sepals (pistillate) 6–7 mm, linear; seeds 2–4 . . . . 4. *Marah guadalupensis*
   4. Sepals (pistillate) 0.4–1 mm, filiform to subulate or deltate, sometimes vestigial; seeds 3–6 or 4–20(–24).
      5. Capsules 4–6.5(–8) cm, surface sparsely to moderately echinate (usually smooth distally), spinules weak, flexible, 3–6 mm; seeds 3–6 . . . . 5. *Marah oregana*
      5. Capsules (5–)8–15(–20) cm, surface densely echinate, spinules rigid, 5–35 mm; seeds 4–20(–24).
         6. Corollas shallowly cupulate to rotate; seeds not flat at one end . . . . 6. *Marah macrocarpa*
         6. Corollas deeply cupulate to campanulate or campanulate-rotate; seeds flat at one end . . . . 7. *Marah horrida*

1. **Marah fabacea** (Naudin) Greene, Leafl. Bot. Observ. Crit. 2: 36. 1910 • California manroot E

Echinocystis fabacea Naudin, Ann. Sci. Nat., Bot., sér. 4, 12: 154, plate 9. 1859; *E. fabacea* var. *inermis* (Congdon) Jepson; *E. inermis* Congdon; *E. scabrida* Eastwood; *Marah fabacea* var. *agrestis* (Greene) Stocking; *M. inermis* (Congdon) Dunn; *Micrampelis fabacea* (Naudin) Greene var. *agrestis* Greene

Leaf blades shallowly 5–7-lobed, 5–10 cm wide, surfaces not glaucous. **Flowers:** sepals (pistillate) vestigial; petals 3–5 mm (pistillate) or 1.5–2.5 mm (staminate), corolla yellowish green to cream-yellow or (especially inland) white, rotate; staminodes absent in pistillate flowers. **Capsules** yellowish green at maturity, globose, 4–5 cm, surface sparsely to densely echinate, spinules rigid or flexible, 4–12 mm. **Seeds** 1–4, oblong-ovoid, ± compressed, 15–20 mm. $2n = 32$.

Flowering Feb–May. Streamsides, washes, coastal strand, rock outcrops, cliff bases, ledges, grasslands, chaparral, oak woodlands, riparian woodlands, open hillsides, roadsides, powerline cuts; 20–1400 m; Calif., Nev.

Marah fabaceae var. *fabacea* was mapped by K. M. Stocking (1955b) as confined to near-coastal localities centering around San Francisco Bay, from Marin to Monterey counties; he recognized var. *agrestis* as the more widely distributed expression of the species. In his view, var. *fabacea* is characterized by fruits with longer (6–12 mm) and rigid spinules and by seeds more numerous (usually four) and commonly laterally flattened. R. A. Schlising (1993) noted that var. *agrestis* intergrades more or less completely with plants identifiable as var. *fabacea* and did not recognize varieties.

2. **Marah gilensis** (Greene) Greene, Leafl. Bot. Observ. Crit. 2: 36. 1910 • Gila manroot

Megarrhiza gilensis Greene, Bull. Torrey Bot. Club 8: 97. 1881

Leaf blades deeply 5–7-lobed, 4–10 cm wide, surfaces not glaucous. **Flowers:** sepals (pistillate) vestigial; petals 3–4 mm (pistillate) or 3–3.5 mm (staminate), corolla white, rotate; staminodes present but very inconspicuous in pistillate flowers. **Capsules** yellowish green, globose to subglobose or short-ellipsoid, 2.5–3 cm, surface densely echinate, spinules rigid, 2–3(–5) mm. **Seeds** 2–4, ovoid to suborbicular, slightly compressed, 10–12 mm. $2n = 30$.

Flowering (Dec–)Feb–Apr. Desert flats and washes, stream beds, canyon bottoms and slopes, desert scrub, chaparral, oak woodlands, riparian zones; (500–)600–1500 m; Ariz., N.Mex.; Mexico (Sonora).

3. **Marah watsonii** (Cogniaux) Greene, Leafl. Bot. Observ. Crit. 2: 36. 1910 • Taw manroot E

Echinocystis watsonii Cogniaux in A. L. P. P. de Candolle and C. de Candolle, Monogr. Phan. 3: 819. 1881

Leaf blades deeply 5-lobed, 3–8 cm wide, surfaces glaucous abaxially. **Flowers:** sepals (pistillate) linear-subulate, 1 mm; petals 4–5 mm (pistillate) or 2.5–3 mm (staminate), corolla white, deeply cupulate to campanulate; staminodes present in pistillate flowers. **Capsules** often striped dark green, globose or depressed-globose to short-ellipsoid, 2–3.5 cm, surface smooth to sparsely echinate or muriculate, spinules weak, 1–2 mm. **Seeds** 1–4, globose, not compressed, 11–14 mm.

Flowering Mar–May(–Jun). Oak and pine-oak grasslands, oak woodlands, yellow pine woodlands, chaparral, rocky slopes and outcrops, roadsides; 100–800(–1300) m; Calif.

Marah watsonii is distributed in about 15 counties in north-central California.

4. **Marah guadalupensis** (S. Watson) Greene, Leafl. Bot. Observ. Crit. 2: 36. 1910 • Island manroot

Megarrhiza guadalupensis S. Watson, Proc. Amer. Acad. Arts 11: 115, 138. 1876; *Marah macrocarpa* (Greene) Greene var. *major* (Dunn) Stocking; *M. major* Dunn

Leaf blades deeply 5-lobed, (10–)15–20(–25) cm wide. **Flowers:** sepals (pistillate) linear, 6–7 mm; petals 8–10 mm (pistillate) or 6–7 mm (staminate), corolla white, shallowly cupulate; staminodes present in pistillate flowers. **Capsules** yellowish green at maturity, ovoid, beaked, 5–7 cm, surface sparsely hirsute-echinate, spinules weak, 1–3 mm. **Seeds** 2–4, orbicular to ellipsoid or oblong, compressed, (22–)28–33 mm.

Flowering Feb–Apr. Washes, stream edges, grassy slopes, beach bluffs, roadsides; 0–200(–500) m; Calif.; Mexico (Baja California).

The nomenclatural type of *Marah guadalupensis* is from Guadalupe Island off the coast of Baja California; the main cluster of populations is on the Channel Islands of California (Santa Barbara County).

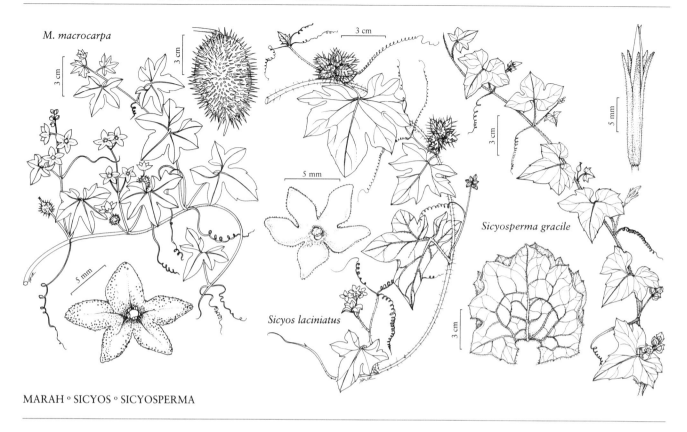

M. macrocarpa

Sicyos laciniatus

Sicyosperma gracile

MARAH ° SICYOS ° SICYOSPERMA

5. Marah oregana (Torrey & A. Gray) Howell, Fl. N.W. Amer., 239. 1898 • Coastal manroot [E]

Sicyos oreganus Torrey & A. Gray, Fl. N. Amer. 1: 542. 1840; *Echinocystis oregana* (Torrey & A. Gray) Cogniaux; *Megarrhiza oregana* (Torrey & A. Gray) S. Watson

Leaf blades usually shallowly, sometimes deeply 5–7-lobed, 8–20(–35) cm wide. **Flowers:** sepals (pistillate) deltate to subulate or filiform, 1 mm; petals 8–10 mm (pistillate) or 5–6 mm (staminate), corolla usually white to cream, rarely greenish, cupulate to cupulate-rotate; staminodes present in pistillate flowers. **Capsules** usually striped dark green at maturity, short-ellipsoid to ovoid, tapered to beak, 4–6.5(–8) cm, surface sparsely to moderately echinate (usually smooth distally), spinules weak, flexible, 3–6 mm. **Seeds** 3–6, orbicular to ellipsoid, compressed, 16–22 mm. $2n = 32$.

Flowering (Feb–)Mar–Jun. Roadsides, disturbed areas, clearings, fields, dunes, open hillsides, streamsides, meadows, thickets, Douglas fir and redwood forests; 0–800(–2000) m; B.C.; Calif., Oreg., Wash.

6. Marah macrocarpa (Greene) Greene, Leafl. Bot. Observ. Crit. 2: 36. 1910 • Large-fruited manroot [F]

Echinocystis macrocarpa Greene, Bull. Calif. Acad. Sci. 1: 188. 1885

Leaf blades deeply 5(–7)-lobed, 5–30 cm wide. **Flowers:** sepals (pistillate) deltate, 0.4–0.6 mm, sometimes vestigial; petals (1–)3–10(–12) mm (pistillate) or 5–8(–10) mm (staminate), corolla white, shallowly cupulate to rotate; staminodes scalelike or absent in pistillate flowers. **Capsules** yellowish green at maturity, short-ellipsoid to broadly ovoid, usually rounded at both ends, sometimes sharply beaked, (5–)8–12 cm, surface densely echinate, spinules rigid, 5–30 mm. **Seeds** 4–20 (–24), usually obovoid to oblong-ellipsoid, sometimes subglobose, not flat at one end, slightly compressed, 15–20 mm. $2n = 32, 64$.

Flowering (Jan–)Mar–May. Pinyon-juniper woodlands, Joshua tree-pinyon transition zones, coastal sage, chaparral, oak woodlands, rocky hillsides, riparian woods and thickets, stream bottoms, disturbed sites, roadsides; 0–1500(–2100) m; Calif.; Mexico (Baja California, Baja California Sur).

K. M. Stocking (1955b) enlarged *Marah macrocarpa* to include *M. micrantha* Dunn [as *M. macrocarpa* var. *micrantha* (Dunn) Stocking], which is known only from

Cedros Island off the Pacific coast of Baja California, outside of Vizcaíno Bay. But the relatively small flowers and seeds of the latter [staminate flowers 3–6(–8) mm diam. versus 8–13 mm diam.; seeds 12–13 mm versus 15–20 mm] and its apparent geographical disjunction suggest that treatment of *M. micrantha* at specific rank is justified.

7. Marah horrida (Congdon) Dunn, Bull. Misc. Inform. Kew 1913: 151. 1913 (as horridus)

• Sierran manroot [E]

Echinocystis horrida Congdon, Erythea 7: 184. 1900

Leaf blades deeply 5–7-lobed, 10–15 cm wide. **Flowers:** sepals (pistillate) filiform, 1 mm; petals 7–9 mm (pistillate) or 5–7 mm (staminate), corolla white, deeply cupulate to campanulate or campanulate-rotate; staminodes present in pistillate flowers. **Capsules** yellowish green to orange-yellow at maturity, short-ellipsoid to oblong, 9–15(–20) cm, surface densely echinate, spinules rigid, 5–35 mm. **Seeds** 6–16(–24), oblong to oblanceoloid or ovoid, flat at one end, slightly compressed, 26–32 mm.

Flowering Feb–Apr(–May). Canyons, open ridges, shrubby and open areas, pine, oak, and pine-oak woodlands, stream sides, roadsides; 300–1200(–1600) m; Calif.

Marah horrida is distributed from Los Angeles County to El Dorado County in the California Sierra Madre.

11. SICYOS Linnaeus, Sp. Pl. 2: 1013. 1753; Gen. Pl. ed. 5, 443. 1754 • Bur cucumber [Greek *sikyos*, cucumber or gourd]

Plants annual, monoecious, climbing or trailing; stems glabrous or hairy, often viscid-pubescent when young; roots fibrous; tendrils 2–5-branched from a common point. **Leaves** sessile or subsessile to petiolate; blade ovate or orbiculate to suborbiculate or reniform, deeply to shallowly palmately 3–5-angular-lobed, lobes triangular to deltate, margins usually serrate to denticulate, surfaces eglandular. **Inflorescences:** staminate flowers 3–22(–34) in axillary racemes or panicles; pistillate flowers 4–16, sessile to subsessile in umbelliform clusters at peduncle apex, from same axils as staminate, peduncles erect at apex; bracts absent. **Flowers:** hypanthium cupulate to shallowly campanulate; sepals 5, ovate to deltate or subulate, linear, linear-triangular, or narrowly triangular; petals 5, connate ¼–½ length, white to greenish white, yellowish green, or yellow, triangular to lanceolate or narrowly lanceolate, 0.5–1.5 mm, glabrous abaxially, often glandular adaxially, corolla campanulate to cupulate. **Staminate flowers:** stamens (2)3(–5); filaments inserted at base of hypanthium, connate 1 mm; thecae connate into head, horseshoe-shaped, connective slightly broadened; pistillodes absent. **Pistillate flowers:** ovary 1-locular, ovoid-fusiform; ovules 1 per locule; styles 1, narrow; stigmas 1, obscurely 2- or 3-lobed; staminodes absent. **Fruits** pepos, dark green to gray at maturity, fusiform to ovoid, beaked or not, dry, thin-walled, echinate or spinulose, usually also with shorter hairs, rarely glabrous, indehiscent. **Seeds** 1, ovoid, compressed, not arillate, margins not differentiated but sometimes with 2 small swellings at base, surface smooth. $x = 12$.

Species ca. 50 (4 in the flora): North America, Mexico, Central America, South America, Pacific Islands, Australia; introduced in Europe, e Asia.

Attribution of *Sicyos parviflorus* A. Gray ex Naudin to the United States has been based on misapplication of the name. The type of *S. parviflorus* was collected from the vicinity of Mexico City, and the species occurs from central Mexico through Central America into South America (R. Lira 2001).

SELECTED REFERENCE Nesom, G. L. 2011. Taxonomy of *Sicyos* (Cucurbitaceae) in the USA. Phytoneuron 2011-15: 1–11.

1. Pepos ovoid-beaked, 9–15 mm; stigmas 3-lobed; e, c North America, including Texas . . . . 1. *Sicyos angulatus*
1. Pepos ovoid, 4–8 mm; stigmas 2- or 3-lobed; Arizona, New Mexico, Texas.
   2. Pepos not echinate. . . . . 2. *Sicyos glaber*
   2. Pepos echinate.
      3. Staminate inflorescences 10–16-flowered, peduncle plus floral axis 40–140 mm; stigmas 3-lobed; mature stems glabrate to sparsely minutely stipitate-glandular; leaf blades deeply lobed, sinuses (1/3–)1/2–2/3 to base, margins not dentate, evenly sharply indurate-apiculate, proximal pair of lateral veins divergent from edge of basal sinus . . . . 3. *Sicyos microphyllus*
      3. Staminate inflorescences 3–10-flowered, peduncle plus floral axis 3–25 mm; stigmas 2-lobed; mature stems glabrescent but remaining villous and stipitate-glandular; leaf blades shallowly lobed to angulate, sinuses 1/5–1/4 to base, margins evenly and shallowly dentate, teeth sharply indurate-apiculate, proximal pair of lateral veins closely bordering edge of basal sinus. . . . . 4. *Sicyos laciniatus*

1. **Sicyos angulatus** Linnaeus, Sp. Pl. 2: 1013. 1753 (as angulata) • One-seed bur or star cucumber, nimble-Kate, sicyos anguleux E

Stems moderately to densely villous-puberulent, hairs glandular-viscid, mixed with stipitate-glandular hairs. **Leaves:** petiole 1–7(–10) cm; blade orbiculate-angulate to broadly ovate-angulate or shallowly (3–)5-lobed, 4–12 × 6–17 cm, terminal lobe deltate-acuminate to ovate-acuminate, basal sinus narrow to broad, margins evenly and minutely green-apiculate-mucronulate, ± ciliate, hairs gland-tipped, surfaces hispidulous-hirsute; proximal pair of lateral veins divergent from edge of basal sinus. **Inflorescences:** staminate 10–21(–34)-flowered, peduncle plus floral axis 30–220 mm; pistillate 8–16-flowered, peduncle 20–50 mm. **Flowers:** staminate: corolla white to greenish white, 4–5 mm, stamens prominently exserted; pistillate: sepals not foliaceous, linear to linear-triangular, 0.5–1 mm, corolla 1–2 mm (essentially without a tube); stigmas 3-lobed. **Pepos** ovoid-beaked, 9–15 mm, echinate, spinules retrorsely barbellate, also densely arachnoid-villous to minutely villosulous, hairs often gland-tipped.

Flowering Jun–Oct(–Nov). Stream banks, alluvial floodplains, disturbed woods, thickets, clearings, vacant lots, fallow fields, railroad banks; 0–300 m; Ont., Que.; Ala., Ark., Conn., Del., D.C., Fla., Ga., Ill., Ind., Iowa, Kans., Ky., La., Maine, Md., Mass., Mich., Minn., Miss., Mo., Nebr., N.H., N.J., N.Y., N.C., N.Dak., Ohio, Okla., Pa., R.I., S.C., S.Dak., Tenn., Tex., Vt., Va., W.Va., Wis.; introduced in Europe (Austria, Czech Republic, England, Finland, France, Hungary, Italy, Russia, Spain), e Asia (Japan, South Korea), Pacific Islands (Taiwan).

In Texas, at the southwestern corner of its range, *Sicyos angulatus* occurs in a narrow band along the eastern margin of the Edwards Plateau, continuing northward into Oklahoma, with a broad gap through eastern Texas and eastward to central Louisiana. The peculiar distribution in Texas appears to be real. Fruits in Texas may average slightly smaller, but there appears to be no other difference between the Texas plants and those elsewhere in the range.

2. **Sicyos glaber** Wooton, Bull. Torrey Bot. Club 25: 310. 1898 • Smooth bur cucumber E

Stems sparsely villous with flattened hairs mixed with glandular-puberulent vestiture, becoming sparsely hirsutulous to glabrous. **Leaves:** petiole 1–3(–4.5) cm; blade pentagonal-angulate to reniform-angulate or shallowly 5-angulate, sinuses 1/4–1/3 to base, (3–)4–7 × 5–9 cm, terminal lobe deltate-acuminate, base concave to semicircular, margins evenly subfoliaceous-apiculate, ciliate, surfaces hirsute-hispidulous; proximal pair of lateral veins closely parallel to slightly divergent from edge of basal sinus. **Inflorescences:** staminate 6–22-flowered, peduncle plus floral axis 15–100(–140) mm; pistillate 4–10-flowered, peduncle 5–8 mm. **Flowers:**

staminate: corolla greenish yellow to yellow, 2–4 mm, stamens slightly exserted; pistillate: sepals not foliaceous, mostly narrowly triangular, 0.5–1.5 mm, corolla 2–3 mm including tube; stigmas 2-lobed. **Pepos** ovoid, 4–6 mm, not echinate, glabrous or sparsely villosulous.

Flowering Aug–Sep. Canyon bottoms, open slopes, talus, dry drainages, sycamore, juniper-mountain mahogany woodlands, chaparral; 1800–2400 m; N.Mex., Tex.

The distribution of *Sicyos glaber* in the United States suggests that it also occurs at least in Chihuahua, Mexico, but apparently it has not been documented from there.

3. **Sicyos microphyllus** Kunth in A. von Humboldt et al., Nov. Gen. Sp. 2(fol.): 95; 2(qto.): 119. 1817 • Little-leaf burr cucumber

Sicyos deppei G. Don

Stems: young sparsely puberulent, stipitate-glandular, often also hirsute-villous, hairs viscid, strongly to weakly glabrescent, becoming glabrate to sparsely minutely stipitate-glandular. **Leaves:** petiole (1–) 2–4(–7) cm; blade orbiculate-pentagonal to deeply 5-lobed, sinuses (⅓–)½–⅔ to base, 4–10 × (4–)6–12 cm, terminal lobe deltate-acuminate to oblong-oblanceolate or obtrullate, basal sinus narrow to broad, margins not dentate, sometimes with few shallow, serrate teeth, evenly sharply indurate-apiculate, barely scabrous, not ciliate, surfaces hirsute to hispid-hirsute; proximal pair of lateral veins divergent from edge of basal sinus. **Inflorescences:** staminate 10–16-flowered, peduncle plus floral axis 40–140 mm; pistillate 4–10-flowered, peduncle (0–)1–5(–20) mm. **Flowers:** staminate: corolla greenish white, 2–3 mm, stamens not exserted; pistillate: sepals not foliaceous, mostly triangular, 0.2–2 mm, corolla 1–2 mm (essentially without a tube), stigmas 3-lobed. **Pepos** ovoid, 5–8 mm, echinate and sparsely, finely villous, spinules retrorsely barbellate.

Flowering Sep–Oct. Rocky slopes and ridges, cliff bases, streamsides, yellow pine woodlands, thickets; 2200–2600 m; N.Mex., Tex.; Mexico (Chihuahua, Coahuila, Durango, Nuevo León, San Luis Potosí, Sinaloa, Sonora, Zacatecas).

4. **Sicyos laciniatus** Linnaeus, Sp. Pl. 2: 1013. 1753 (as laciniata) • Cut-leaf burr cucumber F

Sicyos ampelophyllus Wooton & Standley; *S. laciniatus* var. *subintegrus* Cogniaux

Stems densely villous and stipitate-glandular, glabrescent, remaining pubescent but not glandular. **Leaves:** petiole (1.5–) 3–6(–8) cm; blade broadly ovate-angulate to reniform-angulate or shallowly 5-lobed, sinuses ⅕–¼ to base, 3–5(–8) × 4–11(–13) cm, terminal lobe shallowly deltate to shallowly triangular or widely ovate, acuminate to acute, basal sinus broad-curved to nearly squared, margins evenly and shallowly dentate, teeth sharply indurate-apiculate, surfaces hirsutulous-hispidulous, sometimes abaxially viscid-puberulent and sessile-glandular when very young, glandularity deciduous by maturity; proximal pair of lateral veins closely bordering edge of basal sinus. **Inflorescences:** staminate 3–10-flowered, peduncle plus floral axis 3–25 mm; pistillate 4–12-flowered, peduncle 2–4(–11) mm. **Flowers:** staminate: corolla white, 2 mm, stamens slightly exserted; pistillate: sepals not foliaceous, linear to linear-triangular, 0.2–1.5 mm, corolla 1.5–2 mm including tube, stigmas 2-lobed. **Pepos** ovoid, 5–6 mm, echinate, spinules retrorsely barbellate and glabrous.

Flowering Aug–Oct. Streamsides, flood plains, riparian woods, cottonwood-willow, juniper, oak, pine-oak woods, canyons, rocky slopes, cliff faces and bases; (1000–)1200–2400(–3000) m; Ariz., N.Mex., Tex.; Mexico (Chihuahua, Coahuila, Durango, Nuevo León, San Luis Potosí, Sinaloa, Sonora, Tamaulipas, Zacatecas).

12. SICYOSPERMA A. Gray, Smithsonian Contr. Knowl. 5(6): 62. 1853 • Climbing arrowheads [Greek *sikyos*, cucumber or gourd, and *sperma*, seed, alluding to fruits resembling seeds]

Plants annual, monoecious, climbing; stems glabrous or sparsely puberulent; roots fibrous; tendrils 2-branched. **Leaves:** blade broadly to shallowly triangular, usually shallowly palmately 3(–5)-lobed to hastate, lobes triangular to deltate, margins entire or denticulate, surfaces eglandular. **Inflorescences:** staminate flowers 20–40 in axillary, compound racemes; pistillate flowers solitary, from same axils as staminate; bracts deltate, base cordate, margins loosely enclosing flowers, irregularly sinuate-dentate to incised.   **Flowers:** hypanthium cupulate; sepals 5, filiform-subulate; petals 5, connate ⅓–½ length, white, triangular-lanceolate, 1–1.5 mm, stipitate-glandular along margins, without other hairs, corollas salverform. **Staminate flowers:** stamens 3; filaments inserted at hypanthium base, connate into central column; thecae connate, forming head, sigmoid, connective slightly broadened; pistillodes absent. **Pistillate flowers:** ovary 1-locular, ovoid; ovules 1 per locule; styles 1, narrowly columnar; stigmas 1, capitate, 3-lobed; staminodes absent. **Fruits** pepos, whitish, ellipsoid-ovoid, compressed, dry at maturity, smooth, glabrous, indehiscent. **Seeds** 1, ovoid, compressed, not arillate, margins not differentiated, surface smooth.

Species 1: sw United States, nw Mexico.

1. Sicyosperma gracile A. Gray, Smithsonian Contr. Knowl. 5(6): 62. 1853 [F]

Leaf blades 2–7 cm wide, lobe apex rounded to acute. **Inflorescences:** each flower loosely enclosed within 2 bracts; peduncle hispid, hairs retrorsely curved. **Flowers:** petals apically 2-fid, corolla 2–3 mm diam. **Pepos** 4–5 mm, loosely enclosed by persistent bracts.

Flowering Aug–Oct(–Nov).   North slopes, cliff bases, canyon bottoms, stream sides, riparian scrub, willow-ash, cottonwood-willow, hackberry-acacia, oak woodlands; 1000–1700 m; Ariz., N.Mex.; Mexico (Chihuahua, Sonora).

Sicyosperma gracile is recognized by its annual duration, minute salverform corollas with white, apically bifid petals, each flower enclosed within a pair of basally cordate bracts, and relatively small, 1-seeded, dry, indehiscent, unarmed pepos loosely enclosed by bracts. Specimens sometimes show only staminate inflorescences, but apparently this is an artifact of collection rather than an indication that the plant is unisexual.

13. APODANTHERA   Arnott, J. Bot. (Hooker) 3: 274. 1841   •   Melon loco [Greek *a-*, without, *podos*, foot, and *antheros*, blooming, alluding to sessile anthers]

Plants perennial, monoecious [dioecious], prostrate and trailing; stems annual, strigose; roots tuberous; tendrils unbranched or 2- or 3-branched. **Leaves:** blade reniform to orbiculate-cordate, unlobed or shallowly palmately 5-lobed [3–5-foliolate], lobes rounded, margins dentate, often undulate-crisped, surfaces eglandular. **Inflorescences:** staminate flowers (1)2–5 in racemes [corymbs] from proximal axils; pistillate flowers 5–12 in fascicles in distal axils [solitary]; bracts filiform-subulate or absent. **Flowers:** hypanthium subcylindric to narrowly funnelform; sepals 5, linear-lanceolate; petals 5, distinct, yellow, oblong-obovate to elliptic-oblanceolate, [ovate or ovate-lanceolate], [1–]16–25(–35) mm, glabrate, corolla broadly funnelform-campanulate to rotate. **Staminate flowers:** stamens 3; filaments inserted near hypanthium rim, distinct, nearly vestigial; thecae distinct, oblong to suborbiculate, connective narrow; pistillodes absent. **Pistillate flowers:** ovary 3-locular, ovoid to oblong; ovules ca. 3–35 per locule; styles 1, short-columnar to nearly absent; stigmas 3, linear; staminodes absent. **Fruits** pepos, silvery green to green with darker, raised, broad, longitudinal stripes, subglobose to depressed-globose [ovoid or ellipsoid], smooth or ribbed, indehiscent. **Seeds** [8–]30–80[–100], ellipsoid-obovoid [ovoid to obovoid or broadly ellipsoid], biconvex [compressed], not arillate, margins a broad, flat, light-colored band, surface smooth. $x = 14$.

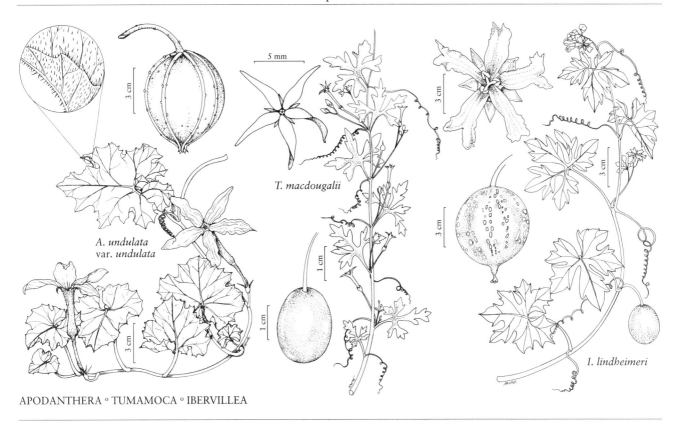

T. macdougalii

A. undulata
var. *undulata*

I. lindheimeri

APODANTHERA ∘ TUMAMOCA ∘ IBERVILLEA

Species ca. 19 (1 in the flora): sw United States, Mexico, Central America, South America.

Plants of *Apodanthera* were reported to be gender-diphasic, switching from staminate to pistillate (V. A. Delesalle 1989). *Apodanthera herrerae* Harms, a species of Andean South America, is grown for its edible tubers.

1. **Apodanthera undulata** A. Gray, Smithsonian Contr. Knowl. 5(6): 60. 1853 F

Varieties 2 (1 in the flora): sw, sc United States, Mexico.

Apodanthera undulata in the flora area is recognized by its reniform, undulate-margined, and white-backed leaves and yellow petals. Variety *australis* McVaugh (*A. aspera* Cogniaux) differs in reduced vesture and pistillate inflorescence; it occurs in Mexico (Aguascalientes, Guanajuato, Hidalgo, Jalisco, Michoacán, Nayarit, Oaxaca, Querétaro, San Luis Potosí, Zacatecas). According to McVaugh, the two varieties are allopatric.

1a. **Apodanthera undulata** A. Gray var. **undulata** F

Vines fetid; branches to 3 m, harshly strigose; tendrils usually branched from 1–5 cm beyond base. **Leaves:** petiole 3–7 cm; blade 8–15 cm wide, surfaces moderately to densely white-sericeous abaxially, glabrous adaxially. **Inflorescences:** staminate flowers in long-pedunculate racemes; pistillate flowers in loose fascicles; peduncle erect, 1–2 cm in flower, 6–9 cm in fruit. **Flowers:** hypanthium 10–18 mm; sepals erect, not recurving. **Pepos** 6–10 cm. **Seeds** light brown, 9–12 mm. **2***n* = 28.

Flowering May–Sep. Roadsides, gravelly stream terraces, interfan alluvium, dunes, flats, rocky slopes, lava beds, grasslands, creosote scrub, Chihuahuan desert scrub of *Flourensia-Larrea*, mesquite, oak chaparral, oak-juniper woodlands; 800–1700 m; Ariz., N.Mex., Tex.; Mexico (Chihuahua, Coahuila, Durango, Sonora).

14. TUMAMOCA Rose, Contr. U.S. Natl. Herb. 16: 21, plate 17. 1912 • Globeberry [Local Native American *Tumamoc*, name for the hill upon which the Carnegie Institute Desert Laboratory is located]

Plants perennial, monoecious and dioecious (developmentally, seasonally), trailing or climbing; stems perennial proximally, annual distally, woody, glabrous [hispid-hirsute or hispidulous]; roots tuberous; tendrils unbranched. **Leaves:** petiolar region an extension of blade midportion; blade ovate to orbiculate, deeply 3-lobed, lobes 1- or 2-lobed, ultimate segments narrow, margins entire, surfaces eglandular at base. **Inflorescences:** staminate flowers 2–19 in axillary, subsessile racemes; pistillate flowers solitary, from same axils as staminate; bracts absent. **Flowers:** hypanthium narrowly tubular-funnelform; sepals 5, deltate to triangular; petals 5, distinct, pale yellow to greenish yellow, narrowly triangular to linear-lanceolate, valvate in bud, 4–6 mm, apex entire, glabrous, corolla salverform. **Staminate flowers:** stamens 3; filaments inserted at hypanthium rim, distinct, nearly vestigial; thecae distinct, oblong, connective narrow, each with glabrous, narrow, exserted, terminal appendage nearly as long as corolla; pistillodes absent. **Pistillate flowers:** ovary 3-locular, ellipsoid; ovules ca. 2–6 per locule; styles 1, narrowly columnar; stigmas 3, coiled; staminodes 3. **Fruits** berrylike, red or yellow, globose, 0.8–1 cm, smooth, glabrous, irregularly dehiscent. **Seeds** 8–20, obovoid to ovoid, compressed, arillate, margins obscure, surface tuberculate-rugose.

Species 2 (1 in the flora): Arizona, n Mexico.

Tumamoca is closely related to *Ibervillea*. Plants of both genera produce orange to red fruits with prominently margined, red-arillate seeds. *Tumamoca* is distinct in its narrowly funnelform hypanthium (versus narrowly campanulate to cylindric in *Ibervillea*), three staminodes (versus five), entire petals (versus apices bifid), interior corolla surfaces glabrous (versus densely pubescent), valvate buds (versus buds with infolded apices), and seeds with roughened surfaces and obscure margins (versus corky-pleated surfaces and raised margins).

Tumamoca mucronata Kearns apparently is known only from the type locality in northwestern Zacatecas, Mexico.

SELECTED REFERENCE Kearns, D. M. 1994. A revision of *Tumamoca* (Cucurbitaceae). Madroño 41: 23–29.

1. Tumamoca macdougalii Rose, Contr. U.S. Natl. Herb. 16: 21, plate 17 [as macdougali]. 1912 • Tumamoc globeberry F

Stems: persistent woody portion 10–15 cm. **Leaf blades** 3–4.5 cm, ultimate segments linear to linear-lanceolate or narrowly ovate, 2–4 cm × 2–10 mm, surfaces hispidulous to pustulate-scabrous abaxially, glabrous adaxially. **Inflorescences:** staminate 2–19-flowered, 3.5–10 cm. **Flowers:** staminate: hypanthium 5–9 mm, petals 4–4.5 mm; pistillate: petals 5–6 mm. **Fruits** 0.8–1 cm diam. **Seeds** 7–8 mm.

Flowering Jun–Oct. Semidesert grasslands, sandy washes and gullies, *Atriplex* flats, Sonoran desert scrub with *Agave, Atriplex, Cercidium, Fouquieria, Jatropha, Lycium, Maytenus, Opuntia, Prosopis,* and *Stenocereus*; 700–1000 m; Ariz.; Mexico (Sonora).

In *Tumamoca macdougalii*, some plants produce all pistillate flowers, some all staminate flowers (dioecy), and some produce both pistillate and staminate (monoecy). Most plants apparently start the early summer as all staminate but only buds are produced, thus the staminate flowers quickly respond if rains come early. Pistillate flowers are not produced until the rains come in season, and then plants may be bisexual. The size of the tuber and the age of the plant may modify this schedule. Smaller plants may not produce pistillate flowers; larger plants may produce all pistillate flowers. (Biological information from Frank Reichenbacher, pers. comm.).

Tumamoca macdouglaii is in the Center for Plant Conservation's National Collection of Endangered Plants.

15. IBERVILLEA Greene, Erythea 3: 75. 1895 • Globeberry [Derivation unknown]

Maximowiczia Cogniaux in A. L. P. P. de Candolle and C. de Candolle, Monogr. Phan. 3: 726. 1881, not Ruprecht 1856 [Magnoliaceae]; *Dieterlea* E. J. Lott

Plants perennial, dioecious, trailing and climbing, often fetid; stems annual or perennial, glabrous [scabridulous]; roots tuberous; tendrils unbranched. **Leaves:** blade ovate to suborbiculate, broadly ovate, or reniform, usually palmately 5-lobed or pedately 3-lobed, rarely unlobed, lobes narrowly oblong to linear or cuneate to flabellate or rhombic-ovate, margins lobulate or coarsely toothed [denticulate], surfaces eglandular. **Inflorescences:** staminate flowers solitary or 4–12(–24) in axillary racemes, racemoid panicles, or corymboid clusters, rarely solitary; pistillate flowers solitary; bracts absent. **Flowers:** hypanthium narrowly campanulate to cylindric [obtriangular]; sepals 5, broadly deltate to triangular; petals 5, distinct, greenish yellow to yellow [whitish], narrowly oblong to linear-oblong, clawed, 3–7 mm, apex shallowly 2-lobed, infolded in bud, densely glandular-pubescent adaxially, corolla salverform. **Staminate flowers:** stamens 3; filaments inserted near hypanthium rim, distinct, nearly vestigial; thecae distinct [weakly connate in *I. fusiformis* and *I. guatemalensis*], linear-oblong, without terminal appendage, connective narrow; pistillodes absent. **Pistillate flowers:** ovary 3(–5)-locular, ellipsoid to fusiform; ovules 2–35 per locule; styles 1, narrowly columnar; stigmas 3, coiled; staminodes absent or present. **Fruits** berrylike, dark orange to red, globose to ellipsoid, smooth, glabrous, irregularly dehiscent. **Seeds** (2–)6–18[–100], irregularly ovoid, compressed, red-arillate, margins prominent, raised, surface corky-pleated [smooth]. $x = 12$.

Species 6 (2 in the flora): sw United States, Mexico, Central America (Guatemala).

Ibervillea was expanded by D. M. Kearns (1994) to include the monospecific Mexican genus *Dieterlea*—as *Ibervillea fusiformis* (E. J. Lott) Kearns. This species had previously been segregated on the basis of its perennial stems, larger, white, nocturnal flowers, connate anthers, four or five placentae, four or five bilobed stigmas, and five staminodes; with incorporation of other species from Mexico and Guatemala, wider variation in *Ibervillea* includes the features of *Dieterlea*.

SELECTED REFERENCE Kearns, D. M. 1994. The genus *Ibervillea* (Cucurbitaceae): An enumeration of the species and two new combinations. Madroño 41: 13–22.

1. Leaf lobes 10–25 mm wide; petals 5–7 mm; fruits 2–3.5 cm . . . . 1. *Ibervillea lindheimeri*
1. Leaf lobes 2–5 mm wide; petals 3–4 mm; fruits 1–1.5 cm . . . . 2. *Ibervillea tenuisecta*

1. Ibervillea lindheimeri (A. Gray) Greene, Erythea 3: 75. 1895 • Lindheimer's globeberry F

Sicydium lindheimeri A. Gray, Boston J. Nat. Hist. 6: 194. 1850; *Ibervillea tenella* (Naudin) Small; *I. tripartita* (Naudin) Greene; *Maximowiczia lindheimeri* (A. Gray) Cogniaux

Stems to 3 m, not glaucous. **Leaves:** petiole 1–3.5 cm; blade usually deeply palmately 5-lobed or pedately 3-lobed, rarely ± unlobed, 6–12 cm wide, often slightly succulent, lobes cuneate to flabellate or rhombic-ovate, 10–25 mm wide, surfaces glabrous or sparsely scabridulous. **Inflorescences:** staminate flowers usually 4–12(–24) in racemes or terminal corymboid clusters 1–3 cm, rarely solitary. **Petals** 5–7 mm. **Fruits** dark orange to red, globose to ellipsoid, 2–3.5 cm. **Seeds** 6–14(–18), 4–6 mm. $2n = 22, 24$.

Flowering Apr–Jul. Thickets, dry woodlands, fencerows, dunes and mudflats, prairies, mesquite, thorn shrublands, hackberry woodlands, scrub oak, oak-juniper, juniper, and oak-hickory woodlands; 0–400(–900) m; N.Mex., Okla., Tex.; Mexico (Coahuila, San Luis Potosí).

2. Ibervillea tenuisecta (A. Gray) Small, Fl. S.E. U.S., 1136. 1903 • Deer-apples

Sicydium lindheimeri A. Gray var. *tenuisectum* A. Gray, Smithsonian Contr. Knowl. 3(5): 75. 1852; *Ibervillea lindheimeri* (A. Gray) Greene var. *tenuisecta* (A. Gray) M. C. Johnston

Stems to 1 m, glaucous. **Leaves:** petiole 1–2 cm; blade deeply pedately 3-lobed, 2–6 cm wide, slightly succulent, lobes narrowly oblong to linear, 2–5 mm wide, surfaces glabrous. **Inflorescences:** staminate flowers 4–7(–20) in racemes 0.5–1.5 cm. **Petals** 3–4 mm. **Fruits** bright red, globose, 1–1.5 cm. **Seeds** (2–)5–9(–12), 5–6 mm.

Flowering Jun–Sep. Rocky hills, draws, arroyos, flats, creosote-blackbrush, Chihuahuan desert scrub, commonly with *Acacia, Chilopsis, Fallugia, Flourensia, Larrea, Opuntia, Parthenium,* and *Prosopis,* mesquite-oak woodlands; (300–)900–1600 m; Ariz., N.Mex., Tex.; Mexico (Chihuahua, Coahuila, Nuevo León, San Luis Potosí, Sonora, Tamaulipas, Zacatecas).

16. CUCUMIS Linnaeus, Sp. Pl. 2: 1011. 1753; Gen. Pl. ed. 5, 442. 1754 • Melon, cucumber [Latin name for cucumber, probably from *curvus*, crooked, alluding to fruit shape] ☐

Cucumella Chiovenda; *Dicoelospermum* C. B. Clarke; *Melo* Miller; *Mukia* Arnott; *Myrmecosicyos* C. Jeffrey; *Oreosyce* Hooker f.

Plants annual [perennial], usually monoecious, rarely dioecious (*C. anguria*) or andromonoecious (*C. melo*), procumbent, trailing, or climbing; stems annual, hispid to hirsute; roots woody or thin [rarely tuberous]; tendrils usually unbranched, rarely unbranched and 2-branched or absent. **Leaves:** blade ovate, palmately 3–6[7]-lobed or unlobed, lobes triangular or elliptic, oblong, ovate, obovate, or spatulate, margins serrate or entire, surfaces eglandular. **Inflorescences:** staminate flowers solitary or 2–10(–18) in axillary racemes, fascicles, or panicles; pistillate flowers usually solitary, rarely 2 or 3 in fascicles, from different axils than staminate, peduncles straight in fruit; bracts absent [rarely present]. **Flowers** unisexual (rarely gynecandrous in *C. metuliferus*); hypanthium infundibular to campanulate, 2–10 mm; sepals (4)5, straight, linear to oblong or narrowly to broadly triangular; petals 5, distinct or basally connate, yellow, elliptic to ovate or obovate, 2–25[–37] mm, glabrous or pubescent, corolla campanulate to infundibular (constricted distally into a neck). **Staminate flowers:** stamens 3; filaments inserted near hypanthium base, distinct; thecae distinct, oblong, replicate (usually 2-folded) [straight to arcuate], connective broadened; pistillodes absent. **Pistillate flowers:** ovary 3–5-locular, ovoid to ellipsoid; ovules 15–150 per locule; styles 1, short-columnar; stigmas 1, sublobate to 3-lobed, lobes sometimes with 1–9 fingerlike projections on margins; staminodes present or rarely absent. **Fruits** pepos, mono- or bicolor, green to yellow, orange, or tan [white, brown, purple] with light or dark green, white, yellow, brown, or orange [purple] longitudinal stripes, usually ellipsoid, globose, or cylindric, rarely obovoid [ovoid, rarely spindle-shaped], smooth, glabrous, surfaces densely echinate, or aculeate to muricate, sometimes netted, warty, scaly, or ridged, indehiscent. **Seeds** 50–500, ellipsoid to ovoid, compressed, not arillate, margins not differentiated, surface smooth. $x = 12$.

Species ca. 55 (6 in the flora): introduced; Asia (China, India, Indonesia, Malaysia), Africa, Atlantic Islands (Cape Verde Islands), Indian Ocean Islands (Aldabra, Madagascar, Seychelles), Pacific Islands (Philippines), Australia; introduced also nearly worldwide.

Two *Cucumis* species are major commercial vegetable crops—*C. melo*, the melons (cantaloupe, honeydew, muskmelon, sugar melon, winter melon) and *C. sativus*, the garden cucumber. *Cucumis anguria* (the West Indian gherkin) and *C. metuliferus* (the kiwano) also are used commercially.

Most *Cucumis* plants in the flora area are escapes from gardens, trash dumps, or picnic discards and do not persist long outside cultivation. The dudaim melon (*C. melo*), paddy melon (*C. myriocarpus*), and a few others possibly have the potential to become aggressively invasive.

C. Jeffrey (1980b) recognized *Cucumis* subg. *Cucumis* (comprising two species, *C. sativus* and the closely similar *C. hystrix* Chakravarty) and subg. *Melo* (Miller) C. Jeffrey. Infrageneric classification has been modified by L. W. D. van Raamsdonk et al. (1989) and J. H. Kirkbride (1993), and based on insights from molecular data, H. Schaefer (2007) divided the genus into two subgenera, subg. *Cucumis* (50 species in five sections) and subg. *Humifructus* H. Schaefer (two species).

C. Jeffrey (1990) circumscribed the subtribe Cucumerinae (tribe Melothriae) to include *Cucumis* and five smaller genera—*Cucumella*, *Dicoelospermum*, *Mukia*, *Myrmecosicyos*, and *Oreosyce*. Molecular studies (A. G. Ghebretinsae et al. 2007; H. Schaefer 2007) indicate that *Cucumis* is paraphyletic without the 19 species of these five genera, and Schaefer formally broadened *Cucumis* to include them.

An Asian origin has been demonstrated for both *Cucumis melo* and *C. sativus* by P. Sebastian et al. (2010), using data from plastid and nuclear DNA. The wild progenitor of *C. melo* occurs in India, and its sister species is *C. picrocarpus* F. Mueller from Australia. The closest relative of *C. sativus* is *C. hystrix* Chakravarty from the eastern Himalayas. The Asian/Australian *Cucumis* clade comprises at least 25 species.

The sister genus to *Cucumis* is *Muellerargia* Cogniaux, with one species in Madagascar (*M. jeffreyana* Keraudren) and one in Indonesia and Australia (*M. timorensis* Cogniaux). Both species are herbaceous trailers or climbers with softly aculeate fruits. The next closest relatives are in African/Asian clades that include the genera *Coccinia* Wight & Arnott, *Neoachmandra* de Wilde & Duyfjes, *Peponium* Engler, and *Zehneria* Endlicher (S. S. Renner and H. Schaefer 2008).

Cucumis sativus (subg. *Cucumis*) is the only species of the genus reported to have a chromosome number of $2n = 14$; the only other species of subg. *Cucumis* (fide J. H. Kirkbride 1993), *C. hystrix*, has been reported to have a number of $2n = 24$ (Chen J. F. et al. 1997). All other species with numbers known have $2n = 24$ or a polyploid level based on $x = 12$. Dysploid numbers of $2n = 20$ and $2n = 22$ have been reported from *C. melo*, which otherwise has $2n = 24$. The chromosome number of *Muellerargia* has not been reported.

Most of the 52 species of *Cucumis* in the sense of Jeffrey are monoecious perennials, and this combination of traits may be the ancestral condition from which an annual habit and dioecy evolved several times. The evolution of smooth fruits from spiny fruits, a traditional key difference in *Cucumis*, and the mode of fruit opening also are much more evolutionarily labile than formerly thought (for details see S. S. Renner et al. 2007).

The monograph by J. H. Kirkbride (1993) has been a primary source in preparing this treatment.

SELECTED REFERENCES Chung, S. M., J. E. Staub, and J. F. Chen. 2006. Molecular phylogeny of *Cucumis* species as revealed by consensus chloroplast SSR marker length and sequence variation. Genome 49: 219–229. Ghebretinsae, A. G., M. Thulin, and J. C. Barber. 2007. Relationships of cucumbers and melons unraveled: Molecular phylogenetics of *Cucumis* and related genera (Benincaseae, Cucurbitaceae). Amer. J. Bot. 94: 1256–1266. Kirkbride, J. H. 1993. Biosystematic Monograph of the Genus *Cucumis* (Cucurbitaceae): Botanical Identification of Cucumbers and Melons. Boone, N.C. Naudin, C. V. 1859. Essais d'une monographie des espèces et des variétés du genre *Cucumis*. Ann. Sci. Nat., Bot., sér. 4, 11: 5–87. Raamsdonk, L. W. D. van, A. P. M. Nijs, and M. C. Jongerius. 1989. Meiotic analyses of *Cucumis* hybrids and an evolutionary evaluation of the genus *Cucumis*. Pl. Syst. Evol. 163: 133–146. Renner, S. S. and H. Schaefer. 2008. Phylogenetics of *Cucumis* (Cucurbitaceae) as understood in 2008. In: M. Pitrat, ed. 2008. Cucurbitaceae 2008: Proceedings of the IXth EUCARPIA meeting on Genetics and

Breeding of Cucurbitaceae, Avignon (France), May 21–24th, 2008. Avignon. Pp. 53–58. Renner, S. S., H. Schaefer, and A. Kocyan. 2007. Phylogenetics of *Cucumis* (Cucurbitaceae): Cucumber (*C. sativus*) belongs in an Asian/Australian clade far from melon (*C. melo*). B. M. C. Evol. Biol. 7: 58. Schaefer, H. 2007. *Cucumis* (Cucurbitaceae) must include *Cucumella, Dicoelospermum, Mukia, Myrmecosicyos,* and *Oreosyce*: A recircumscription based on nuclear and plastid DNA data. Blumea 52: 165–177. Sebastian, P. et al. 2010. Cucumber (*Cucumis sativus*) and melon (*C. melo*) have numerous wild relatives in Asia and Australia, and the sister species of melon is from Australia. Proc. Natl. Acad. Sci. U.S.A. 107: 14269–14273.

1. Pepos netted, ridged, scaly, smooth, or warty.
  2. Leaf lobes triangular; staminate flowers: corolla tube 3.5–5(–6) mm; pistillate flowers: corolla tube 3.5–6.5 mm . . . . 1. *Cucumis sativus*
  2. Leaf lobes elliptic or oblong to ovate; staminate flowers: corolla tube 0.8–2 mm; pistillate flowers: corolla tube 1–1.6(–3) mm . . . . 2. *Cucumis melo*
1. Pepos aculeate, echinate, or muricate (sometimes obscurely in *C. anguria*).
  3. Plants dioecious; pedicels of pistillate flowers distally dilated; staminate flowers 3–10 in racemes . . . . 3. *Cucumis anguria*
  3. Plants monoecious; pedicels of pistillate flowers not distally dilated; staminate flowers 1–7, usually in fascicles or panicles, rarely racemes.
    4. Tendrils antrorsely strigose; petioles proximally retrorsely strigose, medially hirsute, distally antrorsely strigose . . . . 4. *Cucumis myriocarpus*
    4. Tendrils glabrate to hispid or proximally hispidulous, distally glabrous; petioles setose or weakly hispidulous to hispid.
      5. Petioles setose; pistillate flowers: calyx lobes 2–3 mm, corolla lobes sparsely puberulent inside; pepo surface muricate to aculeate; rootstock woody . . . 5. *Cucumis metuliferus*
      5. Petioles weakly hispidulous to hispid; pistillate flowers: calyx lobes 5–6(–11) mm, corolla lobes glabrous inside; pepo surface densely echinate; rootstock not woody . . . . 6. *Cucumis dipsaceus*

1. **Cucumis sativus** Linnaeus, Sp. Pl. 2: 1012. 1753
  • Garden cucumber ☐

Varieties 2 (1 in the flora): introduced; Asia; introduced also nearly worldwide.

Records of cucumber cultivation appear in France in the ninth century, in England in the fourteenth century, and in North America by the mid sixteenth century.

Variety *hardwickii* (Royle) Gabaev (= *Cucumis hardwickii* Royle) occurs in the northwestern Himalayas, the Eastern Ghats and Western Ghats, and the central plateau region of India. J. H. Kirkbride (1993) and C. Jeffrey (1980c, 2001) observed that it can be identified in its usual form but that morphological intergradation argues against its recognition as a distinct evolutionary entity with formal taxonomic status. W. J. J. O. de Wilde and B. E. E. Duyfjes (2010) agreed that *C. hardwickii* is not sharply demarcated from feral forms of *C. sativus* and they reduced its rank to forma within *C. sativus*. On the other hand, *C. hardwickii* is isozymically distinct from *C. sativus*. L. D. Knerr and J. E. Staub (1991) found that it possesses alleles not present in the remainder of their samples of *C. sativus*. V. Meglic et al. (1996) found it to be isozymically distinct from 750 *C. sativus* samples. Without consistent morphological differentiation between *C. hardwickii* and the highly variable domesticated and feral forms of *C. sativus* in the narrow sense, but with an apparent molecular distinction, their recognition at varietal rank seems appropriate. Rank of forma (as interpreted here) would imply that *C. hardwickii* is a populational variant.

1a. **Cucumis sativus** Linnaeus var. **sativus** ☐

Plants: roots thin, without thick, woody rootstock. **Tendrils** hispidulous or antrorsely strigose. **Leaves:** petiole hispid to retrorsely strigose; blade broadly ovate, palmately 5- or 6-lobed, 6–15(–40) × 6–15(–35) cm, length 1–1.2 times width, base cordate, lobes triangular, margins serrate. **Inflorescences:** pedicels of pistillate flowers and fruits cylindric; staminate flowers (1–)3–7 (–10) in fascicles, corolla tube 3.5–5(–6) mm; pistillate flowers: calyx lobes 3–4(–12) mm, petals 10–15(–25) mm, corolla tube 3.5–6.5 mm, glabrous inside. **Pepos** green to yellowish monocolor or with light green to whitish stripes, ellipsoid to cylindric, 5–20(–25) × 3–5 (–12) cm, smooth (ovary and developing fruits weakly aculeate), flesh whitish. **2n** = 14.

Flowering Aug–Nov. Gardens, fields, stream banks, around ponds, trash heaps, sometimes volunteering from past plantings; 10–300(–1500) m; introduced; Ont.; Ariz., Ark., Fla., Ga., Ill., Kans., Ky., La., Mass., Mich., Miss., Mo., N.Y., N.C., Ohio., Pa., S.C., Utah, Va.; Asia (e Himalayas); introduced also in West Indies, Pacific Islands (Hawaii).

2. Cucumis melo Linnaeus, Sp. Pl. 2: 1011. 1753

• Cantaloupe, honeydew, muskmelon [F] [I]

Plants monoecious or andromonoecious, roots thin or with thick, woody rootstock. **Tendrils** sparsely hispid. **Leaves:** petiole hispid to hispidulous or retrorsely strigose; blade broadly ovate, unlobed or palmately 3–5-lobed, 2–14(–26) × 2–15 (–26) cm, length 0.8–1.3 times width, base cordate, lobes elliptic or oblong to ovate, margins entire or weakly serrate. **Inflorescences:** pedicels of pistillate flowers and fruits cylindric; staminate flowers 2–7(–18) in fascicles or panicles, corolla tube 0.8–2 mm; pistillate flowers: calyx lobes 1.5–3(–8) mm, petals 3.5–9(–20) mm, corolla tube 1–1.6(–3) mm, lobes puberulent or glabrous inside. **Pepos** tan, yellowish, lemon yellow, light yellowish green, dark green, light orange, monocolor or orange or brown striped, ellipsoid to subglobose [globose, cylindric, ovoid, or obovoid], 2–12(–100) × 2.5–20+ cm, surface netted, warty, scaly, ridged, or smooth, flesh orange to yellowish, green, or whitish. $2n = 24$.

Subspecies 2 (2 in the flora): introduced; Asia (India); introduced also nearly worldwide.

Only two formal infraspecific taxa within *Cucumis melo* were recognized by J. H. Kirkbride (1993; subsp. *melo* and subsp. *agrestis*), who suggested that other variants should be treated with horticultural names, and classification of melons into two subspecies has been generally supported by molecular phylogenies (A. Stepansky et al. 1999; D. S. Decker et al. 2002).

Wide extremes of variation and horticultural selections exist within *Cucumis melo*, especially as to fruit characters (for example, size, shape, surface features, color, texture, taste, composition). *Cucumis melo* includes feral, wild, and cultivated forms, including dessert melons as well as non-sweet forms that are consumed raw, pickled, or cooked. This has led to a proliferation of names for the variants and various systems of infraspecific classification have been proposed.

A widely used system proposed by C. V. Naudin (1859), dividing *Cucumis melo* into a single wild variety (var. *agrestis*) and six cultivated ones (vars. *cantalupensis, conomon, dudaim, flexuosus, inodorus,* and *momordica*), has been modified and simplified (for example, H. M. Munger and R. W. Robinson 1991) as well as extended into a detailed hierarchical system (I. Grebenscikov 1953; see comments by K. Hammer et al. 1986). An overview of infraspecific nomenclature in *C. melo* was provided by M. Pitrat et al. (2000). This was largely repeated, but adapted to the International Code of Nomenclature for Cultivated Plants, by Y. Burger et al. (2010), who proposed five cultivar groups

in subsp. *agrestis* and 11 cultivar groups in subsp. *melo*; molecular data have not supported their apportionment of the groups among the two subspecies.

Most collections of *Cucumis melo* from the flora area have not been identified to infraspecific rank; for a more accurate assessment of the distribution of subsp. *melo* versus subsp. *agrestis*, all should be reexamined. Presumably, cultivars of each subspecies might be encountered outside of cultivation. As an overview of the diversity of forms and guide to principal names used for melons, a relatively simple taxonomic system is presented here, summarizing the main cultivar groups in each subspecies.

SELECTED REFERENCES Burger, Y. et al. 2010. Genetic diversity of *Cucumis melo*. Hort. Rev. 36: 165–198. Decker, D. S. et al. 2002. The origin and genetic affinities of wild populations of melon (*Cucumis melo*, Cucurbitaceae) in North America. Pl. Syst. Evol. 233: 183–197. Grebenscikov, I. 1953. Die Entwicklung der Melonsystematik. Kulturpflanze 1: 121–138. Hammer, K., P. Hanelt, and P. Perrino. 1986. Carosello and the taxonomy of *Cucumis melo* L. especially of its vegetable races. Kulturpflanze 34: 249–259. Munger, H. M. and R. W. Robinson. 1991. Nomenclature of *Cucumis melo* L. Rep. Cucurbit Genet. Coop. 14: 43–44. Pitrat, M. et al. 2000. Some comments on infraspecific classification of cultivars of melon. Acta Hort. 510: 29–36. Silberstein, L. et al. 1999. Molecular variation in melon (*Cucumis melo* L.) as revealed by RFLP and RAPD markers. Sci. Hort. 79: 101–111. Stepansky, A., I. Kovalski, and R. Perl-Treves. 1999. Intraspecific classification of melons (*Cucumis melo* L.) in view of their phenotypic and molecular variation. Pl. Syst. Evol. 217: 313–332.

1. Hypanthia pilose to lanate, hairs spreading; stems glabrous or sparsely villous; pepos 10–20+ cm diam.. . . . . 2a. *Cucumis melo* subsp. *melo*
1. Hypanthia retrorsely or antrorsely sericeous, hairs appressed; stems retrorsely hispid; pepos 2.5–5 cm diam. . . . . . . 2b. *Cucumis melo* subsp. *agrestis*

2a. Cucumis melo Linnaeus subsp. melo [I]

Plants andromonoecious. **Stems** glabrous or sparsely villous. **Hypanthia** pilose to lanate, hairs spreading. **Pepos** edible, broadly ellipsoid to subglobose, 10–20+ cm diam., rind tan, yellowish, lemon yellow, light yellowish green, dark green, or light orange, smooth, ridged, netted, warty, or scaly; flesh orange to yellowish or green, juicy, sweet.

Flowering Jul–Oct. Gardens, fields, field edges, trash heaps, garbage dumps, lake shores, roadsides, ballast, sometimes volunteering from past plantings; 20–300 m; introduced; Ont.; Ala., Ark., Calif., Conn., Fla., Ga., Ill., Ky., La., Mass., Mich., Miss., Mo., Nev., N.H., N.Mex., N.Y., N.C., Ohio, Okla., Pa., R.I., S.C., Tex., Utah, Va., W.Va.; w Asia; w Africa; introduced also in West Indies.

Plants of subsp. *melo* outside of cultivation in the flora area mostly are the dessert melons; var. *cantalupo* and var. *inodorus*.

Main cultivar groups of subsp. *melo*:

a. *Cucumis melo* var. *cantalupo* Seringe. Asia; pepos medium-large, rind smooth, scaly, or netted, variable in color, aromatic with sweet, juicy flesh, dessert melons; andromonoecious. Formally a nomenclatural synonym of *C. melo* var. *melo*; the name var. *cantalupo* Seringe predates var. *cantalupensis* Naudin, which is often used for this dessert melon.

b. *Cucumis melo* var. *inodorus* Jacquin. Europe (Spain), Asia; pepos large, winter melons, with non-aromatic, non-climacteric and long-storing fruits, rind thick, smooth, wrinkled, or warty, dessert melons (honeydew, winter melon); andromonoecious.

c. *Cucumis melo* var. *melo*. Asia, Africa; pepos large, flesh sweet, rind netted, warty, or scaly, wild form of dessert melons, andromonoecious.

2b. Cucumis melo Linnaeus subsp. **agrestis** (Naudin) Pangalo in P. M. Zhukovsky, Turquie Agric., 534. 1933 • Gulf Coast melon F I

Cucumis melo var. *agrestis* Naudin, Ann. Sci. Nat., Bot., sér. 4, 11: 73. 1859

Plants monoecious or andromonoecious. **Stems** retrorsely hispid. **Hypanthia** retrorsely or antrorsely sericeous, hairs appressed. **Pepos** not edible, broadly ellipsoid to subglobose, 2.5–5 cm diam., rind yellow or yellowish, sometimes orange or brown striped, smooth; flesh whitish.

Flowering Jul–Oct(–Nov). Stream banks, lake shores, marsh edges, hardwood bottomlands, stream beds, sand bars, levees, spoil banks, ditch banks, dunes, clear-cut woods, fallow fields, pastures, soybean, rice, and sugarcane fields, railroad banks, abandoned homesites, vacant lots, roadsides and medians, fencerows, trash dumps, other disturbed areas; 30–100(–1200) m; introduced; Ont.; Ariz., Ark., Calif., Conn., Fla., Ga., La., Miss., N.C., Okla., Tenn., Tex.; c, e Asia; introduced also in Mexico, West Indies, Central America, South America, Europe, se Asia (Indonesia), Africa, Pacific Islands (Marquesas Islands).

Wild populations (subsp. *agrestis*) of *Cucumis melo* in the flora area commonly have been assumed to represent escaped forms of cultivated var. *chito* or var. *dudaim* or, less commonly, of var. *agrestis*. D. S. Decker et al. (2002) showed that populations in the southeastern United States are morphologically and molecularly distinct and appropriately recognized as var. *texanus* (Arkansas, Florida, Mississippi, Texas; Mexico). Variety *texanus* shows the greatest genetic affinities to var. *chito* and

to cultivars from eastern Asia. The wild or cultivated progenitor of subsp. *agrestis* was presumably brought to the Western Hemisphere by humans intentionally (for example, by Asian immigrants) or unintentionally (for example, as seeds mixed with those of other introduced crops), either in pre-Columbian times (given the distinctiveness of genetic differentiation) or post-Columbian.

Variety *dudaim* is encountered outside of cultivation in California, along waterways and irrigation canals, fields, and roadsides. A collection from Sycamore Canyon in Santa Cruz County, Arizona, may be var. *dudaim*: *Thorne 59482* (ARIZ, ASU), described as an apparently naturalized vine on the canyon bottom.

Main cultivar groups of subsp. *agrestis*:

a. *Cucumis melo* var. *agrestis* Naudin. Africa and Asia as weeds; fruits less than 5 cm, inedible with thin mesocarp and tiny seeds; monoecious.

b. *Cucumis melo* var. *chito* (C. Morren) Naudin. Asia, or reportedly of American feral origin (A. Stepansky et al. 1999); fruits plum-sized, aromatic, used as pickles; monoecious. Combined with var. *dudaim* by H. M. Munger and R. W. Robinson (1991).

c. *Cucumis melo* var. *conomon* (Thunberg) Makino. Eastern Asia cultivars; fruits smooth, white-fleshed, with thin rinds, often eaten as pickles; andromonoecious. Includes var. *acidulus* Naudin.

d. *Cucumis melo* var. *dudaim* (Linnaeus) Naudin. Asia (Iran), grown as ornamentals (Dudaim melon, Queen Anne's pocket melon, smellmelon) or for edible fruits; fruits small, aromatic, red or brown striped; andromonoecious.

e. *Cucumis melo* var. *flexuosus* (Linnaeus) Naudin. Middle East, Asia; fruits elongated, non-sweet, eaten immature as cucumbers; usually monoecious; variable in hypanthium vestiture; molecular data place it in subsp. *agrestis*.

f. *Cucumis melo* var. *momordica* (Roxburgh) Cogniaux. Asia (India); pepos large, non-sweet, with thin rind splitting; monoecious.

g. *Cucumis melo* var. *texanus* Naudin. North America; fruits small, rind smooth, yellow; monoecious.

3. Cucumis anguria Linnaeus, Sp. Pl. 2: 1011. 1753 • West Indian or bur gherkin I

Plants dioecious, roots thin, without thick, woody rootstock. **Tendrils** setose. **Leaves:** petiole hispid to setose; blade broadly to narrowly ovate, deeply palmately 3–5-lobed to weakly 3-lobed or nearly unlobed, [sometimes pedately lobed], 3–12(–15) × 2.5–12(–15) cm, length 0.9–1(–2) times width, base cordate, lobes broadly elliptic to obovate, spatulate, or ovate, margins

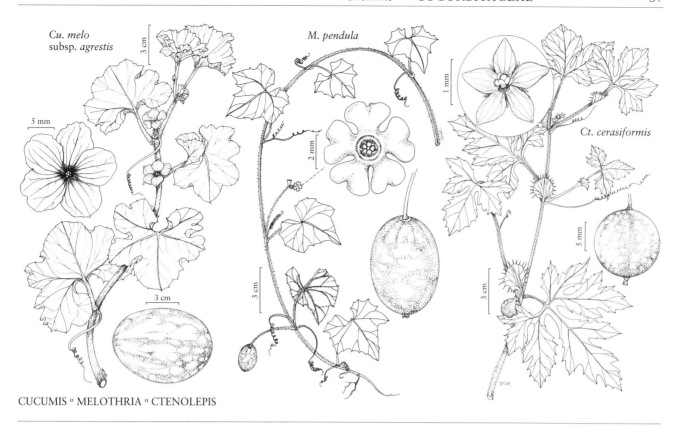

Cu. melo subsp. *agrestis*

M. pendula

Ct. cerasiformis

CUCUMIS ∘ MELOTHRIA ∘ CTENOLEPIS

serrate or entire. **Inflorescences:** pedicels of pistillate flowers distally dilated; staminate flowers 3–10 in racemes, pedunculate or sessile; pistillate flowers: calyx lobes 2–4 mm, petals 6–8 mm, corolla tube 0–1 mm, sparsely puberulent inside. **Pepos** light yellowish green to light yellow, sometimes with light green stripes, usually ellipsoid, rarely obovoid, 2–7 × 1.5–4 cm, surfaces ± aculeate at maturity, flesh greenish white to yellowish green. $2n = 24$.

Varieties 2 (2 in the flora): introduced; Africa, Atlantic Islands (Cape Verde Islands); introduced also in Mexico, West Indies, Central America, South America, Pacific Islands (Marquesas Islands), Australia.

1. Pepos obscurely aculeate, aculei 1–2 mm; leaf blades deeply palmately 3–5-lobed . . . . 3a. *Cucumis anguria* var. *anguria*
1. Pepos prominently aculeate, aculei 4–10(–15) mm; leaf blades deeply palmately 3–5-lobed to weakly 3-lobed or nearly unlobed . . . . 3b. *Cucumis anguria* var. *longaculeatus*

3a. Cucumis anguria Linnaeus var. **anguria** [I]

Leaf blades deeply palmately 3–5-lobed. **Pistillate flowers:** hyaline bristle at apex of hypanthial aculei 2–3 times longer than opaque base. **Pepos** obscurely aculeate, aculei 1–2 mm.

Flowering Jul–Oct. Pastures, old fields, fencerows, roadsides, lake shores, hammocks; 20–500 m; introduced; Calif., Ga., Mass., N.Y.; Africa; Atlantic Islands (Cape Verde Islands); introduced also in Mexico, West Indies, Central America, South America, Pacific Islands (Marquesas Islands), Australia.

3b. Cucumis anguria Linnaeus var. **longaculeatus**
    J. H. Kirkbride, Biosyst. Monogr. Cucumis, 39.
    1993 [I]

Leaf blades deeply palmately 3–5-lobed to weakly 3-lobed or nearly unlobed. **Pistillate flowers:** hyaline bristle at apex of hypanthial aculei 0.5–1 times as long as opaque base. **Pepos** prominently aculeate, aculei 4–10(–15) mm.

Flowering Jul–Oct. Disturbed areas; 0–100(–500) m; introduced; Ala., Fla., Ga., Mont., Tex.; Africa; introduced also in West Indies.

The gherkins are escaped in a few, scattered localities in the United States, mostly in Florida, and it is not even clear that they are naturalized as reproductive populations.

4. Cucumis myriocarpus Naudin, Ann. Sci. Nat., Bot., sér. 4, 11: 22. 1859 • Gooseberry gourd, paddy melon [I]

Plants: rootstock thick, woody. **Tendrils** antrorsely strigose. **Leaves:** petiole proximally retrorsely strigose, medially hirsute, distally antrorsely strigose; blade ovate, palmately 5-lobed, 2.5–9(–16) × 2–8.5 (–13) cm, length 0.8–1.2 times width, base cordate to subcordate, lobes elliptic, margins serrate. **Inflorescences:** pedicels of pistillate flowers and fruits cylindric; staminate flowers 1 or 2–6 in sessile fascicles or panicles; pistillate flowers: calyx lobes 1.5–4.5 mm, petals 2–7 mm, corolla tube 0.3–1.2 mm, puberulent to sparsely puberulent inside. **Pepos** orange to yellow with green to yellow stripes, monocolor or bicolor-striped at maturity, short-ellipsoid to globose, 1.5–3(–5) × 1.5–3 cm, obscurely aculeate, flesh yellowish. $2n = 24$.

Flowering Aug–Nov. Fallow fields, orchards, roadsides, vacant lots, disturbed areas; 50–300 m; introduced; Calif.; Africa; introduced also in Australia.

Cucumis myriocarpus occurs in Fresno, Kings, Santa Barbara, and Tulare counties.

5. Cucumis metuliferus E. Meyer ex Naudin, Ann. Sci. Nat., Bot., sér. 4, 11: 10. 1859 • Horned or jelly melon, kiwano, African horned cucumber, blowfish fruit [I]

Plants: rootstock woody. **Tendrils** glabrate to hispid. **Leaves:** petiole setose; blade ovate, 3-lobate or weakly palmately 3–5-lobed, (4–)6–12(–14) × (3.5–)5.5–10(–12) cm, length 1.2–1.4 times width, base cordate, lobes elliptic to ovate, margins serrate. **Inflorescences:** pedicels of pistillate flowers and fruits cylindric; staminate flowers usually 2–5 in fascicles, rarely solitary; pistillate flowers: calyx lobes 2–3 mm, petals 6–10(–17) mm, corolla tube 1–1.6 mm, lobes sparsely puberulent inside. **Pepos** yellow to yellow-orange, monocolor, cylindric-ellipsoid, 0.6–1.5 × 3–6 cm, surface muricate to aculeate at maturity, spinules thick-based, glabrous, flesh lime green, jellylike. $2n = 24$.

Flowering Jul–Sep. Fences, disturbed sites; 0–200 m; introduced; Ala., Fla.; Africa; introduced also in Australia.

Cucumis metuliferus has been reported as naturalized in Houston County, Alabama, and Pinellas County, Florida.

Cucumis metuliferus is used for food, for garnish, and for decoration, apparently more for the latter. The flavor has been described as bland citrus, bananalike, banana-lime, a banana-cucumber-lemon combination, and a cucumber-zucchini combination. Its success as a cultivar apparently began only in 1982 in New Zealand from particularly colorful selections. It was exported from there to Japan with the trade name Kiwano and now is grown commercially also in California, Chile, and Australia.

6. Cucumis dipsaceus Ehrenberg ex Spach, Hist. Nat. Vég. 6: 211. 1838 • Hedgehog or teasel gourd [I]

Plants: roots thin, without thick, woody rootstock. **Tendrils** proximally hispidulous, distally glabrous. **Leaves:** petiole weakly hispidulous to hispid; blade ovate to broadly ovate, unlobed to 3-lobate, 3–7.5(–12.5) × 2–7(–12) cm, length 1.1–1.5 times width, base cordate, lobes ovate to elliptic, margins serrate or entire. **Inflorescences:** pedicels of pistillate flowers and fruits cylindric; staminate flowers 1 or 2–7, usually in racemoid fascicles, rarely racemes; pistillate flowers: calyx lobes 5–6(–11) mm, petals 6–15 mm, corolla tube 1–1.5 mm, lobes glabrous inside. **Pepos** pale yellow, monocolor, ellipsoid to ellipsoid-cylindric or globose, 3.5–7 × 2.5–4 cm, densely echinate at maturity, spinules narrowly cylindric, mostly obscuring fruit surface, flesh light yellow. $2n = 24$.

Flowering Jul–Sep. Open shrublands, thicket edges, riparian corridors, stream banks, sandy and loamy soil; 50–100 m; introduced; Tex.; Africa; introduced also in Mexico, Pacific Islands (Galapagos Islands, Hawaii), Australia.

Cucumis dipsaceus is documented as adventive in Texas by collections from Hidalgo and Webb counties. It is sometimes cultivated as an ornamental because of its distinctive fruits.

17. MELOTHRIA Linnaeus, Sp. Pl. 1: 35. 1753; Gen. Pl. ed. 5, 21. 1754 • Melonette

[Greek *melothron*, ancient name for some fruiting vine, probably *Bryonia*]

Plants herbs, perennial [annual], monoecious, climbing, trailing, or creeping; stems annual, glabrous or hispid [pilose]; roots tuberous; tendrils unbranched. **Leaves:** blade ovate-reniform, cordate-pentangular, suborbiculate to depressed-ovate, ovate, ovate-triangular, or lanceolate-hastate, subentire or shallowly to deeply palmately 3–5-lobed or -angled, lobes deltate to shallowly triangular, margins denticulate [dentate, shallowly sinuate, or subentire], surfaces eglandular. **Inflorescences:** staminate flowers 2–6 in axillary racemes or corymboid to subumbelloid clusters; pistillate flowers solitary, usually in same axils as staminate; bracts absent. **Flowers:** hypanthium campanulate; sepals 5, triangular to ovate, straight; petals 5, connate ½ length, yellow or orange-yellow [white or pale orange], oblong to ovate-oblong or obovate-oblong, 1.5–2[–3.5] mm, glabrous or villosulous, corolla rotate to campanulate-rotate. **Staminate flowers:** stamens 3; filaments inserted near hypanthium base or near mid tube, distinct; thecae distinct, oblong, connective broadened; pistillodes present, nectariferous. **Pistillate flowers:** ovary 3-locular, globose to ovoid or fusiform; ovules ca. 15–20 per locule; styles 1, short-columnar; stigmas 3, 2-lobed; staminodes 3 or absent. **Fruits** pepos, greenish, often striped or mottled, apparently usually maturing yellow to orange or purplish black, usually ellipsoid to ovoid, sometimes subglobose to globose, (0.8–)1–2.5 cm, smooth, glabrous, indehiscent. **Seeds** 30–60, ovoid to ellipsoid, compressed, not arillate, margins not differentiated, surface smooth, white-sericeous. $x = 12$.

Species 12 (1 in the flora): United States, Mexico, West Indies, Central America, South America; introduced in Asia.

The concept of *Melothria* was considerably narrowed by C. Jeffrey (1962), leaving the genus an entirely New World taxon characterized by long-peduncled fruits and staminate racemes, compressed seeds, and three stamens per flower, two of which are 2-thecous and the other 1-thecous. The segregated (or revived) genera *Mukia* Arnott (four species), *Solena* Loureiro (one species), and *Zehneria* (20–50 species) are from the Old World Tropics.

1. Melothria pendula Linnaeus, Sp. Pl. 1: 35. 1753

• Guadeloupe or creeping cucumber, meloncito [F]

Melothria crassifolia Small; *M. guadalupensis* (Sprengel) Cogniaux; *M. microcarpa* Grisebach; *M. nashii* Small; *M. pendula* var. *aspera* Cogniaux; *M. pendula* var. *chlorocarpa* Cogniaux; *M. pendula* var. *crassifolia* (Small) Cogniaux; *M. pendula* var. *microcarpa* (Grisebach) Cogniaux

Leaf blades 5–10 cm, base cordate, surfaces scabrous to hispid. **Inflorescences:** pistillate flowers on slender peduncles, rarely 2 from single axil. **Flowers:** hypanthium 1–2 mm; petals: apex emarginate to truncate. **Pepos** pendulous on peduncles 15–45 mm. **Seeds** 2–4 mm.

Flowering (Feb–)Mar–Nov. Bottomland forests, riparian thickets, pine-palmetto woods, upland oak woods, marshes and swamps, hammocks, glade margins, dunes and sand flats, pond and lake edges, ditches, roadsides, thickets, fences and fencerows, disturbed areas; 0–200 m; Ala., Ark., Del., D.C., Fla., Ga., Ill., Ind., Kans., Ky., La., Md., Miss., Mo., N.C., Okla., Pa., S.C., Tenn., Tex., Va.; Mexico; West Indies; Central America; South America; introduced in e Asia (Japan), Pacific Islands (Taiwan).

Three varieties have sometimes been recognized within *Melothria pendula*. Variety *aspera* from Alabama and Florida was described by J. K. Small (1933) as distinct in its slightly smaller (12–15 mm versus 10–25 mm) and globose to subglobose (versus ellipsoid to ovoid) pepos; var. *crassifolia* from Florida and Georgia was described as distinct in its trailing or creeping (versus climbing) habit, leaf blades wider than long (versus longer than wide), and mottled green pepos (versus purplish black) at maturity. There appears to be no discernible geographic pattern to any of these features; instead, the variation appears to be populational, especially in habit and pepo and leaf shape. Pepos yellowish to mottled green at maturity (as noted on label data) have been collected over at least all of the Gulf Coast states. The ontogenetic sequence of color change might be useful in interpreting variation patterns but needs further study.

18. CTENOLEPIS   Hooker f. in G. Bentham and J. D. Hooker, Gen. Pl. 1: 832. 1867

• [Greek *ktenos*, comb, and *lepis*, scale, apparently alluding to stiffly spreading cilia on margins of stipuliform bracts]   ⊡

Ctenopsis Naudin, Ann. Sci. Nat., Bot., sér. 5, 6: 12. 1866, not De Notaris 1847 [Poaceae]; *Blastania* Kotschy & Peyritsch

Plants annual, monoecious, prostrate or spreading to climbing; stems glabrous or glabrate; roots not seen; tendrils unbranched. **Leaves:** petiole bracteate basally, bracts prominent, foliaceous, stipuliform, conduplicate; blade suborbiculate to broadly ovate-cordate [triangular], palmately (3–)5–7-lobed, lobes obovate to ovate-oblong or elliptic, 2 lateral lobes deeply cleft, sinuses of main 3 lobes extending completely to base, margins entire or coarsely toothed, surfaces eglandular. **Inflorescences:** staminate flowers (3–)5–10(–20) in axillary racemes; pistillate flowers solitary, in same axils as staminate; bracts filiform [absent]. **Flowers:** hypanthium obconic to cylindric-campanulate; sepals 5, oblong-triangular; petals 5, distinct or nearly so, cream or light yellow to pale greenish, ovate-lanceolate, 1–2 mm, glabrous, corolla rotate. **Staminate flowers:** stamens 3; filaments inserted near hypanthium rim, distinct; thecae distinct, short-oblong, connective narrow; pistillode absent. **Pistillate flowers:** ovary 3-locular, subglobose; ovules 1 per locule; styles 1, columnar; stigmas 3, 2(3)-lobed; staminodes absent. **Fruits** berrylike, red, globose to depressed-globose, broadly ellipsoid-globose [depressed-turbinate], smooth, glabrous, indehiscent. **Seeds** (1)2(3), ovoid to ovoid-pyriform, compressed or planoconvex, not arillate, margins smooth, not bordered, surface smooth. $x = 12$.

Species 2 (1 in the flora): introduced, Maryland; Asia, Africa.

The native range of *Ctenolepis cerasiformis* extends from tropical Africa to India and Pakistan. *Ctenolepis garcinii* (Burman f.) C. B. Clarke is endemic to Sri Lanka and peninsular India. A species sometimes identified as *C. welwitschii* (Hooker f.) Jafri, also with stipuliform bracts, has been maintained in its original position in the monospecific genus *Dactyliandra* Hooker f. (M. M. Bhandari and D. Singh 1964). Both genera are placed in tribe Melothrieae subtribe Trochomeriinae (C. Jeffrey 1990), along with one other genus, *Trochomeria* Hooker f. Differences among these genera—mostly in seed and anther morphology—appear to be no greater than variation within some other genera of the family.

1. **Ctenolepis cerasiformis** (Stocks) Hooker f. in D. Oliver, Fl. Trop. Afr. 2: 558. 1871   Ⓕ ⊡

Zehneria cerasiformis Stocks, Hooker's J. Bot. Kew Gard. Misc. 4: 149. 1852; *Blastania cerasiformis* (Stocks) A. Meeuse; *B. fimbristipula* Kotschy & Peyritsch

Stems glabrous or subscabrous except sparsely hairy at nodes. **Leaves:** petiole 1–5 cm; stipuliform bracts, suborbiculate, 6–15(–20) mm, basally clasping stem, margins stiffly spreading-ciliate; blade broadly ovate to broadly ovate-cordate or suborbiculate, usually distinctly and deeply lobed, 3.5–9 cm, lobes obovate to ovate-oblong or elliptic, base cuneate-attenuate, surfaces scabrous at least on veins and margins. **Floral bracts** 1–3 mm. **Peduncles** 2–7 mm (pistillate), 20–40 mm (staminate). **Flowers:** hypanthium 0.5–1 mm; sepals reflexed, 0.5–1 mm. **Fruits** 1.3–1.5 cm, peduncles 2.5–7 mm. **Seeds** whitish, 7.5–11 mm.

Flowering Aug–Sep.   Chrome ore piles; 0 m; introduced; Md.; Africa; introduced also in Asia.

Ctenolepis cerasiformis was collected by C. F. Reed in 1958 at temporary unloading sites for chrome ore from cargo ships in the Port of Baltimore. The plants are distinctive in their annual duration, deeply lobed leaves, foliaceous stipuliform bracts with stiffly ciliate margins, tiny white flowers, and small, few-seeded, smooth, red, berrylike fruits on relatively short peduncles.

19. **CITRULLUS** Schrader in C. F. Ecklon and K. L. P. Zeyher, Enum. Pl. Afric. Austral. 2: 279. 1836, name conserved • [Generic name *Citrus* and Latin *-ellus*, diminutive, alluding to supposed resemblance of fruits] ①

Anguria Miller, name rejected; *Colocynthis* Miller, name rejected

Plants annual or perennial, monoecious, trailing or climbing; stems annual, villous or pustulate-scabrous to pustulate-hispid and sparsely hirsute; roots fibrous or fleshy to somewhat woody, tuberous; tendrils unbranched or 2- or 3[4]-branched. **Leaves:** blade ovate or elongate-ovate to lanceolate-ovate or ovate-triangular, deeply palmately 3–5(–7)-lobed, lobes oblong to ovate or triangular, each pinnately lobed to shallowly sinuate-lobulate, margins remotely serrate to dentate or denticulate, surfaces eglandular. **Inflorescences:** staminate and pistillate flowers solitary, from different axils; bracts caducous, linear. **Flowers:** hypanthium campanulate; sepals 5, lanceolate to linear-lanceolate [oblong-ovate to deltate]; petals 5, connate in proximal ½, yellow, obovate-oblong to widely oblanceolate [oblong-lanceolate or oblong-ovate], 6–16 mm, pubescent to glabrate, corolla rotate to campanulate. **Staminate flowers:** stamens 3; filaments inserted at hypanthium base, distinct; thecae distinct, replicate, forming a head, connective broad; pistillodes present. **Pistillate flowers:** ovary 6-locular, ellipsoid to obovoid or subglobose; ovules 50–100+ per locule; styles 1, short-columnar, deeply cleft; stigmas 3, 2-lobed; staminodes present. **Fruits** pepos, green, yellow, or variegated green and whitish or green and yellow, ± longitudinally striped, globose or subglobose to ellipsoid, smooth, glabrous, indehiscent. **Seeds** 50–300+ (or many more in some cultivated forms), usually ovoid to ellipsoid or oblong, compressed, not arillate, margins not differentiated, surface smooth or slightly verrucose. *x* = 11.

Species 5 (3 in the flora): introduced; Asia, Africa; introduced widely.

In *Citrullus* the rind (hypanthium/pericarp) may be firm and tough or thin and fragile; it may be durable (persisting intact after maturity) or not. The mesocarp can be fleshy or dry and it varies in color from red or pink to orange, light yellowish orange, yellow, greenish, or whitish.

Citrullus lanatus is commonly established in the flora area as a waif but rarely, if ever, persists more than one season. *Citrullus caffer* is a poorly documented transient, and *C. colocynthis* apparently grows as a weed in crop fields.

The five species of *Citrullus* are *C. lanatus*, the domesticated watermelon; *C. caffer* (*C. lanatus* var. *citroides*), the preserving melon or citron; *C. colocynthis*, the bitter apple; and *C. ecirrhosus* Cogniaux and *C. rehmii* DeWinter, wild species endemic to desert regions of Namibia. *Citrullus caffer*, *C. lanatus*, and *C. rehmii* are monoecious annuals; *C. colocynthis* and *C. ecirrhosus* are perennials. All are native to dry habitats of northern Africa (*C. colocynthis*) or southern Africa (*C. caffer*, *C. ecirrhosus*, *C. lanatus*, and *C. rehmii*), and all are cross-compatible with each other.

The monotypic *Praecitrullus* Pangalo [*P. fistulosus* (Stocks) Pangalo, Indian round gourd, apple gourd, Indian baby pumpkin], a monoecious annual from Afghanistan, India, and Pakistan, was segregated from *Citrullus* and has proved to be more closely related to *Benincasa* Savi (F. Dane and P. Lang 2004; A. Levi and C. E. Thomas 2005; Levi et al. 2010).

Discovery of 5000-year-old *Citrullus lanatus* seeds in Libya (K. Wasylikowa and M. Van der Veen 2004) indicates that its domestication might have occurred in northern Africa. Remains also have been found in tombs and temples in Egypt (ca. 1330 BCE) and Sudan (1500 BCE). Seeds of *C. colocynthis* appear in early Egyptian, Libyan, and near-eastern sites, and it was probably used prior to its domestication (F. Dane and P. Lang 2004; Dane and J. Liu 2007). Watermelons are mentioned in European herbals from the 1500s and by 1625 were widely planted as a minor crop in European gardens. They arrived early in North America with European colonists, appearing in Florida by 1567 and in Massachusetts by 1629.

SELECTED REFERENCES   Aquino Assis, J. G. de et al. 2000. Implications of the introgression between *Citrullus colocynthis* and *C. lanatus* characters in the taxonomy, evolutionary dynamics and breeding of watermelon. Pl. Genet. Resources Newslett. 121: 15–19.   Dane, F. and P. Lang. 2004. Sequence variation at cpDNA regions of watermelon and related wild species: Implications for the evolution of *Citrullus* haplotypes. Amer. J. Bot. 91: 1922–1929.   Dane F., P. Lang, and R. Bakhtiyarova. 2004. Comparative analysis of chloroplast DNA variability in wild and cultivated *Citrullus* species. Theor. Appl. Genet. 108: 958–966.   Dane F. and J. Liu. 2007. Diversity and origin of cultivated and citron type watermelon (*Citrullus lanatus*). Genet. Resources Crop Evol. 54: 1255–1265.   Fursa, T. B. 1972. K sistematike roda *Citrullus* Schrad. (On the taxonomy of genus *Citrullus* Schrad.) Bot. Zhurn. (Moscow & Leningrad) 57: 31–41.   Fursa, T. B. 1981. Intraspecific classification of water-melon under cultivation. Kulturpflanze 29: 297–300.   Jarret, R. L. and M. Newman. 2000. Phylogenetic relationships among species of *Citrullus* and the placement of *C. rehmii* De Winter as determined by internal transcribed spacer (ITS) sequence heterogeneity. Genet. Resources Crop Evol. 47: 215–222.   Levi, A. and C. E. Thomas. 2005. Polymorphisms among chloroplast and mitochondrial genomes of *Citrullus* species and subspecies. Genet. Resources Crop Evol. 52: 609–617.

1. Plants perennial; stems pustulate-scabrous to pustulate-hispid; tendrils usually unbranched, rarely 2-branched; pepos 4–7(–10) cm diam.. . . . . 1. *Citrullus colocynthis*
1. Plants annual; stems villous; tendrils 2- or 3-branched; pepos 12–35+ cm diam.
    2. Leaf blades ovate to lanceolate-ovate or ovate-triangular, mostly 8–20 cm; pepos globose to oblong-ellipsoid, 12–35+ cm diam., rind tough, not durable, mesocarp juicy, red to orange, yellow, or greenish, sweet. . . . . .2. *Citrullus lanatus*
    2. Leaf blades ovate, 3–8 cm; pepos globose to globose-ovoid, 15–25 cm diam., rind tough, durable, mesocarp dry, whitish, bitter . . . . .3. *Citrullus caffer*

1. **Citrullus colocynthis** (Linnaeus) Schrader, Linnaea 12: 414. 1838 • Colocynth, bitter apple, bitter cucumber, vine of Sodom I

Cucumis colocynthis Linnaeus, Sp. Pl. 2: 1011. 1753; *Colocynthis vulgaris* Schrader

Vines perennial. **Stems** trailing, 50–150 cm, pustulate-scabrous to pustulate-hispid and sparsely hirsute with deflexed hairs; roots fleshy to ± woody, tuberous; tendrils usually unbranched, rarely 2-branched. **Leaf blades** elongate-ovate, 1–6(–11) cm, 3–5(–7)-lobed, lobes pinnately shallowly lobed or sinuate, margins serrate to dentate, surfaces pustulate-scabrous. **Flowers:** hypanthium campanulate, 4–8 mm; sepals lanceolate to linear-lanceolate, 5 mm; petals obovate-oblong, 6–10 mm. **Pepos** variegated green and yellow or yellow, green-striped, globose to subglobose, with ± elliptical fissures, 4–7(–10) cm diam.; rind thin, not durable, mesocarp light yellowish orange to pale yellow, dry-spongy, intensely bitter. **Seeds** dark brown to yellowish orange, ovoid-oblong to ellipsoid, 6 mm. *2n* = 22.

Flowering Jun–Sep. Peanut fields, pond dikes in cotton fields, fallow fields, waste ground; 0–100 m; introduced; Calif., Mass., Tex.; Africa; introduced also in Europe, Asia, Pacific Islands, Australia.

Citrullus colocynthis is a traditional food plant in Africa, where it is grown particularly for its edible seeds, which are bitter but nutty-flavored, rich in fat and protein, and eaten whole or used as an oilseed. It has also been a standard cathartic remedy, mostly in combination with other cathartics.

Three main haplotypes within colocynth have been identified via molecular data (F. Dane and P. Lang 2004; Dane and J. Liu 2007; A. Levi and C. E. Thomas 2005).

Colocynth rarely is encountered outside of cultivation. It has been reported as a weed in Texas peanut fields (G. L. Nesom 2011); in Massachusetts, it has been sporadically collected since 1878; in California, it has been collected on a pond dike in an area of cotton fields in Kern County.

2. **Citrullus lanatus** (Thunberg) Matsumura & Nakai, Index Seminum (Tokyo) 30, no. 854. 1916 • Watermelon, sandia F I

Momordica lanata Thunberg, Prodr. Pl. Cap., 13. 1794

Subspecies 2 (1 in the flora): introduced; Asia, Africa; introduced widely.

Subspecies *mucosospermus* Fursa comprises wild, semicultivated, and cultivated forms in western Africa, including the egusi melon, which is cultivated primarily for its oil and protein-rich seeds. Both subspecies have the same cpDNA haplotype (F. Dane and P. Lang 2004).

Citrullus lanatus generally has been regarded to include the citron melon, commonly as var. *citroides*; the latter is treated here as 3. *Citrullus caffer*.

More than a thousand cultivars of *Citrullus lanatus* have been developed, ranging greatly in shape and size,

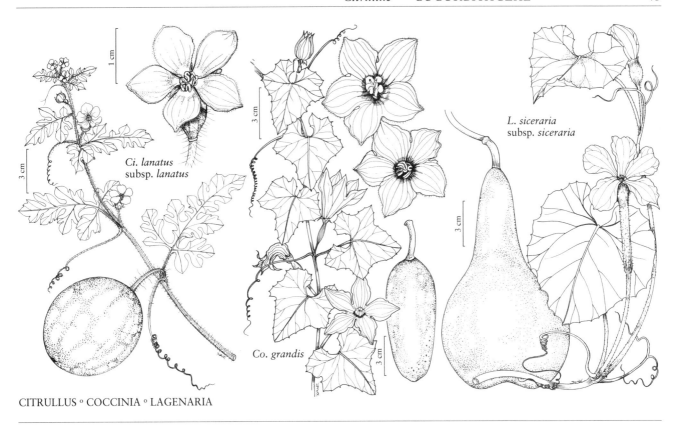

Ci. lanatus
subsp. *lanatus*

L. siceraria
subsp. *siceraria*

Co. grandis

CITRULLUS ∘ COCCINIA ∘ LAGENARIA

from less than a pound to more than 200 pounds, with flesh red, orange, yellow, or greenish. Seedless fruits are from triploid plants produced as hybrids between normal diploids and artificially produced tetraploids. The triploids have sterile pollen and because pollination is required to induce fruit set and enlargement, they must be interplanted with pollenizer diploids.

2a. Citrullus lanatus (Thunberg) Matsumura & Nakai subsp. **lanatus** F I

Citrullus lanatus subsp. *vulgaris* (Schrader) Fursa; *C. vulgaris* Schrader; *Cucurbita citrullus* Linnaeus

Vines annual. **Stems** climbing or trailing, 50–200 cm, villous; roots fibrous; tendrils 2- or 3-branched. **Leaf blades** ovate to lanceolate-ovate or ovate-triangular, mostly 8–20 cm, 3–5-lobed, lobes pinnately shallowly sinuate-lobulate, margins denticulate, surfaces hirsute abaxially, hispid on veins and veinlets, glabrous or scabrous adaxially with translucent dots. **Flowers:** hypanthium broadly campanulate; sepals lanceolate, 3–5 mm; petals obovate-oblong to widely oblanceolate, 7–16 mm. **Pepos** green, mottled with paler green and yellowish to whitish stripes, globose to oblong-ellipsoid, 12–35+ cm diam.; rind tough, not durable, mesocarp red to orange, yellow, or greenish, juicy, sweet. **Seeds** usually black, rarely red or of different shades, ovoid to oblong-ovoid, 7–15 mm. $2n = 22$.

Flowering Jun–Nov. Gardens, fields, vacant lots, trash heaps, dumps, roadsides, clearings in woods, gravel bars, stream banks, riparian thickets, dunes, volunteering from past plantings; 10–400 m; introduced; Ala., Ariz., Ark., Calif., Conn., Fla., Ga., Ill., Ind., Kans., Ky., La., Maine, Md., Mass., Mich., Miss., Mo., Nev., N.J., N.Mex., N.Y., N.C., Ohio, Okla., Pa., R.I., S.C., Tenn., Tex., Utah, Vt., Va., W.Va., Wis.; Asia; Africa; introduced also in Mexico, West Indies, Central America, South America, Europe, Australia.

3. Citrullus caffer Schrader, Linnaea 12: 413. 1838

• Citron, preserving or stock melon I

Citrullus lanatus (Thunberg) Matsumura & Nakai var. *caffer* (Schrader) Mansfeld ex Fursa; *C. lanatus* var. *caffrorum* (Alefeld) Fosberg; *C. lanatus* var. *citroides* (L. H. Bailey) Mansfeld; *C. vulgaris* Schrader var. *citroides* L. H. Bailey

Vines annual. **Stems** climbing or trailing, 50–200 cm, villous; roots fibrous; tendrils 2- or 3-branched. **Leaf blades** ovate, 3–5-lobed, 3–8 cm, lobes pinnately shallowly sinuate-lobulate, margins denticulate, surfaces hirsute abaxially, hispid on veins and veinlets, glabrous or scabrous adaxially with translucent dots. **Flowers:** hypanthium broadly campanulate; sepals lanceolate, 3–5 mm; petals obovate-oblong to widely oblanceolate, 7–16 mm. **Pepos** green, mottled with paler green and yellowish to whitish stripes, globose to globose-ovoid, 15–25 cm diam.; rind tough, durable, mesocarp whitish, dry, bitter. **Seeds** tan or greenish, ovoid to oblong-ovoid, 7–15 mm. $2n = 22$.

Flowering Jun–Oct(–Nov). Sandy fields, pinelands, cotton fields, orange groves, roadsides, railroad banks and tracks, disturbed sites; 50–500 m; introduced; Ariz., Calif., Fla., Nev., N.Mex., Tex.; Africa; introduced also in West Indies.

Records for *Citrullus caffer* outside of cultivation in the flora area are scattered and uncommon. It is called preserving melon because the fruit rind and white flesh are candied or pickled and used in preserves and jellies.

Molecular studies have shown that the cultivated watermelon (*Citrullus lanatus*) and the citron melon represent closely related but distinct lineages (F. Dane and P. Lang 2004; Dane and J. Liu 2007; A. Levi and C. E. Thomas 2005). The two taxa have distinct haplotypes and appear to have evolved from a common ancestor perhaps closely similar to *C. ecirrhosus*. The evolutionary divergence is estimated to have occurred between 0.6 and 0.9 million years ago, long before selection by humans.

Three cpDNA haplotypes exist among the citron melon and the wild tsamma melon of the Kalahari Desert (F. Dane and J. Liu 2007). These three entities apparently have not been unambiguously recognized by formal nomenclature, or at least the genetic affinities of names potentially associated with these haplotypes are not known. The ancestral citron haplotype is known from Swaziland and South Africa; each of the other two ranges across southern Africa. A number of distinct landraces that are cultivated in the Kalahari region (including the tsamma melons) may represent early forms of domestication, as suggested by G. Maggs-Kolling et al. (2000).

20. COCCINIA Wight & Arnott, Prodr. Fl. Ind. Orient. 1: 347. 1834 • Ivy gourd [Latin *coccineus*, scarlet, alluding to bright red mature fruit of *Coccinia grandis*] 1

Plants perennial, dioecious, climbing or trailing; stems annual, glabrous or glabrate [flocculent-arachnoid]; roots tuberous; tendrils unbranched [2-branched]. **Leaves:** blade broadly ovate to rounded-cordate, subreniform, or deltate, unlobed or palmately 5-angular or -lobed, lobes deltate or triangular to broadly angular-elliptic, margins denticulate, adaxial surface with circular, sessile scales [hirsute to hirsutulous], often with glands on both sides of midrib near petiole. **Inflorescences:** staminate flowers solitary, [clustered, racemose, or in spikes], axillary; pistillate flowers solitary, axillary [racemose]; bracts absent. **Flowers:** hypanthium campanulate to turbinate; sepals 5, linear to subulate; petals 5, connate ½ length, bright white, often slightly green-veined [brownish yellow or orange], ovate to ovate-triangular, [8–]15–20[–62] mm, hirtellous or puberulent-hirtellous to glabrate, corolla campanulate. **Staminate flowers:** stamens 3; filaments inserted near hypanthium base, connate; thecae connate into central column and forming central oblong body, sigmoid-triplicate, connective broadened; pistillodes absent. **Pistillate flowers:** ovary 3-locular, ovoid to fusiform; ovules ca. 15–40 per locule; styles 1, narrowly columnar; stigmas 1, 3-lobed; staminodes 3. **Fruits** pepos, usually green with white streaks or lines, sometimes red to scarlet at maturity, broadly cylindric to ellipsoid-cylindric, smooth, glabrous, indehiscent, flesh whitish to greenish. **Seeds** 30–50[–120], asymmetrically pyriform [ovoid or broadly ellipsoid], compressed, arillate, margins thickened or not bordered, surface fibrillose. $x = 12$.

Species ca. 30 (1 in the flora): introduced; s, se Asia (India, Malaysia), Africa; introduced also in Pacific Islands.

1. Coccinia grandis (Linnaeus) Voigt, Hort. Suburb. Calcutt., 59. 1845 • Scarlet gourd [F] [I]

Bryonia grandis Linnaeus, Mant. Pl. 1: 126. 1767

Vines climbing, widely spreading, sometimes prostrate. **Stems** glabrous or glabrate, sometimes rooting at nodes. **Leaves:** petiole 1–5 cm; blade 5–10 × 4–9 cm, base cordate with broad sinus, apex acute, mucronate, adaxial surface with 3–8 glands. **Peduncles** 1–5 cm. **Flowers:** sepals recurved, 2–5 mm; petals 15–20 mm, apices acute to obtuse-apiculate. **Pepos** 2.5–6 cm. **Seeds** 6–8 mm, aril red to red-orange. $2n = 24$.

Flowering May–Nov. Trash dumps, thickets, fencerows, cypress swamps; 0–30 m; introduced; Fla., Tex.; e Africa; introduced also in Asia (China, India, Indonesia, Malaysia, Pakistan, Thailand, Vietnam), Pacific Islands, Australia.

The shoot tips and immature fruits of *Coccinia grandis* are used in Asian and Indian cooking; long-range dispersal is often the result of introduction by humans. It sometimes has been misidentified as *C. cordifolia* (Linnaeus) Cogniaux.

21. LAGENARIA Seringe, Mém. Soc. Phys. Genève 3: 16, 26. 1825 • Bottle gourd [Greek *lagenos*, flask, alluding to shape and use of fruit] [I]

Plants annual [perennial], monoecious [dioecious], scandent or prostrate; stems mostly annual, densely villous to puberulent [sparsely hirsute to villous-hirsute]; taprooted [roots tuberous]; tendrils 2-branched. **Leaves:** petiole with pair of small conical glands at apex; blade broadly reniform or ovate [suborbiculate], ± palmately 3–5(–7)-lobed, lobes triangular to widely obovate, base cordate, margins shallowly sinuate-serrate to dentate or sinuate-dentate, surfaces shortly puberulous or pubescent, eglandular. **Inflorescences:** staminate flowers solitary [racemose], axillary; pistillate flowers solitary, in same axils as staminate; bracts absent. **Flowers:** hypanthium campanulate to funnelform; sepals 5, subulate to triangular or linear; petals 5, distinct, white to cream, obovate to oblong-obovate, (15–)20–25[–45] mm, puberulent or glabrous, corolla broadly campanulate to shallowly cupulate. **Staminate flowers:** stamens 3; filaments inserted in hypanthium tube, distinct; thecae connate, forming a head, triplicate, usually much contorted, connective broad; pistillodes absent. **Pistillate flowers:** ovary 3-locular, subglobose to ellipsoid, ovoid, or cylindric; ovules 30–100 per locule; styles 1, short-columnar; stigmas 3, 2-lobed; staminodes 3. **Fruits** pepos, green to greenish yellow, maturing yellowish or pale brown, commonly mottled or with light green or white longitudinal stripes, subglobose to cylindric, ellipsoid, or lageniform, sometimes 2-ventricose, usually broader distally, smooth, glabrous, indehiscent. **Seeds** 100–300, oblong to ovoid-oblong, compressed, not arillate, with marginal groove (truncate), surface smooth. $x = 11$.

Species 6 (1 in the flora): introduced; Africa, Indian Ocean Islands (Madagascar), Pacific Islands; introduced widely.

SELECTED REFERENCE Clarke, A. C. et al. 2006. Reconstructing the origins and dispersal of the Polynesian bottle gourd (*Lagenaria siceraria*). Molec. Biol. Evol. 23: 893–900.

1. Lagenaria siceraria (Molina) Standley, Publ. Field Mus. Nat. Hist., Bot. Ser. 3: 435. 1930 • Calabash, dipper or birdhouse or Hercules club or Italian edible gourd [F] [I]

Cucurbita siceraria Molina, Sag. Stor. Nat. Chili, 133. 1782

Subspecies 2 (1 in the flora): introduced; Asia, Africa, Indian Ocean Islands (Madagascar), Pacific Islands.

Lagenaria siceraria is broadly cultivated in Asia and locally in the United States for its young and tender fruits, which are eaten as a vegetable. Shells of the dried fruits are often used as water bottles and for making ladles, bowls, pipes, snuff boxes, floats, and horns and other musical instruments. Plants can be confidently identified by their relatively large, solitary, white flowers and pair of glands at the base of the leaf blade.

Subspecies *asiatica* (Kobiakova) Heiser is centered in Asia and the Pacific Islands and is morphologically and molecularly distinct from the primarily African subsp. *siceraria*. The former is characterized by leaves with sharply pointed triangular lobes and serrate margins, larger staminate flowers, and relatively longer seeds.

A wild population of bottle gourd was recently discovered in Zimbabwe, Africa (D. S. Decker et al. 2004), establishing the region of origin for *Lagenaria siceraria*. Compared to the modern wild bottle gourds, all of the archaeological specimens have a distinctly thickened rind, reflecting a loss of natural seed dispersal ability, a significant marker of domestication.

The earliest archaeological evidence for domestication of bottle gourd in Africa is from ca. 4000 BCE. The earliest known archaeological records from Asia are ca. 8000–9000 BCE from China and Japan, and it is probable that initial domestication in Asia (the origin of subsp. *asiatica*) occurred perhaps as much as 2000–3000 years before that, even by 12,000–13,000 BCE.

In the New World, domesticated bottle gourd had a broad distribution by at least 8000 BCE (2700–8100 BCE in eastern North America, 7200–10,000 BCE in southern Mexico, and 8400 BCE in Peru). All archaeological rind fragments of American bottle gourds from pre-European samples are identical in DNA to Asian plants, rather than African ones, and seed morphology is congruent (D. L. Erikson et al. 2005). It seems clear that Asia was the source of the early introduction to the New World.

If the hypothesized time period for earliest Asian domestication of *Lagenaria siceraria* is correct, then the bottle gourd along with the dog were domesticated long before other plants and animals. D. L. Erikson et al. (2005) believed that both of these useful species crossed from Asia to North America with Paleoindian populations as they colonized the New World. Archaeological records of bottle gourd in North America are of subsp. *asiatica*.

After multiple arrivals of Europeans in the New World, cultivars of African origin apparently spread rapidly, and, based on DNA analysis of modern New World races, today have almost completely replaced subsp. *asiatica* in the Western Hemisphere.

1a. **Lagenaria siceraria** (Molina) Standley subsp. **siceraria** F I

Cucurbita lagenaria Linnaeus; *Lagenaria leucantha* Rusby; *L. vulgaris* Seringe

Stems 1–5 m, rooting at nodes. **Leaves:** petiole 3–10(–16) cm; blade 3–25(–40) × 4–25(–40) cm, lobes obscure, rounded, apex apiculate. **Inflorescences:** pistillate peduncles 6–10 cm. **Flowers:** petal apex apiculate, corolla cream to white with darker veins, pale yellow at base. **Pepos** 10–50 cm (to 200 cm in some cultivated forms), [exocarp woody]. **Seeds** slightly tapered, slightly 2-horned on shoulders, with 2 flat facial ridges, 12–22 mm. $2n = 22$.

Flowering Aug–Oct. Gardens, trash heaps, fields, wood edges, railroad banks, roadsides, ditch banks, stream banks, commonly cultivated in home gardens and commercially, abandoned plantings; 20–900 m; introduced; Ala., Ark., Fla., Ga., Ill., Ky., La., Mass., Miss., Mo., N.Y., N.C., Okla., Pa., S.C., Tex., Va.; Asia; Africa; introduced also in West Indies, South America, Europe, Australia.

22. CAYAPONIA Silva Manso, Enum. Subst. Braz., 31. 1836, name conserved

• [Derivation uncertain, perhaps from Caiapó, river or native tribe of Amazonian Brazil]

Plants perennial, usually monoecious, sometimes dioecious, climbing; stems annual or perennial, puberulent or glabrous; roots tuberous; tendrils unbranched or 2[3]-branched. **Leaves:** blade pentagonal, deltate, or ovate [subquadrangular], unlobed or 3–5-lobed, lobes broadly triangular or deltate to ovate-deltate, oblong-oblanceolate, or elliptic, margins denticulate or serrulate, surfaces sometimes with disc-shaped glands abaxially near base or near lobe tips. **Inflorescences:** staminate flowers solitary or sometimes distal on axillary branches with pistillate flowers proximal; pistillate flowers 1–5 in axillary fascicles or proximal on short, leafless, racemoid branches sometimes with staminate flowers distally; bracts filiform or absent. **Flowers:** hypanthium campanulate; sepals 5, deltate to triangular or linear [ovate-lanceolate, oblong-lanceolate, or linear-subulate]; petals 5, connate ½–⅔ length, white, cream, or pale green [orange to yellow], oblong-lanceolate to linear-oblong, 5–10 mm, glabrous or puberulent or tomentose adaxially, corolla rotate to campanulate. **Staminate flowers:** stamens 3; filaments inserted near hypanthium rim, distinct; thecae distinct or connate, forming a head, sigmoid-replicate, connective narrow;

pistillodes with 3-lobed ovary. **Pistillate flowers:** ovary 3-locular (or 1 or 2 by abortion), globose to ovoid or ellipsoid-cylindric; ovules 1–4(–10) per locule; styles 1, narrow; stigmas 1, 3-lobed; staminodes 3, minute. **Fruits** dry pepos or berrylike, red to scarlet, orange, or golden brown [green], mostly ellipsoid-cylindric [to globose], smooth, glabrous, indehiscent. **Seeds** [1–]3–12 [–30], obovoid [ovoid or oblong], subcompressed, not arillate, margins variably differentiated, sometimes thinner or with narrow light-colored stripe, surface smooth.

Species ca. 75 (2 in the flora): se United States, Mexico, West Indies, Central America, South America.

A discussion of plasticity of inflorescence morphology in *Cayaponia* was provided by R. McVaugh (2001b)—a flowering branch apparently constitutes a complex branching system, potentially producing solitary flowers, groups of flowers, or new branches from any node. Although these observations pertain directly only to *C. attenuata* (Hooker & Arnott) Cogniaux, they perhaps also apply to the two species included here.

SELECTED REFERENCE Duchen, P. and S. S. Renner. 2010. The evolution of *Cayaponia* (Cucurbitaceae): Repeated shifts from bat to bee pollination and long-distance dispersal to Africa 2–5 million years ago. Amer. J. Bot. 97: 1129–1141.

1. Petioles glabrous; leaf blades hirsute-hispidulous abaxially, pustulate-scabrous adaxially; fruiting peduncles 8–15(–30) mm . . . . .1. *Cayaponia americana*
1. Petioles usually villous-hirsute, sometimes villous, stipitate-glandular; leaf blades villosulous to hirsutulous; fruiting peduncles 1–4 mm . . . . 2. *Cayaponia quinqueloba*

1. Cayaponia americana (Lamarck) Cogniaux in A. L. P. P. de Candolle and C. de Candolle, Monogr. Phan. 3: 785. 1881 • American melonleaf

Bryonia americana Lamarck in J. Lamarck et al., Encycl. 1: 498. 1785; *Cayaponia americana* var. *angustiloba* Cogniaux; *C. americana* var. *subintegrifolia* Cogniaux

Vines herbaceous. **Stems** sulcate, glabrous; tendrils unbranched. **Leaves:** petiole 1–5 cm, glabrous; blade ovate to deltate or pentagonal, unlobed or shallowly to deeply 3–5-lobed, 5–14 × 5–13 cm, terminal lobe oblong-lanceolate, broadest at base, lateral lobes often sublobed, base cordate to truncate or rounded, margins remotely serrulate or denticulate, surfaces hirsute-hispidulous abaxially (not white-sericeous), pustulate-scabrous adaxially. **Flowers:** sepals deltate to triangular, 1–2 mm; petals pale green to whitish, corolla 6–10 mm; pistillate flowers solitary, axillary, or 3–5 on short racemoid branches; staminate flowers solitary, axillary. **Fruits** usually 1, axillary, or 3–5 on racemoid branches, orange to golden brown, short ellipsoid-cylindric, 1.5–2 cm; peduncle 8–15(–30) mm. **Seeds** usually 3, 10 mm.

Flowering May–Aug. Rockland hammocks, hammock edges and clearings; 0–20 m; Fla.; West Indies.

Cayaponia americana is known in the flora area only from Miami-Dade County. The leaf blades are darker and thicker than those of *C. quinqueloba*, and peduncle length is greater in *C. americana*.

2. Cayaponia quinqueloba (Rafinesque) Shinners, Field & Lab. 25: 32. 1957 • Five-lobe-cucumber E F

Arkezostis quinqueloba Rafinesque, New Fl. 4: 100. 1838; *Bryonia boykinii* Torrey & A. Gray; *Cayaponia boykinii* (Torrey & A. Gray) Cogniaux; *C. grandifolia* (Torrey & A. Gray) Small; *Melothria grandifolia* Torrey & A. Gray

Vines herbaceous. **Stems** sulcate, puberulent; tendrils usually unbranched, sometimes 2-branched. **Leaves:** petiole 15–45 mm, usually villous-hirsute, sometimes villous, moderately to densely short stipitate-glandular; blade ovate to deltate or pentagonal, shallowly to deeply palmately 3–5-lobed, 4–11(–15) × 4–12(–17) cm, terminal lobe ovate-trullate, broadest at base or near middle, sometimes sublobed, base angular-cordate, margins denticulate to serrulate, surfaces sparsely villosulous to hirsutulous (not white-sericeous abaxially). **Flowers:** sepals linear, 1–2 mm; petals white to cream, corolla 6–10 mm; pistillate and staminate flowers 1–3(–5), axillary. **Fruits** 1–3(–5), red to scarlet, ellipsoid-cylindric, 1–1.6(–2) cm; peduncle 1–4 mm. **Seeds** 3–12, 6–8 mm.

Flowering Jun–Oct(–Nov). Swamp forests, floodplain and bottomland woods, clearings in alluvial woods, stream banks, oak-sweetgum-hackberry flatwoods, maple woods, bluffs, hardwood hammocks, roadsides, ditches, spoil banks; 10–200 m; Ala., Ark., Fla., Ga., La., Miss., Mo., Okla., S.C., Tenn., Tex.

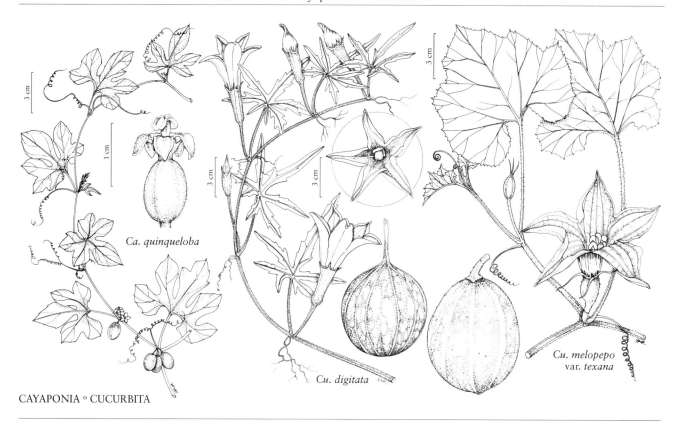

Ca. quinqueloba

Cu. digitata

Cu. melopepo
var. *texana*

CAYAPONIA ° CUCURBITA

Leaves of *Cayaponia quinqueloba* (as in *C. americana* and some other cucurbits) are variable in shape. *Cayaponia grandifolia* has been treated as distinct from *C. quinqueloba* based on extremes of leaf form, *C. grandifolia* with deltate to ovate-deltate lobes broadest at the base (for example, R. K. Godfrey and J. W. Wooten 1981). The range of *C. grandifolia* is described as Arkansas, Louisiana, and Mississippi, sympatric with *C. quinqueloba*; only a single variable species appears to exist.

Most collections of *Cayaponia quinqueloba* show plants with only pistillate flowers, relatively few with both pistillate and staminate flowers (the latter usually relatively less numerous on a given plant).

Apparently pistillate specimens may be from bisexual plants, the sampled portion of the stem lacking staminate flowers.

23. CUCURBITA Linnaeus, Sp. Pl. 2: 1010. 1753; Gen. Pl. ed. 5, 441. 1754 • Squash, gourd [Latin name for gourd, probably from *curvitas*, crookedness, alluding to fruit shape]

Plants sometimes shrublike in cultivated forms of *C. melopepo*, annual or perennial, monoecious, procumbent and trailing or climbing; stems annual, often sulcate or angled, hairy; roots tuberous or fibrous or a taproot; tendrils 2–7-branched or absent. **Leaves:** blade suborbiculate to broadly ovate, ovate-lanceolate, reniform, or triangular, usually deeply to shallowly palmately (3–)5 (–7)-lobed, sometimes unlobed or 2-lobed, lobes depressed-ovate, ovate, broadly or narrowly triangular, or obovate to lanceolate, oblanceolate, oblong-lanceolate, or subrhombic, base truncate to cordate, margins serrate to denticulate or mucronulate, surfaces eglandular or glandular, glands scattered, sessile or stipitate to peltate. **Inflorescences:** staminate flowers solitary [in axillary fascicles]; pistillate flowers solitary, from different axils than staminate; floral bracts absent. **Flowers:** hypanthium campanulate, cylindric, or cupulate; sepals 5, straight, erect, subulate-linear to lanceolate [spatulate]; petals 5, often recurving, connate ½ length, cream or yellow to orange, broadly ovate to oblong-ovate, oblong-elliptic, triangular, triangular-ovate, or lanceolate-ovate, 25–90[–120] mm, pubescent to puberulent, corolla campanulate. **Staminate flowers:** stamens 3; filaments inserted at hypanthium base, distinct or slightly connate; thecae connate, forming central oblong body, sigmoid-flexuous, connective narrow; pistillodes absent. **Pistillate flowers:** ovary 3-locular, (or falsely 5-locular), cylindric to fusiform or ellipsoid, ovoid, globose, pyriform, or lageniform; ovules 30–150 per locule; styles 1, short-columnar; stigmas 3(–5), 2-lobed; staminodes usually present. **Fruits** pepos, orange to brown, yellow, slate blue, grayish, blackish purple, green, white, or cream, often mottled or striped, globose to depressed-globose, cylindric, ellipsoid, ovoid, conic, cushion-shaped, pyriform or lageniform, usually smooth, ribbed, or warty, rarely tuberculate, sometimes longitudinally furrowed, glabrous, indehiscent. **Seeds** (100–)200–500, ovoid to oblong, narrowly to broadly ellipsoid, obovoid, or suborbicular, strongly compressed, not arillate, margins differentiated or not in thickness, texture, and color, surface smooth or slightly rough to punctate-sculptured. $x = 20$.

Species 14–22 (9 in the flora): North America, Mexico, West Indies, Central America, South America; introduced nearly worldwide.

Species of *Cucurbita* have long provided fruits that are staples of the world's diet. Radiocarbon dates for *C. pepo* from Guilá Naquitz Cave in subtropical central Oaxaca, Mexico, document the earliest known cultivation of a squash—between 10,000 to 8000 calendar years ago (B. D. Smith 1997)—predating the appearance of maize, beans, and other directly dated domesticates in the Americas by about 4000 years. The oldest known cultivated maize cobs also have come from Guilá Naquitz Cave, dated about 6250 calendar years ago. Cultivated cucurbits were introduced into the Old World quickly after the discovery of the New World (H. S. Paris et al. 2006), and since the seventeenth century they have spread over the Tropics and subtropics and temperate regions. After its introduction to the Old World, *C. pepo* apparently underwent secondary diversification in Europe and/or Asia Minor (J. R. Harlan 1951).

Cucurbita has been divided into two groups (T. W. Whitaker and W. P. Bemis 1975): the arid-zone perennials with tuberous storage roots, and the more mesophytic annuals or short-lived perennials without storage roots. Four of the domesticated species (*C. argyrosperma* K. Koch, *C. maxima*, *C. moschata*, and *C. pepo*) plus one wild species (*C. ecuadoriensis* Cutler & Whitaker) comprise the mesophytic lineage (O. I. Sanjur et al. 2002). One other among the mesophytic domesticated species, *C. ficifolia*, is outside the mesophytic clade and basal to it.

Uncertainty about the number of species in the genus reflects two areas of subjectivity: some domesticated species are commonly treated as conspecific with a wild species when a derivative-progenitor relationship is indicated—the domesticated *Cucurbita maxima* arose from the wild *C. andreana* Naudin of South America; and, the taxonomic status of some entities is not unambiguously resolved—*C. digitata* was treated by M. Nee (1990) to include four taxa, three of which are sometimes treated as distinct species. In any case, the geography and morphology of all the entities potentially treated at specific rank is fairly well understood. Additionally, Nee has noted that a wild ancestor of *C. moschata*, from reports from Colombia, may remain to be discovered and described; similarly, he has predicted that the wild ancestor of *C. ficifolia* exists somewhere in the Andes.

Four of the domesticated species are treated here because they sometimes are encountered in the flora area outside of cultivation; in most cases, these plants can be characterized as waifs, as they apparently do not maintain themselves in persistent populations for more than one or a few years.

Cucurbita argyrosperma K. Koch (Japanese pie pumpkin, white cushaw) is currently grown in the United States and known to have been in cultivation in eastern North America since pre-Columbian times (G. J. Fritz 1994); it has not been reported to grow outside of cultivation in the flora area.

Identification of species among plants of domesticated *Cucurbita* often is difficult (especially among *C. maxima*, *C. melopepo*, *C. moschata*, and *C. pepo*) because habit, vestiture, leaf shape, and floral characters commonly show parallel variation, and fruit shape and size are highly variable.

SELECTED REFERENCES   Bailey, L. H. 1943. Species of *Cucurbita*. Gentes Herb. 6: 266–322.   Bemis, W. P. and T. W. Whitaker. 1969. The xerophytic *Cucurbita* of northwestern Mexico and southwestern United States. Madroño 20: 33–41.   Decker, D. S., T. W. Walters, and U. Posluszny. 1990. Genealogy and gene flow among annual domesticated species of *Cucurbita*. Canad. J. Bot. 68: 782–789.   Nee, M. 1990. The domestication of *Cucurbita* (Cucurbitaceae). Econ. Bot. 44(3,suppl.): 56–68.   Rhodes, A. M. et al. 1968. Numerical taxonomic study of *Cucurbita*. Brittonia 20: 251–266.   Sanjur, O. I. et al. 2002. Phylogenetic relationships among domesticated and wild species of *Cucurbita* (Cucurbitaceae) inferred from a mitochondrial gene: Implications for crop plant evolution and areas of origin. Proc. Natl. Acad. Sci. U.S.A. 99: 535–540.   Whitaker, T. W. and W. P. Bemis. 1975. Origin and evolution of the cultivated *Cucurbita*. Bull. Torrey Bot. Club 102: 362–368.   Wilson, H. D. 1990. Gene flow in squash species. BioScience 40: 449–455.   Wilson, H. D., J. Doebley, and M. Duvall. 1992. Chloroplast DNA diversity among wild and cultivated members of *Cucurbita* (Cucurbitaceae). Theor. Appl. Genet. 84: 859–865.

1. Leaf blades unlobed or shallowly 2-lobed, longer than broad . . . . 1. *Cucurbita foetidissima*
1. Leaf blades sinuate or deeply to shallowly 3–7-lobed, ± broader than long.
  2. Midveins of leaf lobes adaxially densely hispidulous-strigillose with white hairs; tendrils often gland-tipped; plants perennial; roots tuberous.
    3. Leaf blade sinuses nearly or completely to petiole, lobes narrowly lanceolate to narrowly oblanceolate or oblong-lanceolate . . . . 2. *Cucurbita digitata*
    3. Leaf blade sinuses ½–⅔ to petiole, lobes lanceolate-acuminate to triangular or triangular-lanceolate . . . . 3. *Cucurbita palmata*
  2. Midveins of leaf lobes not adaxially densely hispidulous-strigillose with white hairs; tendrils eglandular; plants usually annual, rarely short-lived perennial; roots taproots or fibrous.
    4. Corollas cream to pale creamy yellow; stems glabrous or glabrate to coarsely hirsute or villous with flattened, vitreous hairs, rarely sparsely short-hispid, without pustulate-based hairs; fruit flesh intensely bitter . . . . 4. *Cucurbita okeechobeensis*
    4. Corollas ± yellow to orange; stems hispid to puberulent, with or without pustulate-based hairs; fruit flesh bitter or not.
      5. Stems hispid with pustulate-based hairs.
        6. Stems hispid with persistent, strongly pustulate-based hairs and hispidulous-hirsutulous understory . . . . 5. *Cucurbita pepo*
        6. Stems densely puberulent to hirsutulous, scattered longer hairs with strongly to weakly pustulate bases absent or present . . . . 6. *Cucurbita melopepo* (in part)
      5. Stems villous to hirsute or puberulent to pubescent, or combination, usually without pustulate-based hairs.
        7. Peduncles in fruit relatively soft, corky-thickened, terete, not prominently ribbed, not abruptly expanded at point of fruit attachment . . . . 7. *Cucurbita maxima*
        7. Peduncles in fruit hardened, woody, 5-ribbed, abruptly expanded or not at point of fruit attachment.
          8. Peduncles in fruit abruptly expanded at point of fruit attachment; sepals narrowly lanceolate; fruit flesh usually yellow to orange to greenish; seeds whitish to cream or light brown, surface ± punctate-sculptured, margins raised-thickened and ± undulate, golden yellow to silvery . . . . 8. *Cucurbita moschata*
          8. Peduncles in fruit slightly expanded or not at point of fruit attachment; sepals linear to linear-lanceolate or subulate-linear; fruit flesh white to whitish or yellow to light orange or greenish; seeds dark brown to black or whitish to cream or tawny, surface ± smooth, margins raised-thickened and smooth, colored as seed surface.
            9. Anther filaments glabrous; fruit flesh usually yellow to light orange to greenish or whitish, bitter or not; seeds whitish to cream or tawny . . . . 6. *Cucurbita melopepo* (in part)
            9. Anther filaments short-villous; fruit flesh usually white, sweet; seeds usually dark brown to black, sometimes whitish . . . . 9. *Cucurbita ficifolia*

1. Cucurbita foetidissima Kunth in A. von Humboldt et al., Nov. Gen. Sp. 2(fol.): 98; 2(qto.): 123. 1817
• Buffalo or foetid or Missouri gourd, calabazilla, chili coyote

Cucurbita perennis (E. James) A. Gray; *Pepo foetidissima* (Kunth) Britton

Plants perennial; roots tuberous. **Stems** prostrate, sometimes rooting adventitiously at nodes, ca. 2–10 m, puberulent to scabrous with pustulate-based hairs; tendrils 3–7-branched 3–6 cm above base, hirsute, eglandular. **Leaves:** petiole 3–12 cm, coarsely hispid to hispidulous with puberulent understory of gland-tipped hairs; blade narrowly triangular or triangular-acuminate to triangular-lanceolate, unlobed or shallowly 2-lobed, (10–)12–30(–40) × (6–)8–20(–30) cm, longer than broad, base hastate-cordate to truncate, margins coarsely and widely mucronulate to denticulate, surfaces densely short-hirsute-pilose abaxially, short strigose-hirsute to hispid-hirsute adaxially, eglandular or minutely sessile-glandular. **Peduncles** in fruit 5-ribbed, slightly expanded or not at point of fruit attachment, hardened, woody. **Flowers:** hypanthium broadly campanulate, 10–18 mm; sepals narrowly lanceolate to linear-oblong or filiform-subulate, 10–25 mm; corolla golden yellow, campanulate to cylindric-campanulate, 6–10 cm; anther filaments usually sparsely short-villous with viscid-glandular hairs; ovary hirsute to short-villous. **Pepos** green with white stripes, white-mottled, evenly yellowish to orange-yellow at full maturity, depressed-globose to globose or oblong-globose, 5–10 cm, smooth. **Seeds** tan to cream or yellowish, oblong-ovate to ovate-elliptic, 9–13 mm, margins thickened-raised, surface smooth. $2n = 40, 42$.

Flowering May–Aug. Sandy fields and hills, sandsage prairies, dunes, gypsum hills, rocky soil, calcareous clay loam, grasslands, mesquite scrub, pinyon-juniper, floodplain woods, vacant lots, roadsides, railroad banks; 100–2000 m; Ariz., Ark., Calif., Colo., Fla., Ill., Ind., Iowa, Kans., Ky., Mich., Mo., Nebr., Nev., N.Mex., Ohio, Okla., Tex., Utah, Va., Wyo.; Mexico (Chihuahua, Coahuila, Nuevo León, San Luis Potosí, Sonora, Zacatecas); introduced in Europe (Germany).

Records of *Cucurbita foetidissima* in localities east of Missouri apparently represent adventives outside of the native range. The single known locality in Florida (Marion County) was where hay from the Midwest was thrown.

Cucurbita foetidissima has been studied as a source of root starch and seed oil and is a potentially productive crop adapted to arid and semiarid regions. The tuberous roots of an individual average 50 kilograms in weight in three to four growing seasons (J. S. DeVeaux and E. B. Shultz 1985).

SELECTED REFERENCE Bemis, W. P. et al. 1978. The feral buffalo gourd, *Cucurbita foetidissima*. Econ. Bot. 32: 87–95.

2. Cucurbita digitata A. Gray, Smithsonian Contr. Knowl. 5(6): 60. 1853 • Finger-leaf or coyote gourd F

Plants perennial; roots tuberous. **Stems** usually sprawling, sometimes climbing, often rooting adventitiously at nodes, to 10 m, sparsely hirsute-strigillose with deflexed hairs to hispid-strigose or hirsute-strigose, muriculate or glabrous on ribs; tendrils 2–5-branched 1–1.5 cm above base, glabrous, often gland-tipped. **Leaves:** petiole 3–4(–8) cm, hispid or hispid and hirsute, often with deflexed hairs; blade depressed-ovate to reniform, palmately 5-lobed, sinuses nearly or completely to petiole, 4–11 × 8–15 cm, usually broader than long, base cordate, lobes narrowly lanceolate to narrowly oblanceolate or oblong-lanceolate, 2–12 cm, margins coarsely toothed or remotely sinuate-dentate to serrate, surfaces hispid to hispidulous, midvein and major veins whitish adaxially, densely hispidulous-strigillose with white hairs, eglandular. **Peduncles** in fruit shallowly 5-ribbed, not abruptly expanded at point of fruit attachment, spongy. **Flowers:** hypanthium cylindric to narrowly campanulate, 2.5–3 mm; sepals linear-subulate, 3–5 mm; corolla bright yellow, narrowly campanulate, 4–7 cm; anther filaments glabrous; ovary villous-hirsute. **Pepos** dark green with 10 whitish stripes and white mottling or yellow at maturity, globose to depressed-globose or oblong-globose, (6–)7–9.5 cm, smooth, rind thin, hard-shelled. **Seeds** dull white, ovate, obtusely pointed, 8–11 mm, margins thickened-raised, surfaces smooth. $2n = 40$.

Flowering (Feb, Apr–)May–Sep(–Oct). *Larrea* desert, often with *Acacia*, *Cercidium*, and *Yucca*, grasslands with *Atriplex*, mesquite, juniper-scrub, oak savannas (rarely), canal banks, stream bottoms and sides, wash and arroyo banks, dry stream channels, alluvial plains, rocky slopes, disturbed sites, roadsides; (100–)300–1500 m; Ariz., Calif., N.Mex., Tex.; Mexico (Baja California, Chihuahua, Sonora).

In Texas, *Cucurbita digitata* is known by a single collection, from southeastern Presidio County in 1975, along gravelly banks of Fresno Creek (M. L. Butterwick 1980).

Leaves of juvenile plants or the first growth of the season of *Cucurbita digitata* produce shorter and broader lobes, but mature leaves are similar to those of *C. palmata* (R. S. Felger 2000). *Cucurbita digitata* and *C. palmata* intergrade where their ranges meet in southwestern Arizona, southeastern California, and northeastern Baja California; with *C. cordata* S. Watson and *C. cylindrata* L. H. Bailey of Baja California, they form a group of closely related, allopatric, mostly intergrading species (W. P. Bemis and T. W. Whitaker 1965, 1969).

3. **Cucurbita palmata** S. Watson, Proc. Amer. Acad. Arts 11: 137. 1876 • Coyote gourd or melon

Cucurbita californica Torrey ex S. Watson

Plants perennial; roots tuberous. **Stems** prostrate, often rooting adventitiously at nodes, to 2 m, villous-hirsute to hispid-hirsute with deflexed hairs, glabrescent and muriculate on angles in age; tendrils mostly 3(–5)-branched 1–1.5 cm above base, retrorsely hispid, gland-tipped. **Leaves:** petiole (2–)3–7 cm, densely hispid-hirsute with short, deflexed hairs mixed with pustulate hairs; blade suborbiculate to depressed-ovate, palmately 5-lobed, sinuses ½–⅔ to petiole, 3–7 × 4–10 cm, usually broader than long or equally so, base cordate, lobes lanceolate-acuminate to triangular or triangular-lanceolate, sometimes irregularly sublobed, margins remotely crenate to remotely serrate-crenate, surfaces appressed-hispid abaxially and muriculate, hirsute-strigillose to hispid-strigose adaxially, midvein and major veins whitish adaxially from densely hispidulous-strigillose with white hairs, eglandular. **Peduncles** in fruit shallowly 5-ribbed, not abruptly expanded at point of fruit attachment, 3–8 cm, spongy. **Flowers:** hypanthium cupulate to campanulate, 2–7(–20) mm; sepals linear-subulate, 2–10 mm; corolla golden yellow to yellow, tubular-campanulate, 2.5–5 cm; anther filaments glabrous; ovary densely pubescent. **Pepos** dull green, narrowly 10-striped and white-mottled, ellipsoid-globose to globose or depressed-globose, 7–10 cm, smooth. **Seeds** dull white, ovate to oblong, 9–14 mm, margins thickened-raised, surfaces smooth or slightly rough. $2n = 40$.

Flowering (Feb–)Apr–Sep(–Oct). Rocky lake shores, washes, stream beds and overflow channels, lava beds, roadsides, waste places, alkali plains, creosote bush scrub, saltbush scrub, grassland-saltbush scrub, annual grasslands, chaparral-desert scrub, upland desert scrub; (-30–)200–1000(–1300) m; Ariz., Calif., Nev., Utah; Mexico (Baja California, Sonora).

Some plants of *Cucurbita palmata* from Baja California and California have uniformly whitish gray, hirsute-strigillose adaxial leaf surfaces; these have been recognized as *C. californica*. Most plants from Arizona and eastern San Bernardino County, California, and some from elsewhere in Baja California and California, have the midrib and major veins whitish like those of *C. digitata*.

4. **Cucurbita okeechobeensis** (Small) L. H. Bailey, Gentes Herbarum 2: 179. 1930 [C]

Pepo okeechobeensis Small, J. New York Bot. Gard. 31: 12, figs. 1, 2. 1930

Subspecies 2 (1 in the flora): United States, e Mexico, West Indies (Hispaniola).

Cucurbita okeechobeensis originally was common in extensive pond-apple (or custard-apple, *Annona glabra*) forests on the southern shore of Lake Okeechobee (J. K. Small 1922). By the time it was described in 1930, most of these forests had been destroyed. Currently *C. okeechobeensis* is known only in two population systems separated by more than 160 km: the shoreline and islands around the northwestern and southern portions of Lake Okeechobee in Glades and Palm Beach counties and both sides of the Saint Johns River from Lake Beresford south to Lake Monroe and along the western side of Lake Jessup along the Volusia–Lake county line and into Seminole County.

Decline and current rarity of *Cucurbita okeechobeensis* have resulted mostly from conversion of swamp forests to agriculture and from water-level management in Lake Okeechobee. Fluctuations in lake level are necessary as high levels facilitate dispersal, and the inundation and withdrawal create open habitats for quick growth of the seedlings. Exotic vegetation around the edges of the lake, especially *Melaleuca*, also is a threat.

Cucurbita okeechobeensis subsp. *martinezii* (L. H. Bailey) Andres & Nabhan ex T. Walters & D. S. Decker (*C. martinezii* L. H. Bailey) is widespread in a narrow strip of eastern Mexico (Chiapas, Oaxaca, Puebla, San Luis Potosí, Tamaulipas, Veracruz), often in riparian vegetation and as a weed in coffee and citrus plantations. Subspecies *okeechobeensis* is morphologically distinct only in its densely pubescent staminate hyphantia and pubescent pistillate ovaries. There is no sterility barrier between the two, and F_1 hybrids are fertile (R. W. Robinson and J. T. Puchalski 1980). There is considerable allelic variation within subsp. *martinezii* (T. C. Andres and G. P. Nabhan 1988) but little or none in subsp. *okeechobeensis*.

R. W. Robinson and J. T. Puchalski (1980) found through isozyme analysis a single allelic difference between the Okeechobee and Martinez gourds. T. W. Walters and D. S. Decker (1991) confirmed the minor allelic difference and estimated the time of divergence as about 450,000 years, with the conclusion that the two were appropriately treated as conspecific. A population of *Cucurbita okeechobeensis* was recently discovered in the Dominican Republic (B. Peguero and F. Jiménez 2005); the authors were unable to assign the plants to either subspecies, and it is unclear whether the population is natural or the result of introduction within the last century (M. Nee, pers. comm.).

The closest relative of *Cucurbita okeechobeensis* and *C. martinezii* appears to be *C. lundelliana* L. H. Bailey (T. C. Andres and G. P. Nabhan 1988), which is geographically isolated in the Yucatán Peninsula but can form artificial hybrids with them. *Cucurbita lundelliana* differs in its orange corollas, less lignified exocarp, and seeds with crenulate margins.

SELECTED REFERENCE   Walters, T. W. and D. S. Decker. 1993. Systematics of the endangered Okeechobee gourd (*Cucurbita okeechobeensis*: Cucurbitaceae). Syst. Bot. 18: 175–187.

4a. Cucurbita okeechobeensis (Small) L. H. Bailey subsp. **okeechobeensis** • Okeechobee gourd [C] [E]

Plants annual; roots fibrous. **Stems** prostrate or climbing, sometimes rooting adventitiously at nodes, to 12 m, usually coarsely hirsute to villous with flattened, vitreous hairs to glabrate or glabrous, rarely sparsely short-hispid, without pustulate-based hairs; tendrils 3-branched 1–3 cm above base, glabrous, eglandular. **Leaves:** petiole 5–15 cm, sparsely hispid to pilose-hirsute or glabrous or glabrate; blade orbiculate to suborbiculate or reniform, shallowly to deeply 5–7-lobed, 4.5–15 × 8–20 cm, usually broader than long or equally so, base cordate, lobes triangular to acuminate-triangular or subrhombic, midveins of leaf lobes not distinctly elongate-whitened, margins closely mucronulate to denticulate, sometimes irregularly toothed, surfaces hispid to hispid-hirsute or villous-hirsute to sparsely hirsute-villous or glabrate, sometimes mottled with silvery-green adaxially, eglandular. **Peduncles** in fruit shallowly 5-ribbed, abruptly expanded at point of fruit attachment, hardened, woody. **Flowers:** hypanthium campanulate, 5–6 mm; sepals lanceolate to subulate, 5–8 mm; corolla cream to pale creamy yellow, campanulate, 6–9 cm (staminate shorter); anther filaments glabrous; ovary densely pubescent. **Pepos** light green with pale white stripes at maturity, green-streaked or flecked with white or dark green, tan to brown when dry, globose to broadly ovoid, (5–)7–9(–11) cm, surface smooth; flesh intensely bitter. **Seeds** whitish to light gray, ovate, 7–12 mm, margins thickened-raised, surfaces smooth. **2*n* = 40.**

Flowering Mar–Nov. Pond-apple swamps, floodplain forests, power-line cuts through cypress-hardwood swamps, hardwood swamp edges, spoil islands, climbing on pond apple, elderberry, and buttonbush, often associated with alligator nests, canal and ditch banks, wet road shoulders; of conservation concern; 0–20 m; Fla.

Subspecies *okeechobeensis* is in the Center for Plant Conservation's National Collection of Endangered Plants.

5. Cucurbita pepo Linnaeus, Sp. Pl. 2: 1010. 1753 [I]

Subspecies 2 (1 in the flora): introduced; Mexico, Central America (Guatemala); introduced also nearly worldwide.

Archaeological and molecular-genetic research, especially data from mitochondrial DNA and RAPD studies (O. I Sanjur et al. 2002; D. S. Decker et al. 2002b) and earlier isozymic and chloroplast DNA studies (for example, Decker et al. 1993), indicates that *Cucurbita pepo* in the broad sense includes two lineages: (1) *C. pepo* in the strict sense, a Mexican lineage of domesticates that differs from plants generally identified previously as *C. pepo* subsp. *ovifera* (here as *C. melopepo*) by a derived molecular feature (a difference in three adjacent base pairs) that occurs also in the *C. moschata* and *C. sororia* L. H. Bailey and *C. argyrosperma* Huber groups, and was shared presumably by the wild ancestor of *C. pepo*, which is unknown and possibly extinct; and (2) *C. melopepo*, a lineage of northeastern Mexico and the eastern Unites States in which the three wild varieties (var. *fraterna*, var. *ozarkana*, var. *texana*) and the domesticated variety (var. *ovifera*) share identical mitochondrial DNA sequences (Sanjur et al.) as well as similarities in isozymes and other kinds of DNA. Domesticates of *C. pepo* and *C. melopepo* are independently derived lineages.

Cucurbita pepo subsp. *gumala* Teppner comprises domesticates from Guatemala and adjacent southern Mexico and apparently is native there (H. Teppner 2000, 2004). The plants have depressed-globose pepos 13–20 cm in diameter with extremely thick rind, ripening orange-yellow, and with orange flesh. Teppner observed that the fruits of subsp. *gumala* are similar to the ancient ones from Guilá Naquitz cave in Oaxaca.

Cultivars of *Cucurbita pepo* with edible pepos have been divided into eight groups (H. S. Paris 1986, 1989; see also E. F. Castetter 1925), based mainly on pepo morphology. Pepos of cultivated forms differ from those of their wild ancestors in their larger size and more variable shape, less durable and more varicolored rinds, and less fibrous, nonbitter flesh.

Plants of *Cucurbita pepo* are likely to be found as non-persistent waifs all over the world, wherever they can be grown in temperate and upland tropical areas.

SELECTED REFERENCES Andres, T. C. and H. B. Tukey. 1995. Complexities in the infraspecific nomenclature of the *Cucurbita pepo* complex. Acta Hort. 413: 65–91. Decker, D. S. 1988. Origin(s), evolution, and systematics of *Cucurbita pepo* (Cucurbitaceae). Econ. Bot. 42: 4–15. Decker, D. S. 1990. Evidence for multiple domestications of *Cucurbita pepo*. In: D. M. Bates et al., eds. 1990. Biology and Utilization of the Cucurbitaceae. Ithaca, N.Y. Pp. 96–101. Decker, D. S. et al. 1993. Isozymic characterization of wild populations of *Cucurbita pepo*. J. Ethnobiol. 13: 55–72. Decker, D. S. et al. 2002b. Diversity in free-living populations of *Cucurbita pepo* (Cucurbitaceae) as assessed by random amplified polymorphic DNA. Syst. Bot. 27: 19–28. Decker, D. S. and H. D. Wilson. 1987. Allozyme variation in the *Cucurbita pepo* complex: *C. pepo* var. *ovifera* vs. *C. texana*. Syst. Bot. 12: 263–273. Paris, H. S. 1986. A proposed subspecific classification for *Cucurbita pepo*. Phytologia 61: 133–138. Paris, H. S. 1989. Historical records, origins, and development of the edible cultivar groups of *Cucurbita pepo* (Cucurbitaceae). Econ. Bot. 43: 423–443. Paris, H. S. 1996. Summer squash: History, diversity, and distribution. HortTechnology 6: 6–13. Paris, H. S. 2001. History of the cultivar-groups of *Cucurbita pepo*. Hort. Rev. 25: 71–170. Teppner, H. 2000. *Cucurbita pepo* (Cucurbitaceae)—History, seed coat types, thin coated seeds and their genetics. Phyton (Horn) 40: 1–42.

5a. Cucurbita pepo Linnaeus subsp. **pepo** • Field or jack-o-lantern pumpkin, cocozelle, vegetable marrow, zucchini, citrouille [I]

Cucurbita aurantia Willdenow; *C. pepo* var. *medullosa* Alefeld

Plants annual; roots taproots or fibrous. **Stems** creeping or climbing, rooting adventitiously at nodes, to 3 m, hispid with persistent, strongly pustulate-based hairs and a hispidulous-hirsutulous understory; tendrils 2–7-branched 1–5 cm above base, hispidulous to hirsutulous, eglandular. **Leaves:** petiole 5–20(–24) cm, pustulate-hispid (pustulate-based) and hispidulous-hirsutulous; blade sometimes white-spotted at vein junctions, broadly ovate-cordate to triangular-cordate or reniform, shallowly to deeply palmately (3–)5–7-lobed, 20–30 × 20–35 cm, usually broader than long or equally so, base cordate, lobes ovate-deltate to obovate or obovate-rhombic, midveins of leaf lobes not distinctly elongate-whitened, margins denticulate to serrate-denticulate, surfaces hirsute, hirsute-strigillose, villous-strigose, or hispidulous-scabrous, eglandular. **Peduncles** in fruit 5-ribbed, not or very gradually expanded at point of fruit attachment, hardened, woody. **Flowers:** hypanthium campanulate, 8–12 mm; sepals linear to subulate-linear, 8–25 mm; corolla yellow to golden yellow or orange, tubular-campanulate, 4–10 cm; anther filaments glabrous; ovary villous. **Pepos** wholly light to dark green, green with white stripes, or minutely cream- or green-speckled to yellow, orange, or bicolored green and yellow, globose or depressed-globose to ovoid, obovoid, ellipsoid-ovoid, broadly ellipsoid, slightly pyriform, cushion-shaped, or cylindric, 5–10(–25) cm, usually smooth or ribbed, rarely verrucose, flesh whitish to yellowish or pale orange, soft, not bitter. **Seeds** whitish to cream or tawny, narrowly to broadly elliptic to obovate, rarely orbiculate, 7–15(–26) mm, margins raised-thickened, surface ± smooth. $2n = 40$.

Flowering Jun–Oct. Vacant lots, trash heaps, roadsides, disturbed sites; 0–300 m; introduced; Ala., Calif., Conn., Kans., Ky., La., Mass., Mich., Nev., N.H., N.Mex., N.Y., N.C., Ohio, Pa., S.C., Tenn., Utah, Va.; Mexico; South America; introduced also in West Indies, Central America, Eurasia, Atlantic Islands.

6. Cucurbita melopepo Linnaeus, Sp. Pl. 2: 1010. 1753 [F]

Plants annual; roots taproots or fibrous. **Stems** creeping or climbing, rooting adventitiously at nodes, sometimes bushy to shrublike in cultivated forms, to 3 m, densely puberulent to hirsutulous, with scattered, longer hairs with strongly to weakly pustulate (many-celled) bases absent or present; tendrils 2–7-branched 2–3 cm above base, hirsutulous to pubescent, eglandular, tendrils sometimes absent in plants with bushy habit. **Leaves:** petiole 4–9(–16) cm, hispidulous to hirsutulous, sometimes minutely stipitate-glandular; blade sometimes white-spotted, broadly ovate-cordate to triangular-cordate or suborbiculate, shallowly to deeply palmately (3–)5–7-lobed, 4–15(–18) × 5–17(–23) cm, ± as broad as long, base cordate, lobes ovate-deltate to obovate or obovate-rhombic, midveins of leaf lobes not distinctly elongate-whitened, margins denticulate to serrate-denticulate, surfaces hispidulous to hirsutulous, eglandular. **Peduncles** in fruit 5-ribbed, slightly expanded at point of fruit attachment, hardened, woody. **Flowers:** hypanthium campanulate, 8–12 mm; sepals linear to subulate-linear, 8–25 mm; corolla yellow to golden yellow or orange, tubular-campanulate, 4–10 cm; anther filaments glabrous; ovary pubescent. **Pepos** wholly light to dark green, green with white stripes, or minutely cream- or green-speckled to slate blue, ivory, yellow, orange, or bicolored green and yellow, globose or depressed-globose to ovoid, obovoid, ellipsoid-ovoid, broadly ellipsoid, slightly pyriform, cushion-shaped, or cylindric, 4–10 cm, smooth, ribbed, or warty, flesh yellow to light orange or greenish or whitish, bitter or not. **Seeds** whitish to cream or tawny, narrowly to broadly elliptic to obovate, rarely orbiculate, 7–15(–26) mm, margins raised-thickened and smooth, surface ± smooth. $2n = 40$.

Subspecies 3 (2 in the flora): United States, ne Mexico; introduced widely.

The wild race of *Cucurbita melopepo* endemic to northeastern Mexico (Nuevo Léon and Tamaulipas) is subsp. *fraterna* (L. H. Bailey) G. L. Nesom, representing a single variety, var. *fraterna* (L. H. Bailey) G. L. Nesom [*C. fraterna* L. H. Bailey; *C. pepo* var. *fraterna* (L. H. Bailey) Filov et al.; *C. pepo* subsp. *fraterna* (L. H. Bailey) Lira, Andres & M. Nee]. These plants grow in lowlands (below 900 m) and are common as weeds in agricultural fields (R. Lira et al. 2009). Variety *fraterna* is indicated to have a sister relationship to the more northern elements of *C. melopepo* (D. S. Decker et al. 2002b).

Evidence indicates that most or all of the domesticated forms of *Cucurbita melopepo* (var. *melopepo*) are derived from var. *ozarkana* (see comments below). For consistency within *Cucurbita* and other genera where wild progenitor/domesticate pairs are identified at subspecific rank, the var. *melopepo* domesticates are recognized also as subsp. *melopepo*, coordinate with *C. melopepo* subsp. *texana* (Scheele) G. L. Nesom, which comprises the two wild varieties of the United States.

The discrete, molecularly delineated lineages within *Cucurbita melopepo* are not so clearly distinguished by morphological features other than mature fruit shape and size. The key below uses features provided in discussions accompanying the molecular studies. Variation in potentially diagnostic features, especially in vesture and fruit coloration, suggests that further study is needed.

1. Pepos solid or striped, multiple colors, including yellow and orange, rind smooth, ribbed, or warty, flesh nonbitter (except for some ornamental gourd cultivars); seed germination within 3–15 days, depending on cultivar. . . . . .6a. *Cucurbita melopepo* subsp. *melopepo*
1. Pepos solid ivory or green-and-white-striped but usually not yellow or orange, rind smooth, flesh almost always bitter; seed germination within 1–7 days . . . . . . 6b. *Cucurbita melopepo* subsp. *texana*

6a. Cucurbita melopepo Linnaeus subsp. **melopepo** [E]

Pepos solid or striped, multiple colors, including yellow and orange, rind smooth, ribbed, or warty, flesh not bitter (except for some ornamental gourd cultivars). **Seeds** germinating within 3–15 days, depending on cultivar.

Variety 1: United States; introduced and cultivated nearly worldwide.

6a.1. Cucurbita melopepo Linnaeus var. **melopepo**
• Scallop, pattypan, crookneck, straightneck, acorn, summer, or spaghetti squash [E]

Cucurbita ovifera Linnaeus; *C. pepo* Linnaeus var. *condensa* L. H. Bailey; *C. pepo* subsp. *ovifera* (Linnaeus) D. S. Decker; *C. pepo* var. *ovifera* (Linnaeus) Alefeld; *C. subverrucosa* Willdenow; *C. verrucosa* Linnaeus

Plants domesticated, sometimes escaping. **Pepos** solid or striped, multiple colors, including yellow and orange, rind smooth, ribbed, or warty, flesh not bitter (except for some ornamental gourd cultivars). **Seeds:** germination within 3–15 days, depending on cultivar.

Flowering Jun–Oct. Vacant lots, trash heaps, roadsides, disturbed sites; 100–200 m; Ark., Calif., Ga., Ill., Ky., La., Miss., Mo., Tex., W.Va.; introduced and cultivated nearly worldwide.

All forms of var. *melopepo* are domesticated; these include most ornamental gourds (for example, Flat-striped, Striped Pear, Bicolor Spoon, Miniature Ball, Mandan, Orange Warted, Warty Hardhead) and the acorn, crookneck, straightneck, scallop, pattypan, summer, and spaghetti squashes. They are hypothesized to be derivatives of var. *ozarkana* (D. S. Decker et al. 1993, 2002b), as var. *texana* has an isozyme pattern distinct from the others (Decker et al. 1993). Wild forms of subsp. *melopepo* once may have been distributed as far as Florida (L. A. Newsom et al. 1993).

Some weedy-habitat populations of *Cucurbita melopepo* in Illinois and Kentucky may have evolved purely as cultivar escapes; these may also have experienced subsequent introgression with other, nearby, cultivated, weedy, or wild populations of var. *ozarkana* (D. S. Decker et al. 2002b and references therein).

6b. Cucurbita melopepo Linnaeus subsp. **texana** (Scheele) G. L. Nesom, Phytoneuron 2011-13: 24. 2011 [E] [F]

Tristemon texanus Scheele, Linnaea 21: 586. 1848 (as texanum); *Cucurbita pepo* Linnaeus subsp. *texana* (Scheele) Filov

Pepos solid ivory or green-and-white-striped but usually not yellow or orange, rind smooth, flesh almost always bitter. **Seeds** germinating within 1–7 days.

Varieties 2 (2 in the flora): c, sc United States.

1. Pepos usually green-and-white-striped, sometimes yellow; seed germination within 3–7 days . . . . .6b.1. *Cucurbita melopepo* var. *texana*
1. Pepos usually solid ivory, sometimes green-and-white-striped; seed germination within 1–4 days . . . . .6b.2. *Cucurbita melopepo* var. *ozarkana*

6b.1. Cucurbita melopepo Linnaeus var. **texana** (Scheele) G. L. Nesom, Phytoneuron 2011-13: 24. 2011 E F

Tristemon texanus Scheele, Linnaea 21: 586. 1848 (as texanum); *Cucurbita pepo* Linnaeus var. *texana* (Scheele) D. S. Decker

Pepos usually green and white striped at maturity, sometimes yellow. **Seeds:** germination within 3–7 days.

Flowering Jun–Nov. Stream beds, lake shores, marsh banks, low woods, dunes, disturbed sandy sites; 0–100 m; Tex.

Isozyme and RAPD data indicate that var. *texana* is limited to south-central Texas (D. S. Decker et al. 1993, 2002b; C. W. Cowan and B. D. Smith 1993). Attribution of var. *texana* to New Mexico apparently has been erroneous. Plants associated with the Mississippi River drainage are var. *ozarkana*, according to the molecular analyses.

Molecular differentiation between var. *ozarkana* and var. *texana* suggests that reproductive isolation has allowed an accumulation of differences; H. Teppner (2004) regarded morphological variability as too overlapping to allow their unarbitrary separation. Molecular data reliably separate them and clearly delimit var. *fraterna* as well; morphology needs to be restudied. Fruits of var. *ozarkana* usually are ivory white at maturity versus striped in var. *texana*, but this is not completely consistent (T. C. Andres and H. B. Tukey 1995; pers. obs.).

6b.2. Cucurbita melopepo Linnaeus var. **ozarkana** (D. S. Decker) G. L. Nesom, Phytoneuron 2011-13: 24. 2011 E

Cucurbita pepo Linnaeus var. *ozarkana* D. S. Decker, J. Ethnobiol. 13: 69. 1993

Pepos usually solid ivory, sometimes green and white striped, usually not yellow or orange. **Seeds:** germination within 1–4 days.

Flowering Jun–Oct. Stream banks, gravel bars, bottomland forests, soybean, corn, and cotton fields, old fields, fencerows, railroad rights-of-way, roadsides, disturbed sites; 20–200 m; Ala., Ark., Ill., Ky., La., Miss., Mo., Okla.

Variety *ozarkana* occurs in the central Mississippi valley and the Ozark Plateau. Populations in Alabama were included in var. *ozarkana* by C. W. Cowan and B. D. Smith (1993), but their evolutionary status was considered uncertain by D. S. Decker et al. (1993).

7. **Cucurbita maxima** Duchesne, Essai Hist. Nat. Courges, 7, 12. 1786 • Winter or Hubbard or blue Hubbard or golden Hubbard or turk's turban or banana or Queensland blue or buttercup or Hokkaido squash, winter marrow, Queensland blue or Atlantic giant or mammoth pumpkin I

Plants annual; roots fibrous. **Stems** climbing, often rooting adventitiously at nodes, 1–5 m, short-pilose to villous or hirsute-villous, usually without pustulate-based hairs; tendrils 2–5-branched ca. 5 cm above base, glabrous, eglandular. **Leaves:** petiole (5–)10–20 cm, villous-hirsute to hispid and sparsely aculeate; blade occasionally with white blotches, ovate to depressed-ovate or reniform, sinuate to shallowly 5–7-lobed, (7–)15–30 × 20–36 cm, broader than long, base cordate, lobes rounded to obtuse, midveins of leaf lobes not distinctly elongate-whitened, margins closely denticulate, surfaces hispid-aculeate primarily on veins, without pustulate-based hairs, eglandular. **Peduncles** in fruit terete, not prominently ribbed, expanded along whole length, not abruptly expanded at point of fruit attachment, relatively soft and corky-thickened. **Flowers:** hypanthium campanulate, 20–25 mm; sepals subulate to linear, 5–20 mm; corolla yellow to orange-yellow, campanulate, 5–7(–8) cm; anther filaments glabrous; ovary pubescent. **Pepos** green to gray-green with cream stripes or mottling, golden yellow to orange, dark purplish green or bluish, blackish purple, or white to grayish, globose to depressed-globose to ovoid or obovoid, oblong-cylindric, or flattened-cylindric, 10–40 cm, smooth, flesh yellow to orange, not bitter. **Seeds** whitish to gray or pale brown, suborbiculate to broadly elliptic or obovate, 12–22(–32) mm, margins raised-thickened or not, sometimes slightly darkened, surfaces smooth or slightly rough. $2n = 40$.

Flowering Jun–Oct. Abandoned agricultural fields, fields, roadsides, disturbed sites, trash heaps; 50–200 m; introduced; Ark., Ga., Maine, Mass., Mich., N.Y., N.C., Ohio, Pa., S.C., Utah, Vt., Va., Wis.; West Indies; South America; introduced also elsewhere in South America (Argentina), Europe (Denmark, England, Germany, Hungary, Spain), Pacific Islands (New Zealand), Australia.

Cucurbita andreana Naudin, a wild species native to Argentina and Bolivia, hybridizes readily with *C. maxima* and is its ancestor (O. I. Sanjur et al. 2002); it sometimes is recognized as *C. maxima* subsp. *andreana* (Naudin) Filov.

Some fruits of *Cucurbita maxima* have a high sugar content and are used for making pies, and they are popular as a soup, especially in Brazil and Africa.

All of the giant pumpkins in weigh-off contests are derived from *Cucurbita maxima*, as are some of the Halloween pumpkins. In 1904, the largest pumpkin was 403 pounds, and winners increased relatively little to 459 pounds in 1980. A rapid increase in size began in 1981, with the champion at 493.5 pounds; from this individual's lineage came seeds for the Atlantic Giant cultivar, which has contributed since to winners burgeoning in size. The first giant pumpkin over 1000 pounds (1061 pounds) was grown in 1996; by 2009 the winner was 1725 pounds and in 2010, 1810 pounds.

8. **Cucurbita moschata** Duchesne, Essai Hist. Nat. Courges, 7, 15. 1786 • Butternut or Tahitian squash, golden cushaw, calabaza, neck or West Indian or Seminole or large cheese or Long Island cheese or Kentucky field or Dickinson pumpkin, Tennessee sweet potato ⒤

Pepo moschata (Duchesne) Britton

Plants annual; roots taproots or fibrous. **Stems** creeping or climbing, rooting adventitiously at nodes, to 10+ m, villous-hirsute with mixture of longer, thick, vitreous hairs with conspicuous cross-walls and puberulent understory of much shorter hairs, without pustulate-based hairs; tendrils 3–5-branched 1.5–8 cm above base, glabrous, eglandular. **Leaves:** petiole 4–25(–40) cm, villous-hirsute with mixture of longer, thick, vitreous hairs with conspicuous cross-walls and puberulent understory of much shorter hairs, without pustulate-based hairs; blade sometimes white-mottled abaxially, suborbiculate to broadly ovate, depressed-ovate, or reniform, shallowly 3–5(–7)-lobed, 5–25 × (8–)10–25(–30) cm, broader than long, base cordate, lobes ovate to broadly triangular or broadly obovate, midveins of leaf lobes not distinctly elongate-whitened, margins closely serrate-denticulate or serrulate-apiculate to denticulate or mucronulate, surfaces densely hirsute to hirsutulous abaxially, less densely hairy adaxially, eglandular. **Peduncles** in fruit 5-ribbed, abruptly expanded at point of fruit attachment, hardened, woody. **Flowers:** hypanthium cupulate, 5–8 mm; sepals narrowly lanceolate, distally foliaceous, 15–25 mm; corolla yellow, tubular-campanulate, 5–7 cm; anther filaments glabrous or sparsely puberulent

at base; ovary pubescent. **Pepos** evenly light or dark green or cream-speckled to evenly light or dark brown, speckled or not, or wholly white, globose or depressed-globose to ovoid, conic, cylindric, pyriform, or lageniform, 10–40(–120) cm, usually smooth or with rounded ribs, rarely with small, raised, wartlike spots, flesh yellow to light or bright orange to greenish, lightly to very sweet. **Seeds** whitish to cream or light brown with golden-yellow to silvery margins, ovate-elliptic to elliptic or obovate, 8–21 mm, margins raised-thickened, ± undulate, surface ± punctate-sculptured. $2n = 40$.

Flowering May–Oct. Oak-pine woods, abandoned agricultural fields, roadsides, disturbed sites; 0–100 m; introduced; Fla., Ga., Ky., La., Miss., N.C., Pa., S.C., Tenn., Tex., Va.; w South America; introduced also in Mexico, West Indies, Central America, elsewhere in South America (French Guiana, Guyana, Surinam), Pacific Islands (Galapagos Islands).

Cucurbita moschata is the primary squash of lowland, humid, tropical and subtropical areas throughout the world. It seems likely that it occurs at least as a waif in more areas than indicated in currently available databases of invasive species.

Fruits of *Cucurbita moschata*, especially the cheese pumpkins, are favorites for making pumpkin pie. Compared to a Halloween "jack-o-lantern" (a "pepo" pumpkin), flesh of a Moschata pumpkin is more richly colored, higher in nutrients and sugars, and has a denser, smoother-grained flesh. The "cheese" name alludes to the pumpkin's resemblance to a wheel of cheddar.

Cucurbita moschata has sometimes been cited as *C. moschata* (Duchesne ex Lamarck) Duchesne ex Poiret, based on *C. pepo* var. *moschata* Duchesne ex Lamarck, but the epithet appeared first at specific rank, slightly earlier than in the work by Lamarck.

The wild ancestor of *Cucurbita moschata* is unknown, but mitochondrial DNA data combined with other information suggest that it will be found in lowland northern South America (O. I. Sanjur et al. 2002).

9. **Cucurbita ficifolia** Bouché, Verh. Vereins. Beförd. Gartenbaues Königl. Preuss. Staaten 12: 205. 1837 • Fig leaf or Malabar gourd, fig leaf or blackseed squash, chilacayote, shark fin melon, cidra, Asian pumpkin, Thai marrow ⒤

Plants annual or short-lived perennial; roots taproots. **Stems** prostrate or climbing, often rooting adventitiously at nodes, to 25 m, usually sparsely hirsute with puberulent understory of gland-tipped hairs, sometimes without hirsute overstory, without pustulate-based hairs; tendrils 3- or 4-branched 3–5 cm above base, glabrous, eglandular. **Leaves:** petiole 8–20 cm, sparsely hirsute to hispid-hirsute, without pustulate-based hairs; blade sometimes mottled with silvery green, suborbiculate to ovate or reniform, palmately 3–5-lobed, 11–16 × 15–26 cm, usually broader than long or equally so, base cordate, sometimes shallowly and irregularly toothed, lobes with sinuses shallow or ± ½ to petiole, obovate to ovate, depressed-ovate, or triangular, midveins of leaf lobes not distinctly elongate-whitened, margins closely mucronulate to denticulate, surfaces short-villous to pubescent, eglandular. **Peduncles** in fruit 5-ribbed, slightly or not expanded at point of fruit attachment, hardened, woody. **Flowers:** hypanthium campanulate, 5–10 mm; sepals linear-lanceolate, not foliaceous, 5–15 mm; corolla yellow to yellow-orange, tubular-campanulate, 6–12 cm (staminate at shorter end of range); anther filaments sparsely short-villous, hairs viscid-glandular; ovary densely pubescent. **Pepos** (1) light or dark green, with or without longitudinal white lines or stripes distally, (2) irregularly, often linearly mottled, green and white, or (3) whitish to cream, globose to broadly ovoid or ovoid-elliptic, (15–)20–50 cm, smooth, flesh usually white, sweet. **Seeds** usually dark brown to black, sometimes whitish, ovate to ovate-elliptic or oblong-elliptic, 15–25 mm, margins raised-thickened and smooth, surfaces smooth. $2n = 40$.

Flowering Jul–Oct. Old habitations, open woods; 100–1500 m; introduced; Calif.; South America; introduced also in Mexico, Central America, Europe (England, Germany, Spain), Asia, Pacific Islands (Galapagos Islands, New Zealand).

Cucurbita ficifolia is included here on the basis of two collections from California: one from Ventura County in 1968 from the site of an old habitation at about 50 m; the other from San Bernardino County in 1948 from a wooded area at about 1400 m.

The wild ancestor of *Cucurbita ficifolia* is not known, but it grows in cool, high-elevation ecological zones and is considered to have originated in South America (M. Nee 1990; O. I. Sanjur et al. 2002), whence the only reliable archaeological records.

SELECTED REFERENCE Andres, T. C. 1990. Biosystematics, theories on the origin, and breeding potential of *Cucurbita ficifolia*. In: D. M. Bates et al., eds. 1990. Biology and Utilization of the Cucurbitaceae. Ithaca, N.Y. Pp. 102–119.

DATISCACEAE Dumortier

• Datisca Family

Walter C. Holmes

Herbs or subshrubs, perennial. **Leaves** alternate, pinnately incised to compound, sometimes simple; stipules absent; petiole present. **Inflorescences** axillary fascicles. **Flowers** usually unisexual (mostly bisexual in *Datisca glomerata*); calyx unequally 4–9-lobed; petals 0; staminate flowers: stamens [6–]8–12[–25]; filaments distinct, very short; pistillate flowers: calyx tube adnate to ovary; ovary inferior, 3–8-carpellate; placentation parietal; ovules 24–64; styles 3, threadlike, deeply forked at apex [subulate or longer with capitate-peltate or clavellate stigmas]; bisexual flowers often with staminodes. **Fruits** capsular, opening apically between styles. **Seeds** 100–300; embryo straight; endosperm little or none.

Genus 1, species 2 (1 in the flora): sw United States, nw Mexico, sw Asia (Mediterranean region to the Himalayas).

1. DATISCA Linnaeus, Sp. Pl. 2: 1037. 1753; Gen. Pl. ed. 5, 460. 1754 • [Roman common name for a species of *Catananche* Linnaeus (Asteraceae); reason for application here obscure]

Tricerastes C. Presl

Stems hollow or pithy. **Flowers:** unisexual or bisexual. **Capsules** 3–5-angled. $x = 11$.

Species 2 (1 in the flora): sw United States, nw Mexico, sw Asia from e Mediterranean region to Himalayas.

Datisca cannabina Linnaeus, akalbir, of southwestern Asia, is occasionally used as an ornamental; it also yields a yellow dye once used for coloring silk.

SELECTED REFERENCE Wolf, D. E. et al. 2001. Sex determination in the androdioecious plant *Datisca glomerata* and its dioecious sister species *D. cannabina*. Genetics 159: 1243–1257.

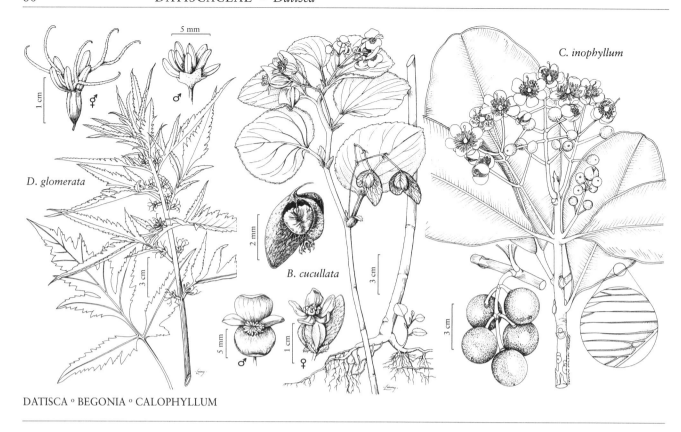

D. glomerata

B. cucullata

C. inophyllum

DATISCA ° BEGONIA ° CALOPHYLLUM

1. Datisca glomerata (C. Presl) Baillon, Hist. Pl. 3: 407. 1871 · Durango root F

Tricerastes glomerata C. Presl, Reliq. Haenk. 2: 88, plate 64. 1835

Plants erect, 1–1.5(–2+) m. **Rootstock:** inner bark yellow. **Stems** 1–30+ from base, glabrous. **Leaves** often appearing subopposite to semiwhorled proximally, gradually smaller distally, simple to asymmetrically laciniate-pinnate or shallowly pinnatifid to incised-trifoliolate, lobes deeply cleft; petiole 2–4 cm; blade lanceolate to ovate, 6–20 × 1–13 cm, margins coarsely serrate, apex acuminate to long-acuminate, surfaces glabrous. **Flowers:** staminate: calyx 2 mm, lobes slender, 0.2–1 mm; stamens borne on rim of calyx, nearly sessile or filaments to 1 mm; anthers yellow, 3–5 mm; bisexual: hypanthium 5–8 mm; calyx lobes narrow, 1–2 mm; ovary usually 3-angled; pistillate: rare, similar to bisexual flowers. **Capsules** ovoid-oblong, 7–11 mm, apex truncate. **Seeds** ellipsoid, 0.8–1 mm, pitted in longitudinal rows.

Flowering late Apr–Aug. Dry streambeds and washes, springs, wet sand, moist areas; 10–2000 m; Calif., Nev.; Mexico (Baja California).

Flowers of *Datisca glomerata* are primarily bisexual, each bearing one to four or more stamens. An occasional plant bears only staminate flowers. Pistillate flowers are rare, and no plants with only these have been observed.

All parts of the plant are reportedly poisonous and have shown some antitumor activity. The species is sometimes cultivated as an unusual ornamental.

The combination *Datisca glomerata* sometimes has been attributed to (C. Presl) Bentham & Hooker f.; it was not validly published by those authors.

BEGONIACEAE C. Agardh

- Begonia Family

Luc Brouillet

Herbs, annual or perennial [canelike shrubs, rarely treelets, rarely climbing, sometimes epiphytic], evergreen [some tuberous species seasonally absent, notably *Hillebrandia*], ± succulent, usually hairy or scaly, sometimes glabrous. **Leaves** alternate in 2 ranks (spirodistichous) [rarely opposite or whorled], usually persistent, rarely deciduous, simple [palmately or pinnately compound], stipulate, petiolate [sessile]; stipules persistent [deciduous]; blade: base asymmetric [subsymmetric], venation palmate [rarely pinnate]. **Inflorescences** axillary cymes [rarely solitary flowers], pedunculate [sessile]. **Flowers** unisexual (staminate flowering first, caducous), pistillate and staminate usually in same inflorescence [rarely in distinct ones], perianth and androecium epigynous, perianth radially [sometimes bilaterally] symmetric, petaloid, white to pink [yellow, orange, red], distinct [rarely ± connate]; staminate flowers: tepals 2–4[–6, rarely more; 5 sepals and 5 petals in *Hillebrandia*], stamens [3–]6–33[–100+], distinct [filaments basally connate], [sometimes collectively bilaterally symmetric]; pistillate flowers: tepals [2–]4 or 5[+] [5 sepals, 5 petals in *Hillebrandia*], ovary inferior [semi-inferior in *Hillebrandia*], [2]3[–9]-carpellate, [1–]3[–6]-winged [horned, ribbed, or angled, rarely not winged], wings unequal, [2]3[–6]-locular [5-locular basally and ± 1-locular distally in *Hillebrandia*], placentation axile [axile and parietal in *Hillebrandia*] (placentae often 2-lobed), styles [2]3[–9; 5 in *Hillebrandia*], distinct [rarely basally connate], stigmatic branches [4–]6[–12+], usually twisted [straight], ovules 15–50 per locule, anatropous, bitegmic, crassinucellate. **Fruits** capsular [baccate], winged [horned, ridged, or angled, rarely not winged, notably in *Hillebrandia*], papery [rarely fleshy], dehiscence loculicidal [indehiscent]. **Seeds** 25–100+, operculate (lid surrounded by collar cells), embryo straight, endosperm insubstantial or absent.

Genera 2, species ca. 1400 (1 genus, 2 species in the flora): introduced; Mexico, West Indies, Central America, South America, s Asia, Africa, Indian Ocean Islands, Pacific Islands (Hawaii).

Molecular phylogenetic and morphologic data show that Begoniaceae comprise two genera: the monotypic, Hawaiian endemic *Hillebrandia* Oliver (*H. sandwicensis* Oliver), and pantropical *Begonia* (W. L. Clement et al. 2004; L. L. Forrest et al. 2005). Although present in New Guinea, the family is natively absent from Australia. The Hawaiian Archipelago is the only group of Pacific Islands where the family is native. The relationships of Begoniaceae are with the Cucurbitales, sister to Datiscaceae in the narrow sense (Angiosperm Phylogeny Group 2009).

1. BEGONIA Linnaeus, Sp. Pl. 2: 1056. 1753; Gen. Pl. ed. 5, 475. 1754 • [For Michel Bégon, 1638–1710, French governor of Haiti and patron of botany] [I]

Plants sometimes rhizomatous [sometimes tuberous]. **Stems** erect or ascending [climbing], reddish [green or brown], simple or branched. **Cymes** [1–]few[–many]-flowered. **Capsules** [2]3[–5+]-locular. *x* = 9.

Species ca. 1400 (2 in the flora): introduced; Mexico, West Indies, Central America, South America, s Asia, Africa, Indian Ocean Islands; pantropical.

Begonia is one of the larger herbaceous pantropical genera. Begonias are widely cultivated as ornamental plants.

Begonia cucullata is usually more robust than *B. hirtella*. The number of flowers per cyme was impossible to establish from the available herbarium specimens, because staminate flowers fall early. In general, *B. cucullata* appears to have more flowers than *B. hirtella*. Flowers of *Begonia* often are described as having tepals, as done here; the staminate flowers sometimes may be described as having sepals and petals, as is the case for the sister genus *Hillebrandia*. Capsules were measured excluding the wings.

1. Perennials, usually glabrous, sometimes sparsely hairy; leaf blades asymmetric, ovate
   to ± reniform, base cuneate on shorter side, usually rounded on longer one . . . . 1. *Begonia cucullata*
1. Annuals, ± densely brownish-villous; leaf blades strongly asymmetric, ± ovate to ± cordate,
   base rounded to shallowly cordate on shorter side, rounded to cordate on longer one. . . . . 2. *Begonia hirtella*

1. Begonia cucullata Willdenow, Sp. Pl. 4: 414. 1805
  • Wax begonia [F] [I]

Plants perennial (rhizomatous), usually glabrous, sometimes sparsely hairy. **Stems** [10–]30–70[–100] cm. **Leaves:** stipules lanceolate to oblong, 7–19 × 3–6 mm; petiole 5–46 mm, glabrous; blade asymmetric, ovate to ± reniform, (28–)46–72[–80] × (28–)33–85 mm, base cuneate on shorter side, usually rounded on longer one, margins not lobed, crenate, teeth apices setose, otherwise eciliate, apex obtuse, surfaces glabrous (or glabrate to sparsely hairy in Alabama specimens). **Peduncles** 22–75 mm (in fruit); bracts lanceolate to ovate. **Flowers** white to pink; staminate: tepals 4, outer 2 (sepals) suborbiculate or reniform, 7–10 mm, inner 2 (petals) narrowly obovate, 5–7 mm; stamens 24–33; pistillate: tepals 4 or 5, obovate, 6–9 mm. **Capsules** 8–15 × 6–12 mm, larger wings deltate-rounded, 10–17 mm wide, smaller 3.5–5 mm wide. **2***n* = 34, 56 (South America).

Flowering spring. Along streams in swamps, floodplain woodlands, cabbage palmetto hummocks, wet ditches, wet and mucky soil, often in shade; 0–10 m; introduced; Ala., Fla., Ga.; South America.

Begonia cucullata is found throughout Florida and recently has been collected in nearby Alabama (Conecuh County). The Alabama plants are typical of this species except that the leaves and petioles are glabrate to sparsely hairy; it may require further study. The distinction between varieties is not always clear (B. G. Schubert 1954), and they are not recognized here. Florida specimens often are treated as var. *hookeri* (de Candolle) L. B. Smith & B. G. Schubert, a name that has been put in synonymy of var. *cucullata* (L. B. Smith et al. 1986). From the morphologic variation observed, it is possible that this species escaped repeatedly from cultivation. It appears to be able to produce abundant seeds; capsules are regularly found on specimens and sometimes are abundant. Known horticulturally as *B. semperflorens*, *B. cucullata* has played an important role in ornamental horticulture.

SELECTED REFERENCE Schubert, B. G. 1954. *Begonia cucullata* and its varieties. Natl. Hort. Mag. 1954(Oct.): 244–248.

2. **Begonia hirtella** Link, Enum. Hort. Berol. Alt. 2: 396. 1822 • Brazilian begonia [1]

Plants annual, ± densely brownish-villous (hairs multicellular). **Stems** 8–15[–90] cm. **Leaves:** stipules lanceolate to narrowly ovate, 5–10 × 2–4 mm; petiole 12–32 mm, ± densely villous; blade strongly asymmetric, ± ovate to ± cordate, (12–)15–90 × (11–)22–70 mm, base rounded to shallowly cordate on shorter side, rounded to cordate on longer one, margins shallowly or not lobed, crenate, ciliate, apex acute to acuminate, surfaces appressed-pilose. **Peduncles** 20–22 mm (in fruit); bracts linear to ovate. **Flowers** pinkish; staminate: tepals 2–4, outer 2 suborbiculate, 2–4 mm, inner 0–2, lanceolate, 4 mm; stamens 6–9[–22]; pistillate: tepals 5, oblong to obovate, 2 mm. **Capsules** 5–10 × 4–6 mm, larger wings deltate-rounded, 6–10 mm wide, smaller 2–5 mm wide.

Flowering spring. Around solution holes of rockland hummocks, greenhouse weeds; 0–10 m; introduced; Fla.; West Indies; South America.

Begonia hirtella is naturalized only in Miami-Dade County; it is also a weed in greenhouses. A specimen similar to *B. hirtella*, collected in Palm Beach County on a floating, rotting log in a cypress strand (*Bradley and Woodmansee 1239*, FTG), is glabrous and cannot be determined in its vegetative condition at the present time; it is unknown whether it has persisted.

CALOPHYLLACEAE J. Agardh

• Alexandrian Laurel Family

Norman K. B. Robson

Trees [shrubs], polygamous or dioecious, glabrous or hairy, hairs unicellular or multicellular, with glandular canals containing latex. **Leaves** opposite [alternate], decussate, simple, petiolate; blade: margins entire, surfaces with ± spherical glands containing secretions other than hypericin and pseudohypericin. **Inflorescences** axillary [terminal], pedicellate. **Flowers** bisexual or unisexual, homostylous; sepals persistent or deciduous, 2–4[–16], distinct or connate, inner pair often petaloid; petals deciduous, [0–]4–6[–8], decussate [contorted]; filaments basally connate [distinct], slender; anthers eglandular [glandular], dehiscing longitudinally; ovary superior, 1–8 -locular; placentation basal; styles 1; stigmas ± expanded to peltate. **Fruits** baccate or drupaceous [capsular]. **Seeds** not arillate; embryo length 1/3+ times seed, cotyledons large.

Genera 14, species ca. 490 (2 genera, 2 species in the flora): introduced, Florida; Mexico, West Indies, Central America, South America, Asia, Africa, Indian Ocean Islands (Madagascar), Pacific Islands, Australia; pantropical.

SELECTED REFERENCE Stevens, P. F. 2007. Clusiaceae-Guttiferae. In: K. Kubitzki et al., eds. The Families and Genera of Vascular Plants. 10+ vols. Berlin etc. Vol. 9, pp. 48–66.

1. Styles slender, lengths ca. 4 times ovary; sepals 4, distinct; leaves without tertiary venation visible between lateral veins . . . . 1. *Calophyllum*, p. 6
1. Styles thick, very short; sepals 2 or 3, completely connate in bud; leaves with tertiary venation prominent between lateral veins . . . . 2. *Mammea*, p. 6.

1. CALOPHYLLUM Linnaeus, Sp. Pl. 1: 513. 1753; Gen. Pl. ed. 5, 229. 1754

• Alexandrian laurel [Greek *kalos*, beautiful, and *phyllon*, leaf] ☐

Trees [shrubs], hairy at least on buds, with stilt or loop roots. **Stems** ± flattened and angled when young. **Leaves:** petiole concave, without adaxial protuberance; blade coriaceous, venation close, apparently unbranched, parallel, uniting to form marginal or submarginal vein and alternating with and usually more prominent than latex canals, without tertiary venation visible between lateral veins. **Inflorescences** racemiform; peduncle and rachis usually present; bracts usually

64

deciduous, small. **Flowers** bisexual; sepals deciduous, 4, distinct, in decussate pairs, outer 1 or 2 pairs rarely much different from rest; stamens obscurely 4-fascicled; style slender, ca. 4 times as long as ovary. **Fruits** drupaceous; pericarp firm or fleshy, smooth when fresh. **Seeds** 1. $x = 16$.

Species ca. 190 (1 in the flora): introduced, Florida; Mexico, West Indies, Central America, South America, Asia, Africa, Indian Ocean Islands (Madagascar), Pacific Islands.

Calophyllum includes about ten species in the American Tropics, the rest in Asia, Africa, Madagascar, and the Pacific.

SELECTED REFERENCE Stevens, P. F. 1980. A revision of the Old World species of *Calophyllum* (Guttiferae). J. Arnold Arbor. 61: 117–699.

1. **Calophyllum inophyllum** Linnaeus, Sp. Pl. 1: 513. 1753 • Mastwood [F] [I]

Trees usually with relatively short trunks and spreading branches, 7–25(–35) m. **Leaves:** petiole broadened and flattened toward blade, 10–25 mm; blade bright green, elliptic-oblong to obovate, 8–20 × 4.5–12 cm, base cuneate to rounded, finally decurrent, margins ± undulate, apex rounded to retuse, lateral veins prominent on both surfaces. **Inflorescences** in distal axils, 3–12-flowered. **Flowers** 2 cm diam.; sepals reflexed, outer pair orbiculate, 7–8 mm, inner pair obovate, ± petaloid, 10 mm; petals 4, white, obovate, 9–12 mm; ovary globose. **Drupes** green, globose to obovoid, 2.5–5 × 2–4 cm; stone subspheric, smooth. $2n = 32$.

Flowering summer. Wetlands, especially coastal; 0–20 m; introduced; Fla.; s Asia (India, Sri Lanka); e Africa; Indian Ocean Islands (Madagascar); Pacific Islands.

Calophyllum inophyllum occurs in southern Florida; it is widely planted and possibly naturalized. The smaller-flowered *C. antillana* Jacquin has also been recorded as seminaturalized on the Keys.

2. MAMMEA Linnaeus, Sp. Pl. 1: 512. 1753; Gen. Pl. ed. 5, 228. 1754 • Mammee- or mamey-apple [West Indian *mammei* altered to Latin *mamma*, breast or teat, alluding to fruit shape] [I]

Trees, glabrous, without stilt or loop roots. **Stems** terete. **Leaves:** petiole canaliculate, without adaxial protuberance; blade coriaceous, venation closely pinnate, without clear submarginal vein, with tertiary venation prominent between lateral veins, with glandular dots in areoles. **Inflorescences** fasciculate or flowers solitary; peduncle absent; bracts absent. **Flowers** bisexual or unisexual, stellate; sepals persistent, 2(3), completely connate in bud, splitting into 2 irregular valves; stamens not fascicled; styles thick, very short. **Fruits** baccate; pericarp firm, rough. **Seeds** 2–4. $x = 16$.

Species ca. 75 (1 in the flora): introduced, Florida; Mexico, West Indies, Central America, South America, Asia, Africa, Indian Ocean Islands (Madagascar), Pacific Islands (New Caledonia), Australia.

Mammea includes two species in Central America, three in Africa, and the rest in Madagascar and Australia, and the Pacific Islands, including New Caledonia.

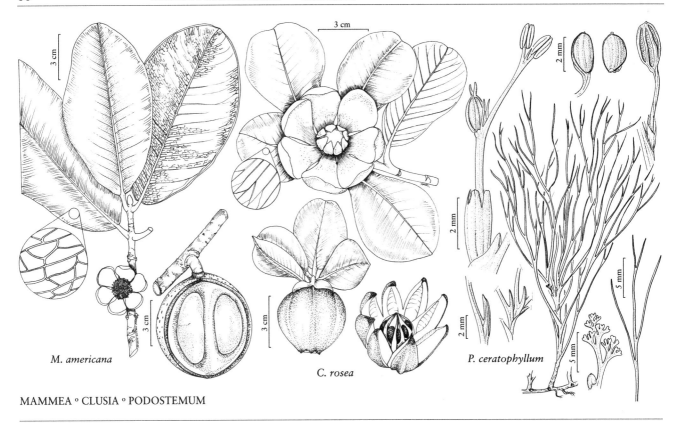

M. americana

C. rosea

P. ceratophyllum

MAMMEA ∘ CLUSIA ∘ PODOSTEMUM

1. Mammea americana Linnaeus, Sp. Pl. 1: 512. 1753 F I

Trees often narrow and densely leafy, to 15(–20) m. **Leaves:** petiole stout, 10–15 mm; blade dark green, elliptic to obovate-elliptic, 9–25 × 5–11 cm, base cuneate to rounded, apex obtuse to rounded or retuse, tertiary reticulation dense. **Inflorescences** 3+-flowered (staminate) or 1-flowered (pistillate or bisexual). **Pedicels** 5–15 mm. **Flowers:** sepals orbiculate-concave, 1–1.5 × 1–1.5 cm; petals 4–6, imbricate, white, orbiculate, 1.5–2.5 cm; filaments basally expanded; ovary pyriform. **Berries** brown, globose or subglobose, 10–15 cm diam.; mesocarp fleshy, soft when ripe; endocarp reddish brown, fibrous. **Seeds** to 7 × 5.5 cm.

Flowering summer–fall (May–Oct). Wetlands; 0–40 m; introduced; Fla.; Mexico; West Indies (Greater Antilles, St. Croix, St. Thomas, Trinidad); Central America.

Mammea americana is widely grown for fruit and occasionally naturalized in southern Florida (Miami-Dade County and the Keys).

CLUSIACEAE Lindley
• Mangosteen Family

Norman K. B. Robson

Shrubs or trees [lianas], evergreen, usually glabrous. **Leaves** decussate [whorled], estipulate; petiole usually present; blade margins entire; glandular canals not containing hypericin or pseudohypericin. **Inflorescences** terminal, dichasial, thyrsoid, or corymbiform or flowers solitary. **Pedicels** absent or relatively stout. **Flowers** unisexual [bisexual], actinomorphic, hypogynous; sepals persistent, [2–]4[–14], imbricate or decussate; petals persistent [deciduous], [3–]6(–8)[–14], without ventral scale; [stamens distinct or fasciculate; filaments nearly as wide as anthers; anthers usually dehiscing longitudinally, sometimes locellate, eglandular]; pistils 1; ovary superior, 3–12-locular; placentation axile; ovules 1–6+ per locule, anatropous, bitegmic, tenuinucellate; styles absent [relatively short]; stigmas [3–]6–9(–12). **Fruits** capsular [berrylike, drupelike], dehiscence septifragal. **Seeds** arillate, embryo usually green, straight, cotyledons minute [absent], endosperm absent.

Genera 15, species ca. 700 (1 in the flora): Florida, Mexico, West Indies, Central America, South America, Asia, Africa, Indian Ocean Islands (Madagascar); pantropical.

SELECTED REFERENCE Stevens, P. F. 2007. Clusiaceae-Guttiferae. In: K. Kubitzki et al., eds. 1990+. The Families and Genera of Vascular Plants. 10+ vols. Berlin etc. Vol. 9, pp. 48–66.

1. CLUSIA Linnaeus, Sp. Pl. 1: 509. 1753; Gen. Pl. ed. 5, 225. 1754 • Balsam- or monkey-apple [For Charles l'Écluse, 1525–1609, Flemish botanist]

Shrubs or trees [lianas], sometimes epiphytic or epilithic with adventitious roots, sometimes polygamodioecious, sap yellowish [white]. **Stems** terete [4-lined]. **Leaves** petiolate [sessile]; petiole planoconvex to terete; blade leathery or fleshy, margins sometimes revolute, venation pinnate, glandular canals crossing veins. **Inflorescences** 1–3-flowered; bracts 2–6. **Flowers** stellate or cupuliform; sepals 4[–6], margins entire, glands linear, often obscure; petals [4–]6(–8) [–10], imbricate or decussate, white, turning pink to brown [red, yellow], margins entire, glands linear, laminar; [staminate flowers: stamens distinct or connate, pistillodes present or absent];

bisexual or pistillate flowers sometimes with ring of ± connate stamens or staminodes or sterile annulus; stigmas radiating. **Capsules** [4–]6–9(–12)-valved, leathery or fleshy. **Seeds:** aril red, basal or ± enveloping seed.

Species 150+ (1 in the flora): Florida, Mexico, West Indies, Central America, South America.

A report of *Clusia flava* Jacquin from the Florida Keys (J. K. Small 1913c) was apparently erroneous.

1. **Clusia rosea** Jacquin, Enum. Syst. Pl., 34. 1760

  • Autograph-tree F

Clusia retusa Poiret

Plants free-standing or epiphytic or epilithic, 0.1–10(–18) m. **Leaves:** petiole 10–20 mm; blade obovate, 7–15(–23) × 6.4–15 cm, base ± cuneate, apex rounded to emarginate. **Inflorescences:** peduncle 2 mm; bracts connate. **Flowers:** staminate unknown; pistillate: sepals in unequal pairs, spatulate to obovate, usually cucullate, to 1.5 × 2 cm; petals obovate-clawed, 3–4 cm, waxy; [staminodes connate, forming resinous cupule]; ovary globose; stigmas 6–9(–12). **Capsules** yellow, flushed red, globose, 5–8 cm diam.

Flowering summer. Disturbed sites, near beaches; 0–10 m; Fla.; Mexico; West Indies; Central America; South America.

Clusia rosea is widely cultivated in Florida and is established in Broward and Miami-Dade counties; it is native in the Keys (Monroe County). In the flora area, staminate flowers are not known; *C. rosea* may be apomictic (B. Maguire 1976).

PODOSTEMACEAE Richard ex Kunth

• Riverweed Family

C. Thomas Philbrick

Garrett E. Crow

Herbs, [annual] perennial, aquatic, attached to rocks and other solid substrata in rapids and waterfalls, submersed in vegetative stage, becoming reproductive as water level drops, exposing plants to air. **Roots** prostrate, elongate, root cap asymmetric. **Stems** often trailing in current, opposite or subopposite, branched or unbranched. **Leaves** distichous [tristichous], stipulate [estipulate], petiolate [sessile]; blade lobed, [pinnate], or dichotomously [to subdichtomously] divided, base sheathing, margins entire. **Inflorescences** solitary flowers [spikelike], enclosed within saclike spathella [absent]. **Flowers:** tepals [2]3[–20], distinct or basally connate; stamens [1]2[–many], restricted to 1 side of flower [1- or 2-whorled or incompletely 1-whorled]; filaments arising from andropodium [individually, not from andropodium], distinct or basally connate; ovary 2[3]-carpellate; stigmas 2[3], apical. **Fruits** capsular, 2-valved [3-valved in *Tristicha*], 1 valve falling away after dehiscence [valves persistent]. **Seeds** 0–40[–numerous] per capsule, orange-brown, minute, ovoid, outer integument expanded and sticky when wet.

Genera ca. 50, species ca. 260 (1 in the flora): c, e North America, Mexico, West Indies, Central America, South America, Africa, se Asia, India, n Australia; pantropical.

Molecular evidence indicates that Podostemaceae is related to Clusiaceae within the Malpighiales (M. H. G. Gustafsson et al. 2002). Structural modifications of vegetative features have led to controversy regarding how best to interpret the structural categories of root, shoot, and leaf in Podostemaceae (for example, I. Jäger-Zürn 2005; R. Rutishauser 1997).

SELECTED REFERENCES Jäger-Zürn, I. 2005. Morphology and morphogenesis of ensiform leaves, syndesmy of shoots and an understanding of the thalloid plant body in species of *Apinagia, Mourera* and *Marathrum* (Podostemaceae). Bot. J. Linn. Soc. 147: 47–71. Les, D. H. et al. 1997c. The phylogenetic placement of river-weeds (Podostemaceae): Insights from *rbc*L sequence data. Aquatic Bot. 57: 5–27. Philbrick, C. T. and A. Novelo R. 1995. New World Podostemaceae: Ecological and evolutionary enigmas. Brittonia 47: 210–222. Rutishauser, R. 1997. Structural and developmental diversity in Podostemaceae (river-weeds). Aquatic Bot. 57: 29–70. van Royen, P. 1951. The Podostemaceae of the New World. Part I. Meded. Bot. Mus. Herb. Rijks Univ. Utrecht, 107: 1–151. van Royen, P. 1953. The Podostemaceae of the New World. Part II. Acta Bot. Neerl. 2: 1–20. van Royen, P. 1954. The Podostemaceae of the New World. Part III. Acta Bot. Neerl. 3: 215–263.

1. PODOSTEMUM Michaux, Fl. Bor.-Amer. 2: 164, plate 44 [as Podostemon]. 1803

• Threadfoot, riverweed, orchid-of-the-waterfall [Greek *podos*, foot, and *stemon*, stamen, alluding to stamens elevated on foot-stalk (andropodium)]

Roots green, photosynthetic; haptera (fleshy disclike or fingerlike outgrowths) arising along flanks of roots serving as anchorage. **Leaves** with 1 or 2 basal sheaths; stipules extensions of boat-shaped leaf base; petiole round to elliptic; blade symmetric, ultimate divisions linear, awl-shaped, or spatulate, apex rounded, blunt, acute, or apiculate. **Inflorescences** apical or lateral, 1–12-flowered, pedicellate; spathella clavate, smooth or papillate, apex rounded or with nipple, rupturing apically into 1–6 irregular segments. **Pedicels** expanded at apex, elongating during anthesis. **Flowers** zygomorphic; tepals scalelike, 2 at base of ovary on either side of andropodium (lateral tepals), shorter than ovary, 1 (andropodial tepal) arising at top of andropodium from fork between 2 stamen filaments; anthers basifixed, quadrangular; thecae equal or inner longer than outer, apices rounded, distinct, dehiscence introrse; ovary oriented obliquely on pedicel, 2-locular, ellipsoidal; ovules 25–200; placentation axile; stigmas distinct, simple, bent toward anthers prior to spathella rupture, divergent and elongating during anthesis. **Capsules** 2-locular, persistent valve oriented obliquely on pedicel, each valve 3-ribbed, suture margins thickened, riblike. **Seeds** 0–42[–203].

Species 11 (1 in the flora): c, e North America, Mexico, West Indies (Dominican Republic), Central America, South America.

SELECTED REFERENCES   Philbrick, C. T. 1984. Aspects of floral biology, breeding system, and seed and seedling biology in *Podostemum ceratophyllum* Michx. (Podostemaceae). Syst. Bot. 9: 166–174. Philbrick, C. T. and G. E. Crow. 1983. Distribution of *Podostemum ceratophyllum* Michx. (Podostemaceae). Rhodora 85: 325–341. Philbrick, C. T. and G. E. Crow. 1992. Isozyme variation and population structure in *Podostemum ceratophyllum* Michx. (Podostemaceae): Implications for colonization of glaciated North America. Aquatic Bot. 43: 311–325. Philbrick, C. T. and A. Novelo R. 2004. Monograph of *Podostemum*. Syst. Bot. Monogr. 70: 1–106. Rutishauser, R. et al. 2003. Developmental morphology of roots and shoots of *Podostemum ceratophyllum* (Podostemaceae–Podostemoideae). Rhodora 105: 337–353.

1. Podostemum ceratophyllum Michaux, Fl. Bor.-Amer. 2: 165, plate 44 [as Podostemon]. 1803 • Podostémon à feuilles cornées F

Lacis ceratophylla (Michaux) Bongard; *Podostemum abrotanoides* Nuttall; *P. ceratophyllum* var. *abrotanoides* (Nuttall) Weddell; *P. ceratophyllum* var. *circumvallatum* P. Royen; *P. ceratophyllum* var. *fluitans* Weddell

Roots flattened or elliptic, 0.3–1.1 mm diam. **Stems** 1.8–9 mm apart, 0.4–300 × 0.5–1.5 mm at base. **Leaves:** stipules caducous (absent from leaves 3–9 nodes back from stem apex), 0.5–3.6 mm, margins entire; petiole arising perpendicular to stem axis or upright, 0.4–59 mm; blade 1–13 times dichotomously divided, 1.7–142 mm, ultimate divisions flattened, spatulate, linear, or awl-shaped, 0.2–40 × 0.05–0.8 mm, central vein faint or absent. **Inflorescences:** spathella 1.5–6.2 × 0.8–1.8 mm. **Pedicels** 0.4–2.9 mm, to 0.6–10.5 mm in anthesis, 0.5–9 mm in fruit. **Flowers:** tepals linear or awl-shaped, apex acute, lateral tepals 0.7–2 mm, andropodial tepals 0.1–1.1 mm; andropodium 0.1–3.3 mm, 0.7–4.3 mm in anthesis; filaments 0.2–0.9 mm, 0.3–2 mm in anthesis; anthers 0.5–1.5 × 0.4–0.8 mm; ovary 0.7–2.6 × 0.4–1.4 mm; stigmas 0.3–1.4 mm, 0.4–1.5 mm in anthesis. **Capsules** 1.4–3.1 × 0.7–1.7 mm. **Seeds** 0.4–0.8 × 0.5–0.8 mm.

Flowering Jun–Sep. Attached to rocks and other solid substrata in swift currents; 0–800 m; N.B., N.S., Ont., P.E.I., Que.; Ala., Ark., Conn., Ga., Ky., La., Maine, Md., Mass., Miss., N.H., N.J., N.Y., N.C., Ohio, Okla., Pa., R.I., S.C., Tenn., Vt., Va., W.Va.; West Indies (Dominican Republic); Central America (Honduras).

Podostemum ceratophyllum is listed as rare, endangered, or of special concern in some regions. Paucity of appropriate habitat and human disturbances (for example, siltation, damming) have contributed to conservation concerns.

HYPERICACEAE Jussieu

• St. John's Wort Family

Norman K. B. Robson

Herbs, annual or perennial, **subshrubs, or shrubs** [**trees**], glabrous or hairy, hairs simple [stellate to dendroid]. **Leaves** opposite [alternate or whorled], simple, estipulate, sessile, subsessile, pseudopetiolate, or petiolate; blade with pellucid glands and/or canals containing essential oils, margins entire [rarely gland-fringed], surfaces with black, reddish, or amber glands containing hypericin and pseudohypericin. **Inflorescences** terminal or axillary, cymose [thyrsoid] or solitary flowers. **Flowers** homostylous [heterostylous]; sepals persistent or deciduous, (3)4 or 5, glanduliferous like leaves; petals persistent or deciduous, 3–5[6], distinct, imbricate or contorted [decussate], orange, pink, or yellow, [white, red], sometimes green or red tinged, [sometimes with adaxial scale], glanduliferous; stamens persistent or deciduous, in 2 whorls, sometimes in fascicles, sometimes reduced to staminodes; filaments distinct or ± connate; anthers 2-locular, dehiscing longitudinally; ovary superior, 2–5-merous; placentation axile to parietal; ovules 1 or 2+ on each placenta, anatropous; styles 2–5, distinct or basally [to completely] connate, elongate; stigmas minute or ± expanded. **Fruits** capsular [baccate], dehiscence septicidal from apex [loculicidal]. **Seeds** sometimes carinate [winged or carunculate]; endosperm absent; embryo straight [curved]; cotyledons 25–40% of total embryo length.

Genera 9, species 700+ (2 genera, 58 species in the flora): nearly worldwide except very cold or very dry regions; almost confined to Tropics except for *Hypericum* and *Triadenum*.

Hypericaceae flowers are always bisexual, the anther and all or part of the filament remain distinct, the cotyledons are distinct and usually 25–40% of the length of the embryo, and specialization has resulted in heterostyly. In Calophyllaceae, the anthers and filaments are distinct, the cotyledons usually form most of the embryo and may be completely connate, and specialization has resulted in dioecism. In Clusiaceae, which are also largely dioecious, distinct anthers are often lacking, the stamens having become more or less connate in masses around the ovary, and the cotyledons are minute or absent.

The dark red compounds hypericin and pseudohypericin, naphtho-dianthrone derivatives that are widespread in Hypericaceae, are contained within black or red gland-dots or -lines in various parts of the plant. These compounds are photosensitizing and lead to eruptions on muzzles of grazing animals. This phenomenon is especially troublesome in dry regions where alternative fodder may be scarce, for example, California, Iraq, Australia, and South Africa. *Hypericum perforatum*, which is native in Iraq and introduced in the other three regions, is the main source of trouble. Reports from Ontario indicate that field workers have experienced severe reactions over prolonged exposure to *H. perforatum*. *Hypericum perforatum* is also used privately (but not prescribed) as a source of an antidepressant; the relevant active ingredients for this treatment are unknown (S. L. Crockett 2003).

SELECTED REFERENCES Adams, W. P. 1973. Clusiaceae of the southeastern United States. J. Elisha Mitchell Sci. Soc. 89: 62–71. Gillett, J. M. and N. K. B. Robson. 1981. The St. John's worts of Canada (Guttiferae). Publ. Bot. (Ottawa) 11: 1–40. Robson, N. K. B. 1977. Studies in the genus *Hypericum* L. (Guttiferae) 1. Infrageneric classification. Bull. Brit. Mus. (Nat. Hist.), Bot. 5: 283–355. Stevens, P. F. 2007b. Hypericaceae. In: K. Kubitzki et al., eds. 1990+. The Families and Genera of Vascular Plants. 10+ vols. Berlin etc. Volume 9, pp. 194–210.

1. Herbs, subshrubs, or shrubs; petals yellow to orange, sometimes red tinged; stamens (5–)10–300(–650), in continuous or interrupted ring or in (3)4 or 5 fascicles; filaments distinct or basally connate, staminode fascicles 0. . . . . .1. *Hypericum*, p. 72
1. Herbs; petals pink or flesh-colored, sometimes green tinged; stamens 9, in 3 fascicles; filaments ⅕–½+ connate, staminode fascicles 3, alternating with stamen fascicles. . . . . .2. *Triadenum*, p. 102

1. HYPERICUM Linnaeus, Sp. Pl. 2: 783. 1753; Gen. Pl. ed. 5, 341. 1754 • St. John's wort, St. Andrew's cross, millepertuis [Greek *hyper*, above, and *ikon*, image, alluding to ancient Greek custom of placing flowers of some species of *Hypericum* above the religious figures to ward off evil spirits]

Ascyrum Linnaeus; *Crookea* Small; *Sanidophyllum* Small; *Sarothra* Linnaeus

Herbs, annual or perennial, **subshrubs, or shrubs [trees],** sometimes rhizomatous, glabrous or hairy, with glandular canals, lacunae, or dots containing resins or waxes (amber), essential oils (pale, translucent), and/or, sometimes, hypericin and pseudohypericin (black or red) in various parts. **Stems:** internodes terete (not lined) or 2-, 4-, or 6-lined at first (lines usually raised), then sometimes becoming angled, terete, or winged; bark smooth or striate, sometimes corky, exfoliating in sheets or strips. **Leaves** sessile, pseudopetiolate, or petiolate; blade relatively broad to narrow, venation pinnate [flabellate] or 1-veined, tertiary veins absent or ± densely reticulate toward margins, glands marginal and/or laminar, pale, red, and/or black, linear to punctiform. **Inflorescences** terminal, cymose, 2+-flowered, or flowers solitary, branching usually dichasial or monochasial, sometimes pseudodichotomous (that is, with 2+-noded branches in distal pair of leaf axils); bracts and bracteoles usually similar to sepals. **Flowers** stellate [cupulate]; sepals persistent or deciduous, (3)4 or 5, distinct or ± connate, margins sometimes glandular-ciliate; petals persistent or deciduous, (3)4 or 5[6], contorted, yellow to orange, sometimes red tinged; stamens persistent or deciduous, (5–)10–300(–650), in continuous or interrupted ring or in (3)4 or 5 fascicles, fascicles distinct or connate, each with 1–60+ stamens; filaments distinct or basally connate; anthers yellow to orange, oblong to ellipsoid, almost isodiametric, sometimes with amber or black gland on connective; staminode fascicles 0[3]; ovary 2–5-merous; placentation axile to parietal; ovules 2+ on each placenta; styles distinct or ± connate basally, spreading to ± appressed. **Capsules** 2–5-valved, sometimes with glandular vittae or vesicles. **Seeds** narrowly cylindric to ellipsoid, sometimes carinate; testa foveolate or reticulate to scalariform [papillose]. x = 12, 9–7, 6 (dihaploid).

Species ca. 490 (54, including 1 hybrid, in the flora): nearly worldwide.

Shrubs with deciduous leaves, petals, and stamens belong to either *Hypericum* sect. *Ascyreia* Choisy (with five stamen fascicles and five styles) or sect. *Androsaemum* (Duhamel) Godron (with five stamen fascicles and three or four styles). These are all introductions, mostly garden escapes. Those in sect. *Ascyreia* include: *Hypericum calycinum* Linnaeus, a low shrub with creeping stolons and flowers 50–95 mm diam. that has been found in California, Oregon, and Washington; *H.* ×*moserianum* Luquet ex André, its hybrid with *H. patulum* Thunberg, a low (sterile?) branching shrub with red anthers; and *H. hookerianum* Wight & Arnott, a shrub to two meters tall with narrow leaves and a dense ring of relatively short stamens, recorded from California (its identity requires confirmation). In sect. *Androsaemum*, *H. androsaemum* is a deciduous shrub with relatively small flowers and baccate fruits that ripen from cherry-red to black; it has been found in British Columbia, California, and Washington.

Introduced herbaceous species with three stamen fascicles and three styles include: *Hypericum hirsutum*, with hairy stems and leaves (Ontario); *H. tetrapterum*, with four-winged internodes and lanceolate sepals (British Columbia and Washington); *H. pulchrum* with cordate leaves and red tinged petals (Newfoundland, St. Pierre and Miquelon); and *H. humifusum* Linnaeus, a procumbent herb with unequal sepals (British Columbia).

SELECTED REFERENCES Adams, W. P. 1957. A revision of the genus *Ascyrum* (Hypericaceae). Rhodora 59: 73–95. Adams, W. P. 1962b. Studies in the Guttiferae. II. Taxonomic and distributional observations in North American taxa. Rhodora 64: 231–242. Adams, W. P. and N. K. B. Robson. 1961. A re-evaluation of the generic status of *Ascyrum* and *Crookea* (Guttiferae). Rhodora 63: 10–16. Crockett, S. L. 2003. Phytochemical and Biosystematic Investigations of New and Old World *Hypericum* Species (Clusiaceae). Ph.D. dissertation. University of Mississippi. Robson, N. K. B. 1981. Studies in the genus *Hypericum* L. (Guttiferae) 2. Characters of the genus. Bull. Brit. Mus. (Nat. Hist.), Bot. 8: 55–226. Robson, N. K. B. 1990. Studies in the genus *Hypericum* L. (Guttiferae) 8. Sections 29. *Brathys* (part 2) and 30. *Trigynobrathys*. Bull. Brit. Mus. (Nat. Hist.), Bot. 20: 1–151. Robson, N. K. B. 1994. Studies in the genus *Hypericum* L. (Guttiferae) 6. Sections 20. *Myriandra* to 28. *Elodes*. Bull. Nat. Hist. Mus. London, Bot. 26: 75–217. Robson, N. K. B. 2001. Studies in the genus *Hypericum* L. (Guttiferae) 4(1). Sections 7. *Roscyna* to 9. *Hypericum* sensu lato (part 1). Bull. Nat. Hist. Mus. London, Bot. 31: 37–88. Robson, N. K. B. 2002. Studies in the genus *Hypericum* L. (Guttiferae) 4(2). Section 9. *Hypericum* sensu lato (part 2): subsection 1. *Hypericum* series 1. *Hypericum*. Bull. Nat. Hist. Mus. London, Bot. 32: 61–123. Robson, N. K. B. 2006. Studies in the genus *Hypericum* (Clusiaceae) 4(3). Section 9. *Hypericum* sensu lato (part 3): subsection 1. *Hypericum* series 2. *Senanensia*, subsection 2. *Erecta* and section 9b. *Graveolentia*. Syst. Biodivers. 4: 19–98. Robson, N. K. B. 2012. Studies in the genus *Hypericum* L. (Hypericaceae) 9. Addenda, corrigenda, keys, lists and general discussion. Phytotaxa 72: 1–111.

1. Herbs, subshrubs, or shrubs, black glands absent; stamens in continuous or interrupted ring or in 4 or 5 barely discernable fascicles, each of 1 or 2 stamens.

    2. Herbs (perennial), subshrubs, or shrubs; petals deciduous; stamens usually persistent, sometimes deciduous, 30–650, in continuous ring or in 4 or 5 barely discernable fascicles; styles ± appressed, bases distinct. . . . . 1a. *Hypericum* sect. *Myriandra*, p. 74

    2. Herbs (annual or perennial); petals persistent; stamens persistent, (5–)10–80, usually in continuous or interrupted ring, sometimes in 5 barely discernable fascicles; styles ± spreading, bases distinct. . . . . 1b. *Hypericum* informal sect. group *Brathys*, p. 88

1. Herbs or shrubs, black and/or red glands usually present throughout, sometimes absent; stamens in 5 fascicles, each of 2+ stamens.

    3. Shrubs; leaves deciduous (base articulated); style bases distinct. . . . . 1c. *Hypericum* sect. *Webbia*, p. 96

    3. Herbs (perennial); leaves persistent or tardily deciduous (base not articulated); style bases ± connate or distinct.

        4. Herbs, black glands absent; flowers 40–70 mm diam.; stamens 150, in 5 fascicles, fascicles usually distinct, rarely 1 pair connate; styles ± appressed, bases ± connate or distinct. . . . . 1d. *Hypericum* sect. *Roscyna*, p. 97

        4. Herbs, black glands usually on leaves, sepals, and petals and, sometimes, on stems and anthers; flowers 6–35 mm diam.; stamens 20–109, in 5 fascicles, fascicles connate (as 2 + 2 + 1); styles spreading, bases distinct . . . .
. . . . .1e. *Hypericum* informal sect. group *Hypericum*, p. 97

1a. HYPERICUM Linnaeus sect. MYRIANDRA (Spach) R. Keller in H. G. A. Engler and K. Prantl, Nat. Pflanzenfam. 95[III,6]: 214. 1893

Myriandra Spach, Ann. Sci. Nat., Bot., sér. 2, 5: 364. 1836

Herbs (perennial), **subshrubs, or shrubs;** black glands absent. **Leaves** deciduous (base articulated) or persistent (base not articulated). **Flowers** 7–45 mm diam.; sepals deciduous or persistent, (2–)4 or 5; petals deciduous, (3)4 or 5; stamens usually persistent, sometimes deciduous, 30–650, in continuous ring or in 4 or 5 barely discernable fascicles, each of 1 or 2 stamens; ovary 2–5-merous; placentation incompletely axile to parietal; styles ± appressed, bases distinct; stigmas minute. **Seeds** ± carinate or not.

Species 29 (28 in the flora): c, e North America, e Mexico, West Indies (Bahamas, Greater Antilles), Bermuda, Central America (Belize).

Hypericum limosum Grisebach is endemic to western Cuba.

1. Sepals enclosing capsule, (2–)4, unequal (outer pair larger) [subsect. *Ascyrum*].
    2. Stamens 30–50; ovaries 2-merous.
        3. Leaf bases without glandlike auricles; inflorescences 1-flowered, pedicels mostly recurved to reflexed in fruit, bracteoles proximal . . . . .27. *Hypericum suffruticosum*
        3. Leaf bases with glandlike auricles; inflorescences 1–12-flowered, pedicels erect in fruit, bracteoles distal . . . . .28. *Hypericum hypericoides*
    2. Stamens 70–100; ovaries 3(4)-merous.
        4. Leaf bases cordate-amplexicaul; inflorescences terminal, branching (at apical node) pseudodichotomous . . . . .25. *Hypericum tetrapetalum*
        4. Leaf bases usually cuneate to subrounded, rarely rounded to slightly subcordate-amplexicaul; inflorescences from 1–4 nodes, branching dichasial or pseudodichotomous.
            5. Leaves: bases without glandlike auricles, margins plane to subrecurved; inflorescences 1–3(–7)-flowered, branching dichasial; outer sepals: apices apiculate or obtuse to rounded. . . . . 24. *Hypericum crux-andreae*
            5. Leaves: bases with glandlike auricles, margins subrecurved to subincrassate; inflorescences 1-flowered, branching pseudodichotomous; outer sepals: apices acute to subacuminate . . . . 26. *Hypericum edisonianum*
1. Sepals not enclosing capsule, usually 5, if 3 or 4, usually subequal.
    6. Leaves: bases usually not articulated, if articulated (*H. myrtifolium*), subcordate-amplexicaul; stamens (all or some) usually persistent (deciduous in *H. apocynifolium*, *H. myrtifolium*, and *H. dolabriforme*).
        7. Shrubs, decumbent, mat-forming, wiry; inflorescences 1(–5)-flowered [subsect. *Pseudobrathydium*]. . . . . .14. *Hypericum buckleyi*
        7. Herbs, subshrubs, or shrubs, erect to ascending or decumbent, not mat-forming, not wiry; inflorescences (1–)3–70-flowered.
            8. Inflorescences widely branched; stamens 120–200 [subsect. *Brathydium*].
                9. Shrubs, erect, unbranched or branched distally, bark corky; leaf blades oblong-ovate to triangular-lanceolate, (5–)7–20 mm wide, bases articulated, subcordate-amplexicaul. . . . . 22. *Hypericum myrtifolium*
                9. Subshrubs, decumbent, ± branching, bark not corky (thin); leaf blades linear-elliptic or linear-oblong to linear, 3–5 mm wide, bases not articulated, narrowly cuneate to rounded. . . . . 23. *Hypericum dolabriforme*
            8. Inflorescences narrowly branched; stamens 30–95 [subsect. *Suturosperma*].
                10. Leaf blades (7–)10–25 mm wide; sepals deciduous or tardily deciduous; placentation incompletely axile.
                    11. Inflorescences terminal, (1–)3–5(–8)-flowered; sepals tardily deciduous, 1.5–2.3 mm wide; capsules 6–15 × 4.5–8 mm. . . . . .15. *Hypericum apocynifolium*
                    11. Inflorescences from 1–3 nodes, the terminal 7–45-flowered; sepals deciduous, 1–1.5 mm wide; capsules 3.5–7 × 3–5 mm. . . . . . . .16. *Hypericum nudiflorum*

10. Leaf blades 1–10(–15) mm wide; sepals persistent; placentation parietal.
    12. Shrubs; leaf blades 1–3 mm wide; inflorescences 1–3-flowered; sepals (3)4(5) . . . . .18. *Hypericum microsepalum*
    12. Herbs, subshrubs, or shrubs; leaf blades (1–)3–15 mm wide; inflorescences usually 7–70-flowered, if flowers 1–6, herbs; sepals usually 5.
        13. Shrubs; capsules ovoid-cylindric to broadly ovoid . . . . . . 17. *Hypericum cistifolium*
        13. Herbs or subshrubs; capsules usually broadly ovoid to depressed-globose, if ovoid-ellipsoid to ellipsoid or globose, rhizomatous herbs.
            14. Herbs or subshrubs, not or rarely rhizomatous; capsules broadly ovoid to depressed-globose; seeds 2–2.7 mm . . . . 19. *Hypericum sphaerocarpum*
            14. Herbs, rhizomatous; capsules ellipsoid to ovoid-ellipsoid or globose; seeds 0.6–0.7 mm.
                15. Herbs, 2–8 dm, relatively stout; leaf blades narrowly oblong or linear to lanceolate or narrowly elliptic . . . . 20. *Hypericum adpressum*
                15. Herbs, 1–3(–5) dm, relatively slender; leaf blades elliptic or oblanceolate to oblong-elliptic . . . . .21. *Hypericum ellipticum*

[6. Shifted to left margin.—Ed.]

6. Leaves: base articulated, not amplexicaul; stamens deciduous [subsect. *Centrosperma*].
  16. Leaf blades usually elliptic, lanceolate, linear, oblanceolate, or oblong, (1–)3–22 mm wide; sepals usually elliptic, linear, oblanceolate, oblong, obovate, ovate, or spatulate.
    17. Inflorescences usually 1–3(–7+)-flowered from apical node, sometimes with single flowers at 1(2) proximal nodes.
      18. Shrubs (6–)10–30 dm; leaf blades 8–22 mm wide, margins plane or subrecurved; sepals 4–10 mm wide; ovaries 3-merous . . . . 1. *Hypericum frondosum*
      18. Shrubs (1.4–)2–6(–10) dm; leaf blades 3–7(–10) mm wide, margins subrecurved to revolute; sepals 1.5–5 mm wide; ovaries (3–)5(6)-merous . . . . .3. *Hypericum kalmianum*
    17. Inflorescences usually (1–)3–25-flowered from apical node, sometimes with triads, dichasia, or flowering branches from 1–4 proximal nodes.
      19. Stem internodes 4- or 6-lined at first; inflorescences cylindric or broadly to narrowly elongate-cylindric with 1–5(–7)-flowered lateral dichasia proximally; internodes 4-lined at first and capsules 7–13 mm or internodes 6-lined at first and capsules 4.5–6 mm.
        20. Leaf blades 30–70 mm, narrowly oblong to narrowly elliptic-oblanceolate; flowers 15–30 mm diam.; placentation incompletely axile; capsules 7–13 mm; seeds 1.5–2 mm . . . . 2. *Hypericum prolificum*
        20. Leaf blades 15–32(–37) mm, narrowly oblong-elliptic or oblanceolate to linear; flowers 9–14 mm diam.; placentation parietal; capsules 4.5–6 mm; seeds 0.7–0.8 mm . . . . .6. *Hypericum galioides*
      19. Stem internodes 4-lined at first; inflorescences shortly and broadly pyramidal or broadly cylindric to globose-cylindric, or obpyramidal with (2–)5–15-flowered lateral dichasia proximally; capsules 5–7 mm.
        21. Sepals: basal veins 3–7; ovaries (3)4- or 5-merous; capsules notably lobed . . . . 4. *Hypericum lobocarpum*
        21. Sepals: basal veins 1–3; ovaries 3- or 4(5)-merous; capsules not or scarcely lobed . . . . 5. *Hypericum densiflorum*
  16. Leaf blades (mature) usually acicular, linear, or linear-subulate, 0.4–1(–1.5) mm wide (juvenile leaf blades sometimes broader); sepals linear to linear-subulate.
    22. Shrubs, usually decumbent and mat-forming, 1–5 dm.
      23. Leaf blades 13–25 mm; sepals (3–)4.5–7 mm . . . . 7. *Hypericum lloydii*
      23. Leaf blades 4–11 mm; sepals 2–4 mm . . . . 8. *Hypericum tenuifolium*
    22. Shrubs, erect, 3–45 dm.

[24. Shifted to left margin.—Ed.]

24. Capsules usually cylindric, rarely narrowly conic or ovoid-conic; bark smooth, not metallic-silvery.
  25. Leaves 9–26 mm; styles shorter than ovaries; capsules (4.5–)5–7 mm . . . . 9. *Hypericum nitidum*
  25. Leaves 6–12 mm; styles longer than ovaries; capsules 3.5–5 mm. . . . . .10. *Hypericum brachyphyllum*
24. Capsules ± narrowly ovoid-conic or pyramidal-ovoid to ovoid or ellipsoid; bark smooth and metallic-silvery or corky to spongy.
  26. Bark smooth and metallic-silvery, exfoliating in thin, curled plates; styles 5 mm; seeds 1–1.6 mm; young stems, leaves, and sepals glaucous. . . . . .11. *Hypericum lissophloeus*
  26. Bark corky to spongy, exfoliating in thin, papery sheets or plates; styles 2.5–4 mm; seeds 0.4–0.8 mm; young stems, leaves, and sepals not glaucous.
    27. Stems: internodes 6-lined at first, soon 2-winged, then terete; bark smooth; terminal inflorescence (3–)7–32-flowered . . . . .12. *Hypericum fasciculatum*
    27. Stems: internodes 4-lined at first, soon 4-angled, then terete; bark striate; terminal inflorescence 1–3-flowered . . . . 13. *Hypericum chapmanii*

1. **Hypericum frondosum** Michaux, Fl. Bor.-Amer. 2: 81. 1803 · Golden St. John's wort E F W

Hypericum amoenum Pursh; *H. splendens* Small

Shrubs, erect, forming rounded bush or treelike, (6–)10–30 dm. **Stems:** internodes 4-lined at first, then 2-lined to terete. **Leaf blades** usually oblong to lanceolate-oblong, sometimes oblanceolate, 25–65 × 8–22 mm, base articulated, broadly to narrowly cuneate, margins plane or subrecurved, apex apiculate-obtuse to rounded, midrib with 10–16 pairs of branches. **Inflorescences** 1–3(–7)-flowered from apical node, sometimes with paired single flowers or triads (3-flowered cymules) or 1–3-flowered branches at proximal node. **Flowers** 24–45 mm diam.; sepals deciduous, not enclosing capsule, (4)5, ovate or oblong to elliptic-spatulate, unequal, 6–14(–20) × 4–10 mm; petals (4)5, golden yellow to orange-yellow, obovate to oblanceolate, 12–25 mm; stamens deciduous, 250–650; ovary 3-merous. **Capsules** ovoid-conic to ovoid-rostrate, 12–15 × 6–8 mm. **Seeds** carinate, 1.5 mm; testa linear-reticulate. **2*n* = 18.**

Flowering summer (Jun–Jul). Dry cedar-glades and barrens on limestone and calcareous shale; 100–500 m; Ala., Conn., Fla., Ga., Ky., La., Mass., Miss., N.Y., N.C., Tenn., Tex.

Hypericum frondosum is endemic to the southwestern end of the Appalachian Range; it is recorded as introduced in Connecticut, Massachusetts, and New York (W. P. Adams 1962). Records from Arkansas and, possibly, South Carolina and Virginia appear to be errors for *H. prolificum.* Although *H. frondosum* is variable over its natural range and approaches *H. prolificum* morphologically in Arkansas, it remains distinct from its immediate relatives. In cultivation, it sometimes hybridizes with *H. prolificum.* Artificial hybrids have been made, as well as artificial tetraploids (O. Myers 1963).

2. **Hypericum prolificum** Linnaeus, Mant. Pl. 1: 106. 1767 · Shrubby St. John's wort E W

Brathys prolifica (Linnaeus) Payer; *Hypericum spathulatum* (Spach) Steudel; *Myriandra ledifolia* Spach; *M. prolifica* (Linnaeus) Spach; *M. prolifica* var. *spathulata* (Spach) K. Koch; *M. spathulata* Spach

Shrubs, erect or ± diffuse, forming rounded or irregular bush, (2–)7.5–15(–20) dm. **Stems:** internodes 4-lined at first, then 2-lined to terete. **Leaf blades** narrowly oblong to narrowly elliptic-oblanceolate, 30–70 × 6–15 mm, base articulated, attenuate to narrowly cuneate, margins plane to recurved, apex rounded-apiculate to acute, midrib with 10–16 pairs of branches. **Inflorescences** cylindric, (1–)3–7(–9)-flowered from apical node, with paired single flowers or triads or 1–3(–7)-flowered branches from 2+ proximal nodes. **Flowers** 15–30 mm diam.; sepals deciduous, not enclosing capsule, 5, elliptic to obovate or oblanceolate-spatulate, unequal or subequal, 4–8 × 1.5–4 mm; petals 5, golden yellow, obovate to oblanceolate-spatulate, 7–15 mm; stamens deciduous, 150–500; ovary 3(–5)-merous, placentation incompletely axile. **Capsules** usually narrowly ovoid-conic to ovoid, rarely ellipsoid, 7–13 × 4–7 mm. **Seeds** carinate, 1.5–2 mm; testa linear-reticulate. **2*n* = 18.**

Flowering summer (Jun–Sep). Rocky slopes, embankments, dry stream bottoms, woodlands (in north), on limestone or granite; 50–600+ m; Ont.; Ala., Ark., Del., D.C., Fla., Ga., Ill., Ind., Iowa, Ky., La., Md., Mich., Miss., Mo., N.J., N.Y., N.C., Ohio, Okla., Pa., S.C., Tenn., Tex., Va., W.Va., Wis.

Hypericum prolificum is variable, the most luxuriant form being found in the southwestern part of its range. Natural hybrids have not been recorded; it hybridizes in gardens with *H. densiflorum* (*H.* ×*arnoldianum* Rehder), *H. frondosum, H. kalmianum,* and *H. lobocarpum* (*H.* ×*dawsonianum* Rehder).

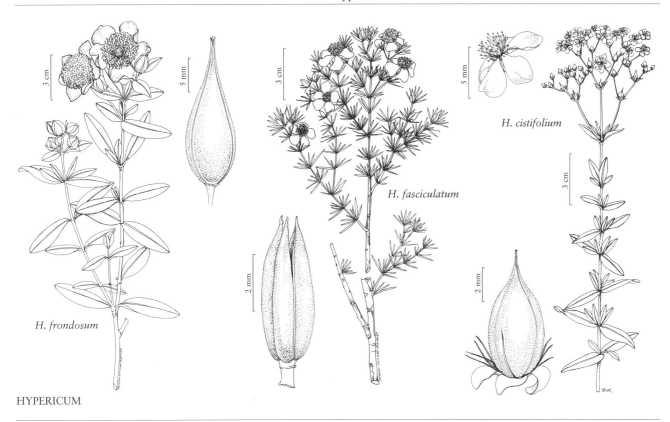

H. frondosum

H. fasciculatum

H. cistifolium

HYPERICUM

3. Hypericum kalmianum Linnaeus, Sp. Pl. 2: 783. 1753 • Kalm's St. John's wort, millepertuis de Kalm E

Norysca kalmiana (Linnaeus) K. Koch

Shrubs, erect, forming slender to rounded or flat-topped bush, (1.4–)2–6(–10) dm. **Stems:** internodes 4-lined at first, then terete. **Leaf blades** narrowly oblong to oblanceolate or linear, (15–)20–45 × 3–7(–10) mm, base articulated, narrowly cuneate to subattenuate, margins subrecurved to revolute, apex rounded to obtuse, midrib with 9–14 pairs of branches. **Inflorescences** usually (1–)3–7(+)-flowered from apical node, rarely with flowers from 1–2 proximal nodes. **Flowers** 20–35 mm diam.; sepals deciduous, not enclosing capsule, (4)5, elliptic or oblong to obovate, subequal, 4–9 × 1.5–5 mm; petals (4)5, golden yellow, obovate to oblong, 8–15 mm; stamens deciduous, 150–200; ovary (3–)5 (6)-merous. **Capsules** narrowly ovoid-conic, 7–11 × 4–7 mm. **Seeds** narrowly carinate, 0.7–1.1 mm; testa subscalariform. $2n = 18$.

Flowering summer (Jul–Aug). Sandy or calcareous dune slacks or swales, rocky shores, plains and low prairies, along streams, sphagnum-sedge swamps; 10–400 m; Ont., Que.; Ill., Ind., Mich., N.Y., Ohio, Wis.

Hypericum kalmianum is a northern derivative of *H. prolificum* with shorter stems, narrower leaves, fewer and larger flowers, and, usually, five styles and placentae. Natural hybrids with *H. prolificum* have been reported from Wisconsin.

4. Hypericum lobocarpum Gattinger ex J. M. Coulter, Bot. Gaz. 11: 275. 1886 • Five-lobe St. John's-wort E

Hypericum densiflorum Pursh var. *lobocarpum* (Gattinger ex J. M. Coulter) Svenson; *H. oklahomense* E. J. Palmer

Shrubs, erect, forming large clumps, 9–15(–20) dm. **Stems:** internodes 4-lined at first, soon 2-lined to terete. **Leaf blades** narrowly oblong to oblanceolate or linear, 35–50 × 3–11 mm, base articulated, narrowly cuneate to attenuate, margins recurved to revolute, apex apiculate-rounded to subacute, midrib with 12–14 pairs of branches. **Inflorescences** globose-cylindric to shortly and broadly pyramidal, 5–25-flowered from apical node, with 3–15-flowered dichasia from 1–3 proximal nodes. **Flowers** 10–15 mm diam.; sepals deciduous, not enclosing capsule, 5, narrowly elliptic to narrowly oblong or oblanceolate-spatulate, subequal or equal, (3.5–)4–4.5 × 0.8–1.5(–2) mm, basal veins 3–7; petals

5, golden yellow, obovate-oblanceolate, 6–7(–8) mm; stamens deciduous, 100–150; ovary (3)4- or 5-merous. **Capsules** narrowly ovoid-conic to ovoid, 5.5–7 × 2.5–3.5 mm, notably lobed. **Seeds** not carinate, 1.2–1.5 mm; testa linear-reticulate. $2n = 18$.

Flowering summer. Rocky stream bottoms and banks, lake margins, swamps and open pine woods; 0–500 m; Ala., Ark., Ill., Ky., La., Miss., Mo., Okla., S.C., Tenn., Tex.

Hypericum lobocarpum usually can be distinguished from *H. densiflorum* by the five-merous ovary; its lobed capsule is a better differentiating characteristic. Hybrid intermediate forms sometimes occur in northwestern Alabama. The South Carolina population (from Allendale County), although well within the area of *H. densiflorum* and well established, has fruits typical of *H. lobocarpum* and appears to be either an outlier or an introduction. *Hypericum ×dawsonianum* Rehder, apparently *H. lobocarpum × prolificum* and intermediate in form between the presumed parents, is known in cultivation only.

5. Hypericum densiflorum Pursh, Fl. Amer. Sept. 2: 376. 1813 • Bushy St. John's wort E

Hypericum glomeratum Small; *H. interior* Small; *H. nothum* Rehder; *H. prolificum* Linnaeus var. *densiflorum* (Pursh) A. Gray

Shrubs, erect, forming slender bush, 6–30 dm. **Stems:** internodes 4-lined at first, soon 2-lined to terete. **Leaf blades** narrowly elliptic-oblong or oblanceolate to linear, 20–45 × 2–7 mm, base articulated, narrowly cuneate to attenuate, margins recurved to revolute, apex apiculate-rounded to subacute, midrib with 14–17 pairs of branches. **Inflorescences** broadly pyramidal to broadly cylindric, 5–25-flowered from apical node, with (2–)5–15-flowered dichasia from 1 or 2 proximal nodes. **Flowers** 10–17(–20) mm diam.; sepals deciduous, not enclosing capsule, 5, narrowly oblong to oblanceolate-spatulate, unequal or subequal, 4–6 × 1–1.5 mm, basal veins 1–3; petals 5, deep golden yellow, obovate-oblanceolate, 6–9 mm; stamens deciduous, 100–150; ovary 3- or 4(5)-merous. **Capsules** narrowly ovoid-conic to cylindric-ovoid, 5–6(–7) × 2–3 mm, not or scarcely lobed. **Seeds** not carinate, 0.8–1.3 mm; testa linear-reticulate. $2n = 18$.

Flowering summer (Jun–Aug). Wet or moist habitats (meadows, lake margins, pinelands, etc.), road embankments, rocky hillsides; 0–1000 m; Ala., Del., Ga., Ky., Md., N.J., N.C., Pa., S.C., Tenn., Va., W.Va.

Hybrid intermediates between *Hypericum densiflorum* and *H. lobocarpum* occur in northwestern Alabama, and a narrow-leaved, small-flowered form

in Tennessee and northern Georgia (*H. interior*) verges toward *H. galioides*. *Hypericum densiflorum* is always distinct from *H. prolificum* in the wild; these species hybridize in gardens. *Hypericum ×arnoldianum* Rehder, known in cultivation only, was thought by Rehder to have the parentage *H. galioides × lobocarpum*; on both morphological and cytological grounds, the conclusion of W. P. Adams (1972) that it was *H. densiflorum × lobocarpum* seems much more likely.

6. Hypericum galioides Lamarck in J. Lamarck et al., Encycl. 4: 161. 1797 E

Brathydium ambiguum (Elliott) K. Koch; *Hypericum ambiguum* Elliott; *H. axillare* Lamarck; *H. galioides* var. *ambiguum* Chapman; *H. galioides* var. *axillare* (Lamarck) Grisebach; *H. michauxii* Poiret; *Myriandra galioides* (Lamarck) Spach; *M. michauxii* (Poiret) Spach

Shrubs, erect, forming rounded clumps, 5–15 dm. **Stems:** internodes 6-lined at first, soon 4-lined, then terete. **Leaf blades** narrowly oblong-elliptic or oblanceolate to linear, 15–32(–37) × 1–7 mm, base articulated, attenuate, margins recurved to revolute, apex rounded to acute, midrib obscurely branched. **Inflorescences** narrowly cylindric, 3–15-flowered from apical node, with (1–)3–5-flowered dichasia from 3 or 4 proximal nodes, sometimes with additional flowering branches. **Flowers** 9–14 mm diam.; sepals deciduous, not enclosing capsule, 5, oblanceolate-spatulate to linear, subequal or equal, 3.5–6.5 × 0.5–1.5 mm; petals 5, bright yellow, obovate-oblanceolate, 5–9 mm; stamens deciduous, 60–120; ovary 3-merous, placentation parietal. **Capsules** narrowly ovoid-conic, 4.5–6 × 2.5–3.5 mm. **Seeds** narrowly carinate, 0.7–0.8 mm; testa finely reticulate. $2n = 18$.

Flowering summer (Jun–Aug). Wet or moist, open habitats (stream banks, flood plains, roadside ditches, low pine forests, etc.), coastal plain; 0–200 m; Ala., Fla., Ga., La., Miss., N.C., S.C., Tex.

The leaves of *Hypericum galioides* vary considerably in width; the lamina is always visible on either side of the midrib.

7. **Hypericum lloydii** (Svenson) W. P. Adams, Contr. Gray Herb. 189: 32. 1962 E

Hypericum galioides Lamarck var. *lloydii* Svenson, Rhodora 54: 207. 1952

Shrubs, decumbent, straggling and rooting, forming low, rounded clumps or mats, 1–5 dm. **Stems:** internodes (4-) 6-lined at first, then terete. **Leaf blades** linear-subulate, 13–25 × 0.5–0.8 mm, base articulated, parallel, margins revolute, apex rounded to retuse, midrib unbranched. **Inflorescences** narrowly pyramidal, 1–3-flowered, with 1–3(–5)-flowered dichasia from to 5 proximal nodes, without additional flowering branches; pedicels 0.5 mm. **Flowers** 12–14 mm diam.; sepals deciduous, not enclosing capsule, 5, linear-subulate, unequal, (3–)4.5–7 × 0.5–0.8 mm; petals 5, golden yellow, oblanceolate-oblong, 5–7.5 mm; stamens deciduous, 100; ovary 3-merous. **Capsules** ovoid, 3–4 × 2–2.5 mm. **Seeds** carinate, 0.7 mm; testa not seen.

Flowering summer (Aug). Dry habitats (pine woods, granite outcrops, roadside embankments), inner coastal plain and foothills; 100–300 m; Ala., Ga., N.C., S.C.

The habit, leaf shape, and drier habitats distinguish *Hypericum lloydii* from *H. galioides*.

8. **Hypericum tenuifolium** Pursh, Fl. Amer. Sept. 2: 377. 1813 E

Hypericum fasciculatum Lamarck var. *laxifolium* Choisy; *H. reductum* (Svenson) W. P. Adams

Shrubs, usually decumbent, not rooting, forming mats, 1–5 dm. **Stems:** internodes 6-lined at first, becoming 4-lined, then terete. **Leaf blades** linear-subulate, 4–11 × 0.4–0.8 mm, base articulated, persistent, parallel or slightly expanded, margins revolute, apex rounded with ± prominent hydathode to long-acuminate, midrib unbranched. **Inflorescences** ± narrowly cylindric, 1–7-flowered, with 1(–3)-flowered dichasia from to 4 proximal nodes, rarely with 1 pair of flowering branches; flowers sessile or nearly so. **Flowers** 10–14 mm diam.; sepals deciduous, not enclosing capsule, 5, linear-subulate, unequal, 2–4 × 0.4–0.5 mm; petals 5, bright yellow, oblanceolate-oblong to obovate, 5–10 mm; stamens deciduous, 50–90; ovary 3-merous. **Capsules** narrowly (sub-) cylindric, (4–)5.7–9.5 × 1.5–2 mm. **Seeds** scarcely carinate, 0.5 mm; testa coarsely reticulate. $2n = 18$.

Flowering spring–mid summer (Apr–Jul). Dry, sandy woods, dunes and dune hollows, coastal plain; 0–200 m; Ala., Fla., Ga., N.C., S.C.

Hypericum tenuifolium differs from *H. galioides* in habit, leaf size, and inflorescence, and in its drier habitat. Its nonrooting stems, longer leaves, and (usually) longer stems distinguish it from *H. lloydii*.

9. **Hypericum nitidum** Lamarck in J. Lamarck et al., Encycl. 4: 160. 1797

Myriandra nitida (Lamarck) Spach

Shrubs, erect, forming dense thickets, 3–45 dm, bark smooth, not metallic-silvery. **Stems:** internodes 4-lined at first, becoming 2-winged, then terete. **Leaf blades** linear or linear-subulate, 9–26 × 0.5–1.5 mm, base articulated, narrowly cuneate or parallel, margins revolute, apex rounded or acute to long-acuminate with prominent hydathode, midrib unbranched. **Inflorescences** narrowly to broadly cylindric, 3–15-flowered, sometimes with 1–3(–7)-flowered dichasia from to 6 proximal nodes, sometimes with 1 or 2 pairs of additional flowering branches. **Flowers** 10–18 mm diam.; sepals deciduous, not enclosing capsule, 5, linear-subulate, unequal or subequal, 3.5–7 × 0.4–0.8 mm; petals 5, yellow, obovate to elliptic-lanceolate, (5–)6–10 mm; stamens deciduous, 50–80(–115); ovary 3-merous; styles shorter than ovaries. **Capsules** narrowly conic to cylindric, (4.5–)5–7 × (1.3–)2–3 mm. **Seeds** scarcely carinate, 0.5 mm; testa finely reticulate.

Subspecies 3 (2 in the flora): se United States, West Indies (Cuba), Central America (Belize).

The two subspecies of *Hypericum nitidum* present in North America apparently remain distinct there; the distinctions are less clear in Cuba, where subsp. *cubense* (Turczaninow) N. Robson is also present.

1. Plants to 45+ dm, with ± numerous, stout, bushy-branched stems from base; leaf blade margins loosely inrolled, apex obtuse to rounded-apiculate; sepal apices shortly apiculate to acute . . . . .9a. *Hypericum nitidum* subsp. *nitidum*
1. Plants 3–10 dm, with ± few, slender, little-branched stems from base; leaf blade margins tightly inrolled, apex acute to long-acuminate; sepal apices acute to long-acuminate. . . . . .9b. *Hypericum nitidum* subsp. *exile*

9a. Hypericum nitidum Lamarck subsp. **nitidum** [E]

Plants with ± numerous, stout, bushy-branched stems from base, to 45 dm. **Leaf blades** 10–18 × 0.5–1.5 mm, subcoriaceous, margins loosely inrolled, apex obtuse to rounded-apiculate. **Inflorescences** with terminal dichasium 3–15-flowered. **Sepal apices** shortly apiculate to acute. **Capsules** cylindric.

Flowering spring–mid summer (Apr–Jul). Moist habitats, open stream banks, pond margins, low pinelands, ditches, coastal plain; 0–300 m; Ala., Fla., Ga., N.C., S.C.

9b. Hypericum nitidum Lamarck subsp. **exile** (W. P. Adams) N. Robson, Bull. Brit. Mus. (Nat. Hist.), Bot. 23: 67. 1993   [C]

Hypericum exile W. P. Adams, Contr. Gray Herb. 189: 33. 1962; *H. galioides* Lamarck var. *cubense* Grisebach

Plants with ± few, slender, little-branched stems from base, 3–10 dm. **Leaf blades** 9–26 × 0.5–0.8 mm, chartaceous, margins tightly inrolled, apex acute to long-acuminate. **Inflorescences** with terminal dichasium 3–7-flowered. **Sepal apices** acute to long-acuminate. **Capsules** cylindric to narrowly conic. $2n = 18$.

Flowering spring (Apr–May). Open pine woods in sand; of conservation concern; 10 m; Fla.; West Indies (w Cuba); Central America (Belize).

10. Hypericum brachyphyllum (Spach) Steudel, Nomencl. Bot. ed. 2, 1: 787. 1840   [E]

Myriandra brachyphylla Spach, Hist. Nat. Vég. 5: 435. 1836

Shrubs, erect, forming round bush, 3–15 dm, bark smooth, not metallic-silvery. **Stems:** internodes sometimes 4-lined at first, soon 2-winged, not terete. **Leaf blades** linear, 6–12 × 0.5–0.7 mm, base articulated, deciduous with leaf, cf. 8. *H. tenuifolium,* parallel, margins revolute, apex rounded-apiculate, midrib unbranched. **Inflorescences** ± narrowly cylindric, 3–15-flowered, with 3–5-flowered dichasia or flowering branches from to 10 proximal nodes. **Flowers** 10–13 mm diam.; sepals deciduous, not enclosing capsule, 5, linear, unequal, 2.5–4.5 × 0.5–1 mm; petals 5, bright yellow, obovate-spatulate, 5–8 mm; stamens deciduous, 40–45; ovary 3-merous; styles longer than ovaries. **Capsules** narrowly cylindric to narrowly ovoid-conic, 3.5–5 × 1.5–2 mm. **Seeds** not carinate, 0.4–0.6 mm; testa finely reticulate. $2n = 18$.

Flowering summer (Jul–Aug). Moist habitats, pine flatwoods, pond margins, borrow pits, swamp woodlands, lowland coastal plain; 0–100 m; Ala., Fla., Ga., La., Miss.

Hypericum brachyphyllum is bushier than *H. nitidum* subsp. *exile,* with relatively small flowers, capsules, and, usually, leaves, and shorter styles. The bushy habit, two-sided shoots, and glossy leaves without prominent base or apical hydathode, as well as the wet habitats, all distinguish it from *H. tenuifolium.*

11. Hypericum lissophloeus W. P. Adams, Contr. Gray Herb. 189: 21. 1962 • Smooth-barked St. John's wort   [C][E]

Shrubs, erect, sparsely branched, forming dense clumps sometimes with prop roots, to 40 dm, bark smooth and metallic-silvery, without laticifers, exfoliating in thin, curled plates. **Stems:** internodes 4-lined at first, soon 4-angled, then terete, dull silvery, glaucous. **Leaf blades** linear-subulate to acicular, (9–)12–17 × 0.5–0.8 mm, glaucous, base articulated, parallel or almost so, margins revolute, apex obtuse to rounded, midrib unbranched. **Inflorescences** narrowly cylindric, 1–3-flowered, usually with paired flowers or triads from to 9 proximal nodes. **Flowers** 20 mm diam.; sepals deciduous, not enclosing capsule, 5, linear-subulate, subequal, 7–8 × 0.5–0.8 mm, glaucous; petals 5, bright yellow, obovate-spatulate, 10–12 mm; stamens deciduous, 170–220; ovary 3-merous; styles 5 mm. **Capsules** narrowly ovoid to ellipsoid, 6–7 × 2.5–3.5 mm. **Seeds** narrowly carinate, 1–1.6 mm; testa coarsely reticulate-sulcate. $2n = 18$.

Flowering summer–early fall (Jun–Oct). Pond and lake margins to 1.5 m deep water; of conservation concern; 0–10 m; Fla.

Hypericum lissophloeus is found in Bay and Washington counties. The larger capsules, one- to three-flowered, lateral inflorescence branches, and smooth-polished, metallic bark (that exfoliates like that of *Betula* species) are among the features that distinguish *H. lissophloeus* from *H. fasciculatum, H. nitidum,* and their allies (*H. brachyphyllum* and *H. chapmanii*).

12. **Hypericum fasciculatum** Lamarck in J. Lamarck et al., Encycl. 4: 160. 1797 • Sandweed E F

Hypericum fasciculatum var. *aspalathoides* Torrey & A. Gray; *H. fulgidum* Rafinesque; *H. galioides* Lamarck var. *fasciculatum* (Lamarck) Svenson

Shrubs, erect, much-branched distally, bushy, not treelike, usually forming mounds, to 15(–30) dm, bark thick, smooth, thin-corky and spongy exfoliating in thin, papery sheets or plates. **Stems:** internodes 6-lined at first, soon 2-winged, then terete, not glaucous. **Leaf blades** linear-subulate, 8–20 × 0.7–1 mm, not glaucous, base articulated, parallel, margins revolute, apex sometimes slightly broadened, midrib unbranched. **Inflorescences** rounded-pyramidal to corymbiform, sometimes intercalary as result of resumed vegetative growth, (3–)7–32-flowered, sometimes with single flowers or 3–5-flowered dichasia from to 3 proximal nodes. **Flowers** 13–16 mm diam.; sepals deciduous, not enclosing capsule, 5, linear-subulate, unequal, (3–)4.5–8(–10) × 0.5 mm, not glaucous; petals 5, bright yellow, obovate-spatulate, 6–9 mm; stamens deciduous, 70–100; ovary 3-merous; styles 2.5–4 mm. **Capsules** ± narrowly ovoid-conic to ovoid-ellipsoid, 5.5 × 2.5–3 mm. **Seeds** not carinate, 0.4 mm; testa finely foveolate-reticulate. $2n = 18$.

Flowering spring–fall (Apr–Nov). Ponds and lake margins, marshes, ditches, coastal plain; 0–500 m; Ala., Fla., Ga., La., Miss., N.C., S.C.

The thick, often spongy, bark, relatively long axillary leaf clusters, paired leaf grooves flanking the midrib abaxially, broader inflorescence, and broader capsules distinguish *Hypericum fasciculatum* (and *H. chapmanii*) from *H. nitidum* and its relatives.

Hypericum aspalathoides Willdenow is an illegitimate name for *H. fasciculatum*.

13. **Hypericum chapmanii** W. P. Adams, Contr. Gray Herb. 189: 22. 1962 E

Hypericum arborescens Chapman, Fl. South. U.S. ed. 2 repr. 2, 680. 1892, not Vahl 1791

Shrubs, erect, treelike, to 40 dm, bark thick-corky, striate, exfoliating in thin, papery sheets or plates. **Stems:** internodes 4-lined at first, soon 4-angled, then terete, not glaucous. **Leaf blades** linear-subulate, 8–25 × 0.5–0.7 mm, not glaucous, base articulated, parallel or slightly expanded, margins revolute, apex acute, midrib unbranched. **Inflorescences** shortly cylindric, 1–3-flowered, often with single flowers or triads from 1 or 2 proximal nodes. **Flowers** 12–15 mm diam.; sepals deciduous, not enclosing capsule, 5, linear-subulate, unequal, 5–7 × 0.5 mm, not glaucous; petals 5, bright yellow, oblong-spatulate, 7–9 mm; stamens deciduous, 75; ovary 3-merous; styles 2.5–4 mm. **Capsules** narrowly pyramidal-ovoid, 6 × 2.4 mm. **Seeds** not carinate, 0.6–0.8 mm; testa finely foveolate-reticulate.

Flowering summer (Jun–Aug). Pond margins, flatwoods, depressions; 0–10 m; Fla.

Hypericum chapmanii differs from *H. fasciculatum* in its taller, single-stemmed habit, thicker stems (to 10–15 cm diameter) with fluted, spongy bark containing large laticifers, and fewer-flowered inflorescences.

Hypericum chapmanii is known from the panhandle of northwestern Florida.

14. **Hypericum buckleyi** M. A. Curtis, Amer. J. Sci. Arts 44: 80. 1843 (as buckleii) • Mountain St. John's wort E

Shrubs, decumbent, spreading and rooting, wiry, branches ascending to erect, forming compact mats, 0.5–4.5 dm. **Stems:** internodes 4-lined. **Leaf blades** oblong or elliptic to obovate, 4–25 × 2–12 mm, base not articulated, cuneate, margins plane, apex rounded, midrib with 2–4 pairs of branches. **Inflorescences** 1(–5)-flowered. **Flowers** 20–25 mm diam.; sepals persistent, not enclosing capsule, 5, broadly elliptic to elliptic-spatulate or obovate, subequal, 4–5 × 2.5–3 mm; petals 5, golden yellow, oblanceolate, 6–10.5 mm; stamens persistent, 100; ovary 3-merous. **Capsules** ovoid to ovoid-cylindric, 8–12 × 5 mm. **Seeds** narrowly to broadly carinate, 1.5–2 mm; testa finely foveolate-reticulate.

Flowering early–mid summer (Jun–Jul). Seepage areas, moist rock crevices, ditches, road embankments; 900–1600 m; Ga., N.C., S.C.

Hypericum buckleyi is found throughout the southern Appalachian Mountains.

The decumbent habit and persistent sepals and stamens distinguish *Hypericum buckleyi* from its nearest relative, *H. prolificum*.

15. Hypericum apocynifolium Small, Bull. Torrey Bot. Club 25: 616. 1898 E

Shrubs, erect, branches ascending, 4–7 dm. **Stems:** internodes narrowly 4-winged at first, then 2-lined. **Leaf blades** oblong to elliptic-oblong, 20–40 × 12–20 mm, base not articulated, broadly cuneate, margins usually plane, rarely recurved, apex rounded to retuse, midrib with 6 pairs of branches. **Inflorescences** terminal, (1–)3–5(–8)-flowered, narrowly branched. **Flowers** 15 mm diam.; sepals tardily deciduous, not enclosing capsule, 5, spatulate to elliptic or ovate, unequal, 3–5 × 1.5–2.3 mm; petals 5, coppery yellow, oblong, 8–10 mm, length 2 times sepals; stamens deciduous, 60–80; ovary 3-merous, placentation incompletely axile. **Capsules** cylindric-conic, 6–15 × 4.5–8 mm. **Seeds** scarcely carinate, 1.8–2 mm; testa finely scalariform-reticulate.

Flowering summer (Jun). Stream banks and moist woods, coastal plain and inland valleys; 10–500 m; Ark., Fla., La., Okla., Tex.

Hypericum apocynifolium has been included in *H. nudiflorum*; it can be distinguished from the latter by the fewer, larger flowers with relatively longer, persistent sepals, the larger, thicker-walled capsules, and the seeds, which are ridged and straight rather than carinate and curved.

A record from Georgia in the Flint River drainage has not been verified.

16. Hypericum nudiflorum Michaux ex Willdenow, Sp. Pl. 3: 1456. 1802 E

Brathydium nudiflorum (Michaux ex Willdenow) K. Koch; *Myriandra nudiflora* (Michaux ex Willdenow) Spach

Subshrubs, erect, usually loosely branched with branches ascending, 5–20 dm. **Stems:** internodes narrowly 4-winged at first, then terete. **Leaf blades** ovate-lanceolate or elliptic to linear-oblong, 30–70 × 7–25 mm, base not articulated, cuneate to subcordate, margins plane, apex obtuse to rounded, midrib with to 6 pairs of branches. **Inflorescences** corymbiform to rounded-pyramidal, 7–45-flowered, narrowly branched, sometimes with 1–7(–40)-flowered dichasia from 1–3 proximal nodes. **Flowers** 15–20 mm diam.; sepals deciduous, not enclosing capsule, 5, oblanceolate-spatulate to narrowly triangular, unequal or subequal,

2–5 × 1–1.5 mm; petals 5, pale or coppery yellow, oblanceolate-oblong to elliptic-oblong, 6–8(–10) × 3–4 mm, length 2–3 times sepals; stamens persistent, 80; ovary 3(4)-merous, placentation incompletely axile. **Capsules** broadly ellipsoid to ovoid-globose, 3.5–7 × 3–5 mm. **Seeds** markedly carinate, 1.5–2 mm; testa ± scalariform-reticulate. $2n = 18$.

Flowering summer (Jun–late Aug). Stream banks, moist woodlands, swamps; 0–1000 m; Ala., Fla., Ga., La., Miss., N.C., S.C., Tenn., Va.

See under 15. *Hypericum apocynifolium* for differences between it and *H. nudiflorum*, which has a more eastern distribution. Records from Texas probably are referable to *H. apocynifolium*.

Hypericum nudiflorum probably is now extirpated in Louisiana.

17. Hypericum cistifolium Lamarck in J. Lamarck et al., Encycl. 4: 158. 1797 E F

Hypericum opacum Torrey & A. Gray; *H. punctulosum* Bertoloni; *H. rosmarinifolium* Lamarck

Shrubs, erect, unbranched or with relatively short branches and sometimes 1 or 2 branches ascending from proximal nodes, 5–13 dm. **Stems:** internodes 4-lined at first, then terete. **Leaf blades** narrowly oblong or narrowly elliptic-oblong to triangular-lanceolate, 15–40 × 2–10 mm, base not articulated, cuneate to subcordate, margins recurved, apex subacute to rounded, midrib with 1 pair of branches. **Inflorescences** corymbiform to cylindric, (7–)15–65-flowered, narrowly branched, sometimes with 3–65-flowered dichasia from 1–2 proximal nodes and relatively short, flowering branches from further 1–4 nodes. **Flowers** 7–12 mm diam.; sepals persistent, not enclosing capsule, 5, obovate or broadly elliptic to oblong, unequal, 2–4 × 1–1.7 mm; petals 5, bright yellow, oblanceolate, 5–8 mm; stamens (some or all) persistent, 30–50; ovary 3-merous, placentation parietal. **Capsules** ovoid-cylindric to broadly ovoid, 4–6 × 3–4 mm. **Seeds** not carinate, 0.6 mm; testa reticulate to linear-foveolate. $2n = 18$.

Flowering spring–early fall (Apr–Oct). Pine flatwoods, margins of bogs, swamps, and marshes, ditches, on sand, coastal plain; 0–300 m; Ala., Fla., Ga., La., Miss., N.C., S.C., Tex.

Hypericum cistifolium is woodier in habit than *H. sphaerocarpum* and has shorter leaves, smaller flowers, narrower sepals, narrower capsules, and smaller seeds.

18. Hypericum microsepalum (Torrey & A. Gray)
A. Gray ex S. Watson, Smithsonian Misc. Collect.
258: 456. 1878 [E]

Ascyrum microsepalum Torrey &
A. Gray, Fl. N. Amer. 1: 157.
1838; *Crookea microsepala*
(Torrey & A. Gray) Small;
Isophyllum drummondii Spach

Shrubs, erect to ascending,
bushy, with ± numerous,
sometimes straggling branches,
1.5–7 dm. **Stems:** internodes
4-lined at first, eventually 2-lined. **Leaf blades** narrowly
oblong or oblanceolate to linear, 5–15 × 1–3 mm,
base not articulated, rounded to cuneate, margins
recurved to subrevolute, apex rounded to obtuse, midrib
with 1–3 pairs of branches. **Inflorescences** rounded-
pyramidal, 1–3-flowered, narrowly branched, sometimes with 1–3-flowered dichasia or branches from to
4 proximal nodes. **Flowers** 15–25 mm diam.; sepals
persistent, not enclosing capsule, (3)4(5), oblong or
elliptic-oblong to linear, subequal or equal, 3–5 × 1–1.4
mm; petals (3)4(5), bright yellow, obovate (larger pair)
to obovate-oblong (smaller pair), 10–12 mm; stamens
persistent, 60–70; ovary 3-merous, placentation
parietal. **Capsules** cylindric-ellipsoid to narrowly
ovoid-conic, 6–8 × 2.5 mm. **Seeds** not carinate,
0.9–1 mm; testa linear-foveolate. $2n = 18$.

Flowering late winter–spring (Feb–May), late fall
(Nov). Low, pine flatwoods, moist to wet, on sand;
0–100 m; Fla., Ga.

Hypericum microsepalum is related to *H. cistifolium*,
not to the other four-petaled species attributed to
Ascyrum, and can be distinguished from it by the bushier
habit and smaller leaves and by the usually four-merous
flowers with larger petals.

Hypericum isophyllum Steudel is an illegitimate name
for *H. microsepalum*.

19. Hypericum sphaerocarpum Michaux, Fl. Bor.-Amer.
2: 78. 1803 [E]

Brathydium chamaenerium Spach;
B. sphaerocarpum (Michaux)
Spach; *Hypericum chamaenerium*
(Spach) Steudel; *H. sphaerocarpum*
var. *turgidum* (Small) Svenson;
H. turgidum Small

Herbs (perennial) **or subshrubs,**
erect or decumbent, not or
rarely rhizomatous, unbranched
or branched proximally, 2–6 dm. **Stems:** internodes
2–4-lined. **Leaf blades** narrowly elliptic or narrowly
oblong to linear, 30–70 × 3–15 mm, base not articulated,
narrowly cuneate to linear, margins plane to revolute,

apex subacute to rounded, midrib with 0–4 pairs of
branches. **Inflorescences** rounded-corymbiform, 7–70-
flowered, narrowly branched, sometimes with dichasia
or branches from to 8 proximal nodes. **Flowers** 10–15
mm diam.; sepals persistent, not enclosing capsule, 5,
broadly ovate to oblong-elliptic, ± unequal, 2.5–5 ×
1.5–3 mm; petals 5, bright yellow, oblanceolate-elliptic
to elliptic, 5–9 mm; stamens persistent, 45–85; ovary
3-merous, placentation parietal. **Capsules** broadly ovoid
to depressed-globose, 4.5–8 × 4–7 mm. **Seeds** carinate,
2–2.7 mm; testa coarsely reticulate.

Flowering summer (Jun–Aug). Rocky outcrops or
embankments, prairies, stream banks, usually wet or
moist, railroad embankments; 500–1000 m; Ont.; Ala.,
Ark., Ill., Ind., Iowa, Kans., Ky., Mich., Miss., Mo.,
Nebr., Ohio, Okla., Tenn., Tex., Wis.

Hypericum sphaerocarpum differs from *H. cistifolium*
and *H. nudiflorum* in its semiherbaceous habit and
more northwestern distribution, as well as in its
combination of relatively long, narrow leaves, persistent
sepals, globose and apiculate to rounded capsules, and
relatively large seeds. The narrow-leaved, bushy form
from eastern parts of the range (var. *turgidum*) merges
with the typical form.

20. Hypericum adpressum W. P. C. Barton, Comp.
Fl. Philadelph. 2: 15. 1818 • Creeping St. John's
wort [E]

Brathydium fastigiatum (Elliott)
K. Koch; *Hypericum adpressum*
var. *fastigiatum* (Elliott) Torrey &
A. Gray; *H. adpressum* var.
spongiosum B. L. Robinson;
H. bonapartee W. P. C. Barton;
H. fastigiatum Elliott; *Myriandra
adpressa* (W. P. C. Barton) K. Koch

Herbs, perennial, erect, with
creeping, rhizomatous, sometimes woody or spongy
base, unbranched or 1(2)-branched, sometimes with
axillary clusters at base, relatively stout, 2–8 dm. **Stems:**
internodes 2-lined. **Leaf blades** narrowly oblong or
linear to lanceolate or narrowly elliptic, 15–80 × 2–10
mm, base not articulated, narrowly cuneate to parallel,
slightly decurrent, margins ± revolute, apex acute,
midrib with 7 pairs of branches. **Inflorescences** rounded-
corymbiform, 13–60-flowered, narrowly branched,
without subsidiary branches. **Flowers** 10–15 mm
diam.; sepals persistent, not enclosing capsule, 5, ovate-
lanceolate to lanceolate, subequal, (2–)4–7 × 1–1.5 mm;
petals 5, bright yellow, obovate-oblanceolate, 6–8 mm;
stamens persistent, 60–80; ovary 3-merous, placentation
parietal. **Capsules** ellipsoid to ovoid-ellipsoid, 3.5–6
× 2–4 mm. **Seeds** slightly carinate, 0.6–0.7 mm; testa
scalariform. $2n = 18$.

Flowering late summer (Jul–Sep). Marshes, pond margins, wet ditches, bogs, coastal plain; 0–1000 m; Conn., Del., Ga., Ill., Ind., Md., Mass., Mo., N.J., N.Y., N.C., Pa., R.I., S.C., Tenn., Va., W.Va.

Hypericum adpressum is more herbaceous and rhizomatous than *H. sphaerocarpum* and has narrower capsules and smaller seeds.

The plants with aerenchymatous tissue in the rhizome (var. *spongiosum*) are not taxonomically distinct but merely the result of a habitat-induced modification.

21. Hypericum ellipticum Hooker, Fl. Bor.-Amer. 1: 110. 1831 • Pale St. John's wort, millepertuis elliptique E

Brathydium canadense Spach; *Hypericum brathydium* Steudel; *H. canadense* Linnaeus var. *oviforme* R. Keller

Herbs, perennial, erect, with creeping, rhizomatous, ± woody base, usually unbranched, sometimes branched proximally, relatively slender, 1.1–3(–5) dm. Stems: internodes 4-lined. Leaf blades broadly to narrowly elliptic or oblanceolate to oblong-elliptic, 11–35 × 3–13 mm, base not articulated, cuneate to shallowly cordate-amplexicaul, margins plane to subrevolute, apex rounded, midrib with 5–7 pairs of branches. Inflorescences corymbiform, (1–)3–15-flowered, narrowly branched, sometimes with branches from 1 or 2 proximal nodes. Flowers 12–15 mm diam.; sepals persistent, not enclosing capsule, (4)5, ± lanceolate to lanceolate-elliptic, ± unequal, 6–7 × 2–3 mm; petals (4–)5, pale yellow, sometimes tinged red, obovate to oblanceolate, 6–8 mm; stamens persistent, 70–95; ovary 3-merous, placentation parietal. Capsules ellipsoid to globose, 4–7 × 3.5–5 mm. Seeds carinate, 0.6–0.7 mm; testa scalariform-reticulate. *2n* = 16, 18.

Flowering summer (Jun–Sep). Stream, lake, and pond margins, wet meadows, swamps; 0–600 m; N.B., Nfld. and Labr. (Nfld.), N.S., Ont., Que.; Conn., Ill., Maine, Md., Mass., Mich., Minn., N.H., N.J., N.Y., Ohio, Pa., R.I., Tenn., Vt., Wash., W.Va., Wis.

Hypericum ellipticum is related to *H. sphaerocarpum*, differing by the shorter, herbaceous, rhizomatous habit, shorter leaves, and smaller seeds. A submerged aquatic form (forma *submersum* Fassett) and one with axillary branches developing after fertilization (forma *foliosum* Marie-Victorin) seem scarcely worth formal recognition. *Hypericum ellipticum* is introduced in Washington.

22. Hypericum myrtifolium Lamarck in J. Lamarck et al., Encycl. 4: 180. 1797 E

Brathydium myrtifolium (Lamarck) K. Koch; *Hypericum glaucum* Michaux; *H. sessiliflorum* Willdenow ex Sprengel; *Myriandra glauca* (Michaux) Spach

Shrubs, erect, with woody caudex, unbranched or branched distally, 3–10 dm, bark on older stems corky. Stems: internodes 4-lined. Leaf blades oblong-ovate to triangular-lanceolate, 8–40 × (5–)7–20 mm, base articulated, subcordate-amplexicaul, margins recurved, apex rounded, midrib with 3 or 4 pairs of branches. Inflorescences hemispheric to ± flat-topped, 7–30-flowered, widely branched, with flowers or flowering branches from to 3 proximal nodes. Flowers 15–25 mm diam.; sepals persistent, not enclosing capsule, 5, ovate to lanceolate, unequal or subequal, 5–8 × 2–4.5 mm; petals 5, bright yellow, obovate to oblong-lanceolate, 8–15 mm; stamens deciduous, 200; ovary 3(4)-merous. Capsules pyramidal-ovoid, 5–6 × 3–4 mm. Seeds narrowly carinate, 1 mm; testa shallowly linear-reticulate. *2n* = 18.

Flowering late spring–summer (May–Jul), sometimes fall. Moist pinewoods, grassy bogs, pond margins, ditches; 0–200 m; Ala., Fla., Ga., Miss., S.C.

Hypericum myrtifolium is related to *H. frondosum*; it differs in its shorter, usually amplexicaul leaves, the widely dichasially branched inflorescences, and persistent sepals.

23. Hypericum dolabriforme Ventenat, Descr. Pl. Nouv., plate 45. 1801 E

Brathydium dolabriforme (Ventenat) Y. Kimura; *Hypericum bissellii* B. L. Robinson; *H. procumbens* Desfontaines ex Willdenow

Subshrubs, decumbent and woody (not rooting) at base, branching at base or throughout, 1.5–5 dm, bark thin. Stems: internodes 4-lined at first, then 2-lined to terete. Leaf blades linear-elliptic or linear-oblong to linear, 20–35 × 3–5 mm (main stem), base not articulated, narrowly cuneate to rounded, margins recurved to revolute, apex obtuse to acute, midrib unbranched. Inflorescences obconic, (1–)3–20-flowered, ± widely branched, sometimes with single flowers at immediately proximal nodes. Flowers 15–20 mm diam.; sepals persistent, not enclosing capsule, 5, ovate-lanceolate to lanceolate, ± foliaceous, unequal, 5–8(–15) × 2–3(–8)mm; petals 5, yellow, curved-dolabriform, 10–13 mm; stamens

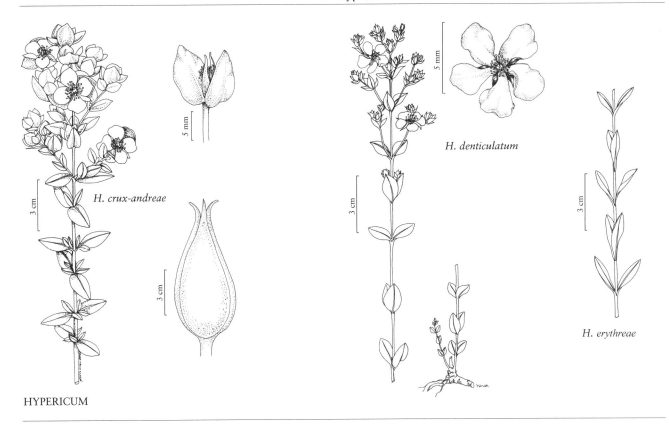

H. crux-andreae

H. denticulatum

H. erythreae

HYPERICUM

deciduous, 120–200; ovary 3-merous. **Capsules** ovoid-conic, rostrate, 4–9 × 3–4 mm. **Seeds** carinate, 1.5–1.8 mm; testa reticulate-scalariform. $2n = 18$.

Flowering summer (Jun–Sep). Limestone outcrops, cedar glades, dry, rocky stream beds; 0–500 m; Conn., Ga., Ind., Ky., Tenn.

Hypericum dolabriforme is superficially similar to *H. sphaerocarpum*, to which W. P. Adams (1962) related it. Apart from the narrow leaves and unequal sepals, it is much nearer morphologically to *H. myrtifolium* (for example, in the wide-spreading inflorescence, numbers of stamens, and ovoid-conic capsules). *Hypericum bissellii*, described from a plant growing in Southington, Connecticut, is unlikely to be indigenous in that state.

24. **Hypericum crux-andreae** (Linnaeus) Crantz, Inst. Rei Herb. 2: 520. 1766 (as crux andreae)

· St. Peter's wort E F

Ascyrum crux-andreae Linnaeus, Sp. Pl. 2: 787. 1753 (as crux andreae); *A. cuneifolium* Chapman; *A. grandiflorum* Rafinesque; *A. simplex* Zeyher ex Turczaninow; *A. stans* var. *obovatum* Chapman ex Torrey & A. Gray; *Hypericoides crux-andreae* (Linnaeus) Poiret; *Hypericum stans* W. P. Adams & N. Robson

Shrubs, usually erect to suberect, rarely decumbent and rooting, usually unbranched, rarely sparsely branched distally, 1–13.5 dm. **Stems:** internodes 2–4-lined at first, then 2-winged. **Leaf blades** usually oblong to elliptic, rarely obovate to oblanceolate or triangular-ovate, 12–36 × 6–16 mm, base articulated, rounded to slightly cordate-amplexicaul, without glandlike auricles, margins plane to subrecurved, apex rounded to obtuse, midrib with to 3 pairs of branches. **Inflorescences** ± narrowly cylindric to narrowly pyramidal, 1–3 (–7)-flowered, branching dichasial, from to 4 proximal nodes. **Flowers** 20–30 mm diam.; sepals persistent, enclosing capsule, 4, unequal, outer broadly ovate to circular, 9–20 × 9–18 mm, apex apiculate or obtuse to rounded, inner narrowly elliptic to lanceolate, 7–14 × 2–4 mm, apex acute to subacute; petals 4, bright yellow, obovate, 11–18 mm; stamens persistent, 80–100; ovary 3(4)-merous. **Capsules** narrowly ellipsoid-ovoid, 7–10 × 5–6.5 mm. **Seeds** not carinate, 0.8 mm; testa shallowly scalariform. $2n = 18$.

Flowering summer–fall. Moist to dry, pine savannas and flatwoods, meadows, bogs, other wet habitats, lake and pond margins; 0–1500 m; Ala., Ark., Del., D.C., Fla., Ga., Ky., La., Md., Miss., N.J., N.Y., N.C., Okla., Pa., S.C., Tenn., Tex., Va.

Hypericum crux-andreae, long known as *Ascyrum stans*, is a derivative of *H. frondosum* in which the tetramerous tendency in the perianth has become fixed. The low, multistemmed form with cuneate leaves, longerpedicellate flowers, and shorter sepals (*A. cuneifolium*, *A. stans* var. *obovatum*) cannot be separated from typical *H. crux-andreae*.

Linnaeus included "*Hypericum ex terra mariana, floribus exiguis luteis*" under his phrase name for *Ascyrum crux-andreae*; that element of the protologue refers to *H. mutilum* Linnaeus; see N. K. B. Robson (1980).

25. Hypericum tetrapetalum Lamarck in J. Lamarck et al., Encycl. 4: 153. 1797

Ascyrum amplexicaule Michaux; *A. cubense* Grisebach; *A. tetrapetalum* (Lamarck) Vail

Herbs (perennial) **or shrubs,** erect, with woody base, usually unbranched, sometimes with divaricate or ascending branches, 2–10 dm. **Stems:** internodes (2–)4-lined at first, then 2-lined to terete. **Leaf blades** oblong-ovate to ovate or triangular-ovate, 5–35 × 4–15 mm, base articulated, cordate-amplexicaul, margins subrecurved, apex apiculate or obtuse to rounded, midrib with 1 pair of branches. **Inflorescences** terminal, 1(–3)-flowered, branching from apical node pseudodichotomous, sometimes with relatively short branches from to 3 proximal nodes. **Flowers** 20–30 mm diam.; sepals persistent, enclosing capsule, 4, unequal, outer broadly ovate, 7–15 × 5.5–10 mm, apex subapiculate to obtuse, inner narrowly lanceolate, 7–15 × 2–3 mm, apex acute; petals 4, bright yellow, obovate-oblong, 10–15 mm; stamens persistent, 100; ovary 3-merous. **Capsules** broadly ellipsoid-ovoid to subglobose, 5–6 × 3.5–4 mm. **Seeds** not carinate, 0.7 mm; testa shallowly scalariform. $2n = 18$.

Flowering winter–spring (Jan–Apr), late summer (Jul–Sep). Moist, low pinelands, ditches; 0–200 m; Ala., Fla., Ga.; West Indies (w Cuba).

Hypericum tetrapetalum differs from *H. crux-andreae* in having broader leaves with strongly cordate-amplexicaul bases and, nearly always, by terminal pseudodichotomous inflorescences.

26. Hypericum edisonianum (Small) W. P. Adams & N. Robson, Rhodora 63: 15. 1961 C E

Ascyrum edisonianum Small, Man. S.E. Fl., 868, 1505. 1933

Shrubs, erect, sometimes unbranched proximal to inflorescence, 3–15 dm. **Stems:** internodes 4–6-lined at first, soon 2-lined. **Leaf blades** elliptic, 15–26 × 5–8(–11) mm, base not articulated, cuneate to subrounded, with glandlike auricles, margins subrecurved to subincrassate, apex obtuse to acute, midrib with to 4

pairs of branches. **Inflorescences** 1-flowered, branching from apical node repeatedly pseudodichotomous, without branches from proximal nodes. **Flowers** 15–20 mm diam.; sepals persistent, enclosing capsule, 4, unequal, outer broadly ovate, 8–17 × 5–9 mm, apex acute to subacuminate, inner linear-lanceolate, 5–6 × 0.6–1.2 mm, apex acuminate; petals 4, bright yellow, obovate, 10–18 mm; stamens persistent, 70–80; ovary 3- or 4-merous. **Capsules** narrowly pyramidal-ovoid, 5–8 × 3–4 mm. **Seeds** not carinate, 0.8 mm; testa reticulate.

Flowering probably year-round. Marshy areas in pine flatwoods, pond margins; of conservation concern; 50 m; Fla.

Hypericum edisonianum differs from *H. crux-andreae* in the smaller, thicker, obtuse to acute leaves with glandlike auricles and the pseudodichotomously branched inflorescence.

27. Hypericum suffruticosum W. B. Adams & N. Robson, Rhodora 63: 15. 1961 (as suffructicosum) E

Ascyrum pumilum Michaux, Fl. Bor.-Amer. 2: 77. 1803, not *Hypericum pumilum* Sessé & Moçiño 1894; *A. pauciflorum* Nuttall 1818, not *H. pauciflorum* Kunth 1822

Shrubs, erect and branched from near base, forming cushion, or decumbent and matted, 0.5–2 dm. **Stems:** internodes 4-lined at first, soon 2-winged. **Leaf blades** elliptic or oblong-linear to narrowly obovate or oblanceolate, 3–10 × 1–3 mm, base not articulated, rounded to cuneate, without glandlike auricles, margins plane to subrecurved, apex obtuse, midrib unbranched. **Inflorescences** 1-flowered, often with pseudodichotomous branches from apical node, without proximal branches; pedicels mostly recurved to reflexed in fruit, bracteoles proximal. **Flowers** 10–15 mm diam.; sepals persistent, enclosing capsule, (2) 4, unequal, outer broadly ovate to broadly elliptic, 4–8 × 4–8 mm, apex obtuse to rounded, inner none or minute; petals 4, pale yellow, narrowly obovate, often unequal, 4–8 mm; stamens persistent, 30; ovary 2-merous. **Capsules** cylindric-ellipsoid, 3–5 × 2–3 mm. **Seeds** scarcely carinate, 1 mm; testa finely reticulate. $2n = 18$.

Flowering spring–early summer (Mar–Jun). Dry, open, sandy pinelands, coastal plain; 0–200 m; Ala., Fla., Ga., La., Miss., N.C., S.C.

Hypericum suffruticosum is near the "cuneifolium" form of *H. crux-andreae*; it differs, among other things, by the two-merous ovary, the relatively small or absent inner sepals, the pedicels recurved or reflexed in fruit, and the cushion or matted habit.

28. Hypericum hypericoides (Linnaeus) Crantz, Inst. Rei Herb. 2: 520. 1766 • St. Andrew's cross

Ascyrum hypericoides Linnaeus, Sp. Pl. 2: 788. 1753

Subshrubs or shrubs, erect or decumbent to prostrate, unbranched or branched, sometimes diffuse and mat-forming, 0.5–3 or 3–15 dm. **Stems:** internodes 2-winged. **Leaf blades** oblanceolate or narrowly oblong or elliptic to linear, 7–25 × 1–8.5 mm, base not articulated, usually cuneate, sometimes rounded, with glandlike auricles, margins subrecurved, apex rounded to obtuse, midrib with 1 or 2 pairs of branches. **Inflorescences** narrowly cylindric to pyramidal, 1–12-flowered from 1–4 nodes, sometimes with branches from to 10 proximal nodes, or branching more elaborate and pseudodichotomous; pedicels erect in fruit, bracteoles distal. **Flowers** 10–20 mm diam.; sepals persistent, enclosing capsule, 4, unequal, outer ovate-suborbiculate to narrowly elliptic, 5–12.5 × 2–13 mm, apex subapiculate to obtuse, inner lanceolate, 1–4 × 2 mm, apex acute; petals 4, bright to pale yellow, obovate to narrowly oblong-elliptic, 6–12 mm; stamens persistent, 40–50; ovary 2-merous. **Capsules** narrowly compressed-ovoid to cylindric-ellipsoid, 5–9 × 2–4 mm. **Seeds** not carinate, 0.6–0.8 mm; testa finely linear-reticulate to linear-foveolate.

Subspecies 3 (2 in the flora): c, e United States, Mexico, West Indies (Bahamas, Greater Antilles), Bermuda, Central America (Guatemala, Honduras).

Hypericum hypericoides can be distinguished from *H. crux-andreae* by the two-merous ovary, narrower leaves, smaller flowers, and more richly-branched stems. It is variable in leaf and sepal shape and size; three subspecies can be recognized.

The erect bushy form (subsp. *hypericoides*) is most widespread and has given rise to a northern diffuse form (subsp. *multicaule*) in the United States and a prostrate form (subsp. *prostratum* N. Robson) in Hispaniola.

1. Plants erect, usually unbranched from base and freely branched well above ground level; leaf blades usually narrowly elliptic to narrowly oblong, broadest near middle . . . . .28a. *Hypericum hypericoides* subsp. *hypericoides*
1. Plants decumbent to prostrate, branching from base; leaf blades usually oblanceolate, broadest beyond middle . . . . 28b. *Hypericum hypericoides* subsp. *multicaule*

28a. Hypericum hypericoides (Linnaeus) Crantz subsp. **hypericoides** E

Ascyrum crux-andreae (Linnaeus) Crantz var. *angustifolium* Nuttall; *A. linifolium* Spach; *A. michauxii* Spach; *A. montanum* Rafinesque; *A. oblongifolium* Spach; *A. plumieri* Bertoloni

Plants erect, usually unbranched from base and freely branched well above ground level, 3–15 dm. **Leaf blades** usually narrowly elliptic to narrowly oblong, 7–25 × 1–8.5 mm, broadest near middle. **Inflorescences:** branching dichasial/lateral to pseudodichotomous. $2n = 18$.

Flowering summer–fall (Jun–Oct). Dry, open, sandy woods to moist, shady, rich woods and thickets; 0–1000 m; Ala., Ark., Del., Fla., Ga., Ill., Ky., La., Md., Miss., Mo., N.J., N.C., Okla., S.C., Tenn., Tex., Va., W.Va.

Subspecies *hypericoides* is probably extirpated in Illinois.

28b. Hypericum hypericoides (Linnaeus) Crantz subsp. **multicaule** (Michaux ex Willdenow) N. Robson, Taxon 29: 273. 1980 E

Ascyrum multicaule Michaux ex Willdenow, Sp. Pl. 3: 1472. 1802; *A. helianthemifolium* Spach; *A. hypericoides* var. *multicaule* (Michaux ex Willdenow) Fernald; *A. spathulatum* Spach; *Hypericum hypericoides* var. *multicaule* (Michaux ex Willdenow) Fosberg; *H. stragulum* W. P. Adams & N. Robson

Plants decumbent to prostrate, branching from base, mat-forming, 0.5–3 dm. **Leaf blades** usually oblanceolate, 10–20 × 3–6 mm, broadest beyond middle. **Inflorescences:** branching dichasial (1–3-flowered) or lateral. $2n = 18$.

Flowering summer (Jun–Aug). Dry, rocky slopes, road embankments, dry to moist, rich woods; 70–900 m; Ala., Ark., Del., D.C., Ga., Ill., Ind., Kans., Ky., La., Md., Mass., Miss., Mo., N.J., N.Y., N.C., Ohio, Okla., Pa., Tenn., Tex., Va., W.Va.

1b. Hypericum Linnaeus informal sect. group Brathys

Brathys Mutis ex Linnaeus f.; *Hypericum* sect. *Brathys* (Mutis ex Linnaeus f.) Choisy; *Hypericum* sect. *Spachium* (R. Keller) N. Robson; *Hypericum* subsect. *Spachium* R. Keller; *Hypericum* sect. *Trigynobrathys* (Y. Kimura) N. Robson; *Sarothra* sect. *Trigynobrathys* Y. Kimura

Herbs, annual or perennial, (stems and leaves glabrous except *H. setosum*); black glands absent. **Leaves** deciduous (base articulated) or persistent (base not articulated). **Flowers** 3–15 mm diam.; sepals persistent, (4)5; petals persistent, (4)5; stamens persistent, (5–)10–80, usually in continuous or interrupted ring, sometimes in 5 barely discernable fascicles, each of 1 or 2 stamens; ovary (2)3(4)-merous; placentation parietal; styles ± spreading, bases distinct; stigmas capitate or clavate. **Seeds** not carinate.

Species 148 (16 in the flora): North America, Mexico, West Indies, Central America, South America, Asia, Africa, Pacific Islands (New Zealand), Australia.

SELECTED REFERENCE   Webb, D. H. 1980. A Biosystematic Study of *Hypericum* Section *Spachium* in Eastern North America. Ph.D. dissertation. University of Tennessee.

1. Herbs annual, wiry; inflorescences: branching mostly monochasial.
   2. Leaf blades linear or linear-subulate to linear-lanceolate, 5–22 mm, margins recurved to revolute; sepals 3–7 mm; capsule lengths 1–1.2 times sepals . . . . 43. *Hypericum drummondii*
   2. Leaf blades narrowly triangular-subulate to linear-subulate, scalelike, 1–4 mm, margins incurved; sepals 1.5–2.5 mm; capsule lengths 2–2.7 times sepals . . . . 44. *Hypericum gentianoides*
1. Herbs annual or perennial, not wiry; inflorescences: branching mostly dichasial.
   3. Stems and leaves scabrous-tomentose to pilose; sepal margins setulose-ciliate. . . . 34. *Hypericum setosum*
   3. Stems and leaves glabrous; sepal margins sometimes ciliate, not setulose-ciliate.
      4. Leaf blades linear-subulate, 0.2–0.3 mm wide, basal veins 1, midrib unbranched
         . . . . 35. *Hypericum cumulicola*
      4. Leaf blades not linear-subulate, (0.5–)2–18 mm wide, basal veins 1–7, midrib with 0–4 pairs of branches.
         5. Leaf blades leathery; petals golden yellow or orange-yellow; stamens (35–) 50–80.
            6. Leaf blades linear-oblong or linear-lanceolate to linear, 0.5–2 mm wide; petals red tinged; stamens 35–50; styles 2.5–5 mm . . . . 41. *Hypericum paucifolium*
            6. Leaf blades elliptic, lanceolate, linear-lanceolate, oblong-elliptic, obovate, or ovate, 3–18 mm wide; petals not red tinged; stamens 50–80; styles 2–4 mm.
               7. Herbs erect or ± spreading, non-aerenchymatous at base; leaves (main stems) longer than internodes.
                  8. Leaf blades (main stem) lanceolate to oblong-elliptic or obovate, 10–30(–50) mm, apex usually acute to acuminate, rarely obtuse; subsidiary inflorescence branches with or without relatively smaller leaves . . . . 32. *Hypericum virgatum*
                  8. Leaf blades (main stem) narrowly lanceolate, 10–40(–55) mm, apex acute to acuminate; subsidiary inflorescence branches with relatively smaller leaves. . . . . 33. *Hypericum radfordiorum*
               7. Herbs erect to ascending, usually ± aerenchymatous (spongiform-thickened) at base; leaves (main stems) mostly shorter than internodes (usually longer in 31. *H. harperi*).
                  9. Herbs 2–7 dm; leaf blades (main stem) usually broadly to narrowly ovate, rarely elliptic or lanceolate . . . . 29. *Hypericum denticulatum*
                  9. Herbs 3–12 dm; leaf blades (main stem) usually lanceolate, sometimes linear-lanceolate or oblong-elliptic, rarely elliptic or ovate.

10. Herbs erect to ascending, branching at base and in inflorescence; leaves strongly ascending to appressed, shorter than internodes, smaller distally . . . . 30. *Hypericum erythreae*
10. Herbs erect, branching (from long-creeping rhizomes) at base and from mid and distal nodes; leaves ascending to deflexed, mostly longer than internodes, not or scarcely smaller distally . . . . .31. *Hypericum harperi*

[5. Shifted to left margin.—Ed.]

5. Leaf blades papery to membranous; petals usually bright, golden, or pale yellow, rarely salmon orange; stamens 5–25.
 11. Herbs decumbent to ascending, forming loose mats; stigmas scarcely capitate . . . . .42. *Hypericum anagalloides*
 11. Herbs usually erect, not forming loose mats; stigmas broadly capitate.
  12. Leaf blades lanceolate, linear, narrowly elliptic, narrowly oblong-elliptic, oblanceolate, oblanceolate-linear, or obovate; capsules broadest proximal to middle.
   13. Leaf blades lanceolate, narrowly oblong-elliptic, or oblanceolate, (2–)6–12 mm wide, basal veins (3–)5–7; inflorescences usually compact . . . . .36. *Hypericum majus*
   13. Leaf blades linear to oblanceolate-linear or (proximal) oblanceolate to obovate, 0.5–5.5 mm wide, basal veins 1–3(–5); inflorescences usually diffuse . . .37. *Hypericum canadense*
  12. Leaf blades usually elliptic, oblong, broadly oblong-elliptic, ovate, ovate-triangular, round, or suborbiculate, rarely lanceolate; capsules usually broadest at or near middle.
   14. Leaf blades (mid and distal) lanceolate-deltate, apex usually subacute; capsules narrowly conic-ellipsoid . . . . 38. *Hypericum gymnanthum*
   14. Leaf blades (mid and distal) elliptic, lanceolate, oblanceolate, oblong, round, suborbiculate, or ovate, apex obtuse to rounded; capsules narrowly ovoid to cylindric-ellipsoid.
    15. Leaf blades bicolor, paler abaxially; inflorescences: bracts linear-subulate . . . . .39. *Hypericum mutilum*
    15. Leaf blades concolor; inflorescences: bracts not linear-subulate . . . . . .40. *Hypericum boreale*

29. Hypericum denticulatum Walter, Fl. Carol., 190. 1788 • Coppery St. John's wort   E F

Hypericum angulosum Michaux ex Willdenow; *H. denticulatum* var. *ovalifolium* (Britton) S. F. Blake; *H. laevigatum* Aiton; *H. virgatum* Lamarck var. *ovalifolium* Britton

Herbs perennial, erect, branching at usually aerenchymatous base and in inflorescence, 2–7 dm. **Stems:** internodes 4-lined. **Leaves** (main stem) spreading to appressed, sessile; blade usually broadly to narrowly ovate, rarely elliptic or lanceolate, 4–20 × 5–15(–18) mm, mostly shorter than internodes, leathery, margins plane, apex acute to subrounded, densely gland-dotted, basal veins 1–5, if 1, midrib with 2 or 3 pairs of branches. **Inflorescences** broadly pyramidal to corymbiform, to 25-flowered, branching mostly dichasial. **Flowers** 5–13 mm diam.;

sepals ovate or lanceolate to elliptic or obovate, subequal, 3–8 × 1.5–4 mm, margins sometimes ciliate, not setulose-ciliate, apex acute; petals orange-yellow, obovate, 5–10 mm; stamens 50–80, irregularly grouped; styles 2–4 mm; stigmas clavate. **Capsules** ovoid to rostrate-subglobose, 3–5 × 2–3 mm. **Seeds** 0.4–0.7 mm; testa obscurely linear-reticulate to finely ribbed-scalariform.

Flowering summer–early fall (Jun–Sep). Wet woods, marshes, bogs; 0–400 m; Ala., Del., Ga., N.J., N.Y., N.C., Pa., S.C., Tenn., Va.

D. H. Webb (1980) regarded the disjunct populations in North Carolina and Tennessee as possible relicts and the Alabama one as due to recent introduction. J. R. Allison (2011) agreed and, in his opinion, the Pennsylvania and Virginia records are historical, and *Hypericum denticulatum* is likely adventive in Georgia.

SELECTED REFERENCE Allison, J. R. 2011. Synopsis of the *Hypericum denticulatum* complex. Castanea 76: 99–115.

30. **Hypericum erythreae** (Spach) Steudel, Nomencl. Bot. ed. 2, 1: 787. 1840 (as erythraeae) E F

Brathys erythreae Spach, Hist. Nat. Vég. 5: 452. 1836

Herbs perennial, erect to ascending, branching at ± aerenchymatous base and in inflorescence, 3–12 dm. **Stems:** internodes 4-lined. **Leaves** ascending to appressed, sessile; blade usually lanceolate, rarely elliptic to ovate, 5–20(–24) × 5–15 mm, shorter than internodes, smaller distally, leathery, margins plane, apex acute, gland-dotted, less conspicuously adaxially, basal veins 1–5, midrib with 2 or 3 pairs of branches. **Inflorescences** corymbiform, to 30-flowered, branching mostly dichasial. **Flowers:** 5–13 mm diam.; sepals lanceolate, unequal, 3.5–4.2(–4.8) mm, margins sometimes ciliate, not setulose-ciliate, apex acute to subacuminate; petals orange-yellow, ± obovate, 5–10 mm; stamens 50–80, filaments basally connate; styles 2–4 mm; stigmas clavate. **Capsules** ovoid, 3–4.5 mm. **Seeds** 0.4–0.7 mm; testa reticulate.

Flowering mid summer–fall (Jul–Oct). Bogs, ditches, coastal plain; 0–300 m; Ga., S.C.

31. **Hypericum harperi** R. Keller, Bot. Jahrb. Syst. 58: 198. 1923 E

Herbs perennial, semiaquatic or aquatic, erect, branching with long-creeping rhizomes at aerenchymatous base and from mid and distal nodes, 3–10 dm. **Stems:** internodes 4-lined. **Leaves** ascending to deflexed, sessile; blade narrowly oblong-elliptic (proximal) or lanceolate to linear-lanceolate, 10–30 × 3–8 mm, mostly longer than internodes, not or scarcely smaller distally, leathery, margins plane, apex acute, basal or near-basal veins 1–3(–5), midrib with 0–2 pairs of branches. **Inflorescences** broadly pyramidal to subcorymbiform, 1(–30)-flowered, branching mostly dichasial. **Flowers** 4–10 mm diam.; sepals lanceolate, usually unequal, 3–5 × 0.8–1 mm, margins sometimes ciliate, not setulose-ciliate, apex acute to acuminate; petals orange-yellow, obovate, 6–10 mm; stamens 50–80, irregularly grouped; styles 2–4 mm; stigmas capitate. **Capsules** ellipsoid to rostrate-subglobose, 3–4.5 × 2–2.5 mm. **Seeds** 0.5–0.6(–0.7) mm; testa obscurely linear-reticulate to irregularly reticulate. **2n** = 24.

Flowering mid–late summer (Jul–Sep). Open *Taxodium* swamps, wet pine barrens; 0–200 m; Ala., Fla., Ga., S.C.

32. **Hypericum virgatum** Lamarck in J. Lamarck et al., Encycl. 4: 158. 1797 E

Brathys lanceolata Spach; *Hypericum acutifolium* Elliott; *H. denticulatum* Walter subsp. *acutifolium* (Elliott) N. Robson; *H. denticulatum* var. *acutifolium* (Elliott) S. F. Blake; *H. denticulatum* var. *recognitum* Fernald & B. G. Schubert; *H. virgatum* var. *acutifolium* (Elliott) J. M. Coulter

Herbs perennial, erect, unbranched or branching at base and in inflorescence, 4–7.5(–10) dm. **Stems:** internodes 4-lined. **Leaves** (main stem) ascending to spreading, sessile; blade lanceolate to oblong-elliptic or obovate, 10–30(–50) × 3–8(–12) mm, on main stem, longer than internodes, leathery, margins plane, apex usually acute to acuminate, rarely obtuse, basal veins 1–3+, midrib with 2 or 3 pairs of inconspicuous branches. **Inflorescences** broadly cylindric to corymbiform, 25–70-flowered, branching mostly dichasial, subsidiary branches with or without relatively smaller leaves. **Flowers** 8–13 mm diam.; sepals narrowly ovate to lanceolate, equal, 3–3.5 × 1.5–2.5 mm, margins sometimes ciliate, not setulose-ciliate, apex acute to acuminate; petals orange-yellow, obovate, 6–10 mm; stamens 50–80, irregularly grouped; styles 2–4 mm; stigmas capitate. **Capsules** ovoid to rostrate-subglobose, 3–5 × 2–3 mm. **Seeds** (0.5–)0.6–0.8(–0.9) mm; testa reticulate. **2n** = 24.

Flowering mid–late summer (Jun–Sep). Dry roadsides, fields, open woods; 0–700 m; Ala., Ark., Fla., Ga., Ill., Ind., Ky., La., Md., Miss., N.C., Ohio, S.C., Tenn., Va., W.Va.

33. **Hypericum radfordiorum** Weakley ex J. R. Allison, Castanea 76: 110, fig. 3. 2011 E

Herbs perennial, erect or ± spreading, branching at base and beyond, (3–)4–7(–7.8) dm. **Stems:** internodes 4-lined. **Leaves** ascending to widely spreading, sessile; blade narrowly lanceolate (linear-lanceolate on axillary branches), 10–40(–55) × 5–10(–13) mm, longer than internodes, leathery, margins plane, apex acute to acuminate, densely gland-dotted, basal vein 1, midrib with inconspicuous branches. **Inflorescences** cylindric, to 40-flowered; branching mostly dichasial, subsidiary branches usually with relatively smaller leaves. **Flowers:** 8–15+ mm diam.; sepals lanceolate, unequal (outer slightly wider than inner), 3–6(–7.5) × 1.1–1.8(–2.2) mm, margins sometimes ciliate, not setulose-ciliate, apex acute or acuminate; petals orange-yellow, obovate, 5–10 mm; stamens 50–80, filaments basally connate; styles 2–4 mm; stigmas capitate. **Capsules** ovoid, 5 × 3–4 mm. **Seeds** 0.4–0.8 mm; testa linear-pitted. $2n = 24$.

Flowering late spring–fall (May–)Jun–Sep(–Oct). Granitic outcrops; 500–800 m; N.C.

34. Hypericum setosum Linnaeus, Sp. Pl. 2: 787. 1753 E

Ascyrum villosum Linnaeus; *Brathys tomentosa* Spach; *Hypericum pilosum* Walter; *H. villosum* (Linnaeus) Crantz

Herbs annual or perennial, erect, usually unbranched proximal to inflorescence, 2–8 dm. **Stems:** internodes 4-lined, scabrous-tomentose to pilose. **Leaves** appressed to ascending, sessile; blade narrowly ovate or lanceolate to narrowly oblong-elliptic (or proximal oblanceolate), 4–15 × 1.5–7 mm, subcoriaceous, margins recurved, apex acute to obtuse, surfaces scabrous-tomentose to pilose, basal veins 1(–5), midrib with 0 or 1 pair of branches. **Inflorescences** cylindric to subcorymbiform, to 30-flowered, branching mostly dichasial. **Flowers** 5–11 mm diam.; sepals ovate to ovate-lanceolate or obovate, subequal, 2.5–5 × 1.5–2.5 mm, margins setulose-ciliate, apex acute; petals 5, deep yellow, obovate, 4–7 mm; stamens (15–)20–40, filaments almost distinct; styles 1.5–2 mm; stigmas ± broadly capitate. **Capsules** ovoid to ellipsoid-subglobose, 3.5–5 × 2–3 mm. **Seeds** 0.4–0.6 mm; testa linear-reticulate. $2n = 12$.

Flowering early–late summer (Jun–Sep). Wet ditches, bogs, savannas, wet pinelands on sandy soil; 0–200 m; Ala., Fla., Ga., La., Miss., N.C., S.C., Tex., Va.

Hypericum setosum is the only American species of *Hypericum* with an indumentum. It is related to *H. virgatum* (*H. denticulatum* subsp. *acutifolium*); in addition to having the indumentum, it is generally smaller and less branched and has a different chromosome number.

35. Hypericum cumulicola (Small) W. P. Adams, Rhodora 64: 234. 1962 C E

Sanidophyllum cumulicola Small, Bull. Torrey Bot. Club 51: 391. 1924

Herbs perennial, erect, branching at or just below ground level and in inflorescence, 2–7.5 dm. **Stems:** internodes 4-lined. **Leaves** appressed, sessile; blade linear-subulate, (1–)2.5–4 × 0.2–0.3 mm, subcoriaceous, margins incurved, apex acute, basal vein 1, midrib unbranched. **Inflorescences**

subcorymbiform, to 13-flowered, branching mostly dichasial. **Flowers** 3–4 mm diam.; sepals ovate to elliptic or narrowly oblong, unequal, 1.5–2 × 0.6–1 mm, margins sometimes ciliate, not setulose-ciliate, apex acute to subacute; petals yellow, obovate-oblong, 3.5–5 mm; stamens 20–25, irregularly arranged; styles 1.5–2 mm; stigmas capitate. **Capsules** narrowly ovoid-conic, subrostrate, 3.5–6 × 1–1.5 mm. **Seeds** 0.5–0.6 mm; testa scalariform-reticulate. $2n = 12$.

Flowering spring–late fall (Mar–Nov). Scrub, on ancient white-sand dunes; of conservation concern; 50 m; Fla.

Hypericum cumulicola is confined to Highlands and Polk counties and its habitat is under threat from bulldozers and citrus groves (D. B. Ward 1980); its nearest relative, with the same chromosome number, appears to be *H. setosum*.

Hypericum cumulicola is in the Center for Plant Conservation's National Collection of Endangered Plants.

36. Hypericum majus (A. Gray) Britton, Mem. Torrey Bot. Club 5: 225. 1894 • Large St. John's wort, millepertuis majeur E

Hypericum canadense Linnaeus var. *majus* A. Gray, Manual ed. 5, 86. 1867 (as major); *H. mutilum* Linnaeus var. *longifolium* R. Keller; *Sarothra major* (A. Gray) Y. Kimura

Herbs perennial, erect, branching at base and in inflorescence, branches relatively few, 0.5–7 dm. **Stems:** internodes 4-angled. **Leaves** spreading, sessile or distal subamplexicaul; blade lanceolate to narrowly oblong-elliptic or (proximal) oblanceolate, 10–45 × (2–)6–12 mm, papery to membranous, margins plane, apex acute to rounded, basal or near-basal veins (3–)5–7, midrib with to 4 pairs of branches. **Inflorescences** corymbiform to cylindric, 3–30-flowered, usually compact, branching mostly dichasial. **Flowers** 6–7 mm diam.; sepals lanceolate to narrowly elliptic, equal, 3.5–6.5 × 0.8–1.5 mm, margins sometimes ciliate, not setulose-ciliate, apex acute; petals golden yellow, sometimes red veined, oblanceolate, 3.5–6 mm; stamens 12–21, obscurely 5-fascicled; styles 0.6–1 mm; stigmas broadly capitate. **Capsules** narrowly conic-ellipsoid, 4–8 × 2.5–3.5 mm, broadest proximal to middle. **Seeds** 0.5–0.7 mm; testa finely linear-scalariform. $2n = 16$.

Flowering summer (Jun–Sep). Fens, marshes, ditches, lake and stream margins, other damp habitats; 0–1200 m; Alta., B.C., Man., N.B., N.S., Ont., P.E.I., Que., Sask.; Colo., Conn., Idaho, Ill., Ind., Iowa, Kans., Maine, Mass., Mich., Minn., Mo., Mont., Nebr., N.H., N.J., N.Y., N.Dak., Ohio, Oreg., Pa., R.I., S.Dak., Vt., Wash., Wis.; introduced in Europe (France, Germany), e Asia (Japan).

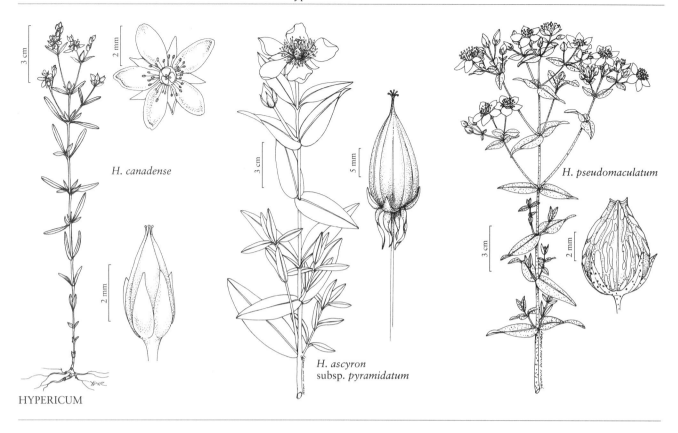

H. canadense

H. ascyron
subsp. *pyramidatum*

H. pseudomaculatum

HYPERICUM

Hypericum majus was the western member of a vicariant species pair, differing from the originally eastern member (*H. canadense*) by the broader leaves, usually more-congested inflorescence, and larger flowers. These species became sympatric in glaciated northeastern North America and now hybridize freely, notably in Wisconsin (F. H. Utech and H. H. Iltis 1970). Hybrids are intermediate in form between the parents and have also been recorded from Connecticut, Massachusetts, and New Hampshire. *Hypericum majus* hybridizes also with *H. mutilum*, with subsp. *mutilum* in Maine, and with subsp. *boreale* in Michigan and Wisconsin.

37. **Hypericum canadense** Linnaeus, Sp. Pl. 2: 785. 1753 • Canadian St. John's wort, millepertuis de Canada E F W

Brathys canadensis (Linnaeus) Spach; *Hypericum canadense* var. *galiiforme* Fernald; *H. canadense* var. *magninsulare* Weatherby; *H. canadense* var. *minimum* Choisy

Herbs annual or perennial, erect, basal branches relatively few or none, usually with strict, ascending branches from distal nodes, 0.3–7.5 dm. **Stems:** internodes 4-angled. **Leaves** erect or spreading, sessile or subsessile; blade linear to oblanceolate-linear or (proximal) oblanceolate to obovate, 6–55 × 0.5–5.5 mm, papery to membranous, margins plane, apex rounded, basal or near-basal veins 1–3(–5), midrib with 1–4 pairs of branches. **Inflorescences** corymbiform to cylindric, usually diffuse, 1–35-flowered, branching mostly dichasial. **Flowers** 5–6 mm diam.; sepals linear-lanceolate to lanceolate, equal, 2.5–4.5 × 0.8–1 mm, margins sometimes ciliate, not setulose-ciliate, apex acute to acuminate; petals golden yellow, sometimes red veined, narrowly obovate to elliptic, 2.5–4 mm; stamens 12–25, obscurely 3–5-fascicled; styles 0.5–0.8 mm; stigmas broadly capitate. **Capsules** narrowly conic to conic-cylindric, 4–6 × 1.5–3 mm, broadest proximal to middle. **Seeds** 0.5–0.7 mm; testa finely linear-scalariform. $2n = 16$.

Flowering summer (Jun–Sep). Fens, marshes, depressions, lake and pond margins; 0–500 m; St. Pierre and Miquelon; N.B., Nfld. and Labr. (Nfld.), N.S., Ont., P.E.I., Que.; Conn., Del., D.C., Fla., Ga., Ill., Ind., Iowa, Ky., Maine, Md., Mass., Mich., Minn., Miss., N.H., N.J., N.Y., N.C., Ohio, Oreg., Pa., R.I., S.C., Tenn., Vt., Va., Wash., W.Va., Wis.; introduced in Europe (Ireland, The Netherlands).

Hypericum canadense is closely related to *H. majus*; it hybridizes with that species and with *H. mutilum*, producing intermediate forms. *Hypericum* ×*dissimulatum* E. P. Bicknell appears to comprise a continuous series of hybrids between *H. canadense* and

H. mutilum or *H. boreale* such that it is not always possible to say which of these species is involved. *Hypericum* ×*dissimulatum* has been recorded from New Brunswick, Newfoundland, Nova Scotia, Ontario, and Quebec and from Connecticut, Maine, Maryland, Massachusetts, New Jersey, New York, North Carolina, Pennsylvania, Rhode Island, and Virginia.

J. Rousseau reduced *Hypericum canadense* var. *minimum* to a form; it does not seem to merit any recognition.

38. Hypericum gymnanthum Engelmann & A. Gray, Boston J. Nat. Hist. 5: 212. 1845 • Small-flowered St. John's wort

Hypericum canadense Linnaeus var. *cardiophyllum* R. Keller; *H. mutilum* Linnaeus var. *gymnanthum* (Engelmann & A. Gray) A. Gray; *Sarothra gymnantha* (Engelmann & A. Gray) Y. Kimura

Herbs annual, usually erect, sometimes shortly decumbent and rooting, basal branches none, rarely with 1–3(–6) pairs of narrowly ascending branches distally, 0.6–7 dm. **Stems:** internodes 4-angled. **Leaves** spreading, sessile or amplexicaul; blade usually ovate-triangular to broadly ovate, rarely oblong, (mid and distal blades lanceolate-deltate), 5–25 × 3–12 mm, papery to membranous, margins plane, apex usually subacute, basal veins (3–)5, midrib usually with 1 or 2 pairs of branches. **Inflorescences** laxly corymbiform to cylindric, (1–)5–65-flowered, branching mostly dichasial. **Flowers** 4.5–7 mm diam.; sepals lanceolate to narrowly ovate, equal, 3–5 × 0.8–1.2 mm, margins sometimes ciliate, not setulose-ciliate, apex acute to acuminate; petals bright yellow, oblanceolate, 2–4 mm; stamens 10–14, scarcely grouped; styles 0.5–0.7 mm; stigmas broadly capitate. **Capsules** narrowly conic-ellipsoid, 3–5 × 1.5–2 mm, usually broadest at or near middle. **Seeds** 0.5–0.6 mm; testa finely linear-scalariform. $2n = 16$.

Flowering summer (Jun–Sep). Bogs, ditches, open and cleared woods, damp habitats; 0–400 m; Ala., Ark., Del., Fla., Ga., Ill., Ind., La., Md., Miss., Mo., N.J., N.C., Ohio, Pa., S.C., Tenn., Tex., Va.; Central America (Guatemala); introduced in Atlantic Islands (Azores).

Hypericum gymnanthum was introduced into Poland; it is now extinct there. It is closely related to *H. mutilum*; it differs from that species in the broader, usually deltate leaves; fewer, stricter branches; no condensed apical stem internode; and larger flowers with lanceolate to ovate sepals.

Hybrids of *Hypericum gymnanthum* with *H. mutilum* have been reported from Maryland, Mississippi, North Carolina, Tennessee, and West Virginia, and, perhaps erroneously, with *H. canadense* from Virginia.

Hypericum gymnanthum has clearly been introduced (recently?) into the Azores, not necessarily by man. Seeds may well have been carried there by birds.

39. Hypericum mutilum Linnaeus, Sp. Pl. 2: 787. 1753 • Dwarf St. John's wort E W

Sarothra mutila (Linnaeus) Y. Kimura

Herbs annual or perennial, usually erect, sometimes decumbent and rooting, basal branches sometimes present, usually with to 10 pairs of spreading branches distal to middle, 0.5–8 dm. **Stems:** internodes 4-angled. **Leaves** spreading, sessile or amplexicaul; blade paler abaxially, ovate, elliptic, or elliptic-oblong (proximal), ovate or suborbiculate to elliptic or lanceolate (mid and distal), 3–27(–40) × 1–15 mm, papery to membranous, margins plane, apex obtuse to rounded, basal veins 3–5, midrib with to 3 pairs of branches. **Inflorescences** cylindric, 5–60-flowered, branching mostly dichasial; bracts linear-subulate. **Flowers** 3–5 mm diam.; sepals lanceolate to linear-lanceolate or elliptic-oblong to oblanceolate, equal or unequal, 2–4.5 × 0.6–1.5 mm, margins sometimes ciliate, not setulose-ciliate, apex usually acute to apiculate, sometimes obtuse to rounded; petals pale yellow, oblong, 1.7–3.5 mm; stamens 5–16, scarcely grouped; styles 0.5 mm; stigmas broadly capitate. **Capsules** narrowly ovoid to cylindric-ellipsoid, 2–5 × 1.6–2.4 mm, usually broadest at or near middle. **Seeds** 0.4–0.7 mm; testa finely linear-scalariform.

Subspecies 2 (2 in the flora): c, e North America; introduced in w North America.

1. Stems: apical internode shorter than adjacent one or almost absent; inflorescences: branching mostly dichasial from 2–10 nodes . . . .
. . . . 39a. *Hypericum mutilum* subsp. *mutilum*
1. Stems: apical internode usually longer than adjacent one; inflorescences: branching from 1–4(–6) nodes, branches diffuse and repeatedly monochasial distally . . . .
. . . . 39b. *Hypericum mutilum* subsp. *latisepalum*

39a. Hypericum mutilum Linnaeus subsp. **mutilum**
    • Millepertuis nain   E

Brathys euphorbioides
(A. St. Hilaire) Spach;
B. quinquenervia (Walter) Spach;
Hypericum euphorbioides
A. St. Hilaire; *H. mutilum* var.
densiflorum Kuntze; *H. mutilum*
var. *foliosissimum* Kuntze;
H. mutilum var. *minimum*
N. Coleman; *H. mutilum* var.
parviflorum Fernald; *H. quinquenervium* Walter

Stems: apical internode shorter than adjacent one or almost absent. **Leaf blades** variable, rarely broadly ovate or suborbiculate. **Inflorescences:** branching from 2–10 nodes, mostly dichasial. **Sepals** not imbricate or broader beyond middle or, if imbricate, leaves not broadly ovate to suborbiculate. $2n = 16$.

Flowering summer–fall (Jun–Oct). Ditches, marshes, lake margins, desiccated temporary pools; 0–400 m; B.C., N.B., N.S., Ont., Que., Sask.; Ala., Ark., Calif., Colo., Conn., Del., D.C., Fla., Ga., Ill., Ind., Iowa, Kans., Ky., La., Maine, Md., Mass., Mich., Miss., Mo., Nebr., N.H., N.J., N.Y., N.C., Ohio, Okla., Oreg., Pa., R.I., S.C., Tenn., Tex., Utah, Vt., Va., Wash., W.Va., Wis.; introduced widely in Central America, South America, Europe, Pacific Islands (Hawaii, New Zealand).

The records of subsp. *mutilum* from British Columbia and Saskatchewan and from California, Colorado, Oregon, and Utah (no details seen) are almost certainly all the result of introduction, as may be some other peripheral records.

39b. Hypericum mutilum Linnaeus subsp. **latisepalum**
    (Fernald) N. Robson, Bull. Brit. Mus. (Nat. Hist.), Bot.
    20: 119. 1990   E

Hypericum mutilum var.
latisepalum Fernald, Rhodora
38: 372. 1936

Stems: apical internode usually longer than adjacent one. **Leaf blades** broadly ovate or broadly oblong-ovate to suborbiculate. **Inflorescences:** branching from 1–4(–6) nodes, branches diffuse and repeatedly monochasial distally. **Sepals** ± imbricate, broader beyond middle.

Flowering late spring–early summer (Apr–Jun). Ditches, marshes, lake margins, desiccated temporary pools; 0–200 m; Fla., Ga., N.J., S.C., Tex.

Fernald recorded subsp. *latisepalum* from New Jersey; D. H. Webb (1980) did not record it north of South Carolina. No specimens have been seen from farther north than Georgia, where it appears to tend morphologically toward subsp. *mutilum*.

40. Hypericum boreale (Britton) E. P. Bicknell, Bull. Torrey Bot. Club 22: 213. 1895 • Northern St. John's wort, millepertuis boréal   E

Hypericum canadense Linnaeus var. *boreale* Britton, Bull. Torrey Bot. Club 18: 365. 1891; *H. mutilum* Linnaeus subsp. *boreale* (Britton) J. M. Gillett; *Sarothra borealis* (Britton) Y. Kimura

Herbs annual or perennial, erect, with decumbent to prostrate, rooting base, usually 2–6-branched, branches spreading or ascending distal to middle, 0.9–3.3 dm. **Stems:** internodes 4-angled, apical internode shorter than adjacent one or almost absent. **Leaves** spreading, sessile; blade (concolor), broadly to narrowly oblong or elliptic to oblanceolate or round, 5–15 × 2–5 mm, papery to membranous, margins plane, apex rounded, basal veins 3–5, midrib branched or not. **Inflorescences** cylindric to rounded-pyramidal, 1–13-flowered, branching mostly dichasial; bracts not linear-subulate. **Flowers** 3–5 mm diam.; sepals usually lanceolate to narrowly oblong, rarely oblanceolate, equal, 2.5 × 0.8–1 mm, margins sometimes ciliate, not setulose-ciliate, apex rounded; petals pale yellow, oblong, 1.7–3.5 mm; stamens 5–16, scarcely grouped; styles 0.5 mm; stigmas broadly capitate. **Capsules** narrowly ovoid to cylindric-ellipsoid, 4–5 × 2–2.5 mm, usually broadest at or near middle. **Seeds** 0.4–0.7 mm; testa finely linear-scalariform. $2n = 16$ ["18"].

Flowering summer–early fall (Jul–Sep). Bogs, poor fens, lake margins, marshes; 0–500 m; St. Pierre and Miquelon; B.C., Man., N.B., Nfld. and Labr. (Nfld.), N.S., Ont., P.E.I., Que.; Conn., Del., Ind., Iowa, Maine, Md., Mass., Mich., Minn., N.H., N.J., N.Y., Oreg., Pa., R.I., Vt., Va., Wash., Wis.

Where *Hypericum boreale* grows submerged, the plants are almost always sterile with elongated stems and suborbiculate leaves (*H. boreale* forma *callitrichoides* Fassett). Such plants intergrade shorewards with typical *H. boreale* (F. H. Utech and H. H. Iltis 1970). All other chromosome counts for *H. mutilum* and its near relatives have given $n = 8$; the count of $2n = 18$ by B. M. Kapoor (1972) must be treated with reserve.

Hypericum mutile var. *boreale* (Britton) E. P. Bicknell is not a validly published name.

The discovery of *Hypericum boreale* near the mouth of the Fraser River at Vancouver in 1989 extends the distribution of this species across Canada almost to the Pacific coast; this occurrence is almost certainly the result of recent introduction.

41. **Hypericum paucifolium** S. Watson, Proc. Amer. Acad. Arts Sci. 25: 143. 1890

Herbs annual or perennial, erect or decumbent at base, not rooting, branches basal, relatively few or none, 1–7 dm. **Stems:** internodes 4-angled. **Leaves** erect to spreading, sessile; blade linear-oblong or linear-lanceolate to linear, 5–33 × 0.5–2 mm, leathery, margins plane to revolute, apex acute, basal veins 1(–3), midrib with (0)1–3 pairs of branches. **Inflorescences** narrowly V-shaped, 1–50-flowered, branching mostly dichasial. **Flowers** 10–15 mm diam.; sepals narrowly oblong to narrowly lanceolate, subequal or unequal, 3–7 × 0.8–1.6 mm, margins sometimes ciliate, not setulose-ciliate, apex acute; petals golden yellow to orange-yellow, red tinged, oblanceolate to obovate, 6–10 mm; stamens 35–50, irregularly 3-fascicled; styles 2.5–5 mm; stigmas broadly capitate. **Capsules** ± narrowly ovoid-conic, 4–9 × 2–4 mm. **Seeds** 0.5–0.7 mm; testa finely linear-scalariform.

Flowering spring. Dry, grassy habitats, roadsides; 0–50 m; Tex.; Mexico.

The relatively long styles distinguish *Hypericum paucifolium* from all other North American species of sect. *Trigynobrathys*.

42. **Hypericum anagalloides** Chamisso & Schlechtendal, Linnaea 3: 127. 1828 · Tinker's penny

Hypericum anagalloides var. *calicifolium* R. Keller; *H. anagalloides* var. *cymosum* R. Keller; *H. anagalloides* var. *nevadense* Greene; *H. anagalloides* var. *pumilum* R. Keller; *H. anagalloides* var. *ramigerum* R. Keller; *H. anagalloides* var. *undulatum* R. Keller; *H. bryophytum* Elmer; *H. tapetoides* A. Nelson

Herbs annual or perennial, decumbent to ascending, with diffusely branching and rooting base, forming loose mats, not usually branched distally, 0.3–1.5 dm. **Stems:** internodes 4-angled. **Leaves** spreading, sessile or subamplexicaul; blade ovate or orbiculate to elliptic or oblanceolate (proximal) or oblong (distal), 3–13 × 1.5–8.5 mm, papery to membranous, margins plane, apex rounded, basal veins 3–5(–7), distally looped, midrib unbranched. **Inflorescences** loosely corymbiform, 1–14-flowered, branching mostly dichasial. **Flowers** 3–5(–8) mm diam.; sepals usually narrowly elliptic-oblong to oblanceolate-spatulate, rarely obovate, unequal, 2–4 × 0.9–2 mm, margins sometimes ciliate, not setulose-ciliate, apex subacute to rounded; petals golden yellow to salmon-orange, oblanceolate, (1.7–)3.5–5 mm; stamens (5–)12–15(–25), separate or obscurely 3-fascicled; styles

0.5–2 mm; stigmas scarcely capitate. **Capsules** ellipsoid to cylindric or subglobose, 2.5–5 × 1.7–2.5 mm. **Seeds** 0.5–0.6 mm; testa linear-scalariform. 2n= 16.

Flowering spring–late summer (May–Sep). Bogs, ditches, lake and stream margins, meadows, other damp habitats; 50–2700 m; B.C.; Ariz., Calif., Idaho, Mont., Nev., Oreg., Utah, Wash.; Mexico (Baja California).

As reflected in the synonymy, *Hypericum anagalloides* is variable; none of the variations merits taxonomic recognition.

43. **Hypericum drummondii** (Greville & Hooker) Torrey & A. Gray, Fl. N. Amer. 1: 165. 1838 [E] [W]

Sarothra drummondii Greville & Hooker, Bot. Misc. 3: 236, plate 107. 1833; *Brathys drummondii* (Greville & Hooker) Spach

Herbs annual, erect, branches strict, in distal ½, 1–8 dm, wiry. **Stems:** internodes 4-lined. **Leaves** erect to suberect, sessile; blade linear or linear-subulate to linear-lanceolate, 5–22 × 0.5–1 mm, subcoriaceous, margins recurved to revolute, apex acute to obtuse, basal vein 1, midrib unbranched. **Inflorescences** narrowly to broadly triangular, 1–12-flowered, branching mostly monochasial. **Flowers** 5–8 mm diam.; sepals narrowly oblong to linear-lanceolate, subequal, 3–7 × 0.7–1.3 mm, apex acute; petals golden yellow to orange-yellow, oblong, 4–7 mm; stamens 10–22, separate or obscurely 3-fascicled; styles (0.5–)0.8–1.5 mm; stigmas broadly capitate. **Capsules** narrowly ovoid to ovoid-cylindric, 3.5–7 × 2.5–3 mm, length 1–1.2 times sepals. **Seeds** 0.9–1.1 mm; testa finely scalariform. 2n = 24.

Flowering summer–early fall (Jul–Sep). Dry, sandy or clay soil in open woods, old fields, waste or rocky places; 0–1100 m; Ala., Ark., Fla., Ga., Ill., Ind., Iowa, Kans., Ky., La., Miss., Mo., N.C., Ohio, Okla., S.C., Tenn., Tex., Va.

Hypericum drummondii is closely related to *H. gentianoides*.

44. **Hypericum gentianoides** (Linnaeus) Britton, Sterns & Poggenburg, Prelim. Cat., 9. 1888 · Orange-grass, pineweed [E] [W]

Sarothra gentianoides Linnaeus, Sp. Pl. 1: 272. 1753; *Brathys gentianoides* (Linnaeus) Spach; *Hypericum nudicaule* Walter

Herbs annual, erect, branches strict, in distal ⅔ or from most nodes, 0.7–6 dm, wiry. **Stems:** internodes 4-lined. **Leaves** appressed, sessile; blade narrowly triangular-subulate to linear-subulate, scalelike, 1–4 × 0.4–0.6 mm, subcoriaceous to chartaceous, margins incurved, apex obtuse to rounded, basal vein 1, midrib unbranched. **Inflorescences** usually pyramidal, 1–24-flowered, branching mostly monochasial. **Flowers** 3–5 mm diam.; sepals lanceolate to narrowly oblong or linear-lanceolate, unequal, 1.5–2.5 × 0.4–0.8 mm, apex acute; petals orange-yellow to golden yellow, oblong, 2–4 mm; stamens 5–11, 5 separate or obscurely 5-grouped; styles 0.8–1.2 mm; stigmas broadly capitate. **Capsules** narrowly cylindric-conic, 4–5 × 1–1.2 mm, length 2–2.7 times sepals. **Seeds** 0.4–0.8 mm; testa markedly ribbed-scalariform. $2n = 24$.

Flowering late spring–fall (May–Oct). Dry, sandy soil in open woods, fields, roadsides, waste or rocky places, tall-grass prairie; 0–500 m; N.S., Ont., P.E.I.; Ala., Ark., Conn., Del., D.C., Fla., Ga., Ill., Ind., Iowa, Ky., La., Maine, Md., Mass., Mich., Miss., Mo., N.H., N.J., N.Y., N.C., Ohio, Okla., Pa., R.I., S.C., Tenn., Tex., Vt., Va., W.Va., Wis.; introduced in West Indies (Dominican Republic), South America (Brazil, Paraguay), Europe (France).

Hypericum gentianoides is smaller in all parts than *H. drummondii* and more branched. The relatively long, narrow capsule is diagnostic, as are the relatively small, often scalelike leaves.

Hypericum sarothra Michaux and *Sarothra hypericoides* Nuttall are illegitimate names that pertain here.

1c. HYPERICUM Linnaeus sect. WEBBIA (Spach) R. Keller in H. G. A. Engler and K. Prantl, Nat. Pflanzenfam. 95[III,6]: 211. 1893 [I]

Webbia Spach, Ann. Sci. Nat., Bot., sér. 2, 5: 356. 1836

Shrubs, black glands absent. **Leaves** deciduous, (base articulated). **Flowers** 20–37 mm diam.; sepals persistent, 5; petals persistent, 5; stamens persistent, 36–75, in 5 fascicles, fascicles connate as 2 + 2 + 1; ovary 3-merous; placentation axile (sometimes incompletely); styles spreading, bases distinct; stigmas subclavate to narrowly capitate. **Seeds** not or slightly carinate.

Species 1: introduced, California; Atlantic Islands (Canary Islands [except Lanzarote and Fuerteventura], Madeira).

45. **Hypericum canariense** Linnaeus, Sp. Pl. 2: 784. 1753 [I] [W]

Hypericum floribundum Aiton; *Webbia canariensis* (Linnaeus) Webb & Berthelot; *W. floribunda* (Aiton) Spach

Shrubs erect, bushy, 10–50 dm. **Stems:** internodes 4-lined at first, then terete. **Leaves** spreading, sessile; blade narrowly elliptic to narrowly elliptic-oblong, 20–70 × 5–15 mm, (proximal usually narrower), base narrowly cuneate to subangustate, margins plane, apex acute to apiculate-obtuse, midrib with 8–12 pairs of branches, tertiary veins densely reticulate toward margins. **Inflorescences** broadly rounded-pyramidal to broadly cylindric, to 30-flowered. **Flowers:** sepals lanceolate, unequal, 3–4.5 × 1–2.2 mm; petals bright yellow, not red tinged, oblanceolate-unguiculate, 12–17 mm; anther gland yellow to orange; styles widely spreading, 8–14 mm. **Capsules** pyramidal-ovoid to ovoid-ellipsoid, 9–12 × 7–8 mm. **Seeds** 1.5–2 mm, narrowly winged; testa linear-reticulate to linear-foveolate. $2n = 40$.

Flowering spring–summer. Disturbed sites; 20–500 m; introduced; Calif.; Atlantic Islands (Canary Islands, Madeira).

Hypericum canariense is established at Montecito and Santa Barbara in the hills (P. A. Munz 1974) and along the coast north of Santa Cruz to San Francisco, at locations in Orange and San Mateo counties, and in the San Diego coast region.

The description above agrees with that for *Hypericum floribundum* regarding sepals lanceolate and acute; in typical *H. canariense* they are oblong-spatulate and rounded. The variation is continuous; only one species is recognized here.

1d. HYPERICUM Linnaeus sect. ROSCYNA (Spach) R. Keller in H. G. A. Engler and K. Prantl, Nat. Pflanzenfam. 95[III,6]: 211. 1893

Roscyna Spach, Ann. Sci. Nat., Bot., sér. 2, 5: 364. 1836

Herbs, perennial, black glands absent. **Leaves** persistent, (base not articulated). **Flowers** 40–70 mm diam.; sepals persistent, 5; petals persistent, 5; stamens persistent, 150, in 5 fascicles, fascicles usually distinct, rarely 1 pair connate; ovary (4)5-merous; placentation axile; styles ± appressed, bases ± connate or distinct; stigmas ± broadly capitate. **Seeds** ± carinate.

Species 2 (1 in the flora): e North America, ne, e Asia.

The other species in the section, *Hypericum przewalskii* Maximowicz, is endemic to China.

46. Hypericum ascyron Linnaeus, Sp. Pl. 2: 783. 1753
• Great or giant St. John's wort F

Subspecies 3 (1 in the flora): e North America, Asia (e China, Japan, Korea, Siberia).

46a. Hypericum ascyron Linnaeus subsp. pyramidatum (Aiton) N. Robson, Bull. Nat. Hist. Mus. London, Bot. 31: 58. 2001 • Millepertuis à grandes fleurs E F

Hypericum pyramidatum Aiton, Hort. Kew. 3: 103. 1789; *H. ascyroides* Willdenow; *H. ascyron* var. *americanum* (Spach) Y. Kimura; *H. bartramium* Miller; *H. macrocarpum* Michaux; *Roscyna americana* Spach

Herbs erect, branched from base and often with strict, narrowly ascending branches, 5–20 dm. **Stems:** internodes 4-lined at first, then 4-angled. **Leaves** spreading, sessile, amplexicaul; blade ovate-lanceolate to lanceolate or oblong, 40–85 × 18–37 mm on main stem, smaller distally and on branches, base rounded to cordate, margins plane, apex usually acute to apiculate, sometimes obtuse, midrib with 4–7 pairs of branches, tertiary veins densely reticulate toward margins. **Inflorescences** cylindric to narrowly pyramidal, 1–35-flowered. **Flowers:** sepals ovate to lanceolate or oblong-elliptic, subequal or equal, 8–13 × 4–8 mm, apex acute to acuminate; petals golden yellow, sometimes red tinged, narrowly obovate to oblanceolate-falcate, 25–40 mm; styles (4)5, 3–7 mm. **Capsules** ovoid, 12–20(–30) × 10–13 mm. **Seeds** 1.5 mm, narrowly winged; testa shallowly linear-reticulate.

Flowering summer (Jun–Aug). Streamsides, roadside ditches, fens, damp meadows; 0–1500 m; Ont., Que.; Conn., Ill., Ind., Iowa, Kans., Md., Mass., Mich., Minn., Mo., N.Y., Ohio, Pa., Vt., Wis.

The American populations of *Hypericum ascyron* sometimes have been treated as distinct from those of eastern Asia, usually as *H. pyramidatum*. These populations can be distinguished by the combination of relatively broad leaves, acute sepals, and, usually, almost distinct styles; they are better treated as subsp. *pyramidatum* (see N. K. B. Robson 2001). The scattered American distribution of *Hypericum ascyron*, which seems to coincide well with earlier Native American campsites, led W. G. Dore (in herbarium notes) to suggest that it was distributed by aborigines for an as yet unknown reason (see J. M. Gillett and N. K. B. Robson 1981).

1e. HYPERICUM Linnaeus informal sect. group HYPERICUM

Hypericum sect. *Concinna* N. Robson; *Hypericum* sect. *Graveolentia* N. Robson

Herbs, perennial, [**subshrubs**]; black glands usually on leaves, sepals, and petals and, sometimes, on stems and anthers. **Leaves** persistent or tardily deciduous, (base not articulated). **Flowers** 6–35 mm diam.; sepals persistent, 5; petals persistent, 5; stamens persistent, 20–109, in 5 fascicles, fascicles connate as 2 + 2 + 1; ovary 3(–4)-merous; placentation axile; styles spreading, bases distinct; stigmas relatively small. **Seeds** not or slightly carinate.

Species 22 (8, including 1 hybrid, in the flora): North America, Mexico, West Indies, Central America, South America.

SELECTED REFERENCE   Culwell, D. E. 1970. A Taxonomic Study of the Section *Hypericum* in the Eastern United States. Ph.D. dissertation. University of North Carolina.

1. Herbs rarely with rooting base, bushy; leaf blades usually conduplicate; sepals markedly imbricate . . . . 47. *Hypericum concinnum*
1. Herbs usually with rooting base (except *H. punctatum*), not bushy; leaf blades not conduplicate; sepals not or scarcely imbricate.
    2. Stems with black glands in raised lines or without black glands; leaf blades: tertiary veins not densely reticulate, black glands intramarginal, (± dense to spaced), and laminar, (relatively few and distal or absent); capsules ovoid, broadly ovoid, narrowly ovoid-pyramidal, or oblanceoloid.
        3. Stems from rooting, creeping base, internodes weakly 2-lined or not lined, without black glands, rarely with reddish glands; petals sometimes red tinged . . . 50. *Hypericum scouleri*
        3. Stems from rooting, base not creeping, internodes 2- or 4-lined, with black glands in lines; petals not red tinged.
            4. Stems: internodes (at least some) 4-lined; sepals broadly ovate to oblong, apex rounded-apiculate to erose-denticulate; capsules with longitudinal vittae (narrow linear glands) . . . . 48. *Hypericum maculatum*
            4. Stems: internodes 2-lined; sepals lanceolate or narrowly oblong to linear, apex acute to aristate; capsules with longitudinal vittae and shorter, oblique vittae . . . . 49. *Hypericum perforatum*
    2. Stems with black glands scattered on or near lines or all over; leaf blades: tertiary veins densely reticulate toward margins, black glands intramarginal (dense) and laminar (scattered); capsules ellipsoid, ovoid, broadly ovoid, or subglobose.
        5. Leaf blade apex usually acute, rarely obtuse to rounded; inflorescence subsidiary branches ascending to widely spreading; anther gland amber or pellucid . . . . 54. *Hypericum pseudomaculatum*
        5. Leaf blade apex usually obtuse, rounded, retuse, or subretuse, rarely acute; inflorescence subsidiary branches narrowly ascending or curved-ascending; anther gland black.
            6. Inflorescences (2–)5–14(–22)-flowered; petals 11–18 mm; styles 5.6–12 mm . . . . 51. *Hypericum graveolens*
            6. Inflorescences (5–)13–600-flowered; petals 3–11 mm; styles 1–5 mm.
                7. Stems with black glands on or near lines or all over; flowers 15–20 mm diam.; capsules with longitudinal vittae . . . . 52. *Hypericum ×mitchellianum*
                7. Stems with black glands scattered; flowers 8–15 mm diam.; capsules with longitudinal vittae or elongate to ovoid vesicles . . . . 53. *Hypericum punctatum*

47. **Hypericum concinnum** Bentham, Pl. Hartw., 300. 1849 • Gold wire  E

Hypericum seleri R. Keller

Herbs erect or ascending, rarely with rooting base, bushy, 1.5–3.3 dm. **Stems:** internodes (at least some) 4-lined, without black glands. **Leaves** spreading, sessile or petiolate (to 0.5 mm); blade narrowly elliptic or narrowly oblong to linear, usually conduplicate, sometimes falcate, 13–22 × 1.5–8 mm, base cuneate, margins plane, apex acute to subacute, midrib with 2–4 pairs of branches, black glands marginal. **Inflorescences** subcapitate to cylindric, 1–7-flowered. **Flowers** 20–35 mm diam.;

sepals markedly imbricate, spreading in fruit, broadly to narrowly ovate, unequal, 6–9 × 2–3 mm, apex acute to acuminate; petals yellow, obovate or oblong-obovate, (10–)12–15 mm; stamens 40–80(–100); anther gland amber; styles 6–9 mm. **Capsules** ovoid, 6–9 × 4–4.5 mm, with longitudinal vittae. **Seeds** not carinate, 1 mm; testa minutely and shallowly pitted. $2n = 16$.

Flowering summer (May–Jul). Dry slopes, chaparral, yellow pine forests; 100–600(–900) m; Calif.

Hypericum concinnum is known from the Sierra Nevada from Mariposa County to Shasta County and the North Coast Ranges from Marin County to Mendocino County. It is isolated, taxonomically and geographically, from its nearest relative, which seems to be the northeastern Asian *H. ascyron* subsp. *gebleri* (Ledebour) N. Robson.

48. Hypericum maculatum Crantz, Stirp. Austr. Fasc. 2: 64. 1763 • Imperforate St. John's wort I

Subspecies 3 (1 in the flora): introduced; Europe, Asia (Siberia).

48a. Hypericum maculatum Crantz subsp. **obtusiusculum** (Tourlet) Hayek, Sched. Fl. Stiriac. 23–24: 27. 1912 I

Hypericum quadrangulum Linnaeus subsp. *obtusiusculum* Tourlet, Bull. Soc. Bot. France 50: 307. 1903

Herbs erect, with rooting, not creeping, base, forming clumps, 1.5–10 dm. **Stems:** internodes (at least some) 4-lined, with black glands in raised lines. **Leaves** spreading, sessile; blade ovate, ovate-lanceolate, oblong, or elliptic, 15–50 × 10–20 mm, base cuneate to rounded, margins plane, apex rounded, midrib with 2 or 3 pairs of branches, tertiary veins not densely reticulate, black glands intramarginal (spaced), pale glands usually sparse or absent. **Inflorescences** subcorymbiform to broadly pyramidal or cylindric, to 40-flowered. **Flowers** 15–25(–30) mm diam.; sepals not imbricate, spreading in fruit, broadly ovate to oblong, unequal or subequal, 4–5 × 2–3.5 mm, apex rounded-apiculate to erose-denticulate; petals golden yellow, obovate-oblong to oblanceolate, 10–15 mm; stamens 30–70; anther gland black; styles 3–4 mm. **Capsules** broadly ovoid, 7–9 × 5–7 mm, with longitudinal vittae. **Seeds** scarcely carinate, 1 mm; testa linear-reticulate. *2n* = 32.

Flowering summer. Waste places, open sites, dry or damp; 0–2500 m; introduced; B.C.; Idaho, Wash.; Europe.

The western European lowland subspecies of *Hypericum maculatum* has been recorded from disturbed habitats near Vancouver and Prince Rupert regions in western British Columbia for more than 50 years. There is no evidence as to whether it has persisted over the whole of this period or has been reintroduced. The mainly eastern European and Siberian subsp. *maculatum* is diploid (*2n* = 16) and has entire sepals, dotted to shortly streaked (not lined) petals, and a narrower angle of branching (ca. 30° rather than ca. 50°).

49. Hypericum perforatum Linnaeus, Sp. Pl. 2: 785. 1753 • Common St. John's wort, millepertuis commun I W

Subspecies 4 (1 in the flora): introduced; Europe, Asia (sw Arabia, China, India, Mongolia, Siberia), n Africa; introduced also widely.

Hypericum perforatum has been introduced into various parts of the world, including North America, where only subsp. *perforatum* is represented. The range of variation in the flora area is less than occurs in Europe, and signs of hybridization that are common there are absent in North America.

49a. Hypericum perforatum Linnaeus subsp. **perforatum** • Klamath weed I W

Hypericum marylandicum Biroli ex Colla

Herbs erect, with rooting, not creeping, base, forming clumps, 2–12 dm. **Stems:** internodes 2-lined, with black glands usually in lines. **Leaves** spreading, sessile or petiolate (to 1 mm); blade oblong or elliptic to lanceolate-elliptic or linear, 7–30 × 2–16 mm, base cuneate to rounded, margins plane or recurved to revolute, apex obtuse or apiculate to rounded, midrib with 2 or 3 pairs of branches, tertiary veins not densely reticulate, black glands intramarginal (spaced) and, usually, laminar (distal). **Inflorescences** subcorymbiform to broadly pyramidal or cylindric, 1–15-flowered. **Flowers** 12–30 mm diam.; sepals not imbricate, spreading in fruit, lanceolate or narrowly oblong to linear, unequal or subequal, 2.5–7 × 0.6–2(–3) mm, apex acute to aristate; petals golden yellow, narrowly obovate to oblanceolate, 7–13 mm; stamens 40–90; anther gland black; styles 4–6 mm. **Capsules** ovoid to narrowly ovoid-pyramidal, 6–10 × 3.5–5 mm, with longitudinal vittae and shorter oblique vittae or vesicles. **Seeds** slightly carinate, 0.8–1.2 mm; testa reticulate-foveolate. *2n* = 32, 48.

Flowering late summer (Jul–Sep). Dry to moist, open to partly shaded, usually disturbed ground; 0–1600 m; introduced; St. Pierre and Miquelon; B.C., Man., N.B., Nfld. and Labr. (Nfld.), N.S., Ont., P.E.I., Que.; Ark., Calif., Colo., Conn., Del., D.C., Idaho, Ill., Ind., Iowa, Kans., Ky., Maine, Md., Mass., Mich., Minn., Mo., Mont., Nebr., Nev., N.H., N.J., N.Mex., N.Y., N.C., N.Dak., Ohio, Okla., Oreg., Pa., R.I., S.C., S.Dak., Tenn., Tex., Vt., Va., Wash., W.Va., Wis., Wyo.; Europe (excluding most of Italy); Asia (e to the Altai Mountains, nw Mongolia).

Subspecies *perforatum* occurs in all of North America from Newfoundland to British Columbia southward, and central Canada. It was introduced into both sides of that continent, and it still appears to be less common in (or absent from) some central and southern regions. The common name Klamath weed alludes to a region of northern California where it occupied over three million acres as an agricultural pest causing photosensitization, spontaneous abortions, and central nervous system damage in stock animals and is also an invasive species altering the composition of natural grasslands. Good control has been attained by beetles, one leaf-eating (*Chrysolina gemellata*) and the other root-boring (*Agrilus hyperici*) (J. K. Holloway 1957).

50. Hypericum scouleri Hooker, Fl. Bor.-Amer. 1: 111. 1831 • Scouler's St. John's wort

Hypericum formosum Kunth subsp. *scouleri* (Hooker) C. L. Hitchcock; *H. formosum* var. *scouleri* (Hooker) J. M. Coulter; *H. nortoniae* M. E. Jones; *H. scouleri* subsp. *nortoniae* (M. E. Jones) J. M. Gillett

Herbs erect or ascending, with rooting, creeping, branching base, 0.5–6.6(–8) dm. **Stems:** internodes usually weakly 2-lined, sometimes not lined, without black glands, rarely with reddish glands. **Leaves** usually spreading, rarely erect, sessile or (proximal) subpetiolate; blade oblong-elliptic or elliptic to triangular-ovate or (proximal) obovate, 12–28(–32) × 6–15(–18) mm, base subcordate to rounded or (proximal) cuneate, margins plane, apex obtuse to rounded, midrib with 4 or 5 pairs of branches, tertiary veins not densely reticulate, black glands intramarginal (± dense) and, rarely, (1 or 2) laminar (distal). **Inflorescences** cylindric to narrowly pyramidal, (1–)8–20-flowered. **Flowers** 6–15(–25) mm diam.; sepals not or scarcely imbricate, erect in fruit, ovate to lanceolate or narrowly oblong, unequal to subequal, 2.5–5.5 × 1–2 mm, apex acute to rounded; petals golden yellow, sometimes red tinged, oblanceolate, 7–12 mm; stamens 50–90(–109); anther gland black; styles 2–8 mm. **Capsules** oblanceoloid, 6–10 × 3.5–6 mm, with longitudinal vittae. **Seeds** not carinate, (0.5–)0.7–0.8 mm; testa linear-reticulate. $2n = 16$.

Flowering summer (Jun–Sep). Wet meadows and banks, coniferous forests, screes, lake margins, marshes, tidal shores; 0–2900 m; Alta., B.C.; Ariz., Calif., Colo., Idaho, Mont., Nev., N.Mex., Oreg., Utah, Wash., Wyo.; Mexico (Baja California, Baja California Sur, Chihuahua, Coahuila, Durango, Sonora, Zacatecas).

Some authors have included *Hypericum scouleri* in central Mexican *H. formosum* Kunth as a synonym, subspecies, or variety; the similarities in sepal form and glandularity between these species are due to convergence. *Hypericum formosum* is related to another

Mexican species (*H. oaxacanum* R. Keller); the affinities of *H. scouleri* are with the *H. attenuatum* group from eastern Asia. *Hypericum scouleri* itself does occur in northern and central Mexico as far south as Michoacán, México, and Hidalgo, where it is known by the synonym *H. simulans* Rose.

J. M. Gillett and N. K. B. Robson (1981) treated the dwarf alpine form of *Hypericum scouleri* as subsp. *nortoniae*. Further work has revealed a range of intermediate forms between the two putative subspecies, which prevents their recognition. Likewise, the differentiation of a southern population (*H. formosum* subsp. *formosum* in the sense of C. L. Hitchcock) from a northern one [*H. formosum* subsp. *scouleri* (Hooker) C. L. Hitchcock] based on the broader, blunter, and less-glandular sepals in the latter, does not appear warranted.

51. Hypericum graveolens Buckley, Amer. J. Sci. Arts 45: 174. 1843 [E]

Herbs erect, with rooting, creeping base, 3–6.5 dm. **Stems:** internodes 4-lined at first, soon 2-lined, with black glands on or near lines. **Leaves** spreading, sessile; blade ovate to oblong or lanceolate, 33–65 × 15–27 mm, base cordate to truncate or broadly cuneate, margins plane, apex obtuse to rounded, midrib with 4 or 5 pairs of branches, tertiary veins densely reticulate toward margins, black glands intramarginal (dense) and, sometimes, laminar (scattered). **Inflorescences** subcorymbiform, (2–)5–14(–22)-flowered, subsidiary branches sometimes narrowly ascending or curved-ascending. **Flowers** 20–25(–30) mm diam.; sepals not imbricate, erect in fruit, lanceolate, subequal, 5–7.5 (–11) × 1–3 mm, apex acute; petals golden yellow, narrowly obovate, 11–18 mm; stamens 50–90(–103); anther gland black; styles 5.6–12 mm. **Capsules** broadly ovoid, 5–8 × 3.5–5 mm, with longitudinal vittae. **Seeds** not carinate, 0.8–1.1 mm; testa linear-reticulate. $2n = 16$.

Flowering summer (Jun–Aug). Open or partly shaded, moist habitats, dry, rocky roadside banks; 1200–2100 m; N.C., Tenn.

The chromosome count of $n = 16$ by Adams (in N. K. B. Robson and W. P. Adams 1968) is now regarded as an error; see D. E. Culwell (1970).

Hypericum graveolens is a relict species with close relatives in Japan; it hybridizes with *H. ×mitchellianum* and, probably, also with *H. punctatum*.

52. **Hypericum ×mitchellianum** Rydberg, Torreya 27: 84, plate 2, figs. 1–6. 1927 (as species) • Blue Ridge St. John's wort E

Herbs erect, with rooting, creeping base, 2–6.5 dm. **Stems:** internodes usually 2-lined, sometimes 4-lined or not lined, with black glands scattered on and near lines or all over. **Leaves** spreading, usually sessile, rarely petiolate (to 0.8 mm); blade ovate-oblong to oblong or elliptic, 30–42(–52) × 8–22 mm, base subcordate to rounded, margins plane, apex usually rounded, rarely obtuse or subretuse, midrib with 4 or 5 pairs of branches, tertiary veins densely reticulate toward margins, black glands intramarginal (dense) and laminar (scattered). **Inflorescences** corymbiform to broadly pyramidal, (5–)13–61(–124)-flowered, subsidiary branches narrowly ascending or curved-ascending. **Flowers** 15–20 mm diam.; sepals not imbricate, erect in fruit, lanceolate to ovate-elliptic or elliptic, subequal, (3–)3.6–4.6(–5.5) × 1–2 mm, apex acute to obtuse; petals golden yellow, narrowly obovate or oblanceolate to elliptic, 6–11 mm; stamens (37–)42–56(–62); anther gland black; styles 1.5–5 mm. **Capsules** ellipsoid to subglobose, 3–7 × 3–4.5 mm, with longitudinal vittae. **Seeds** not carinate, 0.7–0.9 mm; testa not seen. 2*n* = 16.

Flowering summer (Jun–Aug). Open or partly shaded, moist habitats, dry, rocky roadside banks; 1100–1700 m; N.C., Tenn., Va.

Hypericum ×mitchellianum is intermediate in all characters between *H. graveolens* and *H. punctatum* and, like the latter, produces a ring of 16 chromosomes at meiosis (D. E. Culwell 1970). Culwell has shown that it hybridizes with *H. graveolens* in the field and that these species can be crossed artificially. He apparently never suspected that *H. mitchellianum* could itself be a hybrid. Its intermediate morphology and breeding behavior, together with a distribution almost wholly within that of *H. graveolens*, suggests strongly that *H. mitchellianum* is the hybrid *H. graveolens* × *punctatum*, which apparently arose when the area of *H. punctatum* extended into that of *H. graveolens*.

53. **Hypericum punctatum** Lamarck in J. Lamarck et al., Encycl. 4: 164. 1797 • Spotted St. John's wort E W

Hypericum maculatum Walter var. *corymbosum* Kuntze; *H. maculatum* var. *heterophyllum* Kuntze; *H. maculatum* var. *subcordifolium* Kuntze; *H. maculatum* var. *subpetiolatum* E. P. Bicknell ex Britton; *H. micranthum* Choisy; *H. subpetiolatum* (E. P. Bicknell ex Britton) Small

Herbs erect to ascending, with rarely rooting, branching base, 1.3–10.5 dm. **Stems** clustered, internodes not lined, with black glands scattered all over. **Leaves** spreading or ascending, usually sessile, rarely petiolate (to 1 mm); blade elliptic or oblong to lanceolate or triangular-lanceolate, or oblanceolate, 14–40(–58) × 3–17(–22.5) mm, base cordate to narrowly cuneate, margins plane, apex usually rounded to retuse, rarely acute to obtuse, midrib with 3–5 pairs of branches, tertiary veins densely reticulate toward margins, black glands intramarginal (dense) and laminar (scattered). **Inflorescences** subcorymbiform to cylindric, 10–206 (–600)-flowered, subsidiary branches narrowly ascending to curved-ascending. **Flowers** 8–15 mm diam.; sepals not imbricate, erect in fruit, lanceolate or oblong-lanceolate to elliptic or ovate-elliptic, subequal, (1.5–)2–4 × 0.8–1.6 mm, apex acute to rounded; petals pale yellow, oblanceolate to elliptic, 3–6(–9) mm; stamens (20–)30–60; anther gland black; styles 1–4 mm. **Capsules** ovoid to subglobose, 2.5–6 × 2–3.5(–4) mm, with longitudinal vittae or elongate to ovoid vesicles. **Seeds** not carinate, 0.5–0.7 mm; testa linear-reticulate. 2*n* = 14, 16.

Flowering summer (May–Sep). Open or slightly shaded, dry to marshy habitats; 50–1200 m; Nfld. and Labr. (Nfld.), N.S., Ont., Que.; Ala., Ark., Conn., Del., D.C., Fla., Ga., Ill., Ind., Iowa, Kans., Ky., La., Maine, Md., Mass., Mich., Minn., Miss., Mo., Nebr., N.H., N.J., N.Y., N.C., Ohio, Okla., Pa., R.I., S.C., Tenn., Tex., Vt., Va., W.Va., Wis.

Hypericum punctatum has been confused with *H. pseudomaculatum;* it can almost always be distinguished by style length and anther gland (black in *H. punctatum,* amber or pellucid in *H. pseudomaculatum*). D. E. Culwell's (1970) record of one specimen of the latter from Missouri with a black anther gland could refer to a hybrid; he and other authors agree that such plants are rare, despite the considerable overlap in distribution of these species. Culwell remarked on the unexpected fecundity of the experimentally produced cross *H. graveolens* × *punctatum* but nowhere compared the resulting plants with *H. ×mitchellianum,* a probable hybrid with that parentage (see discussion under 52. *H.* ×*mitchellianum*). Both *H. punctatum* and *H.* ×*mitchellianum* (but not *H. graveolens*) display a ring of 16 chromosomes at meiosis, and Culwell suggested that this phenomenon may imply some degree of pseudogamy in the group. C. R. Bell (1965) recorded *n* = 7 chromosomes for this species; his illustration shows *n* = 8.

54. Hypericum pseudomaculatum Bush ex Britton, Man. Fl. N. States, 627. 1901 [E] [F]

Hypericum punctatum Lamarck var. *pseudomaculatum* (Bush ex Britton) Fernald

Herbs erect or ascending to divaricate, with rooting, sometimes creeping, branching base, 4–9.5 dm. **Stems** sometimes clustered, internodes not lined, with black glands scattered all over. **Leaves** usually ascending, sometimes spreading, sessile; blade usually ovate-lanceolate to triangular-lanceolate or narrowly elliptic, rarely ovate, 18–45 × 6–16(–20) mm, base cordate to rounded, margins plane, apex usually acute, rarely obtuse to rounded, midrib with 3 or 4 pairs of branches, tertiary veins densely reticulate toward margins, black glands intramarginal (dense) and laminar (scattered). **Inflorescences** subcorymbiform to broadly pyramidal, 16–164(–280)-flowered, subsidiary branches ascending to widely spreading. **Flowers** 10–20 mm diam.; sepals not imbricate, erect in fruit, lanceolate to ovate or elliptic-oblong, subequal, (3–)3.7–4.9(–6)

mm, apex acute; petals yellow, usually obovate, rarely elliptic, 6–14 mm; stamens 38–61; anther gland amber or pellucid; styles 5.4–8.5 mm. **Capsules** broadly ovoid, 3–6 × 2–4 mm, with longitudinal and lateral vittae or vesicles or only ovoid vesicles (all amber). **Seeds** not or scarcely carinate, 0.6–0.8 mm; testa linear-reticulate. **2n = 16.**

Flowering mid summer (Jun–Jul). Open and partially shaded, dry areas of woods, among rocks, fields, roadsides, well-drained soil; 100–700 m; Ala., Ark., Fla., Ga., Ill., La., Miss., Mo., Okla., S.C., Tenn., Tex.

Hypericum pseudomaculatum has been confused with *H. punctatum*; they are quite distinct and they rarely, if ever, hybridize. The affinities of *H. pseudomaculatum* are with Mexican *H. formosum* Kunth.

J. A. Steyermark (1963) recognized two distinct floral forms in Missouri: forma *pseudomaculatum* with orange-yellow petals and stamen filaments, and forma *flavidum* in which these parts are pale, creamy yellow. The distribution of these two forms within the whole range of the species is unknown.

Excluded Species:

Hypericum elatum Aiton

Hypericum elatum Aiton (a synonym of *H.* ×*inodorum* Miller = *H. androsaemum* Linnaeus × *hircinum* Linnaeus) was wrongly cited as from North America by Aiton and was not conclusively recognized as an Old World taxon until J. M. Coulter (1886) published his account of North American *Hypericum*.

2. TRIADENUM   Rafinesque, Fl. Tellur. 3: 78. 1837 • Marsh St. John's wort [Greek *tri-*, three-, and *aden*, gland, alluding to three glandluliform staminode fascicles]

Gardenia Colden 1756, name rejected, not J. Ellis 1761 [Rubiaceae], name conserved; *Hypericum* Linnaeus sect. *Elodea* Choisy

Herbs, perennial, rhizomatous, glabrous, with glandular canals, lacunae, or dots containing essential oils (pale) in various parts and, sometimes, reddish to purplish gland dots containing hypericin on stems and leaves. **Stems:** internodes with 2 or 4 raised lines at first, then terete (not lined). **Leaves** sessile, subsessile, or petiolate; blade relatively broad, venation pinnate, tertiary veins densely reticulate, glands punctiform, pale (records of black gland dots are probably all due to fungal attack), intramarginal and laminar. **Inflorescences** terminal, sometimes also axillary, cymose, 2–15-flowered, or solitary flower, branching dichasial; bracts and bracteoles relatively small. **Flowers** tubular or campanulate at first, expanding to stellate for short time each day; sepals persistent, 5, distinct or almost so, margins not glandular-ciliate; petals deciduous, 5, partly imbricate or contorted, pink or flesh colored, sometimes green tinged; stamens persistent,

9, in 3 fascicles, each with 3 stamens; filaments of each fascicle ⅕–½+ connate; anthers yellow, isodiametric to oblate or shortly oblong, with amber gland on connective; staminode fascicles 3, alternating with stamen fascicles; ovary 3-merous; placentation axile; ovules relatively numerous on each placenta; styles distinct, spreading. **Capsules** 3-valved, with glandular vittae. **Seeds** narrowly cylindric, carinate; testa reticulate-foveolate. *x* = 19.

Species 6 (4 in the flora): e North America, Asia (e China, India [Assam], Japan, Korea, e Siberia, Taiwan).

Nomenclatural complexities and confusions associated with *Triadenum* were reviewed by N. K. B. Robson (1977). B. R. Ruhfel et al. (2011) concluded from molecular studies that *Triadenum* is part of *Hypericum*. Robson (2012) gave reasons why *Triadenum* is generically distinct.

SELECTED REFERENCE   Gleason, H. A. 1947. Notes on some American plants: *Triadenum*. Phytologia 2: 288–291.

1. Leaves petiolate . . . . 4. *Triadenum walteri*
1. Leaves sessile or (distal) subsessile.
  2. Leaves: gland dots intramarginal. . . . . 3. *Triadenum tubulosum*
  2. Leaves: gland dots laminar and intramarginal.
    3. Sepals 4–7(–8) mm, apex acute to acuminate; styles 2–3.5 mm. . . . . 1. *Triadenum virginicum*
    3. Sepals 2.5–5 mm, apex usually obtuse to rounded, rarely acute; styles 0.5–1.5 mm . . . . 2. *Triadenum fraseri*

1. **Triadenum virginicum** (Linnaeus) Rafinesque, Fl. Tellur. 3: 79. 1837 • Millepertuis de Virginie E F

Hypericum virginicum Linnaeus, Syst. Nat. ed. 10, 2: 1184. 1759; *Elodes campanulata* (Walter) Pursh; *E. campanulata* var. *emarginata* (Lamarck) Pursh; *E. virginica* (Linnaeus) Nuttall; *Gardenia virginica* (Linnaeus) Farwell; *H. campanulatum* Walter; *H. enneandrum* Stokes; *Martia campanulata* (Walter) Sprengel

Herbs erect, 2–7 dm, sometimes with ascending branches distally. **Stems:** internodes 4-lined or 4-angled at first, then terete. **Leaves** sessile, sometimes amplexicaul; blade usually ovate or triangular-ovate to elliptic or oblong, rarely oblanceolate, 20–65 × 10–22(–30) mm, base shallowly cordate, apex usually rounded, rarely obtuse to retuse, gland dots laminar (relatively dense, large) and intramarginal (relatively small). **Inflorescences** laxly cylindric to pyramidal, 3–15-flowered from terminal node, sometimes with subsidiary inflorescences from to 4 proximal nodes and flowering branches from to 6 further nodes. **Flowers** 10–15 mm diam.; sepals oblong to elliptic-oblong or oblong-lanceolate, 4–7(–8) × 1–2 mm, apex acute to acuminate; petals oblong-elliptic to oblanceolate, 6–9(–10) mm; stamen fascicles 4–6.5 mm; filaments ⅕ connate; styles 2–3.5 mm. **Capsules** cylindric to ellipsoid or ovoid-ellipsoid, (8–)9–10(–12) × 4–5 mm, apex acute. **Seeds** 0.5–1.2 mm. *2n* = 38.

Flowering summer–fall (Jul–Oct). Swamps, marshy shores, poor fens; 0–500 m; N.S., Ont.; Ala., Ark., Conn., Del., D.C., Fla., Ga., Ill., Ind., Ky., La., Maine, Md., Mass., Mich., Miss., N.H., N.J., N.Y., N.C., Ohio, Pa., R.I., S.C., Tenn., Tex., Vt., Va., W.Va., Wis.

Triadenum virginicum is the most widely distributed North American species of *Triadenum*. In the southwestern part of its range, the leaves are longer and narrower, approaching those of *T. tubulosum* and *T. walteri*; it is nearly always distinguishable from the northern *T. fraseri* by the sepals and styles (see key).

2. **Triadenum fraseri** (Spach) Gleason, Phytologia 2: 289. 1947 • Millepertuis de Fraser E F

Elodes fraseri Spach, Ann. Sci. Nat., Bot., sér. 2, 5: 168. 1836 (as Elodea); *Hypericum virginicum* Linnaeus var. *fraseri* (Spach) Fernald; *Triadenum virginicum* (Linnaeus) Rafinesque subsp. *fraseri* (Spach) J. M. Gillett; *T. virginicum* var. *fraseri* (Spach) Cooperrider

Herbs erect, 1.5–7.5 dm, sometimes with ascending branches in distal ½+. **Stems** shallowly 4-lined at first, then terete. **Leaves** sessile, sometimes amplexicaul; blade broadly ovate or triangular-ovate to oblong, 15–50(–70) × 10–40(–50) mm, base usually ± shallowly cordate, rarely truncate, apex rounded to retuse, gland dots laminar (relatively dense) and intramarginal.

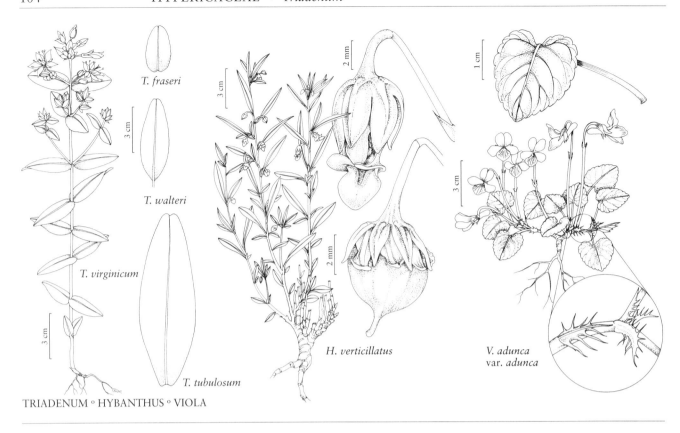

TRIADENUM ∘ HYBANTHUS ∘ VIOLA

Inflorescences spiciform-cylindric to pyramidal, 1–10-flowered from distal 1 or 2 nodes, sometimes with subsidiary inflorescences from to 4 proximal nodes and flowering branches from to 4 further nodes. **Flowers** 8–10 mm diam.; sepals oblong to elliptic-oblong, 2.5–5 × 1–1.5 mm, apex usually obtuse to rounded, rarely acute; petals oblong-obovate to oblanceolate, 5–8 mm; stamen fascicles 3 mm; filaments ½ connate; styles 0.5–1.5 mm. **Capsules** cylindric to ovoid-ellipsoid, 7.5–9 × 3.5–4.5 mm, apex subacute to obtuse. **Seeds** 0.5–11.2 mm. $2n = 38$.

Flowering late summer–early fall (Jul–Sep). Wooded swamps, fens, marshes, lakeshores, organic to silty and sandy substrates, along shores, beaver meadows, poor fens (rarely in true bogs); 0–500 m; St. Pierre and Miquelon; B.C., Man., N.B., Nfld. and Labr., N.S., Ont., P.E.I., Que., Sask.; Conn., Del., Ill., Ind., Iowa, Maine, Md., Mass., Mich., Minn., Mont., Nebr., N.H., N.J., N.Y., N.C., Ohio, Pa., R.I., Tenn., Vt., Va., W.Va., Wis.

According to B. Boivin (1967), *Triadenum fraseri* (as *Hypericum virginicum*) was introduced into British Columbia from eastern Canada in peat. It seems to be almost always distinguishable from *T. virginicum* and to have a distinct habitat; it merits specific rank.

3. **Triadenum tubulosum**   (Walter) Gleason, Phytologia 2: 289. 1947   ☐E ☐F

Hypericum tubulosum Walter, Fl. Carol., 191. 1788; *Elodes drummondii* Spach; *E. pauciflora* Spach; *E. tubulosa* (Walter) Pursh; *H. petiolatum* Linnaeus var. *tubulosum* (Walter) Fernald; *H. walteri* J. F. Gmelin var. *tubulosum* (Walter) Lott; *Triadenum longifolium* Small

Herbs erect, to 10 dm, sometimes branching from near base, sometimes with ascending branches in distal ½+. **Stems:** internodes 2-lined at first, then terete. **Leaves** sessile or (distal) subsessile, rarely amplexicaul; blade narrowly oblong or elliptic to oblanceolate, 50–120 (–150) × 10–40(–50) mm, base rounded to truncate or subcordate, apex rounded or rounded-apiculate to retuse, gland dots intramarginal. **Inflorescences** spiciform-cylindric, 3–7-flowered from terminal node, with sessile or pedunculate inflorescences and flowering branches from to 4 proximal nodes. **Flowers** 15 mm diam.; sepals narrowly oblong, 4–7 × 1–1.5 mm, apex acute to obtuse; petals elliptic-obovate, 5–8 mm; stamen fascicles 4–7 mm; filaments ½+ connate; styles 0.8–1.5 mm. **Capsules** cylindric, 8–12 × 3.5–4 mm, apex obtuse. **Seeds** 0.8–1.2 mm.

Flowering late summer–early fall (Aug–Sep). Swampy or marshy ground in woods; 0–500 m; Ala., Ark., Fla., Ga., Ill., Ind., Ky., La., Miss., Mo., N.C., Ohio, S.C., Tenn., Tex., Va.

The absence of laminar glands in leaves is a more reliable characteristic for differentiating *Triadenum tubulosum* from *T. walteri* than sessile versus petiolate leaves, which are not always quite sessile toward the inflorescence in *T. tubulosum*, or sepals, which can be obtuse in both.

4. Triadenum walteri (J. F. Gmelin) Gleason, Phytologia 2: 289. 1947 E F

Hypericum walteri J. F. Gmelin, Syst. Nat. 2: 1159. 1792 (as Hypericon); *Elodes floribunda* Spach; *E. petiolata* Pursh; *Gardenia petiolata* (Pursh) Farwell; *H. paludosum* Choisy; *H. tubulosum* var. *walteri* (J. F. Gmelin) Lott; *Triadenum petiolatum* (Pursh) Britton

Herbs erect, to 10 dm, sometimes with ascending branches in distal ½+. **Stems:** internodes 4-lined at first, then terete. **Leaves** petiolate, petiole 2–15 mm; blade narrowly elliptic to narrowly oblong, 30–150 × 10–35 mm, base cuneate, apex rounded, gland dots laminar and intramarginal. **Inflorescences** interrupted spiciform-cylindric, 3-flowered from terminal node, sometimes with subsessile flowers or pedunculate triads and/or flowering branches from to 8 proximal nodes. **Flowers** 7–10 mm diam.; sepals narrowly oblong to narrowly elliptic or lanceolate, 3–5 × 1–1.5 mm, apex obtuse to rounded; petals narrowly obovate, 5–7 mm; stamen fascicles 3–5 mm; filaments ½+ connate; styles 0.8–1.2 mm. **Capsules** cylindric, 7–12 × 3–4 mm, apex subacute. **Seeds** 0.8–1.1 mm.

Flowering late summer (Jul–Sep). Swampy or marshy ground in woods, pond or lake margins, on fallen logs; 0–400 m; Ala., Ark., Del., Fla., Ga., Ill., Ind., Ky., La., Md., Miss., Mo., N.J., N.C., Ohio, Okla., S.C., Tenn., Tex., Va., W.Va.

The petiolate leaves usually distinguish *Triadenum walteri* from the other three North American species of *Triadenum* (see note under 3. *T. tubulosum*).

VIOLACEAE Batsch

• Violet Family

R. John Little

Landon E. McKinney†

Herbs, annual or perennial, [**subshrubs, shrubs, lianas, and trees**], glabrous or hairy, hairs simple; taprooted or rhizomatous, sometimes stoloniferous. **Stems** 0–20, prostrate to erect. **Leaves** cauline or basal, (attached directly to rhizome, some *Viola*), alternate (and opposite in *Hybanthus* [and other genera]), simple or compound, stipulate [estipulate], petiolate or sessile; blade unlobed or lobed. **Inflorescences** 1(–4)[5]-flowered, axillary from leaf axils or scapose from rhizomes or stolons (or in racemes of umbels), pedunculate; bracteoles usually present on peduncles, usually alternate. **Flowers** bisexual [unisexual, plants dioecious], perianth and androecium hypogynous, zygomorphic [actinomorphic], chasmogamous, cleistogamous often present, fragrant or odorless; sepals persistent, 5, distinct, ± equal [strongly unequal], with or without basal auricles, margins entire, usually scarious, ciliate or eciliate; petals 5, distinct, unequal, imbricate in bud [convolute], lowermost petal often larger with gibbous or elongated spur; stamens 5, alternate with petals, surrounding ovary, connivent or syngenesious; filaments 0–1 mm, filaments of 2 anterior stamens often with nectaries protruding into spur, anther dehiscence by longitudinal slits; pistil 1, [2]3[5]-carpellate; ovary superior, 1-locular; placentation parietal; ovules [1 or 2] 8–75, anatropous, bitegmic, crassinucellate; styles [0]1, usually enlarged distally, solid or hollow; stigmas 1 [3–5], with or without hairs. **Fruits** capsular [berry, nut], 3-valved, dehiscence loculicidal. **Seeds** [1–](3–)6–75, hard, embryo not developed at time of dispersal, spheroid or ovoid [strongly flattened], glabrous [hairy], some arillate, some with elaiosome [seeds winged in some woody vines].

Genera 23, species 1000–1100 (2 genera, 78 species in the flora): nearly worldwide.

Violaceae are predominantly tropical with nearly worldwide distribution. Most genera are monospecific or oligotypic and are restricted to the New World or Old World Tropics (H. E. Ballard et al. 1998; G. A. Wahlert et al. 2014). Except for *Viola*, *Hybanthus*, and *Rinorea* Aublet, which together account for 98% of all species in the family, most genera are limited to one continent or island system (M. Feng 2005).

Violaceae has been placed in the Violales by most authors (A. Cronquist 1981; R. F. Thorne 1992; A. L. Takhtajan 1997). Based on data from cladistic analyses, it was included in the Malpighiales in 1998 (Angiosperm Phylogeny Group 1998, 2003, 2009).

The Malpighiales clade was first identified by M. W. Chase et al. (1993) in a phylogenetic analysis of nucleotide sequences from the plastid gene *rbc*L (K. J. Wurdack and C. C. Davis 2009). Currently, 35 families are included in Malpighiales (Angiosperm Phylogeny Group 2009). Molecular studies employing multiple gene regions have confirmed the monophyly of Malpighiales, which includes about 16,000 species (Wurdack and Davis). Relationships within Malpighiales remain poorly understood, and it is the most poorly resolved large rosid clade (Wurdack and Davis).

Violaceae were previously organized into three subfamilies: Fusispermoideae Hekking, Leonioideae Melchior, and Violoideae Beilschmied (W. H. A. Hekking 1988; S. A. Hodges et al. 1995). Evidence confirms that *Fusispermum* Cuatrecasas is basal in Violaceae and belongs in the monospecific Fusispermoideae (M. Feng 2005; T. Tokuoka 2008), and Leonioideae should be subsumed in Violoideae (Feng; Feng and H. E. Ballard 2005; Tokuoka). All genera in Violaceae except *Fusispermum* are currently included in the Violoideae. Usually described as having an actinomorphic corolla, *Fusispermum* was reported actually to have weakly zygomorphic calyces and corollas (G. A. Wahlert et al. 2014).

W. H. A. Hekking (1988) divided Violoideae into two tribes, Violeae de Candolle and Rinoreeae Reiche & Taubert. *Viola* and *Hybanthus*, the only two genera in the flora area, are placed in the Violeae.

In a study of Violaceae based on plastid and nuclear DNA sequences (*rbc*L, *atp*B, *mat*K, and 18s rDNA), T. Tokuoka (2008) found that monophyly of the family is strongly supported. A study of 39 species of *Viola* occurring primarily in China using chloroplast sequences *trn*L-*trn*F, *psb*A-*trn*H, *rpL*16, and ITS showed that the supposed subgenus *Viola* is not monophyletic (Liang G. X. and Xing F. W. 2010). Their data imply that 1) erect stems may be more primitive than stolons or rosettes, 2) species with stigmatic beaks might have been derived from ancestors without beaks, 3) simple structure stigmas transformed into complex stigmas then transformed back to simple structures, and 4) long fimbriate margins in stipules and stipules that are one-half to three-fourths adnate may represent distinctive evolutionary trends in sections *Trigonocarpae* Godron and *Adnatae* (W. Becker) Ching J. Wang, respectively.

A study of Violaceae based on plastid DNA sequences showed that most intrafamilial taxa from previous classifications of Violaceae were not supported, that previously unsuspected generic affinities were revealed, and that reliance on floral symmetry (that is, actinomorphy versus zygomorphy) alone provides misleading inferences of relationships and heterogeneous generic circumscriptions (G. A. Wahlert et al. 2014).

SELECTED REFERENCES   Ballard, H. E., J. de Paula-Souza, and G. A. Wahlert. 2014. Violaceae. In: The Families and Genera of Vascular Plants. Vol. 11, pp. 303–322. Berlin, Heidelberg.   Brainerd, E. 1921. Violets of North America. Bull. Vermont Agric. Exp. Sta. 224.   Brizicky, G. K. 1961b. The genera of Violaceae in the southeastern United States. J. Arnold Arbor. 42: 321–333.   Feng, M. 2005. Floral Morphogenesis and Molecular Systematics of the Family Violaceae. Ph.D. dissertation. Ohio University.   Feng, M. and H. E. Ballard. 2005. Molecular systematic, floral developmental and anatomical revelations on generic relationships and evolutionary patterns in the Violaceae. In: International Botanical Congress. 2005. XVII International Botanical Congress, Vienna, Austria, Europe, Austria Center Vienna, 17–23 July 2005. Abstracts. P. 169.   Gershoy, A. 1928. Studies in North American violets. I. General considerations. Bull. Vermont Agric. Exp. Sta. 279.   Hodges, S. A. et al. 1995. Generic relationships in the Violaceae: Data from morphology, anatomy, chromosome numbers and *rbc*L sequences. [Abstract.] Amer. J. Bot. 82(6, suppl.): 136.   McKinney, L. E. and N. H. Russell. 2002. Violaceae of the southeastern United States. Castanea 4: 369–379.   Tokuoka, T. 2008. Molecular phylogenetic analysis of Violaceae (Malpighiales) based on plastid and nuclear DNA sequences. J. Pl. Res. 121: 253–260.   Wahlert, G. A. et al. 2014. Phylogeny of the Violaceae (Malpighiales) inferred from plastid DNA sequences: Implications for generic diversity and intrafamilial classification. Syst. Bot. 39: 239–252.

1. Plants caulescent; sepals not auriculate; upper 2 and lateral 2 petals not showy, 0.5–5 mm; lowest petal showy, narrowed at middle; stamens connate, lowest 2 filaments not spurred with nectary; seeds (3–)6–9 . . . . 1. *Hybanthus*, p. 108
1. Plants caulescent or acaulescent; sepals auriculate; upper 2 and lateral 2 petals showy, 5+ mm; lowest petal showy, not narrowed at middle; stamens connivent, but distinct, lower 2 filaments spurred with nectary that protrudes into petal spur; seeds 6–75 . . . . 2. *Viola*, p. 111

1. HYBANTHUS Jacquin, Enum. Syst. Pl., 2, 17. 1760, name conserved • [Greek *hybos*, hump, and *anthos*, flower, alluding to recurved pedicels]

R. John Little

Cubelium Rafinesque ex Britton & A. Brown; *Ionidium* Ventenat; *Pombalia* Vandelli

Herbs or subshrubs, annual or perennial, [**subshrubs, shrubs**], caulescent, homophyllous, glabrous or hairy. **Stems** persistent, 1–20, usually erect, sometimes suberect or prostrate, leafy, simple or branched, from thick, fleshy or subligneous, branched or unbranched rhizome or taproot. **Leaves** cauline, simple, proximal usually opposite, distal usually alternate, petiolate or sessile; stipules not adnate to petiole, linear-subulate to leaflike, usually shorter than leaves; blade not overlapping basally, linear, lanceolate to oblanceolate, ovate to obovate, or elliptic, surfaces not mottled. **Inflorescences** axillary in leaf axils, 1–3(4)-flowered or in poorly defined racemes; peduncle jointed; bracteoles present or absent. **Flowers:** sepals subequal, not auriculate; upper 2 and lateral 2 petals not showy, 0.5–5 mm, lowest petal showy, larger than others, narrowed at middle, angular-deltate, elliptic, oblong, orbiculate, or ovate, upper and lateral petals usually glabrous, lower sometimes bearded adaxially; spur gibbous; stamens connate, lowest 2 filaments not spurred with nectary; cleistogamous flowers present or absent. **Capsules** ovoid, globose, obtusely trigonous, oblong, or ellipsoid, glabrous. **Seeds** (3–)6–9, globose to slightly flattened, glabrous [hairy], with whitish elaiosome. $x = 4$.

Species 100–115 (5 in the flora): North America, Mexico, West Indies, Central America, South America, Asia, Africa, Australia; tropical and subtropical areas.

In 2010, M. N. Seo et al. concluded that polyploidy likely played a role in speciation in *Hybanthus* and Seo et al. (2011) proposed an infrageneric classification of *Hybanthus* based on foliar micromorphology. Research using *trnL*/*trnL*-F and *rbc*L plastid regions shows that *Hybanthus* is polyphyletic and can be segregated into nine morphologically and biogeographically distinct groups (G. A. Wahlert et al. 2014). Molecular phylogenetic and morphological evidence will be used to circumscribe the nine *Hybanthus* lineages as separate genera (Wahlert et al.). *Hybanthus* would be reduced to two or three species and *H. concolor* would be placed in *Cubelium* (H. E. Ballard, pers. comm.), a position supported in some molecular studies, for example, T. Marcussen et al. (2010). The other four species in the flora area would be placed in *Pombalia* (Ballard, pers. comm.).

The overall outline of the lowest petal (sometimes referred to as the anterior petal) is usually pandurate or panduriform and is one of the more diagnostic characters of *Hybanthus*. The distal limb is usually enlarged and is longer and wider than the upper and lateral petals. The distal limb of some species rolls inward after anthesis.

The method by which capsules dehisce is not discussed in the major references that treat *Hybanthus*. Because of the similarity to *Viola* capsules, it may be assumed that their dehiscence is similar. Although capsules of *Viola* are sometimes described as explosively dehiscent, the capsules of at least some North American species open relatively slowly. As the capsule valves dry out, they contract and squeeze the seeds causing them to be ballistically ejected. J. de Paula-Souza (pers. comm.) reported that Violaceae capsules in South America are not truly explosive, but rather the seeds are expelled as the capsule valves dry out.

SELECTED REFERENCES   Morton, C. V. 1944. Studies of tropical American plants. The genus *Hybanthus* in continental North America. Contr. U.S. Natl. Herb. 29: 74–82.   Morton, C. V. 1971. Some types and range extensions in *Hybanthus* (Violaceae). Phytologia 21: 56–62.

1. Seeds white to cream . . . . .2. *Hybanthus concolor*
1. Seeds dark brown to black with white to gray mottling, brownish black to black, or shiny black.
  2. Leaf margins crenate to serrate.
    3. Leaf blades 1.5–7(–10.5) cm; petioles (3–)4–7 mm; petals white, purple-tinged, or violet, lowest petal 8–12 mm; bracteoles present; Arizona . . . . 1. *Hybanthus attenuatus*
    3. Leaf blades 0.3–3 cm; petioles 0.5–4 mm; petals white, lowest petal 1.5–3.7 mm; bracteoles absent; Georgia, New Jersey. . . . . 4. *Hybanthus parviflorus*
  2. Leaf margins entire or remotely crenulate-dentate.
    4. Lowest petal 5–9.5 mm, limb 4–10 mm wide, with basal purple patch. . . . 3. *Hybanthus linearifolius*
    4. Lowest petal 2.5–6 mm, limb 1–3 mm wide, without basal purple patch. . . 5. *Hybanthus verticillatus*

1. **Hybanthus attenuatus** (Humboldt & Bonpland ex Willdenow) Schulze-Menz, Notizbl. Bot. Gart. Berlin-Dahlem 12: 114. 1934 • Western green-violet

Ionidium attenuatum Humboldt & Bonpland ex Willdenow in J. J. Roemer et al., Syst. Veg. 5: 402. 1819

Plants herbs, annual, 11–46 cm, from taproot. **Stems** 1, erect, often purple mottled, simple to much-branched, glabrous or pilose (especially distal half). **Leaves:** proximal opposite or subopposite, distal mostly alternate, petiolate; stipules linear [narrowly ovate], 1–5 mm, glabrous [ciliate]; petiole (3–)4–7 mm; blade elliptic to lanceolate or oblong-lanceolate, 1.5–7(–10.5) × 0.3–2.5 cm, base attenuate, margins usually serrate, sometimes crenate, ciliate, apex attenuate to acuminate, surfaces coarsely glabrous or coarsely hairy. **Inflorescences** 1- or 2(–4)-flowered (often with 1 or 2 aborting); peduncle erect to horizontal at anthesis, usually erect in fruit, 2–20 mm, puberulent; bracteoles present. **Flowers:** sepals usually appressed to corolla, lanceolate, margins usually ciliate, apex acuminate; petals: upper white proximally, ovate-lanceolate to narrowly oblong, 2.5–3.5 mm, apex lilac to deep reddish violet, deltate, usually glabrous; laterals white or purple-tinged, ovate-lanceolate, inequilateral, ± falcate, 0.5–2 mm, usually glabrous; lowest white or violet with basal purple patch, 8–12 mm, claw 4–5 × 1–2 mm, distal limb broadly angular-deltate or -elliptic, ± flat to convex at anthesis, 4–7 mm wide, apex obtuse to rounded, glabrous or scarcely pubescent on adaxial surface; cleistogamous flowers on short peduncles. **Capsules** ovoid, 4–6 mm. **Seeds** 6, dark brown to black with white to gray mottling, globose or ovoid to ± flattened with angular edges, 1.5–2 mm. $2n = 24, 32$.

Flowering Aug–Oct. Shaded areas, canyons, near water; 1100–1900 m; Ariz.; Mexico; Central America; South America; introduced in Asia (Indonesia).

Hybanthus attenuatus has been reported as a fast-growing, common weed in crops, humid pastures, and waste places in Central America (M. Soerjani et al. 1987). It has been found in rice paddies in central Java (Soerjani et al.; B. M. Waterhouse and A. A. Mitchell 1998).

2. **Hybanthus concolor** (T. F. Forster) Sprengel, Syst. Veg. 1: 805. 1824 • Eastern green-violet [E]

Viola concolor T. F. Forster, Trans. Linn. Soc. 6: 309, plate 28. 1802; *Cubelium concolor* (T. F. Forster) Rafinesque ex Britton & A. Brown

Plants subshrubs, perennial, 30–80(–100) cm, from thick rhizome with fibrous roots. **Stems** 1 or several-clustered, erect, simple, usually hairy, sometimes glabrous or subglabrous. **Leaves:** proximal and distal alternate, petiolate, distalmost sessile or nearly so; stipules narrowly elliptic to linear or lanceolate, (0.5–)4–25(–30) mm, sparsely hirsute; petiole 6–20 mm; blade elliptic or oblanceolate, 6.5–17 × 1.5–6 cm, base attenuate or acuminate, margins entire or sparsely serrate, ciliate or eciliate, apex acute to long-acuminate, surfaces usually strigose and/or sericeous, sometimes glabrous, trichomes scattered throughout or concentrated along outer margins, veins, or petiole. **Inflorescences** 1–3-flowered; peduncle recurved at anthesis, becoming erect, 5–23 mm, pilose; bracteoles present. **Flowers:** sepals often spreading from corolla, linear to linear-lanceolate, margins ciliate or eciliate, apex acute; petals: upper and laterals greenish white, oblong, 3.5–5 mm, margins usually with scattered cilia, apex usually truncate and recurved, glabrous; lowest greenish white, 4–5.5 mm, claw 0–1 × 3–4 mm, usually longer and wider than others, distal limb oblong, ± flat to convex, 3–5 mm wide, apex retuse, sometimes rounded, usually glabrous; cleistogamous flowers, when present, from axils of distal leaves. **Capsules** oblong to ellipsoid, (8–)15–20 mm. **Seeds** 6–9, white to cream, globose to broadly ovoid, (3–)4–7 mm. **2*n*** = 48.

Flowering Apr–early Jul. Rich, limestone soil and talus in woodlands, slopes, ravines; 100–1000 m; Ont.; Ala., Ark., Conn., Del., D.C., Fla., Ga., Ill., Ind., Iowa, Kans., Ky., Md., Mich., Minn., Miss., Mo., N.J., N.Y., N.C., Ohio, Okla., Pa., S.C., Tenn., Vt., Va., W.Va., Wis.

R. E. Brooks and R. L. McGregor (1986) reported that no cleistogamous flowers have been seen on plants of *Hybanthus concolor* collected in the Great Plains. It is believed extinct in Connecticut and is listed as endangered in Florida.

3. **Hybanthus linearifolius** (Vahl) Urban, Symb. Antill. 5: 436. 1908 • Chancleta [I]

Viola linearifolia Vahl in H. West, Bidr. Beskr. Ste Croix, 303. 1793; *V. stricta* (Ventenat) Poiret

Plants herbs, annual or perennial, 10–50 cm, from subligneous rhizome ca. 3 mm diam. **Stems** 1+, erect or ascending, branched, subglabrous or hairy. **Leaves:** proximal usually opposite, distal usually alternate (occasionally opposite), petiolate or sessile; stipules linear-subulate to lanceolate, 0.6–2 mm, usually glabrous; petiole 0–6 mm; blades: proximal obovate or elliptic-ovate, distal oblong-lanceolate to linear-lanceolate, elliptic, or obovate-elliptic, 0.8–5.1 × 0.3–1.8 cm, base attenuate, margins subentire to remotely crenulate-dentate, ciliate, apex acute to obtuse, surfaces glabrate or puberulent. **Inflorescences** 1- or 2-flowered; peduncle ± horizontal or erect at anthesis, horizontal or pendant in fruit, 6–20 mm, puberulent; bracteoles present. **Flowers:** sepals appressed to corolla, lanceolate, ovate-lanceolate, or oblong, margins ciliate, apex acuminate; petals: upper and laterals white, bluish, or purplish white, oblong, 3–5 mm, margins sometimes ciliate, apex deltate, often tinged purple or red, glabrous; lowest white, bluish, or purplish white, with basal purple patch, 5–9.5 mm, claw 3–4 × 1 mm, distal limb deltate to suborbiculate, ± flat to convex at anthesis, 4–10 mm wide, apex retuse or not, surfaces glabrous; presence of cleistogamous flowers not determined. **Capsules** globose or obtusely trigonous, 3–4.5 mm. **Seeds** 3–6,

brownish black to black, ovoid or obovoid, subangular, 1.3–2 mm.

Flowering Jan–Dec. Avocado groves and ruderal areas; 0–10 m; introduced; Fla.; West Indies (Cuba, Dominican Republic, Greater Antilles, Lesser Antilles, Puerto Rico, Virgin Islands).

Hybanthus linearifolius appears to have first been documented in Florida in 2001 by K. A. Bradley from an avocado grove in Miami-Dade County (FTG 140029, 140268). The occurrence was cited by R. P. Wunderlin and B. F. Hansen 2003 (K. A. Bradley, pers. comm.). In 2009, it was collected in Palm Beach County, Florida, and identified at the Division of Plant Industry, FDACS; it was not accessioned (P. J. Anderson, pers. comm.).

Hno. Alain (1985–1997, vol. 3) reported that *Hybanthus linearifolius* occurs as an annual or a perennial; it is unknown if the species behaves as an annual or perennial in Florida.

4. **Hybanthus parviflorus** (Linnaeus f.) Baillon, Traité Bot. Méd. Phan. 2: 841. 1884 [I]

Viola parviflora Linnaeus f., Suppl. Pl., 396. 1782

Plants herbs, annual (or perennial), 30–50 cm, from taproot. **Stems** 1 or several-clustered, erect, suberect, or prostrate, sometimes much-branched from caudex, usually hairy. **Leaves** alternate or opposite, frequently opposite only proximally, petiolate; stipules narrowly lanceolate, 1–2 mm, sparsely ciliate to ciliate; petiole 0.5–4 mm; blade usually elliptic, lanceolate, or oblong, sometimes oblanceolate, widely elliptic, ovate, or obovate, 0.3–3 × 0.2–1.5 cm, base acute, attenuate, rounded, or obtuse, margins serrate, sparsely ciliate to ciliate, apex acute to obtuse, surfaces glabrous or pubescent or puberulent on midrib. **Inflorescences** usually 1-flowered, sometimes in poorly defined racemes; peduncle erect at anthesis, usually pendant in fruit, 7–24 mm, puberulent; bracteoles absent. **Flowers:** sepals appressed to corolla, narrowly lanceolate, rarely ovate to suborbiculate, subfalcate, margins sparsely ciliate to ciliate, apex acuminate, rarely round or obtuse; petals: upper white, oblong, 0.5 mm, margins ciliate, apex rounded, glabrous; laterals white, oblong to lanceolate, falcate, 1.5–1.8 mm, apex rounded or acute, glabrous; lowest white, usually with purplish veins, with basal purple patch, 1.5–3.7 mm, claw 1–1.5 × 0.5–0.7 mm (rarely absent), distal limb ± deltate, ± flat at anthesis, 1.2–1.4 mm wide, margins ciliate, apex retuse, glabrous or adaxial surface sparsely pubescent basally; presence of cleistogamous flowers not determined. **Capsules** ovoid to globose, 3–4 mm. **Seeds** (3–)6–9, black, ovoid, 1.5–1.7 mm. **2*n*** = 12, 24.

Flowering Apr–Oct. Disturbed or grassy areas; 0–10 m; introduced; Ga., N.J.; South America.

Hybanthus parviflorus was first collected in the flora area in 1880 (identified as *Calceolaria glutinosa*) from ship ballast at Communipaw Ferry, New Jersey, and again from ballast (as *Ionidium glutinosum*) at Kaighn Point, Camden, New Jersey, in 1885 (B. E. Wofford et al. 2004). It is unknown whether a 1998 collection from Fort Pulaski, Georgia, is a recent introduction (Wofford et al.).

5. Hybanthus verticillatus (Ortega) Baillon, Hist. Pl. 4: 344. 1873 • Baby-slippers, nodding green-violet F

Viola verticillata Ortega, Nov. Pl. Descr. Dec. 4: 50. 1797; *Hybanthus verticillatus* var. *platyphyllus* (A. Gray) Cory & H. B. Parks

Plants subshrubs, perennial, 10–40 cm, from ligneous rhizome. **Stems** 1–20, clustered, erect, simple or branched from proximal nodes, glabrous or strigose to pilose. **Leaves:** proximal opposite or subopposite, distal usually alternate, petiolate or sessile; stipules linear-subulate and minute to leaflike, 3–40 mm, glabrous or hirsute, gland-tipped; petiole 0–1 mm; blade linear, narrowly elliptic, lanceolate, or oblanceolate, 1–5.5(–6) × 0.1–0.8(–1.1) cm, base attenuate, margins usually entire, ciliate or eciliate, apex acute to acuminate, surfaces glabrous, strigose, or pilose. **Inflorescences** 1-flowered; peduncle usually pendant at anthesis, sometimes horizontal or erect, usually pendant in fruit, 3–14 mm, hirsute to densely puberulent, occasionally glabrous on segment distal to joint; bracteoles present. **Flowers:** sepals appressed to corolla, ovate to lanceolate, margins ciliate or eciliate, apex acute; petals: upper greenish white, cream, or yellowish, with purple tips, oblong, 2–2.5 mm, apex acute to rounded, glabrous; laterals similar to upper except 3 mm; lowest greenish white, cream, or yellowish, sometimes tinged purplish, 2.5–6 mm, claw 2–3 × 0.5 mm, distal limb broadly ovate to orbiculate, ± concave at anthesis, 1–3 mm wide, apex rounded to obtusely angled, adaxial surface usually bearded basally and on claw; presence of cleistogamous flowers not determined. **Capsules** ovoid to globose, 4–7 mm. **Seeds** 6, dark brown to shiny black, subglobose, ± flattened with angular edges, or broadly ovoid, 1.8–2.5(–3) mm. $2n$ = 16, 24, 32.

Flowering Apr–Oct. Dry plains, gravelly soil in prairies, mesas, fields, desert grasslands, pinyon-juniper woodlands, chaparral, rocky slopes, forest edges, oak and mesquite savannahs, riparian habitats; 10–2100 m; Ariz., Colo., Kans., N.Mex., Okla., Tex.; Mexico (Chihuahua, Coahuila, Sonora).

2. VIOLA Linnaeus, Sp. Pl. 2: 933. 1753; Gen. Pl. ed. 5, 402. 1754 • [Classical Latin name derived from Greek *ion*, violet]

R. John Little

Landon E. McKinney†

Chrysion Spach; *Crocion* Nieuwland & Kaczmarek; *Lophion* Spach

Herbs, annual or perennial, caulescent or acaulescent, homophyllous (heterophyllous in *V. palmata*, *V. sagittata*, and *V. septemloba*), hairs concentrated or scattered throughout. **Stems** usually deciduous and withering at end of season, 0–5(–10+), erect to ascending, decumbent, or prostrate, simple, [woody], leafy; from horizontal or vertical, thick, fleshy or subligneous, shallow or deep-seated rhizome (caudex); or from narrow or thick rhizomes; or spreading, thin stolons; or slender taproots in annual species and seedlings; in caulescent species, short, axillary branches are sometimes present on main stems. **Leaves** alternate on caulescent species, simple (compound in *V. beckwithii*, *V. douglasii*, *V. hallii*, *V. sheltonii*, and *V. trinervata*), petiolate; caulescent plants with 0–11(–22) basal leaves per caudex, acaulescent plants with 1–12(–18) leaves per rhizome, prostrate to erect; stipules adnate to petiole or not, not leaflike (sometimes leaflike in *V. lobata*), unlobed, shorter than leaves (except *V. arvensis*, *V. bicolor*, and *V. tricolor*); blade not overlapping basally (except occasionally in *V. blanda* and *V. rotundifolia*), ovate, reniform, deltate, orbiculate, lanceolate, spatulate, or linear, adaxial surface not mottled (except in *V. hastata* and *V. hirsutula*). **Inflorescences** axillary (rarely umbellate in *V. sagittata* forma *umbelliflora*) from distal and proximal stem nodes in caulescent species or scapose from rhizomes or stolons in acaulescent species, 1(–3)[–5]-flowered; peduncles not jointed; bracteoles present. **Flowers:** sepals entire, equal or subequal, margins ciliate or eciliate, auriculate, auricles prominent or not; upper 2 and lateral 2 petals showy, 5+ mm, lowest petal showy, not narrowed at middle of limb; lateral petals and sometimes others bearded proximally with variously shaped hairs; style bearded or beardless; spur gibbous or elongated; stamens connivent, but distinct, forming cone around ovary, not adherent to style, dehiscing introrsely, lower 2 filaments spurred with nectary that protrudes into petal spur; cleistogamous flowers absent or produced in summer, apetalous or petals 2(3) and scarcely developed, stamens 2, usually adherent to style. **Capsules** ovoid, ellipsoid, oblong, spherical, or subglobose, glabrous, puberulent, or tomentose, sometimes muriculate. **Seeds** 6–75, spherical or ovoid, glabrous, often arillate with elaiosome. *x* = 6, 7.

Species 400–600 (73 in the flora): nearly worldwide; temperate regions, also South America, Pacific Islands (Hawaii, Philippines, Taiwan).

The taxonomy of *Viola* is often considered difficult partly because of hybridization; more than 100 named hybrids occur in the flora area. Hybrids among the blue-flowered, acaulescent species in eastern North America and among other species are well known (E. Brainerd 1924; N. H. Russell and M. Cooperrider 1955; G. L. Stebbins et al. 1963; T. S. Cooperrider 1986; L. E. McKinney 1992; H. E. Ballard 1993, 1994; A. Haines 2011).

Other factors contribute to phenotypic variation. The dimensions of leaves and stems often increase substantially between early spring and late summer (D. Klaber 1976). Stem, leaf, and flower size vary with environmental factors such as aspect, light, available soil moisture, and other edaphic conditions (Klaber; L. E. McKinney 1992).

The number of recognized taxa and the ranks attributed to taxa vary among authors (E. Brainerd 1921; N. H. Russell 1965; L. E. McKinney 1992; H. E. Ballard 1994; N. L. Gil-Ad 1997; D. B. Ward 2006).

An effort to deal with the difficulties encountered, specifically with subsect. *Boreali-Americanae* (W. Becker) Gil-Ad, was made by N. L. Gil-Ad (1997, 1998). Much of his research was centered on his belief that hybridization and introgression were extensive and frequent. He investigated both macro- and micro-morphological characters of seed coats and petal hairs. His research was based on a relatively small number of samples; he obtained new and innovative data from which he distinguished orthospecies from possible hybrids and/or introgressants. This work represents a significant contribution to our understanding of some species; we do not share his conclusions about some species, such as *Viola hirsutula* and *V. subsinuata*.

An infrageneric classification for *Viola* is not presented here. Some species complexes have been identified (for example, *V. adunca, canadensis, nuttallii, palustris,* and *purpurea*). W. Becker (1925) provided the first infrageneric classification, which was largely followed by G. K. Brizicky (1961b), a scheme largely based on style morphology. For the most part, this classification continues to be followed (T. Marcussen and T. Karlsson 2010); some of the names used to refer to species complexes (for example, *Nuttallianae*) were published without rank and have been incorrectly assumed to be subsections [for example, J. Clausen (1964); Marcussen and Karlsson]. Chen Y. S. et al. (2007) replaced sections with subgenera.

Branching of stems in perennial caulescent North American *Viola* species is uncommon and has been documented in ten species: *V. adunca, V. canadensis, V. canina, V. douglasii, V. glabella, V. pedunculata, V. pinetorum, V. purpurea, V. quercetorum,* and *V. walteri*. Branching in these species involves the development of one or more relatively short, leafy, axillary shoots on one or more stems. The three annual species in the flora area, *V. arvensis*, *V. bicolor*, and *V. tricolor*, commonly branch from the base of the main stem near or at the crown and from nodes higher on the stem.

The leaves of acaulescent species develop from the rhizome. In homophyllous plants, all leaf blades are lobed from early season through late season; the depth of sinuses depends somewhat on the age of the plant. In heterophyllous plants, the earliest leaf blades are not lobed; later-season blades are lobed. E. Brainerd (1910, 1921) placed great emphasis on these morphological distinctions to differentiate *Viola* taxa. Like Brainerd, L. E. McKinney (1992) found these differences to be reliable taxonomic characters.

The shape, size, position of the stigmatic surface, and degree of bearding vary among species. Differences among style heads have been used in *Viola* classification (W. Becker 1925; J. Clausen 1929) and in keys. In this treatment it is reported whether the style head is bearded or beardless based on reports in the literature and personal observations. Of the 30 acaulescent species in the flora area, one has a bearded style; of the 43 caulescent species, 35 are always bearded, three are bearded or beardless, and five are always beardless.

Of the 73 species of *Viola* in the flora area, 60 are known to produce cleistogamous flowers, nine do not, and the condition in four is unknown. In some species of acaulescent blue violets, the cleistogamous flowers are formed beneath the soil surface or in duff and arise on peduncles as they mature. Some species disperse their cleistogamous seeds in soil or duff. A. Mayers and E. Lord (1983, 1983b) reported that pollen grains in cleistogamous flowers of *V. odorata* germinate in the undehisced anther sacs; the pollen tubes then penetrate the sac and grow toward the stigma. In other groups, observations suggest that pollen is released in proximity to the recurved style (T. Marcussen and T. Karlsson 2010).

Substantial research has been conducted on *Viola* pollination including cleistogamous flowers (A. J. Beattie 1969, 1969b, 1971, 1972, 1974, 1976, 1978; Beattie and D. C. Culver 1979; A. C. Cortés-Palomec and H. E. Ballard 2006; T. M. Culley 2000, 2002; G. Davidse 1976; L. Freitas and M. Sazima 2003; C. M. Herrera 1990; A. Mayers and E. Lord 1983, 1983b). Pollinator rewards available in most *Viola* flowers include nectar and pollen.

Pollinators of violets include bumblebees, honeybees, solitary bees, syrphid flies, butterflies, skippers, hawkmoths, moths, and beeflies. Thrips have been reported in *Viola* flowers (M. S. Baker 1935) and while these insects are often observed with pollen attached to their bodies, there is currently no evidence they play a role in pollination of *Viola*. Due to the high frequency that thrips, aphids, and other minute insects with pollen on their bodies have been observed in *Viola* flowers, it seems probable that these insects sometimes effect pollination.

The three valves of *Viola* capsules usually are thick in perennial species and thin in annual species. The capsules of at least some species open relatively slowly, exposing the seeds. As the valves dry, they contract and squeeze the seeds causing them to be ejected (R. J. Little and G. Leiper 2012). Capsules that disperse seeds ballistically are usually on erect peduncles; capsules that passively release their seeds usually point downward (A. J. Beattie and N. Lyons 1975).

The rhizomes of some western caulescent species sometimes branch underground into two, three, or four caudices. The number of stems and basal leaves reported in descriptions pertains to a single caudex.

Most *Viola* seeds possess an outgrowth (elaiosome), or food body, of variable size that is often attractive to ants. S. Lengyel et al. (2010) estimated that over 70% of Violaceae species are myrmecochorous. Studies have been conducted on various aspects of myrmecochory in *Viola* (A. J. Beattie and N. Lyons 1975; R. Y. Berg 1975; D. C. Culver and Beattie 1978, 1980; Beattie and Culver 1981; G. Matlack 1994).

Violets of horticultural importance include *Viola arvensis* (field or wild pansy), *V. odorata* (English or sweet violet), *V. tricolor* (Johnny-jump-up), and *V. ×wittrockiana* Gams ex Nauenburg & Buttler (garden pansy). Over 120 species of *Viola* are grown as ornamentals (L. Watson and M. J. Dallwitz, http://delta-intkey.com).

E. Brainerd (1908) reported that *Viola chinensis* G. Don (= *V. patrinii* de Candolle ex Gingins) was established at The New York Botanical Garden, at a residence in the District of Columbia, and in his garden in Massachusetts. It appears not to have persisted; it was not mentioned by N. Taylor (1915). A. Haines (2011) noted that reports of *V. chinensis* in New England were based on misidentifications of *V. japonica*.

Native Americans in the United States and Canada used *Viola* species for drugs, dyes, and food (D. E. Moerman 1998). Medicinal uses included pain relief and treatment of ailments including colds and coughs and for dermatological, gastrointestinal, eye, heart, and respiratory problems. Leaves and stems of some species were used as vegetables, usually cooked. Some tribes are reported to have soaked the seeds of corn in an infusion of violet roots before planting to repel insects.

Mature plants are often needed for identification of violets. In preparing specimens of violets, care should be taken to record data on petal and spur colors, and presence and distribution of beards on lateral and other petals, or note if lacking.

Measurements of the lowest petal in the descriptions here include the spur.

When this treatment was being finalized, *Viola calcicola* R. A. McCauley & H. E. Ballard was described as new. Time constraints prevented it from being incorporated. *Viola calcicola* is acaulescent, heterophyllous, has short, vertical rhizomes, nearly white to purple corollas, and occurs only on limestone substrates. It is endemic to the Guadalupe Mountains of Texas and New Mexico.

SELECTED REFERENCES Baird, V. B. 1942. Wild Violets of North America. Berkeley and Los Angeles. Ballard, H. E. 1992. Systematics of *Viola* Section *Viola* in North America North of Mexico. M.S. thesis. Central Michigan University. Ballard, H. E., K. J. Sytsma, and R. R. Kowal. 1998. Shrinking the violets: Phylogenetic relationships of infrageneric groups in *Viola* (Violaceae) based on internal transcribed spacer DNA sequences. Syst. Bot. 23: 439–458. Beattie, A. J. and N. Lyons. 1975. Seed dispersal in *Viola* (Violaceae): Adaptations and strategies. Amer. J. Bot. 62: 714–722. Fabijan, D. M., J. G. Packer, and K. E. Denford. 1987. The taxonomy of the *Viola nuttallii* complex. Canad. J. Bot. 65: 2562–2580. Gil-Ad, N. L. 1997. Systematics of *Viola* L. subsection *Boreali-Americanae*. Boissiera 53: 1–130. Gil-Ad, N. L. 1998. The micro-morphologies of seed coats and petal trichomes of the taxa of *Viola* subsect. *Boreali-Americanae* (Violaceae) and their utility in discerning orthospecies from hybrids. Brittonia 50: 91–121. Klaber, D. 1976. Violets of the United States. South Brunswick, N.J. Marcussen, T. et al. 2012. Inferring species networks from gene trees in high-polyploid North American and Hawaiian violets (*Viola*, Violaceae). Syst. Biol. 61: 107–126. McKinney, L. E. 1992. A Taxonomic Revision of the Acaulescent Blue Violets (*Viola*) of North America. Fort Worth. [Sida Bot. Misc. 7.] Russell, N. H. 1965. Violets (*Viola*) of the central and eastern United States: An introductory survey. Sida 2: 1–113.

1. Plants acaulescent.
  2. Style head bearded; Arizona . . . . .69. *Viola umbraticola*
  2. Style head beardless; not confined to Arizona.
    3. Petals deep lemon-yellow . . . . .56. *Viola rotundifolia*
    3. Petals not lemon yellow.
      4. Leaf blades lobed.
        5. Leaf blades lobes similar in width and shape.
          6. All petals beardless; cleistogamous flowers absent . . . . .43. *Viola pedata*
          6. Lateral petals bearded, lowest sometimes bearded; cleistogamous flowers present . . . . 44. *Viola pedatifida*
        5. Middle and lateral blade lobes differ in width and/or shape.
          7. Mid-season leaf blades incised or lobed at base only . . . . 57. *Viola sagittata* (in part)
          7. Mid-season leaf blades incised or lobed throughout.
            8. Earliest leaf blades lobed (plants homophyllous), similar to mid-season blades.

9. Middle lobes of leaf blades lanceolate or spatulate to narrowly obovate; plants usually glabrous; sepal auricles 2–3 mm . . . 9. *Viola brittoniana*
9. Middle lobes of leaf blades narrowly deltate to narrowly elliptic; plants usually pubescent; sepal auricles 1–2 mm . . .64. *Viola subsinuata*
8. Earliest leaf blades unlobed, or sometimes 3-lobed (plants heterophyllous), mid-season blades lobed.
10. Mid-season leaf blades with 3–5 primary lobes, middle lobe elliptic, ovate, to widely ovate . . . . .41. *Viola palmata* (in part)
10. Mid-season leaf blades with 5–9 primary lobes, middle lobe lanceolate, spatulate, to narrowly ovate.
11. Petioles, leaf surfaces, and peduncles rarely glabrous . . . . .41. *Viola palmata* (in part)
11. Petioles, leaf surfaces, and peduncles rarely pubescent.
12. Earliest leaf blades ± deltate or 3-lobed; lowest petal 10–15 mm; plants of limestone glades and barrens . . . . .17. *Viola egglestonii*
12. Earliest leaf blades ± ovate, sometimes 3-lobed; lowest petal 15–25 mm; plants of sandy, dry or seasonally wet pine or mixed pine-deciduous woods . . . . 60. *Viola septemloba*

[4. Shifted to left margin.—Ed.]

4. Leaf blades unlobed (mid-season incised or lobed at base only in *V. sagittata*).
13. Lowest petal spur 3–10 mm.
14. Capsules puberulent . . . . .39. *Viola odorata*
14. Capsules glabrous.
15. Petioles not winged . . . . .58. *Viola selkirkii*
15. Petioles narrowly winged distally.
16. Leaf base cordate. . . . . 27. *Viola japonica*
16. Leaf base usually truncate, sometimes ± cordate or broadly cuneate . . . 49. *Viola prionantha*
13. Lowest petal spur 1–3 mm.
17. Leaf blades lanceolate, narrowly elliptic, linear, or ovate.
18. Leaf blades lanceolate or narrowly elliptic to nearly linear. . . . . 29. *Viola lanceolata*
18. Leaf blades elliptic to narrowly or broadly ovate . . . . .48. *Viola primulifolia*
17. Leaf blades ovate to broadly ovate, reniform, orbiculate, elliptic, or deltate.
19. Plants stoloniferous.
20. Petals white.
21. Lateral petals usually beardless; leaf blade margins serrate, ciliate or eciliate. . . . . .8. *Viola blanda*
21. Lateral petals usually bearded, rarely beardless; leaf blade margins ± entire or shallowly crenate, eciliate. . . . . 33. *Viola macloskeyi*
20. Petals lilac, pale blue, violet, or deep purple, sometimes nearly white.
22. Bracteoles usually above middle of peduncle in chasmogamous flowers; leaf blade margins crenate, denticulate, or entire, ciliate or eciliate . . . . 18. *Viola epipsila*
22. Bracteoles usually below middle of peduncle; leaf blade margins crenulate, eciliate. . . . . 42. *Viola palustris*
19. Plants not stoloniferous.
23. All petals beardless.
24. Petals light violet on both surfaces; leaf blades ± deltate . . . . 13. *Viola clauseniana*
24. Petals white on both surfaces; leaf blades reniform or ovate to broadly ovate or orbiculate . . . . .53. *Viola renifolia* (in part)
23. Lateral 2 and sometimes also lowest petals bearded.
25. Petals white. . . . . .53. *Viola renifolia* (in part)
25. Petals violet, light to dark blue-violet, lavender-violet, light to deep or dull reddish violet, dark purple-violet or reddish purple, or deep bluish violet, rarely white.

[26. Shifted to left margin.—Ed.]

26. Sepal auricles 2–6 mm.
    27. Leaf blade base reniform to cordate; plants of mesic to wet habitats. . . . . 14. *Viola cucullata*
    27. Leaf blade base truncate, slightly sagittate or hastate, or ± cordate; plants of dry, sandy,
        open woods and thickets. . . . . 57. *Viola sagittata* (in part)
26. Sepal auricles 1–2 mm.
    28. Adaxial leaf surface with silvery strigose patches . . . . 25. *Viola hirsutula*
    28. Adaxial leaf surface without silvery strigose patches.
        29. Petioles densely pubescent.
            30. Leaf blades narrowly ovate to narrowly deltate; plants of wet, rocky shores
                of lakes and streams, meadows . . . . 36. *Viola novae-angliae*
            30. Leaf blades elliptic, ovate, or reniform; plants of sandy pine-oak or pine-oak-
                hickory woods and disturbed ground. . . . . 72. *Viola villosa*
        29. Petioles glabrous or pubescent.
            31. Leaf blades ovate, broadly ovate, or reniform to broadly reniform or
                orbiculate.
                32. Leaf blades sparsely pubescent, rarely glabrous adaxially; lower petal
                    obviously bearded, rarely beardless . . . . 2. *Viola affinis*
                32. Leaf blades glabrous, rarely pubescent; lower petal beardless, rarely
                    lightly bearded. . . . . .34. *Viola missouriensis*
            31. Leaf blades narrowly ovate or narrowly deltate to broadly deltate.
                33. Leaf blades somewhat fleshy, surfaces usually glabrous, grayish green or
                    purplish green abaxially; plants of wet habitats in saturated soil; plants
                    5–15 cm. . . . . .35. *Viola nephrophylla*
                33. Leaf blades not fleshy, surfaces usually pubescent, green abaxially; plants
                    of dry or mesic habitats, not in saturated soil; plants 5–50 cm. . . . . 62. *Viola sororia*
1. Plants caulescent.
    34. Cauline stipules palmately lobed or pinnatifid, equaling leaf blade.
        35. Lateral petals ± equaling or shorter than sepals . . . . .3. *Viola arvensis*
        35. Lateral petals longer than sepals.
            36. Sepal auricles 0.5–2 mm; style head bearded; cleistogamous flowers axillary . . . . 6. *Viola bicolor*
            36. Sepal auricles 2–4 mm; style head beardless; cleistogamous flowers absent . . . . .66. *Viola tricolor*
    34. Cauline stipules unlobed, shorter than leaf blade.
        37. Leaves compound.
            38. Upper 2 and lower 3 petals light golden- to deep lemon-yellow.
                39. Leaf blades ovate, lobes 1–2.5(–5) mm wide; cleistogamous flowers
                    absent; capsules glabrous. . . . . 16. *Viola douglasii*
                39. Leaf blades reniform or ovate to ± orbiculate, lobes 2–10 mm wide;
                    cleistogamous flowers axillary; capsules glabrous or puberulent. . . . . .61. *Viola sheltonii*
            38. Upper 2 petals dark reddish violet, lower 3 lilac, pale yellow, or cream, seldom
                white.
                40. Lower 3 petals pale yellow, cream, or ± white . . . . 23. *Viola hallii*
                40. Lower 3 petals lilac, rarely white.
                    41. Abaxial leaf surface without distinct vein parallel to each margin,
                        margins usually ciliate, surfaces usually puberulent; California,
                        Idaho, Nevada, Oregon, Utah. . . . . .5. *Viola beckwithii*
                    41. Abaxial leaf surface usually with distinct vein parallel to each
                        margin, margins eciliate, surfaces glabrous; Oregon, Washington . . . 67. *Viola trinervata*
        37. Leaves simple.
            42. Plants stoloniferous; stems prostrate, spreading.
                43. Petals lemon-yellow; cauline stipule margins entire or sparingly toothed
                    . . . . .59. *Viola sempervirens*
                43. Petals pale to bluish violet; cauline stipule margins laciniate . . . . .73. *Viola walteri*
            42. Plants not stoloniferous; stems erect, ascending, spreading, decumbent,
                or prostrate (sometimes later reclining to nearly prostrate in *V. adunca*,
                V. howellii).

[44. Shifted to left margin.—Ed.]

44. Petals white or cream adaxially.
    45. Basal leaf blades orbiculate-ovate to deltate, usually shiny, leathery, base cuneate; cauline blade base cuneate. . . . . .15. *Viola cuneata*
    45. Basal leaf blades ovate to reniform or deltate, not shiny or leathery, base cordate, subcordate, rounded, hastate, attenuate (oblique or not), or truncate; cauline blade base cordate to truncate.
        46. Petals white or cream on both surfaces, without yellow patch basally, spur white, 3–6 mm . . . . .63. *Viola striata*
        46. Petals white on adaxial surface, with yellow patch basally, spur white, yellow, or greenish, 1–2.5 mm.
            47. Spurs white, upper 2 petals, sometimes lower 3, usually tinged soft reddish violet, rarely white abaxially . . . . .10. *Viola canadensis*
            47. Spurs yellow or greenish, upper 2 petals, sometimes lower 3, deep reddish violet abaxially. . . . . .38. *Viola ocellata*
44. Petals not white or cream adaxially (sometimes almost white in *V. frank-smithii*; rarely white in *V. adunca* and *V. labradorica*).
    48. Petals blue to gray, violet to pale or soft blue-violet, pale to deep lavender-violet, soft reddish violet, or ± white adaxially.
        49. All petals yellow basally.
            50. Leaf blades broadly reniform to ovate, base cordate; petals soft reddish violet; lowest petal 10–15 mm, spur yellow; Washington . . . . .19. *Viola flettii*
            50. Leaf blades broadly ovate, deltate, or broadly deltate, base cordate to truncate; petals blue to pale violet; lowest petal 5.5–11 mm, spur white to pale violet; Nevada, Utah. . . . . .31. *Viola lithion*
        49. All or just 3 lower petals white basally.
            51. Spurs 10–20 mm; petals beardless; style head beardless . . . . .55. *Viola rostrata*
            51. Spurs 1.6–8 mm; lateral petals sparsely to densely bearded; style head bearded or beardless.
                52. Lateral and upper 2 petals pale purple or almost white adaxially, violet abaxially; Utah . . . . .20. *Viola frank-smithii*
                52. Lateral and upper 2 petals similar color on both surfaces; not limited to Utah.
                    53. Basal leaves absent . . . . .11. *Viola canina*
                    53. Basal leaves present.
                        54. Spurs 2–5 mm, tip straight.
                            55. Sepal margins ciliate or eciliate; style head bearded; seeds light brown. . . . . .26. *Viola howellii*
                            55. Sepal margins eciliate; style head usually beardless, sometimes bearded; seeds dark olive to ± black . . . . 30. *Viola langsdorffii*
                        54. Spurs 3–8 mm, tip straight, curved, or pointed, hooked.
                            56. Sepal auricles enlarged in fruit . . . . .54. *Viola riviniana*
                            56. Sepal auricles not enlarged in fruit.
                                57. Leaf blades usually decurrent on petiole; cauline stipule margins lacerate to laciniate. . . . . 1. *Viola adunca*
                                57. Leaf blades not decurrent on petiole; cauline stipule margins ± entire or laciniate. . . . . .28. *Viola labradorica*
    48. Petals yellow adaxially.
        58. Stems leafless proximally, leafy distally; style head bearded.
            59. Cauline leaf blades widely or narrowly hastate to ovate, unlobed, usually mottled light green adaxially . . . . 24. *Viola hastata*
            59. Cauline leaf blades ovate, reniform, deltate, rhombic, ovate-orbiculate, or reniform-cordate, unlobed or 3–12-lobed, if deltate, not mottled light green adaxially.

60. Petals lemon-yellow on both surfaces; cauline blades unlobed.
    61. Cauline stipules ovate to oblong, margins erose or subserrate, often glandular, apex acute to acuminate; petioles 0.2–2.9 cm; blades ovate to deltate, base cordate to truncate, apex acute; Alberta, British Columbia, Alaska, California, Idaho, Montana, Oregon, Washington . . . . .21.   *Viola glabella*
    61. Cauline stipules ovate, margins entire or coarsely serrate or erose, apex acute; petioles 1–10 cm; blades reniform or ovate to ovate-orbiculate or deltate, base cordate, apex acute to acuminate; c (incl. Wyoming), e North America . . . . 50.   *Viola pubescens*
60. Petals lemon-yellow adaxially, upper 2 and sometimes lateral 2 brownish purple abaxially; cauline blades unlobed or 3–12-lobed.
    62. Stems 1–3; basal leaves 0–2; cauline stipules sometimes ± leaflike; peduncles 2–13 cm; California, Oregon. . . . . 32.   *Viola lobata*
    62. Stems 1(2); basal leaves 0(–2); cauline stipules not leaflike; peduncles 1.5–4 cm; e United States . . . . 68.   *Viola tripartita*
[58. Shifted to left margin.—Ed.]
58. Stems leafy proximally and distally; style head bearded or beardless.
    63. Basal leaves 0; style head bearded.
        64. Petioles glabrous; leaf blades 1.2–2.4 cm, margins entire or with 1–3 crenations on proximal ½, eciliate; Texas . . . . 22.   *Viola guadalupensis*
        64. Petioles usually finely puberulent, sometimes glabrate; leaf blades 1–5.5 cm, margins crenate to serrate, ciliate; California . . . . 45.   *Viola pedunculata*
    63. Basal leaves 1–11; style head bearded or beardless.
        65. Style head beardless . . . . .7.   *Viola biflora*
        65. Style head bearded.
            66. Cleistogamous flowers absent. . . . . .65.   *Viola tomentosa*
            66. Cleistogamous flowers present.
                67. Capsules puberulent.
                    68. Basal blade with prominent whitish veins adaxially; seeds black . . . . 12.   *Viola charlestonensis*
                    68. Basal blade without prominent whitish veins adaxially; seeds light to medium or dark brown or mottled gray and brown.
                        69. Capsules 8–12 mm . . . . 52.   *Viola quercetorum*
                        69. Capsules 3.5–7 mm.
                            70. Cauline blades 2.8–9.6 × 0.3–1.4 cm, length 4–11 times width . . . . .46.   *Viola pinetorum*
                            70. Cauline blades 0.9–5.2 × 0.2–2.9 cm, length 0.8–7.1 times width . . . . .51.   *Viola purpurea*
                67. Capsules glabrous or finely puberulent.
                    71. Capsules ellipsoid to oblong.
                        72. Base of basal and cauline leaf blades cordate; upper 2 petals deep lemon-yellow abaxially. . . . . 40.   *Viola orbiculata*
                        72. Base of basal and cauline leaf blades attenuate to ± truncate or subcordate; upper 2 petals brownish purple abaxially . . . . .47.   *Viola praemorsa*
                    71. Capsules spherical, subglobose, or ovoid.
                        73. Basal leaf base usually truncate, sometimes attenuate . . . . 71.   *Viola vallicola*
                        73. Basal leaf base attenuate (rarely truncate or subcordate in *V. utahensis*).
                            74. Basal blade margins ± coarsely crenate-serrate . . . . .70.   *Viola utahensis*
                            74. Basal blade margins entire or serrulate, sometimes with a few sharp teeth or crenulate.
                                75. Elaiosome not covering funiculus; basal and cauline leaf surfaces glabrous or puberulent on margins or veins . . . . 4.   *Viola bakeri*
                                75. Elaiosome completely covering funiculus; basal and cauline leaf surfaces glabrous or puberulent. . . . . .37.   *Viola nuttallii*

1. **Viola adunca** Smith in A. Rees, Cycl. 37: Viola no. 63. 1817 [E] [F]

Lophion aduncum (Smith) Nieuwland & Lunell; *V. canina* var. *adunca* (Smith) A. Gray

Plants perennial, caulescent, not stoloniferous, 1.8–30(–35) cm. **Stems** 1–5, erect, ascending, or decumbent, sometimes later reclining to nearly prostrate, glabrous or puberulent, on caudex from subligneous rhizome. **Leaves** basal and cauline; basal: 1–4; stipules linear to linear-lanceolate, margins entire or laciniate with gland-tipped projections, apex acute to acuminate; petiole 0.5–13.5 cm, glabrous or puberulent; blade usually ovate or ovate-deltate to ovate-orbiculate, sometimes ± reniform or oblong, 0.5–6.9 × 0.4–5 cm, base cordate, subcordate, truncate, or attenuate, usually decurrent on petiole, margins crenate to crenulate or entire, ciliate or eciliate, apex acute to obtuse, surfaces glabrous or sparsely to densely puberulent; cauline similar to basal except: stipule margins lacerate to laciniate; petiole 0.5–6.5 cm; blade 0.6–5.5 × 0.4–4.7 cm. **Peduncles** 1.7–13.8 cm, glabrous or puberulent. **Flowers:** sepals lanceolate, margins ciliate or eciliate, auricles not enlarged in fruit, 0.5–2 mm; petals light- to deep- to lavender-violet on both surfaces, rarely white, lower 3 usually white basally, dark violet-veined, lateral 2 (and sometimes upper 2) bearded, lowest 7–17(–23) mm, spur purple to violet or white, elongated, 5–7 mm, tip straight or pointed, curved up or lateral; style head sparsely to densely bearded, sometimes beardless; cleistogamous flowers axillary. **Capsules** short-ovoid, 6–11 mm, glabrous. **Seeds** dark brown to olive-black, 1.5–2 mm.

Varieties 2 (2 in the flora): North America.

Viola adunca is polymorphic with over 50 named taxa. G. D. McPherson and J. G. Packer (1974) reported that diploid, triploid, and tetraploid races of *V. adunca* occur in Canada and northwestern United States. They found that diploid and tetraploid chromosome races can be distinguished morphologically based on style beards and on the size of guard cells and pollen grains and recommended taxonomic recognition of both races. In diploid races, the style projections are more or less cylindrical and about one-sixth the width of the style; in tetraploid races they are short-conical or globular, and about one-tenth or less than the width of the style beard.

1. Plants (4–)4.5–30(–35) cm; basal blades 1.3–6.9 × 1.2–5 cm; peduncles 3–13.8 cm; 0–3600 m; widespread in North America . . . . 1a. *Viola adunca* var. *adunca*
1. Plants 1.8–4.5(–6.5) cm; basal blades 0.5–1.7 × 0.4–1.4 cm; peduncles 1.7–5 cm; 2500–3800 m; restricted to Rocky Mountains (Colorado, Montana, Wyoming, British Columbia) .1b. *Viola adunca* var. *bellidifolia*

1a. **Viola adunca** Smith var. **adunca** • Western dog, hooked, early blue, sand, or hookedspur violet, violette à éperon crochu [E] [F]

Viola adunca subsp. *ashtoniae* M. S. Baker; *V. adunca* var. *cascadensis* (M. S. Baker) C. L. Hitchcock; *V. adunca* var. *kirkii* Duran; *V. adunca* subsp. *oxyceras* (S. Watson) Piper; *V. adunca* var. *oxyceras* (S. Watson) Jepson; *V. adunca* subsp. *uncinulata* (Greene) Applegate; *V. adunca* var. *uncinulata* (Greene) C. L. Hitchcock; *V. canina* Linnaeus var. *puberula* S. Watson; *V. cascadensis* M. S. Baker; *V. minima* M. S. Baker; *V. montanensis* Rydberg; *V. oreocallis* Greene; *V. oxyceras* (S. Watson) Greene; *V. sylvestris* Lamarck var. *puberula* E. Sheldon; *V. uncinulata* Greene

Plants erect, decumbent, or prostrate, (4–)4.5–30(–35) cm. **Basal leaves:** blade usually ovate or ovate-deltate, sometimes ± reniform or oblong, 1.3–6.9 × 1.2–5 cm, base cordate, subcordate, truncate, or attenuate, apex acute to obtuse, surfaces glabrous or sparsely to densely puberulent, especially abaxially. **Cauline leaves:** petiole 0.8–6.5 cm; blade 1.1–5.5 × 1–4.7 cm. **Peduncles** 3–13.8 cm, bracteoles alternate or opposite. **Flowers:** sepal margins ciliate or eciliate; lowest petal 9–17(–23) mm; style head bearded or beardless. **Seeds** 1.5–2 mm. $2n = 20, 30, 40$.

Flowering Apr–Aug. Dry to moist meadows, open ground, including lawns, damp banks, openings, meadow edges, rocky areas in coniferous or mixed forests, sandy loam; 0–3600 m; Greenland; Alta., B.C., Man., N.B., Nfld. and Labr., N.W.T., N.S., Nunavut, Ont., Que., Sask., Yukon; Alaska, Ariz., Calif., Colo., Idaho, Iowa, Maine, Mass., Mich., Minn., Mont., Nebr., Nev., N.H., N.Mex., N.Y., N.Dak., Oreg., S.Dak., Utah, Vt., Wash., Wis., Wyo.

Variety *adunca* can be confused with *Viola nephrophylla*, which is acaulescent, and with *V. labradorica*, which lacks decurrent petioles.

H. E. Ballard (1992) noted that eastern specimens of *Viola adunca* were rarely glabrous, but western ones were often glabrous. Dead, elongated stems often persist on caudices.

J. Clausen (1929) noted that although many American botanists had treated *Viola adunca* as closely related to or as a variety of *V. canina*, he thought *V. adunca* should be treated as a subspecies of the European *V. rupestris*.

Variety *adunca* is the food plant for the larvae of the federally listed Behren's silverspot butterfly (*Speyeria zerene behrensii*), Myrtle's silverspot butterfly (*S. z. myrtleae*), and Oregon silverspot butterfly (*S. z. hippolyta*).

Evidence of nectar thievery (presence of holes in the spur) has been observed in populations of var. *adunca* in California and Colorado. Nectar thievery by bumblebees has been reported in *Viola hirta* and *V. riviniana* in Europe (A. J. Beattie 1969b).

Viola cascadensis, from Oregon and adjacent Washington, was described by M. S. Baker (1949b) as having dimorphic growth in which, after chasmogamous flowering, the plant changes from acaulescent to caulescent, forming well-developed, nonpersistent stems with cleistogamous flowers, as well as other differences between it and *V. adunca*. *Viola cascadensis* was treated as a variety of *V. adunca* by C. L. Hitchcock et al. (1955–1969, vol. 3) and was synonymized with *V. adunca* by H. E. Ballard (1992). Study is needed to determine if recognition of *V. cascadensis* is warranted.

1b. Viola adunca Smith var. **bellidifolia** (Greene) H. D. Harrington, Man. Pl. Colorado, 641. 1954 • Violette à feuilles de pâquerette E

Viola bellidifolia Greene, Pittonia 4: 292. 1901

Plants erect, usually appearing small and tufted, 1.8–4.5(–6.5) cm. **Basal leaves:** blade ovate to ovate-orbiculate, 0.5–1.7 × 0.4–1.4 cm, base subcordate, truncate, or attenuate, apex usually obtuse, surfaces usually glabrous. **Cauline leaves:** petiole 0.5–3.8 cm; blade 0.6–1.5 × 0.4–1.4 cm. **Peduncles** 1.7–5 cm, bracteoles usually opposite. **Flowers:** sepal margins usually eciliate; lowest petal 7–13(–14) mm; style head sparsely bearded. **Seeds** 1.5 mm. 2*n* = 20.

Flowering May–Jul. Alpine areas, wet meadows, lake margins; 2500–3800 m; B.C.; Colo., Mont., Wyo.

Variety *bellidifolia* is found in the Rocky Mountains. Although forms transitional with var. *adunca* occur, the diminutive var. *bellidifolia* is quite distinct. V. B. Baird (1942) reported that *Viola bellidifolia* occurs in the Siskiyou Mountains of California; no supporting specimens have been seen.

2. Viola affinis Leconte, Ann. Lyceum Nat. Hist. New York 2: 138. 1826 • Leconte's or sand violet, violette affine E

Viola affinis var. *subarctica* J. Rousseau; *V. crenulata* Greene; *V. sororia* Willdenow subsp. *affinis* (Leconte) R. J. Little; *V. sororia* var. *affinis* (Leconte) L. E. McKinney; *V. subviscosa* Greene; *V. venustula* Greene

Plants perennial, acaulescent, not stoloniferous, 5–15 cm; rhizome slender, becoming thick and fleshy with age. **Leaves** basal, 1–6, ascending to erect; stipules lanceolate, margins entire or fimbriate, apex acute; petiole 2–10 cm, glabrous; blade green abaxially, unlobed, narrowly to broadly ovate or narrowly deltate, 1.5–10 × 1.5–10 cm, not fleshy, base cordate to broadly cordate or almost truncate, margins crenate to serrate, ciliate or eciliate, apex acute to obtuse, surfaces sparsely pubescent adaxially, rarely glabrous. **Peduncles** 3–15 cm, glabrous or pubescent. **Flowers:** sepals lanceolate to ovate, margins ciliate or eciliate, auricles 1–2 mm; petals lavender-violet to dull reddish violet on both surfaces, lower 3 white basally and darker violet-veined, lateral 2 bearded, lowest 10–22 mm, usually obviously bearded, rarely beardless, spur white or same color as petals, gibbous, 2–3 mm; style head beardless; cleistogamous flowers from prostrate to ascending peduncles. **Capsules** often reddish or purplish-flecked or green, ellipsoid, 5–10 mm, glabrous or puberulent. **Seeds** beige, mottled to bronze, 1.5–2.5 mm. 2*n* = 54.

Flowering Apr–Jun. Open or wooded wet areas, meadows, stream banks, thickets, shores of lakes, seasonally dry areas; 100–2000 m; Ont., Que.; Conn., Del., D.C., Ill., Ind., Md., Mass., Mich., Minn., N.H., N.J., N.Y., Ohio, Pa., R.I., Vt., Va., W.Va., Wis.

L. E. McKinney (1992) considered *Viola affinis*, and much of what botanists had called *V. nephrophylla*, to be essentially the same taxon. After studying additional specimens, reviewing literature (H. E. Ballard 1994; A. Haines 2001b), and discussions with others (J. Cayouette, H. E. Ballard, A. Haines, pers. comm.), he chose to maintain these as separate taxa. Reports of *V. affinis* in the Gulf coastal states based on specimens or photographs are usually attributable to *V. missouriensis*.

Viola affinis reportedly hybridizes with *V. hirsutula* (= *V. ×consobrina* House), *V. cucullata* (= *V. ×consocia* House), *V. brittoniana* (= *V. ×davisii* House), *V. sororia* (= *V. ×filicetorum* Greene [as species]), *V. sagittata* var. *sagittata* (= *V. ×hollickii* House), and *V. nephrophylla* (= *V. ×subaffinis* House).

3. Viola arvensis Murray, Prodr. Stirp. Gott., 73. 1770

• Field or wild pansy, European field pansy, violette des champs

Mnemion arvense (Murray) Nieuwland & Kaczmarek; *Viola tricolor* Linnaeus var. *arvensis* (Murray) de Candolle

Plants annual, caulescent, not stoloniferous, 5–35 cm. **Stems** 1–5, usually erect, sometimes prostrate or decumbent, branched, subglabrous or puberulent, clustered on taproot. **Leaves** cauline; stipules palmately lobed, middle lobe oblanceolate, obovate, elliptic, or lanceolate, ± equaling leaf blade, proximal lobes dissected, shorter, margins ciliate, apex acute to obtuse; petiole 0.5–2.3 cm, glabrous or puberulent; blade: proximal ovate to ± oblong, distal narrowly or broadly lanceolate, 0.8–3.4 × 0.3–1.9 cm, base attenuate to ± truncate, margins coarsely crenate-serrate, ciliate or eciliate, apex acute to obtuse, surfaces pubescent abaxially, at least on major veins, glabrous adaxially. **Peduncles** 2–8 cm, glabrous or pubescent. **Flowers:** sepals lanceolate, margins ciliate or eciliate, auricles 2–4 mm; petals white to pale yellow on both surfaces, upper 2 ± violet, lower 3 with yellow basal area, often violet-veined, lateral 2 bearded, ± equaling or shorter than sepals, lowest with dark yellow area basally, 7–15 mm, spur blue-violet to purple, elongated, 3–5 mm; style head bearded; cleistogamous flowers absent. **Capsules** ± spherical, 5–9 mm, glabrous. **Seeds** brown, 1.5–1.9 mm. $2n = 34$.

Flowering May–Jul. Abandoned fields, roadsides, lawns; 0–3000 m; introduced; Greenland; St. Pierre and Miquelon; Alta., B.C., Man., N.B., Nfld. and Labr., N.S., Ont., P.E.I., Que., Sask.; Calif., Conn., Del., D.C., Ga., Idaho, Ill., Ind., Kans., Ky., La., Maine, Md., Mass., Mich., Minn., Miss., Mo., Mont., Nebr., N.H., N.J., N.Y., N.C., Ohio, Oreg., Pa., R.I., S.C., Tenn., Utah, Vt., Va., Wash., W.Va., Wis.; Europe; Asia (Siberia); Africa.

A. R. Clapham et al. (1987) noted that *Viola arvensis* is pollinated by insects and is often selfed; T. Marcussen and T. Karlsson (2010) stated that *V. arvensis* regularly self-pollinates. Roots of *V. arvensis* have the odor of wintergreen when crushed (A. E. Radford et al. 1968; W. J. Hayden and J. Clough 1990). *Viola arvensis* is not vegetatively distinguishable from *V. tricolor* var. *tricolor*.

G. Halliday (pers. comm.) reported that *Viola arvensis* occurs in southwestern Greenland, where it is introduced, and is ephemeral on St. Pierre and Miquelon.

4. Viola bakeri Greene, Pittonia 3: 307. 1898

• Baker's violet [E]

Viola bakeri subsp. *shastensis* M. S. Baker; *V. nuttallii* Pursh var. *bakeri* (Greene) C. L. Hitchcock

Plants perennial, caulescent, not stoloniferous, 3–30 cm. **Stems** 1–4, usually erect, sometimes prostrate or decumbent, leafy proximally and distally, usually puberulent, from usually vertical, subligneous rhizome. **Leaves** basal and cauline; basal: 1–4; stipules adnate to petiole, forming 2 linear-lanceolate wings, margins entire, apex of each wing free, acute to acuminate; petiole 1–15.4 cm, glabrous or puberulent; blade lanceolate, oblanceolate, or elliptic, rarely ovate, 1.8–8.8 × 0.7–3.9 cm, thin, base attenuate, often oblique, margins usually entire, sometimes with a few sharp teeth or crenulate, ciliate, apex acute to obtuse, mucronulate, surfaces glabrous or puberulent on margins or veins; cauline similar to basal except: stipules ovate to lanceolate, margins entire or lacerate, sometimes with glandular projections, apex with 2 or 3 projections; petiole 1.5–7.5 cm; blade 1.9–6.7 × 0.5–1.6 cm. **Peduncles** 1.5–11.6 cm, glabrous or pubescent. **Flowers:** sepals lanceolate, margins eciliate, auricles 0.5–1 mm; petals deep lemon-yellow adaxially, upper 2 often brownish purple abaxially, lower 3 dark brown- to brownish purple-veined, lateral 2 sparsely bearded, lowest 6–14 mm, spur yellow, gibbous, 1–2 mm; style head bearded; cleistogamous flowers axillary. **Capsules** spherical to ovoid, 5–10 mm, usually glabrous, rarely finely puberulent. **Seeds** light to medium brown or dark red-brown, 2.6–3.1 mm, elaiosome not covering funiculus. $2n = 48$.

Flowering May–Jul. Wet and dry places in openings of coniferous forests; 900–3800 m; Calif., Nev., Oreg., Wash.

M. S. Baker (1957) wrote that *Viola bakeri* flowers were often without a brownish tinge on the back. Subsequent collections have shown that the upper two petals of *V. bakeri* are often brownish purple abaxially and thus this characteristic cannot be used to distinguish *V. bakeri* from *V. praemorsa*, as previously done (P. A. Munz 1959; C. L. Hitchcock et al. 1955–1969, vol. 3).

Greene wrote in his description of *Viola bakeri* that the whole plant was glabrous; he did not mention leaf margins. M. S. Baker (1957) examined the type specimen at UC and apparently was the first to document that its leaf margins are entire. He also noted that under magnification the leaves are ciliate and more or less puberulent. D. M. Fabijan et al. (1987) stated that margins were always entire and ciliate and more or less pubescent throughout. Some collections of *V. bakeri* from California have leaves with a sharp point or two on the margin, or crenulations on some portion of the basal or cauline blade margins.

5. **Viola beckwithii** Torrey in E. M. Durand, Pl. Pratten. Calif., 82. 1855 • Beckwith's or Great Basin violet [E]

Viola beckwithii var. *cachensis* C. P. Smith; *V. beckwithii* subsp. *glabrata* M. S. Baker; *V. bonnevillensis* Cottam

Plants perennial, caulescent, not stoloniferous, 2–22 cm. **Stems** 1–3, decumbent, ascending, or erect, ca. ½ subterranean, glabrous or usually puberulent, on single, short, vertical, deep-seated caudex. **Leaves** basal and cauline; basal: 1–6 per caudex, palmately compound, ± 2- or 3-ternate, leaflets 3; stipules adnate to petiole, forming 2 linear-lanceolate wings, margins entire, apex of each wing free, acute; petiole 2–10.5 cm, usually puberulent; blade ovate to deltate, 2.4–5 × 3.5–4.5 cm, base tapered, ultimate leaflets dissected into oblong, elliptic, lanceolate, or oblanceolate lobes 1–7 mm wide, lobe margins entire, usually ciliate, apex acute to obtuse, mucronulate, surfaces usually puberulent, seldom glabrous, abaxial surface without prominent vein parallel to each margin; cauline similar to basal except: stipules linear, unlobed, apex acuminate; petiole 2–5.7 cm, usually puberulent, rarely glabrous; blade 1–2.7 × 1.5–3 cm. **Peduncles** 1.5–10.6 (–15.7) cm, usually puberulent, seldom glabrous. **Flowers:** sepals lanceolate, margins eciliate, auricles 0.1–1 mm; petals dark reddish violet on both surfaces, lower 3 usually lilac, rarely white or whitish, lateral 2 bearded, with yellow patch basally, dark reddish violet-veined, lowest 10–22 mm, with yellow patch, dark reddish violet-veined, spur whitish or yellowish, tinged purple, gibbous, 0.5–2 mm; style head bearded; cleistogamous flowers absent. **Capsules** oblong-ovoid, 7–12 mm, glabrous. **Seeds** brown, 3–4 mm. $2n = 24$.

Flowering Mar–May. Dry or moist places, among shrubs or beneath pines; 900–2700 m; Calif., Idaho, Nev., Oreg., Utah.

In some populations of *Viola beckwithii*, the three lower petals are white with a yellow area proximally (V. B. Baird 1942). Leaves have been described as palmately biternate or triternate (L. Abrams and R. S. Ferris 1923–1960, vol. 3), ternately decompound into linear segments (C. L. Hitchcock et al. 1955–1969, vol. 3), palmately three-parted then bipinnately parted into ultimate linear or spatulate segments (P. A. Munz 1959), and palmately about three times three-parted into linear or spatulate-linear segments (W. L. Jepson 1951). Some populations in northern California are nearly or completely glabrous, which M. S. Baker recognized as var. *glabrata*.

Viola beckwithii is reported to hybridize with *V. utahensis* (G. Davidse 1976). Observed pollinators of *V. beckwithii* in Utah include *Apis mellifera* Linnaeus and *Anthophora ursina* Cresson (Davidse).

When Cottam described *Viola bonnevillensis*, he suggested that it could be a hybrid between *V. beckwithii* and *V. utahensis*, and G. Davidse (1976) concurred. The type specimen of *V. bonnevillensis* (*Cottam 7067*, UT) was examined by R. J. Little. Because no similar forms are known to have been collected since 1939, it is presumed that this taxon is a hybrid.

6. **Viola bicolor** Pursh, Fl. Amer. Sept. 1: 175. 1813 • Wild or field pansy, violette de Rafinesque [E] [F]

Viola kitaibeliana Roemer & Schultes var. *rafinesquei* (Greene) Fernald; *V. rafinesquei* Greene

Plants annual, caulescent, not stoloniferous, 3–25 cm. **Stems** 1–5, prostrate, decumbent, or erect, branched, glabrous or hairy on stem angles only, clustered on taproot. **Leaves** cauline; stipules palmately lobed, middle lobe not leaflike, oblanceolate to spatulate, margins entire, ciliate or eciliate, apex acute to obtuse; petiole 0.5–2 cm, glabrous or sparsely pubescent; proximal blades ± orbiculate, distal blades spatulate to broadly oblanceolate, 0.5–3 × 0.3–1 cm, base attenuate, margins entire or crenate-serrate especially toward apex, eciliate, apex rounded to obtuse, surfaces glabrous or sparsely pubescent. **Peduncles** 1–4.5 cm, glabrous or pubescent. **Flowers:** sepals ovate to lanceolate, margins ciliate or eciliate, auricles 0.5–2 mm; petals white or cream to pale bluish violet on both surfaces, dark purple-veined, lateral 2 longer than sepals, bearded, lowest 8–10 mm, spur white to blue-violet, gibbous, 1–1.5 mm, shorter than or equaling sepal auricles; style head bearded; cleistogamous flowers axillary. **Capsules** ellipsoid to oblong, 4–7 mm, glabrous. **Seeds** beige to bronze, 0.3–1.5 mm. $2n = 34$.

Flowering Mar–May. Prairies, open woodlands, fields, pastures, roadsides, lawns, waste ground; 0–3000 m; Ont., Sask.; Ala., Ariz., Ark., Colo., Conn., Del., D.C., Fla., Ga., Idaho, Ill., Ind., Iowa, Kans., Ky., La., Md., Mass., Miss., Mo., Nebr., N.J., N.Mex., N.Y., N.C., Ohio, Okla., Pa., R.I., S.C., S.Dak., Tenn., Tex., Va., W.Va.

Viola bicolor is the only pansy native to North America (V. B. Baird 1942; J. Clausen et al. 1964; A. E. Radford et al. 1968) and is the only annual *Viola* species that produces cleistogamous flowers (Baird; A. Gershoy 1934). Roots of *V. bicolor* have the odor of wintergreen when crushed (W. J. Hayden and J. Clough 1990).

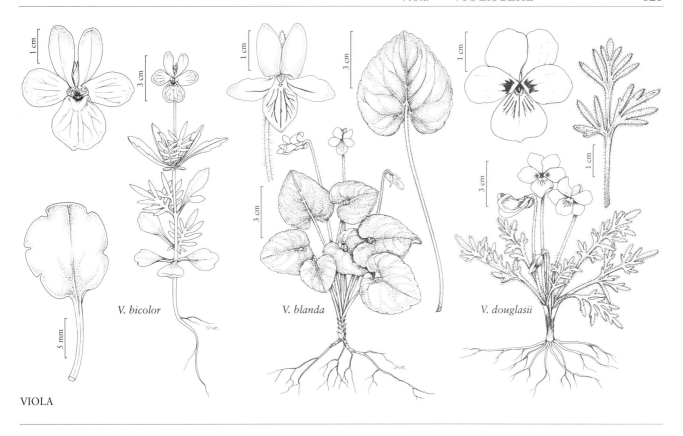

V. bicolor   V. blanda   V. douglasii

VIOLA

---

7. Viola biflora Linnaeus, Sp. Pl. 2: 936. 1753

Chrysion biflorum (Linnaeus) Spach

Plants perennial, caulescent, not stoloniferous, 3–20(–25) cm. **Stems** 1–3+, ascending or erect, leafy proximally and distally, glabrous, on caudex from fleshy rhizome. **Leaves** basal and cauline; basal: 2 or 3(4); stipules ovate-lanceolate, ovate, or oblong, margins entire, apex ± acute; petiole 1.5–15 cm, glabrous; blade broadly reniform to orbiculate, 0.5–4.6 × 0.9–6.4 cm, base cordate, margins crenate to crenate-serrate, ciliate, apex obtuse, rounded, or truncate, rarely with terminal point, abaxial surface sparsely puberulent on veins, adaxial surface glabrate to ± densely puberulent; cauline similar to basal except: stipules lanceolate, ovate, or oblong, margins entire to erose, apex acute to obtuse; petiole 0.3–7(–10) cm; blade sometimes ovate, 0.8–3.7 × 1–4.8 cm. **Peduncles** 2–9 cm, usually glabrous. **Flowers:** sepals with or without purple stripe on both sides of midvein, narrowly lanceolate to oblong, margins ciliate or eciliate, auricles 0.5–1 mm; petals deep lemon-yellow on both surfaces, lower 3 and often upper 2 brownish purple-veined, lateral 2 beardless, lowest 6–15 mm, spur yellow to yellowish green, gibbous, 2–2.5 mm; style head beardless; cleistogamous flowers axillary.

Capsules oblong-ovoid, 3–6 mm, glabrous or sparsely pubescent. **Seeds** purple, sometimes streaked with light and dark brown, 1.5–2.5 mm.

Varieties 5 (2 in the flora): nw North America, Europe, Asia, w Pacific Islands.

Viola biflora may be the most widely distributed species of the genus in the Northern Hemisphere. Although occurring most often in mountainous areas at high elevations, it is known from Alaska near the coast at elevations of approximately 45 m (PNW Herbaria Portal 2010) and has been reported from lowland meadows in Kamchatka (V. B. Baird 1942), and is occasionally found near sea level in exposed rocky habitats on the west coast of the Queen Charlotte Islands (R. A. Calder and R. L. Taylor 1968).

Sometimes described as high latitude, circumpolar, *Viola biflora* also occurs in mid latitudes north of the equator. It is not truly circumpolar; it does not occur in eastern Canada or in Greenland. The deeply cleft style head sets *V. biflora* apart from all other species in North America. V. B. Baird (1942) suggested that the occasional presence of two cleistogamous flowers in the axil of the same leaf may account for the name given it by Linnaeus.

H. N. Ridley (1930) said that in Europe, deer (*Cervus dama* Linnaeus and *Rangifer tarandus* Linnaeus) may play a role in the dispersal of *Viola biflora*; seeds have been recovered from their droppings.

Nonflowering *Viola biflora* can be confused with *V. renifolia* where their ranges overlap in Yukon, Alaska, and Colorado.

1. Lowest petal 6–10 mm; seeds 1.5–2 mm; sepals usually without purple stripe; Yukon, Alaska, Colorado. . . . . 7a. *Viola biflora* var. *biflora*
1. Lowest petal 11–15 mm; seeds 2.5 mm; sepals usually with purple stripe; Queen Charlotte Islands, Vancouver Island, British Columbia . . . . .7b. *Viola biflora* var. *carlottae*

7a. **Viola biflora** Linnaeus var. **biflora** • Arctic yellow or northern yellow or twinflower violet, violette à deux fleurs

Plants 3–20 cm. **Basal leaf blades** 0.5–3[–3.5] × 0.9–3.5[–5] cm. **Flowers:** sepals usually without purple stripe; lowest petal 6–10 mm. **Seeds** 1.5–2 mm. $2n = 12$.

Flowering Apr–Jul. Montane and arctic regions, tundra, mesic alpine meadows, scree slopes, grassy places in shade of alders and willows, lake margins, granite outcrops, moist coniferous forests, along streams and falls in wet moss, alluvial soil, sand; 40–3900 m; Yukon; Alaska, Colo.; c, s Europe; Asia (c, n China, ne India, Indonesia [n Sumatra], Japan, Malaysia, Mongolia, North Korea, Pakistan, Russia, Siberia); Pacific Islands (Philippines, Taiwan).

In North America, var. *biflora* occurs in Alaska (45–3300 m; PNW Herbaria Portal 2010), Yukon (1200–1500 m), and nine counties in central Colorado (1950–3850 m).

7b. **Viola biflora** Linnaeus var. **carlottae** (Calder & Roy L. Taylor) B. Boivin, Naturaliste Canad. 93: 437. 1966 • Queen Charlotte twinflower violet, violette des îles de la Reine-Charlotte E

Viola biflora subsp. *carlottae* Calder & Roy L. Taylor, Canad. J. Bot. 43: 1395. 1965

Plants 5–20(–25) cm. **Basal leaf blades** 1.7–4.6 × 2–6.4 cm. **Flowers:** sepals usually with purple stripe; lowest petal 11–15 mm. **Seeds** 2.5 mm. $2n = 12$.

Flowering Jun–Jul. Moist cliff faces, talus and rock outcrops, meadows and heathlands, rocky-grassy slopes, along streams, montane to alpine; 300–1300 m; B.C.

Variety *carlottae* is known from the Queen Charlotte Islands and the Brooks Peninsula of Vancouver Island, British Columbia. It is reported to hybridize with *Viola glabella* (G. W. Douglas et al. 1998–2002, vol. 5).

One of the characteristics used by Calder and Taylor to distinguish var. *carlottae* was the presence of a prominent purple stripe on the sepals. V. B. Baird (1942) reported that sepals of var. *biflora* plants in Colorado occasionally have one purple line down the center. Purple midveins are also reported on sepals of var. *biflora* in Taiwan (Hsieh C. F. 1977). Some herbarium specimens at UC identified as var. *carlottae* from the Queen Charlotte Islands do not have a purple midvein or stripe on the sepals. The presence of a purple midvein or stripe on the sepals should not by itself be used to make a determination of var. *carlottae* in areas other than the Queen Charlotte Islands and Vancouver Island.

Floras that discuss the capsules of *Viola biflora* report that the surface is glabrous. One exception is G. W. Douglas et al. (1998–2002, vol. 5), where the capsules of var. *carlottae* are described as sparsely short-hairy.

8. **Viola blanda** Willdenow, Hort. Berol. 1(2): plate 24. 1804, name proposed for conservation • Sweet white violet, violette agréable E F

Viola blanda var. *palustriformis* A. Gray; *V. incognita* Brainerd; *V. incognita* var. *forbesii* Brainerd; *V. leconteana* G. Don

Plants perennial, acaulescent, stoloniferous, 3–20 cm; stolons pale, often rooting and leafy at nodes; rhizome short, slender, fleshy. **Leaves** basal, 2–9, prostrate to ascending; stipules linear-lanceolate, margins entire, apex acute; petiole 2–11 cm, usually sparsely pubescent; blade unlobed, reniform to ovate, 2–4 × 2–4 cm, base cordate, lobes often overlapping, margins serrate, ciliate or eciliate, apex rounded or acute to acuminate, surfaces sometimes glabrous, usually sparsely pubescent adaxially. **Peduncles** 3–11 cm, glabrous or pubescent. **Flowers:** sepals lanceolate to ovate, margins mostly eciliate, auricles 1–2 mm; petals white on both surfaces, lower 3 purple-veined, lateral 2 usually beardless, lowest 8–10 mm, spur white, gibbous, 1–2 mm; style head beardless; cleistogamous flowers axillary. **Capsules** ovoid to ellipsoid, 4–6 mm, glabrous. **Seeds** beige to bronze, 1.5–2 mm. $2n = 44, 48$.

Flowering Apr–Jun. Rich woods; 30–2000 m; Alta., Man., N.B., Nfld. and Labr., N.S., Ont., P.E.I., Que., Sask.; Ala., Conn., Del., D.C., Ga., Ill., Ind., Iowa, Ky., Maine, Md., Mass., Mich., Minn., N.H., N.J., N.Y., N.C., N.Dak., Ohio, Pa., R.I., S.C., Tenn., Vt., Va., W.Va., Wis.

Viola blanda occurs in small colonies; individual plants are interconnected by stolons.

Whether to recognize *Viola incognita* at any taxonomic level is currently unresolved. It is said to have pubescent leaf blades, greenish peduncles, nontwisted lateral petals, and a preference for moister habitats. Most of these characters fall within the range of variation observed in *V. blanda*.

N. H. Russell (1965) noted that *Viola incognita* is principally found in glaciated areas whereas *V. blanda* is found in nonglaciated areas. Á. Löve and D. Löve (1982b) and J. M. Canne (1987) reported a chromosome count of $2n = 44$; J. Clausen (1929) and A. Gershoy (1934) reported $2n = 48$. V. B. Baird (1942) reported that *V. blanda* (and *V. incognita*) have fragrant flowers.

9. **Viola brittoniana** Pollard, Bot. Gaz. 26: 332. 1898

• Coast or coastal violet [E]

Viola atlantica Britton, Bull. Torrey Bot. Club 24: 92. 1897, not Pomel 1874; *V. baxteri* House; *V. brittoniana* var. *pectinata* (E. P. Bicknell) Alexander; *V. pectinata* E. P. Bicknell; *V. pedatifida* G. Don subsp. *brittoniana* (Pollard) L. E. McKinney; *V. pedatifida* var. *brittoniana* (Pollard) R. J. Little & L. E. McKinney

Plants perennial, acaulescent, not stoloniferous, 5–30 cm; rhizome thick, fleshy. **Leaves** basal, 5–9, ascending to erect, 5–9-lobed; stipules linear-lanceolate, margins entire, apex obtuse; petiole 3–16 cm, usually glabrous; mid-season blades incised or lobed throughout, earliest blades lobed (plants homophyllous), similar to mid-season blades, middle and lateral blade lobes differ in width and/or shape, middle lobes lanceolate or spatulate to narrowly obovate, lateral lobes lanceolate or spatulate to falcate (each sometimes with deltate to falcate appendages or teeth along margins), 1–7 × 2–8 cm, base truncate to cordate, margins entire, ciliate, apex acute to obtuse, surfaces usually glabrous, rarely with a few strigose hairs concentrated on veins. **Peduncles** 5–18 cm, usually glabrous. **Flowers:** sepals lanceolate to ovate, margins ciliate or eciliate, auricles 2–3 mm; petals light to soft reddish violet on both surfaces, lower 3 white basally, dark violet-veined, lateral 2 bearded, lowest 10–25 mm, sometimes bearded, spur same color as petals, gibbous, 2–3 mm; style head beardless; cleistogamous flowers on ascending to erect peduncles. **Capsules** ellipsoid, 10–15 mm, glabrous. **Seeds** beige, mottled to bronze, 1.5–2.5 mm. $2n = 54$.

Flowering Apr–Jun. Alluvial woods, mesic forests on slopes near streams, wet fields, salt meadows; 0–100 m; Conn., Del., Maine, Md., Mass., N.J., N.Y., N.C., Pa., S.C., Va.

Britton considered *Viola brittoniana* similar to *V. pedatifida*, apparently making him the first to recognize their affinity. Both are homophyllous and similar in other characters (L. E. McKinney 1992).

McKinney recognized *V. brittoniana* as a variety of *V. pedatifida*. Additional evidence (N. L. Gil-Ad 1997; A. Haines 2011) suggests that these taxa are sufficiently distinct to continue recognition as separate species.

Viola pectinata has sharply dentate leaves and is closely related to *V. brittoniana*, usually occurring with it. N. L. Gil-Ad (1997) made a convincing argument to recognize it as a form; A. Haines (2011) recognized it as a species. Others have considered it to be a sporadic form that may be of hybrid origin, or, as N. H. Russell (1965) suggested, is due to genetic dimorphism. Additional studies are needed to determine the status of *V. pectinata*.

A. Haines (2011) reported that the distribution of *Viola brittoniana* in Maine is actually based on the hybrid *V.* ×*insolita* House (*V. brittoniana* × *V. sororia*).

Viola brittoniana reportedly hybridizes with *V. cucullata* (= *V.* ×*notabilis* E. P. Bicknell), *V. sagittata* var. *sagittata* (= *V.* ×*mulfordiae* Pollard), and *V. sororia* (= *V.* ×*insolita*).

10. **Viola canadensis** Linnaeus, Sp. Pl. 2: 936. 1753 [E]

Plants perennial, caulescent, not stoloniferous, 3–46(–60) cm, with branching rhizomes forming colonies or not. **Stems** 1–3(4), usually erect to ascending, glabrous or puberulent, from fleshy or subligneous rhizome. **Leaves** basal and cauline; basal: 1–5; stipules ± oblong, ovate, or lanceolate, margins entire, sometimes glandular, apex acuminate to cuspidate; petiole 1.1–23 cm, glabrous or puberulent; blade ovate to broadly ovate or ovate-reniform, 0.7–12.4 × 0.9–11.1(–12.3) cm, base cordate, subcordate, or truncate, margins crenate, crenulate, or serrulate, ciliate (sometimes only on proximal ½) or eciliate, apex acute to acuminate, surfaces glabrous or puberulent (often only on veins); cauline similar to basal except: stipules also deltate, margins also erose or laciniate, apex acute, long-acuminate to cuspidate, or ± truncate, occasionally 2- or 3-fid; petiole 0.1–6.9(–15.2) cm; blade ovate to deltate, 1.2–7.7 × 0.8–7.8 cm, base cordate to truncate, margins crenate or crenulate to ± serrulate. **Peduncles** 1–6.1 cm, glabrate to puberulent, sometimes glabrous below bracteoles. **Flowers:** sepals lanceolate, margins usually eciliate, auricles 0.5–1.3 mm; petals white adaxially, upper 2 and lower 3 tinged soft reddish violet abaxially, rarely white on both surfaces, all petals usually with yellow patch basally, lower 3 usually purple-veined, lateral 2 bearded, lowest 5.5–20 mm, spur white, gibbous, 1–2 mm; style head bearded; cleistogamous flowers axillary or absent. **Capsules** ovoid to ellipsoid, 2.5–10 mm, sometimes muriculate, glabrous or puberulent. **Seeds** brown to dark brown or purplish black, 1.5–2.5 mm.

Varieties 3 (3 in the flora): North America.

1. Rhizomes branched. . . . .
. . . . .10b. *Viola canadensis* var. *rugulosa*
1. Rhizomes not branched.
　2. Plants 11–46 cm; basal leaf blades 2.4–9.2 cm
　　wide. . . . . 10a. *Viola canadensis* var. *canadensis*
　2. Plants 3–14(–18) cm; basal leaf blades 0.9–3.8
　　cm wide. . . . . 10c. *Viola canadensis* var. *scopulorum*

10a. Viola canadensis Linnaeus var. **canadensis**
　　• Canada or Canada white or tall white violet,
　violette du Canada ☐E

Viola canadensis var. *pubens*
Farwell; *V. canadensis* var.
rydbergii (Greene) House;
V. canadensis var. *scariosa* Porter;
V. geminiflora Greene;
V. muriculata Greene;
V. neomexicana Greene;
V. rydbergii Greene

Plants solitary, 11–46 cm. **Rhizomes** not branched. **Stems** 1 or 2(–4). **Leaves:** basal: petiole 1.2–20 cm; blade ovate to ovate-reniform, 1.4–8 × 2.4–9.2 cm, base cordate, margins crenate; cauline: stipules oblong, ovate, or lanceolate, margins usually entire, apex acute, long-acuminate to cuspidate, or ± truncate; petiole 0.1–6.1(–15.2) cm; blade ovate to deltate, 2–7.1 × 0.9–6.2 cm, base cordate or truncate, margins crenate to crenulate, ciliate or eciliate. **Peduncles** 1–2.6(–6.1) cm, sometimes glabrous below bracteoles. **Flowers:** lowest petal 8–15 mm. **Capsules** 4–8.5 mm, sometimes minutely muricate. $2n = 24$.

Flowering Apr–Oct. Riparian, coniferous or aspen forests, deciduous and mixed forests, cove hardwoods, northern hardwood forests, moist, shaded slopes, sandy, rich, or rocky soil, talus slopes, road cuts; 50–3600 m; N.B., N.W.T., N.S., Ont., Que.; Ala., Ariz., Ark., Colo., Conn., Ga., Idaho, Ill., Ind., Ky., Maine, Md., Mass., Mich., N.H., N.J., N.Y., N.C., Ohio, Pa., R.I., S.C., Tenn., Utah, Vt., Va., W.Va., Wis., Wyo.

H. J. Scoggan (1978–1979) stated that reports of var. *canadensis* from Newfoundland require confirmation.

10b. Viola canadensis Linnaeus var. **rugulosa** (Greene) C. L. Hitchcock in C. L. Hitchcock et al., Vasc. Pl. Pacif. N.W. 3: 442. 1961 • Tall white or rugose violet, violette de l'Ouest ☐E

Viola rugulosa Greene, Pittonia 5: 26. 1902; *Lophion rugulosum* (Greene) Lunell; *V. discurrens* Greene

Plants forming colonies, 10–40 (–60) cm. **Rhizomes** branched. **Stems** 1–3(4). **Leaves:** basal: petiole 4.1–23 cm; blade broadly ovate to ovate-reniform, 1.9–12.4 × 2.1–11.1(–12.3) cm, base cordate, margins crenate, sometimes serrulate proximally; cauline: stipules lanceolate to deltate, margins entire to laciniate, apex acuminate; petiole 0.2–6.9(–15.2) cm; blade ovate, 2.9–7.7 × 1.8–7.8 cm, base cordate to ± truncate, margins crenate to ± serrulate, ciliate. **Peduncles** 1.2–5 cm. **Flowers:** lowest petal 10–20 mm; cleistogamous flowers sometimes absent. **Capsules** 6–10 mm, sometimes muriculate. $2n = 24$.

Flowering May–Aug. Shady areas, woodlands, forest edges, thickets, riparian corridors; 300–2600 m; Alta., B.C., Man., N.W.T., Ont., Sask., Yukon; Alaska, Ariz., Colo., Idaho, Ill., Iowa, Minn., Mont., Nebr., N.Mex., N.Dak., Oreg., S.Dak., Utah, Wash., Wis., Wyo.

Viola discurrens is cited here in synonymy because in his original description, based on a plant from Tennessee, Greene reported that it forms somewhat extensive colonies by a system of loosely connected, slender, horizontal whitish rootstocks. Subsequent authors have referred to these structures as stoloniferous rhizomes, superficial rhizomes, runners, or stolons. Greene based his description of *V. rugulosa* on a plant collected in Minnesota stating that it had a thick, suberect rootstock, but did not say that it had stolons, whitish rootstalks, or stoloniferous rhizomes or that it formed colonies. It is unclear why E. Brainerd (1921) dismissed *V. discurrens* as an "unsatisfactory segregate" and instead assumed that colony-forming plants interconnected by subterranean, branching rhizomes should be termed *V. rugulosa* instead of *V. discurrens*. Most authors have followed Brainerd. W. J. Cody (2000) applied the name *V. canadensis* var. *rydbergii* (Greene) House to rhizomatous *V. canadensis* plants in Yukon; when Greene described *V. rydbergii*, he did not mention rhizomes as he did for *V. discurrens*.

V. B. Baird (1942) stated that *V. rugulosa* lacks cleistogamous flowers; R. E. Brooks and R. L. McGregor (1986) noted that cleistogamous flowers are borne for a short time in axils of distal leaves.

10c. Viola canadensis Linnaeus var. **scopulorum**
A. Gray, Bot. Gaz. 11: 291. 1886 E

Viola canadensis subsp.
scopulorum (A. Gray) House;
V. scopulorum (A. Gray) Greene

Plants solitary, 3–14(–18) cm.
Rhizomes not branched. **Stems**
1–3(4). **Leaves:** basal: petiole
1.1–13.8 cm; blade ovate,
0.7–3.1 × 0.9–3.8 cm, base
cordate to subcordate, margins
crenate, irregularly crenulate, or serrulate; cauline:
stipules oblong to broadly lanceolate, margins entire
to erose, apex acute to long-acuminate, occasionally 2-
or 3-fid or ± truncate; petiole 0.4–6 cm; blade ovate,
1.2–2.8 × 0.8–2.7 cm, base cordate to ± truncate, margins
crenulate to serrulate, ciliate or eciliate. **Peduncles**
1.2–2.9(–4.2) cm, sometimes glabrous below bracteoles.
Flowers: lowest petal 5.5–13 mm. **Capsules** 2.5–5 mm.
2n = 24.

Flowering May–Jul. Moist soil, shaded forests,
stream banks, hillsides; 1500–3500 m; Colo., N.Mex.,
Utah.

E. Brainerd (1921) wrote that after growing a plant of
Viola scopulorum for three years received from the type
locality, he was convinced of its right to specific status,
noting the length of the stem and the size of the capsule,
leaf, and seed were but half that of *V. canadensis.*
V. B. Baird (1942) stated that *V. scopulorum* produces
cleistogamous flowers throughout the growing season.

11. Viola canina Linnaeus, Sp. Pl. 2: 935. 1753
• Heath dog-violet I

Viola canina var. *montana*
(Linnaeus) Fries; *V. montana*
Linnaeus; *V. nemoralis* Kützing

Plants perennial, caulescent,
not stoloniferous, 2–40 cm.
Stems 1–5, ascending to erect,
glabrous, on caudex from
subligneous rhizome. **Leaves**
cauline; stipules linear to
lanceolate, margins ± lacerate to subentire, points often
gland-tipped, apex acute; petiole 0.3–2.5 cm, glabrous;
blade ovate to narrowly ovate, 1.2–5.2 × 1–2.7 cm, base
cordate or deeply cordate to truncate, margins crenate,
eciliate, apex acute to obtuse, abaxial surface glabrous
or sparsely pubescent on veins, adaxial surface glabrous
or sparsely pubescent. **Peduncles** 1–10 cm, glabrous
or sparsely puberulent. **Flowers:** sepals lanceolate,
margins eciliate, auricles 2–3 mm; petals usually blue-
to gray-violet on both surfaces, rarely white, white
basally, lower 3 dark violet-veined, lateral 2 bearded,
lowest 15–25 mm, spur light yellow to light green or

white, gibbous to usually elongated, 3–5 mm; style head
beardless; cleistogamous flowers axillary. **Capsules**
ovoid, 7.5–9 mm, glabrous. **Seeds** light to dark brown;
1.7–2.1 mm. **2n** = 40.

Flowering May–Jul. Sunny to shady, dry to slightly
moist ground, heath lands, grazed or mown grasslands,
dunes, rock ledges, open woodlands, disturbed areas
(roadsides, railway banks, gravel pits), littoral vegetation
of lakes and streams; 20–300 m; introduced; Greenland;
Europe; Asia.

Viola canina was possibly introduced in Greenland
(T. Marcussen and T. Karlsson et al. 2010).

Hybrids of *Viola canina* are known with the
European species *V. mirabilis* Linnaeus, *V. pumila*
Chaix, *V. reichenbachiana* Jordan ex Boreau,
V. riviniana Reichenbach, *V. rupestris* F. W. Schmidt
(both subspecies), *V. stagnina* Kitaibel, and *V. uliginosa*
Besser.

12. Viola charlestonensis M. S. Baker & J. C. Clausen,
Madroño 8: 58. 1945 • Charleston Mountain or
Charleston violet E

Viola purpurea Kellogg var.
charlestonensis (M. S. Baker &
J. C. Clausen) S. L. Welsh &
Reveal

Plants perennial, caulescent, not
stoloniferous, 6–15 cm. **Stems**
1–3, prostrate, decumbent,
or erect, leafy proximally and
distally, ½–⅔ subterranean,
glabrous or puberulent, on caudex from usually
vertical, subligneous rhizome. **Leaves** basal and cauline;
basal: 1–3; stipules adnate to petiole, forming 2 linear-
lanceolate wings, margins entire or sparingly lacerate,
apex of each wing free, acuminate; petiole 3.5–13.5 cm,
densely short-puberulent; blade purplish abaxially (often
dark purple-veined), grayish adaxially with prominent
whitish veins (from dense hairs), usually orbiculate
to broadly ovate, sometimes reniform, thick, 1–3.5 ×
1.1–3.3 cm, base attenuate or truncate, margins entire,
ciliate mostly on proximal half of blade, apex acute to
obtuse, mucronulate, surfaces densely short-puberulent;
cauline similar to basal except: stipules deltate to
lanceolate, apex acute; petiole 1.9–3.4 cm; blade ovate
or elliptic to deltate, 0.7–3.1 × 0.6–2.2 cm, length
0.7–1.9 times width, base usually attenuate, sometimes
subcordate, margins entire, ciliate. **Peduncles** 1.7–6.6
cm, pubescent. **Flowers:** sepals linear-lanceolate, margins
ciliate or eciliate, auricles 0.5–1 mm; petals deep lemon-
yellow adaxially, upper 2 usually conspicuously reddish
brown to brownish purple abaxially, lateral 2 streaked
or solid reddish brown, lower 3 and sometimes upper 2
dark brown-veined proximally, lateral 2 bearded, lowest
8–13 mm, spur usually reddish brown, sometimes

yellowish, gibbous, 0.4–2 mm, glabrous or scabrous abaxially; style head bearded; cleistogamous flowers axillary. **Capsules** spherical, 4.5–9 mm, puberulent. **Seeds** black, 3.4–3.5 mm. $2n = 12$.

Flowering May–Jun. On limestone hills, slopes, and dry washes beneath *Pinus monophylla*, *P. ponderosa*, *Juniperus osteosperma*, and/or *Cercocarpus* spp.; 2000–2900 m; Nev., Utah.

Viola charlestonensis is known only from the Spring Mountains (previously called Charleston Mountains) in Nevada and Zion National Park, Utah. M. S. Baker (in I. W. Clokey 1945) stated that E. C. Jaeger reportedly collected it at Jacob's Pool, Arizona, in July 1926. The location of this single Arizona collection may be an error (R. J. Little 2001).

M. S. Baker and J. Clausen (in I. W. Clokey 1945) stated that *Viola charlestonensis* is the only species in Becker's *Nuttallianae* group with the spur pubescent on the exterior. In some populations in the Spring Mountains, Nevada, the spur and the midvein on the abaxial surface of the lowest petal to more or less the middle of the lowest petal are covered more or less densely with short hairs. Scattered hairs are also present on the abaxial surface of the lateral and upper petals. In other populations in the Spring Mountains, short hairs are mostly absent on the spur, lowest petal, and abaxial surfaces of lateral and upper petals.

M. S. Baker (in I. W. Clokey 1945) commented that he observed numerous sterile flowers and relatively few mature capsules and seeds of *Viola charlestonensis* plants when he visited the Spring Mountains in June 1937. Similar observations were made of *V. charlestonensis* plants in the Spring Mountains at one location in 2009 and of three other locations in 2010.

13. **Viola clauseniana** M. S. Baker, Madroño 4: 194. 1938 • Clausen's violet C E

Plants perennial, acaulescent, not stoloniferous, 3–20 cm; rhizome short, slender, fleshy. **Leaves** basal, 2–10, often prostrate, sometimes ascending; stipules narrowly lanceolate, margins faintly glandular-toothed, apex acute or obtuse; petiole 5–11 cm, glabrous; blade unlobed, ± deltate, 3–5 × 4–5 cm, base ± truncate, margins serrate, ciliate or eciliate, apex obtuse, surfaces usually glabrous, sometimes sparsely pubescent abaxially. **Peduncles** 8–14 cm, glabrous. **Flowers:** sepals lanceolate to ovate, margins eciliate or ciliate around auricles, auricles 1–2 mm; petals light violet on both surfaces, lower 3 purple-veined, all beardless, lowest petal 8–20 mm, spur white, gibbous, 2–3 mm; style head beardless; cleistogamous flowers on prostrate to ascending peduncles. **Capsules** oblong, 8–10 mm, glabrous. **Seeds** dark brown to black, 1.2–2 mm. $2n = 44$.

Flowering Apr–Jun. Hanging gardens, seeps, springs, shady areas; of conservation concern; 1600 m; Utah.

Viola clauseniana is endemic to Zion National Park, Washington County. M. S. Baker reported that its seeds were minutely roughed, a characteristic not recorded for other *Viola* species. *Viola clauseniana* was originally thought to be closely related to the acaulescent blue violets, most notably *V. nephrophylla* (S. L. Welsh et al. 1987; L. E. McKinney 1992). After contemplating the $2n = 44$ chromosome count obtained by J. Clausen (1964), H. E. Ballard (pers. comm.) suggested that *V. clauseniana* might be more closely related with the stemless white violets (for example, *V. blanda*) than with *V. nephrophylla*. *Viola clauseniana* was considered a distinct species by N. H. Holmgren (2005d) and T. Marcussen and T. Karlsson (2010).

14. **Viola cucullata** Aiton, Hort. Kew. 3: 288. 1789 • Marsh blue or northern bog or marsh or blue marsh violet, violette cucullée E

Viola cucullata var. *leptosepala* (Greene) W. Stone; *V. cucullata* var. *macrotis* (Greene) W. Stone; *V. cucullata* var. *microtitis* Brainerd; *V. dicksonii* Greene; *V. leptosepala* Greene; *V. macrotis* Greene

Plants perennial, acaulescent, not stoloniferous, 5–20 cm; rhizome thick, fleshy. **Leaves** basal, 4–8, ascending to erect; stipules linear-lanceolate, margins entire, apex acute; petiole 6–20 cm, usually glabrous; blade unlobed, ovate to reniform, 1.5–11 × 1–9 cm, base reniform to cordate, margins crenate to serrate, ciliate, apex acute or obtuse to slightly caudate, surfaces glabrous, rarely pubescent. **Peduncles** 7–25 cm, glabrous or pubescent. **Flowers:** sepals lanceolate, margins ciliate or eciliate, auricles 3–6 mm; petals light to dark blue-violet to violet on both surfaces, upper 2 and lateral 2 darker basally, lateral 2 densely bearded, lowest white basally, dark purple-veined, 9–13 mm, spur color same as petals, gibbous, 2–3 mm; style head beardless; cleistogamous flowers on erect peduncles. **Capsules** ellipsoid to ± ovoid, 10–15 mm, glabrous. **Seeds** beige, mottled to bronze, 1–2 mm. $2n = 54$.

Flowering Apr–Jun. Open or wooded wet areas, seeps, springs, swamps, marshes, streamsides; 0–2000 m; St. Pierre and Miquelon; N.B., Nfld. and Labr. (Nfld.), N.S., Ont., P.E.I., Que.; Ark., Conn., Del., D.C., Ga., Ill., Ind., Iowa, Ky., Maine, Md., Mass., Mich., Minn., Miss., Mo., N.H., N.J., N.Y., N.C., Ohio, Pa., R.I., S.C., Tenn., Vt., Va., W.Va., Wis.

Viola cucullata reportedly hybridizes with *V. sororia* (= *V.* ×*bissellii* House), *V. brittoniana* (= *V.* ×*notabilis* E. P. Bicknell), and *V. sagittata* var. *sagittata* (= *V.* ×*porteriana* House).

15. **Viola cuneata** S. Watson, Proc. Amer. Acad. Arts 14: 290. 1879 • Northern two-eyed or wedge-leaved violet E

Plants perennial, caulescent, not stoloniferous, 2–25 cm. Stems 1–3, usually erect, sometimes prostrate or ascending, glabrous, from shallow, fleshy rhizome or deep-seated caudex with fleshy roots. Leaves basal and cauline; basal: 2–6; stipules lanceolate, margins entire or gland-toothed, apex acute; petiole 4.5–9.8 cm, glabrous; blade purple-veined, orbiculate-ovate to deltate, 1–4 × 2.7–3.8 cm, usually shiny, leathery, base cuneate, margins serrate, eciliate, apex acute, mucronulate, surfaces glabrous; cauline similar to basal except: stipules lanceolate to ovate, margins entire or erose to lacerate, with or without gland-tipped processes, apex acute to acuminate; petiole 0.5–20 cm; blade usually rhombic, sometimes orbiculate, 0.9–2.6 × 0.7–1.8 cm, margins crenate to ± serrate, apex acute to obtuse. Peduncles 1–10.5 cm, glabrous. Flowers: sepals lanceolate, margins eciliate, auricles 0.5–1 mm; petals deep reddish violet abaxially, white adaxially, upper 2 sometimes with purple patch basally, lateral 2 with purple patch basally beyond smaller yellow area, usually bearded, sometimes beardless, lowest purple-veined with yellow area basally, 8–14 mm, spur yellow, gibbous, 1–2 mm; style head bearded; cleistogamous flowers axillary. Capsules ± spherical, 5–9 mm, glabrous. Seeds deep brown-purple, 2.1–3 mm.

Flowering Mar–Sep. Open pine or oak forests, often serpentine soil; 600–2200 m; Calif., Oreg.

Viola cuneata often occurs on serpentine-derived soil and is closely related to *V. flettii*, *V. hallii*, and *V. ocellata* (V. B. Baird 1942). Originally described as beardless by Watson, the lateral petals of flowers in the same population can be prominently bearded or essentially beardless. The cauline blades of *V. cuneata* are often vertical, especially early in season.

16. **Viola douglasii** Steudel, Nomencl. Bot. ed. 2, 2. 771. 1841 • Douglas's or Douglas's golden violet F

Viola chrysantha Hooker, Icon. Pl. 1: plate 49. 1836, not Schrader ex Reichenbach 1830

Plants perennial, caulescent, not stoloniferous, 3–20 cm. Stems 1–3, decumbent or ascending to erect, ca. ½ subterranean, glabrous or puberulent, from single, short, vertical, deep-seated caudex. Leaves basal and cauline; basal: 1–6, bipinnately compound, leaflets 3–5; stipules adnate to petiole forming 2 linear-lanceolate wings, margins entire, apex of each wing free, acute to acuminate; petiole 5–6.8 cm, glabrous or puberulent; blade ovate, 3.5–5 × 2.4–3.5 cm, base tapered, leaflets 3–5-lobed, lobes linear, narrowly elliptic, or oblong, 1–2.5(–5) mm wide, margins entire, usually densely ciliate, apex acute to obtuse, mucronulate, surfaces glabrous or puberulent; cauline similar to basal except: stipules ovate to linear-lanceolate, unlobed, margins entire or toothed, apex acute; petiole 0.9–4 cm; blade 1.1–4.1 × 1–3.6 cm. Peduncles 2–12.5 cm, glabrous or puberulent. Flowers: sepals lanceolate, margins ciliate, auricles 0.5–1.5 mm; petals light golden yellow adaxially, upper 2 dark brown to ± black abaxially, lower 3 dark brown-veined, lateral 2 bearded, lowest 8–21 mm, spur dark greenish to dark brown, gibbous, 1.5–2 mm; style head bearded; cleistogamous flowers absent. Capsules spherical to oblong, 5–12 mm, glabrous. Seeds light brown, 2.8–3.3 mm. $2n = 24, 48$.

Flowering Feb–Jul. Vernally moist grassy slopes and flats, often serpentine soil (except Oregon); 20–2300 m; Calif., Oreg.; Mexico (Baja California).

Viola douglasii is tetraploid ($n = 12$) south of, and octoploid ($n = 24$) north of, San Francisco Bay, California. It forms sterile hybrids with *V. quercetorum* (J. Clausen 1964). V. B. Baird (1936) described *V. douglasii* × *purpurea*, which Clausen later said was actually *V. quercetorum*, not described at the time of publication by Baird.

17. **Viola egglestonii** Brainerd, Bull. Torrey Bot. Club 37: 526, plates 34, 35. 1910 • Cedar glade or glade violet E

Viola septemloba Leconte subsp. *egglestonii* (Brainerd) L. E. McKinney

Plants perennial, acaulescent, not stoloniferous, 5–20 cm; rhizome thick, fleshy. Leaves basal, 3, prostrate to ascending; stipules linear-lanceolate, margins entire, apex acute; petiole 1.5–7 cm, usually glabrous; earliest leaf blades ± deltate or 3-lobed, mid-season blades 5–9-lobed, 1–9 × 1–10 cm, base truncate to cuneate, middle lobe lanceolate or spatulate to narrowly ovate, lateral lobes lanceolate or spatulate to falcate, margins serrate, sometimes with deltate or falcate appendages or teeth, ciliate, apex acute, surfaces usually glabrous, rarely pubescent. Peduncles 2–15 cm, usually glabrous, rarely pubescent. Flowers: sepals lanceolate to ovate, margins ciliate or eciliate, auricles 0.5–1 mm; petals light to dark blue-violet on both surfaces, lower 3 and sometimes upper 2 white basally, lower 3 darker violet-veined, lateral 2 densely bearded, lowest 10–15 mm, spur usually lilac, gibbous, 2–3 mm; style head beardless; cleistogamous flowers on prostrate peduncles. **Capsules** ellipsoid, 11–14 mm, glabrous. **Seeds** beige, mottled to bronze, 2–3 mm. $2n = 54$.

Flowering Mar–May. Limestone glades and barrens; 100–200 m; Ala., Ga., Ind., Ky., Tenn.

N. H. Russell (1965) was the first to consider *Viola egglestonii* similar to *V. septemloba*. After analyzing leaf-blade lobing patterns, L. E. McKinney considered it a subspecies of *V. septemloba*. Although similarities exist between these taxa, they are maintained as distinct species.

18. **Viola epipsila** Ledebour, Index Seminum (Dorpat) 1820: 5. 1820

Varieties 2 (1 in the flora): North America, Europe, Asia.

18a. **Viola epipsila** Ledebour var. **repens** (W. Becker) R. J. Little, J. Bot. Res. Inst. Texas 4: 225. 2010
• Dwarf marsh or marsh or northern marsh violet, violette rampante

Viola epipsila subsp. *repens* W. Becker, Beih. Bot. Centralbl. 34(2): 406. 1917, based on *V. repens* Turczaninow ex Trautvetter & C. A. Meyer in A. T. von Middendorff, Reise Siber. 1(2,3): 18. 1856, not Schweinitz 1822; *V. achyrophora* Greene

Plants perennial, acaulescent, stoloniferous, 0.5–20 cm; stolons pale, often rooting and leafy at nodes; rhizome slender, fleshy. **Leaves** basal, 1–4, ascending to erect; stipules linear-lanceolate, margins entire, apex acute; petiole 1–5 cm, glabrous or pubescent; blade unlobed, broadly ovate, reniform, or orbiculate, 1.5–2.5(–4.5) × 2–3(–4) cm, base cordate, margins crenate, denticulate, or entire, ciliate or eciliate, apex rounded to obtuse, sometimes acute, surfaces glabrous or sparsely pubescent throughout or abaxially. **Peduncles** 2–7 cm, glabrous or pubescent, bracteoles usually above middle in chasmogamous flowers. **Flowers:** sepals ovate-lanceolate, margins ciliate or eciliate, auricles 1–2 mm; petals lilac, violet, or deep purple on both surfaces, sometimes nearly white, lowest dark violet-veined, lateral 2 bearded or beardless, lowest 12–15 mm, spur same color as petals, gibbous, 1–2 mm; style head beardless; cleistogamous flowers on ascending peduncles. **Capsules** ellipsoid, 5–10 mm, glabrous. **Seeds** beige, 1.4–1.6 mm. $2n = 24, 36$.

Flowering May–Aug. Swamps, bogs, marshes, along streams, mossy wet banks, cool, wet places; 10–2000 m; Alta., B.C., Man., N.W.T., Nunavut, Ont., Yukon; Alaska, Colo., Mich.; Europe; Asia.

Viola epipsila occurs in small colonies; individual plants are interconnected by stolons.

Variety *repens* differs from var. *epipsila* by its smaller size, glabrous abaxial leaf surfaces (V. B. Baird 1942), acute leaves, and larger flowers (M. Sorsa 1968). When flowering, usually only one or two leaves, rarely more, are present per caudex.

M. Sorsa (1968) noted that diploid plants in British Columbia and Washington differed morphologically from diploid plants in Alaska. J. G. Packer in G. W. Douglas et al. (1989–1994, vol. 3; 1998–2002, vol. 5) wrote that var. *repens* (as subsp. *repens*) ($2n = 24$) and *Viola palustris* ($2n = 48$) intergrade in British Columbia and are often difficult to separate.

The presence of bracteoles above or below the middle of the peduncle has been used to differentiate *Viola epipsila* from *V. palustris* (V. B. Baird 1942; M. Sorsa 1968). B. A. Bennett (pers. comm.) reported that both varieties of *V. epipsila* at ALA collected in Alaska had bracteoles mostly above, but some at or below the middle of the peduncle. T. Marcussen and T. Karlsson (2010) stated that bracteoles of *V. palustris* occur in the middle third of the peduncle.

E. Hultén (1941–1950, vol. 7, 1968) considered *Viola palustris* synonymous with *V. epipsila* var. *repens* (as subsp. *repens*). He produced maps (Hultén 1968; Hultén and M. Fries 1986) showing the distribution of the taxon extending into the lower 48 states. Variety *repens* was first reported in the United States by H. E. Ballard (1985) stating that it had been documented in California, South Dakota, and Utah, apparently based on the 1968 map by Hultén, as well as in Michigan, where it is listed as endangered. At present, there is no evidence that var. *repens* occurs in California; M. Sorsa (1968) did not state that var. *repens* occurs in California and recently wrote that she could not say she had identified var. *repens* there (pers. comm.). S. L. Welsh et al. (1987) and N. H. Holmgren (2005d) did not recognize it as occurring in Utah.

The overall situation regarding the *Viola palustris* complex in the northwestern United States and southwestern Canada is complicated by the apparent presence of hybrids between *V. epipsila* var. *repens* and *V. palustris*, as commonly occurs in Europe and northern Asia (M. Sorsa 1968), and possibly with other species in the complex. Sorsa commented that the distinctly acute leaves of var. *repens* set it apart from *V. palustris*. A review of on-line images of var. *repens* at ALA (http://arctos.database.museum/SpecimenResults.cfm) of specimens collected in Alaska and scans of specimens collected by B. A. Bennett in Yukon show a range of apices from rounded to acute. Because meiosis and the number of chromosomes of *V. achyrophora* were identical to those of var. *repens*, Sorsa considered *V. achyrophora* to be a more northerly form of var. *repens*. The *V. palustris* complex in western North America, which includes *V. blanda*, *V. epipsila* var. *repens*, *V. macloskeyi*, *V. palustris*, and *V. renifolia*, is in need of detailed study.

19. Viola flettii Piper, Erythea 6: 69. 1898 • Olympic or rock violet E

Plants perennial, caulescent, not stoloniferous, 3–15 cm. **Stems** 1–3, ascending to erect, mostly glabrous, on caudex from fleshy rhizome. **Leaves** basal and cauline; basal: 1–3; stipules linear-lanceolate, margins entire or with glandular processes, apex acuminate; petiole 1.5–9.7 cm, mostly glabrous; blade purple-tinted and -veined, broadly reniform to ovate, 0.9–2.4 × 1.2–4 cm, base cordate, margins finely crenate-serrate, eciliate, apex acute to obtuse, surfaces glabrous or sparsely pubescent along veins adaxially; cauline similar to basal except: stipules ovate to lanceolate, margins entire or shallowly laciniate; petiole 0.7–5.9 cm, usually glabrous; blade 0.8–2.1 × 1.2–3.1 cm. **Peduncles** 1.8–7.1 cm, usually glabrous. **Flowers:** sepals lanceolate, margins eciliate, auricles 0.5–1.5 mm; petals soft reddish violet on both surfaces, all with yellow area basally, lower 3 dark violet-veined, lateral 2 bearded, lowest with white around yellow area, 10–15 mm, spur yellow, gibbous, 0.5–2 mm; style head bearded; cleistogamous flowers axillary. **Capsules** ± spherical, 5–9 mm, glabrous. **Seeds** dark brown to brownish purple, 2.5–3 mm.

Flowering Jun–Aug. Alpine and subalpine rock crevices, vertical faces, talus slopes; 1100–2000 m; Wash.

Viola flettii is endemic to the Olympic Mountains of northwestern Washington. C. S. McCreary (2005) noted that although morphologically and ecologically distinct, *V. cuneata*, *V. flettii*, and *V. ocellata* are closely related.

20. Viola frank-smithii N. H. Holmgren, Brittonia 44: 303, fig. 1E–H. 1992 • Frank Smith violet C E

Plants perennial, caulescent, not stoloniferous, 1.5–12.5 cm. **Stems** 1–4, decumbent or ascending to erect, glabrous, on caudex from fleshy rhizome. **Leaves** basal and cauline; basal: 2–3; stipules lanceolate, margins fimbriate, apex acute; petiole 1.5–6(–14) cm, glabrous; blade broadly ovate to deltate, 1.3–2.9(–4.1) × 1.3–2.9 cm, base cordate to truncate, margins crenate, eciliate, apex obtuse to rounded, surfaces glabrous; cauline similar to basal except: petiole 1.3–6.5 cm; blade 1.1–2.2 × 0.7–1.6 cm. **Peduncles** 2.5–7(–11.2) cm, glabrous. **Flowers:** sepals lanceolate to linear-lanceolate, margins eciliate, auricles 0.5–1.5 mm; all petals usually pale purple, sometimes almost white adaxially, lateral 2 and lowest purple-veined, lateral and upper 2 violet abaxially, white basally, lateral 2 bearded, lowest 10–16(–18) mm, spur greenish to pale lime green, gibbous, 1.6–3 mm; style head bearded; cleistogamous flowers unknown. **Capsules** ovoid, 3–7 mm, glabrous. **Seeds** pale yellow, 2.4–2.8 mm.

Flowering May–Jul. Cracks, crevices, and narrow ledges of vertical limestone and dolomite rock faces, humid, shady places; of conservation concern; 1600–2100 m; Utah.

Viola frank-smithii is endemic to Logan Canyon, Cache County. N. H. Holmgren (2005d) wrote that it likely has its ancestral ties with *V. adunca*.

21. Viola glabella Nuttall in J. Torrey and A. Gray, Fl. N. Amer. 1: 142. 1838 • Stream or smooth yellow or yellow wood or wood or smooth yellow woodland violet, violette glabre F

Viola californica M. S. Baker; *V. canadensis* Linnaeus var. *sitchensis* Bongard ex Ledebour; *V. glabella* var. *remotifolia* Suksdorf

Plants perennial, caulescent, not stoloniferous, 3–38 cm. **Stems** 1–3, erect or prostrate, leafless proximally, leafy distally, glabrous or finely puberulent, on caudex from fleshy rhizome. **Leaves** basal and cauline; basal: 0–4(–7); stipules ovate to obovate, margins entire, crenate, or serrate, usually glandular, apex acute to obtuse; petiole 7–27.5 cm, glabrous or puberulent; blade usually reniform to ovate, sometimes orbiculate, 3.3–8.5 × 2–9.3 cm, base cordate, margins crenate to serrate, ciliate or eciliate, apex acute to obtuse, mucronulate, surfaces glabrous or finely puberulent; cauline similar to basal except: usually restricted to distal ends of naked stems; stipules ovate to oblong, margins erose or subserrate, often glandular, apex acute to acuminate; petiole 0.2–2.9 cm, glabrous or finely puberulent; blade ovate to deltate, 1.4–5.7 × 0.8–4.7 cm, base cordate to truncate, margins crenate to ± serrate, ciliate (sometimes limited to proximal half), apex acute. **Peduncles** 2–8 cm, glabrous or pubescent. **Flowers:** sepals linear-lanceolate, margins eciliate, auricles 0.5–1 mm; petals deep lemon-yellow on both surfaces, lower 3 and sometimes upper 2 brownish purple-veined, lateral 2 bearded, lowest 6–18 mm, spur yellow to greenish, gibbous, 0.5–2 mm; style head bearded; cleistogamous flowers axillary. **Capsules** ovoid to ellipsoid, 7–13 mm, glabrous. **Seeds** light to dark brown, shiny, 2–2.2 mm. $2n = 24$.

Flowering Mar–Aug. Damp, wet, or shady places in forests, stream banks; 0–2600 m; Alta., B.C.; Alaska, Calif., Idaho, Mont., Oreg., Wash.; Asia.

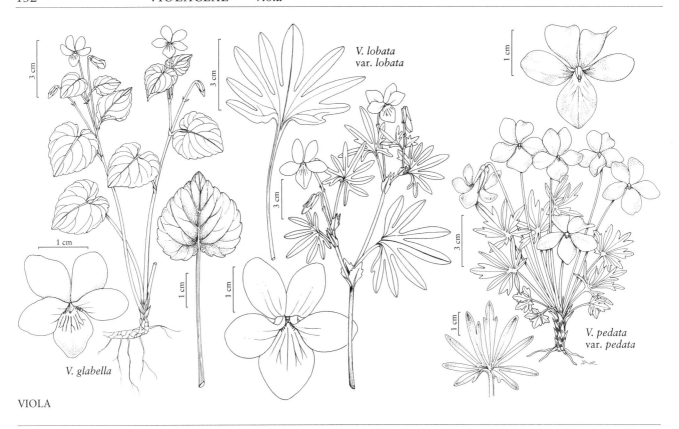

VIOLA

Viola glabella is similar in overall appearance to *V. pubescens*. The cauline leaves are sometimes described as appearing only near the apex of the stem (P. A. Munz 1959), but often a leaf occurs from a node on the stem below the apex.

Viola californica was described from collections made in Humboldt and Trinity counties, California, and may have arisen as a hybrid between *V. glabella* and *V. lobata* var. *integrifolia*. It is apparently limited to *Abies concolor* forests above 1520 m (M. S. Baker 1953). Study of *V. californica* is needed to determine if taxonomic recognition is warranted.

22. Viola guadalupensis A. M. Powell & Wauer, Sida 14: 1, fig. 1. 1990 • Guadalupe Mountains violet Ⓒ Ⓔ

Plants perennial, caulescent, not stoloniferous, 1–10 cm. **Stems** 1–5, decumbent to erect, leafy proximally and distally, glabrous, on caudex from fleshy rhizome. **Leaves** cauline; stipules lanceolate to ovate- or oblong-lanceolate or linear, margins sparingly glandular-fimbriate, apex acute; petiole 2–6 cm, glabrous; blade ovate to ovate-deltate or ovate-lanceolate, 1.2–2.4 × 0.7–1.3 cm, base broadly cuneate to rounded or truncate, margins entire or with 1–3 crenations on proximal ½, eciliate, apex acute to rounded, surfaces glabrous, sometimes with a few short hairs on veins abaxially. **Peduncles** 3.5–6 cm, glabrous. **Flowers:** sepals linear to linear-lanceolate, margins eciliate, auricles 0.5–1.5 mm; petals deep lemon-yellow adaxially, upper 2 reddish brown abaxially, lateral 2 and lowest dark brown-veined basally, lateral 2 bearded, lowest 7–11 mm, spur yellow, gibbous, 1–1.4 mm; style head bearded; cleistogamous flowers absent. **Capsules** ovoid, 3–4.5 mm, glabrous. **Seeds** light brown, ± 2 mm. **2n** = 24.

Flowering May. Openings and narrow ledges on limestone rock faces; of conservation concern; 2600 m; Tex.

Viola guadalupensis is known only from the eastern rim of the Guadalupe Mountains in Culberson County. Powell and Wauer noted that it is the only yellow-flowered violet known in the Guadalupe Mountains and appears to be related to *V. nuttallii* and *V. vallicola*. K. W. Allred (2008) stated that a report of this species in New Mexico by J. T. Kartesz and C. A. Meacham (1999) requires verification. K. Haskins (pers. comm.) reported that experiments are currently being conducted to propagate plants of *V. guadalupensis* via cell tissue culture.

Chloroplast (*trn*L-F spacer) and low-copy nuclear gene (*GPI*) phylogenies indicate that *Viola guadalupensis* is an alloploid that originated through hybridization between an unidentified member of subsect. *Canadenses* (the paternal parent) and a member of the *V. nuttallii* complex (the maternal parent), of sect. *Chamaemelanium* (T. Marcussen et al. 2011). Evidence reported by these authors from a fossil-calibrated relaxed clock dating analysis showed the estimated maximum age of *V. guadalupensis* to be (5.7–)8.6(–11.6) million years.

23. **Viola hallii** A. Gray, Proc. Amer. Acad. Arts 8: 377. 1872 • Hall's violet, wild pansy E

Plants perennial, caulescent, not stoloniferous, 5–22 cm. **Stems** 1–3, decumbent or ascending to erect, ca. ½ subterranean, glabrous, clustered on single, short, vertical, deep-seated caudex. **Leaves** basal and cauline; basal: 1–4, palmately compound, ± 2- or 3-ternate, leaflets 3; stipules adnate to petiole, forming 2 linear-lanceolate wings, margins entire, apex of each wing free, acute; petiole 5–8 cm, glabrous; blade ovate to deltate, 2.8–6 × 2.6–6.5 cm, ± coriaceous, base tapered, ultimate lobes narrowly elliptic, lanceolate, or oblanceolate, 1–7 mm wide, margins entire, ciliate or eciliate, apex acute, mucronulate, surfaces glabrous; cauline similar to basal except: stipules usually lanceolate, sometimes broadly ovate, ± leaflike, unlobed, margins toothed; petiole 1.3–6 cm; blade 2–4.8 × 1.2–5.5 cm. **Peduncles** 2.5–11 cm, glabrous. **Flowers:** sepals lanceolate to ovate, margins ciliate, auricles 0.5–1 mm; petals: upper 2 almost black abaxially, dark reddish violet adaxially, lower 3 pale yellow, cream, or ± white, lateral 2 bearded, with deep yellow to orange patch basally, dark reddish violet-veined, lowest with deep yellow to orange patch basally, dark reddish violet-veined, 5–18 mm, spur yellow, gibbous, 0.5–2 mm; style head bearded; cleistogamous flowers absent. **Capsules** ellipsoid, 4–12 mm, glabrous. **Seeds** light brown, shiny, 3.2–3.5 mm. $2n = 60, 72$.

Flowering Apr–Jul. Open forests, grassy hills, chaparral, frequently serpentine or gravelly soil; 500–2100 m; Calif., Oreg.

Viola hallii was discovered on the grounds of Willamette University in Salem, Oregon, by Elihu Hall, a professor at that institution (V. B. Baird 1942). Leaves of *V. hallii* are similar to those of *V. beckwithii*.

24. **Viola hastata** Michaux, Fl. Bor.-Amer. 2: 149. 1803 • Halberd-leaved or halberdleaf yellow violet E

Plants perennial, caulescent, not stoloniferous, 5–30 cm. **Stems** 1(2), erect, leafless proximally, leafy distally, glabrous or sparsely puberulent, from fleshy rhizome. **Leaves** basal and cauline; basal: 0 or 1; stipules ovate to lanceolate, margins laciniate with gland-tipped projections, apex usually acuminate; petiole 3–11.5 cm, usually glabrous; blade sometimes gray-green abaxially, usually mottled light green adaxially, widely or narrowly hastate to ovate, 2.5–4.4 × 1.8–3.2 cm, base hastate to cordate or truncate, margins serrate or crenate, ciliate or eciliate, apex acute, surfaces usually glabrous, sometimes scabrous; cauline similar to basal except: leaves usually restricted to distal ends of naked stems; petiole 1–2.5 cm; blade 2–10 × 1.1–4.5 cm. **Peduncles** 1–5.3 cm, usually glabrous. **Flowers:** sepals lanceolate to ovate, margins mostly eciliate, auricles 0.5–1 mm; petals lemon-yellow usually on both surfaces, lower 3 and sometimes upper 2 brownish purple-veined, lateral 2 bearded, lowest 9.5–12 mm, spur yellow to greenish, gibbous, 0.5–2 mm; style head bearded; cleistogamous flowers axillary. **Capsules** ovoid to ellipsoid, 6–8 mm, glabrous. **Seeds** beige to bronze, 2–2.5 mm. $2n = 12$.

Flowering late Mar–May. Rich woods, chiefly mountains and piedmont; 50–2000 m; Ala., Ga., Ky., Md., N.Y., N.C., Ohio, Pa., S.C., Tenn., Va., W.Va.

V. B. Baird (1942) commented that *Viola hastata* is more closely related to *V. glabella* than to any eastern species.

25. **Viola hirsutula** Brainerd, Rhodora 9: 98. 1907 • Southern woodland violet E

Plants perennial, acaulescent, not stoloniferous, 2–15 cm; rhizome thick, fleshy. **Leaves** basal, 2–4, prostrate or nearly so; stipules linear-lanceolate, margins entire, apex acute; petiole 1–10 cm, usually glabrous; blade unlobed, reniform to ovate, 1–6 × 1–5 cm, base cordate, margins serrate, ciliate or eciliate, apex rounded, adaxial leaf surface with silvery strigose patches. **Peduncles** 2–12 cm, glabrous. **Flowers:** sepals lanceolate to ovate, margins ciliate or eciliate, auricles 1–2 mm; petals light to deep reddish violet to blue-violet on both surfaces, 3 lower whitish basally, dark violet-veined, lateral 2 densely bearded, lowest 11–17 mm, spur violet, gibbous, 2–3 mm; style head beardless; cleistogamous flowers on prostrate to ascending peduncles. **Capsules** ellipsoid, 8–12 mm, glabrous. **Seeds** beige, mottled to bronze, 1–1.5 mm. $2n = 54$.

Flowering Mar–Jun. Dry to mesic deciduous woodlands; 100–1000 m; Ala., Conn., Del., D.C., Fla., Ga., Ind., Ky., Md., Miss., N.J., N.Y., N.C., Ohio, Pa., S.C., Tenn., Va., W.Va.

Viola hirsutula exhibits little phenotypic plasticity. Although considered by N. L. Gil-Ad (1997) to be a putative hybrid, he did not identify the potential parents and chose not to recognize it. A. Cronquist in H. A. Gleason and Cronquist (1991) considered *V. hirsutula* a form of *V. villosa*. *Viola hirsutula* is considered one of the more distinct species throughout its range by McKinney because of the pattern of silvery strigose patches on adaxial leaf surfaces and a more prostrate growth habit than similar species. He believes *V. hirsutula* is one of the least variable species in the flora. *Viola hirsutula* appears to occur much less frequently than herbarium collections suggest. Some herbarium specimens are misidentified and represent depauperate plants, usually of *V. sororia*. These specimens are often hairy on the adaxial leaf surfaces but lack the silvery patches.

Viola hirsutula reportedly hybridizes with *V. sororia* [= *V. ×cordifolia* (Nuttall) Schweinitz] and *V. sagittata* var. *sagittata* (= *V. ×redacta* House).

26. **Viola howellii** A. Gray, Proc. Amer. Acad. Arts 22: 308. 1887 • Howell's violet, violette de Howell [E]

Plants perennial, caulescent, not stoloniferous, 2–44 cm. **Stems** 1–4, ascending to erect (sometimes later reclining to nearly prostrate), glabrous or sparsely pubescent, on caudex from subligneous rhizome. **Leaves** basal and cauline; basal: 1–6; stipules linear-lanceolate to oblong, margins mostly entire or laciniate with gland-tipped projections, apex acute to acuminate; petiole 4–15 cm, glabrous or sparsely puberulent; blade ovate to reniform, 1.9–6.8 × 2.1–6.4 cm, thin, base cordate, margins crenate, ciliate, apex acute to usually obtuse, surfaces glabrous or sparsely puberulent; cauline similar to basal except: stipules linear-lanceolate or deltate to oblong, apex acute or long-acuminate; petiole 1–4.5 cm; blade 2–5.1 × 1.2–5.8 cm. **Peduncles** 2.8–17.8 cm, glabrous or sparsely pubescent. **Flowers:** sepals lanceolate, margins ciliate or eciliate, auricles 1–2 mm; petals violet to soft blue-violet on both surfaces, whitish basally, lower 3 dark violet-veined, lateral 2 densely bearded, lowest 14–23 mm, spur blue-violet to whitish, gibbous to elongated, 2.4–5 mm, usually less than ½ lowest petal, tip straight; style head bearded; cleistogamous flowers axillary. **Capsules** ellipsoid, 7–11 mm, glabrous. **Seeds** light brown, 2.1–2.8 mm. $2n = 40, 80$.

Flowering Apr–Jul. Moist, shady areas, coniferous forests, prairies, along streams; 50–1500 m; B.C.; Calif., Oreg., Wash.

Viola howellii is similar to *V. adunca* and usually occupies moister habitats along the western side of the Cascade Mountains (H. E. Ballard 1992). The first documented record of this species in California was recently reported from Siskiyou County (R. J. Little 2010).

27. **Viola japonica** Langsdorff ex Gingins in A. P. de Candolle and A. L. P. P. de Candolle, Prodr. 1: 295. 1824 • Japanese violet [I]

Plants perennial, acaulescent, not stoloniferous, 3–10 cm; rhizome thick, fleshy. **Leaves** basal, ca. 5, ascending to erect; stipules ± oblong, 2-fid, proximal margins entire, distal ± serrate, apex acuminate; petiole narrowly winged distally, 1–14 cm, usually glabrous; blade unlobed, broadly ovate or triangular-ovate, 3–8 × 3–5.5 cm, base cordate, margins crenate, usually eciliate, apex acute or ± obtuse, surfaces sparsely puberulent. **Peduncles** 3–6 cm, glabrous or pubescent, bracteoles near middle. **Flowers:** sepals broadly lanceolate, margins eciliate, auricles 6–8 mm; petals light violet or whitish violet on both surfaces, lowest 3 occasionally white basally, often dark violet-veined, lateral 2 sparsely bearded or beardless, lowest 17–20 mm, spur pale to dark violet, elongated, 5–10 mm; style head beardless; cleistogamous flowers present. **Capsules** ellipsoid, 8–10 mm, glabrous. **Seeds** unknown. $2n = 48$.

Flowering Apr–Jun. Gardens and ruderal areas; 10–50 m; introduced; Mass.; Asia (China, Japan, Korea).

A. Haines (2011) reported that *Viola japonica* occurs in Massachusetts. He noted that reports of *V. chinensis* D. Don in New England were misidentified and are *V. japonica*. Reports of *V. patrinii* in District of Columbia, Maryland, and New York may also be *V. japonica*.

28. **Viola labradorica** Schrank, Denkschr. Königl.-Baier. Bot. Ges. Regensburg 1(2): 12. 1818 • American dog or Labrador or alpine violet, violette du Labrador [E]

Viola adunca Smith var. *minor* (Hooker) Fernald; *V. canina* Linnaeus var. *muhlenbergii* (Torrey) A. Gray; *V. conspersa* Reichenbach; *V. leucopetala* Greene; *V. muhlenbergiana* Gingins; *V. muhlenbergiana* var. *minor* Hooker; *V. muhlenbergii* Torrey

Plants perennial, caulescent, not stoloniferous, 5–20 cm. **Stems** 1–5, erect (later reclining), glabrous, on caudex from subligneous rhizome. **Leaves** basal and cauline; basal: 1–5; stipules mostly lanceolate-linear, margins usually laciniate with gland-tipped projections, apex acute; petiole 1–9.5 cm, glabrous or puberulent; blade reniform or ovate to deltate, 1.4–5.5 × 1.8–5.3 cm, base cordate, not decurrent on petiole, margins crenate or serrate, mostly eciliate, apex acute to obtuse, surfaces usually glabrous, sometimes pubescent adaxially near margin; cauline similar to basal except: stipules lanceolate to ovate, margins ± entire or laciniate, projections often gland-tipped, apex acuminate; petiole 0.5–3.6 cm, glabrous; blade narrowly ovate to orbiculate, 1.8–2.9 × 1.5–3.1 cm, base deeply cordate to truncate, margins regularly crenate, apex obtuse to broadly rounded or apiculate. **Peduncles** 5–9 cm, glabrous or pubescent. **Flowers:** sepals lanceolate, margins ciliate or eciliate, auricles not enlarged in fruit, 1–1.5 mm; petals usually lavender-violet to violet on both surfaces, rarely white, lower 3 white basally, darker violet-veined, lateral 2 sparsely bearded, lowest 9–16 mm, spur white to pinkish violet, elongated, 4–8 mm, tip usually straight, sometimes curved up; style head beardless; cleistogamous flowers unknown. **Capsules** ovoid to ellipsoid, 5–7 mm, glabrous. **Seeds** beige to bronze, 1.5–2 mm. $2n = 20$.

Flowering Apr–Aug. Rich, mesic to wet woods, swamps, bogs; 50–3000 m; Greenland; St. Pierre and Miquelon; Man., N.B., Nfld. and Labr., N.W.T., N.S., Ont., P.E.I., Que., Sask.; Ala., Colo., Conn., Del., D.C., Ga., Ill., Ind., Ky., Maine, Md., Mass., Mich., Minn., N.H., N.J., N.Y., N.C., N.Dak., Ohio, Pa., R.I., S.C., Tenn., Vt., Va., W.Va., Wis.

H. E. Ballard (1994) made a compelling case for including *Viola conspersa* within *V. labradorica*, which is followed here. Ballard (1992) did not recognize *V. labradorica* as occurring in Colorado. Several western herbaria have collections identified as *V. labradorica* (ASC, CS, KHD, RM, RMBL; http://swbiodiversity.org/seinet/index.pho). Labels on some collections at KHD imply that *V. labradorica* is considered synonymous with *V. adunca* var. *bellidifolia*.

Viola labradorica reportedly hybridizes with *V. adunca*, *V. rostrata* (= *V.* ×*malteana* House), and *V. striata* (= *V.* ×*eclipes* H. E. Ballard), and less frequently with *V. walteri* var. *appalachiensis* and var. *walteri*.

29. **Viola lanceolata** Linnaeus, Sp. Pl. 2: 934. 1753 • Lance-leaved or bog white violet, violette lancéolée [E]

Viola lanceolata subsp. *vittata* (Greene) N. Russell; *V. lanceolata* var. *vittata* (Greene) Weatherby & Griscom; *V. vittata* Greene

Plants perennial, acaulescent, stoloniferous, 5–30 cm; stolons pale, often rooting and leafy at nodes; rhizome slender, fleshy. **Leaves** basal, 2–6(–9), ascending to erect; stipules linear-lanceolate, margins entire or irregularly lacerate (at least distally), apex acute; petiole 2–12 cm, glabrous or pubescent; blade unlobed, lanceolate or narrowly elliptic to nearly linear, 2.5–12 × 0.7–2.5 cm, longer than wide, base attenuate, margins serrate, mostly eciliate, apex acute, mucronulate, surfaces glabrous. **Peduncles** 2–17 cm, usually pubescent. **Flowers:** sepals ovate to lanceolate, margins mostly eciliate, auricles 1–2 mm; petals white on both surfaces, lowest and sometimes lateral 2 purple-veined, lateral 2 sparsely bearded or beardless, lowest 7–12 mm, spur white, gibbous, 1–2 mm; style head beardless; cleistogamous flowers on prostrate to ascending peduncles. **Capsules** ellipsoid, 5–8 mm, glabrous. **Seeds** beige to bronze, 1.5–2.5 mm. $2n = 24$.

Flowering Mar–May. Open to semi-open wet areas, bogs, meadows, pond and lake shores, stream banks, seasonally inundated depressions; 0–2000 m; St. Pierre and Miquelon; B.C., N.B., Nfld. and Labr. (Nfld.), N.S., Ont., P.E.I., Que.; Ala., Ark., Conn., Del., D.C., Fla., Ga., Ill., Ind., Iowa, Ky., La., Maine, Md., Mass., Mich., Minn., Miss., Mo., Nebr., N.H., N.J., N.Y., N.C., Ohio, Okla., Oreg., Pa., R.I., S.C., Tenn., Tex., Vt., Va., Wash., W.Va., Wis.

Viola lanceolata occurs in small colonies; individual plants are interconnected by stolons.

Viola lanceolata subsp. *vittata* was recognized by N. H. Russell (1955) based on its linear leaf blade shape. The range in leaf blade shape appears to have no distinct line of demarcation. Some believe that leaf shape differences and near restriction to the coastal plain support recognition at some level. Studies are necessary to resolve this issue.

J. H. Shultze (1946) reported that *Viola lanceolata* was introduced into Washington in the early 1900s primarily from Cape Cod and Wisconsin as a result of importation of cranberry vines. In British Columbia, it is known from Lulu Island, where it was introduced from eastern North America (G. W. Douglas et al. 1998–2002, vol. 5).

Viola lanceolata reportedly hybridizes with *V. primulifolia* var. *primulifolia* (= *V.* ×*modesta* House) and *V. macloskeyi* (= *V.* ×*sublanceolata* House).

30. **Viola langsdorffii** Fischer ex Gingins in A. P. de Candolle and A. L. P. P. de Candolle, Prodr. 1: 296. 1824 (as langsdorfii) • Alaska or Langsdorff's violet, violette de Langsdorff

Viola simulata M. S. Baker; *V. superba* M. S. Baker

Plants perennial, caulescent, not stoloniferous, 2–30 cm. **Stems** 1–3, ascending to erect, usually glabrous, from shallow, fleshy rhizome. **Leaves** basal and cauline; basal: 2 or 3; stipules broadly ovate to deltate or long-lanceolate, margins often glandular (glands sessile or stalked), apex acute; petiole 0.8–21 cm, glabrous or sparsely pubescent; blade ovate to reniform, 0.9–5.8 × 1–6 cm, base cordate to ± truncate, margins crenate to crenate-serrulate, usually eciliate, apex rounded or acute to usually obtuse, surfaces glabrous or sparsely pubescent on adaxial veins, seldom on abaxial veins; cauline similar to basal except: stipules broadly ovate to lanceolate-oblong or deltate, apex acute to acuminate; petiole 2.2–12.1 cm; blade 1.9–4.2 × 1.8–5.9 cm. **Peduncles** 2.2–20.7 cm, glabrous or sparsely pubescent. **Flowers:** sepals lanceolate to ovate, margins eciliate, auricles 0.5–1.5 mm; petals light to deep violet on both surfaces, lower 3 white basally and light to deep violet-veined, lateral 2 bearded, lowest 12–24 mm, spur white to violet, gibbous to elongated, 2–5 mm, usually less than ½ lowest petal, tip straight; style head usually beardless, sometimes bearded; cleistogamous flowers axillary. **Capsules** ovoid to oblong, 7.5–13 mm, glabrous. **Seeds** dark olive to ± black, 2.5–2.8 mm. $2n = 12$.

Flowering Apr–Aug. Swamps, bogs, fens, wet meadows, stream banks, rocky slopes, usually near coast; 0–1500 m; B.C., Yukon; Alaska, Calif., Oreg., Wash.; Asia.

Viola simulata and *V. superba* are closely related to *V. langsdorffii* (L. Abrams and R. S. Ferris 1923–1960, vol. 3). They were treated by C. L. Hitchcock et al. (1955–1969, vol. 3) as synonymous with *V. langsdorffii* pending further study.

E. Brainerd (1921) stated that *Viola langsdorffii* presents an interesting transition between the caulescent and acaulescent violets. Plants of the species are reported from Winchester Grade in Idaho (*Gail s.n.*, 1931, ID); verification is needed. *Viola langsdorffii* is reported to have fleshy rhizomes. Plants in Del Norte County, California, have fleshy rhizomes near the soil surface, but these are connected to deeper, subligneous rhizomes.

The style in some populations of *Viola langsdorffii* is reported to be sometimes bearded (G. W. Douglas et al. 1998–2002, vol. 5; W. J. Cody 2000). Illustrations in V. B. Baird (1942; location of plant unknown) and M. S. Baker (1936; a plant from Alaska) show a beardless style; L. Abrams and R. S. Ferris (1923–1960, vol. 3) and C. L. Hitchcock et al. (1955–1969, vol. 3) reported that the style head was beardless. Styles are beardless in plants occurring in Del Norte County, California.

31. **Viola lithion** N. H. Holmgren & P. K. Holmgren, Brittonia 44: 300, fig. 1A–D. 1992 • Rock violet
C E

Plants perennial, caulescent, not stoloniferous, 5–15 cm. **Stems** 1–3, ascending to erect, glabrous, on caudex from fleshy rhizome. **Leaves** basal and cauline; basal: 1–4; stipules unknown; petiole 1–10 cm, glabrous, sometimes finely puberulent; blade broadly ovate or deltate to broadly deltate, 1–2.5(–2.9) × 0.6–2.2(–2.6) cm, base usually cordate to truncate, rarely rounded, margins crenate-dentate, eciliate, apex acute, surfaces glabrous; cauline similar to basal except: stipules lanceolate, margins usually fimbriate-toothed, sometimes entire, apex attenuate or acute; petiole 1.1–3.7 cm, sometimes finely puberulent; blade ovate to deltate, 0.7–2.2 × 0.4–1.2 cm, base sometimes rounded on distal blades. **Peduncles** 3–6(–10) cm, glabrous. **Flowers:** sepals lanceolate, margins eciliate, auricles 0.5–1 mm; petals blue to pale violet on both surfaces with yellow area basally, lower 3 purple-veined, lowest with yellow area, lateral 2 bearded, lowest 5.5–11 mm, spur white to pale violet, gibbous, 0.5–1.3 mm; style head bearded; cleistogamous flowers unknown. **Capsules** subglobose, ca. 5 mm, glabrous. **Seeds** dark brown, ca. 1.8 mm.

Flowering Jun–Aug. Seasonally wet cracks and crevices, narrow ledges of rock outcrops, shaded northeast-facing avalanche chutes, cirque headwalls, subalpine conifer zone; of conservation concern; 2300–3100 m; Nev., Utah.

Viola lithion is known only from the White Pine Range in Nevada and the Pilot Range straddling the Nevada-Utah border. It is related to *V. canadensis* and *V. flettii*.

32. Viola lobata Bentham, Pl. Hartw., 298. 1849 [F]

Plants perennial, caulescent, not stoloniferous, 5–46 cm. **Stems** 1–3, erect, leafless proximally, leafy distally, glabrous or puberulent, from subligneous rhizome. **Leaves** basal and cauline; basal: 0–2, unlobed or palmately 3–11-lobed; stipules ovate to lanceolate, margins ± entire or serrate, apex acute; petiole 5–27 cm, glabrous or puberulent; blade sometimes glaucous, deltate to reniform, 3.5–8.5 × 4.5–13.5 cm, base cordate, truncate, or attenuate, margins on unlobed leaves coarsely dentate-serrate, margins on lobed leaves usually entire, sometimes few-toothed, ciliate (sometimes only proximal ½) or eciliate, apex acute to obtuse, mucronulate, surfaces glabrous or puberulent; cauline similar to basal except: distal on naked stems, unlobed and divided leaves can occur on same plant, if divided, palmately 3–12-lobed; stipules sometimes large and ± leaflike, margins entire, lacerate, or laciniate, sometimes with gland-tipped projections, apex acute to acuminate; petiole 0.2–8.8 cm; blade reniform to reniform-cordate, ± ovate, deltate, or rhombic, 1.5–5.5 × 1.4–10 cm, base cordate, subcordate, truncate, or attenuate, margins entire, crenate-serrate, or dentate, or coarsely lacerate to deeply serrate, often entire distally, ciliate or rarely eciliate, apex acute to obtuse, often long-tapered, mucronulate. **Peduncles** 2–13 cm, glabrous or pubescent. **Flowers:** sepals lanceolate, margins ciliate or eciliate, auricles 0.1–1 mm; petals deep lemon-yellow adaxially, usually upper 2 and sometimes lateral 2 brownish purple abaxially, lower 3 and sometimes upper 2 brownish purple-veined basally, lateral 2 bearded, lowest 8–19 mm, spur yellow to greenish, gibbous, 0.5–2 mm; style head bearded; cleistogamous flowers axillary. **Capsules** ellipsoid-ovoid, 6–16 mm, glabrous. **Seeds** light brown, blotched or streaked with brown, shiny, 2.1–2.7 mm. $2n$ = 12.

Varieties 2 (2 in the flora): w United States, nw Mexico.

V. B. Baird (1942) noted that *Viola lobata* was more closely related to *V. tripartita* than to any western *Viola*.

1. Cauline leaf blades usually deeply divided, seldom unlobed, reniform, ± ovate, or deltate, if divided, palmately 3–12-lobed, lobe margins usually entire.................32a. *Viola lobata* var. *lobata*
1. Cauline leaf blades unlobed, deltate to rhombic, ovate, or reniform-cordate, margins ± serrate, crenate-serrate, or dentate to deeply serrate or coarsely lacerate, often entire distally........ .... 32b. *Viola lobata* var. *integrifolia*

32a. Viola lobata Bentham var. **lobata** • Pine or yellow wood violet [F]

Viola lobata subsp. *psychodes* (Greene) Munz; *V. sequoiensis* Kellogg

Leaves: basal blades unlobed or palmately lobed; cauline blades usually deeply divided, seldom unlobed, if divided, palmately 3–12-lobed, reniform, ± ovate, or deltate, usually wider than long, lobe margins usually entire, apex acute to obtuse.

Flowering Apr–Aug. Dry, shaded or open forests; 100–2300 m; Calif., Oreg.; Mexico (Baja California).

32b. Viola lobata Bentham var. **integrifolia** S. Watson in W. H. Brewer et al., Bot. California 1: 57. 1876 [E]

Viola lobata subsp. *integrifolia* (S. Watson) R. J. Little

Leaves: basal blades unlobed; cauline blades unlobed, deltate to rhombic, ovate, or reniform-cordate, usually longer than wide, margins ± serrate, crenate-serrate, or dentate to deeply serrate or coarsely and irregularly lacerate, often entire distally, apex acute, often long-tapered.

Flowering Mar–Aug. Dry, shaded or open forests; 300–2100 m; Calif., Oreg.

Variety *integrifolia* is considered uncommon; it occurs from southwestern Oregon to southern California and is possibly more common in Oregon than California. Plants in the same population may have subserrate, deeply serrate, or irregularly lacerate leaves.

33. Viola macloskeyi F. E. Lloyd, Erythea 3: 74. 1895 • Macloskey's or northern white or small white or smooth white or wild white violet, violette pâle [E]

Viola blanda Willdenow subsp. *macloskeyi* (F. E. Lloyd) A. E. Murray; *V. blanda* var. *macloskeyi* (F. E. Lloyd) Jepson; *V. macloskeyi* subsp. *pallens* (Banks ex Gingins) M. S. Baker; *V. macloskeyi* var. *pallens* (Banks ex Gingins) C. L. Hitchcock; *V. pallens* (Banks ex Gingins) Brainerd; *V. pallens* subsp. *macloskeyi* (F. E. Lloyd) M. S. Baker; *V. pallens* var. *subreptans* J. Rousseau; *V. rotundifolia* Michaux var. *pallens* Banks ex Gingins

Plants perennial, acaulescent, stoloniferous, 2–10 cm; stolons pale, often rooting and leafy at nodes; rhizome slender, fleshy. **Leaves** basal, 2–6, ascending to erect; stipules ovate to linear-lanceolate, margins entire or glandular-toothed, apex acute; petiole 1–10 cm, strigose; blade unlobed, reniform to ovate, 1–6.5 × 1–5.5 cm, base broadly or shallowly cordate, margins ± entire or shallowly crenate, eciliate, apex rounded to acute, surfaces usually glabrous, abaxial surfaces sometimes puberulent on proximal ½. **Peduncles** 2.5–11(–21) cm, usually glabrous, sometimes villous. **Flowers:** sepals lanceolate to ovate, margins eciliate, auricles 0.5–2 mm; petals white on both surfaces, lower 3 purple-veined, lateral 2 bearded, rarely beardless, lowest 6–12 mm, spur white, gibbous, 1–2.5 mm; style head beardless; cleistogamous flowers on ascending peduncles. **Capsules** ovoid, 5–9 mm, glabrous. **Seeds** beige to bronze, 1–1.5 mm. *2n* = 24.

Flowering Mar–Sep. Bogs, wet meadows, seeps, lake margins, stream banks, flood plains, swampy woods, mesic roadside depressions, often among mosses; 0–3600 m; St. Pierre and Miquelon; Alta., B.C., Man., N.B., Nfld. and Labr., N.W.T., N.S., Nunavut, Ont., P.E.I., Que., Sask.; Ariz., Calif., Colo., Conn., Del., Ga., Idaho, Ill., Ind., Iowa, Maine, Md., Mass., Mich., Minn., Mo., Mont., Nev., N.H., N.J., N.Y., N.C., N.Dak., Ohio, Oreg., Pa., R.I., S.C., S.Dak., Tenn., Vt., Va., Wash., W.Va., Wis., Wyo.

Viola macloskeyi was described by Banks in 1824 as *V. rotundifolia* var. *pallens*, from specimens collected by Banks in Labrador. E. Brainerd made the combination *V. pallens* in 1905 and later (1924) detailed its troublesome nomenclatural history.

Viola macloskeyi was described in 1895 by F. E. Lloyd based on specimens collected at the base of Mount Hood, Oregon. M. S. Baker (1953) placed *V. macloskeyi* as a subspecies of *V. pallens*. He later (1953b) corrected that change based on priority and thus the taxon became known as *V. macloskeyi* subsp. *pallens*. N. H. Russell (1955) maintained that status, separating subspp. *macloskeyi* and *pallens* on the wider, spreading basal leaf lobes of subsp. *pallens*. McKinney, after examining specimens of both taxa, including types, concluded that the differences have been exaggerated and fall within the range of variation of a single species.

Viola macloskeyi occurs in small colonies; individual plants are interconnected by stolons.

In parts of its range, *Viola macloskeyi* often shares habitat with *V. renifolia*; it may be difficult to distinguish the two. The stolons of *V. macloskeyi* are useful for identification during summer months; *V. renifolia* plants lack stolons. Patterns of indument can be useful; *V. renifolia* is usually hairy and *V. macloskeyi* is usually glabrous (petioles sometimes pubescent).

Viola macloskeyi reportedly hybridizes with *V. primulifolia* var. *primulifolia* (= *V. ×mollicula* House).

34. Viola missouriensis Greene, Pittonia 4: 141. 1900   • Missouri violet [E]

Viola candidula Nieuwland; *V. langloisii* Greene; *V. lucidifolia* Newbro; *V. sororia* Willdenow var. *missouriensis* (Greene) L. E. McKinney

Plants perennial, acaulescent, not stoloniferous, 5–50 cm; rhizomes thick, fleshy. **Leaves** basal, 1–8, ascending to erect; stipules linear-lanceolate to broadly lanceolate, margins entire, sometimes distally glandular, apex acute; petiole 5–20 cm, glabrous; blade green abaxially, unlobed, usually narrowly to broadly deltate, 1.5–12 × 1.5–10 cm, not fleshy, base cordate or broadly cordate to truncate, margins ± crenate to serrate, ciliate or eciliate, apex acute to acuminate, surfaces glabrous, rarely pubescent. **Peduncles** 3–25 cm, usually glabrous. **Flowers:** sepals lanceolate to ovate, margins ciliate or eciliate, auricles 1–2 mm; petals light to dark blue-violet, lowest and sometimes lateral 2 purple-veined, lateral 2 bearded, lowest beardless, rarely lightly bearded, 15–25 mm, spur same color as petals, gibbous, 2–3 mm; style head beardless; cleistogamous flowers on prostrate to ascending peduncles. **Capsules** ellipsoid, 5–12 mm, glabrous. **Seeds** beige, mottled to bronze, 1.5–2.5 mm. *2n* = 54.

Flowering Mar–May. Swamps, thickets, stream banks, alluvial woods; 50–2000 m; Ala., Ark., Del., D.C., Fla., Ga., Ill., Ind., Iowa, Kans., La., Md., Minn., Miss., Mo., Nebr., N.J., N.Mex., N.C., Okla., S.C., S.Dak., Tenn., Tex., Va., Wis.

What many have treated as *Viola affinis*, especially in the southern Gulf coastal states, is likely to be *V. missouriensis*. N. H. Russell (1965) considered the two as likely part of a species complex inhabiting alluvial woods and wet areas and exhibiting the typical deltate leaf blade shape. L. E. McKinney (1992) considered *V. missouriensis* a variety of *V. sororia*. *Viola missouriensis* appears to have a closer affinity to *V. affinis*, as Russell suggested; current evidence suggests maintaining *V. missouriensis* as a separate species.

35. Viola nephrophylla Greene, Pittonia 3: 144. 1896
• Northern bog violet, violette néphrophylle E

Viola arizonica Greene; *V. austiniae* Greene; *V. cognata* Greene; *V. crassula* Greene; *V. galacifolia* Greene; *V. lunellii* Greene; *V. mccabeiana* M. S. Baker; *V. nephrophylla* var. *arizonica* (Greene) Kearney & Peebles; *V. nephrophylla* var. *cognata* (Greene) C. L. Hitchcock; *V. peramoena* Greene; *V. pratincola* Greene; *V. prionosepala* Greene; *V. retusa* Greene; *V. subjuncta* Greene; *V. vagula* Greene

Plants perennial, acaulescent, not stoloniferous, 5–15 cm; rhizome slender, becoming thick and fleshy with age. **Leaves** basal, 4–7, ascending to erect; stipules lanceolate, margins entire or fimbriate, apex acute; petiole 2–25 cm, usually glabrous; blade usually grayish green or purplish green abaxially, unlobed, ovate, reniform, or broadly reniform to orbiculate, 1–7 × 1–7 cm, somewhat fleshy, base broadly cordate or reniform to ± truncate, margins crenate to serrate, ciliate or eciliate, apex acute to obtuse or rounded, surfaces usually glabrous, rarely sparsely pubescent. **Peduncles** 3–25 cm, usually glabrous. **Flowers:** sepals ovate, margins usually eciliate, auricles 1–2 mm; petals deep bluish violet on both surfaces, lower 3 white basally and darker violet-veined, lateral 2 bearded, upper 2 sometimes sparsely bearded, lowest densely bearded or beardless, 10–28 mm, spur same color as petals, gibbous, 2–3 mm; style head beardless; cleistogamous flowers on erect to ascending peduncles. **Capsules** ovoid, 5–10 mm, glabrous. **Seeds** beige to brown or dark brown, 1.5–2.5 mm. $2n = 54$.

Flowering Jan–Sep. Wet habitats in saturated soil in prairies, pastures, bogs, fens, sedge meadows, rocky shores of lakes and streams, limestone outcrops, gravelly calcareous stream beds; 100–3000 m; Alta., B.C., Man., N.B., Nfld. and Labr. (Nfld.), N.W.T., N.S., Nunavut, Ont., P.E.I., Que., Sask., Yukon; Ariz., Ark., Calif., Colo., Conn., Idaho, Ill., Iowa, Kans., Maine, Mass., Mich., Minn., Mont., Nebr., Nev., N.H., N.Mex., N.Y., N.Dak., Ohio, Oreg., Pa., S.Dak., Tex., Utah, Vt., Wash., W.Va., Wis., Wyo.

Viola nephrophylla has the widest distribution in North America of any of the acaulescent blue violets.

There is some question whether the species now known as *Viola nephrophylla* fits with the protologue and type specimen(s). McKinney examined the type designated by Greene; it appears that what most botanists have called *V. nephrophylla* may be an undescribed taxon. N. L. Gil-Ad (1997) believed Greene's type contained plants that were genetically impure based on seed coat micromorphology and suggested the specimens might be a mixture of both orthospecies and introgressants.

Viola mccabeiana was described by M. S. Baker in 1940 from collections made in British Columbia. Later, Baker (1949b) discussed the reasons why the name *V. mccabeiana* needed to be abandoned. C. L. Hitchcock et al. (1955–1969, vol. 3) treated *V. mccabeiana* as synonymous with *V. nephrophylla* var. *cognata*. While *V. mccabeiana* is recognized in Canada (L. Brouillet et al., http://data.canadensys.net/vascan/), we believe it is best included within *V. nephrophylla*.

Viola nephrophylla reportedly hybridizes with *V. cucullata* (= *V. ×insessa* House), *V. sororia* (= *V. ×napae* House), and *V. affinis* (= *V. ×subaffinis* House).

36. Viola novae-angliae House, Rhodora 6: 226, plate 59. 1904 • New England blue violet E

Viola septentrionalis Greene var. *grisea* Fernald; *V. sororia* Willdenow var. *grisea* (Fernald) L. E. McKinney; *V. sororia* var. *novae-angliae* (House) L. E. McKinney

Plants perennial, acaulescent, not stoloniferous, 5–50 cm; rhizomes thick, fleshy. **Leaves** basal, 2–8, ascending to erect; stipules linear-lanceolate to broadly lanceolate, margins entire, sometimes distally glandular, apex acute; petiole 5–25 cm, densely pubescent; blade unlobed, narrowly ovate to narrowly deltate, 3.5–7 × 2–5 cm, base cordate, margins uniformly crenate to serrate, ciliate or eciliate, apex acute, surfaces usually glabrous, rarely pubescent. **Peduncles** 3–25 cm, glabrous or pubescent. **Flowers:** sepals lanceolate to ovate, margins ciliate or eciliate, auricles 1–2 mm; petals light to dark blue- or dark purple-violet or reddish purple, lowest and sometimes lateral 2 purple-veined, lateral 2 bearded, lowest bearded or beardless, 15–25 mm, spur same color as petals, gibbous, 2–3 mm; style head beardless; cleistogamous flowers on prostrate to ascending peduncles. **Capsules** ellipsoid, 5–12 mm, glabrous. **Seeds** beige, mottled to bronze, 1.5–2.5 mm. $2n = 54$.

Flowering Apr–Jun. Gravelly, wet, rocky shores of lakes and streams, meadows; 0–100 m; Man., N.B., Ont., Que.; Maine, Mich., Minn., Wis.

L. E. McKinney (1992) considered *Viola novae-angliae* to be a variety of *V. sororia*; N. L. Gil-Ad (1997) treated it as a distinct species. H. E. Ballard and S. C. Gawler (1994) also considered *V. novae-angliae* distinct and suggested it might be a hybrid between *V. sagittata* and *V. sororia*. *Viola novae-angliae* appears to have a close affinity with *V. sororia*.

37. Viola nuttallii Pursh, Fl. Amer. Sept. 1: 174. 1813
    • Nuttall's or Nuttall's yellow or yellow prairie violet,
violette de Nuttall E

Crocion nuttallii (Pursh)
Nieuwland & Lunell

Plants perennial, caulescent,
not stoloniferous, 2–27 cm.
Stems 1–6, ascending to erect,
leafy proximally and distally,
ca. ½ subterranean, puberulent,
on caudex from usually vertical,
subligneous rhizome.   **Leaves**
basal and cauline; basal: 1–6; stipules adnate to petiole,
forming 2 linear-lanceolate wings, margins entire,
apex of each wing free, acute, few-toothed or lobed;
petiole 2–17 cm, glabrous or minutely puberulent;
blade lanceolate, ovate, or elliptic, 1–9 × 0.6–2.5 cm,
base attenuate, margins entire or serrulate, sometimes
sinuate, ciliate, apex acute to obtuse, mucronulate,
surfaces glabrous or puberulent; cauline similar to basal
except: stipules adnate to or free from petiole, linear to
linear-lanceolate or linear-oblong, sometimes leaflike,
margins usually entire, rarely laciniate or glandular, apex
acute to acuminate; petiole 2–7 cm; blade 1.4–7.2(–10)
× 1.1–2.3 cm, length 1.3–4.4 times width, apex acute.
Peduncles 3–13 cm, glabrous or puberulent. **Flowers:**
sepals linear-lanceolate, margins eciliate, auricles 0.5–1
mm; petals deep lemon-yellow adaxially or on both
surfaces, upper 2 often brownish purple abaxially,
lower 3 dark brown- to brownish purple-veined, lateral
2 sparsely bearded, lowest 6–13 mm, spur yellow,
gibbous, 0.5–1.5 mm; style head bearded; cleistogamous
flowers axillary. **Capsules** subglobose to ovoid, 4–10
mm, usually glabrous, rarely finely puberulent. **Seeds**
medium brown, 2–3.2 mm, elaiosome extending over
⅓ length of seed and completely covering funiculus.
2n = 24.

Flowering Apr–Jun. Sagebrush flats, prairie grasslands, dry stream banks, juniper woodlands, scree
slopes; 400–2600 m; Alta., Man., Sask.; Ariz., Colo.,
Idaho, Kans., Minn., Mont., Nebr., N.Mex., N.Dak.,
S.Dak., Utah, Wyo.

D. M. Fabijan et al. (1987) stated that *Viola nuttallii*
showed no evidence of hybridization even when growing
with *V. vallicola* east of the Rocky Mountains.

38. Viola ocellata Torrey & A. Gray, Fl. N. Amer. 1: 142.
1838 • Western heart's ease or two-eyed or southern
two-eyed violet E

Plants perennial, caulescent, not
stoloniferous, 2–37 cm. **Stems**
1–3,   ascending   to   erect,
± glabrous or usually puberulent,
on caudex from shallow, fleshy
rhizome or deep-seated caudex
with fleshy roots. **Leaves** basal
and cauline; basal: 1–6; stipules
deltate, margins laciniate, apex
usually long-acuminate; petiole 3.7–10 cm, puberulent;
blade ovate, deltate, or subreniform, 1–6 × 1.2–4 cm,
base usually cordate to subcordate, margins crenate,
usually ciliate, apex acute, mucronulate, surfaces usually
sparsely puberulent, glabrous early; cauline similar to
basal except: stipules lanceolate, margins ± fimbriate,
sometimes stipitate-glandular, sometimes entire, erose,
or ± laciniate; petiole 0.4–9 cm; blade ovate to deltate,
1.6–4.4 × 1.1–3.6 cm, base subcordate to truncate,
margins crenate to ± serrate, ciliate, apex acute to obtuse.
Peduncles 1–10 cm, puberulent.   **Flowers:** sepals
lanceolate, margins ciliate, auricles 0.5–1.5 mm; petals
white adaxially, upper 2 and sometimes lower 3 deep
reddish violet abaxially, all with yellow area basally,
lateral 2 with purple patch basally distal to smaller
yellow area, bearded, lowest with yellow patch basally,
purple-veined, 5–15 mm, spur yellow or greenish,
gibbous, 1–2.5 mm; style head bearded; cleistogamous
flowers axillary. **Capsules** spherical to spherical-ovoid,
5–11 mm, minutely scabrous. **Seeds** brown-purple, ca.
2 mm. **2n** = 12.

Flowering Mar–Jul.   Rocky areas, grassy banks,
thickets, often on serpentine soil; 100–1100 m; Calif.,
Oreg.

39. Viola odorata Linnaeus, Sp. Pl. 2: 934. 1753
    • English or sweet violet, violette odorante I

Plants perennial, acaulescent,
stoloniferous, 4–12 cm; stolons
green, often rooting at nodes
and   forming   leafy   rosettes;
rooted rosettes often develop
into erect, rhizomatous caudex
from which new stolons are
produced; rhizome thick, fleshy.
Leaves basal (and from stolons),
5–10, ascending to erect; stipules lanceolate to linear-
lanceolate, margins fimbriate, projections gland-tipped,
apex acute; petiole 2–17 cm, puberulent; blade unlobed,
ovate to orbiculate, 1.5–7 × 1.5–5 cm, base cordate,
margins crenate, ciliate, apex obtuse to rounded, surfaces
puberulent. **Peduncles** 4–15 cm, puberulent. **Flowers:**
sepals narrow to broadly lanceolate, margins ciliate,

auricles 3–4 mm; petals deep to pale blue-violet, pale blue, or white on both surfaces, usually white basally, lateral 2 sparsely to densely bearded, lowest usually purple-veined, 12–22 mm, spur usually same color as petals, elongated, 5–7 mm; style head beardless; cleistogamous flowers on prostrate or ascending peduncles. **Capsules** sometimes purple-flecked, ovoid, 5–8 mm, puberulent. **Seeds** brown, 3–4 mm. $2n = 20$.

Flowering Jan–May. Lawns, roadsides, clearings, riparian habitats, parks, urban areas; 0–1700 m; introduced; B.C., N.S., Ont., Que.; Calif., Conn., Idaho, Ill., Maine, Mass., Mich., N.Y., N.C., Ohio, Oreg., Pa., R.I., Utah, Wash., Wis.; Eurasia; introduced also in Australia.

Viola odorata occurs in small colonies; individual plants are interconnected by stolons.

The flowers of *Viola odorata* are noted for their fragrance; some plants are more fragrant than others. It is native to Eurasia and assumed to be introduced in North America where it is usually found in areas associated with human habitation, including parks, lawns, and roadsides. A substantial industry revolved around the commercial production of violets in England, France, and the United States from prior to 1895 and into the 1900s (R. E. Coombs 2003). *Viola odorata* is sometimes found in remote locations not easily explained by anthropogenic influence, for example, Clearwater Mountains, Idaho. It is available through the nursery trade and is cultivated as a garden plant and occasionally reported as an escape. *Viola odorata* is grown in southern France for essential oils used in perfumes, flavorings, and toiletries, and also for the production of the sweet, violet-colored liqueur called *parfait amour* (V. H. Heywood 1978).

40. **Viola orbiculata** (A. Gray) Geyer ex B. D. Jackson in B. D. Jackson et al., Index Kew. 2: 1208. 1895

• Round-leaved or round-leaved wood or western round-leaved or evergreen violet, violette orbiculaire E

Viola sarmentosa Douglas var. *orbiculata* A. Gray in A. Gray et al., Syn. Fl. N. Amer. 1(1,1): 199. 1895; *V. sempervirens* Greene var. *orbiculata* (A. Gray) J. K. Henry

Plants perennial, caulescent, not stoloniferous, 5–9 cm. **Stems** 1–5, prostrate or erect, leafy proximally and distally, usually glabrous, from current and/or previous year's growth, on usually vertical, fleshy rhizome. **Leaves** basal and cauline; basal: 3–11; stipules deltate to lanceolate, margins entire, apex acute to acuminate; petiole 2.1–10.5 cm, glabrous or pubescent; blade usually orbiculate to broadly ovate, sometimes reniform, 1.4–5.3 × 1.4–5.3 cm, base cordate, margins crenate to serrulate-crenulate, eciliate or sparingly ciliate, apex usually obtuse, rarely acute, abaxial surface usually glabrous, adaxial surface sparsely pubescent; cauline similar to basal except: stipules lanceolate to ovate, margins usually entire, ciliate; petiole 0.7–1 cm, glabrous or pubescent; blade 1.2–2 × 1.1–1.4 cm. **Peduncles** 2.7–5.6 cm, glabrous. **Flowers:** sepals narrow to broadly lanceolate, margins eciliate, auricles 0.5–1.5 mm; upper 2 and lateral 2 petals deep lemon-yellow on both surfaces, lower 3 and sometimes upper 2 brownish purple-veined, lateral 2 usually bearded, lowest 8–17 mm, spur usually yellow, sometimes whitish, gibbous, 1.5–2.5 mm; style head bearded; cleistogamous flowers axillary. **Capsules** ellipsoid, 6–8 mm, glabrous. **Seeds** light to dark brown, sometimes mottled with white, 2–2.3 mm. $2n = 24$.

Flowering May–Aug. Alpine and montane slopes, moist montane coniferous forests, canyons, meadows, lake margins; 700–1700 m; Alta., B.C.; Idaho, Mont., Oreg., Wash., Wyo.

During winter, the basal leaves of *Viola orbiculata* are appressed to the ground under the weight of snow. The leaves overwinter and are often partly or entirely green after the snow has melted. Mature plants possess three to 11 basal leaves, some derived from previous years and others from the current year's growth. Occasional plants are found with short, thin stems rooted at the tip and producing a rosette of leaves and stems, showing the close relationship with *V. sempervirens*.

41. **Viola palmata** Linnaeus, Sp. Pl. 2: 933. 1753

• Three-lobed violet E

Viola cucullata Aiton var. *palmata* (Linnaeus) A. Gray

Plants perennial, acaulescent, not stoloniferous, 6–50 cm; rhizome thick, fleshy. **Leaves** basal, 2 or 3, ascending to erect; stipules linear-lanceolate, margins entire, apex acute; petiole 1–20 cm, glabrous or pubescent; earliest leaf blades unlobed, mid-season blades 3–9-lobed, lobes sometimes appearing petiolate and sometimes further lobed; earliest leaf blades reniform to ovate, mid-season blades with middle lobes usually ovate or elliptic to widely obovate, sometimes narrowly elliptic, narrowly ovate, lanceolate, or spatulate to narrowly obovate, lateral lobes elliptic, obdeltate, or spatulate to falcate, smaller lobes similar, 1–14 × 1–10 cm, base truncate to cordate, margins serrate or entire, usually ciliate, apex acute, rounded, blunt, or obtuse, surfaces glabrous or pubescent throughout or along veins. **Peduncles** 3–13 cm, glabrous or pubescent. **Flowers:** sepals lanceolate, margins ciliate or eciliate, auricles 0.5–1 mm; petals violet on both surfaces, lower 3 white basally, lower 3 and upper 2 sometimes purple-veined, lateral 2 bearded, spur sometimes bearded, lowest 15–25 mm, spur white, gibbous, 2–3 mm; style head beardless; cleistogamous flowers on prostrate to ascending peduncles. **Capsules** ellipsoid, 5–15 mm, glabrous. **Seeds** beige, mottled to bronze, 1.5–2.5 mm. $2n = 54$.

Varieties 2 (2 in the flora): c, e North America.

N. L. Gil-Ad (1997) presented evidence based on seed coat micromorphology that some plants known as *Viola palmata* are hybrids between plants with lobed and unlobed leaves and believed the type specimen to be hybrid in origin. He chose not to recognize the name *V. palmata*, choosing instead to recognize *V. triloba*. If the type specimen does represent a hybrid, Gil-Ad would be correct in recognizing *V. triloba*. Because the purity of the type specimen cannot be determined, *V. palmata* is the most appropriate name for this taxon. We acknowledge the presence of hybrids between lobed and unlobed plants; such hybrids obscure lines of demarcation. Some herbarium specimens may represent such hybrids, but most do not.

L. E. McKinney (1992) described the nomenclatural history of *Viola palmata*. Homophylly versus heterophylly differentiates *V. palmata* from *V. subsinuata*, the homophyllous violet most often called *V. palmata*. The taxon described by Linnaeus was heterophyllous, with some undivided leaf blades.

1. Petioles, leaf surfaces, and peduncles usually pubescent, rarely glabrous; apex of middle leaf lobe acute; plants of dry to mesic habitats. . . . . .41a. *Viola palmata* var. *palmata*
1. Petioles, leaf surfaces, and peduncles usually glabrous, rarely pubescent; apex of middle leaf lobe rounded, blunt, or obtuse; plants of wet ground . . . . 41b. *Viola palmata* var. *heterophylla*

41a. Viola palmata Linnaeus var. **palmata** • Wood or palmate-leaved violet, violette palmée E

Viola congener Leconte; *V. falcata* Greene; *V. lovelliana* Brainerd; *V. palmata* var. *dilatata* Elliott; *V. palmata* var. *triloba* (Schweinitz) Gingins; *V. stoneana* House; *V. triloba* Schweinitz; *V. triloba* var. *dilatata* (Elliott) Brainerd; *V. viarum* Pollard

Plants usually pubescent, rarely glabrous. **Leaves** 3–9-lobed, middle lobes ovate, widely ovate, elliptic to widely obovate, sometimes lanceolate, or spatulate to narrowly ovate, apex acute, surfaces pubescent throughout or concentrated along veins or on petioles and peduncles.

Flowering Mar–Jun. Dry to mesic woods, thickets, disturbed ground; 0–2500 m; Ont.; Ala., Ark., Conn., Del., D.C., Fla., Ga., Ill., Ind., Iowa, Kans., Ky., La., Maine, Md., Mass., Mich., Miss., Mo., Nebr., N.J., N.Mex., N.C., N.Dak., Ohio, Okla., Pa., R.I., S.C., Tenn., Tex., Vt., Va., W.Va.

41b. Viola palmata Linnaeus var. **heterophylla** Elliott, Sketch. Bot. S. Carolina 1: 300. 1817 • Salad violet E

Viola esculenta Elliott ex Greene

Plants usually glabrous, rarely pubescent. **Leaves** 3–5-lobed, middle lobes ovate, widely ovate, or elliptic to widely obovate, apex rounded, blunt, or obtuse.

Flowering Mar–Jun. Blackwater flood plains and other wet ground along streams, swamps; 0–100 m; Ala., Fla., Ga., La., Miss., N.C., S.C., Tex., Va.

L. E. McKinney (1992) did not recognize var. *heterophylla*; D. B. Ward (2006) provided convincing evidence to recognize it, but he chose to do so as var. *esculenta* (Elliott ex Greene) D. B. Ward, an illegitimate name.

42. Viola palustris Linnaeus, Sp. Pl. 2: 934. 1753 • Northern marsh or marsh or swamp or alpine marsh violet, violette des marais

Viola palustris subsp. *brevipes* M. S. Baker; *V. palustris* var. *brevipes* (M. S. Baker) R. J. Davis; *V. palustris* var. *leimonia* J. K. Henry; *V. palustris* var. *pensylvanica* Gingins; *V. palustris* subsp. *pubifolia* Kuta

Plants perennial, acaulescent, stoloniferous, 3–21.5 cm; stolons pale, often rooting and leafy at nodes; rhizome slender, fleshy. **Leaves** basal, 2–4, ascending to erect; stipules linear-lanceolate, margins entire, apex acute; petiole 1–17 cm, glabrous; blade unlobed, reniform or ovate to orbiculate, 0.5–6.4 × 0.5–5.5 cm, ± as long as wide, base cordate, margins crenulate, eciliate, apex obtuse to acuminate, surfaces glabrous. **Peduncles** 2–20.7 cm, glabrous, bracteoles usually below middle in chasmogamous flowers. **Flowers:** sepals ovate or lanceolate, margins eciliate, auricles 1–2 mm; petals lilac, pale blue, or pale violet on both surfaces, sometimes nearly white and tinged with lilac, lowest and sometimes lateral 2 dark violet-veined, lateral 2 bearded or beardless, lowest 8–16 mm, spur usually same color as petals, gibbous, 1–3 mm; style head beardless; cleistogamous flowers on ascending peduncles. **Capsules** ellipsoid, 6–10 mm, glabrous. **Seeds** dark brown, 1–2 mm. $2n = 48$.

Flowering Apr–Jul. Marshes, swamps, fens, wet forests, stream banks, brushy places, lowlands to subalpine zones; 0–1800 m; Greenland; Alta., B.C., Man., Nfld. and Labr., N.W.T., Nunavut, Ont., Que., Sask.; Calif., Colo., Idaho, Maine, Mont., Nev., N.H., Oreg., S.Dak., Utah, Wash., Wyo.; Eurasia.

Viola palustris occurs in small colonies; individual plants are interconnected by stolons. When flowering, usually three or more leaves are present per caudex.

M. Sorsa (1968) noted that in western North America *Viola palustris* specimens differ from European *V. palustris* plants and often show variation in character combinations. She stated (pers. comm.) that North American *V. palustris* is not a uniform entity.

E. Hultén (1968) synonymized *Viola palustris* with *V. epipsila*, as did W. J. Cody (2000). G. W. Douglas et al. (1998–2002, vol. 5) recognized both species. We assume that *V. palustris* does not occur in Alaska; verification is needed. See additional discussion under 18a. *V. epipsila* var. *repens*.

M. Sorsa (1968) indicated that *Viola palustris* var. *brevipes* has often been confused with *V. blanda* and that var. *brevipes* may have arisen from hybridization between a tetraploid and a diploid plant. Listed here in synonymy, var. *brevipes* may warrant recognition.

Reports of *Viola palustris* from one location in Coconino County, Arizona, first collected in 1962, likely refer to *V. macloskeyi*.

Viola palustris hybridizes with *V. epipsila* var. *repens* (M. Sorsa 1968; T. Marcussen and T. Karlsson 2010).

43. Viola pedata Linnaeus, Sp. Pl. 2: 933. 1753 E F

Plants perennial, acaulescent, not stoloniferous, 5–30 cm; rhizome thick, fleshy. **Leaves** basal, 4–10, ascending to erect, deeply divided; stipules linear-lanceolate, margins entire, lacerate, or shallowly divided, apex acute; petiole 2–12 cm, glabrous or sparsely puberulent; blade 3–9(10)-lobed, lobes similar in width and shape, spatulate, lanceolate, ± linear, deltate, or ovate, 1–4 × 1–4 cm, base attenuate or broadly cordate to cuneate, margins entire, ciliate or eciliate, apex rounded to usually acute, surfaces usually glabrous, sometimes pubescent on abaxial veins. **Peduncles** 5–12 cm, glabrous or pubescent. **Flowers:** sepals lanceolate, margins mostly ciliate, at least proximally, auricles 1–2 mm; petals uniformly light to dark blue-violet on both surfaces or upper 2 darker adaxially, sometimes white, upper and lateral 2 often darker basally, lowest, seldom others, dark violet-veined, all beardless, lowest white basally, 12–24 mm, spur white, gibbous, 2–3 mm; style head beardless; cleistogamous flowers absent. **Capsules** ellipsoid, 6–10 mm, glabrous. **Seeds** beige, mottled to brown, 1.4–3 mm. $2n$ = 56.

Varieties 2 (2 in the flora): e North America.

1. Leaf blades 7–9-lobed, lobes spatulate, lanceolate, or ± linear, sometimes with narrowly deltate to falcate appendages toward apex . . . . 43a. *Viola pedata* var. *pedata*
1. Leaf blades 3–5(–10)-lobed, lobes deltate or ovate . . . . 43b. *Viola pedata* var. *ranunculifolia*

43a. Viola pedata Linnaeus var. **pedata** • Bird's foot or bird-foot violet, violette pédalée E F

Viola ampliata Greene; *V. digitata* Pursh; *V. pedata* var. *atropurpurea* Gingins; *V. pedata* var. *inornata* Greene; *V. pedata* var. *lineariloba* Gingins; *V. redunca* House

Leaf blades reniform or ovate to elliptic, 7–9-lobed, lobes spatulate, lanceolate, or ± linear, sometimes with narrowly deltate to falcate appendages toward apex, base broadly cordate to cuneate, apex acute, mucronulate.

Flowering Mar–Jun. Sandy, open woods, fields, and rights-of-way; 0–2000 m; Ont.; Ala., Ark., Conn., Del., D.C., Ga., Ill., Ind., Iowa, Kans., Ky., La., Md., Mass., Mich., Minn., Miss., Mo., Nebr., N.H., N.J., N.Y., N.C., Ohio, Okla., Pa., R.I., S.C., Tenn., Tex., Va., W.Va., Wis.

The orange tips of the stamens of var. *pedata* are exserted.

43b. Viola pedata Linnaeus var. **ranunculifolia** Gingins in A. P. de Candolle and A. L. P. P. de Candolle, Prodr. 1: 291. 1824 E

Viola ranunculifolia Jussieu ex Poiret

Leaf blades ovate to narrowly elliptic, 3–5(–10)-lobed, lobes deltate or ovate, base cuneate, apex rounded to acute.

Flowering Mar–Jun. Sandy, open woods; 50–200 m; N.C., S.C.

Variety *ranunculifolia* is apparently endemic to the Fall Line Sandhill physiographic province of the Carolinas.

44. Viola pedatifida G. Don, Gen. Hist. 1: 320. 1831
• Prairie or crowfoot or larkspur violet, violette pédatifide [E]

Viola delphiniifolia Nuttall; *V. palmata* Linnaeus var. *pedatifida* (G. Don) Cronquist

Plants perennial, acaulescent, not stoloniferous, 5–30 cm; rhizome thick, fleshy. **Leaves** basal, 2–11, ascending to erect, 5–9-lobed; stipules linear-lanceolate, margins entire, apex acute; petiole 3–16 cm, pubescent; blade similar in width and shape, lobes lanceolate, spatulate, falcate, or linear, 1–7 × 2–8 cm, base truncate to reniform, margins entire, ciliate, apex acute to obtuse, mucronulate, surfaces pubescent, hairs sometimes concentrated on veins. **Peduncles** 5–18 cm, glabrous or pubescent. **Flowers:** sepals lanceolate to ovate, margins ciliate or eciliate, auricles 1–2 mm; petals light to soft reddish violet on both surfaces, lower 3 white basally, dark violet-veined, lateral 2 and lowest usually bearded, lowest 10–25 mm, spur same color as petals, gibbous, 2–3 mm; style head beardless; cleistogamous flowers on ascending to erect peduncles. **Capsules** ellipsoid, 10–15 mm, glabrous. **Seeds** beige, mottled to bronze, 1.5–2.5 mm. $2n = 54$.

Flowering Mar–Jun. Prairies, grasslands, disturbed ground, dry gravelly hills; 500–1000 m; Alta., Man., Ont., Sask.; Ariz., Ark., Colo., Ill., Ind., Iowa, Kans., Mich., Minn., Mo., Mont., Nebr., N.Mex., N.Dak., Ohio, Okla., S.Dak., Va., Wis., Wyo.

Viola pedatifida was reported historically from, and recently rediscovered in, the Appalachian shale barrens of Virginia (T. Wieboldt, pers. comm.).

Viola pedatifida reportedly hybridizes with *V. sororia* [= *V.* ×*bernardii* (Greene) Greene].

45. Viola pedunculata Torrey & A. Gray, Fl. N. Amer. 1: 141. 1838 • California golden violet, Johnny-jump-up, wild pansy

Viola pedunculata subsp. *tenuifolia* M. S. Baker & J. C. Clausen

Plants perennial, caulescent, not stoloniferous, 5–39 cm. **Stems** 1–10+, decumbent, ascending, or erect, leafy proximally and distally, glabrous or puberulent, from shallow to deep-seated, enlarged rhizome with fleshy to subligneous roots. **Leaves** cauline; stipules ovate, linear-lanceolate, or oblanceolate, sometimes leaflike, margins entire or glandular-toothed, apex acute to acuminate; petiole 2.7–7.2 cm, usually finely puberulent, sometimes glabrate; blade deltate to ovate, 1–5.5 × 1–5.5 cm, base truncate, subcordate, or attenuate, margins crenate to serrate, ciliate, apex acute to obtuse, surfaces subglabrous or sparsely puberulent. **Peduncles** 2.9–20 cm, sparsely to densely puberulent. **Flowers:** sepals lanceolate to ovate, margins ciliate or eciliate, auricles 1–3 mm; petals golden yellow adaxially, upper 2 reddish brown abaxially, lower 3 dark brown-veined, lateral 2 bearded, lowest 10–20 mm, spur dark reddish brown, gibbous, 2–4 mm; style head bearded; cleistogamous flowers absent. **Capsules** ellipsoid, 5–11 mm, glabrous. **Seeds** dark brown or black, shiny, 2.7 mm. $2n = 12$.

Flowering Feb–Apr. Open, grassy coastal and inland slopes and hillsides, usually in full sun, chaparral, foothill and oak woodlands; 0–1000 m; Calif.; Mexico (Baja California).

The stems of *Viola pedunculata* arise from an enlarged, subterranean, spongy or fibrous rhizome. Often, these rhizome structures are deep seated; it is unknown how they get so deeply buried. The anther appendages of *V. pedunculata* are hairy distally, a characteristic not known to occur in other members of the *V. purpurea* complex.

Larvae of the federally listed Callippe silverspot butterfly [*Speyeria callippe* (Boisduval) *callippe*] feed only on *Viola pedunculata*.

Plants with leaves reported to be smaller, thinner, deltate, mostly longer than wide, with yellow petals (versus orange for *Viola pedunculata* var. *pedunculata*), style 2.1 mm (versus 2.9 mm for var. *pedunculata*), from the Pinnacles region in San Benito County, have been called subsp. *tenuifolia*.

46. Viola pinetorum Greene, Pittonia 2: 14. 1889 [E]

Viola purpurea Kellogg var. *pinetorum* (Greene) Greene

Plants perennial, caulescent, not stoloniferous, 3–18 cm, cespitose or not. **Stems** 1–3, prostrate or erect, leafy proximally and distally, puberulent or canescent to gray-tomentose, sometimes glabrous, on caudex from subligneous rhizome. **Leaves** basal and cauline; basal: 1–4; stipules adnate to petiole forming 2 linear-lanceolate wings, margins entire or laciniate, apex of each wing free, tips usually filamentous; petiole 2.3–9.5 cm, puberulent or canescent; blade purple-tinted abaxially or not, usually linear to narrowly lanceolate, oblanceolate or obovate, or lanceolate-elliptic, rarely ovate, 1.3–5 × 0.3–2.5 cm, base attenuate, margins usually lacerate, dentate, or serrate, sometimes entire, usually undulate, ciliate, apex acute, mucronulate, surfaces puberulent to canescent or gray-tomentose; cauline similar to basal except: stipules lanceolate, oblanceolate, or linear-oblong, margins entire or lacerate, apex acute to acuminate; petiole 0.9–8.3 cm; blade 2.8–9.6 × 0.3–1.4 cm, length 4–11 times width. **Peduncles** 2.9–11.5 cm, puberulent or canescent. **Flowers:** sepals lanceolate, margins ciliate, auricles 0.5–1 mm; petals deep lemon-yellow adaxially, upper 2 red- to purple-brown abaxially, lower 3 dark brown-veined, lateral 2 bearded, lowest 5–12 mm, spur color same as petals, gibbous, 1.5–3 mm; style head bearded; cleistogamous flowers axillary. **Capsules** ovoid, 3.5–7 mm, puberulent. **Seeds** medium to dark brown, 2–3.5 mm. $2n = 12$.

Varieties 2 (2 in the flora): California.

Flowers of *Viola pinetorum* have been observed to close up in late afternoon then fully reopen the following morning.

Although E. O. Wooton and P. C. Standley (1915) reported *Viola pinetorum* from New Mexico, the plant was probably *V. nuttallii*. K. W. Allred (2008) noted that *V. pinetorum* occurs in California; he did not recognize it in New Mexico.

1. Plants 6.5–18 cm, not cespitose; basal leaf blades 0.7–2.5 cm wide, surfaces puberulent; peduncles 3.4–11.5 cm . . . . . 46a. *Viola pinetorum* var. *pinetorum*
1. Plants 3–7(–9) cm, usually cespitose; basal leaf blades 0.3–1 cm wide, surfaces canescent, sometimes appearing gray-tomentose; peduncles 2.9–6(–7) cm. . . . . 46b. *Viola pinetorum* var. *grisea*

46a. Viola pinetorum Greene var. pinetorum

• Mountain yellow violet E

Plants 6.5–18 cm, not cespitose. **Stems** usually puberulent, sometimes glabrous. **Leaves:** basal: petiole (3.2–)4–9.5 cm; blade usually purple-tinted abaxially, linear or lanceolate-elliptic, rarely ovate, 1.3–5 × 0.7–2.5 cm, margins usually irregularly lacerate, dentate, or serrate, sometimes entire, undulate, surfaces puberulent; cauline similar to basal except: petiole 1.7–8.3 cm; blade (2.8–)3.5–9.6 × 0.3–1.4 cm. **Peduncles** 3.4–11.5 cm. **Flowers:** lowest petal 6–12 mm.

Flowering late May–late Jul. Montane slopes, often in moist, eroding soil, often beneath pines and firs in coniferous forests; 1400–3000 m; Calif.

Viola pinetorum was described from collections made by E. L. Greene in the Tehachapi Mountains of southern California. There are similarities between the elongated leaves of var. *pinetorum* and those of *V. purpurea* var. *mesophyta*. The leaves of var. *pinetorum* are often twice as long as those of var. *mesophyta* and margins are usually undulate.

Variety *pinetorum* has been observed to be pollinated by two unidentified species of solitary bees and by an unidentified species of skipper (Hesperiidae).

46b. Viola pinetorum Greene var. grisea (Jepson)

R. J. Little, J. Bot. Res. Inst. Texas 5: 633. 2011

• Gray-leaved violet E

Viola purpurea Kellogg var. *grisea* Jepson, Fl. Calif. 2: 521. 1936; *V. pinetorum* subsp. *grisea* (Jepson) R. J. Little

Plants 3–7(–9) cm, usually cespitose. **Stems** canescent to gray-tomentose. **Leaves:** basal: petiole 2.3–5.5(–7.5) cm; blade usually not purple-tinted abaxially, usually linear to narrowly lanceolate, sometimes oblanceolate or obovate, 1.7–4 × 0.3–1 cm, margins irregularly serrate or lacerate, surfaces canescent, sometimes appearing gray-tomentose; cauline similar to basal except: petiole 0.9–4.4 cm, canescent; blade 1.5–4 × 0.3–0.5(–0.9) cm. **Peduncles** 2.9–6(–7) cm. **Flowers:** lowest petal 5–9 mm.

Flowering Jun–Jul. Montane slopes and peaks, often in moist, eroding soil, alpine zones; 2000–3100 m; Calif.

47. Viola praemorsa Douglas ex Lindley, Edwards's Bot. Reg. 15: plate 1254. 1829 E F

Viola nuttallii Pursh subsp. *praemorsa* (Douglas ex Lindley) Piper; *V. nuttallii* var. *praemorsa* (Douglas ex Lindley) S. Watson

Plants perennial, caulescent, not stoloniferous, 5.5–36.5 cm. **Stems** 1–3(–5), prostrate, decumbent, or erect, leafy proximally and distally, glabrous or puberulent, on caudex from usually vertical, subligneous rhizome. **Leaves** basal and cauline; basal: 1–5; stipules adnate to proximal ⅓ of petiole, forming 2 narrow, linear-lanceolate wings, margins entire, apex of each wing free, acute; petiole 2.6–19.2 cm, glabrous or densely puberulent; blade usually elliptic to ovate, sometimes oblong-lanceolate to ± orbiculate, 1.7–14(–17) × 0.8–5.3(–6.7) cm, base attenuate to ± truncate or subcordate, often oblique, margins usually crenate, serrulate, or serrate, sometimes entire, ciliate or eciliate, apex acute to obtuse, surfaces glabrous or densely puberulent; cauline similar to basal except: stipules lanceolate to ovate, margins entire or toothed, with or without gland-tipped projections, apex acute to acuminate; petiole 1.3–16.2 cm, glabrous or puberulent; blade 2.3–11(–14.8) × 1.1–3.6(–5.5) cm, length 1.1–6.5 times width. **Peduncles** 4.4–27 cm, glabrous or puberulent. **Flowers:** sepals lanceolate, margins eciliate, auricles 1–2 mm; petals deep lemon-yellow adaxially, upper 2 and sometimes lateral 2, brownish purple abaxially, lower 3 brownish purple-veined, lateral 2

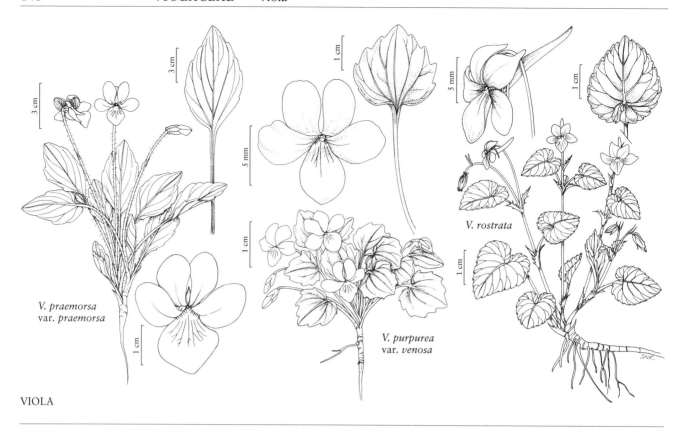

V. praemorsa
var. *praemorsa*

V. rostrata

V. purpurea
var. *venosa*

VIOLA

sparsely bearded, lowest 11–19 mm, spur yellow or pale green, gibbous, 0.5–3 mm; style head bearded; clcistogamous flowcrs axillary. **Capsules** ellipsoid to oblong, 6–14 mm, glabrous or finely puberulent. **Seeds** medium to dark brown or red-brown, 2–3 mm, elaiosome completely covering funiculus. $2n$ = 36, 48.

Varieties 3 (3 in the flora): w North America.

Viola praemorsa is a complex group that appears to be evolving. Sympatric populations and the similarity of their flowers provide opportunities for hybridization and introgression (D. M. Fabijan et al. 1987). Some botanists have suggested that all taxa in the *V. nuttallii* complex, which includes *V. bakeri*, *V. nuttallii*, *V. praemorsa*, *V. tomentosa*, and *V. vallicola*, should be treated as varieties of *V. nuttallii*, as C. L. Hitchcock et al. (1955–1969, vol. 3) did, on the basis that these taxa are more or less sympatric in range, intergrade with one another, and possess no distinctive gross morphological features by which they can be consistently recognized. S. L. Welsh et al. (1987) treated *V. praemorsa* and *V. linguifolia* Nuttall as synonymous with *V. nuttallii*, stating that *V. nuttallii* (and related taxa) are acaulescent to short-caulescent. Although some plants have short internodes, all plants in the *V. nuttallii* complex are caulescent. There is complexity and taxonomic difficulties inherent in the *V. nuttallii* complex; patterns of morphology, cytology, and leaf flavonoid chemistry provide a basis for recognizing infrataxa. Fabijan et al. conducted the most extensive study of the complex to date; their treatment of *V. praemorsa* is followed here.

1. Basal and cauline leaf blade bases ± truncate to subcordate; basal leaf blades 1.7–5.9(–7.3) cm, length 1.1–3 times width; plants 5.5–21.5 cm. . . .
. . . . 47a. *Viola praemorsa* var. *praemorsa*
1. Basal and cauline leaf blade bases attenuate; basal leaf blades 2.7–17 cm, length 1.7–6.5 times width; plants 10.5–36.5 cm.
  2. Basal leaf blades (3.4–)5.3–17 cm, usually ovate to elliptic, sometimes oblong-lanceolate, length 1.8–6.5 times width; peduncles 5.9–27 cm . . . . 47b. *Viola praemorsa* var. *flavovirens*
  2. Basal leaf blades 2.7–8.5 cm, usually ovate to elliptic, sometimes oblong-lanceolate, length 1.7–3.4 times width; peduncles 5–15 cm . . . .
. . . . 47c. *Viola praemorsa* var. *linguifolia*

47a. **Viola praemorsa** Douglas ex Lindley var. **praemorsa** • Astoria or yellow montane or hairy violet, violette jaune des monts E F

Plants 5.5–21.5 cm. **Leaves:** basal: petiole 2.6–13.8 cm; blade usually narrowly to broadly ovate, sometimes oblong-lanceolate to ± orbiculate or ± elliptic, 1.7–5.9(–7.3) × 0.8–3.3 cm, length 1.1–3 times width, base ± truncate to subcordate, margins ± crenate to serrate or entire; cauline similar to basal except: petiole 1.7–11.5 cm; blade 2.3–5.8(–6.7) × 1.1–3.5 cm. **Peduncles** 4.4–16.5 cm. **Lowest petal** 12–15(–18) mm. **Capsules** 6–11 mm. **Seeds** dark brown. $2n = 36$.

Flowering Mar–Jul. Vernally moist slopes and banks, meadows, grassy areas, open forests, sagebrush, shade or open areas; 100–2500 m; B.C.; Calif., Oreg., Wash.

In British Columbia, var. *praemorsa* is known only from southeastern Vancouver Island and adjacent Saltspring Island (G. W. Douglas and M. Ryan 1998).

47b. **Viola praemorsa** Douglas ex Lindley var. **flavovirens** (Pollard) R. J. Little, J. Bot. Res. Inst. Texas 4: 225. 2010 E

Viola flavovirens Pollard, Bull. Torrey Bot. Club 24: 405. 1897; *V. praemorsa* subsp. *flavovirens* (Pollard) Fabijan

Plants (11–)14–36.5 cm. **Leaves:** basal: petiole (5–)7.5–19 cm; blade usually elliptic to ovate, sometimes oblong-lanceolate, (3.4–)5.3–17 × 1.1–6.7 cm, length 1.8–6.5 times width, base attenuate, margins crenate, serrate, or serrulate; cauline similar to basal except: petiole 1.3–16.2 cm; blade 3–11(–14.8) × 1.4–3.6(–5.5) cm. **Peduncles** 5.9–27 cm. **Lowest petal** 11–15(–19) mm. **Capsules** 6–14 mm. **Seeds** medium brown. $2n = 36, 48$.

Flowering Apr–Jun. Vernally moist slopes and banks, grassy areas, rocky stream banks; 500–1600 m; Idaho, Mont., Utah, Wash., Wyo.

Populations of var. *flavovirens* in Idaho County, Idaho, have cauline leaves longer (14.8 versus 8 cm) and wider (5.5 versus 3.1 cm) than the maximum lengths and widths cited by D. M. Fabijan et al. (1987).

47c. **Viola praemorsa** Douglas ex Lindley var. **linguifolia** (Nuttall) M. Peck, Man. Pl. Oregon, 486. 1941 • Broad-leaved yellow prairie violet, violette à feuilles ligulées E

Viola linguifolia Nuttall in J. Torrey and A. Gray, Fl. N. Amer. 1: 141. 1838 (as linguaefolia); *V. erectifolia* A. Nelson; *V. nuttallii* Pursh var. *linguifolia* (Nuttall) Piper; *V. praemorsa* var. *altior* Blankinship; *V. praemorsa* subsp. *arida* M. S. Baker; *V. praemorsa* subsp. *linguifolia* (Nuttall) M. S. Baker; *V. praemorsa* subsp. *major* (Hooker) M. S. Baker; *V. praemorsa* var. *major* (Hooker) M. Peck; *V. praemorsa* subsp. *oregona* M. S. Baker

Plants 10.5–30 cm. **Leaves:** basal: petiole 7.2–19.2 cm; blade usually ovate to elliptic, sometimes oblong-lanceolate, 2.7–8.5 × 1.6–3.7 cm, length 1.7–3.4 times width, base attenuate, margins usually crenate, serrate, or serrulate, sometimes entire; cauline similar to basal except: petiole 3.8–11.8 cm; blade 3.3–8 × 1.4–3.5 cm. **Peduncles** 5–15 cm. **Lowest petal** 14–15 mm. **Capsules** 6–12 mm. **Seeds** brown to red-brown. $2n = 36, 48$.

Flowering May–Aug. Vernally moist slopes and banks, grassy areas; 700–3500 m; Alta.; Calif., Colo., Idaho, Mont., Nev., Oreg., Utah, Wash., Wyo.

The leaves of var. *linguifolia* are sometimes described as tongue-shaped (M. S. Baker 1957; D. M. Fabijan et al. 1987). Fabijan et al. commented that the distribution of var. *linguifolia* overlaps with *Viola bakeri* in central California, Oregon, and Washington; with var. *flavovirens* in Idaho, Washington, and Wyoming; and with *V. tomentosa* in the Sierra Nevada. She stated that the morphological variants of var. *linguifolia* observed during her study may be symptomatic of these sympatric distributions, which lead to the possibilities of hybridization and/or introgression. She described the cauline leaves of var. *linguifolia* as ovate to narrowly ovate. As a possible consequence of hybridization and/or introgression, differences in leaf outline have been observed among plants in the same population. In a population of var. *linguifolia* in Modoc County, California, some plants had ovate to narrowly ovate leaf blades, as described by Fabijan et al.; leaf blades on other plants in the same population were elliptic.

Viola praemorsa var. *altior* was treated by M. S. Baker (1957) and D. M. Fabijan et al. (1987) as a synonym of *V. praemorsa* subsp. *linguifolia*. Variety *altior*, described from Montana, is recognized in the Intermountain Flora as occurring in Utah (N. H. Holmgren 2005d). Because the original description of var. *altior* was meager and no specimens or the type have been examined, it is retained here as a synonym of var. *linguifolia* pending further study.

M. S. Baker (1957) noted that in comparison with *Viola nuttallii*, which grows in open, sunny areas, var. *linguifolia* is usually found in the shade of trees or shrubs.

Variety *linguifolia* is reported to hybridize with *V. utahensis* (G. Davidse 1976).

48. Viola primulifolia Linnaeus, Sp. Pl. 2: 934. 175 $\boxed{\text{E}}$

Plants perennial, acaulescent, stoloniferous, (3–)5–20(–36) cm; stolons pale, often rooting and leafy at nodes; rhizome thick or slender, fleshy. Leaves basal, 4–8, prostrate to ascending; stipules linear-lanceolate, margins ± crenate-serrate, sometimes glandular, apex acute; petiole (1–)3–13(–29) cm, glabrous or pubescent; blade unlobed, elliptic to narrowly or broadly ovate, (1.5–)3–7(–9) × 1–3(–3.5) cm, longer than wide, base broadly cordate to attenuate, rarely ± truncate, margins crenulate to serrulate, sometimes glandular, mostly eciliate, apex acute to rounded, mucronulate, surfaces glabrous or sparsely pubescent. Peduncles (3–)6–18(–28) cm, glabrous. Flowers: sepals lanceolate to ovate, margins usually eciliate, auricles 1–2 mm; petals white on both surfaces, lower 3 purple-veined, lateral 2 bearded or beardless, lowest 9–14(–16) mm, spur white, gibbous, 1–2 mm; style head beardless; cleistogamous flowers present or absent. Capsules ellipsoid, (5–)8–9 mm, glabrous. Seeds beige to bronze, 1.5–2 mm. $2n = 24$.

Varieties 2 (2 in the flora): United States.

Viola primulifolia occurs in small colonies; individual plants are interconnected by stolons.

1. Leaf blades elliptic to widely ovate, base broadly cordate to attenuate, surfaces sparsely pubescent, rarely glabrous; cleistogamous flowers present; c, e United States. . . . . 48a. *Viola primulifolia* var. *primulifolia*
1. Leaf blades elliptic to narrowly ovate, base usually attenuate, rarely ± truncate, surfaces glabrous, rarely pubescent; cleistogamous flowers absent; nw California, sw Oregon. . . . . 48b. *Viola primulifolia* var. *occidentalis*

48a. Viola primulifolia Linnaeus var. primulifolia

• Primrose-leaved violet $\boxed{\text{E}}$

Viola primulifolia var. *acuta* (Bigelow) Torrey & A. Gray; *V. primulifolia* subsp. *villosa* (Eaton) N. Russell; *V. primulifolia* var. *villosa* Eaton

Leaf blades elliptic, ovate, or widely ovate, base broadly cordate to attenuate, surfaces sparsely pubescent, rarely glabrous. Cleistogamous flowers present.

Flowering Mar–May. Marshes, wet meadows and other open to semi-open wet areas and depressions in mesic forests, hardwood-pine swamp forests, white cedar swamps, often in *Sphagnum* or other mosses; 100–2500 m; Ala., Ark., Conn., Del., D.C., Fla., Ga., Ill., Ind., Ky., La., Maine, Md., Mass., Mich., Minn., Miss., N.H., N.J., N.Y., N.C., Ohio, Okla., Pa., R.I., S.C., Tenn., Tex., Vt., Va., W.Va.

Some workers have suggested that var. *primulifolia* is a hybrid between *Viola lanceolata* and either *V. macloskeyi* or *V. blanda*. The distributions of *V. macloskeyi* and *V. blanda* overlap with *V. lanceolata* and var. *primulifolia* in parts of the northeast United States; var. *primulifolia* has its greatest distribution along the southern Atlantic coast throughout much of the southeast where neither *V. macloskeyi* nor *V. blanda* occur, except in the southern Appalachians.

Variety *primulifolia* reportedly hybridizes with *Viola lanceolata* (= *V.* ×*modesta* House) and *V. macloskeyi* (= *V.* ×*mollicula* House).

48b. Viola primulifolia Linnaeus var. occidentalis

A. Gray, Bot. Gaz. 11: 255. 1886 • Western bog or western white bog violet $\boxed{\text{E}}$

Viola lanceolata Linnaeus subsp. *occidentalis* (A. Gray) N. Russell; *V. occidentalis* (A. Gray) Howell; *V. primulifolia* subsp. *occidentalis* (A. Gray) L. E. McKinney & R. J. Little

Leaf blades elliptic to narrowly ovate, base usually attenuate, rarely ± truncate, surfaces glabrous, rarely pubescent. Cleistogamous flowers absent.

Flowering Mar–May. Perennially wet places, marshes, bogs; 100–500 m; Calif., Oreg.

Variety *occidentalis* is found in northwestern California and southwestern Oregon, often with *Darlingtonia californica* (V. B. Baird 1942; R. J. Little 1993, 2012).

49. Viola prionantha Bunge, Enum. Pl. China Bor., 8. 1833 • Japanese violet [I]

Plants perennial, acaulescent, not stoloniferous, 3–20 cm; rhizome stout, fleshy. **Leaves** basal, 6–22, ascending to erect; stipules linear-lanceolate, margins remotely denticulate, apex acuminate; petiole 1–13 cm, narrowly winged distally, glabrous or puberulent; blade unlobed, oblong-ovate, ovate-lanceolate, or narrowly ovate, 1–4.5(–10) × 0.6–2(–4) cm, base usually truncate, sometimes ± cordate or broadly cuneate, margins crenulate, ciliate or eciliate, apex obtuse or ± acute, surfaces glabrous or puberulent. **Peduncles** 2.5–6 cm, glabrous, bracteoles near middle. **Flowers:** sepals lanceolate or ovate-lanceolate, margins ciliate or eciliate, auricles 1–2 mm; petals light violet or purple on both surfaces, rarely all white, lower 3 dark violet-veined, lateral 2 beardless or sparsely bearded, lowest 14–25 mm, spur pale to dark violet, elongated, 6–9 mm, tip hooked up; style head beardless; cleistogamous flowers present. **Capsules** narrowly ellipsoid, 5–12 mm, glabrous. **Seeds** dark brown, ca. 2 mm. $2n = 48$.

Flowering Apr–May. Lawns, roads, roadsides; 200–900 m; introduced; Kans., Nebr.; Asia (China, Korea, Russia).

In 1952, plants of *Viola* initially identified as *V. patrinii* were collected and reported by C. T. Rogerson to be established in lawns on the campus of Kansas State University, Manhattan, Kansas, where they are still found. S. B. Rolfsmeier recently determined that these plants are *V. prionantha*. R. B. Kaul reported that *V. prionantha* occurs along roadsides in Kansas and is also established in the flood plain of the North Platte River in Nebraska, where it was first collected by him in 1992.

50. Viola pubescens Aiton, Hort. Kew. 3: 290. 1789 [E]

Crocion pubescens (Aiton) Nieuwland & Kaczmarek

Plants perennial, caulescent, not stoloniferous, 10–45 cm. **Stems** 1–3+, ascending to erect, leafless proximally, leafy distally, glabrous or puberulent, from subligneous rhizome. **Leaves** basal and cauline; basal: 0–4; stipules ovate to lanceolate, margins entire, coarsely toothed, or crenate, apex acute or obtuse; petiole 2.7–19.6 cm, glabrous or puberulent; blade reniform to ovate, 1.5–5.8 × 1–6.8 cm, base usually narrowly or broadly cordate, rarely truncate, margins crenate to serrate, ciliate or eciliate, apex acute to long-acuminate, surfaces glabrous or puberulent, sometimes sparsely strigose; cauline similar to basal except: leaves usually at apex of stem, sometimes with a leaf proximal to apex; stipules ovate, margins entire, coarsely serrate, or erose, apex acute; petiole 1–10 cm; blade reniform or ovate to ovate-orbiculate or deltate, 1.4–10.9 × 1–12 cm, base cordate, margins ± crenate or ± serrate, apex acute to acuminate. **Peduncles** 5–8 cm, glabrous or puberulent. **Flowers:** sepals lanceolate to oblong-lanceolate, margins ciliate or eciliate, auricles 0.1–0.5 mm; petals lemon-yellow on both surfaces, lower 3 brownish purple-veined, lateral 2 and sometimes others, bearded, lowest 8–18 mm, spur yellow to greenish, gibbous, 1–2.5 mm; style head bearded; cleistogamous flowers axillary. **Capsules** ellipsoid, 8–12(–20) mm, glabrous or tomentose. **Seeds** brown, 1.8–3 mm. $2n = 12$.

Varieties 2 (2 in the flora): c, e North America.

Viola pubescens is similar in overall appearance to *V. glabella*.

1. Stems 1 or 2, puberulent; basal leaves 0(1), surfaces usually puberulent . . . . 50a. *Viola pubescens* var. *pubescens*
1. Stems (2)3+, glabrous or puberulent; basal leaves 0–4, surfaces glabrous, sometimes sparsely strigose . . . . 50b. *Viola pubescens* var. *scabriuscula*

50a. Viola pubescens Aiton var. **pubescens** • Downy yellow or downy or hairy yellow or yellow violet, violette pubescente [E]

Viola pensylvanica Michaux; *V. pubescens* var. *eriocarpa* Nuttall; *V. pubescens* var. *peckii* House

Stems 1 or 2, erect, 10–45 cm, puberulent. **Basal leaves** 0(1), surfaces usually puberulent; stipules ovate, apex obtuse.

Flowering Apr–Jun. Rich woodlands, deciduous, mixed, and coniferous riverine forests; 30–1500 m; Ont., Que.; Ala., Ark., Conn., Del., D.C., Ill., Ind., Iowa, Kans., Ky., La., Maine, Md., Mass., Mich., Minn., Mo., Nebr., N.H., N.J., N.Y., N.C., N.Dak., Ohio, Okla., Pa., R.I., S.Dak., Tenn., Tex., Vt., Va., W.Va., Wis., Wyo.

Reports of var. *pubescens* in Manitoba (L. Brouillet et al., http://data.canadensys.net/vascan) are considered doubtful and those from Nunavut are considered unlikely (M. Oldham, pers. comm.).

50b. Viola pubescens Aiton var. **scabriuscula** Torrey & A. Gray, Fl. N. Amer. 1: 142. 1838 • Smooth yellow violet, violette scabre E

Viola eriocarpa Schweinitz

Stems (2)3+, ascending to erect, 10–30 cm, glabrous or puberulent. **Basal leaves** 0–4, surfaces glabrous, sometimes sparsely strigose; stipules lanceolate, apex acute.

Flowering Mar–May. Rich woodlands, mesic forests, alluvial forests; 50–1500 m; Man., N.B., N.S., Ont., P.E.I., Que., Sask.; Ala., Ark., Conn., Del., Ga., Ill., Ind., Kans., Ky., Maine, Md., Mass., Mich., Minn., Miss., Mo., Nebr., N.H., N.J., N.Y., N.C., N.Dak., Ohio, Okla., Pa., S.C., S.Dak., Tenn., Tex., Vt., Va., W.Va., Wis.

51. Viola purpurea Kellogg, Pacific (San Francisco), 2 Feb. 1855: unnumb. 1855 F

Plants perennial, caulescent, not stoloniferous, 1.5–25 cm. **Stems** 1–5(–7), decumbent or spreading to erect, leafy proximally and distally, ± glabrous, puberulent, canescent, or tomentose, on caudex from subligneous rhizome. **Leaves** basal and cauline; basal: 1–6; stipules adnate to petiole, forming 2 linear, membranous wings, wing margins entire or laciniate, each wing with lanceolate to ± deltate projection, margins entire or laciniate, apex acute to long-acuminate; petiole 1.8–14.5 cm, puberulent to tomentose; blade purplish, purple-tinted, or gray-green abaxially, gray, green, or gray-green adaxially, sometimes shiny adaxially, ovate, orbiculate, oblong, deltate, or lanceolate, 0.8–5.3 × 0.4–4.1 cm, often fleshy, base cordate, subcordate, truncate, or attenuate, oblique or not, margins usually ± crenate, serrate, dentate, or coarsely or irregularly repand-dentate, sometimes entire, usually ciliate, apex acute to obtuse or rounded, surfaces glabrous, puberulent, or tomentose; cauline similar to basal except: stipules linear, lanceolate, oblanceolate, or ± oblong to ovate, margins entire, lacerate, or laciniate, usually ciliate, apex sometimes divided into 2 or 3 filiform processes or obtuse; petiole 0.3–19.7 cm, glabrous or puberulent; blade ovate, oblong, elliptic, deltate, or lanceolate, 0.9–5.2 × 0.2–2.9 cm, length 0.8–7.1 times width, margins crenate, serrate, dentate, repand-denticulate, undulate-denticulate, sinuate, undulate, or entire, abaxial surface puberulent, canescent, or tomentose, adaxial surface glabrous, sparsely pubescent, puberulent, canescent, or tomentose. **Peduncles** 1–14 cm, puberulent

or tomentose, sometimes glabrous above bracteoles. **Flowers:** sepals lanceolate, margins ciliate, auricles 0.5–1.5 mm; petals deep lemon-yellow adaxially, upper 2 and sometimes lateral 2 reddish brown to brownish purple abaxially, lower 3 dark brown-veined, lateral 2 sparsely to densely bearded, lowest 6–14 mm, spur yellow to reddish brown, gibbous, 1.5–2 mm; style head bearded; cleistogamous flowers axillary. **Capsules** ovoid to ± spherical, 4–7 mm, puberulent. **Seeds** light to dark brown or mottled gray and brown, 2–3.1 mm. $2n = 12$.

Varieties 7 (7 in the flora): w North America, nw Mexico.

Varieties of *Viola purpurea* are variable and intergrade. All are found in California; six occur in other western states, one in Mexico, and one in British Columbia. Mature plants are needed for determination.

1. Basal and cauline leaf blade surfaces tomentose .... 51b. *Viola purpurea* var. *aurea*
1. Basal and cauline leaf blade surfaces ± glabrous, canescent, or puberulent.
  2. Stems mostly buried, not much elongated by end of season; plants 1.5–9(–12) cm.
    3. Cauline leaf margins usually entire, sometimes sinuate; basal leaf margins ± crenate to irregularly repand-dentate or entire ...... 51d. *Viola purpurea* var. *integrifolia*
    3. Cauline leaf margins usually coarsely crenate or dentate, sometimes ± serrate or ± entire; basal leaf margins coarse-serrate or irregularly dentate or crenate with 2–4 rounded teeth per side .... 51g. *Viola purpurea* var. *venosa*
  2. Stems usually not buried, usually elongated by end of season; plants 3–25 cm.
    4. Basal leaf base cordate or truncate .... 51c. *Viola purpurea* var. *dimorpha*
    4. Basal leaf base usually attenuate, sometimes subcordate or truncate.
      5. Cauline leaf blade length 3.2–7.1 times width, margins usually undulate-denticulate, sometimes entire........ .... 51e. *Viola purpurea* var. *mesophyta*
      5. Cauline leaf blade length 1–3 times width, margins with 3 or 4(5) pointed or rounded teeth per side or crenate-serrate.
        6. Basal leaf blade margins irregularly crenate, without pointed or rounded teeth; adaxial surface often shiny .... 51a. *Viola purpurea* var. *purpurea*
        6. Basal leaf blade margins dentate-serrate with 4 or 5(6) pointed or rounded teeth per side; adaxial surface not shiny ....51f. *Viola purpurea* var. *mohavensis*

51a. Viola purpurea Kellogg var. **purpurea**

• Mountain violet

Plants 3–25 cm. **Stems** spreading to erect, usually not buried, usually elongated by end of season, ± glabrous or puberulent. **Leaves:** basal: 1–5; petiole 4–11 cm, puberulent; blade purple-tinted abaxially, green adaxially, often shiny, ± orbiculate to widely ovate, 1.6–4.6(–5.3) × 1.6–4.1 cm, ± fleshy, base usually attenuate, sometimes subcordate or truncate, margins irregularly crenate, apex acute to obtuse, abaxial surface glabrous or puberulent, hairs sometimes only on veins, adaxial surface usually glabrous, sometimes sparsely puberulent; cauline: petiole 1.5–19.7 cm, usually puberulent; blade ovate, deltate, or lanceolate, 1.2–4.5 × 0.8–2.5 cm, length 1–2.3 times width, base usually attenuate, sometimes ± cordate or truncate, margins crenate-serrate, abaxial surface puberulent (or along veins), adaxial surface glabrous or sparsely pubescent. **Peduncles** 4.5–10 cm, usually puberulent, sometimes glabrous. **Lowest petals** 10–12 mm. **Capsules** 5–6 mm. **Seeds** dark brown, 2.1–2.3 mm.

Flowering Mar–Jul. In openings or beneath shrubs, usually in yellow pine (*Pinus ponderosa*) forests or higher; 200–2900 m; Calif., Oreg.; Mexico (Baja California).

51b. Viola purpurea Kellogg var. **aurea** (Kellogg) M. S. Baker ex Jepson, Fl. Calif. 2: 521. 1936

• Golden violet [E]

Viola aurea Kellogg, Proc. Calif. Acad. Sci. 2: 185, fig. 54. 1862; *V. purpurea* subsp. *aurea* (Kellogg) J. C. Clausen

Plants 4–12 cm. **Stems** erect or decumbent, mostly buried, not much elongated by end of season, canescent to tomentose. **Leaves:** basal: 1–6; petiole 4.6–9.5 cm, tomentose; blade purplish abaxially, gray adaxially, ovate, oblong, or orbiculate, 1.1–5 × 1–3.4 cm, base attenuate, often oblique, margins coarsely repand-dentate to irregularly crenate or shallowly and irregularly serrate, apex acute to obtuse, surfaces tomentose; cauline: petiole 1.5–5.5 cm, tomentose; blade ovate to ovate-lanceolate, 1.3–3.7 × 0.7–2 cm, length 0.8–1.4 times width, base attenuate, sometimes oblique, margins usually irregularly serrate, sometimes entire and undulate, apex usually acute, surfaces tomentose. **Peduncles** 2.4–10.5 cm, tomentose. **Lowest petals** 8–13 mm. **Capsules** 4–7 mm. **Seeds** medium brown, 2.5–2.9 mm.

Flowering Apr–Jun. Pinyon-juniper woodlands, sagebrush (Great Basin); 1000–2300 m; Calif., Nev.

M. S. Baker initially considered *Viola aurea* a subspecies of *V. purpurea*; he did not publish the transfer (J. Clausen 1964b). Later (1953), he considered it a distinct species. G. L. Stebbins et al. (1963) stated that chemical and morphological evidence supported the 1953 decision by Baker to maintain *V. aurea* as a species distinct from *V. purpurea*. Clausen (1964) disagreed with the conclusions of Stebbins et al. and (1964b) treated the taxon as *V. purpurea* subsp. *aurea*.

Flowers of var. *aurea* have been observed to close up in late afternoon, then fully reopen the following morning.

51c. Viola purpurea Kellogg var. **dimorpha** (M. S. Baker & J. C. Clausen) J. T. Howell, Mentzelia 1: 8. 1976 • Dimorphic mountain violet [E]

Viola purpurea subsp. *dimorpha* M. S. Baker & J. C. Clausen, Madroño 10: 122, plates 4 [right lower center], 8, fig. 8. 1949; *V. purpurea* subsp. *geophyta* M. S. Baker & J. C. Clausen

Plants 3–25 cm. **Stems** spreading to erect, usually not buried, usually elongated by end of season, puberulent. **Leaves:** basal: 1–6; petiole 1.8–8.5 cm, puberulent; blade usually purple-tinted abaxially, green adaxially, ± orbiculate to ovate, 0.8–3 × 0.5–2.2 cm, base cordate or truncate, margins irregularly dentate or crenate with 3 or 4 rounded and/or pointed teeth per side, apex obtuse to rounded, surfaces puberulent; cauline: petiole 0.5–4.1 cm, puberulent; blade ovate to lanceolate to ± oblong, 1.2–2.7 × 0.7–1.9 cm, length 0.8–2.5 times width, base attenuate, oblique or not, subcordate, or truncate, margins entire, crenate, serrate, or repand-denticulate, apex acute to obtuse, surfaces canescent. **Peduncles** 4.6–6 cm, puberulent. **Lowest petals** 6–11 mm. **Capsules** 5–6.5 mm. **Seeds** dark brown, 2.2–2.9 mm.

Flowering May–Jul. Pine, fir, cedar forests; 1200–2500 m; Calif., Oreg.

51d. Viola purpurea Kellogg var. **integrifolia**
(M. S. Baker & J. C. Clausen) J. T. Howell, Mentzelia
1: 8. 1976 　E

Viola purpurea subsp. *integrifolia*
M. S. Baker & J. C. Clausen,
Madroño 10: 118, plate 4 [left
lower center]. 1949

Plants 1.5–9(–12) cm. **Stems**
erect, mostly buried, not much
elongated by end of season,
puberulent. **Leaves:** basal: 1–4;
petiole 1.9–7.5 cm, puberulent;
blade purple-tinted abaxially, green adaxially, lanceolate
or ovate to ± orbiculate, 0.8–3.1 × 0.4–2.3 cm, ± fleshy,
base cordate, subcordate, or truncate, margins ± crenate
to irregularly repand-dentate or entire, apex obtuse,
abaxial surface puberulent, adaxial surface sparsely
puberulent; cauline: petiole 1.3–3 cm, puberulent;
blade ovate to oblong-lanceolate, 1.2–3.1 × 0.3–1.4 cm,
length 1.2–5.7 times width, base truncate to attenuate,
margins usually entire, sometimes sinuate, apex acute to
obtuse, surfaces similar to basal blades. **Peduncles** 1–9
cm, puberulent. **Lowest petals** 8–11 mm. **Capsules** 5–7
mm. **Seeds** dark brown, 2–2.9 mm.

Flowering May–Aug. Dry, red fir (*Abies magnifica*
in California) and pine forests to timberline, sandy,
gravelly, or rocky soil, sometimes on serpentine;
1200–2600 m; Calif., Nev., Oreg.

51e. Viola purpurea Kellogg var. **mesophyta**
(M. S. Baker & J. C. Clausen) J. T. Howell, Mentzelia
1: 8. 1976 　E

Viola purpurea subsp. *mesophyta*
M. S. Baker & J. C. Clausen,
Madroño 10: 114, plate 4 [left
upper center]. 1949

Plants 9–18.5 cm. **Stems**
usually erect, usually not buried,
elongated by end of season,
usually puberulent, sometimes
± glabrous. **Leaves:** basal:
1–5; petiole 2.8–13.8 cm, puberulent; blade purple-
tinted abaxially, green adaxially, oblong, ovate-oblong,
or lanceolate, 1.7–4.2 × 0.5–2.3 cm, base attenuate,
usually oblique, margins usually irregularly repand-
dentate to -serrate, sometimes ± entire, apex acute to
obtuse, surfaces puberulent; cauline: petiole 0.3–12.3
cm, puberulent; blade lanceolate to ovate- or oblong-
lanceolate, 1.5–5.2 × 0.2–1.5 cm, length 3.2–7.1 times
width, base attenuate, usually oblique, margins usually
undulate-denticulate, sharp-angled, sometimes entire,
apex acute to obtuse, surfaces puberulent. **Peduncles**
2.6–6.7 cm, glabrous or usually puberulent. **Lowest
petals** 7–10 mm. **Capsules** 4–5.5 mm. **Seeds** mottled
gray and brown, 2.3–2.9 mm.

Flowering May–Aug. Damp, shady areas in lodgepole
pine, red and white fir forests; 1400–3600 m; Calif.

Variety *mesophyta* was reported from Oregon by
M. E. Peck (1941), but is not currently known there
(H. L. Chambers in T. Cook and S. Sundberg, http://
www.oregonflora.org/checklist.php).

51f. Viola purpurea Kellogg var. **mohavensis**
(M. S. Baker & J. C. Clausen) J. T. Howell, Mentzelia
1: 8. 1976 　E

Viola aurea Kellogg subsp.
mohavensis M. S. Baker &
J. C. Clausen, Madroño 12: 9,
fig. 1. 1953; *V. purpurea* subsp.
mohavensis (M. S. Baker &
J. C. Clausen) J. C. Clausen

Plants 5–24 cm. **Stems** spreading
to erect, usually not buried,
usually elongated by end of
season, ± glabrous or usually puberulent. **Leaves:** basal:
1–5; petiole 4.5–14.5 cm, puberulent; blade gray-green
to purple-tinted abaxially, green or gray-green adaxially,
ovate, ± orbiculate, or ± deltate, 1–4 × 1–3.5 cm, base
usually attenuate, margins dentate-serrate with 4 or
5(6) pointed or rounded teeth per side, apex obtuse,
abaxial surface glabrous or puberulent, adaxial surface
glabrous or finely puberulent, not shiny; cauline: petiole
0.8–11 cm, glabrous or puberulent; blade ovate, elliptic,
or lanceolate, 1.5–3.7 × 0.5–2.5 cm, length 1–3 times
width, base attenuate, margins with 3 or 4(5) pointed
or rounded teeth per side, apex acute, abaxial surface
puberulent, adaxial surface usually glabrous, sometimes
puberulent. **Peduncles** 1.7–14 cm, glabrous or usually
puberulent. **Lowest petal** 10–14 mm. **Capsules** 5–7
mm. **Seeds** light brown, 2.7–3.1 mm.

Flowering Mar–Jul. Desert scrub, sagebrush, dry
areas in yellow pine forests; 900–2600 m; Ariz., Calif.,
Nev.

Variety *mohavensis* is variable; cauline leaf margins
in some populations are similar to basal blades. The
basal leaves of plants from Long Valley, Mono County,
California, are more or less truncate or subcordate.
In 1953, M. S. Baker changed his previous position
and decided to treat *Viola aurea* as a separate species
rather than as a subspecies of *V. purpurea*, and to treat
mohavensis as a subspecies of *V. aurea*. G. L. Stebbins
et al. (1963) stated that additional study may reveal that
aurea and *mohavensis* might be better treated as species
separate from each other and from *V. purpurea*.

M. S. Baker (1953) noted that a form of var. *mohavensis*
found in Mono and Inyo counties, California, is much
greener in aspect and lacks microscopic pubescence.
G. L. Stebbins et al. (1963) wrote that these taxa
appeared to be a variable assemblage perhaps of forms
transitional between *Viola aurea* and subspecies of
V. purpurea other than subsp. *purpurea*.

51g. Viola purpurea Kellogg var. **venosa** (S. Watson) Brainerd, Bull. Vermont Agric. Exp. Sta. 224: 111. 1921 • Purple-marked yellow violet, violette veinée E F

Viola nuttallii Pursh var. *venosa* S. Watson, Botany (Fortieth Parallel), 35. 1871; *V. atriplicifolia* Greene; *V. aurea* Kellogg var. *venosa* (S. Watson) S. Watson; *V. purpurea* subsp. *atriplicifolia* (Greene) M. S. Baker & J. C. Clausen; *V. purpurea* subsp. *venosa* (S. Watson) M. S. Baker & J. C. Clausen

Plants 3–8.5(–12) cm. **Stems** decumbent or erect, mostly buried, not much elongated by end of season, ± glabrous or puberulent. **Leaves:** basal: 1–5; petiole 5–10.3 cm, puberulent; blade purple-tinted abaxially, green adaxially, often shiny, ± orbiculate to ovate, 0.8–3.3 × 0.5–3.6 cm, ± fleshy, base truncate or ± cordate, oblique or not, margins usually irregularly dentate or crenate with 2–4 rounded teeth per side, sometimes coarse-serrate, apex acute to obtuse, abaxial surface puberulent, adaxial surface glabrous, glabrate, or sparsely puberulent; cauline: petiole 0.8–6 cm, puberulent; blade ovate, ± oblong or lanceolate, 0.9–2.7 × 0.6–2.9 cm, length 0.9–2.3 times width, margins usually coarsely crenate or dentate, sometimes ± serrate or ± entire, abaxial surface puberulent, adaxial surface glabrous (except along veins) to sparsely puberulent. **Peduncles** 2.9–7 cm, glabrous or puberulent. **Lowest petal** 6–14 mm. **Capsules** 4–5.5 mm. **Seeds** medium brown, 2–2.6 mm.

Flowering Apr–Sep. Pine forests, deserts, gravelly plains, edges of wet meadows, grassy or rocky slopes, shaded or exposed areas, ridges, dry to moist soil, near snowdrifts; 1300–3400 m; B.C.; Ariz., Calif., Colo., Idaho, Mont., Nev., Oreg., Utah, Wash., Wyo.

Variety *venosa* is the most wide-ranging member of the *Viola purpurea* complex. It has been reported to hybridize with *V. utahensis* in Idaho, Nevada, and Utah (G. Davidse 1976).

52. Viola quercetorum M. S. Baker & J. C. Clausen, Leafl. W. Bot. 5: 101. 1948 • Oakwoods violet E

Viola purpurea Kellogg subsp. *quercetorum* (M. S. Baker & J. C. Clausen) R. J. Little

Plants perennial, caulescent, not stoloniferous, 4–25(–34.5) cm. **Stems** 1–5, spreading to erect, leafy proximally and distally, usually elongated by end of season, puberulent to canescent, on caudex from subligneous rhizome. **Leaves** basal and cauline; basal: 1–6; stipules adnate

to petiole, forming 2 linear, membranous wings, each wing with lanceolate to ± deltate projection, margins entire or laciniate, apex usually long-acuminate or divided into narrow, filiform processes; petiole 1.9–9.5 cm, puberulent; blade usually grayish green to whitish, sometimes ± purple-tinted abaxially, green to grayish green adaxially, ± ovate to ± orbiculate, deltate, pandurate, or broadly lanceolate, 1–5.3 × 1–3.5 cm, base truncate to ± cordate or attenuate, often oblique, margins often undulate-sinuate, ± crenate to ± serrate, ciliate or eciliate, apex obtuse or ± acute, surfaces finely to densely puberulent; cauline similar to basal except: stipules lanceolate, often very unequal at same node, margins lacerate to laciniate, apex acute to acuminate; petiole 1.3–5.3 cm; blade ovate, deltate, lanceolate, or ± rhombic, 1.4–3.8 × 0.8–2 cm, length 0.7–1.9 times width, base usually attenuate, sometimes truncate to ± cordate, margins crenate to serrate, undulate-sinuate, apex acute to obtuse. **Peduncles** 2.7–17 cm, puberulent. **Flowers:** sepals lanceolate, margins ciliate, auricles 1–2 mm; petals deep lemon-yellow adaxially, upper 2 and sometimes lateral 2 reddish brown to brownish purple abaxially, lower 3 dark brown-veined, lateral 2 sparsely to densely bearded, lowest 10–16 mm, spur yellow to reddish brown, gibbous, 1–2 mm; style head bearded; cleistogamous flowers axillary. **Capsules** ovoid to ellipsoid, 8–12 mm, puberulent. **Seeds** medium brown, 2.7 mm. $2n = 24$.

Flowering Feb–Jul. Dry, grassy or brushy slopes, chaparral, in or lower than yellow pine forests; 300–2000 m; Calif., Oreg.

Herbarium specimens of *Viola quercetorum* and *V. purpurea* var. *purpurea* can be difficult to distinguish; pressing obliterates the undulate leaf margins of *V. quercetorum*. In Oregon, most collections of *V. quercetorum* are from yellow pine forests (A. Liston, pers. comm.).

Viola quercetorum hybridizes with *V. douglasii* (J. Clausen 1964).

53. Viola renifolia A. Gray, Proc. Amer. Acad. Arts 8: 288. 1870 • Kidney-leaved or kidney-shaped or white violet, violette réniforme E

Viola blanda Willdenow var. *renifolia* (A. Gray) A. Gray; *V. brainerdii* Greene; *V. mistassinica* Greene; *V. renifolia* var. *brainerdii* (Greene) Fernald

Plants perennial, acaulescent, not stoloniferous, 5–30 cm; rhizome thick, fleshy. **Leaves** basal, 1–5, prostrate to ascending; stipules linear-lanceolate, margins entire to sparsely laciniate, apex acute; petiole 3–10 cm, strigose, sericeous, or villous, occasionally glabrous; blade unlobed, reniform or ovate to broadly ovate or orbiculate, 1.5–3.5 × 2–5 cm, base cordate to broadly cordate, margins serrate-crenate, ciliate or eciliate, apex acute, obtuse, or rounded, occasionally acuminate, surfaces usually sparsely to densely strigose, sericeous, or villous throughout or along veins, sometimes glabrous. **Peduncles** 3–8 cm, puberulent. **Flowers:** sepals lanceolate to ovate, margins usually eciliate, auricles 1–2 mm; petals white on both surfaces, lower 3 purple-veined, all beardless or lower 3 sparsely bearded, lowest 8–10 mm, spur white, gibbous, 2–3 mm; style head beardless; cleistogamous flowers on prostrate to ascending peduncles. **Capsules** ovoid to ellipsoid, 5–8 mm, glabrous. **Seeds** mottled beige to bronze, 1.5–2.2 mm. $2n = 24$.

Flowering Apr–Jun. Moist, often shaded alluvial or upland forests, shrub thickets, stream banks, swamp forests, bogs, fens; 200–3000 m; Alta., B.C., Man., N.B., Nfld. and Labr., N.W.T., N.S., Nunavut, Ont., P.E.I., Que., Sask., Yukon; Alaska, Colo., Conn., Idaho, Iowa, Maine, Mass., Mich., Minn., Mont., N.H., N.Y., Pa., S.Dak., Vt., Wash., Wis., Wyo.

Nonflowering plants of *Viola renifolia* and *V. epipsila* appear similar. The abaxial leaf surfaces of *V. renifolia* usually have a few short, straight hairs on the main veins; leaves of *V. epipsila* are usually glabrous (C. Parker, pers. comm.). *Viola renifolia* can appear similar to *V. macloskeyi*. V. B. Baird (1942) wrote that *V. renifolia* sometimes produces ascending stems. H. A. Gleason and A. Cronquist (1991) noted that if stolons were present, they were short and racemelike with cleistogamous flowers. The ascending stems and stolons mentioned by these authors may or may not be different phases of the same structure.

54. **Viola riviniana** Reichenbach, Iconogr. Bot. Pl. Crit. 1: 81, plate 95. 1823 · Dog or wood violet [I]

Plants perennial, caulescent, not stoloniferous, 1.8–30 cm. **Stems** 1–5, erect, ascending, or decumbent, glabrous or puberulent, on caudex from subligneous rhizome. **Leaves** basal and cauline; basal: 3–5; stipules subulate or lanceolate, margins fimbriate, apex acute to acuminate; petiole 2.5–12 cm, usually glabrous, rarely puberulent; blade ovate to reniform, 2.3–5.5 × 2.6–5.3 cm, base cordate to deeply cordate, margins crenate, eciliate, apex acute or obtuse, abaxial surface glabrous or sparsely puberulent, adaxial surface usually sparsely puberulent, sometimes glabrous; cauline similar to basal except: stipules ovate or lanceolate, margins fimbriate, apex long-acuminate; petiole 1.3–6 cm; blade ovate or broadly ovate, 1.8–5 × 1.5–4.2 cm, apex acute. **Peduncles** 2.8–10 cm, glabrous, sometimes puberulent. **Flowers:** sepals lanceolate, margins eciliate; auricles 1.8–2.5 mm

(enlarged in fruit); petals violet on both surfaces, lower 3 usually white basally, rarely violet, dark violet-veined, lateral 2 bearded, lowest 13–17 mm, spur white, rarely violet, elongated, 5–7 mm, tip straight or hooked, blunt; style head beardless (surface papillose); cleistogamous flowers axillary. **Capsules** ovoid or oblong, 8–12 mm, glabrous. **Seeds** pale to light brown, 1.8–2.1 mm. $2n = 40$.

Flowering Mar–Sep. Roadsides and trails, sidewalk cracks, parks; 0–200 m; introduced; B.C.; Calif., Oreg., Wash.; Eurasia; nw Africa; introduced also in Australia.

Viola riviniana has become established in several states on the Pacific Coast. It is cultivated and sold through nurseries in the United States. In the nursery trade in California and elsewhere, it is incorrectly referred to as *V. labradorica* 'Purpurea.' In some plants shoots arise from adventitious buds on the roots (A. R. Clapham et al. 1987; T. Marcussen and T. Karlsson 2010).

55. **Viola rostrata** Pursh, Fl. Amer. Sept. 1: 174. 1813 · Long-spurred violet, violette rostrée [F]

Lophion rostratum (Pursh) Nieuwland & Kaczmarek

Plants perennial, caulescent, not stoloniferous, 5–20 cm. **Stems** 1–7, ascending to erect (often declining during anthesis), glabrous, on caudex from fleshy rhizome. **Leaves** basal and cauline; basal: 1–5; stipules lanceolate, margins laciniate, apex acute; petiole 1–9.6 cm, glabrous; blade sometimes purple-spotted abaxially and/or adaxially, reniform to ovate, 1–4.5 × 1–4 cm, base broadly cordate to cordate, margins crenate to serrate, mostly eciliate, apex obtuse to acute, surfaces usually pubescent, mostly adaxially toward base, sometimes glabrous; cauline similar to basal except: petiole 0.4–4 cm; distal cauline blades ovate to deltate, 1–4.5 × 1–4 cm, base cordate, apex acuminate to acute. **Peduncles** 5–9 cm, usually glabrous. **Flowers:** sepals lanceolate, margins eciliate, auricles 1–2 mm; petals pale lavender-violet on both surfaces, all white basally, lower 3 purple-black-veined, all beardless, lowest 8–20 mm, spur white, purple, or lavender-tinged, elongated, 10–20 mm; style head beardless; cleistogamous flowers axillary. **Capsules** ellipsoid, 4–6 mm, glabrous. **Seeds** beige to bronze, 1.3–2 mm. $2n = 20$.

Flowering Apr–May. Rich, mesic to dry, well-drained woodlands, mountains; 200–1800 m; Ont., Que.; Ala., Conn., Ga., Ind., Ky., Md., Mass., Mich., N.J., N.Y., N.C., Ohio, Pa., S.C., Tenn., Vt., Va., W.Va., Wis.; e Asia (Japan).

Viola rostrata has the longest spur of any North American *Viola* species.

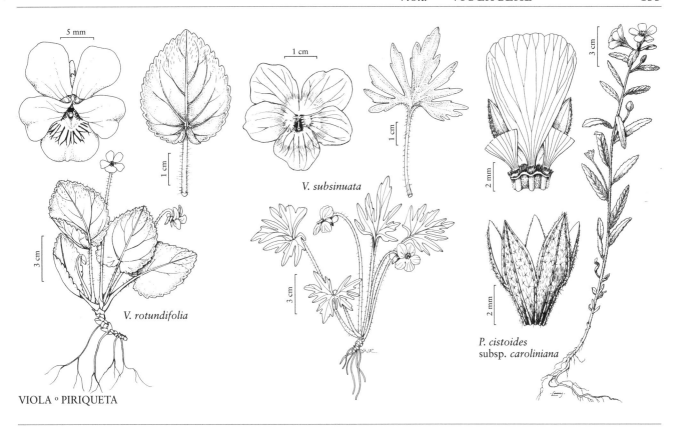

V. subsinuata

V. rotundifolia

P. cistoides
subsp. caroliniana

VIOLA ° PIRIQUETA

Viola rostrata reportedly hybridizes with *V. labradorica* (= *V. ×malteana* House) and *V. striata* (= *V. ×brauniae* Grover ex Cooperrider).

56. Viola rotundifolia Michaux, Fl. Bor.-Amer. 2: 150. 1803 • Early yellow or roundleaf yellow violet, violette à feuilles rondes E F

Plants perennial, acaulescent, not stoloniferous, 1–20 cm; rhizome thick, fleshy. **Leaves** basal, 2–5, prostrate to ascending, often overlapping basally; stipules linear-lanceolate, margins entire, apex acute; petiole 2–8 cm, pubescent; blade unlobed, orbiculate, reniform, or ovate, 2–12 × 1.5–9 cm, base cordate, margins crenate to serrate, sometimes glandular, ciliate or eciliate, apex rounded to acute, surfaces usually pubescent throughout or concentrated proximally on both surfaces. **Peduncles** 1.5–7 cm, usually pubescent. **Flowers:** sepals lanceolate to ovate, margins ciliate or eciliate, auricles 0.5–1 mm; petals deep lemon-yellow on both surfaces, lower 3 brownish purple-veined, lateral 2 bearded, lowest 8–11 mm, spur yellow, gibbous, 1–2 mm; style head beardless; cleistogamous flowers on prostrate or partially subterranean rhizomes or on racemelike, nonrooting, and usually leafless branches growing from rhizome apex. **Capsules** ellipsoid, 5–10 mm, glabrous. **Seeds** beige, 1–2 mm. 2*n* = 12.

Flowering Mar–May. Rich montane forests and other mesic woodlands; 200–2000 m; Ont., Que.; Conn., Del., Ga., Ky., Maine, Md., Mass., N.H., N.Y., N.C., Ohio, Pa., R.I., S.C., Tenn., Vt., Va., W.Va.

N. H. Russell (1955b) stated that *Viola rotundifolia* is a primitive member of *Viola* and probably one of the ancestral species of stemmed yellow violets of North America. Russell (1965) stated that morphologically, *V. rotundifolia* is one of the most invariable violets and suggested that its nearest relative is *V. orbiculata*.

57. Viola sagittata Aiton, Hort. Kew. 3: 287. 1789 E

Plants perennial, acaulescent, not stoloniferous, 5–50 cm; rhizome thick, fleshy. **Leaves** basal, 4–8, ascending to erect; stipules linear-lanceolate, margins entire, apex acute; petiole 2–13 cm, glabrous or pubescent; middle and lateral blade lobes differ in width and/ or shape, earliest blades ovate to elliptic, mid-season blades ovate, elliptic, or narrowly elliptic to narrowly deltate, 1–8 × 1–5 cm, incised or lobed at base only, base sagittate or hastate, truncate, attenuate, or ± cordate, margins crenate or serrate, ciliate or eciliate, apex acute, surfaces glabrous or pubescent. **Peduncles** 3–15 cm, glabrous or pubescent. **Flowers:** sepals lanceolate to ovate, margins ciliate or eciliate, auricles 2–3 mm; petals light to dark violet on both surfaces, lower 3 white basally, lowest dark violet-veined, lateral 2 densely bearded, spur on lowest petal occasionally bearded, lowest 10–15 mm, spur light to dark violet, gibbous, 2–3 mm; style head beardless; cleistogamous flowers on prostrate, ascending, or erect peduncles. **Capsules** ellipsoid, 10–14 mm, glabrous. **Seeds** beige, mottled to bronze, 1.3–2.5 mm.

Varieties 2 (2 in the flora): c, e North America.

1. Mid-season leaf blades: base sagittate, hastate, or ± cordate, surfaces glabrous or sparsely pubescent; sepal margins eciliate . . . .
. . . . 57a. *Viola sagittata* var. *sagittata*
1. Mid-season leaf blades: base truncate, attenuate, slightly sagittate or hastate, or ± cordate, surfaces usually densely pubescent; sepal margins ciliate .57b. *Viola sagittata* var. *ovata*

57a. Viola sagittata Aiton var. **sagittata** • Arrowhead or arrow-leaved violet, violette sagittée [E]

Viola emarginata (Nuttall) Leconte; *V. sagittata* var. *emarginata* Nuttall; *V. sagittata* var. *subsagittata* (Greene) Pollard; *V. subsagittata* Greene

Leaves: earliest blades ovate to elliptic; mid-season blades ovate or narrowly elliptic to narrowly deltate, 1.5–7 × 1–5 cm, base sagittate, hastate, or ± cordate, margins serrate, ciliate or eciliate, surfaces glabrous or sparsely pubescent. **Sepal margins** eciliate. **Cleistogamous flowers** on ascending to erect peduncles. **2***n* = 54.

Flowering Mar–May. Sandy, open woods, fields, disturbed ground, roadsides, powerline rights-of-way; 50–2500 m; N.S., Ont., Que.; Ala., Ark., Conn., Del., D.C., Ga., Ill., Ind., Iowa, Kans., Ky., La., Maine, Md., Mass., Mich., Minn., Miss., Mo., N.J., N.Y., N.C., Ohio, Okla., Pa., R.I., S.C., Tenn., Tex., Va., W.Va., Wis.

Although exhibiting variation in leaf-blade shape, var. *sagittata* is usually easily distinguished. There have been misconceptions regarding the identity of *Viola emarginata*, a species thought for a time to be different from *V. sagittata*. The type specimen of *V. emarginata* is conspecific with that of *V. sagittata*; E. L. Greene (1898b), apparently not having seen the type, portrayed *V. emarginata* with a somewhat deltate leaf form. N. H. Russell and A. C. Risser (1960) suggested that most collections representing *V. emarginata* were hybrids.

Variety *sagittata* reportedly hybridizes with *Viola cucullata* (= *V.* ×*porteriana* Pollard), *V. hirsutula* (= *V.* ×*redacta* House), *V. brittoniana* (= *V.* ×*mulfordiae* Pollard), *V. sagittata* var. *ovata* (= *V.* ×*abundans* House), and *V. sororia* (= *V.* ×*conjugens* Greene).

57b. Viola sagittata Aiton var. **ovata** (Nuttall) Torrey & A. Gray, Fl. N. Amer. 1: 138. 1838 • Northern downy or sand violet, violette à feuilles frangées [E]

Viola ovata Nuttall, Gen. N. Amer. Pl. 1: 148. 1818; *V. amorphophylla* Pollard; *V. fimbriatula* Smith; *V. ovata* var. *hicksii* (Pollard) Pollard; *V. sagittata* var. *hicksii* Pollard

Leaves: earliest blades ovate to widely ovate; mid-season blades ovate to elliptic, 1–8 × 1–4.5 cm, base truncate, attenuate, slightly sagittate or hastate, or ± cordate, margins crenate or serrate, ciliate, surfaces usually densely pubescent. **Sepal margins** ciliate. **Cleistogamous flowers** on prostrate to ascending peduncles. **2***n* = 54.

Flowering Apr–Jun. Dry, open woods and thickets, disturbed ground, roadsides, powerline rights-of-way; 10–1000 m; N.B., N.S., Ont., P.E.I., Que.; Ala., Conn., Del., D.C., Ill., Iowa, Ky., Maine, Md., Mass., Mich., N.H., N.J., N.Y., N.C., Ohio, Pa., R.I., Tenn., Vt., Va., W.Va., Wis.

Variety *ovata* was known as *Viola fimbriatula* by those who considered it a distinct species. Torrey and Gray were the first to recognize its affinity with *V. sagittata*. The early leaf-blade structure is similar in var. *ovata* and var. *sagittata*, and both have prominent sepal auricles on cleistogamous flowers (L. E. McKinney 1992).

Fernald described *Viola fimbriatula* forma *umbelliflora*, an apparently rare taxon from Halifax, Nova Scotia, in which flowers occur in umbels of three. H. J. Scoggan (1978–1979) included forma *umbelliflora* under *V. sagittata*. Study is needed of this taxon because umbels are otherwise unknown in North American species of *Viola*. Several species in the Hawaiian Islands produce umbellate inflorescences with up to five flowers (W. L. Wagner et al. 1990).

Variety *ovata* reportedly hybridizes with *Viola sagittata* var. *sagittata* (= *V.* ×*abundans* House).

58. Viola selkirkii Pursh ex Goldie, Edinburgh Philos. J. 6: 324. 1822 • Great-spurred or long-spurred or Selkirk's violet, violette de Selkirk

Plants perennial, acaulescent, not stoloniferous, 4–15 cm; rhizome slender, not fleshy. **Leaves** basal, 2–12, prostrate to ascending; stipules linear-lanceolate, margins entire, apex acute; petiole 1.5–7 cm, not winged, glabrous or pubescent; blade unlobed, usually ovate, rarely orbiculate, 1–5 × 1–5 cm, base cordate, margins crenate to crenulate or serrate, eciliate, apex rounded to acute, surfaces glabrous or sometimes pubescent abaxially, strigose adaxially. **Peduncles** 3–6 cm, glabrous or pubescent. **Flowers:** sepals lanceolate to ovate-lanceolate, margins eciliate, auricles 1–2 mm; petals light violet on both surfaces, lower 3 white basally and dark violet-veined, lateral 2 beardless, lowest 8–13 mm, spur pale to dark violet, elongated, 4–7 mm; style head beardless; cleistogamous flowers on prostrate to ascending peduncles. **Capsules** ovoid to ellipsoid, 4–8 mm, glabrous. **Seeds** brown, 1–2 mm. $2n = 24$.

Flowering May–Jun. Wet to moist places, thickets, mixed or coniferous woods; 200–3000 m; Greenland; Alta., B.C., Man., N.B., Nfld. and Labr., N.S., Ont., Que., Sask., Yukon; Alaska, Colo., Maine, Mass., Mich., Minn., N.H., N.Y., Pa., S.Dak., Vt., Wash., Wis.; Eurasia.

Viola selkirkii occurs on the southwestern coast of Greenland, north to 63°N (L. Brouillet, pers. comm.). Presence of *V. selkirkii* in the Northwest Territories and Nunavut is considered doubtful (L. Brouillet et al., http://canadensys.net/vascan). K. W. Allred (2008) said that *V. selkirkii* was considered by W. C. Martin and C. R. Hutchins (1980) to be expected in New Mexico.

59. Viola sempervirens Greene, Pittonia 4: 8. 1899 • Evergreen or redwood violet, violette toujours verte E

Viola sarmentosa Douglas in W. J. Hooker, Fl. Bor.-Amer. 1: 80. 1830, not M. Bieberstein 1808; *V. sempervirens* subsp. *orbiculoides* M. S. Baker

Plants perennial, caulescent, stoloniferous, 10–30 cm; stolons green or reddish, leafy, sometimes rooting at nodes, becoming lignified in age. **Stems** 1–5, prostrate, spreading, glabrous or sparsely puberulent, from current and/or previous year's growth, on usually vertical, fleshy rhizome, rooting and forming rosettes at or near tip; rooted rosettes often develop into an erect, fleshy caudex from which new stems are produced. **Leaves**
evergreen, basal and cauline; basal: 1–6(–10); stipules deltate to ovate or linear-lanceolate, margins entire or glandular-toothed, apex acute to long-acuminate; petiole 2–16 cm, glabrous; blade often purple-spotted abaxially and/or adaxially, orbiculate to ovate, 1–4.5 × 2–3.9 cm, base cordate to truncate, margins crenate, eciliate, apex blunt to obtuse, mucronulate, surfaces glabrous or with scattered bristles on one or both surfaces; cauline similar to basal except: stipules deltate to lanceolate, margins entire or sparingly toothed; petiole 0.3–3 cm; blade 1.2–2.2 × 1.2–2 cm. **Peduncles** 5–10 cm, glabrous. **Flowers:** sepals lanceolate to ovate-lanceolate, margins eciliate, auricles 1–2 mm; petals lemon-yellow on both surfaces, lower 3 and sometimes upper 2 brownish purple-veined, lateral 2 bearded, lowest 8–17 mm, spur yellow or whitish, gibbous, 1–2.5 mm; style head bearded; cleistogamous flowers axillary. **Capsules** mottled with purple, spherical to ovoid, 5–8 mm, glabrous. **Seeds** brown, tinged purple, 2–2.5 mm. $2n = 24, 48$.

Flowering Jan–Jul. Redwood forests, other coastal forests, Douglas fir, other coniferous forests; 30–1400 m; B.C.; Alaska, Calif., Idaho, Oreg., Wash.

In California, *Viola sempervirens* occurs in shaded redwood forests and other coastal forest habitats. In Oregon and Washington, it occurs in Douglas fir and other coniferous forests, where it can form mats (clones) one meter or more in diameter; its prostrate, spreading growth habit is similar to *V. walteri*. The leafy stems of *V. sempervirens* are similar to the leafy stolons of *V. odorata*.

60. Viola septemloba Leconte, Ann. Lyceum Nat. Hist. New York 2: 141. 1826 • Southern coastal violet E

Plants perennial, acaulescent, not stoloniferous, 5–30 cm; rhizome thick, fleshy. **Leaves** basal, 5 or 6, prostrate to ascending; stipules linear-lanceolate, margins entire, apex acute; petiole 1.5–7 cm, usually glabrous; earliest leaf blades ± ovate, sometimes 3-lobed, mid-season blades 7–9-lobed, 1–9 × 1–10 cm, base broadly cordate to cordate, middle lobes narrowly elliptic, lanceolate, spatulate, or obovate, (rarely linear), lateral lobes lanceolate or spatulate to falcate, margins usually entire, sometimes serrate, sometimes with narrowly deltate or falcate appendages or teeth, ciliate or eciliate, apex acute to mucronulate, surfaces usually glabrous. **Peduncles** 2–20 cm, usually glabrous. **Flowers:** sepals lanceolate to ovate, margins ciliate or eciliate, auricles 0.5–1 mm; petals light to dark blue-violet on both surfaces, lower 3 and sometimes upper 2 white basally, lower 3 darker violet-veined, lateral 2 densely bearded, spur sometimes bearded, lowest 15–25 mm, spur usually lilac, sometimes whitish, gibbous, 2–3 mm;

style head beardless; cleistogamous flowers on ascending to erect peduncles. **Capsules** ellipsoid, 11–14 mm, glabrous. **Seeds** beige, mottled to bronze, 2–3 mm. $2n = 54$.

Flowering Mar–May. Sandy, dry or seasonally wet pine or mixed pine/deciduous woods; 0–200 m; Ala., Fla., Ga., La., Miss., N.C., S.C., Tex.

For years *Viola septemloba*, a heterophyllous species, was either ignored or included in *V. palmata*. C. L. Pollard (1898) and E. Brainerd (1910, 1921) treated it as *V. insignis* Pollard, a later homonym.

61. Viola sheltonii Torrey in War Department [U.S.], Pacif. Railr. Rep. 4(5): 67, plate 2. 1857 • Shelton's violet E

Viola sheltonii var. *biternata* (Greene) A. Nelson

Plants perennial, caulescent, not stoloniferous, 3–27 cm. **Stems** 1–3, prostrate, decumbent, or erect, glabrous or sparsely puberulent, from short, often vertical, deep-seated or usually shallow, subligneous rhizome. **Leaves** basal and cauline; basal: 1–3, palmately compound, leaflets 3; stipules lanceolate-ovate, margins laciniate with gland-tipped projections, apex acute to acuminate; petiole 8.6–21 cm, glabrous or sparsely puberulent; blade reniform or ovate to ± orbiculate, 2–7 × 2–11 cm, coriaceous, base tapered, each leaflet cleft or dissected into 3 ± obovate lobes, each lobe further divided into 2 or 3 oblanceolate, pandurate, spatulate, oblong, lanceolate, or elliptic lobes 2–10 mm wide, margins entire, ciliate or eciliate, apex acute to obtuse, mucronulate, surfaces glabrous or sparsely puberulent; cauline similar to basal except: stipules ovate to lanceolate, unlobed, margin projections gland-tipped or eglandular, apex long-acuminate; petiole 5.5–12 cm; blade 1.2–6.3 × 1.2–10.5 cm. **Peduncles** 5–19 cm, glabrous or sparsely puberulent. **Flowers:** sepals lanceolate, margins ciliate or eciliate, auricles 0.5–1 mm; petals deep lemon-yellow adaxially, upper 2 dark brown to brownish purple abaxially, lower 3 and sometimes upper 2 brownish purple-veined, lateral 2 bearded or beardless, lowest 7–18 mm, spur yellowish with brownish purple specks, gibbous, 1–2 mm; style head bearded; cleistogamous flowers axillary. **Capsules** oblong to ovoid, 6–8 mm, glabrous or puberulent. **Seeds** brownish, shiny, ca. 2.5 mm. $2n = 12$.

Flowering Mar–Jul. Red fir, yellow pine, mixed evergreen, chaparral, oak woodlands, rich or gravelly soil; 800–2500 m; Calif., Colo., Idaho, Oreg., Wash.

The cleistogamous flowers of *Viola sheltonii* are borne on long, prostrate peduncles usually buried in duff around the plant. Mature cleistogamous capsules are usually hidden and the dehisced seeds remain close to the parent plant. Some populations of *V. sheltonii* produce only cleistogamous flowers (D. Klaber 1976).

62. Viola sororia Willdenow, Hort. Berol. 1: plate 72. 1806 • Common or downy or woolly blue violet, violette parente E

Viola chalcosperma Brainerd; *V. floridana* Brainerd; *V. latiuscula* Greene; *V. palmata* Linnaeus var. *sororia* (Willdenow) Pollard; *V. papilionaceae* Pursh; *V. priceana* Pollard; *V. rosacea* Brainerd; *V. septentrionalis* Greene; *V. wilmattiae* Pollard & Cockerell

Plants perennial, acaulescent, not stoloniferous, 5–50 cm; rhizomes thick, fleshy. **Leaves** basal, 1–8, ascending to erect; stipules linear-lanceolate to broadly lanceolate, margins entire, sometimes glandular distally, apex acute; petiole 2–25 cm, pubescent or glabrous; blade green abaxially, unlobed, ovate or broadly ovate to reniform, 2–5 × 2–10 cm, not fleshy, base cordate, margins crenate to serrate, ciliate or eciliate, apex acute to obtuse or rounded, surfaces usually pubescent, rarely glabrous. **Peduncles** 3–25 cm, glabrous or sparsely pubescent. **Flowers:** sepals lanceolate to ovate, margins ciliate or eciliate, auricles 1–2 mm; petals light to dark blue- or dark purple-violet, reddish purple, or rarely white on both surfaces, usually white basally, lowest and sometimes lateral 2 purple-veined, lateral 2 bearded, lowest bearded or beardless, 15–25 mm, spur same color as petals, gibbous, 2–3 mm; style head beardless; cleistogamous flowers on prostrate to ascending peduncles. **Capsules** ellipsoid, 5–12 mm, glabrous. **Seeds** beige, mottled to bronze, 1.5–2.5 mm. $2n = 54$.

Flowering Mar–Jun. Dry to mesic habitats in woods, thickets, stream banks, moist prairies, pastures, disturbed ground, not in saturated soil; 0–3000 m; B.C., Man., N.B., Nfld. and Labr. (Nfld.), N.S., Ont., P.E.I., Que., Sask.; Ala., Ark., Colo., Conn., Del., D.C., Fla., Ga., Ill., Ind., Iowa, Kans., Ky., La., Maine, Md., Mass., Mich., Minn., Miss., Mo., Mont., Nebr., N.H., N.J., N.Mex., N.Y., N.C., N.Dak., Ohio, Okla., Pa., R.I., S.C., S.Dak., Tenn., Tex., Utah, Vt., Va., W.Va., Wis.

Viola sororia is similar to *V. palmata* in the high degree of phenotypic plasticity. The petal color is usually light to dark blue-violet. In *V. sororia* forma *priceana* (Pollard) Cooperrider (the Confederate violet) petals are grayish white with violet veins. *Viola septentrionalis* was recognized by N. L. Gil-Ad (1997) based on what he considered distinct capsule and seed morphology. His assumptions were based on one to four specimens. A. Haines (2011) alluded to the variability of *V. sororia* and indicated that a more northern form has often been called *V. septentrionalis*. He suggested that range-wide patterns of variation make it difficult or impractical to separate. Currently, there is no valid reason to recognize *V. septentrionalis*.

Viola floridana was recognized by D. B. Ward (2006) as distinct from *V. sororia* based on being glabrous except for scattered hairs on petioles.

Viola sororia reportedly hybridizes with *V. cucullata* (= *V.* ×*bissellii* House), *V. hirsutula* [= *V.* ×*cordifolia* (Nuttall) Schweinitz], *V. brittoniana* (= *V.* ×*insolita* House), *V. pedatifida* [= *V.* ×*bernardii* (Greene) Greene], and *V. sagittata* var. *sagittata* (= *V.* ×*conjugens* Greene).

63. Viola striata Aiton, Hort. Kew. 3: 290. 1789
• Cream or cream-white or pale or striped or striped cream violet, violette striée E

Lophion striatum (Aiton) Nieuwland & Kaczmarek; *Viola conspersa* Reichenbach var. *masonii* Farwell

Plants perennial, caulescent, not stoloniferous, 10–60 cm. **Stems** 1–4, ascending to erect (often declining during anthesis), glabrous or pubescent, on caudex from fleshy rhizome. **Leaves** basal and cauline; basal: 2–6; stipules lanceolate to narrowly deltate, margins laciniate, apex acute; petiole 3–6 cm, glabrous or puberulent; blade ovate to reniform, 2–7 × 1–2.5 cm, base cordate, margins crenate to serrate, ciliate or eciliate, apex obtuse to acute, surfaces glabrous or pubescent; cauline similar to basal except: stipules lanceolate, margins laciniate; petiole 3–7 cm; distal blades ovate to deltate, 1–6 × 1–4 cm, base cordate, apex acuminate to acute. **Peduncles** 5–12 cm, glabrous or pubescent. **Flowers:** sepals lanceolate, margins ciliate or eciliate, auricles 2–3 mm; petals white or cream on both surfaces, without yellow patch basally, lowest and usually lateral 2 purple-veined, lateral 2 and sometimes all densely bearded, lowest 10–18 mm, spur white, gibbous to elongated, 3–6 mm; style head bearded; cleistogamous flowers axillary. **Capsules** ellipsoid, 6–7 mm, glabrous. **Seeds** beige to bronze, 1.5–3 mm. $2n = 20$.

Flowering Mar–Jun. Riparian or alluvial woods, flood plains in silty loam, meadows; 40–1000 m; Ont.; Ala., Ark., Conn., Del., D.C., Ga., Ill., Ind., Iowa, Ky., Md., Mass., Mich., Mo., N.H., N.J., N.Y., N.C., Ohio, Okla., Pa., S.C., Tenn., Va., W.Va., Wis.

Dead, long-persistent stems of *Viola striata* are often present the following season. In flower, the plants are often mistaken for *V. canadensis* because the petals are whitish (H. E. Ballard 1992).

Viola striata reportedly hybridizes with *V. labradorica* (= *V.* ×*eclipes* H. E. Ballard), *V. rostrata* (= *V.* ×*brauniae* Grover ex Cooperrider), *V. walteri* var. *appalachiensis* (= *V.* ×*wujekii* H. E. Ballard), and *V. walteri* var. *walteri* (= *V.* ×*cooperrideri* H. E. Ballard).

64. Viola subsinuata (Greene) Greene, Pittonia 4: 4. 1899 • Wood violet E F

Viola emarginata (Nuttall) Leconte var. *subsinuata* Greene, Pittonia 3: 313. 1898; *V. palmata* Linnaeus var. *angelliae* (Pollard) W. Stone

Plants perennial, acaulescent, not stoloniferous, 10–30 cm; rhizome thick, fleshy. **Leaves** basal, 2–11, ascending to erect; stipules linear-lanceolate, margins entire, apex acute; petiole 5–25 cm, glabrous or pubescent; mid-season leaf blades incised or lobed throughout, earliest leaf blades lobed (plants homophyllous), similar to mid-season blades, blade 5–9 (–16)-lobed, sinuses usually narrower, shallower toward leaf base, middle and lateral blade lobes differ in width and/or shape, middle lobes narrowly deltate to narrowly elliptic, lateral lobes narrowly elliptic, lanceolate, or falcate, 1–11 × 1–12 cm, base truncate to cordate, margins entire or crenate, ciliate or eciliate, apex acute, mucronulate, surfaces usually pubescent throughout or on veins. **Peduncles** 5–15 cm, glabrous or pubescent. **Flowers:** sepals lanceolate to ovate, margins ciliate or eciliate, auricles 1–2 mm; petals light to dark blue-violet on both surfaces, lower 3 and upper 2 sometimes purple-veined, lateral 2 bearded, lowest sometimes bearded, 15–25 mm, spur color same as petals, gibbous, 2–3 mm; style head beardless; cleistogamous flowers on prostrate to ascending peduncles. **Capsules** ellipsoid, 8–12 mm, glabrous. **Seeds** beige, mottled to bronze, 1.5–2.5 mm. $2n = 54$.

Flowering Apr–Jun. Rich woods; 100–3000 m; Ont.; Conn., Del., D.C., Ga., Ill., Ky., Md., Mass., Mich., N.J., N.Y., N.C., Ohio, Pa., R.I., S.C., Tenn., Vt., Va., W.Va., Wis.

L. E. McKinney (1992) described the misconceptions surrounding *Viola subsinuata* that many came to think of as the heterophyllous *V. palmata*, whereas *V. subsinuata* is homophyllous. As E. Brainerd (1910, 1921) pointed out, heterophylly versus homophylly is an important and steadfast character difference in acaulescent *Viola* species. N. L. Gil-Ad (1997) chose not to recognize *V. subsinuata*, suggesting that it represented a hybrid or introgressant. H. E. Ballard (2000) and A. Haines (2011) both recognized *V. subsinuata* as a species.

Viola subsinuata is a state historical species in Rhode Island, where it was last documented in 1941 (R. W. Enser, http://rinhs.org/wp-content/uploads/ri_rare_plants_2007.pdf).

65. Viola tomentosa M. S. Baker & J. C. Clausen, Leafl. W. Bot. 5: 142. 1949 • Woolly or felt-leaved or feltleaf violet E

Plants perennial, caulescent, not stoloniferous, 7–10 cm. **Stems** 1–3(–5), prostrate or decumbent to erect, leafy proximally and distally, densely white-tomentose, from usually vertical, subligneous rhizome. **Leaves** basal and cauline; basal: 1–6; stipules linear to broadly ovate-oblong, margins entire, sometimes with scattered glandular hairs, apex acute to obtuse; petiole 2–6 cm, densely white-tomentose; blade ± oblanceolate to elliptic, 1.5–5 × 1.4–2.1 cm, base attenuate, usually oblique, margins usually entire, rarely crenate distally, ciliate, apex acute to usually obtuse, mucronulate, surfaces densely white-tomentose abaxially, strigose adaxially; cauline similar to basal except: stipules ovate, lanceolate, oblanceolate, or oblong, margins entire or toothed, densely ciliate with white hairs; petiole 1.5–3.5 cm; blade 1.8–4 × 0.6–1.1 cm. **Peduncles** 1–4 cm, densely white-tomentose. **Flowers:** sepals lanceolate, margins ciliate-tomentose, auricles 0.5–1 mm; petals deep lemon-yellow adaxially, upper 2 often brownish purple abaxially, lower 3 dark brown- to brownish purple-veined, lateral 2 bearded, lowest 6–11 mm, spur yellow, gibbous, 0.5–1.5 mm; style head bearded; cleistogamous flowers absent. **Capsules** ± spherical, 4–5 mm, densely white-tomentose. **Seeds** brown with lighter brown mottling, 2.5–2.8 mm. $2n = 12$.

Flowering May–Aug. Dry, gravelly places, open ponderosa, Jeffrey, lodgepole pine forests; 1500–2000 m; Calif.

Viola tomentosa occurs in El Dorado, Nevada, Placer, Plumas, and Sierra counties. M. S. Baker (1949) reported that nearly every leaf axil of *V. tomentosa* produces a flower bud and that these buds produce chasmogamous flowers late in season instead of cleistogamous flowers, as do other members of the *V. nuttallii* complex.

Viola tomentosa hybridizes with *V. purpurea*; the hybrids appeared sterile (M. S. Baker 1949). J. Clausen (1964) reported a putative hybrid between *V. tomentosa* and *V. sheltonii* from one location in Sierra County.

66. Viola tricolor Linnaeus, Sp. Pl. 2: 935. 1753 • Johnny-jump-up, pansy, wild pansy, heart's ease, ladies delight, violette tricolore I

Varieties 3 (1 in the flora): introduced; Europe, Asia.

66a. Viola tricolor Linnaeus var. **tricolor** I

Plants usually annual, rarely perennial, caulescent, not stoloniferous, 3–45 cm. **Stems** 1–9, prostrate, decumbent, or erect, simple or branched, subglabrous or puberulent, clustered on taproot. **Leaves** cauline; stipules palmately lobed or pinnatifid, middle lobe oblanceolate, obovate, elliptic, or lanceolate, sometimes ± equaling leaf blade, margins entire or usually crenate, ciliate, apex acute to obtuse; petiole 1–3.3 cm, glabrous or puberulent; blade ovate to ± oblong (distal blades lanceolate), 1.5–2.8 × 0.8–2.3 cm, base cordate, truncate, or attenuate, margins coarsely crenate-serrate, ciliate or eciliate, apex acute to obtuse, surfaces pubescent abaxially at least on major veins, glabrous adaxially. **Peduncles** 3–12.5 cm, usually glabrous. **Flowers:** sepals deltate-lanceolate to linear-lanceolate, margins usually ciliate, auricles 2–4 mm; petals: upper 2 commonly violet, lateral white, and lowest yellow or white on both surfaces with darker yellow patch basally, variants include lateral 2 violet distally or all petals violet, white, yellow, or red, lateral 2 densely bearded, lowest and lateral 2 dark violet- to brownish purple-veined, lateral 2 longer than sepals, lowest 10–15 mm, spur usually yellow to violet, rarely white, elongated, 4–5 mm; style head beardless; cleistogamous flowers absent. **Capsules** subglobose to ovoid, 6–8 mm, glabrous. **Seeds** tan, 1.5–2 mm. $2n = 26$.

Flowering Apr–Sep. Cultivated and waste ground, roadsides, lawns, urban parks; 10–1000 m; introduced; St. Pierre and Miquelon; Alta., B.C., Man., N.B., Nfld. and Labr. (Nfld.), N.S., Ont., P.E.I., Que., Sask., Yukon; Ala., Alaska, Ark., Calif., Conn., Del., D.C., Fla., Ga., Idaho, Ill., Ind., Iowa, Kans., Ky., La., Maine, Md., Mass., Mich., Minn., Miss., Mo., Nebr., N.H., N.J., N.Y., N.C., Ohio, Okla., Oreg., Pa., R.I., S.C., Tex., Utah, Vt., Va., Wash., Wis., Wyo.; Europe; Asia (Siberia).

Variety *tricolor* was introduced from Europe and is not vegetatively distinguishable from *Viola arvensis*. In Europe, the variety is pollinated mainly by long-tongued bees (A. R. Clapham et al. 1987).

67. **Viola trinervata** (Howell) Howell ex A. Gray, Bot. Gaz. 11: 290. 1886 • Rainier or sagebrush or three-nerved violet E

Viola beckwithii Torrey var. *trinervata* Howell, Bot. Gaz. 8: 207. 1883

Plants perennial, caulescent, not stoloniferous, 5–15 cm. **Stems** 1–4, decumbent, ascending, or erect, ca. ½ subterranean, glabrous, from single, vertical, deep-seated caudex. **Leaves** basal and cauline; basal: 1–7, palmately compound, leaflets 3–5; stipules adnate to petiole, forming 2 linear-lanceolate wings, margins entire, apex of each wing free, acute; petiole 4.5–15 cm, glabrous; blade reniform or ovate to ± orbiculate, 2–5 × 2.5–5 cm, coriaceous, base tapered, leaflets cleft or dissected into 2 or 3 elliptic, lanceolate, or oblanceolate lobes 2–7 mm wide, margins usually entire, eciliate, apex acute, mucronulate, surfaces glabrous (± glaucous), abaxial surface usually with prominent vein parallel to each margin; cauline similar to basal except: stipules lanceolate, unlobed; petiole 1–5.5 cm; blade 1–3 × 2–4.5 cm. **Peduncles** 1.1–7 cm, glabrous. **Flowers:** sepals lanceolate, margins eciliate, auricles 0–1 mm; petals: upper 2 often overlapping, dark reddish violet on both surfaces, lower 3 lilac, rarely white, lateral 2 bearded, with yellow patch basally and reddish violet patch distal to yellow patch, lowest 9–15 mm with yellow patch, dark reddish violet-veined, spur yellow, gibbous, 0.6–1.5 mm; style head bearded; cleistogamous flowers absent. **Capsules** ovoid, 7–12 mm, glabrous. **Seeds** tan, 3.2–4.5 mm.

Flowering Mar–May. Sagebrush flats, dry, rocky hillsides, usually in gravelly soil; 400–1200 m; Oreg., Wash.

In some populations of *Viola trinervata* the lower three petals are white with a yellow area proximally (V. B. Baird 1942).

68. **Viola tripartita** Elliott, Sketch Bot. S. Carolina 1: 302. 1817 • Three-parted or Piedmont or threepart violet E

Viola hastata Michaux var. *glaberrima* Gingins; *V. hastata* var. *tripartita* (Elliott) A. Gray; *V. tripartita* var. *glaberrima* (Gingins) R. M. Harper

Plants perennial, caulescent, not stoloniferous, 10–40 cm. **Stems** 1(2), erect, leafless proximally, leafy distally, glabrous or puberulent, from subligneous rhizome. **Leaves** basal and cauline; basal: 0(–2); stipules ovate, not leaflike, margins entire, apex acute to acuminate, surfaces

glabrous or puberulent; petiole 9–11.5 cm, glabrous or puberulent; blade unlobed, ovate, or 3–5-lobed, 4–5 × 1–5 cm, base cordate, margins entire or crenate-serrate, ciliate or eciliate, apex acute, surfaces glabrous or ± puberulent; cauline similar to basal except: restricted to distal ends of stems; stipules ovate to oblong; petiole 0.7–7.2 cm, glabrous or puberulent; blade unlobed, ovate or deltate, or 3-lobed (if 3-lobed, lateral lobes falcate, middle rhombic, longer than others, lobes may appear petiolate; unlobed and 3-lobed leaves can occur on same plant), 1–6 × 0.5–5.5 cm, base truncate to cuneate, margins serrate, ciliate or eciliate, surfaces glabrous or pubescent. **Peduncles** 1.5–4 cm, glabrous or pubescent. **Flowers:** sepals lanceolate to ovate, margins ciliate or eciliate, auricles 0.1–0.5 mm; petals lemon-yellow adaxially, upper 2, rarely others, brownish purple abaxially, lowest and usually lateral 2 brownish purple-veined, lateral 2 and sometimes lowest bearded, lowest 10–18 mm, spur yellow, gibbous, 0.5–2 mm; style head bearded; cleistogamous flowers axillary. **Capsules** ovoid to ellipsoid, 9–12 mm, glabrous. **Seeds** beige, bronze, or brown, 2.4–3 mm. $2n = 12$.

Flowering Mar–May. Rich woods; 50–1600 m; Ala., Fla., Ga., Ky., Miss., N.C., Ohio, Pa., S.C., Tenn., Va., W.Va.

Some authors recognize two varieties of *Viola tripartita* based on lobed versus unlobed leaves. F. L. Lévesque and P. M. Dansereau (1966) suggested that leaf variation is the only character difference between vars. *tripartita* and *glaberrima*. N. H. Russell (1965) stated that *V. tripartita* plants with lobed and unlobed leaves are sympatric and frequently intergrade and did not recognize them as distinct. The situation with two leaf forms in *V. tripartita* is similar to *V. lobata*, which also has two leaf forms.

69. **Viola umbraticola** Kunth in A. von Humboldt et al., Nov. Gen. Sp. 5(fol.): 289; 5(qto.): 370. 1823

Varieties 2 (1 in the flora): Arizona, n Mexico.

69a. **Viola umbraticola** Kunth var. **glaberrima** W. Becker, Repert. Spec. Nov. Regni Veg. 20: 7. 1924 • Ponderosa violet

Plants perennial, acaulescent, not stoloniferous, 5–11 cm; rhizome thick, subligneous. **Leaves** basal, 3–14, prostrate to ascending; stipules linear-lanceolate, margins laciniate, apex long-acuminate; petiole 1–7.8 cm, puberulent; blade unlobed, ovate to oblong, 1–3.8 × 1–3.6 cm, base truncate to cordate, margins crenate to serrate, eciliate, apex obtuse to acute, surfaces glabrous. **Peduncles** 3–9 cm, glabrous. **Flowers:** sepals linear-lanceolate, margins eciliate, auricles 1–1.5 mm; petals blue to violet on both surfaces, white at base, lower 3 dark violet-veined, lateral 2 bearded, lowest 9–16 mm, spur white, gibbous to elongated, 2.5–4 mm, tip straight or hooked up; style head bearded; cleistogamous flowers unknown. **Capsules** ellipsoid, 4–11 mm, glabrous. **Seeds** unknown.

Flowering May–Aug.   Mixed oak and pine and coniferous forests, bases of shrubs or pine trees in dense duff; 1600–2400 m; Ariz.; Mexico (Chihuahua, Durango, Sonora).

The mature rhizomes of var. *glaberrima* occur to 15 cm beneath the soil surface and eventually become lignified and woody. Variety *glaberrima* differs from var. *umbraticola* in being glabrous except for hairy stipule apices.

In Arizona, var. *glaberrima* is known from the higher elevations of the Chiricahua, Santa Catalina, and Santa Rita mountain ranges. It was first collected in the United States by J. J. Thornber in 1913 in the Santa Catalina Mountains, Pima County. *Viola umbraticola* has no close relatives in the United States or in Mexico (M. S. Baker 1949).

70. **Viola utahensis** M. S. Baker & J. C. Clausen, Leafl. W. Bot. 5: 145. 1949   [E]

Plants perennial, caulescent, not stoloniferous, 5–20 cm. **Stems** 1–4, prostrate, decumbent, or erect, leafy proximally and distally, glabrate or sparsely puberulent, on caudex from subligneous rhizome. **Leaves** basal and cauline; basal: 1–3; stipules adnate to petiole, forming 2 linear-lanceolate wings, margins entire or laciniate with glandular hairs, apex of each wing free, acute, sometimes divided into several filiform processes; petiole 2–8(–12) cm, glabrate or sparsely puberulent; blade usually broadly ovate, sometimes ± orbiculate, 1–4.5 × 0.7–3.4 cm, base usually attenuate, rarely truncate or subcordate, margins ± coarsely crenate-serrate, ciliate, apex obtuse, surfaces glabrate or sparsely puberulent; cauline similar to basal except: stipules lanceolate or linear-lanceolate to linear, margins ± entire or lacerate to laciniate with sparse gland-tipped projections, apex acute to acuminate, often divided into 2–4 filiform processes; petiole 3–8.7 cm; blade ovate to oblong-ovate or oblong-lanceolate, 0.7–1.6 × 2.6–3.7 cm, margins shallowly serrate, irregularly crenate, or ± entire, apex acute to obtuse. **Peduncles** 4–10 cm, glabrate or sparsely puberulent. **Flowers:** sepals lanceolate, margins ciliate or eciliate, auricles 1–1.5 mm; petals deep lemon-yellow adaxially, upper 2 and sometimes lateral 2 reddish brown to brownish purple abaxially, lower 3 and sometimes upper 2 dark brown-

veined, lateral 2 sparsely bearded, lowest 10–13 mm, spur yellow, gibbous, 0.5–1.5 mm; style head bearded; cleistogamous flowers axillary. **Capsules** ± spherical, 6–7 mm, glabrous or finely puberulent. **Seeds** brown, ca. 3 mm. $2n = 24$.

Flowering May.   Open sagebrush areas, semidesert slopes, grasslands, open forests; 1200–2600 m; Colo., Idaho, Nev., Utah, Wyo.

Viola utahensis occurs only in areas where its closest relatives, *V. vallicola* and *V. purpurea* var. *venosa*, are also found, and it appears to have originated from the crossing of those taxa followed by a doubling of the chromosomes (M. S. Baker 1949). The elaiosome of *V. utahensis* seeds has the same general appearance as those of other members of *V. purpurea* complex (except *V. purpurea* var. *venosa*), covering about one-third the length of the seed. G. Davidse (1976) found that tetraploid members of Clausen's "subsect." *Purpurea* (*V. utahensis*) have serrate leaf margins; diploid plants (*V. purpurea*) have crenate leaf margins. He also found that the size of *V. utahensis* plants is correlated with elevation, the largest plants at low elevations and the smallest at higher elevations.

G. Davidse (1976) reported putative hybrids between *Viola utahensis* and *V. praemorsa* subsp. *major* (= var. *linguifolia*), *V. purpurea* var. *venosa*, and *V. beckwithii*.

Observed pollinators of *Viola utahensis* in Utah include the bees *Apis mellifera* Linnaeus and *Anthophora ursina* Cresson (G. Davidse 1976). In other populations, flies in the genera *Eristalis* Latreille and *Bombylius* Linnaeus were observed indiscriminately visiting the yellow-flowered species *V. praemorsa* subsp. *major*, *V. utahensis*, and *V. vallicola* (Davidse).

71. **Viola vallicola** A. Nelson, Bull. Torrey Bot. Club 26: 128. 1899   [E]

Crocion vallicola (A. Nelson) Nieuwland & Lunell; *Viola nuttallii* Pursh subsp. *vallicola* (A. Nelson) Roy L. Taylor & MacBryde; *V. nuttallii* var. *vallicola* (A. Nelson) H. St. John; *V. physalodes* Greene; *V. russellii* B. Boivin; *V. subsagittifolia* Suksdorf

Plants perennial, caulescent, not stoloniferous, 2–18 cm. **Stems** 1–5, decumbent or ascending to erect, leafy proximally and distally, ca. ½ subterranean, glabrate to puberulent, on caudex from usually vertical, subligneous rhizome. **Leaves** basal and cauline; basal: 1–6; stipules adnate to petiole, forming 2 linear-lanceolate wings, margins entire, apex of each wing free, sometimes divided into filiform processes; petiole 3–10 cm, glabrous or puberulent; blade ovate to oblong-ovate,

1.5–4.3 × 0.9–1.1 cm, base usually truncate, sometimes attenuate, margins entire or serrulate, sometimes sinuate, ciliate (sometimes only on proximal ½ of leaf), apex acute to obtuse, surfaces glabrous or puberulent; cauline similar to basal except: stipules linear to linear-lanceolate, margins ± entire, apex acute; petiole 1.5–9.2 cm; blade ovate or lanceolate to elliptic, 2.3–4 × 1–2 cm, length ca. 2.2 times width, base usually ± truncate to subcordate, sometimes attenuate on new leaves, margins sinuate, apex acute. **Peduncles** 3–11.5 cm, glabrous or puberulent. **Flowers:** sepals lanceolate, margins eciliate, auricles 0.5–1 mm; petals deep lemon-yellow adaxially or on both surfaces, upper 2 often brownish purple abaxially, lower 3 dark brown- to brownish purple-veined, lateral 2 sparsely bearded, lowest 9–14 mm, spur yellow, gibbous, 0.8–1.2 mm; style head bearded; cleistogamous flowers axillary. **Capsules** spherical, ca. 5 mm, glabrous or finely puberulent. **Seeds** tan, 2.1–2.2 mm, elaiosome extending beyond and covering funiculus. $2n = 12$.

Flowering late Mar–early Jul. Sagebrush flats, prairie grasslands, open forests, juniper woodlands; 400–2800 m; B.C., Sask.; Colo., Kans., Mont., N.Dak., S.Dak., Wyo.

D. M. Fabijan et al. (1987) distinguished two varieties of *Viola vallicola* based on geographic location and type of leaf flavonoids: var. *major* (Hooker) Fabijan, occurring west of the Continental Divide with leaf flavonoids primarily kaempferol derivatives; and var. *vallicola* occurring east of the Continental Divide with leaf flavonoids all apigenin derivatives. Fabijan et al. reported that var. *major* (valley violet) occurs in sagebrush flats, prairie grasslands, open forests, juniper woodlands, 400–2800 m, in Alberta, Colorado, Idaho, Montana, Nevada, Oregon, Utah, and Washington.

M. S. Baker (1957) noted that there is only a tendency for leaves of *Viola vallicola* to be wide with truncate bases; on some plants only a few leaves are truncate. Images of type specimens at NY show truncate leaf bases for most basal and cauline leaves, with only some cauline leaves being attenuate. The key to *V. vallicola* in D. M. Fabijan et al. (1987) described the cauline leaf bases as truncate with some later cauline leaves becoming cuneate. G. Davidse (1976) stated that the *V. vallicola* plants he studied had cordate to truncate leaf bases; he made no distinction between basal and cauline leaves. Baker stated that *V. vallicola* may have given rise to *V. nuttallii* through a doubling of its chromosomes.

Because intermediate leaf forms are found in areas where *Viola vallicola* and *V. nuttallii* are sympatric, some question the specific status of *V. vallicola*. No hybrids involving *V. nuttallii* and *V. vallicola* are known (D. M. Fabijan et al. 1987).

V. Harms (pers. comm.) reported that *Viola vallicola* is frequent in southern Saskatchewan grasslands, usually occurring with *V. nuttallii* and appearing the more frequent of the two.

Observed pollinators of *Viola vallicola* in the intermountain region include flies in the genera *Bombylius* Linnaeus and *Eristalis* Latreille (G. Davidse 1976).

72. Viola villosa Walter, Fl. Carol., 219. 1788
  • Southern woolly violet E

Viola alabamensis Pollard

Plants perennial, acaulescent, not stoloniferous, 5–15 cm; rhizome thick, fleshy. **Leaves** basal, 4–9, prostrate to ascending; stipules linear-lanceolate, margins entire, apex acute; petiole 3–10 cm, densely pubescent; blade unlobed, reniform or ovate to elliptic, 1–8 × 1–5.5 cm, base cordate, margins serrate, ciliate, apex rounded to acute, mucronulate, surfaces densely pubescent. **Peduncles** 4–10 cm, puberulent. **Flowers:** sepals lanceolate to ovate, margins ciliate, auricles 1–2 mm; petals light to dark blue-violet on both surfaces, lower 3 white basally and dark violet-veined, lateral 2 bearded, spur sometimes bearded, lowest 10–20 mm, spur usually white, gibbous, 2–3 mm; style head beardless; cleistogamous flowers on ascending to erect peduncles. **Capsules** ellipsoid, 6–10 mm, glabrous. **Seeds** beige, mottled to bronze, or dark brown, 1.5–2 mm. $2n = 54$.

Flowering Apr–Jun. Sandy, pine-oak and pine-oak-hickory woods and disturbed ground; 10–300 m; Ala., Ark., Fla., Ga., La., Miss., N.C., Okla., S.C., Tex.

Much of the foliage of *Viola villosa* remains green throughout the winter (V. B. Baird 1942).

73. Viola walteri House, Torreya 6: 172. 1906 E

Plants perennial, caulescent, stoloniferous, 5–19 cm; stolons green or reddish, leafy, sometimes rooting at nodes, becoming lignified in age. **Stems** 1–5, prostrate, spreading, finely puberulent, from current and/or previous year's growth, on usually vertical, fleshy rhizome, rooting and forming rosettes at or near tip; rooted rosettes often develop into an erect, fleshy caudex from which new stems are produced. **Leaves** basal and cauline; basal: 3–6; stipules lanceolate, margins laciniate, projections often long-filamentous, apex long-acuminate; petiole 2.3–7.3 cm, glabrous or pubescent; blade often purple-spotted abaxially and/or adaxially, ovate to reniform, 1.2–5 × 1.6–3.6 cm, base deeply to broadly cordate, margins crenulate to serrate, ciliate or eciliate, apex rounded to obtuse, surfaces glabrous or pubescent; cauline similar to basal except: stipules ovate to lanceolate, margins laciniate; petiole 1–3.5 cm; blade 1.3–2.9 × 1.4–3.2 cm. **Peduncles** 5–9.6 cm, glabrous or pubescent. **Flowers:** sepals lanceolate to ovate, margins mostly eciliate, auricles 0.5–1 mm; petals pale to bluish violet on both surfaces, lower 3 white basally and darker violet-veined, lateral 2 and often upper 2 and lowest bearded, lowest 15–18 mm, spur white, gibbous to usually elongated, 3–5 mm; style head bearded; cleistogamous flowers axillary. **Capsules** ovoid to ellipsoid, 5–7 mm, glabrous. **Seeds** brown, 1–1.5 mm. $2n = 20$.

Varieties 2 (2 in the flora): e United States.

1. Leaf blades pubescent, hairs scattered throughout or concentrated along veins, adaxial surfaces, or toward margins; stipule margin processes ½+ stipule length. . . . . 73a. *Viola walteri* var. *walteri*
1. Leaf blades glabrous or pubescent, hairs usually confined to adaxial surfaces; stipule margin processes to ½ stipule length. . . . . 73b. *Viola walteri* var. *appalachiensis*

73a. Viola walteri House var. **walteri** · Walter's or prostrate blue violet   E

Stipule margin processes ½+ stipule length. **Leaf blade surfaces** pubescent, hairs scattered throughout or concentrated along veins, adaxial surfaces, or toward margins. $2n = 20$.

Flowering Mar–May. Rich woodlands, upper flood plains, rocky ledges; 100–2000 m; Ala., Ark., Fla., Ga., Ky., La., Miss., N.C., Ohio, S.C., Tenn., Tex., Va.

Variety *walteri* shows an interesting transition between caulescent and acaulescent violets; its growth habit is similar to *Viola odorata* (A. E. Radford et al. 1968). D. Klaber (1976) reported that *V. walteri* forms mats to 0.6 m in diameter.

Viola walteri and *V. sempervirens* have a similar growth habit, the abaxial and adaxial leaf surfaces of their mature leaves are often distinctly purple-dotted, and their stems persist through the winter.

Variety *walteri* is reported to hybridize with *Viola striata* (= *V.* ×*cooperrideri* H. E. Ballard).

73b. Viola walteri House var. **appalachiensis** (L. K. Henry) L. E. McKinney ex S. P. Grund & B. L. Isaac, Castanea 72: 59. 2007 · Appalachian blue or Appalachian violet   E

Viola appalachiensis L. K. Henry, Castanea 18: 131. 1953, based on *V. allegheniensis* L. K. Henry, Castanea 18: 53, plate 2. 1953, not *V. alleghanensis* Roemer & Schultes 1819

Stipule margin processes to ½ stipule length. **Leaf blade surfaces** glabrous, if pubescent, hairs usually confined to adaxial surface. $2n = 20$.

Flowering Apr–Jun. Mesic woodlands, open woodlands on serpentine, flood plains, edges of bogs and swamps; 500–1100 m; Md., N.C., Pa., W.Va.

L. E. McKinney (1986) considered the differences between *Viola walteri* and *V. appalachiensis* to be minor, choosing not to recognize the latter. H. E. Ballard and D. E. Wujek (1994) provided evidence to continue recognition of *V. appalachiensis*.

Variety *appalachiensis* is reported to hybridize with *Viola striata* (= *V.* ×*wujekii* H. E. Ballard).

TURNERACEAE Kunth ex de Candolle

• Turnera Family

María Mercedes Arbo

Herbs, subshrubs, or shrubs [trees], perennial, often rhizomatous. **Stems** usually branched, hairs simple and/or porrect-stellate [stellate] and, sometimes, glandular hairs. **Leaves** alternate, sessile or petiolate; stipules present or absent; blade margins crenate or serrate [entire]. **Inflorescences** solitary flowers [cymose or racemose], ephemeral, sometimes epiphyllous; prophylls 2. **Flowers** homostylous or distylous; sepals 5, connate; petals 5, claw adnate to calyx forming 10-veined floral tube; stamens 5; filaments partially adnate to floral tube; anthers 4-locular, dehiscence introrse, longitudinal; pistils 1, 3-carpellate; ovary superior or half-inferior, 1-locular; ovules 1–120, anatropous; placentation parietal; styles 3, connivent, filiform; stigmas 3, penicillate. **Fruits** capsular, 3-valved, dehiscence loculicidal. **Seeds** obovoid, straight or curved; aril inserted around hilum, lobed, plump, membranous when dry; endosperm fleshy; embryo straight.

Genera 10, species 226 (2 genera, 4 species in the flora): sc, se United States, Mexico, West Indies, Central America, South America, Africa, Indian Ocean Islands (Madagascar, Mascarene Islands).

The close relationships of Turneraceae with Passifloraceae, Malesherbiaceae, and Violaceae have long been recognized, especially by the presence of cyclopentenoid cyanogenic glycosides and cyclopentenyl fatty acids. These families have traditionally been positioned in Parietales or Violales with other taxa that have parietal placentation; analyses of DNA sequence data indicate that only a subset of the taxa with parietal placentation are closely related. The group that includes Turneraceae is embedded within Malpighiales (O. I. Nandi et al. 1998; V. Savolainen et al. 2000b; D. E. Soltis et al. 2000; M. W. Chase et al. 2002).

SELECTED REFERENCES Brizicky, G. K. 1961. The genera of Turneraceae and Passifloraceae in the southeastern United States. J. Arnold Arbor. 42: 204–218. Urban, I. 1883. Monographie der Familie der Turneraceen. Jahrb. Königl. Bot. Gart. Berlin 2: 81–148.

1. Hairs porrect-stellate and simple, seldom absent, glandular hairs absent; leaves without nectaries; peduncle free; corona present. . . . . 1. *Piriqueta*, p. 166
1. Hairs glandular and simple, glandular hairs microcapitate or sessile-capitate; leaves often with nectaries; peduncle free or adnate to petiole (flowers epiphyllous); corona absent. . . . 2. *Turnera*, p. 167

1. PIRIQUETA   Aublet, Hist. Pl. Guiane 1: 298, plate 117. 1775   •   [Common name in Guiana]

Herbs [shrubs], hairs porrect-stellate and simple, seldom absent, glandular hairs absent [present]. **Leaves** petiolate or sessile, without nectaries; stipules reduced or absent. **Inflorescences** axillary; peduncle free; prophylls inconspicuous. **Pedicels** articulate. **Flowers** distylous [homostylous]; sepals ⅕–½ connate; petals yellow or whitish; corona inserted at throat on sepals and petals, annular, fimbriate; filaments often with nectaries; anthers dorsifixed. **Capsules** smooth [granulate/tuberculate]. **Seeds** straight [curved], reticulate; aril relatively narrow [wide]. *x* = 7.

Species 44 (1 in the flora): se United States, Mexico, West Indies, Central America, South America, s Africa.

SELECTED REFERENCES   Arbo, M. M. 1995. Turneraceae Parte I. *Piriqueta*. In: Organization for Flora Neotropica. 1968+. Flora Neotropica. 109+ nos. New York. No. 67. Maskas, S. D. and M. B. Cruzan. 2000. Patterns of intraspecific diversification in the *Piriqueta caroliniana* complex in southeastern North America and the Bahamas. Evolution 54: 815–827. Ornduff, R. 1970c. Relationships in the *Piriqueta caroliniana–P. cistoides* complex (Turneraceae). J. Arnold Arbor. 51: 492–498.

1. Piriqueta cistoides   (Linnaeus) Grisebach, Fl. Brit. W. I., 298. 1860   F

Turnera cistoides Linnaeus, Sp. Pl. ed. 2, 1: 387. 1762

Subspecies 2 (1 in the flora): se United States, West Indies, South America.

Within *Piriqueta*, *P. cistoides* has the widest distribution. Both subspecies show wide morphological variation; subsp. *cistoides* is homostylous and self compatible; subsp. *caroliniana* is distylous and self incompatible. In Cuba, Dominican Republic, and northern South America, intermediate specimens (10% of the total) cannot be assigned morphologically to one subspecies or the other. R. Ornduff (1970c) made crosses between and among the subspecies; he found no reproductive barriers among the different morphs of subsp. *caroliniana*; the fertility of the hybrids between both subspecies was higher than that of the hybrids among populations of subsp. *cistoides*. Subspecies *cistoides* has been reported from Georgia; the author has not seen specimens.

1a. Piriqueta cistoides   (Linnaeus) Grisebach subsp. **caroliniana**   (Walter) Arbo, Ann. Missouri Bot. Gard. 77: 351. 1990   •   Pitted stripeseed   F

Waltheria caroliniana Walter, Fl. Carol., 175. 1788; *Piriqueta caroliniana* (Walter) Urban; *P. caroliniana* var. *tomentosa* Urban; *P. caroliniana* var. *viridis* (Small) G. S. Torrey; *P. glabra* Chapman; *P. viridis* Small

Plants 10–60 cm. **Leaves:** petiole to 2(–8) mm; blade linear to elliptic or obovate, 8–70 × 1.5–22 mm, base cuneate to rounded, apex obtuse or acute, surfaces tomentose or glabrous. **Peduncles and pedicels** erect-ascending, combined length 6.5–22 mm. **Flowers** not self compatible; calyx 4.7–11 mm, tube 1.3–3 mm; petals 8–24 × 3–15 mm; stamens 3–5 mm in long-styled flowers, 5–9 mm in short-styled flowers; styles often forked, 3.5–6 mm in long-styled flowers, 1.3–3 mm in short-styled flowers. **Capsules** ovoid, 4–8 mm. **Seeds** 12–30, dark, 1.5–2.1 × 0.9–1.2 mm; aril unilateral, 1–2 mm. *2n* = 14.

Flowering and fruiting year-round. Marshes to dry pinelands and sand scrub, dry or scrubby flatwoods, sandhills, roadsides, other ruderal areas, sandy soil; 0–60 m; Ala., Fla., Ga., S.C.; West Indies; South America.

Some authors consider subsp. *caroliniana* to be a species complex composed of multiple morphotypes that hybridize along a wide zone in Florida; North American populations would have been derived from three independent migrations from the northern Bahamas (M. B. Cruzan 2005; S. D. Maskas and Cruzan 2000; K. McBreen and Cruzan 2004; Jiongqion Wang and Cruzan 1998). The morphotypes differ mainly in leaf shape and indument; leaf traits and plasticity responses toward narrower shape and higher trichome density are associated with increased drought conditions (J. J. Picotte et al. 2009).

2. TURNERA Linnaeus, Sp. Pl. 1: 271. 1753; Gen. Pl. ed. 5, 131. 1754 • Turnera [For William Turner, 1515–1568, English botanist]

Herbs, subshrubs, or shrubs [trees], hairs glandular and simple [stellate], glandular hairs sessile-capitate or microcapitate. **Leaves** petiolate or sessile, often with nectaries; stipules present or absent. **Inflorescences** axillary; peduncle free or adnate to petiole (flowers epiphyllous); prophylls persistent. **Pedicels** absent. **Flowers** mostly distylous; sepals at least ⅓ connate; petals yellow or white [salmon, pink, orange, or red], sometimes with dark basal spot; corona absent; filaments often with nectaries; anthers dorsifixed or basifixed. **Capsules** granulose, rugose, or verrucose [smooth]. **Seeds** reticulate [striate]. *x* = 5, 7, (13).

Species 142 (3 in the flora): sc, se United States, Mexico, West Indies, Central America, South America, Africa.

I. Urban (1883) divided *Turnera* into nine series, and M. M. Arbo (2008) added two. Phylogenetic studies (S. Truyens et al. 2005; Arbo and S. M. Espert 2009) indicate that the genus is monophyletic.

Turnera subulata Smith, with a dark basal spot in the yellow petals, has been collected twice, probably as a garden escape, in Miami-Dade County, southern Florida.

SELECTED REFERENCES   Arbo, M. M. 2000. Estudios sistemáticos en *Turnera* (Turneraceae). II. Series *Annulares, Capitatae, Microphyllae y Papilliferae*. Bonplandia (Corrientes) 10: 1–82. Arbo, M. M. 2005. Estudios sistemáticos en *Turnera* (Turneraceae). III. Series *Anomalae y Turnera*. Bonplandia (Corrientes) 14: 115–318.   Arbo, M. M. 2008. Estudios sistemáticos en *Turnera* (Turneraceae). IV. Series *Leiocarpae, Conciliatae y Sessilifoliae*. Bonplandia (Corrientes) 17: 107–334.   Arbo, M. M. and S. M. Espert. 2009. Morphology, phylogeny and biogeography of *Turnera* L. (Turneraceae). Taxon 58: 457–467.   Arbo, M. M. and S. M. Mazza. 2011. The major diversity centre for neotropical Turneraceae. Syst. Biodivers. 9: 203–210.   Truyens, S., M. M. Arbo, and J. S. Shore. 2005. Phylogenetic relationships, chromosome and breeding system evolution in *Turnera* (Turneraceae): Inferences from ITS sequence data. Amer. J. Bot. 92: 1749–1758.

1. Leaves: petiole without nectaries, blade margins revolute; inflorescences 0.5–0.8 cm, not epiphyllous; filaments adnate for 0.2 mm to base of floral tube, anther pockets absent; anthers dorsifixed . . . . .1. *Turnera diffusa*
1. Leaves: petiole with nectaries, blade margins not revolute; inflorescences 2–4 cm, epiphyllous; proximal part of filaments adnate by margins to floral tube forming 5 nectar pockets; anthers basifixed.
   2. Flowers distylous; petals white . . . . 2. *Turnera coerulea*
   2. Flowers homostylous; petals yellow . . . . 3. *Turnera ulmifolia*

1. Turnera diffusa Willdenow in J. J. Roemer et al., Syst. Veg. 6: 679. 1820 [F]

Varieties 2 (1 in the flora): Texas, Mexico, West Indies, Central America, South America (ne Brazil).

Turnera diffusa is used extensively as an anticough, diuretic, and aphrodisiac agent. It has antibacterial activity against the most common gastrointestinal diseases of Mexico (T. Hernández et al. 2003). Variety *aphrodisiaca* (Ward) Urban, with glabrous leaves, grows only in Mexico and West Indies.

1a. Turnera diffusa Willdenow var. diffusa
• Damiana, yerba de venado [F]

Shrubs 0.3–1.2[–2] m, hairs simple and sessile-capitate glandular. **Leaves** aromatic; petiole absent or to 3.5(–7) mm, without nectaries; blade obovate or ovate, 5–30 × 2–13 mm, base cuneate or attenuate, margins revolute, crenate-serrate, apex obtuse, surfaces tomentose to pilose abaxially, often glabrate adaxially. **Inflorescences** not epiphyllous, 0.5–0.8 cm; peduncle inserted at foliar base, 0.3–1(–2) mm; prophylls inserted at peduncle apex, ovate-lanceolate, 2.5–7 × 0.4–2.5 mm, without nectaries. **Flowers** distylous; calyx 3–8 mm, tube 2–5 mm; petals usually yellow, rarely whitish, 4–7(–9) × 1.5–4(–7) mm; filaments adnate 0.2 mm to floral tube

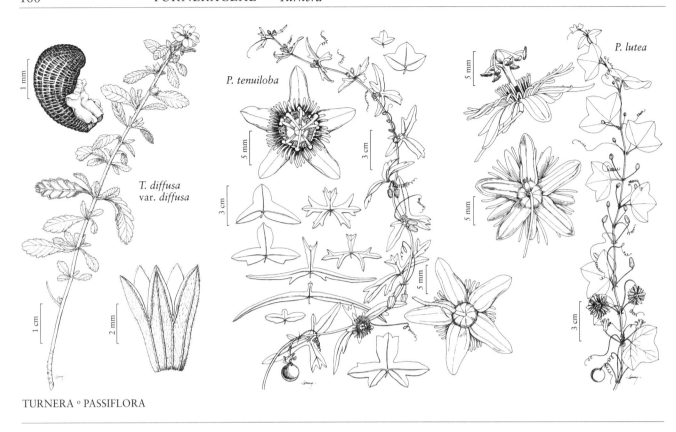

T. diffusa
var. diffusa

P. tenuiloba

P. lutea

TURNERA ° PASSIFLORA

from base, 3–3.5 mm in long-styled flowers, (2–)4.5–7.5 mm in short-styled flowers, nectaries absent; anthers dorsifixed; styles 2–4.5 mm in long-styled flowers, 1–2 mm in short-styled flowers. **Capsules** ovoid, 2.5–4.5 × 2–3 mm, rugose or verrucose. **Seeds** blackish, 1.5–2.2 × 0.7–1.1 mm; epidermis smooth; aril 0.9–1.5 mm. $2n = 14$.

Flowering and fruiting Sep–Mar. Gravelly, brushy hills, sandy soil; 50–100 m; Tex.; Mexico; West Indies; Central America; South America (ne Brazil).

Variety *diffusa* is known from southern Texas, mainly in counties along the lower Rio Grande.

2. **Turnera coerulea** de Candolle in A. P. de Candolle and A. L. P. P. de Candolle, Prodr. 3: 346. 1828

Turnera ulmifolia Linnaeus var. *coerulea* (A. P. de Candolle) Urban

Varieties 2 (1 in the flora): Texas, Mexico, South America (Bolivia, Brazil, Guyana).

Variety *surinamensis* (Urban) Arbo & Fernández grows in Mexico and South America (Brazil, Guyana, Suriname).

2a. **Turnera coerulea** de Candolle var. **coerulea**

Herbs or shrubs, 0.2[–1.5] m, hairs simple, glandular hairs microcapitate. **Leaves** not aromatic; petiole 3–4 mm, with 2 apical nectaries; blade narrowly ovate or elliptic, 19–31 × 5–9 mm, base cuneate or attenuate, margins not revolute, serrulate, apex acute, surfaces pilose. **Inflorescences** epiphyllous, 2 cm; peduncle adnate to petiole; prophylls inserted at receptacle base, lanceolate or subulate, 6–7 × 0.5–1 mm, without nectaries. **Flowers** distylous; calyx 15 mm, tube 6.5 mm; petals white, sometimes veins purple at base, 15 × 10 mm; proximal part of filaments adnate along margins to floral tube forming 5 nectar pockets, 6 mm in long-styled flowers, 10 mm in short-styled flowers; anthers basifixed; styles 7 mm in long-styled flowers, 5 mm in short-styled flowers. **Capsules** globose, 2–4 mm, verrucose. **Seeds** blackish, 2 × 1 mm; epidermis smooth; aril unilateral, 1.5 mm. $2n = 10, 20$.

Flowering and fruiting Feb. Sunny, open places, dry or humid, often on rocky or sandy soil; 0–10 m; Tex.; Mexico; South America (Bolivia, Brazil, Guyana).

Variety *coerulea* was collected once near Rockport, Aransas County; perhaps it has not persisted.

3. Turnera ulmifolia Linnaeus, Sp. Pl. 1: 271. 1753 [I]

Varieties 2 (1 in the flora): introduced; s Mexico, West Indies, Bermuda, Central America; introduced also in tropical Asia, Africa, Indian Ocean Islands (including Madagascar), Pacific Islands, Australia.

Both varieties of *Turnera ulmifolia* are homostylous with var. *acuta* (Sprengel) Urban occurring in Bermuda and the West Indies. Both are used as medicinal plants in the West Indies.

I. Urban (1883) described 12 varieties in *Turnera ulmifolia*. Experimental studies have demonstrated that this species as he circumscribed it is a polyploid complex in which most of the taxa deserve species status (M. M. Arbo and Fernández, Aveliano 1987; Fernández 1997; Fernández and Arbo 2000; J. S. Shore and S. C. H. Barrett 1985; V. G. Solís Neffa and Fernández 2000).

3a. Turnera ulmifolia Linnaeus var. **ulmifolia**

• Yellow alder, ramgoat dashalong, cat bush [I]

Turnera angustifolia Miller; *T. ulmifolia* var. *angustifolia* (Miller) Willdenow

Subshrubs or shrubs, 0.5–2 m; hairs simple and glandular, glandular hairs microcapitate. **Leaves** not aromatic; petiole (1.5–)4–27 mm, with 2 apical nectaries; blade elliptic to ovate, 3.5–15 × 1.3–5.7 cm, base attenuate to cuneate, margins not revolute, serrate, apex acute or acuminate, surfaces pilose. **Inflorescences** epiphyllous, 2–4 cm; peduncle adnate to petiole; prophylls inserted at calyx base, ovate, 6–29 × 3–12 mm, often with nectaries. **Flowers** homostylous; calyx 14–30 mm, tube 5–16 mm; petals bright yellow, 14–30 × 7–20 mm; proximal part of filaments adnate by their margins to floral tube forming 5 nectar pockets, 11–21 mm; anthers basifixed; styles 8.5–21 mm. **Capsules** globose, 3–8 mm, granulose. **Seeds** blackish, 2.1–3 × 0.8–1.1 mm; epidermis smooth; aril unilateral, 1.8–3 mm. $2n = 30$.

Flowering and fruiting Mar–Dec. Sheltered locations; 0–10 m; introduced; Fla., La.; s Mexico; West Indies; Central America; introduced also in Asia, Africa, Indian Ocean Islands (including Madagascar), Pacific Islands, Australia.

Variety *ulmifolia* is often planted and has probably escaped from cultivation in various locations.

PASSIFLORACEAE Jussieu ex Roussel

• Passionflower Family

Douglas H. Goldman

John M. MacDougal

Vines [shrubs, trees], perennial [rarely annual], woody or herbaceous, with [without] tendrils; axils with multiple axillary buds, primary axillary bud often developing into inflorescence or tendril; bark smooth to rough or corky. **Leaves** alternate, simple [rarely compound], often bearing extrafloral nectary glands on petiole or blade; stipules persistent or deciduous, setaceous to leaflike; blade pinnately to palmately [pedately] veined, unlobed or palmately [pedately] lobed, margins entire or serrate. **Inflorescences** axillary, solitary or paired flowers [cymose]; bracts absent or setaceous to leaflike or pinnatifid. **Flowers** usually actinomorphic [zygomorphic]; perianth sometimes connate basally into floral tube or cup; sepals [3–]5[–8]; petals as many as sepals, sometimes absent, corona present at base of perianth near outer edge of floral tube as 1–7[–ca. 15] series of filaments or outgrowths, sometimes membranous; extrastaminal nectary disc often present; stamens [4]5[–ca. 25], usually borne on androgynophore [hypanthium]; ovary superior, [2]3[–5]-carpellate, 1-locular, usually borne on androgynophore; placentation parietal; styles distinct [variously connate]; stigmas capitate or clavate to reniform, sometimes 2-lobed [fimbriate]. **Fruits** baccate or capsular. **Seeds** (1–)3–ca. 200, arillate, usually compressed, surface usually pitted to reticulate or grooved.

Genera 17, species ca. 750 (1 genus, 18 species in the flora): nearly worldwide; primarily in tropical regions, especially South America and Africa.

Passifloraceae have been historically assumed to be closely related to Achariaceae, Caricaceae, Cucurbitaceae, Flacourtiaceae, Malesherbiaceae, and Turneraceae (A. Cronquist 1981); they are currently considered to be most closely related to the latter two families, seated within Malpighiales (D. E. Soltis et al. 2000; M. W. Chase et al. 2002; Angiosperm Phylogeny Group 2003; N. Korotkova et al. 2009). All three families have cyanogenic compounds, flowers with a hypanthium often not bearing the stamens, a floral corona (although uncommon in Turneraceae), tricarpellate ovaries with three elongate styles, parietal placentation, and five stamens (Cronquist; Angiosperm Phylogeny Group). Because of the close relationship of these three families, they have recently been treated as an expanded Passifloraceae (Chase et al.; Angiosperm Phylogeny Group). Passifloraceae are still considered relatively closely related to members of the now ex-Flacourtiaceae, particularly Salicaceae in the broad sense (Chase et al.), which includes the apparently corona-bearing *Abatia* Ruiz & Pavón (coronal filaments in *Abatia*, unlike those in Passifloraceae, are clearly staminal appendages; see A. Bernhard 1999), although the members of Flacourtiaceae that have cyanogenic compounds are possibly more distantly related to Passifloraceae and are now placed in a greatly expanded Achariaceae (Chase et al.). The family description provided above represents Passifloraceae in the narrow sense; Malesherbiaceae and Turneraceae are not included.

Passifloraceae are divided into two tribes: Passifloreae de Candolle, containing genera with tendril-bearing vines and scandent shrubs (or rarely small trees without tendrils), and Paropsieae de Candolle, containing genera without tendrils that are shrubby or arborescent (W. J. J. O. de Wilde 1971). Paropsieae have been assigned to Flacourtiaceae or are seen as transitional between Flacourtiaceae and Passifloraceae (de Wilde). Morphological evidence supports the inclusion of Paropsieae within Passifloraceae (E. S. Ayensu and W. L. Stern 1964), further supported by recent molecular phylogenetic evidence (D. E. Soltis et al. 2000; M. W. Chase et al. 2002). Beyond this, intergeneric relationships are largely unknown within Passifloraceae. The tribes may not be distinct as currently defined and some genera are probably derived from within *Passiflora* [that is, *Hollrungia* K. Schumann and *Tetrapathaea* (de Candolle) Reichenbach; S. E. Krosnick and J. V. Freudenstein 2005].

The floral corona of Passifloraceae is of uncertain origin, considered derived from the perianth (V. Puri 1948) or consisting of staminodes (P. K. Endress 1996). It has an unusual pattern of development, inward from the outer and inner margins of a ringlike primordium; its development is delayed until well after the fertile stamens begin to form; and the arrangement and number of coronal filaments is not consistent with that of the fertile stamens. There is little support for the corona being derived from perianths or staminodes (A. Bernhard 1999).

SELECTED REFERENCES de Wilde, W. J. J. O. 1971. The systematic position of the Paropsieae, in particular the genus *Ancistrothyrsus*, and a key to the genera of Passifloraceae. Blumea 14: 99–104. de Wilde, W. J. J. O. 1974. The genera of tribe Passifloreae (Passifloraceae) with special reference to flower morphology. Blumea 22: 37–50. Feuillet, C. and J. M. MacDougal. 2007. Passifloraceae. In: K. Kubitzki et al., eds. 1990+. The Families and Genera of Vascular Plants. 10+ vols. Berlin. Vol. 9, pp. 270–281. Killip, E. P. 1938. The American Species of Passifloraceae. Publ. Field Mus. Nat. Hist., Bot. Ser. 19.

1. PASSIFLORA Linnaeus, Sp. Pl. 2: 955. 1753; Gen. Pl. ed. 5, 410. 1754, name conserved
  • Passionflower [Latin *passio*, passion or suffering, and *flos*, flower, alluding to floral morphology perceived to symbolize Christ's crucifixion]

Vines, glabrous or densely hairy, sometimes glandular. **Stems** terete to angled; tendrils simple [branched]. **Leaves** petiolate; stipules leaflike to minutely setaceous, margins entire, serrate, or deeply cleft, sometimes glandular; blade (2)3(–9)-lobed or unlobed, base cuneate to cordate or rarely peltate, surfaces sometimes glandular, glands or nectaries associated with marginal teeth or abaxially near margins or between primary veins. **Inflorescences** solitary or paired flowers, simple or many-branched [pedunculate cyme with central pedicel often as aborted tendril], secondary inflorescences sometimes present as condensed, axillary (terminal) shoots; bracts 0 or (1–)3, scattered to whorled, margins sometimes glandular. **Flowers** bisexual or sometimes functionally unisexual [staminate]; hypanthium flattened to cuplike or tubular; sepals sometimes with subapical setose to leaflike projection; stamens 5[8], usually alternate with petals, borne on short to elongate androgynophore; anthers dorsifixed, versatile; ovary 3[–5]-carpellate, borne at tip of androgynophore. **Fruits** baccate [capsular or capsulelike berries]. $x = 6$.

Species ca. 550 (18 in the flora): North America, Mexico, Central America, West Indies, South America, Australasia; warm-temperate to tropical areas; some species introduced in the tropics worldwide.

Passiflora species have been arranged under a variety of infrageneric systems, with the 22 subgeneric system by E. P. Killip (1938) the most commonly used until recently. The most recent system (C. Feuillet and J. M. MacDougal 2003) recognizes only four subgenera: *Astrophea* (de Candolle) Masters, *Decaloba* (de Candolle) Reichenbach, *Deidamioides* (Harms) Killip, and *Passiflora*. The flora area includes species of only the two largest subgenera: *Passiflora*, with relatively large flowers and fruits and a predominant chromosome number of $n = 9$, and *Decaloba*, with relatively small flowers and fruits and a predominant chromosome number of $n = 6$. Recent molecular-based phylogenetic studies have investigated the relationships and circumscriptions of infrageneric groups, generally supporting the Feuillet and MacDougal subgeneric classification (for example, V. C. Muschner et al. 2003; R. Yockteng and S. Nadot 2004; A. K. Hansen et al. 2006; S. E. Krosnick et al. 2013).

Primarily outcrossers, *Passiflora* flowers attract pollinators by fragrance, showiness, or nectar secreted within the hypanthium. While collecting nectar, visitors inadvertently collect pollen on their backs or heads from the downward-facing anthers and deposit the pollen on stigmas of the reflexed styles. The styles move over time, beginning erect and flexing downward when receptive, then erect again when no longer receptive. *Passiflora* species are pollinated principally by insects (generally flying Hymenoptera), which seem to prefer scented species. However, species with long-tubular, scentless, red flowers (especially common in the Andes) are hummingbird-pollinated, and bat pollination is also known (T. Ulmer and J. M. MacDougal 2004). Our native species have a conspicuous corona, are scented, and are insect-pollinated. *Passiflora incarnata* is pollinated primarily by carpenter bees (Hymenoptera, Anthophoridae; C. M. McGuire 1999). *Passiflora lutea* is pollinated by various hymenoptera, including wasps (MacDougal 1983), and a ground-nesting bee, *Anthemurgus passiflorae* (Hymenoptera, Andrenidae), that pollinates only this species of *Passiflora* (and no other plant species) and with which it shares a similar geographic distribution (C. D. Michener et al. 1994; J. L. Neff and J. G. Rozen 1995; Neff 2003).

Passiflora fruits are generally adapted to animal consumption for seed dispersal. The relatively large, fragrant, and often yellowish fruits of subgenus *Passiflora* are probably mammal-dispersed; the relatively small, often purplish berries with occasionally brightly colored arils of subgenus *Decaloba* are probably bird-dispersed (J. M. MacDougal 1983).

Some butterflies of the family Nymphalidae (Nymphalinae, Heliconiinae) are obligate-associates of *Passiflora*, using plants as larval hosts (L. E. Gilbert 1982). The larvae sequester and metabolize plant toxins for their own defense against predation, even into adulthood, which has resulted in complicated Batesian and Muellerian mimicry networks of Heliconiinae species. *Passiflora*-Heliconiinae coevolution has been a subject of interest (W. W. Benson et al. 1975). *Passiflora* defenses include extreme leaf polymorphism (Benson et al.), defensive trichomes (Gilbert; J. M. MacDougal 1994), extrafloral nectaries (Benson et al.), and butterfly egg-mimicry (Gilbert; K. S. Williams and Gilbert 1981). Some *Passiflora* species exhibit striking variation in leaf shape and color (variegation) within individual plants, providing camouflage or denying predators a search image. Within the flora area, leaf-shape polymorphism is best exemplified in the closely related *P. pallida* and *P. tenuiloba*; color variation is most strongly expressed in *P. mexicana* and *P. pallens*. Extrafloral nectaries occur most commonly on petioles, stipules, leaf margins and abaxial surfaces, and floral bracts, attracting potential defenders such as predatory or parasitic Hymenoptera; see J. A. Scott (1986) for a survey of Heliconiinae associated with *Passiflora* in North America.

Fifteenth-century Spanish explorers were the first Europeans to encounter *Passiflora*. Some saw religious symbolism in its floral morphology, signifying various aspects of Christ's crucifixion (E. E. Kugler and L. A. King 2004), the three styles representing the nails, the corona representing the crown of thorns, and red coloration (if present) representing the blood of Christ. This symbolism led to its cultivation in Europe in the early seventeenth century; *P. incarnata* was probably the first species grown there.

Several species are cultivated for their edible fruits, including *Passiflora edulis* Sims, *P. laurifolia* Linnaeus, *P. quadrangularis* Linnaeus, *P. tarminiana*, *P. tripartita* (Jussieu) Poiret var. *mollissima* (Kunth) Holm-Nielsen & P. Jørgensen, and *P. vitifolia* Kunth (T. Ulmer and J. M. MacDougal 2004). However, because the genus is generally cyanogenic, immature fruits of all species are probably poisonous and should not be consumed.

Although *Passiflora* is commonly considered tropical, the flora area has several native cold-tolerant species. Many of these invest heavily in below-ground structures, having rhizomes or root-suckers (for example, *P. affinis*, *P. filipes*, *P. foetida* var. *gossypiifolia*, *P. incarnata*, *P. lutea*, *P. mexicana*, and *P. tenuiloba*).

Passiflora species are horticulturally desirable because of the unusual appearance and attractiveness of their flowers and leaves. Because of their popularity, it is expected that the number of species naturalized in the flora area will increase. In the flora area, *P. arida*, *P. caerulea*, and *P. tarminiana* are well-established escapes that are also popular in horticulture. However, there are currently far fewer species of *Passiflora* introduced and established in the flora area than have been reported (J. T. Kartesz 1994), some cultivated only (D. G. Burch et al. 1988). Species long-persisting in old plantings, "naturalizing" merely by root-suckering, or historical adventives that are long-absent or poorly documented, are not formally covered in this treatment, for example, *P.* ×*belotii* Pepin (*P. alata* Curtis × *P. caerulea*; R. P. Wunderlin and B. F. Hansen 2003), *P. edulis* (Wunderlin and Hansen), *P. gracilis* (E. P. Killip 1938), *P.* 'Incense' (*P. incarnata* × *P. cincinnata*), *P. manicata* (C. F. Smith 1976), *P. miniata* (Wunderlin and Hansen, identified as *P. coccinea*), *P. mixta* Linnaeus, and *P. morifolia* Masters (Killip; A. E. Radford et al. 1968; syn. *P. warmingii*). *Passiflora* ×*belotii* does not form fruit, and it is unlikely to naturalize except by rooting at the nodes of fallen branches or possibly by suckering. J. C. Estill and M. B. Cruzan (2001) mentioned that *P. morifolia* is endemic to South Carolina, whereas it is actually native only to Central and South America. Currently, any minimally "naturalized" species or hybrids listed above that might be encountered are in California or Florida, specifically *P.* ×*belotii*, *P. edulis*, *P.* 'Incense', *P. miniata*, and *P. mixta*. In the key below, *P.* ×*belotii* will key to *P. caerulea*, *P. edulis* and *P.* 'Incense' to *P. incarnata*, *P. miniata* to *P. multiflora*, and *P. mixta* to *P. tarminiana*. For broad and well-illustrated treatments of the genus, see T. Ulmer and J. M. MacDougal (2004), and J. Vanderplank (2000).

SELECTED REFERENCES Feuillet, C. and J. M. MacDougal. 2003. A new infrageneric classification of *Passiflora* L. (Passifloraceae). Passiflora 13: 34–35, 37–38. Hansen, A. K. et al. 2006. Phylogenetic relationships and chromosome number evolution in *Passiflora*. Syst. Bot. 31: 138–150. Krosnick, S. E. et al. 2013. New insights into the evolution of *Passiflora* subgenus *Decaloba* (Passifloraceae): Phylogenetic relationships and morphological synapomorphies. Syst. Bot. 38: 692–713. Muschner, V. C. et al. 2003. A first molecular phylogenetic analysis of *Passiflora* (Passifloraceae). Amer. J. Bot. 90: 1229–1238. Ulmer, T. and J. M. MacDougal. 2004. *Passiflora*: Passionflowers of the World. Portland. Vanderplank, J. 2000. Passion Flowers, ed. 3. Cambridge, Mass.

1. Primary leaf lobes 2 or 3, often appearing 2-lobed (middle lobe absent), middle lobe much shorter than lateral lobes, usually symmetric, or strongly asymmetrically 2- or 3-lobed with 1 lateral lobe greatly reduced or absent.
    2. Leaf blades strongly asymmetric and 2- or 3-lobed, with 1 of the 2 primary lateral lobes greatly reduced or absent . . . . 5. *Passiflora pallida* (in part)
    2. Leaf blades roughly symmetric, 2- or 3-lobed, both primary lateral lobes present.

3. Leaf blades with conspicuous nectaries on abaxial surface, usually forming lines; sepals 13–17 mm.
    4. Stems minutely puberulent; abaxial leaf nectaries usually in lines extending to leaf lobes, at least on flowering stems; fine leaf veins prominently raised abaxially; corona filaments laterally flattened, green basally, yellow apically; se Florida . . . . 1. *Passiflora biflora*
    4. Stems glabrous; abaxial leaf nectaries in lines rarely extending to leaf lobes, at least on flowering stems; fine leaf veins weakly to moderately raised abaxially; corona filaments terete, red, becoming purple; Arizona . . . . 2. *Passiflora mexicana*
3. Leaf blades without conspicuous nectaries on abaxial surface, or if present, not forming lines and always near leaf margins; sepals 6–13 mm.
    5. Leaf blades densely soft-hairy, lobes unlobed, abaxial nectaries absent; petioles eglandular; se Florida . . . . 3. *Passiflora sexflora*
    5. Leaf blades glabrous or subglabrous to short-hairy, not soft-hairy, primary 3 lobes often lobed, abaxial nectaries present or absent; petioles with cuplike glands; Texas . . . . 4. *Passiflora tenuiloba* (in part)
1. Primary leaf lobes 3+, or 0, not appearing 2-lobed, middle primary lobe ± as long as or longer than lateral 2 primary lobes, blade usually symmetric.
    6. Stipules pectinate, with usually prominent gland-tipped hairs or bristles; leaves often pungent; floral bracts pinnatifid, often with glandular bristles or hairs.
        7. Plants densely woolly-hairy; stipules, blades, and floral bracts with obscurely glandular to eglandular bristles or hairs; leaf blades deeply (3–)5–7(–9)-lobed . . . . .6. *Passiflora arida*
        7. Plants glabrous or hairy but not densely woolly; stipules, blades, and floral bracts with obvious gland-tipped bristles or hairs; leaf blades moderately to deeply 3(–5)-lobed.
            8. Leaf margins sharply dentate; Arizona . . . . 7. *Passiflora arizonica*
            8. Leaf margins serrate to nearly entire; Florida, Texas.
                9. Fruits bright red to crimson; leaves not or weakly pungent, glabrous or hairy, blade 3–10(–12) × 3–8(–10) cm; Florida, Texas . . . . .8. *Passiflora ciliata*
                9. Fruits green to yellow-green; leaves pungent, densely hairy, blade 3–5(–8) × 2.5–5.5(–7) cm; Texas. . . . . 9. *Passiflora foetida*
    6. Stipules not pectinate, glabrous or hairy but without gland-tipped bristles or hairs; leaves usually not pungent; floral bracts absent, or present and not pinnatifid but with margins entire or serrate, dentate, or erose, eglandular to sessile-glandular.
      10. Leaves soft-hairy abaxially.
        11. Leaf blades deeply 3-lobed, margins serrate; stipules subreniform, often leaflike, 2–3 mm wide; petals 40–54 mm, floral tube 60–80 mm . . . . . . 10. *Passiflora tarminiana*
        11. Leaf blades unlobed to rarely obscurely 3–5-lobed, margins entire; stipules linear-setaceous, 0.5 mm wide; petals 4–5 mm, floral tube absent . . . . . . 11. *Passiflora multiflora*
      10. Leaves glabrous or hairy to hispid abaxially, but not soft-hairy.
        12. Petioles eglandular.
          13. Leaf blades with abaxial nectaries; floral bracts 1–3 mm; sepals 10–16 mm, outer corona filaments apically clavate; c Texas. . . . . .12. *Passiflora affinis*
          13. Leaf blades without abaxial nectaries; floral bracts absent; sepals 6–10 mm, outer corona filaments not apically clavate; s Texas or widespread.
             14. Fine veins on abaxial leaf surface raised (especially in dried specimens), leaf blade 2–10(–15) cm, middle lobe usually ⅓–⅔ blade length; widespread (excluding southernmost Texas) . . . . .13. *Passiflora lutea*
             14. Fine veins on abaxial leaf surface not raised, leaf blade 1–3(–5) cm, middle lobe usually to ¼ blade length; southernmost Texas . . . . 14. *Passiflora filipes*
        12. Petioles glandular.
          15. Stems and leaves uncinate-hairy. . . . . 15. *Passiflora bryonioides*
          15. Stems and leaves not uncinate-hairy.
             16. Stipules subreniform, 5–14 mm wide.
                17. Leaf blades shallowly 3-lobed; young stems terete; Florida. . . . 16. *Passiflora pallens*
                17. Leaf blades deeply 3–9-lobed; young stems angular; California . . . . 17. *Passiflora caerulea*

[16. Shifted to left margin.—Ed.]

16. Stipules linear-subulate or linear-setaceous, 0.5–1 mm wide.

    18. Leaf blades 3(–5)-lobed, margins serrate; sepals 20–30 mm, petals 25–35 mm; fruits 30–60 mm, green to pale yellow; bark not corky . . . . 18. *Passiflora incarnata*

    18. Leaf blades unlobed or 3–9-lobed, margins entire; sepals 4–10 mm, petals absent; fruits 5–13 mm, dark blue to black; bark corky or not.

        19. Bark not corky; leaf blades as wide as to usually wider than long, 3–9-lobed, lobes acute, abaxial nectaries present or absent, petiole glands cuplike; Texas. . . . . 4. *Passiflora tenuiloba* (in part)

        19. Bark with corky ridges or wings; leaf blades as long as to usually longer than wide, unlobed or 3-lobed, lobes rounded to acute, abaxial nectaries absent, petiole glands clavate; Florida, Texas. . . . . 5. *Passiflora pallida* (in part)

1. **Passiflora biflora** Lamarck in J. Lamarck et al., Encycl. 3: 36. 1789 [I]

Stems angular, minutely puberulent. **Leaves** not pungent, minutely puberulent; stipules linear, 2–5 × 0.5–1 mm, eglandular; petiole eglandular; blade roughly symmetric, 2–7 × 3–10 cm, shallowly to deeply 2-lobed, margins entire; abaxial fine veins prominently raised, abaxial nectaries not along leaf margins, usually in lines extending into leaf lobes on at least flowering stems. **Floral bracts** setaceous, 1–3 × 0.5 mm, margins entire, eglandular. **Flowers:** floral tube cuplike, 1–3 mm deep; sepals white, 13–17 × 4–6 mm; petals white, 9–13 × 4–6 mm; corona filament whorls 2, outer filaments green basally, yellow apically, linear-spatulate, laterally flattened, 5–8 mm. **Berries** blue-black, ovoid to ellipsoid, 20–30 × 15–20 mm. $2n = 12$.

Flowering Jun–Aug. Intact or disturbed tropical woodlands in loamy soil over limestone; 0–10 m; introduced; Fla.; Mexico; West Indies (Bahamas); Central America; South America (Colombia, Ecuador, Venezuela).

In the flora area, *Passiflora biflora* occurs only in southeastern Florida, where it is locally abundant and often an aggressive weed.

2. **Passiflora mexicana** Jussieu, Ann. Mus. Natl. Hist. Nat. 6: 108, plate 38, fig. 2. 1805

Stems subangular, glabrous. **Leaves** not pungent, glabrous; stipules linear-subulate, 1.5–2.5 × 0.5 mm, eglandular; petiole eglandular; blade roughly symmetric, 1.5–7(–15) × 2.5–8(–14) cm, moderately to deeply 2-lobed, margins entire; abaxial fine veins weakly to moderately raised, abaxial nectaries not along leaf margins, usually in 2 lines and rarely extending into leaf

lobes at least on flowering stems. **Floral bracts** linear-subulate, 1–4 × 0.5 mm, margins entire, eglandular. **Flowers:** floral tube absent; sepals green, 13–17 × 4–6 mm; petals green, 3–4 × 1 mm; corona filament whorls 2, outer filaments red, becoming purple, linear, terete, 5–12 mm. **Berries** purple-black, globose to subellipsoid, 8–16 × 7–14 mm.

Flowering Jul–Sep. Riparian woodlands, semiarid shrublands; 600–1400 m; Ariz.; Mexico.

The leaf shape of *Passiflora mexicana* varies greatly within individual plants. Lateral leaf lobes are much shorter and apically rounded on slower-growing or flowering branches, but are relatively long and truncate on vigorously-growing, nonflowering branches, suckers, and young plants. A striking feature of this species is that as the flowers become less receptive to pollination the coronal filaments and limen (disc at base of the androgynophore) dramatically change color; the corona from red to purple, the limen from orange to yellow or white. The typically fetid flowers of this species may be wasp-pollinated (J. M. MacDougal and R. McVaugh 2001).

Passiflora mexicana may consist of a small complex of species.

3. **Passiflora sexflora** Jussieu, Ann. Mus. Natl. Hist. Nat. 6: 110, plate 37, fig. 1. 1805 • Goatsfoot

Stems terete to slightly flattened, densely soft-hairy. **Leaves** not pungent, densely soft-hairy; stipules linear-setaceous, 2–5 × 0.5 mm, eglandular; petiole eglandular; blade roughly symmetric, 2–10 × 2–15 cm, 2- or 3-lobed, middle lobe shorter than lateral lobes, lobes unlobed, margins entire; abaxial fine veins prominently raised but obscured by pubescence, abaxial nectaries absent. **Floral bracts** linear-subulate to setaceous, 2–6 × 0.5–1 mm, margins entire or incised, eglandular. **Flowers:** floral tube absent; sepals greenish white, 9–13 × 2–3 mm; petals white, 5–9 × 1–2 mm; corona filament whorls 2, outer filaments purple basally, white apically, linear-filiform, terete, 5–9 mm. **Berries** blue-black, ovoid, ellipsoid, or subglobose to dorsiventrally compressed, 5–10 × 5–8 mm.

Flowering Oct–Mar. Margins of, and sunny gaps within mesic, tropical woodlands over oölitic limestone; 0–10 m; Fla.; Mexico; West Indies (Cuba, Hispaniola, Jamaica, Puerto Rico); Central America; South America (Colombia, Ecuador).

Found in the flora area only in extreme southeastern Florida, *Passiflora sexflora* responds vigorously to moderate disturbance of its forested habitats, with rapid seed germination and plant growth in treefalls or other canopy openings (J. Blakley, pers. comm.). There have been successful attempts at reintroducing it into the wild in Florida (J. Possley et al. 2007), where it is listed as endangered.

4. **Passiflora tenuiloba** Engelmann, Boston J. Nat. Hist. 6: 192. 1850 [F]

Stems terete, sparsely hairy, (bark not corky). **Leaves** not pungent, glabrous or subglabrous to short-hairy, not softhairy; stipules linear-subulate, 2–5 × 0.5–1 mm, eglandular; petiole glandular, glands cuplike; blade roughly symmetric, 2–7 × 2.5–16 cm, as wide as to usually wider than long, 3–9-lobed, primary lobes 3 and often further lobed, lobes acute, middle primary lobe much shorter than to ± as long as lateral 2 primary lobes, margins entire; abaxial fine veins moderately to prominently raised, abaxial nectaries usually absent, sometimes present near leaf margins but not forming lines. **Floral bracts** obscure, linear-subulate, 1 × 0.2–0.5 mm, margins entire, eglandular. **Flowers:** floral tube absent; sepals green, 6–10 × 2–4 mm; petals absent; corona filament whorls 2, outer filaments purple basally, yellow apically, or yellow throughout, linear, terete to slightly transversely compressed, 3–7 mm. **Berries** dark blue to black, globose to dorsiventrally compressed, 6–12 × 6–12 mm. $2n = 24$.

Flowering Apr–Aug(–Dec). Open oak-juniper or coastal woodlands and savannas, grasslands and semideserts, primarily over limestone; 30–900 m; Tex.; Mexico (Coahuila, Nuevo León, Tamaulipas).

E. P. Killip (1938) and W. C. Martin and C. R. Hutchins (1980) incorrectly indicated that *Passiflora tenuiloba* is native to New Mexico. This is based upon the misinterpretation of confusing labels on old specimens (see D. H. Goldman 2004).

Passiflora tenuiloba has the most variable leaves of any species in the genus. They vary considerably in texture, color, indument, lobe length and number, and petiole-gland size and shape, not only between populations but also within individual plants.

5. **Passiflora pallida** Linnaeus, Sp. Pl. 2: 955. 1753
   • Corkystem passionflower

Stems terete, glabrous or sparsely hairy, (bark with corky ridges or wings). **Leaves** not pungent, glabrous or sparsely hairy; stipules linear-subulate, 2–5 × 0.5 mm, eglandular; petiole glandular, glands clavate; blade usually symmetric, (1–)3–10(–17) × (0.5–)2–6(–14) cm, as long as to usually longer than wide, unlobed or deeply 3-lobed, middle lobe longer than lateral lobes, or asymmetrically 2- or 3-lobed with 1 lateral lobe greatly reduced or absent, lobes rounded to acute, margins entire; abaxial fine veins moderately to prominently raised, abaxial nectaries absent. **Floral bracts** absent or minute. **Flowers:** floral tube absent; sepals green, often becoming purple, 4–8 × 2–3 mm; petals absent; corona filament whorls 2, outer filaments green, green-white, or purple basally, yellow apically, linear, terete, 2–4 mm. **Berries** blue-black, globose to ovoid, 5–13 × 5–10 mm. $2n = 24$.

Flowering year-round, primarily Oct–Dec. Subtropical to tropical woodlands, shrublands, pine forests, scrub and disturbed areas, in rocky, loamy to sandy soil, often calcareous; 0–60 m; Fla., Tex.; Mexico; West Indies (Bahamas, Greater and Lesser Antilles); Bermuda; South America (Venezuela); introduced in s Asia, Indian Ocean Islands, w Pacific Islands, n Australia.

The leaves of *Passiflora pallida* show variation in lobe number, from juvenile to adult plants, and within and among individuals, with both unlobed and three-lobed leaf forms most common. In Texas, leaves on mature plants are always symmetrically three-lobed; in Florida, leaves can be unlobed, symmetrically three-lobed, or asymmetrically two- to three-lobed. *Passiflora pallida* is also self-compatible, an unusual feature in the genus.

In peninsular Florida, *Passiflora pallida* is widespread and locally common to occasionally weedy. In Texas, it is relatively uncommon, known only from the extreme southernmost part of the state and from a single disjunct population about 150 miles northward in Refugio County (S. R. Hill 1981).

Traditionally synonymized under *Passiflora suberosa* in the flora area (for example, D. S. Correll and M. C. Johnston 1970; R. L. Hammer 2002; R. P. Wunderlin and B. F. Hansen 2003), *P. pallida* is closely related to the larger-flowered *P. suberosa* and commonly is treated as conspecific with it worldwide. The hypanthium of *P. pallida* generally is 3–4 mm in diameter, with inner coronal filaments usually less than 1.5 mm and outer filaments less than 4 mm, whereas in *P. suberosa* the hypanthium generally is 4–8.8 mm in diameter, with inner coronal filaments usually 1.5–3.9 mm and outer filaments (2.5–)3–8.1 mm. *Passiflora suberosa* is native in Mexico, the West Indies, and Central and South America, and has been widely introduced in tropical regions (weedy in Hawaii), but does not occur in the flora area. In regions where the two species naturally co-occur, *P. pallida* generally is found at lower elevations than *P. suberosa*. These two species and close relatives were reviewed by K. E. Porter-Utley (2003).

6. **Passiflora arida** (Masters & Rose) Killip, J. Wash. Acad. Sci. 12: 256. 1922 [1]

Passiflora foetida Linnaeus var. *arida* Masters & Rose, Contr. U.S. Natl. Herb. 5: 182. 1899

Stems terete, densely woolly-hairy. **Leaves** pungent, densely woolly-hairy, with obscurely glandular or eglandular bristles; stipules pectinate, 2–5 × 3–8 mm, with obscurely glandular or eglandular bristles or hairs; petiole with obscurely glandular or eglandular bristles or hairs; blade roughly symmetric, 3–6 × 2–7 cm, deeply (3–)5–7(–9)-lobed, middle lobe longer than lateral lobes, margins entire or serrate; abaxial fine veins weakly to moderately raised but obscured by pubescence, abaxial nectaries absent. **Floral bracts** pinnatifid, 16–28 × 14–28 mm, margins entire or dentate, with obscurely glandular or eglandular bristles or hairs. **Flowers:** floral tube cuplike, 3–5 mm deep; sepals white, 13–23 × 5–9 mm; petals white, 14–20 × 5–8 mm; corona filament whorls 5 or 6, outer 2 whorls purple to magenta basally, white medially, pale pink apically, linear, terete, 10–17 mm. **Berries** green to pale yellow-green, ovoid, 2–3 × 15–25 mm.

Flowering Jul–Aug. Deserts and desert grasslands, urban areas and disturbed sites, usually near adequate water run-off; 300–1200 m; introduced; Ariz.; Mexico (Baja California, Baja California Sur, Sonora).

In the flora area, *Passiflora arida* is known only from the vicinity of Tucson, where it is introduced. Although native to desert and semidesert areas, in cultivation, this species readily adapts to mesic conditions, grows aggressively, is self-pollinating, and exhibits rapid seed germination, suggesting that it could become a problematic, widespread weed (D. H. Goldman 2003). In recent years it has been distributed in the horticultural trade and among enthusiasts, misidentified as *P. foetida* var. *arizonica*. In greenhouse cultivation this plant can produce leaves less densely hairy and much larger than those described above.

7. **Passiflora arizonica** (Killip) D. H. Goldman, Madroño 50: 249. 2004

Passiflora foetida Linnaeus var. *arizonica* Killip, Publ. Field Mus. Nat. Hist., Bot. Ser. 19: 490. 1938

Stems terete, densely hairy. **Leaves** pungent, densely hairy, glandular-ciliate; stipules pectinate, 1–8 × 1–4 mm, with glandular bristles or hairs; petiole with glandular bristles or hairs; blade roughly symmetric, 1.5–5 × 1–7 cm, moderately to deeply 3–5-lobed, middle lobe as long as or longer than lateral lobes, margins sharply dentate; abaxial fine veins weakly to moderately raised, abaxial nectaries absent. **Floral bracts** pinnatifid, 15–35 × 10–28 mm, margins sharply dentate, with glandular bristles or hairs. **Flowers:** floral tube cuplike, 5–7 mm deep; sepals white, 17–38 × 6–9 mm; petals white, 16–30 × 6–12 mm; corona filament whorls 5 or 6, outer 2 whorls white basally, pale purple apically, linear, terete to transversely compressed, 9–25 mm. **Berries** green to yellow-green, ovoid, 20–35 × 18–30 mm.

Flowering Jun–Sep. Rocky, igneous slopes in semidesert grasslands and oak savannas; 1000–1800 m; Ariz.; Mexico (Sonora).

Passiflora arizonica is known only from portions of Pima and Santa Cruz counties, and eastern Sonora, Mexico, and has been confused with *P. arida* (D. H. Goldman 2003). It flowers during the summer rainy season, usually August and September. Unlike most other members of the genus, particularly those of sect. *Dysosmia*, to which it belongs, *P. arizonica* flowers in the evening, closing around midnight. The fragrant flowers have a deep floral cup, and may be pollinated by nocturnal moths.

8. **Passiflora ciliata** Aiton, Hort. Kew. 3: 310. 1789 [1]

Passiflora foetida Linnaeus var. *ciliata* (Aiton) Masters

Stems terete to longitudinally ridged, glabrous or densely hairy. **Leaves** not or weakly pungent, glabrous or densely hairy, glandular-ciliate; stipules pectinate, 5–6 × 2–4 mm, with glandular bristles or hairs; petiole with glandular bristles or hairs; blade roughly symmetric, 3–10(–12) × 3–8(–10) cm, moderately 3(–5)-lobed, middle lobe much longer than lateral lobes, margins weakly serrulate to nearly entire; abaxial fine veins weakly raised, abaxial nectaries absent. **Floral bracts** pinnatifid, 20–30 × 15–25 mm, margins serrate to nearly entire, with glandular bristles or hairs. **Flowers:** floral tube cuplike, 4–6 mm deep; sepals white, 18–25 × 7–9 mm; petals purple to white, 20–25 × 8–10 mm;

corona filament whorls 5 or 6, outer 2 whorls purple basally, white medially, pale purple apically, linear, terete to transversely compressed, 16–20 mm. **Berries** bright red to crimson, ovoid to broadly ellipsoid, 30–35 × 20–25 mm.

Flowering Aug–Dec. Warm-temperate to subtropical woodlands, shrublands, or disturbed areas, in moist to dry, loamy to sandy soil; 10–200 m; introduced; Fla., Tex.; Mexico; West Indies; Central America.

Passiflora ciliata is relatively uncommon and apparently introduced in the flora area and probably is spreading. Nearly all plants in Florida (for example, R. P. Wunderlin and B. F. Hansen 2003) and some in Texas identified as *P. foetida* are actually *P. ciliata*. E. P. Killip (1938) noted the presence of *P. foetida* var. *riparia* (C. Wright ex Grisebach) Killip in Florida, which is a synonym of *P. ciliata*.

Plants of this species within the flora area can be either entirely glabrous or densely hairy throughout; otherwise, they are essentially identical morphologically. The description provided here best reflects its variation within the flora area, and the species is more variable outside of the flora area. We treat this species broadly to include many members of sect. *Dysosmia* that have mature leaves unscented to weakly pungent when bruised and that have bright red to scarlet fruits, characteristics that seem to be consistently associated.

9. Passiflora foetida Linnaeus, Sp. Pl. 2: 959. 1753 [W]

Varieties ca. 30 (1 in the flora): Texas, Mexico, West Indies, Central America, South America; introduced in tropical Asia, Africa, Pacific Islands, Australia.

9a. Passiflora foetida Linnaeus var. **gossypiifolia** (Desvaux ex Hamilton) Masters, Trans. Linn. Soc. London 27: 631. 1871 [W]

Passiflora gossypiifolia Desvaux ex Hamilton, Prodr. Pl. Ind. Occid., 48. 1825 (as gossypifolia)

Stems terete, densely hairy. **Leaves** pungent, densely hairy, glandular-ciliate; stipules pectinate, 2–5 × 1–3 mm, with glandular bristles or hairs; petiole with glandular bristles or hairs; blade roughly symmetric, 3–5(–8) × 2.5–5.5 (–7) cm, moderately to deeply 3-lobed, middle lobe as long as or longer than lateral lobes, margins serrate to nearly entire; abaxial fine veins weakly to moderately raised, abaxial nectaries absent. **Floral bracts** pinnatifid, 20–30 × 20–25 mm, margins serrate to nearly entire, with glandular bristles or hairs. **Flowers:** floral tube cuplike, 3–5 mm deep; sepals white, 15–25 × 5–7 mm; petals white, 15–23 × 5–7 mm; corona filament whorls 5 or 6, outer 2 whorls purple to violet, linear, terete to

transversely compressed, 10–17 mm. **Berries** green to yellow-green, ovoid to subglobose, 15–25 × 15–25 mm.

Flowering May–Aug(–Dec). Woodlands and shrublands, usually subtropical, in moist to dry, loamy to sandy soil; 10–200(–300) m; Tex.; Mexico; West Indies; Central America; South America.

Variety *gossypiifolia* is perhaps the most naturally widespread member of this diverse species and exhibits a large amount of vegetative morphological variation as well (E. P. Killip 1938). Varieties of *Passiflora foetida* and some related species share the common features of mature leaves that are pungent when bruised and fruits that are green to yellow. The taxonomy of this species and other members of sect. *Dysosmia* is complicated and requires much more research.

Plants with small, orange fruit, possibly of var. *hispida* (de Candolle ex Triana & Planchon) Killip ex Gleason, recently have been found naturalized in the city of Coconut Creek, Broward County, Florida, although efforts to eradicate it are being implemented.

Passiflora foetida is suspected of insectivory. The glandular trichomes have the ability to trap and kill small invertebrates, with the potential for enzymatic absorption (T. R. Radhamani et al. 1995).

10. Passiflora tarminiana Coppens & V. E. Barney, Novon 11: 9, figs. 1, 3. 2001 • Banana poka [I]

Stems terete, densely hairy. **Leaves** not pungent, densely soft-hairy abaxially, sparsely hairy adaxially; stipules subreniform, often leaflike, 4–7 × 2–3 mm, eglandular; petiole glandular, glands emergent protuberances; blade roughly symmetric, 5.5–16(–28) × 7–16(–29) cm, deeply 3-lobed, middle lobe as long as or longer than lateral lobes, margins serrate; abaxial fine veins prominently raised, abaxial nectaries absent. **Floral bracts** leaflike, 25–50 × 20–30 mm, margins entire, eglandular. **Flowers:** floral tube elongate, 60–80 mm deep; sepals pink, 45–60 × 12–25 mm; petals pink, 40–54 × 15–20 mm; corona filament whorls 1, filaments tuberculate knobs, purple basally, white apically, 1–2 mm. **Berries** yellow to orange-yellow, oblong to ellipsoid-fusiform, 100–140 × 35–45 mm.

Flowering Jun–Sep(–Dec). Pine or oak woodlands and woodland edges; 0–100 m; introduced; Calif.; South America (Colombia, Ecuador, Peru, Venezuela).

Passiflora tarminiana is sparingly naturalized in the eastern San Francisco Bay area (F. Hrusa et al. 2002) and southward along the coast to San Luis Obispo County, in areas of minimal summer drought.

This species was recently described, and is commonly confused with *Passiflora mollissima* (Kunth) L. H. Bailey [now usually recognized as *P. tripartita* var. *mollissima*

(Kunth) Holm-Nielsen & P. Jørgensen]. Many reports of *P. mollissima* in agricultural, horticultural, and weed-science literature actually apply to *P. tarminiana*. An attractive plant with large, edible fruits (T. Ulmer and J. M. MacDougal 2004), it is an extremely aggressive weed in Hawaii (A. M. La Rosa 1984, as *P. mollissima*) and other areas where it has been introduced in the Old World Tropics and Subtropics. The species is unlikely to become a widespread weed in the continental United States because it cannot survive frost nor occasional desiccation.

A similar, closely related species, *Passiflora mixta* Linnaeus f., is a rare escape in San Francisco, California; it can be distinguished from *P. tarminiana* by its angular young stems, persistent stipules (deciduous in *P. tarminiana*), and a floral tube 80–110 mm deep, 1.6–2.6 times the sepal length (1.3–1.6 times in *P. tarminiana*).

11. Passiflora multiflora Linnaeus, Sp. Pl. 2: 956. 1753

• Many-flowered passionflower

Stems terete, becoming coarsely fluted when old, densely hairy. Leaves not pungent, densely soft-hairy; stipules linear-setaceous, 3–5 × 0.5 mm, eglandular; petiole glandular, glands small, emergent protuberances; blade roughly symmetric, 4–10(–13) × 1.5–4(–6) cm, unlobed to rarely obscurely 3–5-lobed, middle lobe much longer than lateral lobes, margins entire; abaxial fine veins prominently raised, abaxial nectaries absent or sometimes present near leaf margins. **Floral bracts** obscure, ovate-lanceolate, 1–2 × 0.5–1 mm, margins entire, eglandular. **Flowers:** floral tube absent; sepals green-white, 5–6 × 2–3 mm; petals white, 4–5 × 1–1.5 mm; corona filament whorls 2, outer filaments white basally, yellow apically, filiform, terete, clavate, 3–5 mm. **Berries** dark blue, globose, 5–8 × 5–8 mm.

Flowering Nov–Mar. Tropical woodlands primarily over coral-reef limestone; 0–10 m; Fla.; West Indies (Bahamas, Cuba, Hispaniola, Puerto Rico).

Passiflora multiflora can produce hundreds to thousands of flowers at once, often in several fragrant flushes per season. It has perhaps the smallest flowers of any passionflower, and the androgynophore is nearly absent. A glabrous form of this otherwise densely hairy species has been collected outside the flora area.

Passiflora miniata Vanderplank (often misidentified as *P. coccinea* Aublet) will key to *P. multiflora* based upon superficial similarity of their leaves. However, *P. miniata* has serrated leaf margins and large, red flowers.

Passiflora multiflora is listed as endangered in Florida.

12. Passiflora affinis Engelmann, Boston J. Nat. Hist. 6.: 233. 1850

Stems terete, glabrous, minutely puberulent to scabrous when young. Leaves not pungent, glabrous or minutely puberulent; stipules linear-setaceous, 1–2 × 0.5 mm, eglandular; petiole eglandular; blade roughly symmetric, 1–8(–10) × 1.5–10 (–14) cm, shallowly to deeply 3(–5)-lobed, middle lobe as long as or longer than lateral lobes, margins entire; abaxial fine veins weakly to moderately raised, abaxial nectaries circular, usually in 2 short lines or also scattered near leaf margins. **Floral bracts** linear-subulate, 1–3 × 0.5 mm, margins entire, eglandular. **Flowers:** floral tube absent; sepals pale green to white, 10–16 × 2–4 mm; petals pale green to white, 6–13 × 1–2 mm; corona filament whorls 2, outer filaments purple basally, white medially, green apically, linear-filiform, terete, apically clavate, 9–18 mm. **Berries** purple-black, globose to ovoid, 10–15 × 10 mm.

Flowering (May–)Jun–Oct. Oak-juniper woodlands, shrublands, and savannas, in moist to dry, loamy soil over limestone; 100–800 m; Tex.; Mexico (Coahuila, Nuevo León, Tamaulipas).

Passiflora affinis is similar to *P. lutea* in leaf shape and flower appearance. However, *P. affinis* has nectaries on the abaxial surface of the leaves, which are absent in *P. lutea*. *Passiflora affinis* also has floral bracts, and flowers greater than 25 mm in diameter with clavate to capitate, sinuous outer corona filaments more than 10 mm long. *Passiflora lutea* lacks floral bracts and has flowers less than 25 mm in diameter with apically unornamented, typically straight outer corona filaments that are usually less than 10 mm long.

In the flora area *Passiflora affinis* is restricted to central Texas, where its leaves are often appropriately shaped like cowboy hats. E. P. Killip (1938) erroneously suggested that it is native to New Mexico, based upon the misinterpretation of herbarium label data (D. H. Goldman 2004).

13. Passiflora lutea Linnaeus, Sp. Pl. 2: 958. 1753

• Yellow passionflower [E] [F] [W]

Passiflora lutea var. *glabriflora* Fernald

Stems terete to subangular, glabrous or sparsely hairy, sometimes densely hairy when young, (above-ground stems annual). Leaves not pungent, glabrous or minutely hairy; stipules linear-lanceolate, falcate, 2–5 × 0.5–1 mm, eglandular; petiole eglandular; blade roughly symmetric, 2–10(–15) × 1–8(–12) cm, shallowly to rarely deeply 3(–5)-lobed, middle lobe slightly shorter to longer than lateral lobes (⅓–⅔ blade length), margins entire; abaxial fine veins prominently raised (especially in dried specimens), abaxial nectaries absent. **Floral bracts** absent. **Flowers:** floral tube absent; sepals green, 6–10 × 2–3 mm; petals pale green-yellow, 3–7 × 1 mm; corona filament whorls 2, outer filaments pale green to white basally, pale yellow apically, linear, terete, 5–10 mm. **Berries** purple-black, globose to ovoid, 8–15 × 8–15 mm. $2n = 24, 84$.

Flowering (Mar–)Jun–Sep(–Oct). Mesic, open woodlands and forest margins; 0–600(–1100) m; Ala., Ark., Del., D.C., Fla., Ga., Ill., Ind., Kans., Ky., La., Md., Miss., Mo., N.C., Ohio, Okla., Pa., S.C., Tenn., Tex., Va., W.Va.

Perhaps the most widespread species of the genus endemic in the flora area, *Passiflora lutea* is probably becoming rarer at its geographic limits because of habitat loss. It is absent from southernmost Texas, and is therefore not sympatric with the similar *P. filipes*.

Passiflora lutea is our only native species, and one of very few in the genus, that has hypogeal seed germination. It is also extensively rhizomatous, forming rhizomes even as small seedlings. Very cold-hardy, it is able to thrive in cultivation far north of its natural range. The plants are shade-tolerant, but those growing in the open can occasionally form dense masses. The leaves often turn bright yellow in autumn.

14. Passiflora filipes Bentham, Pl. Hartw., 118. 1843

Stems terete, glabrous. **Leaves** not pungent, glabrous; stipules linear-lanceolate, falcate, 2–4 × 0.5–1 mm, eglandular; petiole eglandular; blade roughly symmetric, 1–3(–5) × 1.5–6(–8) cm, shallowly 3-lobed, middle lobe ± as long as lateral lobes (usually to ¼ blade length), margins entire; abaxial fine veins not raised (especially in dried specimens), abaxial nectaries absent. **Floral bracts** absent. **Flowers:** floral tube absent; sepals pale green, 7–10 × 2–3 mm; petals pale green, 3–5 × 1 mm; corona filament whorls 2, outer filaments white basally, pale yellow apically, filiform, terete, 6–7 mm. **Berries** purple-black, globose, 5–10 × 5–10 mm.

Flowering Oct–Feb. Subtropical woodlands, in moist, loamy soil; 10 m; Tex.; Mexico; Central America; South America (Ecuador, Venezuela).

Passiflora filipes and *P. lutea* are morphologically similar and closely related (E. P. Killip 1938), yet they do not occur sympatrically. In the flora area, *P. filipes* is restricted to southernmost Texas; the southern range limit of *P. lutea* is at least 150 km north of the range of *P. filipes*. In addition to the relative lengths of middle leaf lobes, seed surface features distinguish these two

species, with sharply and coarsely foveate seeds in *P. filipes* versus the more delicately and transversely sulcate seeds of *P. lutea*.

15. Passiflora bryonioides Kunth in A. von Humboldt et al., Nov. Gen. Sp. 2(fol.): 111; 2(qto.): 140. 1817

Stems terete, uncinate-hairy. **Leaves** not pungent, uncinate-hairy; stipules ovate-lanceolate, falcate, 2–7 × 1–4 mm, margins entire, eglandular; petiole glandular, glands clavate; blade roughly symmetric, 2–7(–10) × 2–7(–11) cm, deeply (3–)5–7 (–9)-lobed, middle lobe as long as or longer than lateral lobes, margins serrate; abaxial fine veins prominently raised, abaxial nectaries absent or sometimes present along lateral veins. **Floral bracts** linear-subulate, 2–7 × 0.5 mm, margins entire, eglandular. **Flowers:** floral tube cuplike, 1–2 mm deep; sepals white, 15–20 × 5–8 mm; petals white, 8–11 × 2–3 mm; corona filament whorls 1, filaments purple basally, yellow-white apically, linear, terete, sometimes clavate, 5–11 mm. **Berries** green to green-white, ovoid-fusiform, 35–45 × 20–25 mm. $2n = 12$.

Flowering May–Sep. Semidesert grasslands, oak savannas; 1000–1300 m; Ariz.; Mexico.

J. M. Coulter (1890, 1891) reported *Passiflora bryonioides* (as the synonym *P. inamoena* A. Gray) from Hidalgo, Texas; that specimen may now be lost or it may have been misidentified. The nearest to southernmost Texas this species is confirmed to occur is about 500 km to the southwest (J. M. MacDougal 1994).

16. Passiflora pallens Poeppig ex Masters in C. F. P. von Martius et al., Fl. Bras. 13(1): 567, plate 128, fig. 4. 1872

Stems terete when young, glabrous. **Leaves** weakly to moderately pungent, glabrous; stipules subreniform, 10–20 × 5–14 mm, eglandular; petiole glandular, glands clavate; blade roughly symmetric, 1.5–6 × 2.5–9 cm, shallowly 3-lobed, middle lobe as long as or longer than lateral lobes, margins serrate basally; abaxial fine veins moderately raised, abaxial nectaries absent. **Floral bracts** ovate, 10–20 × 9–12 mm, margins basally serrate to glandular-serrate. **Flowers:** floral tube cuplike, 3–5 mm deep; sepals white, 30–35 × 7–12 mm; petals white, 20–30 × 6–10 mm; corona filament whorls 4, outer filaments green basally, white apically, with alternating lines of purple, linear, terete, 7–15 mm. **Berries** yellow to yellow-orange, ovoid, 30–50 × 25–35 mm.

Flowering Apr–Dec. Margins of and sunny gaps within mesic to wet tropical to subtropical woodlands; 0–10 m; Fla.; West Indies (Cuba, Hispaniola).

Leaf and stipule variegation, although rare within subg. *Passiflora*, is occasionally found in *Passiflora pallens*, and may serve to camouflage plants from predators in the dappled shade of their habitats. Listed as endangered in Florida, in the flora area it is found only in the southernmost part of the state.

17. Passiflora caerulea Linnaeus, Sp. Pl. 2: 959. 1753

• Blue passionflower I

Stems angular when young, glabrous. Leaves weakly pungent, glabrous; stipules subreniform, 10–20 × 5–10 mm, glandular-serrate; petiole glandular, glands clavate; blade roughly symmetric, 3–10(–16) × 4–11(–14) cm, deeply (3–)5–7 (–9)-lobed, middle lobe as long as or longer than lateral lobes, margins entire but often serrate basally on lobes; abaxial fine veins prominently raised, abaxial nectaries scattered along margins. Floral bracts ovate to ovate-oblong, 15–25 × 10–15 mm, margins entire or weakly serrate, eglandular. Flowers: floral tube cuplike, 4–5 mm deep; sepals white, 20–35 × 10–15 mm; petals white, 20–40 × 10–15 mm; corona filament whorls 4, outer filaments dark purple basally, white medially, purple apically (rarely entirely white), linear, terete to slightly flattened, 10–20 mm. Berries yellow-orange to orange, ovoid to ellipsoid, 30–50 × 30–35 mm. **2n** = 18.

Flowering Mar–Jun. Disturbed areas, open woodlands, chaparral; 0–400 m; introduced; Calif.; South America (Argentina, Brazil, Paraguay, Uruguay).

Passiflora caerulea is cultivated widely in the flora area but naturalized only in the Los Angeles metropolitan area (see F. Hrusa et al. 2002). It is possibly introduced in the Mule Mountains of southeastern Arizona (J. Koweek, pers. comm.), although this has not been confirmed. It was reported also from Utah by S. L. Welsh et al. (2003), although as "cultivated...long persisting," suggesting that it is not actually naturalized there. This species is cold-hardy and can be cultivated in gardens in relatively cold regions (at least USDA plant hardiness Zone 6), but it is unlikely to flower in such areas because of relatively short growing seasons, although it still makes an interesting foliage plant. Even in the absence of sexual reproduction, this species can persist and even spread locally by root suckering.

The artificial hybrid *Passiflora ×belotii* Pepin will key to *P. caerulea* in this treatment. However, the leaves of *P. ×belotii* are consistently three-lobed, unlike the primarily five- to seven-lobed leaves of *P. caerulea*.

18. Passiflora incarnata Linnaeus, Sp. Pl. 2: 959. 1753, name conserved • Maypop E F W

Stems terete, short-hairy, (bark not corky, above-ground stems annual). Leaves weakly to moderately pungent, sparsely to moderately hairy; stipules linear-setaceous, 3–5 × 0.5 mm, eglandular; petiole glandular, glands emergent protuberances; blade roughly symmetric, 4–12(–22) × 4–12(–30) cm, deeply 3(–5)-lobed, middle lobe as long as or longer than lateral lobes, margins serrate; abaxial fine veins prominently raised, abaxial nectaries absent. Floral bracts ovate, 3–8 × 2.5–4 mm, margins erose apically, basally glandular. Flowers: floral tube cuplike, 2–4 mm deep; sepals violet or pink to white, 20–30 × 10–15 mm; petals violet or pink to white, 25–35 × 10–15 mm; corona filament whorls 5–7, outer filaments with several concentric rings of violet, pink, and white (rarely entirely white), linear, terete, 30–50 mm. Berries green to pale yellow, ovoid to globose, 30–60 × 30–50 mm. **2n** = 18, 36.

Flowering (Mar–)Apr–Oct(–Nov). Open woodlands, savannas, prairies, dunes, cultivated ground and disturbed areas, in clayey, loamy, or sandy, dry to mesic, and often poor soil; 0–400(–1000) m; Ala., Ark., Del., D.C., Fla., Ga., Ill., Ind., Kans., Ky., La., Md., Miss., Mo., N.J., N.C., Ohio, Okla., S.C., Tenn., Tex., Va., W.Va.

Passiflora incarnata probably has been expanding its range due to its preference for open, disturbed areas, possibly since prior to European settlement (K. J. Gremillion 1989). The species is probably native no farther north than southern Illinois and Ohio, central or southern Virginia, and central West Virginia (C. Frye and B. McAvoy, pers. comm.); it is very cold-hardy and is introduced sporadically northwards (for example, G. Moore 1989; G. J. Wilder and M. R. McCombs 2002). This species can spread over large areas in well-drained soil, primarily by suckers from deep, far-spreading roots, and has been considered a weed of croplands (W. C. Muenscher 1980). Despite its aggressiveness, it is considered rare in some parts of its range (Indiana, Ohio).

Passiflora incarnata is used for its sedative properties (J. A. Beutler and A. DerMarderosian 2005; A. Chevallier 1996). The species was possibly cultivated by Native Americans, or at least exploited by them for its edible fruit, for over 3000 years prior to European settlement in North America (K. J. Gremillion 1989). The Cherokee used it for skin infections and earaches, as a liver tonic, and for weaning babies, and the Houma used it as a blood tonic (D. E. Moerman 1986). Fruits of *P. incarnata* vary in palatability, more flavorful ones suggesting potential value as a fruit crop (C. M. McGuire 1999).

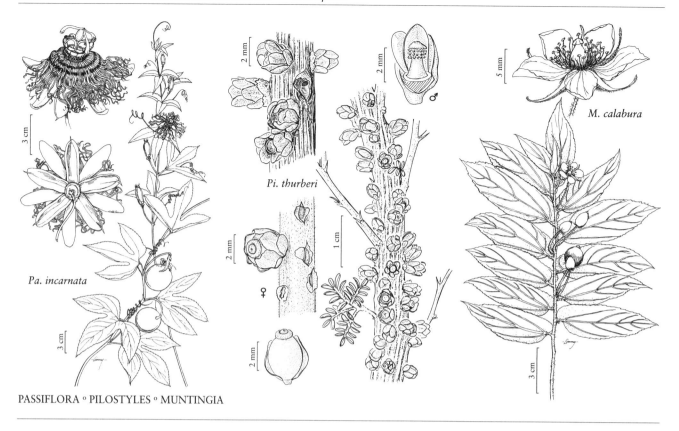

Pi. thurberi

Pa. incarnata

M. calabura

PASSIFLORA ° PILOSTYLES ° MUNTINGIA

A similar species, *Passiflora edulis* Sims, has been sparingly naturalized in southern Florida (R. P. Wunderlin and B. F. Hansen 2003), although such material has not been collected there since the 1960s. It is closely related to *P. incarnata* (V. C. Muschner et al. 2003) and will key to this species in this treatment. *Passiflora edulis* differs from *P. incarnata* by its larger stipules, at least 10 mm; leaf margins generally more coarsely serrate; larger, leaflike floral bracts, at least 17 × 8 mm; and a broadly and abruptly expanded androgynophore base (gradually expanded in *P. incarnata*). *Passiflora* 'Incense', an artificial hybrid between *P. incarnata* and the South American *P. cincinnata* Masters, also will key to *P. incarnata* in this treatment. It differs from the latter by consistently having five-lobed leaves with lobes that are much-narrowed basally, and larger floral bracts, at least 15 × 10 mm.

APODANTHACEAE Tieghem ex Takhtajan

George Yatskievych

Plants apparently dioecious stem parasites, achlorophyllous; roots not produced. **Stems** not apparent, almost entirely endophytic in host. **Leaves** 2–6, usually in 2(3) series, highly reduced, scalelike, appearing bracteal. **Inflorescences** solitary flowers terminating short branches. **Flowers** unisexual, radially symmetric, ± epigynous; sepals 4 or 5, often connate basally, sometimes petaloid; petals absent; stamens in staminate flowers usually numerous, sessile, 1-locular; anthers connate to distal portion of central column, not produced in pistillate flowers, column expanded at apex, sometimes into knob or disc; pistils in staminate flowers with only stylar column present, in pistillate flowers ovary partially inferior, indistinctly 4- or 5-carpellate, 1-locular; placentation parietal; ovules numerous. **Fruits** berrylike capsules, ± fleshy. **Seeds** numerous, minute.

Genera 2, species 10 or 11 (1 in the flora): sw United States, Mexico, West Indies, Central America, South America, sw Asia, Africa, Australia; mostly in tropical areas.

The genera of Apodanthaceae as here circumscribed have been included previously in Rafflesiaceae. Molecular studies have provided evidence that Rafflesiaceae as traditionally circumscribed (A. Cronquist 1981; W. Meijer 1993) are polyphyletic and include three or four independent lineages (T. J. Barkman et al. 2004; D. L. Nickrent et al., www.biomedcentral.com/1471-2148-4-40; C. C. Davis et al. 2007). The Indo-Malesian genera *Rafflesia* R. Brown, *Rhizanthes* Dumortier, and *Sapria* Griffith (Rafflesiaceae in the narrow sense, 20–25 species, stem and root parasites of Vitaceae), which are characterized by very large flowers, currently are thought to be related to Euphorbiaceae (Malpighiales). The two (or more) species of *Mitrastemon* Makino (Mitrastemonaceae, root parasites of Fagaceae), with a disjunct distribution in the Neotropics plus eastern and southeastern Asia, are now placed with the Ericales. The two genera of Cytinaceae, the neotropical *Bdallophytum* Eichler (two species, root parasites of Burseraceae) and *Cytinus* Linnaeus (about eight species, root parasites on various taxa), appear to be related to the Malvales. A. Blarer et al. (2004) thought that the affinities of Apodanthaceae were either with the Malvales or Cucurbitales; more recent molecular analyses have provided some support for the latter placement (N. Filipowicz and S. S. Renner 2010; H. Schaefer and Renner 2011b).

SELECTED REFERENCES Bellot, S. and S. S. Renner. 2014. The systematics of the worldwide endoparasite family Apodanthaceae (Cucurbitales), with a key, a map, and color photos of most species. PhytoKeys 36: 41–57. Filipowicz, N., and S. S. Renner. 2010. The worldwide holoparasitic Apodanthaceae confidently placed in the Cucurbitales by nuclear and mitochondrial gene trees. B. M. C. Evol. Biol. 10: 219.

1. PILOSTYLES   Guillemin, Ann. Sci. Nat., Bot., sér. 2, 2: 21, plate 1. 1834   •   [Greek *pilos*, cap, and *stylos*, pillar or column, alluding to style terminated by caplike stigma]

Leaves reduced to imbricate bracts subtending flowers. **Inflorescences** solitary flowers produced endogenously and rupturing through host stems. **Flowers** with ringed nectary at base of sepals; sepals 4 or 5, free; stylar column expanded at apex into convex, knoblike disc, this sometimes fringed or papillose; staminate flowers with 1–3 series of anthers near column apex, under margin of disc; pistillate flowers with stigmatic areas in ring along or under margin of disc. **Capsules** irregularly dehiscent, perianth and column remaining essentially unchanged after flowering. **Seeds** 0.2–0.3 mm; embryos relatively undifferentiated.

Species 9 or 10 (1 in the flora): sw United States, Mexico, Central America, South America, sw Asia, Africa, Australia.

Only the flowers and bracts of *Pilostyles* are visible on the surface of the host stems. It is not known definitely whether species of this genus are monoecious or dioecious. They are parasitic on woody legumes (Fabaceae in the broad sense).

1. **Pilostyles thurberi** A. Gray, Pl. Nov. Thurb., 326.
   1854 [F]

Pilostyles covillei Rose

Bracts 4–7, imbricate, circular to ovate, 1–1.5 mm. **Flowers** brown or maroon, 1.5–2 mm; sepals similar to bracts; stylar column with apex expanded to 1 mm diam., papillose along margins; anthers 15–21, in ring of ca. 3 rows; ovules mostly 120–200. **Seeds** mostly 120–200. $2n = 60 + 0–1$B.

Flowering Jan–May. Open desert scrub; 100–1000 m; Ariz., Calif., Nev., N.Mex., Tex.; Mexico.

Documented hosts from the flora area are *Dalea formosa, D. frutescens, Psorothamnus emoryi,* and *P. polydenius* (Fabaceae). Other species in these genera apparently serve as hosts in Mexico. The taxonomic limits of *Pilostyles thurberi* are poorly understood. It is difficult to circumscribe the host range of this taxon.

MUNTINGIACEAE   C. Bayer, M. W. Chase & M. F. Fay

John L. Strother

Shrubs or trees, to 12 m. **Leaves** alternate (distichous), petiolate, stipitate; stipules subulate or filiform [absent or peltate discs]; blade palmately veined, seldom lobed, base often asymmetric, margins serrate, surfaces hairy, hairs usually mixed: unicellular and multicellular, simple with some setiform and some glandular, branched, and stellate, often ± tangled, forming tomentum. **Inflorescences:** flowers solitary or in clusters of 2 or 3+, usually supra-axillary; involucel absent [bracteoles ca. 15, filiform]. **Flowers:** sepals caducous [persistent], (4)5(–7), valvate, basally distinct or weakly connate; petals caducous, (4)5(–7), distinct; nectaries absent; stamens 10–75+, filaments distinct or bases connate; ovary superior [inferior], 5–7-carpellate; styles 1[0]; stigmas 5–7, ± decurrent. **Fruits** baccate, ± spheric. **Seeds** [25–]100–200+.

Genera 3, species 3 (1 in the flora): introduced, Florida, Mexico, West Indies, Central America, South America; introduced also in Old World.

Plants included in Muntingiaceae (in the sense of C. Bayer et al. 1998) have been treated in Eleocarpaceae, Flacourtiaceae, or Tiliaceae. *Dicraspidia* Standley (Central America and Colombia) and *Neotessmannia* Burret (Peru) are relatively poorly known.

SELECTED REFERENCE Bayer, C., M. W. Chase, and M. F. Fay. 1998. Muntingiaceae, a new family of dicotyledons with malvalean affinities. Taxon 47: 37–42.

1. MUNTINGIA   Linnaeus, Sp. Pl. 1: 509. 1753; Gen. Pl. ed. 5, 225. 1754

• [For Abraham Munting, 1626–1683, Dutch botanist]   ⊡

Leaves: blade lanceolate to lanceolate-linear, marginal teeth irregular, abaxial indument more persistent, denser than adaxial. **Flowers:** sepals lanceolate-attenuate, base navicular; petals imbricate, white or pinkish [yellow], crumpled in bud, obovate and ± clawed or spatulate; ovary obscurely stipitate, subtended by ring of setiform hairs. **Berries** red [yellow]. **Seeds** yellowish, plumply lenticular. $x = 15$.

Species 1: introduced, Florida, Mexico, West Indies, Central America, South America; introduced also in Old World Tropics.

Muntingia is widely grown in warm to hot climates for fruit, fiber, and firewood.

1. **Muntingia calabura** Linnaeus, Sp. Pl. 1: 509. 1753
   • Calabura, Jamaica cherry [F] [I]

Leaves: petiole 2–5 mm; blade 60–150 × 20–50 mm. **Pedicels** 5–20(–35) mm. **Flowers:** 8–12+ mm; petals 12–20 mm. **Berries** 10–15 mm diam. **Seeds** 0.4–0.5 × 0.2–0.3 mm. *2n* = 28 (Costa Rica), 30 (India).

Flowering ± year-round. Disturbed, nonsalty sites; 10+ m; introduced; Fla.; Mexico; West Indies; Central America; South America; also introduced in Old World Tropics.

Muntingia calabura has been reported as a spontaneous weed in commercial greenhouses in California. It was evidently brought into California with coco fiber used in hydroponics installations (F. Hrusa et al. 2002). Fruits of *M. calabura* are reputed to be prized by bats, birds, children, and fish.

MALVACEAE Salisbury

• Mallow Family

Margaret M. Hanes

Herbs, subshrubs, shrubs, or trees, usually stellate-hairy. **Leaves** alternate, usually spiral, sometimes distichous (Malvoideae), usually petiolate, sometimes subsessile or sessile (Malvoideae), stipulate (usually well developed), simple (compound in *Abelmoschus*); blade unlobed or palmately lobed, palmately veined. **Inflorescences** axillary, terminal, or leaf-opposed. **Flowers** bisexual or unisexual, usually actinomorphic; involucel (epicalyx) immediately subtending calyx, sometimes absent; sepals usually caducous or persistent, sometimes deciduous (Malvoideae, Sterculioideae), (4)5(–8), distinct or connate; petals 4 or 5 (absent in Bombacoideae and Sterculioideae, rarely absent in Grewioideae); nectaries glandular hairs on adaxial base of sepals, petals, or androgynophores, sometimes absent; androgynophore present or absent; stamens [4]5–100[–1500], usually in antipetalous groups; filaments distinct or connate, often forming conspicuous staminal tube; ovary superior; carpels usually same number as sepals, distinct or connate, sessile or on androgynophore; ovules (1)2–many per ovary. **Fruits** usually capsules, sometimes follicles, schizocarps, berries, or nuts. **Seeds:** cotyledons usually folded, endosperm absent or sparse to copious.

Genera ca. 240, species ca. 4350 (52 genera, 250 species in the flora): nearly worldwide.

Malvaceae comprise taxa traditionally separated among four families: Malvaceae, Bombacaceae Kunth, Sterculiaceae Ventenat, and Tiliaceae Jussieu. Morphological characters distinguishing these previously recognized families are notoriously ambiguous and/or absent. Multiple tribes and genera (for example, *Fremontodendron*) have been transferred between families as taxonomic boundaries changed throughout history. Molecular phylogenies indicate that only one of the four previously recognized families (Malvaceae in the narrow sense) forms a monophyletic group (C. Bayer et al. 1999; W. S. Alverson et al. 1999) and the monophyly of an expanded familial circumscription, including all four previously accepted families, is well documented (W. S. Judd and S. R. Manchester 1997; Alverson et al. 1998; Bayer et al.). Based on morphological, molecular, and biogeographic data, Malvaceae now include nine subfamilies (Bayer et al.; Alverson et al. 1999), six of which are represented in the flora area; Malvoideae and Bombacoideae form a monophyletic group that is part of a larger clade that includes some traditional components of Tiliaceae and Sterculiaceae (now Tilioideae and Sterculioideae). A second clade contains Grewioideae and Byttneriodeae. Not included in the flora are members of subfamilies Brownlowioideae Burret, Dombeyoideae Beilschmied, and Helicteroideae Meisner.

Malvaceae range widely in inflorescence structure; all members share a basic repeating bicolor unit (a terminal flower and three bracts or an epicalyx; C. Bayer 1999). Floral nectaries in the family are composed of dense multicellular, glandular hairs on sepals, petals, or androgynophore (S. Vogel 2000). These nectaries provide nectar to a broad range of pollinators; some flowers are pollinated abiotically. Most malvaceous tissues contain large amounts of mucilage; seeds contain cyclopropanoid fatty acids. Tile cells (empty, upright ray cells in wood; M. M. Chattaway 1933) are restricted to Malvaceae but are not present in every taxon.

Representatives of Malvaceae are present on every continent except Antarctica; diversity increases in warmer regions. A majority of the genera in Malvoideae and Bombacoideae are native to the New World (P. A. Fryxell 1997); members of Byttnerioideae, Grewioideae, and Sterculioideae are more evenly distributed throughout the Tropics. Tilioideae are restricted to the Northern Hemisphere (C. Bayer and K. Kubitzki 2003).

Malvaceae have an extensive pollen fossil record, and a majority of the subfamilies are represented in the Paleocene or Eocene. *Tilia* (Tilioideae) fossil pollen and leaves are present in western North American temperate forests (where it is now absent) from the mid Eocene (S. R. Manchester 1994; H. W. Meyer and Manchester 1997). North American fossil pollen deposits of Bombacoideae are plentiful in the Cretaceous (W. Krutzsch 1989; B. E. Pfeil et al. 2002 and references therein).

The hairy seed coat of cotton [*Gossypium* (Malvoideae)] is the most economically valuable product in the family and the historical and evolutionary importance of its domestication is well documented. The seeds of cacao [*Theobroma cacao* (Sterculioideae)] are the basis of chocolate. Okra [*Abelmoschus esculentus* (Malvoideae)] is a major vegetable crop in the southeastern United States. *Tilia* (Tilioideae) trees are planted throughout temperate regions to beautify streets and parks. The marshmallow, *Althaea officinalis* (Malvoideae), is a perennial herb found in northeastern North America and Europe; the mucilage from its roots was used to make the original marshmallow.

SELECTED REFERENCES　Bayer, C. et al. 1999. Support for an expanded family concept of Malvaceae with a recircumscribed order Malvales: A combined analysis of plastid *atp*B and *rbc*L DNA sequences. Bot. J. Linn. Soc. 129: 267–303. Bayer, C. and K. Kubitzki. 2003. Malvaceae. In: K. Kubitzki et al., eds. 1990+. The Families and Genera of Vascular Plants. 10+ vols. Berlin etc. Vol. 5, pp. 225–311. Brizicky, G. K. 1966. The genera of Sterculiaceae in the southeastern United States. J. Arnold Arbor. 47: 60–74. Judd, W. S. and S. R. Manchester. 1997. Circumscription of Malvaceae (Malvales) as determined by a preliminary cladistic analysis of morphological, anatomical, palynological, and chemical characters. Brittonia 49: 348–405. Whetstone, R. D. 1983. The Sterculiaceae in the flora of the southeastern United States. Sida 10: 15–23.

Key to Subfamilies of Malvaceae

1. Gynoecium apocarpous; petals absent; epicalyx absent; androgynophore present; flowers functionally unisexual . . . . a. Sterculioideae, p. 189
1. Gynoecium syncarpous; petals present or absent; epicalyx present or absent; androgynophore present or absent; flowers usually bisexual.
　　2. Petals absent . . . . b. Bombacoideae, p. 191
　　2. Petals usually present.
　　　　3. Epicalyx usually absent.
　　　　　　4. Androgynophore absent; nectaries on sepals . . . . c. Tilioideae, p. 193
　　　　　　4. Androgynophore present or absent; nectaries on petals or androgynophore
　　　　　　. . . . . d. Grewioideae, p. 197
　　　　3. Epicalyx usually present, rarely absent.
　　　　　　5. Staminodes usually present; anthers 2- or 3-thecate . . . . .e. Byttnerioideae, p. 202
　　　　　　5. Staminodes absent or relatively small; anthers 1-thecate. . . . . f. Malvoideae, p. 215

a. Malvaceae Jussieu subfam. Sterculioideae Burnett, Outlines Bot., 821, 1119. 1835 (as Sterculidae) ☐

Margaret M. Hanes

Trees [shrubs]. Leaf blades unlobed or lobed, ultimate margins entire or serrate. **Inflorescences** axillary or terminal, paniculate [racemose]. **Flowers** usually functionally unisexual, rarely bisexual; epicalyx absent; sepals persistent or late-deciduous, (4)5(–8), connate, petaloid, nectaries usually at base on adaxial surface, sometimes absent; petals absent; androgynophore present; stamens [4–]10–30, basally connate; anthers 2-thecate; staminodes absent; gynoecium apocarpous. **Fruits** follicles [nuts]. **Seeds** 1–22[–144], glabrous or hairy.

Genera ca. 12, species ca. 400 (2 genera, 2 species in the flora): introduced; Asia, Pacific Islands, Australia; pantropical.

Genera in Sterculioideae historically have been recognized as a morphological group (H. W. Schott and S. L. Endlicher 1832; A. L. Takhtajan 1997; P. Wilkie et al. 2006) due to the presence of mostly unisexual flowers with androgynophores and without petals. Individually, these elements are found throughout Malvaceae; their combination is exclusive to Sterculioideae. Generic relationships within Sterculioideae are complicated. *Firmiana* is closely related to *Hildegardia* Schott & Endlicher and, as currently circumscribed, is not monophyletic (Wilkie et al.).

SELECTED REFERENCE Wilkie, P. et al. 2006. Phylogenetic relationships within the subfamily Sterculioideae (Malvaceae/Sterculiaceae-Sterculieae) using the chloroplast gene *ndh*F. Syst. Bot. 31: 160–170.

1. Follicles woody; seeds hairy. . . . . 1. *Brachychiton*, p. 189
1. Follicles chartaceous; seeds glabrous . . . . .2. *Firmiana*, p. 190

1. BRACHYCHITON Schott & Endlicher, Melet. Bot., 34. 1832 • [Greek *brachys*, short, and *chiton*, tunic, evidently alluding to covering of short hairs on seeds] ☐

John L. Strother

Trees [shrubs], trunks often swollen proximally, plants monoecious. **Leaves:** stipules caducous [persistent]; blade often lobed, base usually cuneate to truncate or cordate, margins usually entire or serrate, palmately or pinnately veined. **Flowers** functionally unisexual; sepals ± valvate, basally connate; nectaries 0 or 10–16; stamens (staminodes in pistillate flowers) [10–]25–30; anthers often crowded; ovary stalked, carpels 5, distinct; styles 5, connate; stigmas 5. **Follicles** woody, outsides glabrous, insides stellate-hairy. **Seeds** hairy. $x = 20$.

Species 31 (1 in the flora): introduced, California; Pacific Islands (Papua New Guinea), Australia.

SELECTED REFERENCE Guymer, G. P. 1988. A taxonomic revision of *Brachychiton* (Sterculiaceae). Austral. Syst. Bot. 1: 199–323.

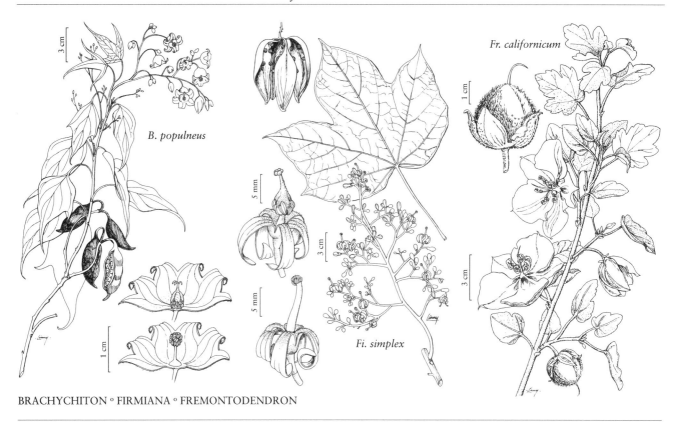

BRACHYCHITON ° FIRMIANA ° FREMONTODENDRON

1. Brachychiton populneus (Schott & Endlicher)
R. Brown, Pterocymbium, 234. 1844 [F] [I]

Poecilodermis populnea Schott &
Endlicher, Melet. Bot., 33. 1832

Plants mostly 6–20 m. **Leaf blades** mostly lanceolate-ovate to lanceolate, sometimes 1 or 2 (–5)-lobed, 5–16 cm, base mostly cuneate, sometimes truncate or cordate, apex apiculate to cirrhous, surfaces glabrous. **Panicles** 10–60-flowered. **Sepals** abaxially cream to greenish, adaxially usually green and marked with red, rarely mostly red, forming perianths 10–16 × 13–16 mm. **Follicles** stalked, ellipsoid to ovoid, becoming navicular on dehiscence, mostly 2–7 cm. **Seeds** 2–22[–140]. $2n = 40$ (Australia).

Flowering (Jan–)May–Jul(–Nov). Disturbed sites; 200–500 m; introduced; Calif.; Australia.

Brachychiton populneus may be marginally established in the flora area or may persist outside cultivation only as abandoned plantings.

2. FIRMIANA Marsili, Saggi Sci. Lett. Accad. Padova 1: 115. 1786 • [For Carlo Giuseppe, Conte di Firmian, 1717–1782, Austrian statesman and Governor-General of Lombardy] [I]

Laurence J. Dorr

Trees [shrubs], deciduous. **Leaves:** stipules caducous; blade usually lobed, base often cordate, palmately 5–7-veined. **Flowers** unisexual or rarely bisexual (plants monoecious or polygamous); sepals 5, ± connate into tube. **Staminate flowers:** stamens [10–]15; anthers shortly stipitate or sessile, 2-celled, dehiscing longitudinally; pistillode present. **Pistillate flowers:** ovary 5-locular, pyriform [ovoid or globose], base ringed with [10–]15 indehiscent anthers; ovules 2–6 per locule; styles 5, basally connate; stigmas 5. **Follicles** stipitate, chartaceous, foliaceous, dehiscent long before maturity and exposing seeds adhering to margins. **Seeds** globose, smooth, wrinkled when dry, glabrous; endosperm abundant; embryo straight. $x = 10$.

Species ca. 12 (1 in the flora): introduced; Asia, Pacific Islands.

A. J. G. H. Kostermans (1954, 1956, 1957) used follicle dehiscence to separate *Firmiana* from the closely related *Hildegardia*; follicles are dehiscent in the former and indehiscent in the latter. However, molecular evidence suggests that fruit dehiscence is not a defining generic character, and it now appears that the species of *Firmiana* and *Hildegardia* with deeply lobed calyces form a clade (P. Wilkie et al. 2006). This would tend to support the practice in some Asian floras of recognizing the species of *Firmiana* with infundibuliform calyces as *Erythropsis* Lindley ex Schott & Endlicher. Although generic boundaries are poorly defined, and with better molecular sampling probably will change, *Firmiana* is the oldest generic name available for these three related taxa and likely will continue to be applied to the species found in our flora.

SELECTED REFERENCES Kostermans, A. J. G. H. 1956. The genus *Firmiana* Marsili (Stercul.). Pengum. Balai Besar Penjel. Kehut. Indonesia 54: 3–33. Kostermans, A. J. G. H. 1957. The genus *Firmiana* Marsili (Sterculiaceae). Reinwardtia 4: 281–310. Kostermans, A. J. G. H. 1989. Notes on *Firmiana* Marsili (Sterculiaceae). Blumea 34: 117–118. Ridley, H. N. 1934. *Firmiana* and *Erythropsis*. Bull. Misc. Inform. Kew 1934: 214–217.

1. **Firmiana simplex** (Linnaeus) W. Wight, U.S.D.A. Bur. Pl. Industr. Bull. 142: 67. 1909 • Chinese parasol or bottle tree, Japanese varnish tree, phoenix tree, sycamore-leaf sterculia, varnish tree [F] [I]

Hibiscus simplex Linnaeus, Sp. Pl. ed. 2, 2: 977. 1763, name and type proposed for conservation; *Firmiana platanifolia* (Linnaeus f.) Schott & Endlicher

Trees single or multi-stemmed, 10–15(–20) m; bark green with paler greenish white vertical stripes, becoming gray or chalky white, smooth. **Leaves:** petiole 15–30(–40) cm; blade palmately 3–5-lobed, rarely unlobed, broadly ovate, lobes often constricted at base, (8–)12–40 × (12–)20–50 cm, membranous, apex acute or acuminate, surfaces abaxially minutely stellate-puberulent with domatia in axils of primary and secondary veins, adaxially glabrous. **Inflorescences** terminal, erect, paniculate, 20–50 × 20–100 cm, often borne on leafy branches; bracts caducous. **Pedicels** articulated, 2–3(–5) mm. **Flowers:** calyx 7–9(–10) mm, divided nearly to base, tube cupuliform, lobes reflexed, lanceolate-oblong, abaxially puberulent, adaxially villous in throat proximally, androgynophore exserted, glabrous. **Staminate flowers:** androgynophore to 2 cm; anthers irregularly fascicled; pistillode obscured by anthers. **Pistillate flowers:** androgynophore to 0.5 cm; ovary densely stellate-pubescent. **Follicles** pendant, ovate-lanceolate, 6–11 × 2.5–4.5 cm, abaxially stellate-puberulent or glabrous. **Seeds** (1)2–4, 6–7 mm diam. $2n = 40$.

Flowering May–Aug; fruiting Jun–Oct. Roadsides, thickets, mixed deciduous woods; 0–300 m; introduced; Ala., Ark., D.C., Fla., Ga., La., Miss., N.C., S.C., Tenn., Tex., Va.; Asia (China); introduced also in Europe, e Asia (Japan).

Firmiana simplex has long been cultivated in Japan and was brought to Europe by the mid eighteenth century. André Michaux reportedly introduced the species to North America before 1796 through his garden in Charleston, South Carolina (J. P. F. Deleuze 1804), and it is grown today as an ornamental or street tree in at least 20 states. The species is considered an invasive tree in the southeastern states, but only a few occurrences of its being naturalized have been well documented.

b. MALVACEAE Jussieu subfam. BOMBACOIDEAE Burnett, Outlines Bot., 816, 818, 1094, 1119. 1835 (as Bombacidae)

Margaret M. Hanes

Shrubs or trees. Leaf blades lobed [digitate], ultimate margins serrate or entire. **Inflorescences** leaf-opposed [axillary, terminal], solitary flowers [cymose or paniculate]. **Flowers** bisexual;

epicalyx present [absent], 3-lobed; sepals persistent or late-caducous, 5, connate ½ length, petaloid, nectaries usually at base; petals absent [present, basally connate]; androgynophore absent; stamens 5[–1500], basally connate into tube; anthers 1-thecate; staminodes absent; gynoecium syncarpous. **Fruits** capsules, rarely indehiscent. **Seeds** 8–15[–many], glabrous or hairy.

Genera 27, species ca. 250 (1 genus, 3 species in the flora): sw United States, Mexico, West Indies, Central America, South America, Asia, Africa, Indian Ocean Islands, Pacific Islands, Australia; especially New World Tropics.

The phylogenetic placement of *Fremontodendron* has long challenged botanists (W. M. Kelman 1991); it has been affiliated with Sterculiaceae/ioideae (J. Hutchinson 1964–1967), Malvoideae (W. S. Alverson et al. 1999), and Bombacoideae (C. Bayer et al. 1999). It is closely related to the Mexican monospecific *Chiranthodendron* Larreátegui, and D. A. Baum et al. (2004) suggested (with minimal support) that the two genera are sister to the Malvatheca clade (Malvoideae and Bombacoideae).

3. FREMONTODENDRON   Coville, Contr. U.S. Natl. Herb. 4: 74. 1893   •   [For John Charles Frémont, 1813–1890, U.S. military explorer and politician, and Greek *dendron*, tree]

John L. Strother

Fremontia Torrey, Proc. Amer. Assoc. Advancem. Sci. 4: 191. 1851, not Torrey 1843 [Chenopodiaceae]

Shrubs or trees to 100 dm. **Leaves** stipulate (stipules caducous); blade usually palmately lobed, base cuneate to rounded or cordate, surfaces hairy, hairs mostly stellate, (1–)3–7-veined from base. **Inflorescences** solitary flowers, (usually supra-axillary). **Flowers:** epicalyx caducous; sepals coppery, orange, or yellow, sometimes with red, weakly imbricate or valvate, adaxially glandular or pitted near base, apex often apiculate to attenuate; filament bases connate to ½ their lengths; ovary sessile, carpels (4)5, connate; styles 1; stigmas 1. **Capsules** 4- or 5-locular, usually densely hairy abaxially. **Seeds** 2 or 3 per locule, often carunculate.

Species 3 (3 in the flora): sw United States, nw Mexico.

SELECTED REFERENCES Kelman, W. M. 1991. A revision of *Fremontodendron* (Sterculiaceae). Syst. Bot. 16: 3–20. Kelman, W. M. et al. 2006. Genetic relationships among *Fremontodendron* (Sterculiaceae) populations of the central Sierra Nevada foothills of California. Madroño 53: 380–387.

1. Plants decumbent, to 10+ dm; pedicels 10–27 mm; sepals usually coppery to orange, rarely yellow . . . . 1. *Fremontodendron decumbens*
1. Plants erect, mostly 15–100 dm; pedicels 4–18 mm; sepals usually orange to yellow.
    2. Leaf blade bases cuneate to rounded or weakly cordate; sepals usually yellow, lobes sometimes red at bases and margins, gland/pit borders usually setose; seeds usually carunculate, hairy . . . . 2. *Fremontodendron californicum*
    2. Leaf blade bases notably cordate; sepals usually orange, lobes sometimes reddish at bases and along veins, gland/pit borders rarely setose; seeds rarely carunculate, usually glabrous . . . . 3. *Fremontodendron mexicanum*

1. Fremontodendron decumbens R. M. Lloyd, Brittonia 17: 382, figs. 1–4. 1965 [C][E]

Fremontodendron californicum (Torrey) Coville subsp. *decumbens* (R. M. Lloyd) Munz

Plants decumbent, to 10+ dm, usually broader than high. **Leaf blades** 15–95 × 9–61 mm, base moderately to notably cordate. **Pedicels** 10–27 mm. **Flowers** 20–50 mm diam. (pressed); sepals usually coppery to orange, rarely yellow, gland/pit borders setose. **Seeds** carunculate, hairy. $2n$ = ca. 98.

Flowering May. On rocks in pine woodlands, chaparral borders; of conservation concern; 400–800 m; Calif.

Fremontodendron decumbens is known with certainty only from El Dorado County.

Differences between plants treated here as *Fremontodendron decumbens* and *F. californicum* are primarily in habit, pedicel length, and sepal color (W. M. Kelman 1991). The two may be taxonomically indistinct at any rank.

Fremontodendron decumbens is in the Center for Plant Conservation's National Collection of Endangered Plants.

2. Fremontodendron californicum (Torrey) Coville, Contr. U.S. Natl. Herb. 4: 74. 1893 [F][I]

Fremontia californica Torrey, Proc. Amer. Assoc. Advancem. Sci. 4: 191. 1851; *F. californica* subsp. *crassifolia* (Eastwood) Abrams; *F. californica* subsp. *napensis* (Eastwood) Munz; *F. californica* var. *napensis* (Eastwood) McMinn; *Fremontodendron californicum* subsp. *crassifolium* (Eastwood) J. H. Thomas; *F. californicum* subsp. *napense* (Eastwood) Munz; *F. californicum* subsp. *obispoense* (Eastwood) Munz

Plants erect, 15–100 dm. **Leaf blades** 5–113 × 4–65 mm, base cuneate to rounded or weakly cordate. **Pedicels** 4–18 mm. **Flowers** 23–76(–100) mm diam. (pressed); sepals usually yellow, lobes sometimes red at bases and margins, gland/pit borders usually setose. **Seeds** usually carunculate, hairy. $2n$ = 40.

Flowering Mar–Jul(–Dec). Chaparral, pinyon scrublands, sagebrush scrublands; (10–)400–2300 m; introduced; Ariz., Calif.; Mexico (Baja California).

Morphologically distinctive local and/or regional populations of *Fremontodendron californicum* have been treated as subspecies and/or varieties; they intergrade across the range of the species.

3. Fremontodendron mexicanum Davidson, Bull. S. Calif. Acad. Sci. 16: 50. 1917 [C]

Fremontia mexicana (Davidson) J. F. Macbride

Plants erect, 15–60 dm. **Leaf blades** 15–100 × 10–70 mm, base notably cordate. **Pedicels** 5–15 mm. **Flowers** 45–85 mm diam. (pressed); sepals usually orange, lobes sometimes reddish at bases and along veins, gland/pit borders rarely setose. **Seeds** rarely carunculate, usually glabrous.

Flowering Mar–Jul. Canyon slopes, chaparral; of conservation concern; 100–700 m; Calif.; Mexico (Baja California).

Fremontodendron mexicanum is in the Center for Plant Conservation's National Collection of Endangered Plants.

---

c. MALVACEAE Jussieu subfam. TILIOIDEAE Arnott, Botany, 100. 1832 (as Tilieae)

Margaret M. Hanes

[Shrubs] trees. **Leaf blades** usually unlobed, margins serrate, usually with domatia. **Inflorescences** axillary, usually dichasia, 3–24[–80]-flowered, rarely solitary flowers, associated with winged bract in *Tilia*. **Flowers** bisexual; epicalyx absent; sepals caducous, 5[–7], distinct, not petaloid, nectaries basal, adaxial; petals 5, distinct, base not clawed; androgynophore absent; stamens

[15–]40–60, ± distinct or in fascicles of 4 or 5; anthers 2-thecate; staminodes basally fused with stamens and petals absent; gynoecium syncarpous. **Fruits** capsules or nuts, indehiscent. **Seeds** 1–3, glabrous or hairy.

Genera 2 or 3, species ca. 40 (1 species in the flora): North America, Europe, Asia; mostly temperate areas.

Tilioideae represent a narrow and fairly atypical circumscription of the former family Tiliaceae. The subfamily is unusual in Malvaceae in being restricted to the Northern Hemisphere.

4. TILIA Linnaeus, Sp. Pl. 1: 514. 1753; Gen. Pl. ed. 5, 230. 1754 • Basswood, lime tree, linden, tilleul [Classical Latin name]

John L. Strother

Trees to 150(–300+) dm. **Leaves** deciduous, distichous; stipules caducous; petiole not winged; blade mostly ovate to orbiculate, base cordate to truncate, usually asymmetric, sometimes oblique, margins serrate, teeth usually gland-tipped, apex apiculate to acuminate, surfaces glabrate or notably hairy, hairs mostly simple, forked, fascicled, stellate, and/or glandular, glandular hairs minute, venation palmate, primary veins (3–)5(–7+). **Inflorescences** usually (2)3–24 [–80]-flowered dichasia, rarely solitary flowers, each inflorescence associated with an elliptic, oblanceolate, oblong, or spatulate bract. **Pedicels** 2-bracteolate at base, bracteoles caducous. **Flowers** actinomorphic; sepals valvate, bases sometimes coherent, abaxially hairy [glabrous], adaxially usually densely pilose to villous; petals cream to yellow, lanceolate to oblanceolate; stamens usually in 5 fascicles [+ 5 separate stamens]; staminodes [0–]5, similar to, smaller than, petals; ovary sessile, (3–)5-carpellate, ultimately 1-celled; styles 1 per ovary; stigmas 5-lobed. **Fruits** nutlike, indehiscent. $x = 41$.

Species ca. 20 (1 in the flora): North America, Mexico, Europe, Asia; mostly temperate areas.

Tilia (Tiliaceae) has been moved into Malvaceae in a broad sense (see Angiosperm Phylogeny Group 1998, 2003, 2009, and http://www.mobot.org/MOBOT/Research/APweb/welcome.html).

The name basswood evidently derives from bast (the inner bark), which is a source of fibers used in making cords and ropes from tilias. The names linden and lime tree may be English derivatives from German (linde baum). Tilias are commonly planted as shade trees along paths and streets; a potential problem with such plantings is aphid rain of honeydew, usually followed by sooty molds. Some living tilias are thought to be more than 300, possibly 1000+, years old and some have trunk diameters to two or more meters. Tilia wood is fine-grained and has long been prized by carvers. Flowers, fruits, and wood from tilias are used by herbalists and in perfumery. Nectar from tilia flowers is used by bees in making honey; some reports suggest the nectar may sometimes be poisonous to bees.

For native tilias in the flora area, 50+ specific and infraspecific basionyms have been published; the proposed taxa have been distinguished largely on differences in indument on leaves. Leaf surfaces that look to be glabrous at magnifications of 1–10× are usually found, at 30× or higher magnification, to have minute glandular hairs (0.05–0.1 mm); such surfaces are called glabrate here. More readily seen hairs on tilia leaves are simple, forked, fascicled, or stellate (see J. W. Hardin 1990). Indument on leaves of flowering shoots of tilias may differ from tree to tree within a population, may differ within trees on shoots in full sun versus relatively shaded shoots, and may differ on a single leaf from early season to late season. Leaves on vegetative sucker shoots from at and near bases of tilia trunks are usually larger and thinner than leaves on flowering shoots. H. Kurz and R. K. Godfrey (1962) noted in their review of the taxonomy of tilias that a "student of *Tilia*" identified as distinct species individual specimens from different parts of a single tree.

W. C. Ashby (1964) studied leaf induments and other traits in specimens from wild-collected tilias and from tilias grown from seed in common gardens. Ashby concluded that native tilias in an area ranging from North Dakota to Illinois and thence to Massachusetts and North Carolina belong to one species.

L. G. Hickok and J. C. Anway (1972) sampled 102 populations of tilias from two Canadian provinces and 17 of the United States for 32 morphological characters (151 specimens) and for flavonoid chemistry (92 leaf samples). They noted geographical trends in induments and flavonoid patterns; they did not find justification for recognizing more than one native species of *Tilia* in the flora area.

[Editorial note added in proof: Molecular data support the existence of two discrete genetic lineages of North American *Tilia*. One includes most Mexican haplotypes and those from the southern Appalachians, while the second includes most exemplars from the United States and two haplotypes from eastern Mexico (D. McCarthy 2012).]

Non-native tilias that may be encountered in the flora area outside of cultivation as ruderals or abandoned plantings include: *Tilia cordata* Miller (leaf blades orbiculate, 4–7 cm, abaxially glabrate but for tufts of hairs at vein axils; fruits not notably ribbed; European), *T. platyphyllos* Scopoli (leaf blades ovate to orbiculate, 7–10 cm, abaxially sparsely stellate-hairy and with tufts of hairs at vein axils; fruits notably 3–5-ribbed; European), and *T. tomentosa* Moench (including *T. petiolaris* de Candolle with pendulous branches; leaf blades suborbiculate, 7–13 cm, abaxially usually densely white stellate-hairy; fruits weakly five-angled; Eurasian). *Tilia* ×*europaea* Linnaeus evidently applies to hybrids derived from crosses between *T. cordata* and *T. platyphyllos*.

Tilia glabra Ventenat is a superfluous name; *T. americana* Linnaeus was cited as a synonym. *Tilia michauxii* Nuttall was a provisional name and was not validly published. *Tilia pubescens* Aiton is a superfluous name; *T. caroliniana* Miller was cited as a synonym.

SELECTED REFERENCES Ashby, W. C. 1964. A note on basswood nomenclature. Castanea 29: 109–116. Hardin, J. W. 1990. Variation patterns and recognition of varieties of *Tilia americana* s.l. Syst. Bot. 15: 33–48. Hickok, L. G. and J. C. Anway. 1972. A morphological and chemical analysis of geographical variation in *Tilia* L. of eastern North America. Brittonia 21: 2–8. Jones, G. N. 1968. Taxonomy of American species of linden (*Tilia*). Illinois Biol. Monogr. 39: 1–156. McCarthy, D. 2012. Systematics and Phylogeography of the Genus *Tilia* in North America. Ph.D. dissertation. University of Illinois, Chicago. Pigott, D. 2012. Lime-trees and Basswoods: A Biological Monograph of the Genus *Tilia*. Cambridge and New York.

1. Tilia americana Linnaeus, Sp. Pl. 1: 514. 1753 F

Tilia americana var. *caroliniana* (Miller) Castiglioni; *T. americana* var. *heterophylla* (Ventenat) Loudon; *T. americana* var. *neglecta* (Spach) Fosberg; *T. australis* Small; *T. caroliniana* Miller; *T. caroliniana* subsp. *floridana* (Small) A. E. Murray; *T. caroliniana* subsp. *heterophylla* (Ventenat) Pigott; *T. caroliniana* subsp. *occidentalis* (Rose) Pigott; *T. eburnea* Ashe; *T. floridana* Small; *T. georgiana* Sargent; *T. heterophylla* Ventenat; *T. lasioclada* Sargent; *T. leucocarpa* Ashe; *T. littoralis* Sargent; *T. michauxii* Nuttall ex Sargent; *T. monticola* Sargent; *T. neglecta* Spach; *T. porracea* Ashe; *T.* ×*stellata* Hartig; *T. truncata* Spach

Trees sometimes multitrunked. **Leaf blades** (on flowering shoots) 5–15(–20) × 5–12+ cm, abaxial surfaces initially glabrate but for tufts of simple, forked, or fascicled hairs at some vein axils, or initially densely to sparsely stellate-hairy, then sometimes glabrescent, with or without tufts of simple, forked, or fascicled hairs at some vein axils, adaxial surfaces usually glabrous or glabrate. **Inflorescences:** bracts 7–15 cm, notably reticulate-veined, notably hairy, glabrescent, or glabrate; peduncle diverging from near or beyond midlength of bract. **Pedicels** weakly clavate, 4–15+ mm, notably hairy, glabrescent, or glabrate. **Flowers:** sepals 4–6(–9+) mm; petals 5–9(–11) mm; staminodes 4–7(–10) mm. **Fruits** ellipsoid to globose, 5–10 mm diam. $2n = 82$.

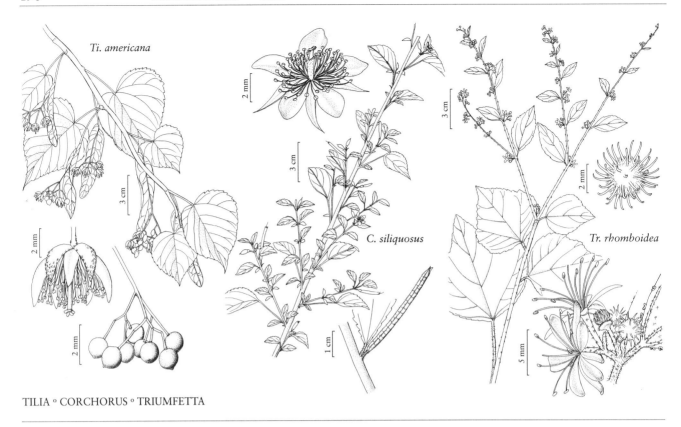

TILIA ∘ CORCHORUS ∘ TRIUMFETTA

Flowering (Apr–)Jun–Jul. Forests, stream and lake shores; 10–800 m; Man., N.B., Ont., Que., Sask.; Ala., Ark., Conn., Del., D.C., Fla., Ga., Ill., Ind., Iowa, Kans., Ky., La., Maine, Md., Mass., Mich., Minn., Miss., Mo., Nebr., N.H., N.J., N.Y., N.C., N.Dak., Ohio, Okla., Pa., R.I., S.C., S.Dak., Tenn., Tex., Vt., Va., W.Va., Wis.; Mexico.

In preparing this treatment, I tried to see merit in taxonomies in which more than one *Tilia* species native to the flora area are recognized. In the end, my experiences with specimens and my attempts at using taxonomies that purport to delineate distinct species and/or infraspecific taxa of *Tilia* in the flora area led me to agree with W. C. Ashby (1964), J. W. Hardin (1990), L. G. Hickok and J. C. Anway (1972), and H. Kurz and R. K. Godfrey (1962) that taxonomic recognition of more than one native *Tilia* species in the flora area is not tenable. I also concluded that recognition of infraspecific taxa within that one species in the flora area is not tenable.

If *Tilia americana* is partitioned (for example, by J. W. Hardin 1990), infraspecific taxa in the flora area may be characterized like so:

Variety *americana* has abaxial surfaces of leaves of flowering shoots initially glabrate but for tufts of simple, forked, and/or fascicled hairs at some vein axils.

Variety *caroliniana* has abaxial surfaces of leaves of flowering shoots initially sparsely to moderately stellate-hairy and with simple, forked, and/or fascicled hairs scattered and/or in tufts at some vein axils; attachments of the stellate hairs are usually fragile and the stellate hairs become sparser as seasons progress.

Variety *heterophylla* has abaxial surfaces of leaves of flowering shoots initially moderately to densely stellate-hairy and with simple, forked, and/or fascicled hairs scattered and/or in tufts at some vein axils; attachments of the stellate hairs are usually firm, and the stellate hairs usually persist as seasons progress.

J. W. Hardin (1990) included *Tilia mexicana* Schlechtendal, native to Mexico (at least 14 states), in the circumscription of *T. americana* adopted here and treated it as *T. americana* var. *mexicana* (Schlechtendal) Hardin.

D. Pigott (2012) treated native tilias of North America as: *Tilia americana* var. *americana*, *T. americana* var. *neglecta*, *T. caroliniana* subsp. *caroliniana*, *T. caroliniana* subsp. *floridana* (extending into Mexico), *T. caroliniana* subsp. *heterophylla*, *T. caroliniana* subsp. *occidentalis* (Mexico), and *T.* ×*stellata*.

d. MALVACEAE Jussieu subfam. GREWIOIDEAE Dippel, Handb. Laubholzk. 3: 56. 1893 (as Grewieae)

Margaret M. Hanes

Herbs, subshrubs, or shrubs [trees]. Leaf blades usually unlobed, margins crenate to dentate. **Inflorescences** axillary, terminal, or leaf-opposed, fasciculate or cymose or solitary flowers. **Flowers** bisexual [plants rarely gynodioecious]; epicalyx absent; sepals usually persistent, 4 or 5, distinct, not petaloid, appendaged near apex [not appendaged]; petals (0–)4 or 5, distinct, base sometimes clawed, base with or without nectaries; androgynophore present or absent, base with or without nectaries; stamens [4]5–100, distinct; anthers 2-thecate; staminodes absent [outermost stamen sometimes sterile]; gynoecium syncarpous. **Fruits** capsules, dehiscence loculicidal. **Seeds** 2–150, glabrous.

Genera 25, species ca. 700 (2 genera, 8 species in the flora): United States, Mexico, West Indies, Central America, South America, Asia, Africa, Indian Ocean Islands, Pacific Islands, Australia; mostly tropical.

A large portion of former Tiliaceae is now included in Grewioideae; the remaining members of the former Tiliaceae have been transferred to other subfamilies in Malvaceae or to other families (C. Bayer et al. 1999). Grewioideae are sister to Byttnerioideae; together they represent the earliest branching subfamilies in the family (Bayer et al.; R. Nyffeler et al. 2005). Relationships within the morphologically diverse Grewioideae have been difficult to resolve (Bayer et al.; W. S. Alverson et al. 1999). U. Brunken and A. N. Muellner (2012) provided a new tribal classification for Grewioideae and evidence that morphological characters traditionally used to delimit taxonomic groups in Grewioideae (for example, the presence of an androgynophore) have arisen multiple times independently and should not be used as synapomorphies to define groups.

SELECTED REFERENCE Brunken, U. and A. N. Muellner. 2012. A new tribal classification of Grewioideae (Malvaceae) based on morphological and molecular phylogenetic evidence. Syst. Bot. 37: 699–711.

1. Capsules glabrous or hairy, without uncinate spines . . . . 5. *Corchorus*, p. 197
1. Capsules glabrous or hairy, with uncinate spines . . . . 6. *Triumfetta*, p. 200

5. CORCHORUS Linnaeus, Sp. Pl. 1: 529. 1753; Gen. Pl. ed. 5, 234. 1754 • Jute [Greek *kore*, eye pupil, and *koreo*, to purge or clear, alluding to use of leaves]

Guy L. Nesom

Herbs, annual or perennial, [**subshrubs**], **or shrubs,** taprooted. **Stems** erect to ascending or decumbent, usually unbranched or relatively few-branched, hairy, hairs simple or stellate. **Leaves** petiolate; stipules caducous to subpersistent, filiform; blade unlobed, oblong to ovate or lanceolate, herbaceous, margins serrate, [serrulate], or crenate, surfaces glabrous or hairy, hairs simple or stellate, palmately 3–7-veined. **Inflorescences** leaf-opposed, solitary flowers or fasciculate or cymose, 2–8-flowered. **Flowers:** sepals [4]5, narrowly lanceolate to oblong or obovate, apically awned or not, glabrous or hairy; petals [4]5, yellow, obovate to oblanceolate, glands absent; stamens [4–]10–70[–100], on androgynophore; ovary 2–4[–10]-locular; ovules 2–50 per locule; styles 1, simple, short-cylindric; stigmas peltate or discoid, usually irregularly crenulate or lobulate. **Capsules** usually cylindric to short-ellipsoid, rarely subglobose, 2–4[–10]-valved, glabrous or hairy, dehiscence loculicidal. **Seeds** 4–10[–30] per locule, angular, smooth or pitted. $x = 7$.

Species 90–100 (5 in the flora): United States, Mexico, West Indies, Central America, South America, se Asia, Africa, Pacific Islands, Australia.

Corchorus capsularis Linnaeus and *C. olitorius* Linnaeus have been domesticated for production of bast fibers and are the commercially important cultivated species. *Corchorus aestuans* Linnaeus also is used and cultivated for fiber production, although the fibers are weaker than those of the others. Fibers of other species (for example, *C. tridens* Linnaeus) also are used locally. *Corchorus* leaves (*C. capsularis* in China and Japan; *C. olitorius* in southern Asia, the Middle East, and northern Africa) are commonly eaten as leafy vegetables and in stews. Within *C. olitorius*, the fiber types and the vegetable types comprise two separate cultivar groups, the former with a little-branched habit.

Generic boundaries among the species of *Corchorus* probably will be modified. The monospecific *Oceanopapaver* Guillaumin is part of a clade with the endemic Malagasy genus *Pseudocorchorus* Capuron and species of *Corchorus* with stellate vestiture (C. Tirel et al. 1996; B. A. Whitlock et al. 2003).

1. Stems, leaves, and fruits densely tawny-pubescent, hairs stellate; capsules subglobose to short-ellipsoid . . . . .1. *Corchorus hirsutus*
1. Stems, leaves, and fruits glabrous or sparsely hairy, hairs simple; capsules cylindric.
   2. Capsules 3(4)-valved, each valve terminated by bifurcate awn.
      3. Capsules 4–5 mm diam., wing-angled, awns 1.5–3 mm; pedicels 2–3 mm; sepals awned. . . . . 2. *Corchorus aestuans*
      3. Capsules 1.5–2 mm diam., not wing-angled, awns 1.5–2 mm; pedicels 0–1 mm; sepals not awned. . . . . .3. *Corchorus tridens*
   2. Capsules 2-valved, awns none or not bifurcate.
      4. Stems puberulent in lines, hairs retrorse, blunt-tipped; capsules terete, sparsely to moderately strigose to strigulose, distally abruptly constricted to beaklike apex, awnless; sepals hirsute; stamens 15–25 . . . . 4. *Corchorus hirtus*
      4. Stems short-pilose in 1(2) lines, hairs erect, sharp-tipped; capsules flattened, minutely hirtellous to hispidulous, distally subtruncate, each valve 2-awned; sepals glabrous; stamens 50–70. . . . . 5. *Corchorus siliquosus*

1. Corchorus hirsutus Linnaeus, Sp. Pl. 1: 530. 1753

   • Jackswitch, woolly corchorus, mallet, cadillo   [I]

Plants shrubs. **Stems** erect to semiprostrate, mostly unbranched, 5–10(–15) dm, densely tawny-pubescent, hairs stellate. **Leaves** persistent; petiole 2–8 mm; blade oblong to oblong-elliptic or elliptic-lanceolate, 1–3.5(–6) cm, base rounded-truncate, margins crenate, proximal teeth not prolonged, apex obtuse to acute, surfaces densely tawny-pubescent, hairs stellate.

Inflorescences fasciculate, 2–8-flowered. **Pedicels** 5–7 mm. **Flowers:** sepals linear-oblong, 4–6 mm, not awned, stellate-pubescent; petals 5–6 mm; stamens 50–60. **Capsules** subglobose to short-ellipsoid, not wing-angled, 4-valved, without awns, 7–9(–12) × 7–8 mm, densely tawny-pubescent, hairs stellate. $2n = 14$.

Flowering Jul–Oct. Lake edges, disturbed pine rocklands; 0–50 m; introduced; Fla.; West Indies; introduced also in Mexico, Central America, South America, Africa.

Corchorus hirsutus is known only by collections from Hillsborough and Miami-Dade counties.

2. Corchorus aestuans Linnaeus, Syst. Nat. ed. 10, 2: 1079. 1759 • East Indian or West African mallow I W

Corchorus acutangulus Lamarck

Plants herbs, annual or short-lived perennial. **Stems** erect to decumbent or ascending, 2–13 (–20) dm, puberulent in lines, hairs simple, arched, down-turned to irregularly oriented. **Leaves:** petiole 5–30(–36) mm; blade ovate or broadly ovate to ovate-elliptic, 2–8 cm, base rounded-truncate, margins serrate, one or both of basal pair of teeth sometimes prolonged into caudate-setaceous point 3–5 mm, apex short-acuminate or acute, surfaces sparsely pilose to strigose on veins and lamina. **Inflorescences** usually solitary flowers, sometimes fasciculate or cymose, 2- or 3-flowered. **Pedicels** 2–3 mm. **Flowers:** sepals narrowly oblong, 3–5 mm, awned, glabrous; petals 4–6 mm; stamens 10–20. **Capsules** cylindric, wing-angled, wings to 2 mm wide, 3(4)[5]-valved, each valve terminated by bifurcate awn 1.5–3 mm, 15–25(–40) × 4–5 mm, glabrous. $2n = 14$.

Flowering Jul–Sep(–Oct). Roadsides, fallow fields, vacant lots, abandoned road beds, concrete cracks, edges of scrub oak thickets and pine-oak woodlands, lake edges, shady woodlands; 0–50 m; introduced; Ala., Fla., Ga., La.; Asia (India); Africa; introduced also in West Indies, Central America, South America, elsewhere in Asia (Japan, Pakistan, Singapore), Pacific Islands (Papua New Guinea), Australia.

Apparently the first and perhaps only record of *Corchorus aestuans* for Alabama is an 1891 collection from ballast ground in Mobile County.

3. Corchorus tridens Linnaeus, Mant. Pl. 2: 566. 1771 • Horn-fruited jute I

Plants herbs, annual. **Stems** erect to ascending, 3–6 dm, mostly glabrous. **Leaves:** petiole 5–15(–25) mm; blade oblong-lanceolate to elliptic-obovate, linear-lanceolate, or narrowly elliptic, 1.5–9(–12) cm, base rounded, margins crenate to serrate, each of proximal pair of teeth usually prolonged into caudate-setaceous point 3–5 mm, apex acute to acuminate, surfaces glabrous or sparsely hirsute on veins. **Inflorescences** solitary flowers or fasciculate or cymose, 2 or 3(4)-flowered. **Pedicels** 0–1 mm. **Flowers:** sepals linear-oblong to narrowly elliptic or narrowly obovate, 3–5 mm, not awned, glabrous; petals 3–5 mm; stamens 10–15(–20). **Capsules** cylindric, terete, not wing-angled, 3-valved, each valve terminated by bifurcate awn 1.5–2 mm, 20–40 × 1.5–2 mm, glabrous. $2n = 14$.

Flowering Mar–Oct. Disturbed sites; 0–100 m; introduced; N.J., Pa.; s Asia (India, Pakistan); introduced also in Africa, Australia.

Corchorus tridens is included for New Jersey and Pennsylvania based on documentation posted on the PLANTS website. No information is available on the abundance, but presumably it is rare or a waif, as other accounts do not include the species as naturalized in the flora area.

4. Corchorus hirtus Linnaeus, Sp. Pl. ed. 2, 1: 747. 1762 • Red or Orinoco jute I

Corchorus hirtus var. *glabellus* A. Gray; *C. hirtus* var. *orinocensis* (Kunth) K. Schumann; *C. orinocensis* Kunth

Plants herbs, annual. **Stems** erect, 5–10(–25) dm, puberulent in lines, hairs retrorse, blunt-tipped. **Leaves:** petiole 2–20 mm; blade narrowly lanceolate to elliptic-lanceolate, 2–7 cm, base rounded to obtuse, margins crenate-serrate, proximal teeth not prolonged, apex acute, surfaces glabrous or sparsely hirsute-strigose on veins. **Inflorescences** usually solitary flowers, sometimes fasciculate or cymose, 2- or 3-flowered. **Pedicels** 3–4 mm. **Flowers:** sepals narrowly oblong-lanceolate, 3–5(–7) mm, not awned, hirsute; petals 4–6 mm; stamens 15–25. **Capsules** cylindric, terete, not wing-angled, 2-valved, distally abruptly constricted to beaklike apex, without awns, (20–)25–45(–60) × 2 mm, sparsely to moderately strigose to strigulose. $2n = 28$.

Flowering Jun–Nov. Hammock edges, fallow fields, remnant prairie railroad rights-of-way, wet-weather ponds, lake, stream, and marsh edges, levees, floodplains, roadsides, ditches, grazed pastures, desert grasslands; 0–1500 m; introduced; Ala., Ariz., Fla., Miss., Tex.; Mexico; West Indies; Central America; South America; introduced also in Europe (France), Asia (Afghanistan, Thailand, Turkey), Africa, w Pacific Islands (Philippines), Australia.

In Florida, *Corchorus hirtus* occurs at 0–50 m, in Texas 20–200 m, in Arizona 1200–1500 m.

Attributions of *Corchorus hirtus* to Alabama mostly have been based on garden-grown plants from Mobile and Tuscaloosa counties. A single collection of naturalized plants has been seen: Mobile Co., Dauphin Island, pine forest, 17 Oct 1973, *Taylor and Taylor 15196* (BRIT). In Mississippi, it is known from collections in 1985 and 1986 from adjacent Issaquena and Sharkey counties (BRIT, VDB; these collections previously have been identified as *C. siliquosus*).

5. Corchorus siliquosus Linnaeus, Sp. Pl. 1: 529. 1753
  • Slippery burr, broomweed F

Plants herbs, annual. **Stems** erect, 3–25 dm, short-pilose in 1(2) lines, hairs erect, dense, sharp-pointed. **Leaves:** petiole 5–10 mm; blade elliptic-ovate, 0.5–2.5(–4) cm, base rounded-truncate, margins serrate, proximal teeth not prolonged, apex acute, surfaces glabrous, adaxial surface minutely pustulate-roughened. **Inflorescences** usually solitary flowers, sometimes fasciculate or cymose, 2- or 3-flowered. **Pedicels** 4–7 mm. **Flowers:** sepals narrowly oblong-lanceolate, 4–5 mm, not awned, glabrous; petals 4–6 mm; stamens 50–70. **Capsules** cylindric, flattened, not wing-angled, subtruncate distally, 2-valved, each valve with 2 awns 1.5–3 mm, 20–30(–60) × 2–3 mm, minutely hirtellous to hispidulous. $2n = 28$.

Flowering Mar–Oct(–Dec). Canal banks, glade edges, hammocks, hammock and swamp edges, beaches, open pine woodlands with limestone outcrops and solution holes, rock outcrops, roadsides; 0–50 m; Fla.; Mexico; West Indies; Central America; South America.

Corchorus siliquosus is native in Collier, Miami-Dade, and Monroe counties. Reports of it from Alabama have been based on a garden-grown plant from Mobile County, and attribution to Mississippi has been based on misidentifications of *C. hirtus*.

6. TRIUMFETTA Linnaeus, Sp. Pl. 1: 444. 1753; Gen. Pl. ed. 5, 203. 1754 • Burrbark [For Giovanni Battista Triumfetti, 1658–1708, Italian botanist, director of the botanical garden in Rome]

Guy L. Nesom

Bartramia Linnaeus, name rejected

Herbs, annual or perennial, **subshrubs, [trees],** taprooted. **Stems** erect, usually much-branched, hairy, hairs stellate or 1-rayed and apparently simple. **Leaves** petiolate; stipules subpersistent, linear-lanceolate to lanceolate; blade ovate or broadly ovate to oblong-elliptic, lanceolate-elliptic, or ovate-elliptic, rhombic-ovate, or suborbiculate, often palmately 3–5-lobed, herbaceous, margins serrate to dentate, surfaces usually stellate-pubescent, palmately 3–7-veined. **Inflorescences** axillary or terminal, cymose [solitary flowers or fasciculate], 2 or 3(–5)-flowered, simple or ultimately 1–4-branched, clusters leaf-opposed at axils, sometimes appearing subspicate. **Flowers:** sepals 5, narrowly oblong to linear [triangular-lanceolate], valvate, often cucullate, commonly with hornlike subapical apiculum, stellate-pubescent to glabrate; petals 5 [absent or relatively small], yellow, narrowly oblanceolate to oblong or linear-oblong, usually with basal, adaxial glands; stamens 5–25[–60], on androgynophore; staminodes 0; ovary 2–4[5]-locular; ovules 1 or 2 per locule; styles 1, simple, short-cylindric; stigmas 2–5-lobed. **Capsules** globose to ovoid-globose or ovoid, 2 or 3[–6]-valved, spiny, spines uncinate [straight], glabrous or hairy, indehiscent [dehiscence loculicidal]. **Seeds** 1 or 2 per locule, irregularly ovoid to obovoid or pyriform, smooth. $x = 8$.

Species 100–150 (3 in the flora): se United States, Mexico, West Indies, Central America, South America, Asia, Africa.

Plants of *Triumfetta* vary in habit, in leaf size, shape, and vesture, and in some floral features. Leaves are shed or smaller during stress (R. G. Collevatti et al. 1998). Fruits usually are necessary for unambiguous identification.

SELECTED REFERENCES Fryxell, P. A. 1998. A synopsis of the neotropical species of *Triumfetta* (Tiliaceae). In: P. Matthew and M. Sivadasan, eds. 1998. Diversity and Taxonomy of Tropical Flowering Plants. Calicut. Pp. 167–192. Lay, K. K. 1950. The American species of *Triumfetta* L. Ann. Missouri Bot. Gard. 37: 315–395.

1. Capsules: spines glabrate, capsule surfaces densely tomentose-pilose . . . . 2. *Triumfetta rhomboidea*
1. Capsules: spines retrorsely barbed or densely hirsute on 1 side, capsule surfaces glabrous or sparsely minutely hirtellous or densely tomentose.
 2. Herbs; capsules ovoid, densely tomentose, spines densely hirsute on 1 side; leaf surfaces: abaxial moderately hairy with mix of stellate hairs (laminae) and simple hairs (veins) . . . . 1. *Triumfetta pentandra*
 2. Subshrubs; capsules globose to slightly ovoid, glabrous or sparsely minutely hirtellous, spines retrorsely barbed; leaf surfaces: abaxial (veins and laminae) densely stellate-pubescent . . . . 3. *Triumfetta semitriloba*

1. **Triumfetta pentandra** A. Richard in J. B. A. Guillemin et al., Fl. Seneg. Tent., 93, plate 19. 1831 • Five stamen burrbark ⬚I

Herbs, annual. **Stems** erect, branched, 2.5–6 dm, stellate-pubescent. **Leaves:** petiole 1–5 cm; proximal blades rhombic-ovate, palmately 3-lobed, 4.5–9 cm, distal ovate-lanceolate, not lobed, base cuneate to obtuse, margins unequally coarsely serrate, apex acute to acuminate, surfaces: abaxial moderately hairy with mix of stellate hairs (laminae) and simple hairs (veins), adaxial sparingly simple-hairy, 5-veined from base. **Inflorescences** mostly 2 or 3 per axil, often subspicate; peduncle 1–2 mm. **Pedicels** 0.5–1 mm. **Flowers:** sepals narrowly oblong, subapically appendaged, 2–3 mm, sparsely stellate abaxially; stamens 5(–10); ovaries 2-locular. **Capsules** ovoid, 2.5–3 mm, surface densely tomentose; spines uncinate, densely hirsute on 1 side.

Flowering Aug–Sep. Disturbed sites, roadsides, pastures, ditch banks; 10–50 m; introduced; Fla.; Africa; introduced also in South America, Asia (India, Pakistan, Sri Lanka, Taiwan), Atlantic Islands (Cape Verde Islands), Australia.

Triumfetta pentandra is known from Baker and DeSoto counties.

2. **Triumfetta rhomboidea** Jacquin, Enum. Syst. Pl., 22. 1760 • Diamond burrbark ⬚F ⬚I ⬚W

Bartramia indica Linnaeus; *Triumfetta angulata* Lamarck; *T. velutina* Vahl

Herbs, annual. **Stems** erect, branched, 3–18 dm, stellate-pubescent to glabrate. **Leaves:** petiole 1–5(–7) cm; proximal blades broadly ovate-orbiculate, rhombic, elliptic, or broadly ovate, usually palmately 3-lobed, 3–9(–15) cm, distal ovate to ovate-lanceolate or oblong-lanceolate, not lobed, base broadly cuneate or rounded to cordate or truncate, margins irregularly serrate, apex acute, surfaces: abaxial densely stellate-pubescent, adaxial sparsely stellate-pubescent, 3–5-veined from base. **Inflorescences** 3–5(6) per axil, often subspicate; peduncle 1–3 mm. **Pedicels** 0.5–2 mm. **Flowers:** sepals narrowly oblong to linear-oblong, subapically appendaged, 4–5.5 mm, stellate abaxially; stamens 10–15; ovaries 3- or 4-locular. **Capsules** globose to ovoid-globose, 2.5–3 mm, surface densely tomentose-pilose; spines uncinate, glabrate, sparsely and minutely stipitate-glandular. $2n = 32$.

Flowering Oct–Dec. Roadsides, disturbed shrubby areas; 10–50 m; introduced; Ala., Fla.; West Indies; introduced also in South America, Asia, Africa.

The type of *Triumfetta indica* Lamarck is not conspecific with that of *Bartramia indica* and blocks transfer of *B. indica* to *Triumfetta. Triumfetta bartramia* Linnaeus, which pertains here, is an illegitimate, superfluous name, because Linnaeus cited the earlier *B. indica.*

Triumfetta rhomboidea is known from Florida only by collections from Broward, DeSoto, and Jackson counties, and from Alabama only by a collection from Houston County.

3. **Triumfetta semitriloba** Jacquin, Enum. Syst. Pl., 22. 1760 • Mosote, burweed, Sacramento burrbark ⬚W

Subshrubs. Stems erect, branched, 3–17 dm, stellate-pubescent, sometimes with simple hairs intermixed. **Leaves:** petiole 0.5–6(–9) cm; proximal blades broadly ovate to rhombic-ovate, sometimes obscurely 3-lobed, 3–8 cm, distal usually oblong, base rounded, rarely shallowly cordate, margins irregularly serrate-dentate, apex acuminate, surfaces: abaxial (veins and lamina) densely and uniformly stellate-pubescent, 5–7-veined from base. **Inflorescences** 2 or 3 per axil, usually foliar-bracteate; peduncle 1–2(–3) mm. **Pedicels** 2–3 mm. **Flowers:** sepals linear, subapically appendaged, 6 mm, stellate abaxially; stamens 15–25; ovaries 3(4)-locular. **Capsules** globose to slightly ovoid, 3–4 mm, surface glabrous or sparsely minutely hirtellous and sparsely and inconspicuously glandular with minute viscid hairs; spines uncinate, retrorsely barbed. $2n = 32$.

Flowering May–Dec. Hammocks and hammock margins, mulberry woodlands, thicket edges, disturbed sites, old groves; Fla., Ga.; Mexico; West Indies; Central America; South America; introduced in Asia (Taiwan), w Pacific Islands (Hawaii, Micronesia, Philippines), Australia.

Triumfetta semitriloba is known from the southern counties of Florida, north to Manatee and Okeechobee counties, and in Georgia only from Baker and Camden counties.

SELECTED REFERENCE Leitão C. A. E. et al. 2005. Anatomy of the floral, bract, and foliar nectaries of *Triumfetta semitriloba* (Tiliaceae). Canad. J. Bot. 83: 279–286.

e. MALVACEAE Jussieu subfam. BYTTNERIOIDEAE Burnett, Outlines Bot., 821, 1119. 1835 (as Buttneridae)

Margaret M. Hanes

Herbs, subshrubs, shrubs [trees]. **Leaf blades** usually unlobed, rarely lobed (*Hermannia*), margins serrate, dentate, or entire. **Inflorescences** axillary, terminal, or leaf-opposed, usually umbellate, cymose, or paniculate, rarely solitary flowers. **Flowers** bisexual; epicalyx present or absent; sepals usually persistent, 5, distinct or basally connate, rarely petaloid, nectaries usually absent; petals 5, distinct, base sometimes clawed; androgynophore usually absent; stamens 5[–10], connate into tube, opposite petals; anthers 2- or 3-thecate; staminodes antisepalous or absent (*Hermannia* and *Waltheria*); gynoecium syncarpous. **Fruits** capsules or schizocarps, dehiscence loculicidal or septicidal. **Seeds** (1–)5–10, usually glabrous.

Genera 26, species ca. 650 (4 genera, 16 species in the flora): North America, Mexico, West Indies, Central America, South America, Asia, Afric, Australia; pantropical.

Byttnerioideae comprise five tribes previously ascribed to the family Sterculiaceae (B. A. Whitlock et al. 2001). It is sister to Grewioideae; together they represent the earliest branching taxa in the family (C. Bayer et al. 1999; R. Nyffeler et al. 2005).

SELECTED REFERENCE Whitlock, B. A., C. Bayer and D. A. Baum. 2001. Phylogenetic relationships and floral evolution of the Byttnerioideae ("Sterculiaceae" or Malvaceae s.l.) based on sequences of the chloroplast gene, *ndh*F. Syst. Bot. 26: 420–437.

1. Anthers 3-thecate; staminodes present; sepals distinct or connate . . . . 7. *Ayenia*, p. 202
1. Anthers 2-thecate; staminodes usually absent; sepals connate.
  2. Ovules 4–14 per locule . . . . 8. *Hermannia*, p. 207
  2. Ovules 2 per locule.
    3. Ovaries 5-locular . . . . 9. *Melochia*, p. 210
    3. Ovaries 1-locular . . . . 10. *Waltheria*, p. 212

7. AYENIA Linnaeus, Kongl. Svenska Vetensk. Acad. Handl. 17: 23, plate 2. 1756 • [For Louis de Noailles, 1713–1793, first Duc d'Ayen]

Laurence J. Dorr

Nephropetalum B. L. Robinson & Greenman

Subshrubs or shrubs [trees], erect, ascending, or decumbent; taprooted. **Stems** unarmed, hairy [glabrous], hairs simple, [bifurcate], fasciculate, or stellate, [glandular]. **Leaves** petiolate; stipules persistent or caducous; blade: margins serrate, dentate, or crenate [entire], palmately 3–7-veined from base. **Inflorescences** axillary or terminal, cymose, 1–11-flowered; epicalyx absent. **Flowers** bisexual [rarely unisexual or cleistogamous]; sepals basally connate, basally abaxially glandular-hairy; petals clawed, claws filiform, lamina rhombic, deltate, reniform, or ± triangular, base divided into 2 lobes or not, lobe margins entire or erose, apex notched, entire, or with 2 widely spaced teeth, surfaces glabrous or hairy, abaxial surface appendaged or not, appendage ± filiform, cylindric, or clavate, inserted in middle of lamina distally, petals inflexed toward center of flower and connivent to apex of staminal tube such that when viewed from above corolla resembles a disc with staminodes, style, and stigma, if exserted, protruding from center; androgynophore present [absent]; staminodes 5, connate into cylindric or ± campanulate tube; anthers 3-thecate, sessile, subsessile, or filaments distinct, each anther covered by petal lamina; gynoecium syncarpous, 5-carpellate; ovary 5-locular, globose or pyriform, shortly stipitate, minutely tuberculate or mammillate; placentation axile; ovules 2 per locule, 1 aborting; styles simple, included or exserted; stigmas capitate, 5-lobed, lobes rounded, minute or indistinct. **Fruits** capsules, often pendulous, oblate or subspheric, 5-locular, dehiscence septicidal and then loculicidal, hairy [glabrous], prickled. **Seeds** 1 per locule, ovoid, smooth, wrinkled, or tuberculate; endosperm absent; cotyledons leafy, folded and rolled around hypocotylar axis. $x = 10$.

Species ca. 80 (7 in the flora): United States, Mexico, West Indies, Central America, South America (to Argentina and Paraguay).

The name *Ayenia pusilla* Linnaeus has been misapplied to some species found in the flora area. C. L. Cristóbal (1960) selected the South American element in the Linnaean protologue of this name as lectotype; the name is now correctly applied only to a species in Ecuador and Peru.

Molecular data (B. A. Whitlock and A. M. Hale 2011) indicate that a monophyletic *Ayenia* is embedded within a paraphyletic *Byttneria* Loefling and that the monospecific, Brazilian endemic *Rayleya* Cristóbal is sister to a combined clade of *Ayenia* and *Byttneria*. The taxonomic implications of these relationships have not been fully resolved. A relatively small proportion of species in this *Ayenia/Byttneria/Rayleya* clade have been sampled, and the African genus *Megatritheca* Cristóbal, which shows a combination of morphological features usually associated with either *Ayenia* or *Byttneria*, has not been included in molecular analyses.

The structure of *Ayenia* flowers is complicated, and the terminology applied to the various parts has not been consistent (see L. J. Dorr 1996 and included references). In particular, the terms claw and lamina as applied to petals have been used to describe different structures. Here the term claw is used in the traditional topographical sense as the narrowed base of a petal and lamina is applied to the expanded, rhombic, deltate, or reniform portion of the petal, which may be lobed or not, and which may have an abaxial appendage (sometimes termed a gland or ligule) or not. W. Leinfellner (1960), using comparative and developmental data, argued that the rhombic, deltate, or reniform portion of the petal (lamina) is homologous with the distal part of the claw and that the lamina is reduced to an appendage or absent.

SELECTED REFERENCES Cristóbal, C. L. 1960. Revisión del género *Ayenia* (Sterculiaceae). Opera Lilloana 4: 1–230. Whitlock, B. A. and A. M. Hale. 2011. The phylogeny of *Ayenia*, *Byttneria*, and *Rayleya* (Malvaceae s.l.) and its implications for the evolution of growth forms. Syst. Bot. 36: 129–136.

1. Petals: base of lamina lobed, apex entire or with 2 widely spaced teeth (not notched), abaxial appendage absent.
    2. Cymes borne on short shoots (brachyblasts); sepals not reflexed at anthesis; petal lamina apex with 2 widely spaced teeth; stamen filaments present . . . . .6. *Ayenia microphylla*
    2. Cymes not borne on short shoots (brachyblasts); sepals reflexed at anthesis; petal lamina apex entire; stamen filaments absent or nearly so.
        3. Leaf blades broadly ovate to ovate-lanceolate, slightly 3–5-lobed; petal lobe margins erose; androgynophore 0.5–1 mm; Arizona. . . . . 4. *Ayenia jaliscana*
        3. Leaf blades ovate to ovate-lanceolate, unlobed; petal lobe margins not erose; androgynophore 0.2–0.3 mm; Texas. . . . . 5. *Ayenia limitaris*

1. Petals: base of lamina attenuate on claw, apex notched, abaxial appendage present.
    4. Sepals caducous (not present in young fruit); capsule prickles 0.1–0.3 mm . . . . 1. *Ayenia compacta*
    4. Sepals persistent; capsule prickles 0.3–1.5 mm.
        5. Leaf blades: proximals and distals similar in shape and size, apex cuspidate; petiole (0.1–)0.2–0.5 cm; Florida . . . . 2. *Ayenia euphrasiifolia*
        5. Leaf blades: proximals and distals dissimilar in shape and size (sometimes only slightly), apex acute or subacute; petiole 0.4–1(–1.5) cm; Arizona, New Mexico, Texas.
            6. Leaves: blades of proximals ovate to orbiculate, distals oblong to ovate-lanceolate or linear, base rounded to truncate, surfaces usually stellate-puberulent, sometimes glabrescent . . . . 3. *Ayenia filiformis*
            6. Leaves: blades of proximals broadly ovate to orbiculate, distals ovate to oblong-ovate or oblong-lanceolate, base cordate, surfaces moderately to sparingly hirsute, hairs usually simple, bifurcate, and fasciculate, sometimes also stellate . . . . 7. *Ayenia pilosa*

1. **Ayenia compacta** Rose, Contr. U.S. Natl. Herb. 8: 321. 1905 • California ayenia

Ayenia californica Jepson

Subshrubs, erect or decumbent, (0.1–)0.5–0.8(–1) m. **Stems** hairy, hairs stellate, arms appressed. **Leaves:** petiole 0.5 (–0.8) cm; blades of proximal leaves orbiculate to broadly ovate, 0.3–0.5 × 0.3–0.5 cm, distal ovate to ovate-lanceolate, unlobed, 1–1.5 × 0.5–0.9 cm, base rounded to slightly cordate, margins serrate, ciliate, apex obtuse to subacute, 3(–5)-veined from base, surfaces stellate-hairy. **Cymes** axillary, not borne on short shoots (brachyblasts), 1(2)-flowered; peduncle 1–2.5 mm. **Flowers:** sepals caducous, not present in young fruit, not reflexed at anthesis, ovate-lanceolate, 1.5–3 mm, sparingly stellate-pubescent abaxially; petal claws 2–3 mm, lamina rhombic, 1–1.5 × 1.5–3 mm, base attenuate on claw, margins entire, apex notched, surfaces hairy abaxially, hairs simple, multicellular, abaxial appendage cylindric, 0.2–0.3 mm; androgynophore 1–1.5 mm; stamen filaments present; stigmas scarcely exserted. **Capsules** subglobose, (3.5–)4.5–5 × 3–4.5 mm, sparingly stellate-pubescent, prickles 0.1–0.3 mm. **Seeds** 2–2.5(–2.9) mm, slightly to densely tuberculate. $2n = 20$.

Flowering and fruiting spring–early fall. Rocky slopes, gravelly or sandy washes, dry canyons; 100–1200 m; Ariz., Calif.; Mexico (Baja California, Baja California Sur, Sonora).

In the United States, *Ayenia compacta* is known from southernmost California and Arizona; in Mexico, it is found on the Baja California peninsula and in Sonora on islands in the Gulf of California.

2. **Ayenia euphrasiifolia** Grisebach, Cat. Pl. Cub., 29. 1866 (as euphrasifolia) • Dwarf or eyebright ayenia

Ayenia tenuicaulis Urban

Subshrubs, decumbent to ascending, 0.1–0.4 m. **Stems** hairy, hairs mostly simple, often retrorse, sometimes also fasciculate and/or stellate. **Leaves:** petiole (0.1–)0.2–0.5 cm; blade orbiculate to ovate, unlobed, 0.4–1.4 × 0.1–0.7 cm, proximal and distal similar in shape and size, base truncate to cordate or rounded, margins serrate, each tooth terminating in 1 or 2 bristles (bristles often absent on older leaves), apex cuspidate, 3–5-veined from base, surfaces: abaxial sparingly hairy, hairs simple, bifurcate, fasciculate, and/or stellate, adaxial glabrescent. **Cymes** axillary, not borne on short shoots (brachyblasts), 1(2)-flowered; peduncle 2–3 mm. **Pedicels** 2–3 mm. **Flowers:** sepals persistent, not reflexed at anthesis, lanceolate to elliptic, 1.8–2 mm, sparingly hairy abaxially, hairs simple, bifurcate, and fasciculate; petal claws 2–3 mm, lamina rhombic to ± triangular, 1 × 1 mm, base attenuate on claw, margins entire, apex notched, surfaces slightly hairy abaxially, hairs minute, simple, or glandlike, abaxial appendage cylindric to ± clavate, 0.5 mm; androgynophore (1–)2.2–3 mm; stamen filaments present; stigmas exserted. **Capsules** oblate to subspheric, 4–5 × 4–5 mm, sparingly stellate-pubescent, prickles 0.3–0.5 mm. **Seeds** 2 mm, densely tuberculate.

Flowering and fruiting year-round. Marl over limestone, pinelands, sandy scrub, rocky flats, waste places; 0–10 m; Fla.; West Indies (Bahamas, Cuba).

In Florida, *Ayenia euphrasiifolia* is found in the Florida Keys, south and east of Lake Okeechobee, and near Tampa Bay on the mainland.

Three subspecies, two of them endemic, are recognized in Cuba but are not recognized here; they are based on rather weak morphological characters (habit, vesture,

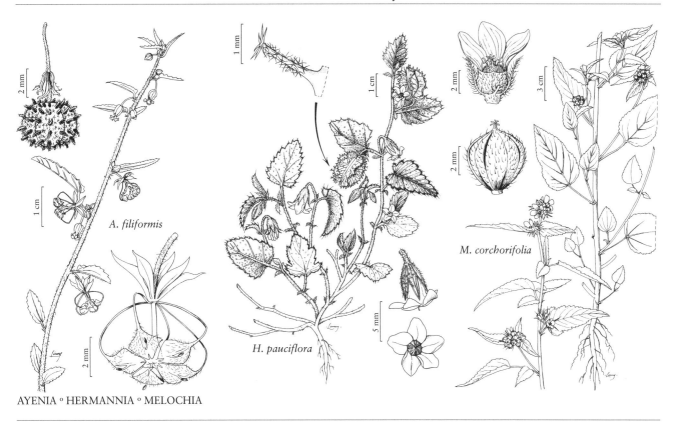

A. filiformis

H. pauciflora

M. corchorifolia

AYENIA ° HERMANNIA ° MELOCHIA

and leaf texture) ostensibly correlated with vegetation type and soil substrate. If subspecies are accepted, plants in the flora area are subsp. *euphrasiifolia*.

3. **Ayenia filiformis** S. Watson, Proc. Amer. Acad. Arts 24: 42. 1889 • Trans-Pecos ayenia F

Ayenia cuneata Brandegee; *A. reflexa* Brandegee

Subshrubs, decumbent or erect, 0.2–0.5(–0.9) m. **Stems** hairy, hairs simple and retrorse or simple and stellate. **Leaves:** petiole 0.4–1(–1.5) cm; blades of proximal leaves ovate to orbiculate, 0.5–2(–3.2) × 0.2–1 (–1.5) cm, distal oblong-ovate to ovate-lanceolate or linear, unlobed, 1–4.3(–7) × 0.2–1(–1.5) cm, base rounded to truncate, margins serrate to doubly serrate or dentate, sometimes ciliate, apex subacute, 3(–5)-veined from base, surfaces usually stellate-puberulent, sometimes glabrescent. **Cymes** axillary, not borne on short shoots (brachyblasts), 1–3(–11)-flowered; peduncle 2–4(–7) mm. **Pedicels** 2–4(–7) mm. **Flowers:** sepals persistent, not reflexed at anthesis, ovate-lanceolate, 1–3 mm, stellate-hairy abaxially; petal claws 2–2.5(–6) mm, lamina subtriangular or rhombic, 1–2.5 × 1–2 mm, base ± attenuate on claw, margins entire, apex notched, surfaces hairy abaxially, hairs simple, multicellular,

abaxial appendage filiform to slightly clavate, 0.7 mm; androgynophore 2–3 mm; stamen filaments present; stigmas slightly exserted. **Capsules** oblate, 2–4 × 5 mm, densely stellate-hairy, prickles 0.5–1 mm. **Seeds** 2–3 mm, tuberculate.

Flowering and fruiting spring–early fall. On limestone and granite soil, steep rocky slopes, canyons, sandy washes; 1000–1500 m; Ariz., N.Mex., Tex.; Mexico (Chihuahua, Sinaloa, Sonora, Zacatecas).

In Texas, *Ayenia filiformis* is known from the trans-Pecos region.

4. **Ayenia jaliscana** S. Watson, Proc. Amer. Acad. Arts 26: 133. 1891

Shrubs, erect, to 2 m. **Stems** densely stellate-hairy to glabrate, hairs simple, fasciculate, and/or stellate, sometimes minutely stellate, not appressed. **Leaves:** petiole (0.4–)2–3.5 cm; blade broadly ovate to ovate-lanceolate, slightly 3–5-lobed, 2.5–11 × (1.5–)2.5–7.5 cm, noticeably reduced in size distally on stem, base truncate to slightly cordate, margins dentate-serrate, apex acute to attenuate, 5–7-veined from base, surfaces: abaxial densely to sparingly hairy, hairs simple, bifurcate, and/or stellate, not appressed, adaxial sparingly hairy to glabrescent, hairs simple, bifurcate, and/or stellate, not appressed. **Cymes** axillary or terminal, not borne on short shoots (brachyblasts), 3–5-flowered; peduncle 10–20 mm. **Pedicels** 5–10 mm. **Flowers:** sepals caducous, not present in young fruit, reflexed at anthesis, narrowly ovate-lanceolate, 3–3.5 mm, sparingly stellate-hairy abaxially; petal claws 2–2.5 mm, lamina ± rhombic, 1 × 1–1.5 mm, base deeply lobed, lobes oblong, margins erose, apex entire, glabrous, abaxial appendage absent; androgynophore 0.5–1 mm; stamen filaments absent; stigmas slightly exserted. **Capsules** oblate to subspheric, 6 × 7 mm, densely short stellate-pubescent, prickles 0.5–1 mm. **Seeds** 4 mm, tuberculate.

Flowering and fruiting late summer–early winter. Rocky slopes, canyon bottoms, desert grasslands; 1100–1200 m; Ariz.; Mexico (Baja California, Baja California Sur, Chihuahua, Sinaloa, Sonora).

5. **Ayenia limitaris** Cristóbal, Opera Lilloana 4: 61, fig. 17. 1960 • Texas or Rio Grande ayenia, Tamaulipan kidney petal, kidney petal [C]

Nephropetalum pringlei B. L. Robinson & Greenman 1896, not *Ayenia pringlei* Cristóbal 1960

Shrubs, erect, to 1.5 m. **Stems** hairy to glabrate, hairs minutely stellate, ± appressed. **Leaves:** petiole to 2.5(–4) cm; blade ovate to ovate-lanceolate, unlobed, (3–)5–12.5 × (1.7–)3.2–7.5 cm, noticeably reduced in size distally on stem, base rounded to cordate, margins irregularly dentate to dentate-crenate, apex acute to acuminate, 7-veined from base, surfaces: abaxial tomentulose, adaxial sparingly stellate-pubescent. **Cymes** axillary, not borne on short shoots (brachyblasts), 2 or 3(4)-flowered; peduncle to 10 mm. **Pedicels** 3–7 mm. **Flowers:** sepals caducous, not persistent in fruit, reflexed at anthesis, ovate-lanceolate, 3 mm, stellate-pubescent abaxially; petal claws 2.5 mm, lamina reniform, 1 × 1.5 mm, base deeply lobed, lobes ovate-rounded, margins ± entire, apex entire, surfaces glabrous, abaxial appendage absent; androgynophore 0.2–0.3 mm; stamen filaments absent or nearly so; stigmas slightly exserted. **Capsules** oblate, 6–7 × 8–10 mm, densely short stellate-pubescent, prickles 1–1.5 mm. **Seeds** 3–5 mm, tuberculate.

Flowering summer; fruiting summer–fall. Subtropical thorn woodlands or tall shrublands, calcareous or fine sandy loam to heavy clay alluvial soil; of conservation concern; 0–30 m; Tex.; Mexico (Coahuila, Tamaulipas).

Ayenia limitaris was designated an endangered species by the U.S. Fish and Wildlife Service in 1994. In the flora area, the species is known only from near the mouth of the Rio Grande.

Ayenia limitaris is in the Center for Plant Conservation's National Collection of Endangered Plants.

6. **Ayenia microphylla** A. Gray, Smithsonian Contr. Knowl. 3(5): 24. 1852 • Dense or little-leaf ayenia

Shrubs, erect or spreading, 0.2–0.6 m. **Stems** hairy, hairs stellate, arms appressed. **Leaves:** petiole 0.4–1.2 cm; blades of proximal leaves orbiculate to suborbiculate, 0.5 × 0.4 cm, distal ovate to narrowly ovate, unlobed, 0.5–2.1(–3) × 0.4–1.8 (–2) cm, base rounded or obscurely subcordate, margins serrate, stellate-hairy (not ciliate), apex acute to rounded, 3(–5)-veined from base, surfaces minutely, densely stellate-hairy. **Cymes** axillary, borne on short shoots (brachyblasts), 1- or 2-flowered; peduncle to 4 mm. **Pedicels** to 4 mm. **Flowers:** sepals caducous, not reflexed at anthesis, ovate-lanceolate, 2.5–3.2 mm, densely stellate-pubescent abaxially; petal claws 4 mm, lamina reniform-deltate, 1.4–1.6 × 1.5 mm, base lobed, lobes ± rectangular, margins entire, apex with 2 widely spaced teeth, surfaces glabrous, abaxial appendage absent; androgynophore 0.5–1 mm; stamen filaments present; stigmas scarcely exserted. **Capsules** subspheric, 4–5 × 4–5 mm, puberulent to densely stellate-pubescent, prickles 0.5 mm. **Seeds** 3 mm, tuberculate.

Flowering and fruiting spring–early fall. Dry limestone, igneous rocky slopes; 600–1400 m; Ariz., N.Mex., Tex.; Mexico (Coahuila).

Ayenia microphylla can be distinguished from other *Ayenia* species in the flora area by the presence of short shoots (brachyblasts) from which leaves and inflorescences emerge. In Texas, the species is known from the trans-Pecos region.

7. **Ayenia pilosa** Cristóbal, Opera Lilloana 4: 185, fig. 65. 1960 • Dwarf ayenia

Subshrubs, decumbent, 0.1–0.2 (–0.3) m. **Stems** hairy, hairs simple and retrorse, or simple, fasciculate, and stellate. **Leaves:** petiole 0.5–1 cm; blades of proximal leaves broadly ovate to orbiculate, 0.5–1.3 × 0.5–1.1 cm, distal ovate to oblong-ovate or oblong-lanceolate, unlobed, 0.5–2(–3.5) × 0.4–1.4(–1.7) cm, base cordate, margins serrate to doubly serrate, ciliate with 1+ bristles per tooth, apex acute, 3–5-veined from base, surfaces moderately to sparingly hirsute, hairs usually simple, bifurcate, and fasciculate, sometimes also stellate. **Cymes** axillary, not borne on short shoots (brachyblasts), (1)2- or 3-flowered; peduncle 2–3 mm. **Pedicels** 1.5–2 mm. **Flowers:** sepals persistent, not reflexed at anthesis, ovate, 2 mm, sparingly stellate-hairy abaxially; petal claws 4 mm, lamina rhombic, 1 × 1 mm, base attenuate on claw, margins entire, apex notched, surfaces sparingly hairy abaxially, hairs simple, multicellular, abaxial appendage cylindric to slightly clavate, 0.5 mm; androgynophore 1.5 mm; stamen filaments present; stigmas slightly exserted. **Capsules** subspheric, 3–3.5 × 3.5 mm, sparingly stellate-hairy, prickles 0.5 mm. **Seeds** 2–2.5 mm, sparingly tuberculate.

Flowering and fruiting year-round. Edges of thickets; 600–1200 m; N.Mex., Tex.; Mexico (Coahuila, San Luis Potosí, Tamaulipas).

Leaf shape and vesture are the principal characters used to distinguish *Ayenia pilosa* from *A. filiformis*. Where they co-occur in southwestern Texas there are intermediates in which the leaves are oblong-ovate as in *A. pilosa* but the vesture is more like that of *A. filiformis*.

8. HERMANNIA Linnaeus, Sp. Pl. 2: 673. 1753; Gen. Pl. ed. 5, 304. 1754 • Burstworts [For Paul Hermann, 1646–1695, German-born Dutch botanist and explorer]

Janice G. Saunders

Subshrubs, [shrubs], prostrate to erect; taprooted. **Stems** hairy, hairs usually stellate, sometimes intermixed with capitate-glandular and subsessile glandular hairs. **Leaves** petiolate; stipules deciduous, foliaceous, narrowly dimidiate-lanceolate or narrowly dimidiate-ovate, triangular, margins simple-bristled; blade usually unlobed, rarely lobed, margins dentate or serrate. **Inflorescence subunits** axillary, cymose, 1–7-flowered, epicalyx absent; bracts persistent. **Pedicels** reflexed. **Flowers** sweet-smelling, pendulous, cernuous, approach herkogamous; sepals persistent, not accrescent in fruit, basally connate ¼–½ length, nectaries present; petals late-deciduous, distinct, not cucullate, involutely clawed, narrowly obovate or pandurate-concave, 6–12 mm, subglabrous or distally glandular-capitate abaxially; androgynophore absent; gynophore present; stamens persistent, opposite petals, included; staminodes 0; filaments ligulate, very compressed, adnate to petal base and gynophore or ovary base, distally free, incurved, not abruptly dilated, expanded region narrowly oblong from base to above anther base, apex acuminate or acute, glabrous; anthers 2-thecate, lanceolate, [1–]2–3.5[–10] mm, inflexed, connivent to style, longitudinally dehiscent; thecae with rim ciliate from simple hairs, apex acuminate, slightly twisted, gland at apex only (*H. texana*) or also at theca base (*H. pauciflora*); gynoecium syncarpous, 5-carpellate, stipitate, 5-angled, locules opposite sepals; ovary 5-locular; ovules 4–14 per locule, ascending or horizontal, anatropous or amphitropous; styles persistent, 5, shortly exserted, presumably connate at anthesis (connate, distinct, or partially distinct dried), filiform; stigmas inconspicuous, terete and 1-dentate (acute) or filiform and often few-minutely papillate at apex, rarely truncate, inconspicuous. **Fruits** capsules, 5-locular, in apical view 5-angled, 5-lobed, parted between angles, in lateral view emarginate at apex, margins curved, stipitate, valve margins dark-rimmed, dentate, teeth terminated by hairy tubercles not elsewhere on fruit (*H. pauciflora*) or hairy processes (*H. texana*), denser on valves, stellate-pubescent. **Seeds** 0–8 per locule, brown, crescentiform-reniform, chalazal end wider, other end acute, large-pitted; elaiosome conspicuous, white; endosperm present; embryo curved, chlorophyllous; cotyledons flat, narrowly elliptic or oblong-elliptic. *x* = 6.

Species ca. 180 (2 in the flora): sw, sc United States, Mexico, Central America (Guatemala), w Asia (Saudi Arabia), Africa, Australia; subtropical and tropical areas.

Hermannia is primarily southern African, with other species found scattered outside Africa, notably former *Gilesia biniflora* F. Mueller from Australia, distinguished by totally free filaments. Only four species are exclusively from the Americas, found in Mexico and across its national borders. One species extends north to Arizona, and one extends north to Texas. Two are found outside the flora area: *H. palmeri* Rose from rocky granitic hillsides, sandy mesas, and coastal dunes of Baja California Sur and *H. inflata* Link & Otto from tropical dry deciduous forests in mountains of southern Mexico and northern Guatemala.

I. C. Verdoorn (1980) placed the only two fringed-capsuled species in southern Africa in subg. *Hermannia*, the basal subgenus, and she related them to the two American species that are fringed-capsuled: *H. texana* (Texas and Mexico) and *H. palmeri* (Baja California Sur). Both species from the flora area would lie also within subg. *Hermannia* in her key to taxa by possession of "narrowly oblong stamen filaments with the expanded portion of the filaments overlapping the anther bases."

The coherence of styles and connivance of anthers at anthesis (M. Jenny 1985, 1988), nectaries (S. Vogel 2000), and ovule/seed number per carpel need further study in the field to investigate possible cryptic floral differentiation between plants of *Hermannia* that would indicate the initial stages of the well-developed floral and pollen distylous dimorphisms of *Melochia* and *Waltheria*.

SELECTED REFERENCE   Rose, J. N. 1897. A synopsis of American species of *Hermannia*. Contr. U.S. Natl. Herb. 5: 130–131.

1. Styles 3.9–5.8 mm; leaf blades olive, with terminal trichome at each tooth apex never simple, rather stellate or stellate-bristled and slightly exserted from stellate trichomes of tooth sides; distal petal lamina densely capitate-glandular abaxially; capsules: endocarp corneous, rigid, valve margins fringed, processes filiform, flexible, 0.9–3.2 mm; terminal trichome of fringe process with trichome rays ascending, fine, and only slightly firmer or slightly longer than trichome rays of hairs below process apex; seeds: elaiosome narrowly transversely rhombic . . . . 1. *Hermannia texana*
1. Styles 1.7–3 mm; leaf blades red to red-rimmed, with terminal trichome at each tooth apex with 1 simple thin seta or seta 2-rayed exserted to 2 times length of stellate trichomes of tooth sides; distal petal lamina subglabrous abaxially; capsules: endocarp chartaceous, slightly sclerified, flexible, valve margins tuberculate, tubercles rigid, 0.2–1.8 mm; terminal trichome of tubercle apex with trichome rays usually planar (spreading); seeds: elaiosome narrowly transversely elliptic . . . . 2. *Hermannia pauciflora*

1. **Hermannia texana** A. Gray, Gen. Amer. Bor. 2: 88, plate 135. 1849 • Texas burstwort

Plants 50–70 cm, densely stellate-tomentose to glabrescent. **Stems** usually ascending to erect, sometimes decumbent, stellate-tomentose, inconspicuously glandular. **Leaves:** stipule triangular, 3–4 × 1–2 mm, adaxially both stellate and simple hairs present; petiole 12–27 × 0.8–2 mm, eglandular; blade olive, unlobed, sometimes 1 or 2(–6)-lobuled, elliptic to oblate, narrowly ovate to widely ovate, oblong-ovate, orbiculate, or depressed-obovate, 1–6 × 1–6 cm, base usually rounded or truncate, rarely obtuse, subcordate to cordate, margins dentate to sinuate-dentate, with terminal trichome of each tooth apex never simple, rather stellate or stellate-bristled and slightly exserted from stellate trichomes of tooth sides, costal vein obscure, apex mostly broadly rounded, truncate-rounded, rounded-obtuse, or broadly obtuse, rarely acute. **Inflorescence subunits** at leaf axils, 1–7-flowered. **Pedicels** 1–3 mm.

Flowers 7–12 × 3.2–5.6 mm fresh; calyx 6.5–8.5 × 4–6.4 mm, stellate-hispidulous or stellate-tomentose and minutely glandular abaxially, with tufted trichomes at base adaxially over globular nectaries between hairless nectiferous bands, tube 2.5–4.5 × 5–7 mm, lobes widely deltate, 2–3 mm wide, margins simple-setose, apex acuminate, setose, adaxially villous; corolla urn- or barrel-shaped, 8–9 mm curved, 10–12 mm flat, aperture terminal, never opening wider, petals erect, never spreading, convolute, yellow, becoming yellow-orange to red-orange, pandurate if spread out, 8.4–12 × 3.4–5.9 mm, base clawed, claw distinct, narrowly obtriangular, 4–5.2 × 1.6–2 mm, lamina widely concave between sides (fresh), widely obovate if flattened, distally densely capitate-glandular abaxially; stamens 6.2–9 mm; filaments dark yellow, edges pale yellow, midvein prominent abaxially; stamen tube connate, 1.5–2.5 × 0.7–1.5 mm; free filaments 3.2–3.8 × 0.5–0.6 mm; anthers 2.6–3.5 × 0.6–0.9 mm; theca: base rounded, eglandular, apex acuminate, stellate-bristled, with conspicuous gland; ovary 1.7–2.3 × 1.8–2.4 mm, lanate; gynophore 1.8–2.5 mm; ovules 6–14 per locule; styles connate completely or to 1 mm below apex in fresh and rehydrated specimens, 3.9–5.8 mm, glabrous or scant simple-hirtellous near ovary; stigmas erect or ascending, apex acute, not papillate. **Capsules** broadly depressed-ovoid to broadly ovoid or broadly ellipsoid, 7–12 mm, delicately stellate-tomentose, sparsely fringed (over nerves) and valve margins densely fringed, fringe processes flexible, filiform, 0.9–3.2 mm, tomentose along length below apex and at apex, trichome rays of terminal trichome of each fringe process ascending, fine, only slightly firmer or slightly longer than trichome rays of proximal hairs below apex; stipe 1.8–3 mm; exocarp membranaceous, mesocarp chartaceous, endocarp corneous, rigid, not fissured; calyx tube 3.4–5.9 mm wide in fruit; dehiscence loculicidal. **Seeds** 0–6 per cell, 1.3–1.7 × 1.3–1.6 × 0.6–0.7 mm, sinus obtuse, shallowly triangular, covered by narrowly transversely rhombic elaiosome; embryo ± V-shaped, cotyledons narrowly elliptic, papillose and simple-hirsutulous adaxially. $2n$ = ca. 24.

Flowering and fruiting Mar–Nov. Upland prairies, juniper-oak woodlands, Tamaulipan scrub or chaparral, salt-marsh borders, coastal dunes, on granite, limestone, or dry, calcareous soil, clay-loams or black silt, rocky hillsides, streambeds, canyons; 20–700 m; Tex.; Mexico (Coahuila, Nuevo León, Tamaulipas).

Hermannia texana is found in the Texas upland prairies and coastal plain regions. It occurs usually in scattered colonies and sometimes is locally common. Plants are eaten by deer, sheep, goats, and cows.

Leaves of *Hermannia texana* are variable in shape and size. A. Gray (1850) later corrected the corolla color to cinnabar, and the corolla attitude to be convolute and erect, never spreading as originally figured and described.

Fresh flower styles are usually coherent completely or to 1 mm below the recurved apex. The flowers attract halictid and anthophorid bees. The large bee *Centris atripes* Mocsary is probably an effective pollinator (Saunders, unpubl.).

2. **Hermannia pauciflora** S. Watson, Proc. Amer. Acad. Arts 17: 368. 1882 • Santa Catalina burstwort, hierba del soldado C F

Plants 5–34 cm, sparsely [densely] stellate-pubescent to glabrescent and densely to sparsely capitate-glandular. **Stems** usually decumbent, rarely prostrate or erect, stellate-pubescent, conspicuously glandular. **Leaves:** stipule dimidiate-narrowly ovate or dimidiate-narrowly lanceolate, 2.5–5 × 0.5–2 mm, adaxially with simple, not stellate, hairs present; petiole 2–17 × 0.2–0.8 mm, glandular; blade red to red-rimmed adaxially,

usually unlobed, rarely 2–4-lobuled basally, usually narrowly ovate to deltate, rarely widely ovate, 1–4 × 1–3 cm, base often rounded or truncate, rarely subcordate, margins usually dentate, sometimes serrate, terminal trichome at each tooth apex with 1 simple thin seta or setae 2-rayed, exserted to 2 times length of stellate trichomes of tooth sides, costal vein prominent adaxially, apex usually narrowly rounded or obtuse-rounded, rarely acute. **Inflorescence subunits** at leaf axils, 1- or 2-flowered, cymes probably strictly 2-flowered rather than solitary. **Pedicels** 1–3 mm. **Flowers** 6 × 6–12 mm fresh (when fully open), 6–10 mm wide dried; calyx 3.9–7 × 3.6–7.2 mm, stellate-hirsute, capitate-glandular hairs (0.4 mm) abaxially, adaxially glabrous within proximally, nectaries globular at calyx base and stamen base, tube 1–3 × 2.7–4.3 mm, lobes narrowly triangular, 1–2 mm wide, margins setose, apex acute, bristled, adaxially villulose, margin and apex trichome rays 1–3; corolla campanulate-rotate, aperture at midlength (fresh), petals spreading from erect base, yellow, orange basally, narrowly obovate, 6–8.2 × 2–3.6 mm, base clawed, claw not very distinct, narrowly obtriangular, 2.5–3.5 × 1–1.3 mm; lamina flat, slightly reflexed apically, abaxially subglabrous, papillose, not densely capitate-glandular, with a few minute simple or 2-rayed hairs, sometimes also a few minute capitate-glandular hairs; stamens 4.6–6 mm; filaments red, edges yellow, midvein prominent abaxially; free filaments 2.2–3.2 × 0.3–0.5 mm; pseudotube yellow, 1–2 mm, segments free, imbricate laterally above base and distally adnate by midvein to ovary base, globular nectaries at base; anthers 2–2.8 × 0.5–0.6 mm, theca apex acute, singly simple-setulose at tip; thecal glands inconspicuous at apex, conspicuous at apiculate base; ovary 2.5–3.4 × 1.8–3 mm, hirsute; gynophore 0.9–2 mm; ovules 4–10 per locule; styles connate completely, or distally or basally or completely distinct but contiguous in rehydrated specimens, or very distally distinct fresh, usually reflexed at apex, hirsutulous ⅔ of length, 1.7–3 mm; stigmas minutely recurved lobules, apices truncate, usually introrsely minutely few-papillate, rarely epapillate. **Capsules** stipitate, ellipsoid to oblong or obovoid, 10–20 mm, sparsely to densely stellate-pubescent; stipe 0.9–2 mm; valve margins undulate, exclusively and densely tuberculate, tubercles rigid, stout, columnar, 0.2–1.8 mm, stellate-pubescent along length, trichome at tubercle apex stellate-hispidulous, its trichome rays in one plane (spreading), compressed; fruit wall with exocarp membranaceous, mesocarp chartaceous, endocarp chartaceous, slightly sclerified, flexible, multifissured, calyx tube 1.7–3.5 mm wide in fruit; dehiscence loculicidal. **Seeds** 0–8 per cell, 1.4–1.9 × 1.3–1.6 × 0.8–1 mm, sinus linear-oblong, covered by narrowly transversely elliptic elaiosome; embryo arcuate (C-shaped); cotyledons oblong-elliptic, papillose, trichomes absent adaxially.

Flowering and fruiting year-round. Desert scrub or grasslands, hillsides, rocky slopes, rock crevices, rocky soil, soil pockets, canyon bottoms, washes, rock outcrops, limestone, granitic, or talus soil; of conservation concern; 700–1300 m; Ariz.; Mexico (San Luis Potosí, Sonora, Tamaulipas).

Hermannia pauciflora is found in upland Sonoran desert scrub, in palo verde and saguaro cactus association of the Santa Catalina, Tucson, and Waterman mountains, and the San Pedro River floodplain in the Charleston Hills. Mexican collections are mostly from Sonora.

The flower illustration in H. W. Rickett (1966–1973, vol. 4, part 2, page 342) depicts petals erect, agreeing with the original description by S. Watson, and J. N. Rose (1897) described petals as "at most spreading"; photos sent by P. Jenkins (pers. comm.) show that distal petals above the pentagonal throat at midlength are almost flat when flowers are fully open, and anthers are included. Jenkins also reported the flowers to be attractive to cactus bees.

9. MELOCHIA Linnaeus, Sp. Pl. 2: 674 [as 774]. 1753; Gen. Pl. ed. 5, 304. 1754

• Broom-wood [Arabic *melóchich*, name for *Corchorus olitorius* Linnaeus, a salad plant in the East]

Aaron Goldberg†

Moluchia Medikus; *Physodium* C. Presl

Herbs, annual or perennial, **subshrubs, or shrubs** [trees], erect, spreading, or decumbent, taprooted. **Stems** unarmed, hairy, hairs usually simple and/or stellate, sometimes also glandular. **Leaves** petiolate; stipules usually deciduous, linear-lanceolate; blade: margins serrate, surfaces glabrous or subglabrous to tomentose, hairs usually simple and/or stellate. **Inflorescences** axillary, leaf-opposed, or terminal, umbels or spikelike or cymose glomerules, (2–)5–25-flowered; epicalyx sometimes present; bracteoles immediately subtending flower or at base of pedicel. **Flowers:** dimorphic (pistil and stamens of unequal lengths varying inversely in different plants), or monomorphic (pistil and stamens of equal lengths); sepals basally connate; petals sometimes with tuft of hair between blade and claw; androgynophore rarely present; stamens and carpels opposite petals; stamens connate at least basally; staminodes rarely present; anthers 2-thecate; ovary 5-locular; ovules 2 per locule; styles 5, distinct or basally connate, filiform; stigmas included, decurrent on style, papillose. **Fruits** capsules, pyramidal and 5-winged, or schizocarps, subglobular and not winged, dehiscence loculicidal, sometimes also septicidal, fruit then falling apart. **Seeds** 1 or 2 per locule, obovoid, with 2 flat sides and 1 rounded, smooth; endosperm present, fleshy; embryo straight; cotyledon flat. $x = 7, 9, 10$.

Species ca. 60 (4 in the flora): c, se United States, Mexico, West Indies, Central America, South America, Asia, Africa, Pacific Islands, Australia; tropical and subtropical areas.

SELECTED REFERENCE Goldberg, A. 1967. The genus *Melochia* L. (Sterculiaceae). Contr. U.S. Natl. Herb. 34: 191–363.

1. Inflorescences: bracteoles at base of pedicel; fruits pyramidal, 5-winged, 7–14 mm diam.
    2. Leaf blades glabrous or glabrate, hairs mostly simple; inflorescences usually leaf-opposed; fruits often purple-blotched . . . . 1. *Melochia pyramidata*
    2. Leaf blades tomentulose and canescent, hairs stellate; inflorescences usually axillary; fruits not purple-blotched . . . . 2. *Melochia tomentosa*
1. Inflorescences: bracteoles immediately subtending flower; fruits subglobose, not winged, 2.5–6 mm diam.
    3. Inflorescences spikelike; petals 8–11 mm; flowers dimorphic (pistil and stamens of unequal lengths varying inversely in different plants); fruits capsules . . . . 3. *Melochia villosa*
    3. Inflorescences terminal and axillary, cymose glomerules; petals 3–6 mm; flowers monomorphic (pistil and stamens of equal lengths); fruits schizocarps . . . . 4. *Melochia corchorifolia*

1. **Melochia pyramidata** Linnaeus, Sp. Pl. 2: 674 [as 774]. 1753 • Pyramid bush

Moluchia pyramidata (Linnaeus) Britton

Varieties 2 (1 in the flora): sc, se United States, Mexico, West Indies, Central America, South America.

1a. **Melochia pyramidata** Linnaeus var. **pyramidata**

Sida sabeana Buckley

Herbs, annual or perennial, **subshrubs, or shrubs,** 0.1–2 m, sometimes taprooted. **Leaves:** petiole 0.5–3 cm; blade lanceolate to ovate, 1–6 × 0.5–3 cm, surfaces glabrous or glabrate, hairs mostly simple. **Inflorescences** usually leaf-opposed, umbellate, 2–25-flowered; peduncle 1–12 mm; bracteoles 3, at base of pedicel. **Pedicels** 1–5 mm. **Flowers** dimorphic; calyx 3.5–5 mm, teeth acuminate; petals purple, pink, or violet, sometimes yellow proximally, 6.5–8 × 1.5–2.5 mm; longistylous form: stamens 3.5–4 mm, pistil 5–6 mm; brevistylous form: stamens 5–5.3 mm, pistil 3.5–4 mm. **Fruits** capsules, often purple-blotched, 5-winged, short-beaked, pyramidal, 7–9 × 8–14 mm, dehiscence loculicidal. **Seeds** 1 or 2 per locule.

Flowering and fruiting Apr–Nov. Low, disturbed ground; 0–1600 m; Fla., La., Tex.; Mexico; West Indies; Central America; South America.

2. **Melochia tomentosa** Linnaeus, Syst. Nat. ed. 10, 2: 1140. 1759 • Woolly pyramid bush

Moluchia tomentosa (Linnaeus) Britton

Varieties 4 (1 in the flora): sc, se United States, Mexico, West Indies, Central America, South America.

2a. **Melochia tomentosa** Linnaeus var. **tomentosa**

Herbs, perennial, **subshrubs, or shrubs,** to 2.5 m, sometimes taprooted. **Leaves:** petiole 0.3–2.5 cm; blade ovate to lanceolate, 1.5–6 × 0.8–4 cm, surfaces tomentulose and canescent, hairs stellate. **Inflorescences** usually axillary, sometimes terminal or leaf-opposed, cymose and umbellate, umbels 3–5-flowered; peduncle 6–8 mm; bracteoles 3, at base of pedicel. **Pedicels** 1–5 mm. **Flowers** dimorphic; calyx 5–7 × 4–6 mm, teeth deltate to ovate, acute or acuminate; petals purple, pink, or violet, (6–)8–11 × 2–3 mm; longistylous form: stamens 4–6.5 mm, pistil 7.5–10.5 mm; brevistylous form: stamens 9–9.5 mm, pistil 6 mm. **Fruits** capsules, not

purple-blotched, 5-winged, long-beaked, pyramidal, 6–9 × 7–9 mm, dehiscence loculicidal. **Seeds** 1 or 2 per locule. **2n** = 18.

Flowering spring–fall. Dry, open woodlands, brushlands, pinelands; 0–1300 m; Fla., Tex.; Mexico; Central America (Belize, El Salvador, Guatemala, Honduras, Nicaragua); West Indies; South America (Brazil, Colombia, Suriname, Venezuela).

3. **Melochia villosa** (Miller) Fawcett & Rendle, Fl. Jamaica 5: 165. 1926

Sida villosa Miller, Gard. Dict. ed. 8, Sida no. 6. 1768; *Melochia hirsuta* Cavanilles; *M. hirsuta* var. *glabrescens* (C. Presl) A. Gray; *M. serrata* (Ventenat) A. St. Hilaire & Naudin; *Riedlea glabrescens* (C. Presl) Small; *R. hirsuta* (Cavanilles) A. de Candolle; *R. serrata* Ventenat

Herbs, perennial, **rarely shrubs,** 0.3–2 m, taprooted. **Stems** erect or spreading, usually branched. **Leaves:** petiole 0.1–0.8 cm; blade lanceolate to ovate, 1–5 × 0.5–2 cm, surfaces villous, sometimes also sericeous or glabrate, hairs simple, appressed. **Inflorescences** usually terminal, sometimes axillary, spikelike, 1–3 (–12 subterminal) flowers per node; bracteoles 2 or 3 immediately subtending flower, relatively long. **Flowers** dimorphic; calyx 4–5 mm, teeth acute; petals pink, purple, or violet, 8–11 × 2–3 mm; longistylous form: stamens 4–6 mm, filaments completely connate, pistil 7–8 mm; brevistylous form: stamens 5–8 mm, pistil 4–5 mm. **Fruits** capsules, brown, subglobose, not winged, not beaked, obscurely obtusely 5-angled, 2–3.5 × 2.5–4 mm, dehiscence loculicidal, tardily septicidal, fruit falling apart. **Seeds** usually 1 per locule.

Flowering year-round. Wet or dry pinelands and savanna; 0–2300 m; Fla., Ga., La.; Mexico; West Indies (Greater Antilles); South America.

4. **Melochia corchorifolia** Linnaeus, Sp. Pl. 2: 675. 1753 • Chocolate weed, hirsute melochia, redweed

F I W

Riedlea corchorifolia (Linnaeus) de Candolle

Herbs, annual or perennial, **or subshrubs,** to 2 m, taprooted. **Leaves:** petiole 0.3–5.5 cm; blade ovate to lanceolate, 1.3–9 × 1–7 cm, glabrate, hairs simple on veins. **Inflorescences** terminal, cymose glomerules, 10–25-flowered; bracteoles 3 or 4, immediately subtending flower. **Flowers** monomorphic; calyx 2–2.5 mm, teeth acute, distant and sinuses between them rounded; petals purple, pink, or white with yellow base, 3–6 × 1–1.5 mm; stamens and pistil equal, 2.5–3.2 mm. **Fruits** schizocarps, white, pink, green, purplish, or black, subglobose, not winged, not beaked, 4–5 × 5–6 mm, dehiscence loculicidal and then septicidal, fruit falling apart. **Seeds** 1 per locule. $2n$ = 36, 46.

Flowering and fruiting (spring–)summer–fall. Agricultural lands, open pinelands, meadows, waste ground, wet or dry soil; 0–300(–1000) m; introduced; Ala., Ark., Fla., Ga., La., Miss., N.C., S.C., Tex.; Asia; Pacific Islands (East Indies, Papua New Guinea, Philippines); Australia (Queensland); introduced also in Central America (Panama), South America (Brazil), Africa.

10. WALTHERIA Linnaeus, Sp. Pl. 2: 673. 1753; Gen. Pl. ed. 5. 304. 1754 • [For Augustin Friedrich Walther, 1688–1746, German physician, anatomist, and botanist at Leipzig University]

Janice G. Saunders

Subshrubs [herbs, shrubs], prostrate to erect, taprooted. **Stems** unarmed, hairy, hairs stellate and/ or simple, sometimes also glandular. **Leaves** petiolate; stipules deciduous, narrowly triangular or linear-triangular; blade: margins dentate, serrate, or crenate. **Inflorescences** axillary, cincinni in double or compound dichasial clusters, glomerulate, or paniculiform, 5–40-flowered; epicalyx absent; bracts 4, unilateral subtending floral pair, unequal or subequal. **Flowers** sweet-smelling, sessile or subsessile, homostylous [mostly distylous]; sepals connate ½–⅔ length, nectary present; petals late-deciduous, pale yellow, bright yellow, or yellow-orange, usually darker in corolla throat, spatulate, clawed [rarely not clawed], lamina obovate, oblong, oblanceolate, or obtriangular; androgynophore absent or present, to 0.7 mm; stamens connate most of length [connate basally to entirely]; anthers 2-thecate, oblong or ovate; staminodes 0; ovary 1-locular, obovoid or obconic, stipitate or not; ovules 2 per locule; styles included, cylindric; stigmas 10–40-branched, apex slightly exserted above stamen apices and corolla. **Fruits** capsules, 1-locular, obconic, obovoid, or oblique, apically hairy, dehiscence partially or completely loculicidal. **Seeds** 1(2) per locule, brown or black, slightly laterally compressed, obovoid, obconic, or oblique, smooth or minutely granulate; endosperm present; cotyledons oblate [widely elliptic to elliptic, circular]. x = 5, 6, 7, 10, 13.

Species ca. 60 (3 in the flora): sw, se United States, Mexico, West Indies, Bermuda, Central America, South America, Asia, Africa, Atlantic Islands, Indian Ocean Islands, Pacific Islands, Australia; tropical and subtropical areas.

Waltheria albicans Turczaninow and *W. rotundifolia* Schrank are found in Mexico, and apparent hybrids of them with *W. indica* in the broad sense occur in the United States.

SELECTED REFERENCES Köhler, E. 1971. Zur Pollenmorphologie der Gattung *Waltheria* L. (Sterculiaceae). Feddes Repert. 82: 125–153. Köhler, E. 1976. Pollen dimorphism and heterostyly in the genus *Waltheria* L. (Sterculiaceae). In: I. K. Ferguson and I. Muller, eds. The Evolutionary Significance of the Exine. London. Pp. 147–161. Saunders, J. G. 1995. Systematics and Evolution of *Waltheria* (Sterculiaceae). Ph.D. dissertation. University of Texas.

1. Capsules: endocarp corneous, to 1.3 mm from apex; bracts unequal, major bract elliptic, oblong, ovate, or obovate, minor bracts lanceolate; major bract apex obtuse or rounded and (2)3-cuspidate or 3-dentate distally . . . . .1. *Waltheria detonsa*
1. Capsules: endocarp corneous, to 1.6+ mm from apex, (sometimes thin at base and along valve margins); bracts subequal, elliptic, lanceolate, or linear, apex acute, entire distally.

[2. Shifted to left margin.—Ed.]

2. Calyx tube 1.7–2.5 mm; stem nodes knobby, joints extended 0.8–1 mm; leaf surface glabrate, puberulent, drying brownish olive, dark brown, bronze, or coppery; anthers ovate dehisced; styles densely white-sericeous; capsules with dense white puberulent line at apical valve margin . . . . .2. *Waltheria bahamensis*

2. Calyx tube 2.5–2.9 mm; stem nodes even or extended to 0.5 mm; leaf surface tomentose or pubescent, drying olive; anthers oblong dehisced; styles hirsutulous distal to base; capsules without white puberulent line at apical valve margin. . . . . .3. *Waltheria indica*

1. **Waltheria detonsa** A. Gray, Smithsonian Contr. Knowl. 5(6): 24. 1853 [F]

Plants to 5.2 dm, finely minutely ash-gray tomentose, overlayer sparsely pubescent, hirsutulous. **Stems** mostly 1–2 mm wide, nodes slightly extended to 0.7 mm. **Leaves:** stipules narrowly triangular, 3–4 mm, apex geniculate or curved; petiole 0.6–0.8 mm wide; blade discolorous, abaxially gray-olive, adaxially darker, reddish brown, ovate-elliptic, to 6 × 4.5 cm, slightly resinous, base widely cuneate, truncate, or rounded, margins crenate-dentate or crenate-denticulate, apex widely obtuse, rounded, surfaces: abaxial sparsely to densely tomentose, adaxial densely tomentulose, trichome rays 0.2–0.4 mm. **Inflorescences** lax, paniculiform, lateral branches to 15 cm, clusters 5 or 6, oblong or deltate, with 1 terminal flower, the 2 or more dichasia (cincinni) arising from accessory shoots below it; bracteoles free, bracts unequal, major bract of primary cyme elliptic, oblong, ovate, or obovate, 0.8–2.2 mm wide, apex obtuse or rounded, (2)3-cuspidate or 3-dentate, sharply pointed "cuspidate" incised lobes linear, narrowly triangular, or triangular, unequal, to 1.7 m long; minor bracts lanceolate. **Flowers** sessile; calyx 3.5–4.8 mm, densely to sparsely tomentulose, and hirsutulous and sparsely hirsute, tube 1.8–2.4 mm, lobes 1.5–2.4 mm; petals bright yellow proximally, pale yellow distally, obtriangular, 3.7–4 × 1–1.4 mm, apex truncate, minutely puberulent across apical margin, trichome rays 1 or 2, abaxial surface minutely stellate-puberulent, stellate trichome rays dense, 8–10, adaxial surface sparsely pilose or villous, claw adherent for (0.4–)0.8–1 mm to stamen tube base; stamens 2.6–3 mm, tube 2.2 mm, red-papillose distally; anthers (0.7–)0.9–1 mm, base sagittate; pistil 2.6–3.6 mm; ovary velutinous-hirsute apically; styles 1.3 × 0.1 mm, lateral, stellate-hirsutulous; stigmas 10–26-branched, plumose to distended plumose, 1.2 × 0.9 mm, column 0.6–1.2 mm, branches 0.3–0.7 mm. **Capsules** 2.3–2.8 × 1.6–1.8 mm, transverse-truncate to declining-truncate apically, without dense white puberulent line at apical valve margin, usually with oblong sessile glandular hairs near base, walls membranous most of length; endocarp corneous to 1.3 mm from apex, saddle-shaped in lateral view; capsule dehiscence partially loculicidal across apex and upper sides of valve suture. **Seeds** dark chocolate brown with light brown zone, obovoid-obconic, 1.6–2.3 × 1–1.4 × 1.2–1.3 mm, apex broadly rounded, nearly truncate in lateral view, smooth, apex with very fine raised concolorous line.

Flowering and fruiting late spring–fall. Canyons, exposed boulders, granitic slopes, foothills; 900–1400 (–1500) m; Ariz.; Mexico (Baja California Sur, Chihuahua, Jalisco, Sinaloa, Sonora, Zacatecas).

In Arizona, *Waltheria detonsa* occurs in the Baboquivari, Las Guijas, and Santa Catalina mountains, Buenos Aires National Wildlife Refuge, Coronado National Forest, Huachuca foothills (Fort Huachuca), and, presumably, still at the type locality in Santa Cruz County on Sonoita Creek, near Deserted Rancho.

Some authors have placed *Waltheria detonsa* under *W. indica* or its synonym *W. americana*, or misidentified it as such. Specimens of *W. detonsa* have also been misidentified as *W. acapulcensis* Rose, *W. albicans* Turczaninow, *W. paniculata* Bentham, or *W. preslii* Walpers.

The flowers of *Waltheria detonsa* usually have the stigma exceeding the stamens by 0.5–1 mm. I have seen a distylous thrum-flowered plant with separation between stamens and stigma from Mexico in Sinaloa (*Lamb 327*, US), and a possible thrum from Chihuahua (*Palmer 20*, NY).

2. **Waltheria bahamensis** Britton, Torreya 3: 105. 1903

Plants to 6(–10) dm, resinous, appearing subglabrous but microscopically densely sessile-glandular, usually sparsely and very minutely stellate-puberulent. **Stems** 1–3 mm wide, nodes knobby, joints exserted 0.8–1 mm. **Leaves:** stipules linear-triangular, to 5 mm; petiole 1–12(–20) × 0.4–0.6 mm, with low tubercles; blade concolor, dark brown, bronze, or coppery, drying brownish olive, usually broadly oblong to oblong, ovate-oblong, or obovate-oblong, sometimes ovate, to 5 × 3 cm, resinous, base widely cuneate to slightly cordate, margins sharply dentate or sharply crenate-dentate,

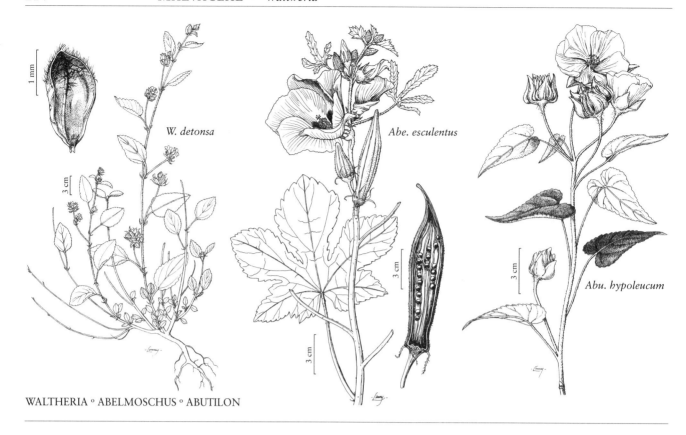

W. detonsa

Abe. esculentus

Abu. hypoleucum

WALTHERIA ° ABELMOSCHUS ° ABUTILON

revolute, apex obtuse to rounded, surfaces glabrate, slightly scabrous, puberulent, and/or sparsely hispidulous with rays planar, almost lepidote, at times moderately dense finely pubescent, trichome rays 0.1–0.5 mm, densely subsessile-glandular, costal vein thick. **Inflorescences** subsessile glomerules on ultimate 5–7 (–12) nodes; peduncle 1 × 0.8 mm; glomerules compact, 0.4–0.9 × 0.5–1.2 cm, 6–12-flowered, composed of terminal flower subtended by accessory shoots with simple elongate cincinnate dichasia; bracteoles free, bracts subequal, narrowly elliptic, lanceolate, or linear, exterior lanceolate, interior linear, wider bracts 0.4–1.4 mm wide, apex acute, entire distally. **Flowers** sessile; calyx 2–4.8 mm, hirsutulous and hirsute, tube 1.7–2.5 mm, lobes 1–2 mm; petals bright yellow-orange, oblanceolate, obovate, or oblong, 2.5–4.8 × 0.7–1.3 mm, abaxially glabrous, adaxially subglabrous, densely papillose, with a few pilulose hairs, margins glabrous or very minutely stellate-puberulent, claw adherent for 0.3–0.7 mm to stamen tube base; stamens 1.8–3.1 mm, uneven, tube at apex incised, arcuate, or truncate, yellow-papillose; anthers 0.6–1 mm, base sagittate, ovate-dehisced; pistil 2.6–3.5 mm; ovary sericeous apically; styles 1–1.5 × 0.1–0.2 mm, lateral, densely white-sericeous; stigmas 20–40-branched, penicillate or distended-penicillate, obconic when dried, 0.8–1.4 × 1.2–1.5 mm, column 0.3–0.4 mm, branches 0.6–1 mm. **Capsules** obconic, transverse-truncate at apex, 1.7–2.4 × 1.5–1.7 mm, with dense white puberulent line at apical valve margin; walls with endocarp corneous 1.6+ mm from apex, sometimes thin at base and along valve margins; dehiscence usually 2-valvate, valve halves completely separate. **Seeds** dull black, obovoid, 1.8–2.2 × 1.1–1.5 × 1.1 mm, apex with finely keeled linear low brown ridge, surface generally smooth but slightly pusticulate.

Flowering and fruiting year-round. Dry, pineland regions (Big Pine Key), pine barrens, hammocks, everglades, savannas, coastal rocky ground, sandy beaches; 0 m; Fla.; West Indies (Bahamas).

Waltheria bahamensis has been considered endemic to the Bahamas. Plants of *W. bahamensis* are found in Florida from Camp Jackson, Frostproof, Miami, Pine Crest, and Sanibel Island.

Some Florida collections previously called *Waltheria indica* are attributable to *W. bahamensis* or *W. bahamensis* × *W. glabra*; other specimens are hybrids of *W. bahamensis* and *W. indica* (J. G. Saunders 1995).

3. Waltheria indica Linnaeus, Sp. Pl. 2: 673. 1753

• Sleepy morning, basora prieta, hierba del soldado, malva del monte, malva, malva blanca, uhaloa W

Waltheria americana Linnaeus

Plants branched or nearly simple, to 7–12 dm, indument variable, tomentose to short-woolly, sparsely to densely pubescent. **Stems** to 8 mm wide, nodes even or extended to 0.5 mm. **Leaves:** stipules linear-triangular, 3–6 mm; petiole 2.5–5 cm × 1–4 mm; blade concolorous or discolorous, drying olive, ovate-oblong to lanceolate, to 7(–10) × 5 cm, base subcordate to obtuse, margins crenate-serrate, serrate, irregularly serrate, or crenate-dentate, apex narrowly rounded, obtuse, or subacute, surfaces tomentose or pubescent, trichome rays 0.4–1 mm, and sparsely hispidulous, slightly scabrous or not. **Inflorescences** subsessile, or on short peduncle-like lateral branches to 3.6 cm, glomerules 5–20 along stems, compact, often dense, sympodial, at times lax clusters with terminal flower and subtending dichasia; bracteoles free or with 1 adherent to stipule forming an elliptic 2-cuspidate foliole at glomerule base, bracts subequal, lanceolate or linear, 0.3–1.5 mm wide, apex acute, entire. **Flowers** sessile or subsessile; calyx 4.2–5.8 mm, tomentose, hirsute, tube 2.5–2.9 mm, lobes 1.8–2.5 × 0.9–1 mm; petals yellow, bright yellow, or yellow-orange, narrowly obovate or oblong, 3.3–5.8 × 0.8–1.6 mm, abaxially glabrous or stellate-pilose, 5-rayed, adaxially pilose at midlength, apex rounded, at times truncate or almost emarginate, glabrous or with few simple or 2-rayed hairs, claw adherent for 0.2–1.2 mm to stamen tube base; stamens 2.1–4.2 mm, tube 1.7–3.7 mm, glabrous or yellow-papillose; anthers 0.6–0.9 mm, base 2-apiculate, oblong-dehisced; pistil 2.5–4.7 mm; ovary setose apically; styles 0.9–1.7 × 0.1–0.3 mm, lateral or excentric, broader and stellate-hirsute distal to base; stigmas 12-branched, plumose (1.2–1.4 × 0.5–0.8 mm, column 0.9–1.2 mm, branches to 0.7 mm) or dissolute-penicillate (0.6–1.2 × 0.7–1.6 mm, column 0.4–0.7 mm, branches to 1.1 mm). **Capsules** oblique, obconic, or obovoid, 2.1–3 × 1.5–2 mm, walls with endocarp corneous for 1+ mm from apex, sometimes merely membranous along one or both valve margins laterally; partially apically dehiscent to 2-valvate. **Seeds** dark brown, obovoid or obconic, 2–2.2 × 1.2–1.5 × 1–1.3 mm, smooth. $2n$ = 24, 26 (India), 40 (Africa).

Flowering and fruiting year-round. Ruderal communities, disturbed scrublands, scrub hummocks, pine-oak sandhills, dunes, open savannas, dry coastal hummocks, sandy or shell soil; 0–1000 m; Ala., Ariz., Fla., Tex.; Mexico; West Indies; Bermuda; Central America; South America; Asia; Africa; Atlantic Islands; Indian Ocean Islands; Pacific Islands; Australia.

Robert Brown (in J. H. Tuckey 1818) was the first to unequivocally synonymize the names *Waltheria americana* and *W. indica* published by Linnaeus in the same work. Brown adopted the name *W. indica* for the combined species and thus gave priority to that name.

Waltheria indica in the narrow sense is globally in need of revision, being very complex and problematic. Heterostyly has been reported for some populations of *W. indica* in India (B. Bahadur et al. 1996). Flora area populations have the pistil subequal (0.2–0.5 mm longer) and more or less homostylous, or up to 1 mm longer than stamens, more like monomorphic pin floral morphology. Multiple introductions are apparent from the variability seen within the flora area. Apparent hybrids of *W. indica* with other species are found in Arizona, Florida, and Texas.

f.   MALVACEAE   Jussieu subfam. MALVOIDEAE   Burnett, Outlines Bot., 816, 1094, 1118. 1835 (as Malvidae)

Paul A. Fryxell†

Steven R. Hill

Herbs, subshrubs, shrubs, or trees, erect or procumbent, often stellate-hairy, sometimes with glandular hairs, generally mucilaginous, inner bark tough-fibrous. **Leaves** usually cauline, alternate, also distichous in *Krapovickasia, Malvella, Meximalva,* and some species of *Sida,* subsessile or sessile in some *Herissantia, Horsfordia, Sida, Sidastrum,* and *Sphaeralcea;* blade usually symmetric, asymmetric in *Malvella,* sometimes asymmetric in *Pavonia,* usually unlobed, sometimes palmately lobed or dissected, margins serrate or entire, primary veins often palmate (except in *Lagunaria*), without nectaries (except in *Gossypium, Hibiscus, Talipariti, Thespesia,* and *Urena*). **Inflorescences** terminal or axillary, umbels, spikes, racemes, or panicles or solitary flowers; bracts usually present. **Flowers** bisexual or unisexual, some species with staminate or pistillate flowers, plants usually hermaphroditic, some dioecious, some monoecious; involucel (epicalyx) present or absent; sepals usually persistent (deciduous in *Abelmoschus*), 5, ± connate; petals 5, usually distinct, adnate to staminal column and connate to each other at base, falling together, without clawlike appendage; androgynophore absent; androecium monadelphous, in more than 1 concentric series in *Sidalcea*, staminal column sometimes toothed at apex, stamens 5–many, filaments connate; anthers 1-thecate; staminodes absent or 5 teeth at apex of staminal column; gynoecium syncarpous, ovary superior, 3–40-carpellate; styles 1, branched or unbranched; stigmas truncate, capitate, linear, or filiform, 1–2 times number of carpels. **Fruits** usually schizocarps with 5–many wedge-shaped mericarps or folliclelike segments, or capsules with 3–5 cells, rarely berries (in *Malvaviscus*), carpels sometimes with internal protrusion dividing carpel into 2 cells, dehiscence loculicidal, rarely indehiscent. **Seeds** 2–30, often reniform, glabrous or hairy.

Genera ca. 110, species ca. 1800 (42 genera, 220 species in the flora): nearly worldwide; mostly in tropical areas.

Some ornamental malvaceous trees have been introduced to California, none of which is known in the flora area outside of cultivation. These include the Mexican *Robinsonella cordata* Rose & Baker f. and, from Australia and New Zealand, *Hoheria glabrata* Sprague, *Plagianthus divaricatus* J. R. Forster & G. Forster, and *P. regius* (Poiteau) Hochreutiner. *Malope trifida* Cavanilles from the Mediterranean region, and the subshrubs *Alyogyne* Alefeld from Australia, species of *Anisodontea* C. Presl from South Africa, and *Lavatera* species from the Mediterranean are sometimes grown as ornamentals; they are not naturalized in the flora area.

SELECTED REFERENCES Fryxell, P. A. 1978. Neotropical segregates from *Sida* L. (Malvaceae). Brittonia 30: 447–462. Fryxell, P. A. 1979. The Natural History of the Cotton Tribe. College Station. Fryxell, P. A. 1988. Malvaceae of Mexico. Syst. Bot. Monogr. 25. Fryxell, P. A. 1997. The American genera of Malvaceae—II. Brittonia 49: 204–269.

1. Involucel usually absent, sometimes present in *Callirhoë*, *Malachra*, *Malvella*, and *Sidalcea*, often deciduous in *Sphaeralcea*; fruits schizocarps.
  2. Inflorescences subtended by boat-shaped sessile or subsessile bracts . . . . 33. *Malachra*, p. 279
  2. Inflorescences not subtended by boat-shaped sessile or subsessile bracts.
    3. Stigmas linear or filiform.
      4. Herbs, (1–)2.2(–3) m; inflorescences terminal panicles; flowers unisexual, plants dioecious; corollas white . . . . .41. *Napaea*, p. 305
      4. Herbs or subshrubs, usually less than 2 m; inflorescences axillary solitary flowers or terminal, open or spicate racemes or panicles; flowers bisexual or rarely unisexual, plants not dioecious; corollas usually not white.
        5. Anthers borne ± evenly along staminal column; staminal columns not comprised of concentric series . . . . 20. *Callirhoë* (in part), p. 240
        5. Anthers borne distally on staminal column, staminal columns comprised of concentric inner and outer series of filaments . . . . 46. *Sidalcea* (in part), p. 319
    3. Stigmas capitate.
      6. Mericarps 1- or 2-celled; distally dehiscent; seeds 1–3 per mericarp.
        7. Calyces subequal to or longer than fruits; seeds 2 or 3 per mericarp.
          8. Calyces usually 8–13 mm; mericarps with prominent, medial constriction resulting in 2 cells . . . . .14. *Allowissadula*, p. 229
          8. Calyces 4–8 mm; mericarps with obscure medial constriction (except *P. umbellatum*) . . . . 43. *Pseudabutilon*, p. 308
        7. Calyces usually shorter than fruits; seeds 1–3 per mericarp.
          9. Styles 3–6-branched; petals 3–5 mm . . . . .52. *Wissadula*, p. 374
          9. Styles 6–12-branched; petals 6–21 mm.

10. Mericarps with medial constriction, distal cell unwinged; seeds
1 per mericarp; Texas . . . . 18. *Batesimalva*, p. 238
10. Mericarps without medial constriction, distal cell apically
winged; seeds 2 or 3 per mericarp; Arizona, California. . . . . . 27. *Horsfordia*, p. 267

[6. Shifted to left margin.—Ed.]

6. Mericarps 1-celled; distally dehiscent, partially dehiscent, or indehiscent; seeds usually 1
per mericarp (3–6 in *Abutilon*; 2–6 in *Herissantia*).
11. Calyces completely enclosing fruits or leaves maplelike.
12. Corollas white; calyces not completely enclosing fruits; leaves maplelike; east of
Mississippi River. . . . . 45. *Sida* (in part), p. 310
12. Corollas yellow or white, fading rose; calyces completely enclosing fruits; leaves
not maplelike; west of Mississippi River.
13. Corollas white, fading rose; calyces brownish-membranous at maturity;
mericarps fragile-walled, unornamented. 30. *Krapovickasia*, p. 274
13. Corollas yellow; calyces green-membranous at maturity; mericarps indurate,
laterally reticulate-walled, with horizontal obtuse rostrum. . . . . 44. *Rhynchosida*, p. 309
11. Calyces closely subtending, not enclosing fruits (fruits closely invested by calyx tube in
Fryxellia); leaves seldom maplelike (*Anoda*; *Sida hermaphrodita*).
14. Fruits spheric, inflated, not indurate, setose, reflexed . . . . 25. *Herissantia*, p. 251
14. Fruits not both spheric and inflated (can appear somewhat inflated in *Fryxellia* but
not spheric or pendent), usually indurate, hairy, seldom setose, usually erect.
15. Mericarps 3–6-seeded. . . . . .12. *Abutilon*, p. 220
15. Mericarps 1-seeded.
16. Leaf blades cuneate at base, otherwise broadly oblanceolate, 0.5–1.5 cm;
mericarps 5; usually saline habitats . . . . .19. *Billieturnera*, p. 239
16. Leaf blades variously shaped at base, ovate, reniform, oblong, elliptic,
lanceolate, or linear, usually 1.5+ cm; mericarps 5+; seldom saline habitats
(except *Malvella*).
17. Mericarps: lateral walls evanescent, spur rarely absent . . . . 16. *Anoda*, p. 234
17. Mericarps: lateral walls persistent (firm or indurate), spur usually
absent (except *Fryxellia*).
18. Plants cespitose; mericarps with dorsal spur and endoglossum
. . . . .23. *Fryxellia*, p. 248
18. Plants not cespitose; mericarps without dorsal spur (sometimes
with apical spines) or endoglossum.
19. Corollas purple, 3–6 mm; calyces 2–4 mm, not costate;
pedicels capillary.
20. Mericarps 7 or 8, lateral walls prominently reticulate;
leaves distichous; styles 7 or 8, purple; corollas rotate;
inflorescences solitary flowers . . . . 39. *Meximalva*, p. 302
20. Mericarps 5, lateral walls smooth or weakly reticulate;
leaves not distichous; styles 5, pallid; corollas reflexed;
inflorescences terminal panicles . . . . .47. *Sidastrum*, p. 356
19. Corollas yellow or yellowish, usually 6+ mm; calyces 4–10
mm, costate or not; pedicels not capillary.
21. Leaf blades asymmetric; stems prostrate; indument
sometimes ± lepidote; often saline habitats. . . . 38. *Malvella* (in part), p. 301
21. Leaf blades symmetric; stems usually erect, seldom
prostrate; indument never lepidote; not saline habitats.
22. Herbage prominently viscid; calyces divided nearly
to base, not costate; fruit walls papery . . . . 17. *Bastardia*, p. 237
22. Herbage seldom viscid; calyces usually ½-divided,
usually costate; fruit walls indurate . . . . 45. *Sida* (in part), p. 310

1. Involucel usually present, sometimes absent; fruits schizocarps or capsules.
    23. Ovaries 3–5-carpellate; fruits usually capsules.
        24. Trees; leaves: venation pinnate; involucellar bractlets basally connate . . . . .31. *Lagunaria*, p. 276
        24. Trees, subshrubs, or herbs; leaves: venation palmate; involucellar bractlets usually distinct (except *Abelmoschus*, *Talipariti*, and *Urena*), not enclosing bud.
            25. Involucellar bractlets 3, distinct.
                26. Capsules ovoid or subglobose to oblong, dehiscent; shrubs or trees. . .24. *Gossypium*, p. 250
                26. Capsules oblate, indehiscent; trees . . . . .50. *Thespesia*, p. 372
            25. Involucellar bractlets 4+, distinct or ± connate.
                27. Calyces deciduous, spathaceous . . . . 11. *Abelmoschus*, p. 219
                27. Calyces persistent, not spathaceous.
                    28. Fruits ovoid, subglobose, or spheroid, not fleshy; seeds 2–20 per locule.
                        29. Calyces gland-dotted; ovaries 3–5-carpellate; styles usually connate to apex; stigmas 3–5, decurrent . . . . .21. *Cienfuegosia*, p. 245
                        29. Calyces not gland-dotted; ovaries 5-carpellate; styles proximally connate, distally 5-fid; stigmas 5, capitate to discoid.
                            30. Stipules persistent or caducous, not enlarged in bud; relatively dry habitats or freshwater swamps .26. *Hibiscus*, p. 252
                            30. Stipules caducous, enlarged in bud, leaving annular scars; estuarine habitats . . . . 49. *Talipariti*, p. 370
                    28. Fruits oblate, rarely fleshy, seeds 1 per locule.
                        31. Fruits 5-angled, sections dehiscent; foliage scabrid; styles 5-fid . . . . .29. *Kosteletzkya*, p. 272
                        31. Fruits not angled, indehiscent; foliage glabrous or hairy, seldom scabrid; styles 10-fid.
                            32. Leaves with abaxial nectaries; involucellar bractlets 5 . . . . . . .51. *Urena*, p. 373
                            32. Leaves without nectaries; involucellar bractlets 5–9.
                                33. Petals basally auriculate, usually red; fruits fleshy, red; stamens and stigmas usually exserted . . . . 37. *Malvaviscus*, p. 298
                                33. Petals usually not auriculate, lavender, pink, or yellow; fruits dry, not colored; stamens and stigmas usually included . . . . 42. *Pavonia*, p. 305
    23. Ovaries (5)6–36-carpellate; fruits schizocarps.
        34. Involucellar bractlets 6–12.
            35. Petals 3+ cm; mericarps 2-celled, proximal cell fertile, distal cell sterile; staminal columns 5-angled, anthers pale yellow . . . . 13. *Alcea*, p. 227
            35. Petals (0.9–)1–2 cm; mericarps 1-celled; staminal columns cylindric, anthers purple or pale pink to almost white or yellow . . . . .15. *Althaea*, p. 231
        34. Involucellar bractlets (0)1–3.
            36. Leaf blades asymmetric; herbage hairy, hairs mixed stellate and lepidote . . . . .38. *Malvella* (in part), p. 301
            36. Leaf blades symmetric; herbage usually stellate-hairy, glabrate, or glabrous, sometimes some hairs simple, seldom lepidote.
                37. Mericarps 2-celled; seeds 2 per mericarp; corollas salmon-orange; mericarps drying black; stems prostrate to ascending . . . . .40. *Modiola*, p. 303
                37. Mericarps 1- or 2-celled, seeds 1–4 per mericarp; corollas sometimes salmon-orange; mericarps drying black, brown, or tan; stems usually erect, sometimes ascending, decumbent, or prostrate.
                    38. Corollas yellow or yellow-orange, without darker veins . . . . 36. *Malvastrum*, p. 293
                    38. Corollas usually white, pinkish, pink, magenta, rose-purple, rose-pink, purple, mauve, orange, red-orange, or red, sometimes salmon-orange, sometimes with darker veins.

[39. Shifted to left margin.—Ed.]

11. ABELMOSCHUS Medikus, Malvenfam., 45. 1787 • [Presumably Arabic *habb-el-misk*, musk seed, alluding to scented seeds] ☐

David M. Bates

Herbs, annual [perennial], [**subshrubs**], glabrate to densely hairy, often harshly so. **Stems** erect, glabrous or sparsely hairy, sometimes glandular-viscid. **Leaves** appearing palmately compound; stipules deciduous, filiform to falcate; blade broadly ovate to orbiculate, palmately 3-, 5-, or 7-angled to deeply divided, base ± hastate, margins usually toothed or lobate. **Inflorescences** axillary, solitary flowers or racemes; involucellar bractlets persistent or caducous, 5–10, usually connate basally. **Flowers:** calyx deciduous, not accrescent, ribbed, not inflated, spathaceous, 5-toothed at apex, splitting longitudinally, falling with corolla and staminal column; corolla ± funnelform, yellow to white [pinkish], purple or maroon at base; staminal column included; style 5-branched; stigmas peltate. **Fruits** capsules, erect, 5-locular, cylindric [ovoid], angled, bristly, glabrescent. **Seeds** many, obovoid-reniform, mostly striate-hairy. x = 29, 30, 33, 34, 36.

Species 6+ (1 in the flora): introduced, Florida; s, se Asia, Africa, sw Pacific Islands, Australia; introduced nearly worldwide, often becoming naturalized, especially in tropical and subtropical regions.

Abelmoschus is a segregate of *Hibiscus*; it is unique in calyx characters among others. Conservative interpretations of *Abelmoschus* recognize six species; many others have been named.

In addition to *Abelmoschus esculentus*, *A. manihot* (Linnaeus) Medikus and *A. moschatus* Medikus are cultivated as ornamentals in North America and may occasionally escape. The leaves of *Abelmoschus* are edible; the seeds of *A. moschatus* have a musk odor and yield ambrette, an oil used in perfumery.

SELECTED REFERENCE Bates, D. M. 1968. Notes on cultivated Malvaceae. 2. *Abelmoschus*. Baileya 16: 99–112.

1. **Abelmoschus esculentus** (Linnaeus) Moench, Methodus, 617. 1794 • Okra, gumbo, lady's finger F I W

Hibiscus esculentus Linnaeus, Sp. Pl. 2: 696. 1753

Plants 1–2 m. **Stems** often red blotched, coarse. **Leaf blades** scarcely lobed to palmately divided, 10–25 cm, ± broader than long. **Pedicels** not articulated, stout; involucellar bractlets linear, to 2.5 cm. **Flowers:** corolla to 8 cm diam.; staminal column anther-bearing from near base, apex 5-toothed. **Capsules** cylindric, slightly 5-angled, beaked, 8–30 cm. $2n$ = 72, 108, 118, 120, 122, 130, 132, 144.

Flowering spring–fall. Fertile, well-drained soil with ample moisture, waste places; 0–30 m; introduced; Fla., Ga., La., Miss., N.C., S.C., Va.; s Asia; Africa; sw Pacific Islands; introduced also in Mexico and elsewhere nearly worldwide.

Abelmoschus esculentus is a cultigen, apparently domesticated in India for the edible, unripe, succulent, mucilaginous young capsule and edible leaves; it may escape from commercial and garden cultivation and sometimes persist in waste places.

12. ABUTILON Miller, Gard. Dict. Abr. ed. 4, vol. 1. 1754 • Indian-mallow [Probably Arabic *abu*, father of, and Persian *tula* or *tulha*, mallow]

Paul A. Fryxell†

Steven R. Hill

Subshrubs, shrubs, or herbs. Stems erect, sometimes trailing (*A. parvulum*) or procumbent or ascending (*A. wrightii*), glabrescent or pubescent, sometimes viscid (*A. hirtum*, *A. reventum*, *A. trisulcatum*). **Leaves:** stipules usually persistent, subulate, lanceolate, or filiform; blade elliptic, ovate, [cordiform], sometimes shallowly lobed, but not maplelike [sometimes maplelike], base often cordate, margins usually crenate or serrate, [rarely subentire], pubescent. **Inflorescences** axillary, solitary flowers or cymose, racemose, or paniculate, [sometimes umbellate]; involucel absent. **Flowers:** calyx not accrescent (except *A. hulseanum*, *A. hypoleucum*, *A. palmeri*, and *A. wrightii*), not inflated, not completely enclosing fruit, lobes not ribbed, lanceolate, ovate, cordate, or acuminate; corolla usually yellow or orange, less often pinkish, sometimes with dark red center; staminal column included or exserted; ovules 3(–6) per carpel; style 5–25-branched; stigmas sometimes black, capitate. **Fruits** schizocarps, erect, not inflated, globose, ovoid, oblate, cask-shaped, or cylindric, usually not indurate, variably hairy but not setose; mericarps 5–25, 1-celled follicle, adherent to adjacent mericarps and persistent on their axes, without dorsal spur, apex usually acute or acuminate to spinescent, sometimes rounded or obtuse, abaxially dehiscent. **Seeds** usually 3–6 per mericarp, usually turbinate, puberulent or scabridulous. x = 7, 8.

Species ca. 160 (18 in the flora): United States, Mexico, West Indies, Central America, South America, Asia, Africa, Australia.

Abutilon is mostly tropical and subtropical with relatively few species reaching into fully temperate climates.

The South American native *Abutilon megapotamicum* (A. Sprengel) A. Saint-Hilaire & Naudin (= *A. vexillarium* E. Morren) is sometimes grown as a basket plant in colder regions and perhaps in the open in frost-free areas; it is not naturalized in the flora area. Another South American species, *A. striatum* Dickson ex Lindley [= *A. pictum* (Gillies ex Hooker) Walpers], has been widely introduced elsewhere as an ornamental and is naturalized in some tropical countries. This species is grown in North American gardens as a perennial in warmer climates and an annual in colder ones, but is not known to be naturalized in the flora area. *Abutilon indicum* (Linnaeus) Sweet has been said to be naturalized in southern Florida (L. H. Bailey et al. 1976); its occurrence in the flora area has not been substantiated. This species has an Indo-Australian origin (J. van Borssum Waalkes 1966) and is naturalized in the West Indies. *Abutilon grandifolium* (Willdenow) Sweet is sometimes cultivated and may escape.

Outside North America, plants of *Abutilon* may be arborescent, and their corollas may be of other colors although yellow or yellow-orange predominates.

SELECTED REFERENCES Fryxell, J. E. 1983. A revision of *Abutilon* sect. *Oligocarpae* (Malvaceae), including a new species from Mexico. Madroño 30: 84–92. Fryxell, P. A. 2002. An *Abutilon* nomenclator. Lundellia 5: 79–118. Kearney, T. H. 1955. A tentative key to the North American species of *Abutilon* Miller. Leafl. W. Bot. 7: 241–254.

1. Styles 5-branched.
    2. Stems trailing; leaf blades sparsely pubescent, surface visible; flowers solitary; petals pinkish, without dark center . . . . .13. *Abutilon parvulum*
    2. Stems erect; leaf blades puberulent or tomentulose, surface obscured by pubescence; flowers sometimes solitary, usually in panicles; petals yellow or pink, sometimes with dark center.
        3. Calyces 5–8 mm; corollas rotate, petals 9–15 mm, yellow, without dark center; leaf blade margins sharply serrate.
            4. Calyces 5–6 mm, lobes fully reflexed in fruit; petals 10–15 mm; inflorescences open panicles or solitary flowers . . . . .3. *Abutilon coahuilae*
            4. Calyces 6–8 mm, lobes erect in fruit; petals 9–15 mm; inflorescences compact panicles . . . . 9. *Abutilon malacum*
        3. Calyces 3–5 mm; corollas often reflexed, petals 4–6 mm, yellow or pink, often with dark center; leaf blade margins irregularly crenate-serrate.
            5. Young stems terete, minutely tomentulose; fruits cask-shaped, not medially constricted . . . . 8. *Abutilon incanum*
            5. Young stems 3-sulcate, often viscid in inflorescence; fruits subcylindric, medially constricted . . . . 17. *Abutilon trisulcatum*
1. Styles 6–25-branched.
    6. Stems pubescent and with simple hairs 2–5 mm.
        7. Stems viscid; corollas orange-yellow with dark red center; styles 20–25-branched . . . . .5. *Abutilon hirtum*
        7. Stems not viscid; corollas pale yellow or yellow-orange without red center; styles 6–12-branched.
            8. Petals 20 mm, yellowish, fading pinkish; seeds 4–6 per mericarp . . . . .6. *Abutilon hulseanum*
            8. Petals 5–12 mm, yellowish or yellow-orange, not fading pinkish; seeds 3 per mericarp.
                9. Leaf blades 10–20 cm; inflorescences paniculate; petals 5–8 mm; styles 8–10-branched. . . . . 10. *Abutilon mollicomum*
                9. Leaf blades 2.5–6.5 cm; inflorescences solitary flowers; petals 8–12 mm; styles 6–8-branched . . . . 12. *Abutilon parishii*

[6. Shifted to left margin.—Ed.]

6. Stems stellate-pubescent to glabrate or obscurely glandular, simple hairs absent or to 2 mm.

    10. Apices of mericarps spinose. . . . . 16.   *Abutilon theophrasti*

    10. Apices of mericarps not spinose (obtuse, rounded, acute, acuminate, or apiculate).

        11. Plants procumbent or ascending, to 0.5 m; leaf blades 1.5–4 cm, ± as long as wide, markedly discolorous; styles 6–9-branched. . . . . 18.   *Abutilon wrightii*

        11. Plants erect, 0.5–2 m; leaf blades 2–12(–20) cm, often longer than wide, concolorous or discolorous; styles 6–15-branched.

            12. Styles 13–15-branched; calyx lobes broadly cordate, often 8–20+ mm wide, accrescent in fruit. . . . . .7.   *Abutilon hypoleucum*

            12. Styles 6–12-branched; calyx lobes cordate, lanceolate-ovate, or acuminate, to 8 mm wide, little if at all accrescent.

                13. Calyces 3–6 mm.

                    14. Styles 6–9-branched; mericarp apices acute or apiculate; stems stellate-tomentulose; leaf blades ± concolorous. .4.   *Abutilon fruticosum*

                    14. Styles 10-branched; mericarp apices rounded or obtuse; stems minutely glandular-pubescent to glabrate; leaf blades strongly discolorous . . . . .15.   *Abutilon reventum*

                13. Calyces 8–15(–20) mm.

                  15. Leaf blades roughly pubescent; styles 8–10-branched.

                    16. Stems without glandular hairs; seeds reticulately scabridulous; calyx basally truncate. . . . . .1.   *Abutilon abutiloides*

                    16. Stems with glandular and stellate hairs intermixed; seeds uniformly puberulent; calyx basally cuneate. . . . . .2.   *Abutilon berlandieri*

                15. Leaf blades softly tomentose; styles 10–12-branched.

                  17. Leaf blades ± as wide as long; petals 20–25 mm; stems stellate-pubescent and sometimes with simple hairs 1–2 mm . . . . 11.   *Abutilon palmeri*

                  17. Leaf blades ca. 2 times as long as wide; petals 10–18 mm; stems stellate-pubescent . . . . 14.   *Abutilon permolle*

---

1. **Abutilon abutiloides** (Jacquin) Garcke ex Hochreutiner, Annuaire Conserv. Jard. Bot. Genève 6: 22. 1902

Sida abutiloides Jacquin, Observ. Bot. 1: 17, plate 7. 1764; *Abutilon lignosum* (Cavanilles) G. Don

Subshrubs, to 1.5 m. **Stems** erect, roughly stellate-pubescent, without glandular hairs. **Leaves:** stipules subulate, 6–8 mm; petiole usually shorter than blade; blade slightly discolorous, ovate, 2–10 cm, longer than wide, base truncate to cordate, margins obscurely to prominently crenate-serrate, apex usually acuminate, surfaces roughly pubescent. **Inflorescences** solitary flowers or racemose panicles. **Flowers:** calyx 9–12 mm, basally truncate, lobes basally overlapping, erect in fruit, acuminate, to 8 mm wide; corolla rotate, orange-yellow throughout, petals 10–12 mm; staminal column pubescent; style 8–10-branched. **Schizocarps** ± ovoid, 9–14 × 10–16 mm; mericarps: apex acuminate, surface uniformly stellate-pubescent and with simple hairs on abaxial margins. **Seeds** 3 per mericarp, 3 mm, reticulately scabridulous. *2n* = 28.

Flowering year-round. Open, arid habitats; 0–1000 m; Ariz., Tex.; Mexico; West Indies (Bahamas, Cuba, Hispaniola, Jamaica).

Abutilon abutiloides has been confused with *A. californicum* Bentham and *A. berlandieri*; the three species are distinct (P. A. Fryxell 1988). It is found in extreme southern Texas and more commonly in west-central and southeastern Arizona.

2. **Abutilon berlandieri** A. Gray ex S. Watson, Proc. Amer. Acad. Arts 20: 358. 1885

Shrubs, 0.5–1.5 m. **Stems** erect, roughly stellate-pubescent with minute intermingled glandular hairs. **Leaves:** stipules subulate, 3–10 mm; petiole shorter than blade; blade slightly discolorous, ovate, 2.5–10 cm, slightly longer than wide, base truncate to cordate, margins obscurely crenate to serrate, apex acute, sometimes acuminate, surfaces roughly pubescent. **Inflorescences** solitary flowers or racemes or panicles. **Flowers:** calyx 8–13 mm, basally cuneate, lobes basally overlapping, erect in fruit, acuminate, to 8 mm wide; corolla orange-yellow throughout, petals 10 mm; staminal column pubescent; style 8–10-branched. **Schizocarps** ± ovoid, 10–13 × 10–13 mm; mericarps: apex acuminate, surface with both stellate and glandular hairs. **Seeds** 3 per mericarp, 3 mm, uniformly puberulent. $2n = 42$.

Flowering year-round. Open, arid habitats; 0–200 m; Tex.; Mexico (Coahuila, Nuevo León, Tamaulipas).

Abutilon berlandieri is found in southern Texas and is scarce elsewhere.

3. Abutilon coahuilae Kearney, Madroño 12: 115. 1953

Subshrubs, to 1 m. **Stems** erect, densely tomentulose, without simple hairs. **Leaves:** stipules subulate, 3–5 mm; petiole more than ½ as long as blade; blade nearly concolorous, ovate, 4–6 cm, ± as long as wide, base truncate to cordate, margins sharply serrulate, apex acute, surfaces tomentulose, obscured by pubescence. **Inflorescences** solitary flowers or open, terminal panicles. **Flowers:** calyx 5–6 mm, lobes not overlapping, fully reflexed in fruit, lanceolate; corolla rotate, yellow throughout, petals 10–15 mm; staminal column with few stellate hairs apically; style 5-branched. **Schizocarps** ± cylindric, 8–10 × 6–8 mm; mericarps: apex long-acuminate, surface stellate-pubescent. **Seeds** 3 per mericarp, 2 mm, puberulent.

Flowering year-round. Open, arid habitats; 1000–1800 m; Ariz.; Mexico (Coahuila, Sonora).

Abutilon coahuilae is known from Pima and Pinal counties.

4. Abutilon fruticosum Guillemin & Perrottet in J. B. A. Guillemin et al., Fl. Seneg. Tent. 1: 70. 1831
• Pelotazo

Abutilon texense Torrey & A. Gray

Subshrubs, 1–1.5 m. **Stems** erect, stellate-tomentulose, without simple hairs. **Leaves:** stipules subulate, 2 mm; petiole shorter than blade; blade ± concolorous, ovate, 2–10 cm (often smaller), somewhat longer than wide, base cordate, margins irregularly serrate, apex acute, surfaces minutely stellate-pubescent. **Inflorescences** solitary flowers or terminal panicles. **Flowers:** calyx 3–5 mm, lobes not overlapping, reflexed in fruit, lanceolate-ovate, to 8 mm wide; corolla yellow throughout, petals 5–10 mm; staminal column glabrous; style 6–9-branched. **Schizocarps** cask-shaped, 8–9 × 8–9 mm; mericarps: apex acute or apiculate, surface tomentulose. **Seeds** 3 per mericarp, 2 mm, puberulent but appearing glabrous. $2n = 14$.

Flowering year-round. Open, arid habitats; 0–2000 m; Ark., N.Mex., Okla., Tex.; Mexico (Chihuahua, Coahuila, Nuevo León, San Luis Potosí, Tamaulipas, Zacatecas); se Europe; n Africa.

Abutilon fruticosum is thought to be native to the New World; it also occurs disjunctly in northern Africa and the Levant countries. It is widespread in southwestern Texas, scarce in Oklahoma and Arkansas, and is known from Chaves and Lincoln counties, New Mexico.

5. Abutilon hirtum (Lamarck) Sweet, Hort. Brit., 53. 1826 □

Sida hirta Lamarck in J. Lamarck et al., Encycl. 1: 7. 1783

Herbs, 1 m. **Stems** erect, puberulent and with simple hairs 2–5 mm, viscid. **Leaves:** stipules recurved, lanceolate, 7–9 mm; petiole as long as or longer than blade; blade somewhat discolorous, ovate to suborbiculate, 5–7 cm, base cordate, margins finely serrate, apex acuminate, surfaces softly tomentose. **Inflorescences** solitary flowers or terminal panicles. **Flowers:** calyx 12–17 mm, lobes not overlapping, erect, ovate; corolla orange-yellow with dark red center, petals 18–20 mm; staminal column pubescent; style 20–25-branched. **Schizocarps** oblate, 12–14 × 20 mm; mericarps: apex obtuse to acute, surface stellate-hirsute. **Seeds** 3 per mericarp, 2.4–2.8 mm, minutely scabridulous. $2n = 42$.

Flowering year-round. Disturbed sites; 0–100 m; introduced; Fla.; Asia; Africa; Australia; introduced also in Mexico, West Indies (Cuba, Lesser Antilles, Puerto Rico), Central America, South America (Peru, Venezuela).

Abutilon hirtum has been found in Alachua, Lake, Miami-Dade, and Monroe counties, primarily in the Florida Keys. It is native in tropical parts of Africa, Asia, and Australia.

6. Abutilon hulseanum (Torrey & A. Gray) Torrey ex Baker f., J. Bot. 30: 328. 1892

Sida hulseana Torrey & A. Gray, Fl. N. Amer. 1: 233. 1838

Herbs or subshrubs, perennial, 1–2 m. **Stems** erect, stellate-tomentulose and with simple hairs 2–4 mm. **Leaves:** stipules filiform, 8 mm; petiole subequal to blade; blade ± discolorous, ovate, mostly 6–10 cm, longer than wide, base cordate, margins crenate, apex rounded-acute, surfaces softly tomentulose. **Inflorescences** solitary flowers. **Flowers:** calyx 12–15 mm, lobes basally overlapping, erect, cordate, accrescent to 15–20 mm;

corolla yellowish throughout, fading pinkish, petals 20 mm; staminal column glabrous; style 12-branched. **Schizocarps** ± oblate, 12–15 × 20–25 mm; mericarps: apex apiculate, surface prominently hirsute, hairs 1–2 mm. **Seeds** 4–6 per mericarp, 2 mm, puberulent. $2n = 14$.

Flowering winter–spring. Roadsides, disturbed sites, secondary vegetation; 0–100 m; Fla., La., Tex.; Mexico; West Indies.

Plants of *Abutilon hulseanum* are sometimes misidentified as *A. pauciflorum* A. Saint-Hilaire, which is known only from Bolivia, Brazil, Paraguay, and Peru.

7. **Abutilon hypoleucum** A. Gray, Smithsonian Contr. Knowl. 3(5): 20. 1852 F

Shrubs, 0.5–1.5 m. **Stems** erect, sparsely to densely stellate-pubescent, sometimes also with few simple hairs to 2 mm. **Leaves:** stipules subulate, 2–4 mm; petiole ½ to as long as blade; blade markedly discolorous, narrowly ovate, 5–11 cm, to 2 times as long as wide, base cordate, margins crenate, apex acute or slightly acuminate, surfaces densely soft-tomentose abaxially, sparsely scabridulous adaxially. **Inflorescences** solitary flowers. **Flowers:** calyx 10–15 mm, lobes basally overlapping, erect, broadly cordate, accrescent to 25 mm, to 20 mm wide; corolla yellow or yellow-orange throughout, petals 15–25 mm; staminal column glabrous; style 13–15-branched. **Schizocarps** ± ovoid, 15 × 15 mm; mericarps: apex acuminate, surface stellate-pubescent. **Seeds** 3 per mericarp, 2.5 mm, minutely scabridulous. $2n = 14$.

Flowering spring–fall. Dry shrublands, roadsides, disturbed sites; 0–1300 m; Tex.; Mexico.

Abutilon hypoleucum is limited to the Rio Grande plains and valley.

8. **Abutilon incanum** (Link) Sweet, Hort. Brit., 53. 1826
• Pelotazo chico

Sida incana Link, Enum. Hort. Berol. Alt. 2: 204. 1822

Subshrubs, 1–2 m. **Stems** erect, terete when young, minutely tomentulose. **Leaves:** stipules subulate, 3 mm; petiole ca. ½ as long as blade; blade concolorous, ovate, to 6 cm, longer than wide, base cordate, margins irregularly serrulate or crenate, apex acute or acuminate, surfaces densely tomentulose, obscured by pubescence. **Inflorescences** solitary flowers or in open panicles. **Flowers:** calyx 3–5 mm, lobes not overlapping, spreading or somewhat reflexed in fruit, lanceolate-

ovate; corolla reflexed, yellow or pink with dark red center, petals 4–6 mm; staminal column pubescent; style 5-branched. **Schizocarps** cask-shaped, not medially constricted, 7–9 × 6 mm; mericarps: apex acute or apiculate, surface tomentulose. **Seeds** 3 per mericarp, 2 mm, puberulent but appearing glabrous. $2n = 14$.

Flowering spring–fall. Open, arid habitats, hills and plains; 0–1000 m; Ariz., Colo., N.Mex.; Mexico (Baja California, Sinaloa, Sonora); Pacific Islands (Hawaii).

Abutilon incanum is found in Baja California, Sinaloa, Sonora, adjacent Arizona, and disjunctly in Hawaii, where it is considered to be native.

9. **Abutilon malacum** S. Watson, Proc. Amer. Acad. Arts 21: 446. 1886

Herbs or subshrubs, usually perennial, to 1 m. **Stems** erect, minutely stellate-tomentulose, hairs yellowish. **Leaves:** stipules subulate, 5–9 mm; petiole ½ to as long as blade; blade concolorous, suborbiculate to ovate, 3–7 cm, ± as long as wide, base cordate, margins sharply serrate, apex acuminate, surfaces minutely tomentulose, obscured by pubescence. **Inflorescences** terminal, compact panicles. **Flowers:** calyx 6–8 mm, lobes not overlapping, erect in fruit, lanceolate-ovate; corolla rotate, yellow throughout, petals 9–15 mm; staminal column pubescent; style 5-branched. **Schizocarps** ± cylindric, 6–7 × 6–7 mm; mericarps: apex usually acute, surface coarsely stellate-pubescent. **Seeds** 3 per mericarp, 2 mm, puberulent but appearing glabrous. $2n = 14$.

Flowering summer. Arid habitats, hillsides, plains, canyons; 300–1500 m; Ariz., N.Mex., Tex.; Mexico (Chihuahua, Coahuila, Durango, San Luis Potosí, Sonora, Zacatecas).

Abutilon malacum is found in southwestern Arizona, southwestern New Mexico, and the Big Bend area of Texas.

10. **Abutilon mollicomum** (Willdenow) Sweet, Hort. Brit., 54. 1826 • Pintapán cimarrón F

Sida mollicoma Willdenow, Enum. Pl., 725. 1809

Shrubs, 1–2 m. **Stems** erect, prominently hirsute, with simple hairs 2–4 mm. **Leaves:** stipules subulate, 5–9 mm; petiole subequal to blade, prominently hirsute; blade markedly discolorous, broadly ovate or 3–5-angulate, mostly 10–20 cm, ± as long as wide, base cordate, margins irregularly crenate-serrate, apex acuminate, surfaces minutely stellate-pubescent.

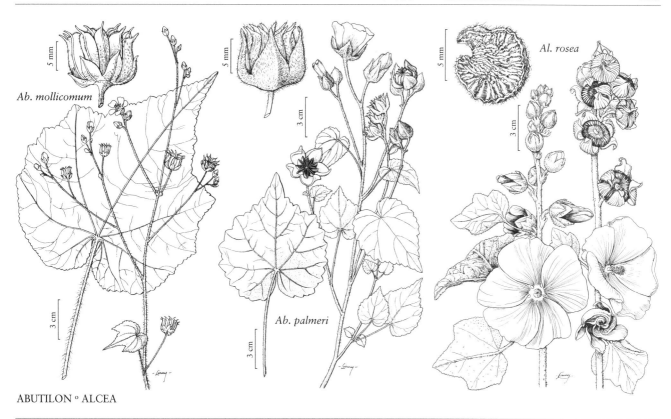

Ab. mollicomum

Ab. palmeri

Al. rosea

ABUTILON ° ALCEA

Inflorescences terminal panicles. **Flowers:** calyx 4–6 mm, lobes not overlapping, erect in fruit, ovate; corolla yellowish throughout, petals 5–8 mm; staminal column glabrous; style 8–10-branched. **Schizocarps** ovoid, 8–10 × 8–10 mm; mericarps: apex short-apiculate, stellate-pubescent. **Seeds** 3 per mericarp, 2 mm, minutely verruculate. $2n = 14$.

Flowering late summer. Open desert habitats, along water courses; 500–2200 m; Ariz., N.Mex., Tex.; Mexico (Chihuahua, Guerrero, Oaxaca, Sonora, Veracruz).

Abutilon mollicomum occurs principally in Chihuahua and Sonora and in adjacent Arizona and New Mexico.

11. Abutilon palmeri A. Gray, Proc. Amer. Acad. Arts 8: 289. 1870 F

Shrubs, to 1.5(–2) m. **Stems** erect, ± stellate-pubescent, sometimes also with simple hairs 1–2 mm. **Leaves:** stipules filiform, 6–9 mm; petiole variable, usually subequal to blade; blade nearly concolorous, broadly ovate or weakly 3-lobed, 4–8 cm, ± as long as wide, base cordate, margins dentate, apex acuminate, surfaces softly tomentose. **Inflorescences** solitary flowers or terminal panicles. **Flowers:** calyx 9–15 mm, lobes accrescent to 20 mm, basally overlapping, not reflexed in fruit, cordate, to 8 mm wide; corolla yellow-orange

throughout, petals 20–25 mm; staminal column glabrous; style 10-branched. **Schizocarps** broadly ovoid, 10–12 × 15–16 mm; mericarps: apex acute or apiculate, densely hirsute. **Seeds** 3 per mericarp, 3 mm, puberulent.

Flowering late winter–spring. Desert habitats; 0–1000 m; Ariz., Calif.; Mexico (Baja California, Sinaloa, Sonora, Tamaulipas).

Abutilon palmeri has become popular in cultivation. It is distributed from the Sonoran Desert to San Diego County, in the San Jacinto Mountains, and in the southern counties of Arizona.

12. Abutilon parishii S. Watson, Proc. Amer. Acad. Arts 20: 357. 1885 C

Subshrubs, 1 m. **Stems** erect, stellate-pubescent and with ± retrorse, simple hairs 2 mm. **Leaves:** stipules caducous, subulate, 5 mm; petiole equaling or to 2 times length of blades of lower leaves; blade discolorous, ovate, 2.5–6.5 cm, longer than wide, base cordate, margins coarsely dentate, apex acute, abaxial surface softly stellate-pubescent, adaxial surface with appressed simple and bifurcate hairs 1 mm. **Inflorescences** solitary flowers. **Flowers:** calyx 6–8 mm, lobes not overlapping, erect in fruit, ovate; corolla yellow-orange throughout, not fading pinkish, petals 8–12 mm; style 6–8-branched. **Schizocarps** ± ovoid, 10 × 10 mm; mericarps: apex spinose, spines 2–3 mm, stellate-pubescent and ciliate on keel, hairs 1.5 mm. **Seeds** 3 per mericarp, 2.5 mm, minutely scabridulous. $2n = 14$.

Flowering late summer.   Mountain slopes, desert scrublands; of conservation concern; 1000 m; Ariz.; Mexico (Sonora).

Abutilon parishii is a regional endemic, found north to south in Arizona.

Abutilon parishii is in the Center for Plant Conservation's National Collection of Endangered Plants.

13. Abutilon parvulum A. Gray, Smithsonian Contr.
Knowl. 3(5): 21. 1852

Herbs or subshrubs, perennial, usually to 0.5 m. **Stems** trailing, minutely stellate-pubescent and sometimes with simple hairs 1 mm. **Leaves:** stipules subulate, 2–4 mm; petiole ½ to as long as blade; blade concolorous, broadly ovate, to 5 cm, length ± equaling width, base cordate, margins coarsely dentate, apex acute, surfaces sparsely pubescent, visible through pubescence. **Inflorescences** axillary, solitary flowers. **Flowers:** calyx 3–5 mm, lobes not overlapping, reflexed in fruit, ovate; corolla ± pinkish throughout, petals 4–7 mm; staminal column glabrous; style 5-branched. **Schizocarps** ± ovoid, 8–9 × 8–9 mm; mericarps: apex acute or apiculate, minutely stellate-pubescent. **Seeds** 3 per mericarp, 2 mm, puberulent. $2n = 14$.

Flowering spring–fall.   Dry, open habitats; 500–1500 m; Ariz., Calif., Colo., Nev., N.Mex., Tex., Utah; Mexico (Baja California, Chihuahua, Coahuila, Sonora, Tamaulipas).

Abutilon parvulum is an inconspicuous plant and thus not often collected. It has trailing branches, uncommon in *Abutilon*.

14. Abutilon permolle (Willdenow) Sweet, Hort. Brit.,
53. 1826 • Velvety abutilon

Sida permollis Willdenow, Enum. Pl., 723. 1809

Shrubs, 1–2 m. **Stems** erect, softly stellate-pubescent, without simple hairs. **Leaves:** stipules lanceolate, 5–7 mm; petiole ½ to as long as blade; blade discolorous, ovate, 7–12 cm, ca. 2 times as long as wide, base deeply cordate, margins serrate, apex acuminate, surfaces softly tomentose. **Inflorescences** solitary flowers

or terminal panicles. **Flowers:** calyx 10–15 mm, lobes basally overlapping, erect in fruit, lanceolate-ovate, to 8 mm wide; corolla yellow or yellow-orange throughout, petals 10–18 mm; staminal column glabrous; style 10–12-branched. **Schizocarps** ± ovoid, 8–10 × 8–10 mm; mericarps: apex acute, hirsute. **Seeds** 3 per mericarp, 2 mm, papillate.

Flowering year-round.   Coastal areas, disturbed sites, often on limestone soil; 0–100 m; Fla.; Mexico; West Indies; Central America.

Abutilon permolle is restricted primarily to the four southernmost counties of Florida with an additional collection from Manatee County.

15. Abutilon reventum S. Watson, Proc. Amer. Acad.
Arts 21: 418. 1886

Herbs, annual, **or subshrubs,** to 2 m. **Stems** erect, usually reddish or purplish, minutely glandular-pubescent to glabrate, without simple hairs, obscurely viscid. **Leaves:** stipules lost early and seldom observed, subulate; petiole subequal to blade; blade markedly discolorous, broadly round-ovate, 8–10(–20) cm, ± as long as wide, base deeply cordate, margins obscurely serrulate, apex acuminate, surfaces softly pubescent. **Inflorescences** terminal panicles. **Flowers:** calyx 3–6 mm, lobes not overlapping, not reflexed in fruit, lanceolate-ovate, to 8 mm wide; corolla pale yellow throughout, petals 9–15 mm; staminal column glabrous; style 10-branched. **Schizocarps** globose, 7–10 mm; mericarps: apex rounded or obtuse, stellate-tomentose. **Seeds** 3 per mericarp, 2.5–3 mm, sparsely reticulately scabridulous.

Flowering late summer.   Arid hillsides, along water courses; 1000–1500 m; Ariz.; Mexico.

Abutilon reventum ranges in western Mexico from Chihuahua and Sonora to Oaxaca. It is also found at adjacent sites in Pima County, southern Arizona.

16. Abutilon theophrasti Medikus, Malvenfam., 28.
1787 • Butterprint, China jute, chingma, Indian hemp, pie-marker, velvetleaf [I] [W]

Sida abutilon Linnaeus, Sp. Pl. 2: 685. 1753

Herbs, annual, to 1+ m. **Stems** erect, stellate-tomentose, without simple hairs. **Leaves:** stipules lanceolate; petiole subequal to blade; blade concolorous, broadly ovate to suborbiculate, 8–15 cm, ± as long as wide, base cordate, margins crenulate, apex acuminate, surfaces softly pubescent. **Inflorescences** usually solitary flowers, sometimes cymose or racemose. **Flowers:** calyx 10 mm, lobes not overlapping, erect in fruit, ovate; corolla pale yellow throughout, petals 8–13 mm; staminal column glabrous; style 13–15-branched. **Schizocarps** broadly ovoid, 15 × 20 mm; mericarps: apex spinose, spines divergent, 3–6 mm, hirsute. **Seeds** 3 per mericarp, 3–4 mm, minutely puberulent. $2n = 84$.

Flowering summer–fall. Soybean, corn, and cotton fields, naturalized in disturbed sites; 0–1400 m; introduced; Ont.; Ala., Ariz., Ark., Calif., Colo., Conn., Del., Fla., Ga., Idaho, Ill., Ind., Iowa, Kans., Ky., La., Maine, Md., Mass., Mich., Minn., Miss., Mo., Mont., Nebr., Nev., N.H., N.J., N.Mex., N.Y., N.C., N.Dak., Ohio, Okla., Oreg., Pa., R.I., S.C., S.Dak., Tenn., Tex., Utah, Vt., Va., Wash., W.Va., Wis., Wyo.; Europe; Asia.

The history of the introduction of *Abutilon theophrasti* to North America was recounted by N. R. Spencer (1984). It can be abundant locally, thriving when rich cultivated soils are disturbed, especially in the midwestern region. Interference with crops has been extensive.

17. Abutilon trisulcatum (Jacquin) Urban, Repert. Spec. Nov. Regni Veg. 16: 32. 1919

Sida trisulcata Jacquin, Enum. Syst. Pl., 26. 1760; *Abutilon nealleyi* J. M. Coulter; *A. ramosissimum* C. Presl; *A. triquetrum* Sweet

Herbs, annual, sometimes perennial, **or subshrubs,** 1–2.5 m. **Stems** erect, prominently 3-sulcate when young, minutely puberulent, usually viscid (especially in inflorescence). **Leaves:** stipules subulate, 2–4 mm; petiole usually shorter than blade; blade slightly discolorous, ovate, 6–11 cm, longer than wide, base cordate, margins crenulate-serrulate, apex acuminate, surfaces tomentulose, obscured by pubescence. **Inflorescences** terminal, open panicles. **Flowers:** calyx 3–4 mm, lobes not overlapping, erect in fruit, lanceolate-ovate; corolla sometimes reflexed, yellow, often with dark red center, petals 4–6 mm; staminal column glabrous; style 5-branched. **Schizocarps** subcylindric, usually medially constricted, 6–8 × 5–8 mm; mericarps: apex apiculate, minutely puberulent. **Seeds** 3 per mericarp, 2 mm, minutely pubescent. $2n = 14$.

Flowering winter–spring. Dry shrublands, disturbed vegetation, roadsides; 0–300 m; Tex.; Mexico; West Indies; Central America.

Abutilon trisulcatum occurs from Texas to Nicaragua and in the West Indies. It is a common roadside weed in most of Mexico, and has been found in southern Texas in Cameron and Hidalgo counties.

18. Abutilon wrightii A. Gray, Boston J. Nat. Hist. 6: 162. 1850

Subshrubs, to 0.5 m. **Stems** procumbent or ascending, sometimes purplish, minutely stellate-pubescent, usually with simple hairs 1 mm. **Leaves:** stipules filiform, 3–4 mm; petiole often exceeding blade; blade markedly discolorous, ovate, 1.5–4 cm, ± as long as wide, base deeply cordate, margins dentate, apex acute to obtuse, abaxial surface with soft, densely matted pubescence, adaxial surface sparsely pubescent. **Inflorescences** solitary flowers. **Flowers:** calyx 10–15 (–20) mm, lobes accrescent, basally overlapping, cordate; corolla pale yellow throughout, petals 14–18 mm; staminal column glabrous; style 6–9-branched. **Schizocarps** ± ovoid, 10 × 11 mm; mericarps: apically acuminate, densely hirsute. **Seeds** 3 per mericarp, 2.6 mm, muriculate. $2n = 14$.

Flowering spring–fall. Dry, open habitats, roadsides; 100–300 m; N.Mex., Tex.; Mexico (Chihuahua, Coahuila, Nuevo León, San Luis Potosí, Tamaulipas).

Abutilon wrightii is found in Texas primarily in the Big Bend area and scattered locations from Cameron and Travis counties. It has been reported in New Mexico from Eddy and Otero counties and likely elsewhere in the extreme southeast.

13. **ALCEA** Linnaeus, Sp. Pl. 2: 687. 1753; Gen. Pl. ed. 5, 307. 1754 • Hollyhock [Greek *alkea*, a kind of mallow] ⊡

Steven R. Hill

Herbs, [annual], biennial or perennial, stellate-hairy to pilose or hirsute or glabrous, [sometimes with some long, simple hairs, sometimes glabrate]. **Stems** erect, usually simple. **Leaves:** stipules persistent or caducous, ovate [unlobed] or 2–4-fid, sparsely to densely stellate-pilose; blade orbiculate, angled, weakly lobed or deeply palmately parted, base cordate, cuneate, or truncate, margins crenate-serrate, apex acute to obtuse. **Inflorescences** terminal and/or axillary, usually unbranched, racemes, often with 1–5-flowered axillary fascicles, elongate, flowers axillary, solitary or fascicled; involucellar bractlets persistent, attached to apex of pedicel, connate basally, 6 or 7[–9]-parted, stellate-hairy. **Flowers:** calyx usually accrescent, not inflated, lobes slightly or conspicuously striate, lanceolate, margins entire, apex obtuse to acuminate, densely stellate-pilose-hairy; corolla rotate, white, pink, red, purple, or yellow, darker or paler basally, base densely white-pilose-hairy; staminal column exserted, 5-angled, anthers crowded, pale yellow, glabrous; ovary [15–]20–40-carpellate; ovules 1 per carpel; style [15–]20–40-branched (equaling number of locules); stigmas decurrent, filiform. **Fruits** schizocarps, erect, not inflated, disc-shaped, dry, central axis equaling or shorter than mericarps, indehiscent; mericarps [15–]20–40, 2-celled (proximal cell 1-seeded, distal cell sterile), laterally compressed and reniform-circular with prominent ventral notch, smooth to wrinkled, hairy [glabrous]. **Seeds** 1 per mericarp, brown, reniform, glabrous or minutely hairy. $x = 21$ [$n = 13, 21$].

Species ca. 70 (2 in the flora): introduced; s Europe, Asia (Mediterranean region to c Asia).

A few species of *Alcea* are widely cultivated and both species in the flora area are escapes from cultivation. Various authors have treated some of these taxa within *Althaea*, but this disagrees with Linnaeus's concepts of the two genera as quite distinct. The primary difference is that *Alcea* has a two-chambered mericarp (the upper chamber being empty and vestigial) and yellowish anthers, and *Althaea* has a one-chambered mericarp and purple or brownish purple anthers. Current treatments consistently accept the two genera as distinct; see M. E. Uzunhisarcikii and M. Vural (2012) for a discussion of the two genera and their circumscriptions.

Alcea biennis Winterl occasionally is planted and rarely is found as an escape. It differs from the two species treated here by its white to pink corolla with a pale yellow to greenish center and by its generally more deeply lobed petals that are usually more separated and less overlapping. Its involucel is more than one-half as long as the calyx, sometimes equal in length, the sepals are conspicuously striate, the pedicel is 1–25 mm long, and the mericarps are conspicuously winged; its leaves are inconspicuously lobed or merely angled.

1. Leaf blades angled or shallowly lobed or rarely more deeply; flowers usually white, pink, red, or purple, rarely yellow, not drying greenish; involucellar bractlets ½+ calyx length. . . . . . . .1. *Alcea rosea*
1. Leaf blades lobed usually halfway or more to midrib, often figlike; flowers pale yellow, usually drying greenish; involucellar bractlets usually ½–⅔ calyx length . . . . .2. *Alcea rugosa*

1. **Alcea rosea** Linnaeus, Sp. Pl. 2: 687. 1753
   • Common hollyhock, rose trémière   F I W

Alcea ficifolia Linnaeus; *A. glabrata* Alefeld; *Althaea ficifolia* (Linnaeus) Cavanilles; *A. mexicana* Kunze; *A. rosea* (Linnaeus) Cavanilles; *A. rosea* var. *sinensis* (Cavanilles) S. Y. Hu; *A. sinensis* Cavanilles

Plants 1–2.5+ m, roughly stellate-hairy to hirsute. **Leaves:** stipules ovate, 8 mm, apically 3-lobed; petiole equaling or longer than blade; blade suborbiculate to 5–7-angled or shallowly triangular-lobed, sometimes more deeply lobed, 7–16 cm, margins crenate. **Inflorescences** simple, flowers solitary or fascicled, subsessile; bracts leaflike. **Pedicels** 5 mm, to 8–10 mm in fruit; involucellar bractlets ovate-triangular, 6- or 7-parted, ½+ calyx length. **Flowers:** calyx campanulate, rough-tomentose; corolla sometimes doubled, 8–10 cm diam., petals usually white, pink, red, purple, or black-purple, rarely yellow, subentire to shallowly notched, narrowed to base, 5 cm; staminal column 2 cm, bearing anthers at apex; anthers distinctly 5-rowed viewed from above; style 20–40-branched, puberulent. **Schizocarps** enclosed by calyx, 2 cm diam., hairs somewhat harsh and irritating; mericarps 20–40, channeled and winged dorsally, 6–8 mm. **Seeds** tuberculate or not, often minutely hairy. $2n = 42$.

Flowering May–Oct; fruiting Jun–Oct. Disturbed sites, roadsides, vacant lots; 0–3000 m; introduced; N.B., Ont., Que.; Ariz., Calif., Colo., Conn., Del., D.C., Ga., Idaho, Ill., Ind., Iowa, Kans., Ky., Md., Mass., Mich., Minn., Miss., Mo., Mont., Nebr., Nev., N.J., N.Mex., N.Y., N.C., N.Dak., Ohio, Oreg., Pa., R.I., S.Dak., Tenn., Tex., Utah, Vt., Va., Wash., W.Va., Wis., Wyo.; Asia (China); introduced also nearly worldwide.

Alcea rosea is a showy and popular ornamental that is essentially cosmopolitan in cultivation. The species is thought to have originated in the southwestern provinces of China but is apparently not known in the wild. It occasionally escapes and naturalizes in disturbed temperate areas nearly worldwide. However, it is often difficult to determine if a given specimen was cultivated or an established adventive. Plants with more deeply lobed leaves and rose-pink flowers have been called *A. ficifolia*; plants in cultivation under this name are most likely a mix of *A. rosea* and *A. rugosa* or of hybrid origin.

2. **Alcea rugosa** Alefeld, Oesterr. Bot. Z. 12: 254. 1862

• Russian hollyhock [1]

Alcea novopokrovski Iljin; *A. taurica* Iljin; *Althaea rugosa* (Alefeld) Litvinov

Plants 1.5–2.5 m, stellate-hairy. **Leaves:** stipules ovate, 6–11 mm, apically 2–4-lobed; petiole equaling or longer than blade; blade ovate to ovate-orbiculate, usually 5–7-lobed ½ or more to midrib, often figlike, 7–15 cm, margins crenate. **Inflorescences** simple, flowers solitary, subsessile; bracts leaflike. **Pedicels** 6 mm, to 8–10 mm in fruit; involucellar bractlets 6-parted, usually ½–⅔ calyx length. **Flowers:** calyx hairy; corolla not doubled, 8–10 cm diam., petals pale yellow, usually drying greenish, notched, narrowed to base, 5 cm; staminal column bearing anthers at apex; anthers 5-rowed when viewed from above; style 20–40-branched. **Schizocarps** enclosed by calyx, 2 cm diam.; mericarps (20–)30(–40), channeled and winged dorsally, 4–6 mm. **Seeds** tuberculate or not, glabrous or minutely hairy axially. $2n = 42$.

Flowering May–Oct; fruiting Jun–Oct. Disturbed sites, roadsides, vacant lots; 0–300 m; introduced; Md., Wis.; e Europe; w Asia; introduced widely.

Alcea rugosa is a showy ornamental that is becoming more frequent in cultivation because it is more resistant to hollyhock rust (*Puccinia malvacearum*) than *A. rosea*. The species is native in temperate eastern Europe in Ukraine and the Russian Federation, and to western Asia in Armenia, Azerbaijan, and Georgia. It occasionally escapes and naturalizes in disturbed temperate areas worldwide.

14. ALLOWISSADULA D. M. Bates, Gentes Herbarum 11: 337, figs. 3–5. 1978

• [Greek *allo-*, different, and genus *Wissadula*]

David M. Bates

Subshrubs, herbage usually hairy, often viscid. **Stems** erect or spreading. **Leaves:** stipules caducous, filiform; blade ovate to subround, unlobed or 3-lobed, base cordate [obtuse or subtruncate], margins toothed. **Inflorescences** axillary and solitary flowers or cymes or terminal and racemes or panicles; involucellar bractlets absent. **Flowers:** calyx [5–]8–13.5[–14] mm, not or slightly accrescent in fruit, not inflated, lobes smooth to strongly ribbed, ovate to narrowly to broadly triangular, apex apiculate to caudate; corolla campanulate, yellow to orange-yellow [white] (sometimes fading rose); staminal column ± included, anther-bearing at apex; filaments terminal and subterminal; ovary 5-carpellate; ovules 3 per carpel, pendulous or horizontally pendulous, 2 collateral in upper carpel cell, 1 in lower carpel cell; style 5-branched; stigmas capitate. **Fruits** schizocarps (± functionally capsular), erect, not inflated, turbinate, 5-lobed, starlike in cross section, papery to indurate, hairy; mericarps 5, 2-celled, lower cell ± trapezoidal, 1-seeded, upper cell angular-orbiculate, collaterally 2-seeded, cells sometimes separated by endoglossum. **Seeds** obovoid-reniform, glabrous. $x = 8$.

Species 9 (2 in the flora): sc United States, Mexico.

1. Herbage viscid, glandular hairs present; mericarps moderately constricted, endoglossum absent . . . . 1. *Allowissadula holosericea*
1. Herbage not viscid, glandular hairs absent; mericarps strongly constricted, endoglossum present . . . . 2. *Allowissadula lozanoi*

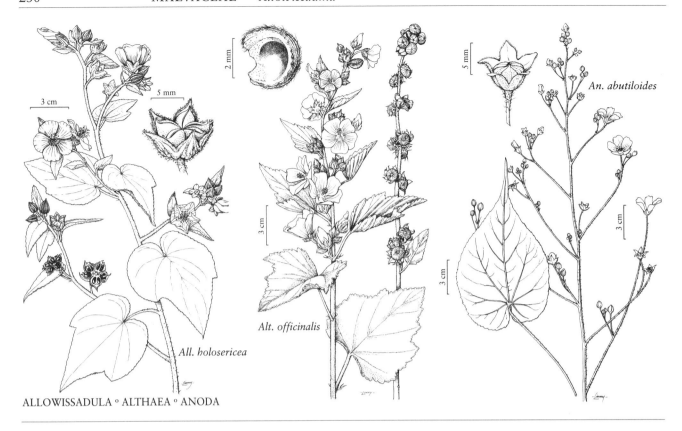

An. abutiloides

Alt. officinalis

All. holosericea

ALLOWISSADULA ∘ ALTHAEA ∘ ANODA

1. Allowissadula holosericea (Scheele) D. M. Bates, Gentes Herbarum 11: 340. 1978 [F]

Abutilon holosericeum Scheele, Linnaea 21: 471. 1848; *A. marshii* Standley; *Wissadula holosericea* (Scheele) Garcke; *W. insignis* R. E. Fries

Plants 1–2 m; herbage sparsely to densely, sometimes velvety tawny, stellate-hairy, glandular. **Leaf blades** ovate to subspheric, unlobed or acutely 3-lobed, 5–10(–30) cm, base closed-cordate. **Inflorescences** terminal, usually open panicles or cymes, individual units 2+-flowered, sometimes solitary flowers. **Flowers:** calyx campanulate, smooth or ribbed, (7–)8–13.5 mm, subequal or slightly exceeding fruits, lobes ovate to triangular, (3–)4–6.5 × (3–)4–5.5 mm; corolla orange-yellow, (12–)15–25 mm; staminal column 10 mm, including filaments. **Mericarps** moderately dorsolaterally constricted, 7–9 mm including dorso-apical apiculus or spine to 1.5(–2) mm, stellate- and glandular-hairy, endoglossum absent. **Seeds** reddish brown to blackish, 2.3–3 mm. $2n$ = 16.

Flowering May–Nov. Open habitats, roadsides; 100–1600 m; N.Mex., Tex.; Mexico (Coahuila, Nuevo León, San Luis Potosí, Tamaulipas).

In the flora area, *Allowissadula holosericea* ranges from the Edwards Plateau to the trans-Pecos mountains and westward; in Mexico, it occurs north of the Sierra Madre Oriental and Sierra de Tamaulipas.

The name *Abutilon velutinum* A. Gray 1849 (not G. Don 1831), which pertains here, is illegitimate.

2. Allowissadula lozanoi (Rose) D. M. Bates, Gentes Herbarum 11: 347. 1978 (as lozanii)

Wissadula lozanoi Rose, Contr. U.S. Natl. Herb. 10: 124, plate 41. 1906 (as lozani); *Pseudabutilon lozanoi* (Rose) R. E. Fries

Plants 1–1.5 m; herbage densely, closely tawny-yellowish, roughly stellate-hairy, not glandular. **Leaf blades** ovate to transversely ovate, unlobed or with low hiplike lobe on either side, 5–13 cm, base mostly open-cordate. **Inflorescences** usually dense or open cymes, sometimes solitary flowers. **Flowers:** calyx campanulate, densely tawny-hairy, strongly 10-ribbed at base, 8–12 mm, subequal to fruits, lobes keeled, ovate to deltate-ovate, (3–)4–8 × 4–6 mm; corolla yellow to orange-yellow, 13–20 mm; staminal column to 12 mm including filaments. **Mericarps** strongly dorsolaterally constricted, 7.5–10 mm including dorso-apical apiculus or spur 0.5–2 mm, densely tawny stellate-hairy, endoglossum narrow, stout, spearlike, nearly reaching ventral suture. **Seeds** grayish with reddish-brown raphe, 2.7–3 mm. $2n = 16$.

Flowering May–Oct. Mostly clay soil, open scrub or waste places; 10–200 m; Tex.; Mexico (Nuevo León, Tamaulipas).

Allowissadula lozanoi occurs east and south of the Edwards Plateau through the southern Texas coastal plains into northeastern Mexico.

15. **ALTHAEA** Linnaeus, Sp. Pl. 2: 686. 1753; Gen. Pl. ed. 5, 307. 1754 • Marshmallow, guimauve [For Althaea, wife of King Oeneus of Aetolia or Calydon] ☐

Steven R. Hill

Herbs, annual or perennial, usually hairy, hairs stellate or simple, not viscid. **Stems** erect to decumbent. **Leaves:** stipules persistent or caducous, subulate or linear to lanceolate, simple to 2- or 3-fid; blade reniform, deltate-ovate, or ovate, bluntly or acutely lobed or palmately parted, base cuneate, truncate, obtuse, or cordate, margins crenate to coarsely dentate or serrate. **Inflorescences** axillary, solitary flowers or in 2–4-flowered fascicles, these concentrated on ends of stems and branches; involucel present, bractlets persistent, 6–9(–12), connate basally. **Flowers:** calyx not or slightly accrescent, not inflated, campanulate, 5-lobed, lobes connate basally, not ribbed, ovate or lanceolate, apex obtuse, acuminate, or apiculate; corolla rotate to wide-campanulate, white or pink to lilac-purple or bluish, petals asymmetric, broadly obovate, 1–2 cm, margins entire or notched; staminal column exserted, cylindric viewed from above, glabrous or hairy, anthers purple or pale pink to nearly white (yellow); ovary [8–]10–20 [–24]-carpellate; ovules 1 per carpel; style low-branched and appearing to be [8–]10–20[–24]-branched; stigmas decurrent on upper inner surface, linear. **Fruits** dry schizocarpic achenes, erect, not inflated, reniform to orbicular, rounded to angled, unwinged, not or slightly indurate, stellate-pilose-hairy or glabrous, indehiscent; mericarps [8–]10–20[–24], 1-celled. **Seeds** 1 per mericarp, reniform-round, minutely rugulose, glabrous. $x = 21$.

Species 6–12 (4 in the flora): introduced; Europe, w, c Asia (Mediterranean region).

Two of the species of *Althaea* in the flora area occur in cultivation and rarely escape. The perennial species generally have stellate pubescence and the annual species generally have simple hairs; both have pink to purple anthers; the annuals are thought by some to be better placed within the genus *Malva*. *Althaea* is very similar to *Alcea*, and the latter has been combined with it by various authors. Hybrids between the two (×*Alcathaea* Hinsley) have been described, as have hybrids between *Malva* and *Althaea* (×*Malvalthaea* Iljin). See discussion under 13. *Alcea* for more information.

1. Annuals, hispid-hirsute; petals to 1.5 times as long as calyx; staminal columns glabrous; anthers pale pink to almost white (yellow). . . . . .3. *Althaea hirsuta*
1. Perennials, stellate-hairy; petals (1.1–)2–3 times as long as calyx; staminal columns clavate-hairy or glabrous; anthers purple.
   2. Leaves unlobed or at most 3(–5)-lobed ½ to midrib; mericarps densely stellate-hairy
     . . . . .4. *Althaea officinalis*
   2. Leaves palmately divided ½+ to midrib, 5-lobed; mericarps glabrous or stellate-hairy.
     3. Peduncles many-flowered; mericarps stellate-pilose-hairy . . . . .1. *Althaea armeniaca*
     3. Peduncles 2- or 3-flowered; mericarps glabrous . . . . .2. *Althaea cannabina*

1. Althaea armeniaca Tenore, Index Seminum (Naples) 1839: 11. 1839 • Armenian marshmallow, guimauve arménienne [I]

Althaea micrantha Wiesbaur ex Borbás; *A. officinalis* Linnaeus var. *pseudoarmeniaca* Polgár

Herbs perennial, 0.4–2 m. **Stems** erect, branched distally, softly stellate-tomentose. **Leaves:** stipule somewhat persistent, linear, simple, 2–5 mm, white pilose-hairy; petiole 0.5–5 cm, stellate-hairy; blade orbiculate, palmately divided, 5-lobed, 5–7 × 6–8 cm, lobes lanceolate, 0.5–5 × 0.3–3.5 cm, base truncate, margins irregularly serrate, apex obtuse, upper leaves reduced, orbiculate-triangular, 3–5-lobed, 1–6 × 0.6–5 cm, middle lobe longer than others, surfaces stellate-hairy. **Inflorescences** 1–4-flowered fascicles in leaf axils or peduncles many-flowered; peduncles 0.5–3 cm. **Pedicels** 0.1–5 cm, densely stellate-hairy; involucellar bractlets 6–10, linear-lanceolate, free portions 2–6 × 0.5–2 mm, ½+ as long as calyx, margins entire, apex acuminate, densely stellate-hairy. **Flowers:** calyx 8–10 mm, not or only slightly accrescent, lobes lanceolate, margins entire, apex acuminate, stellate-hairy; petals lilac-purple, 9–15 × 4–6 mm, 1.1–1.5(–3) times as long as calyx, margins entire, apex obtuse or slightly notched; staminal column 1–3 mm, glabrous or rarely sparsely papillose-hairy; anthers in upper ½, purple; style 10–20-branched. **Fruits** partially concealed by incurved calyx lobes, 7 mm diam.; mericarps 10–20, erect, brown, unwinged, orbiculate, rugose, stellate pilose-hairy, sides glabrous. **Seeds** brown, reniform-round, 1–2 × 2–2.5 mm, glabrous. $2n = 84$.

Flowering summer–fall. Wet or moist ditches, disturbed areas, stream drainages; 300–400 m; introduced; S.Dak.; w Asia (n Iran, s Russia, Turkey, Turkistan).

Althaea armeniaca is very similar to *A. officinalis* and is known to hybridize with it. *Althaea armeniaca* differs primarily by the more deeply five-lobed leaves rather than the three in *A. officinalis* and can be distinguished from *A. cannabina* by its pubescent fruits. Only a single population is known from the flora area, found in Hutchinson County, near Lonetree Creek west of Olivet. The species dominated a fence row at the edge of a field and was locally common in scattered clumps over an area of about three miles.

2. Althaea cannabina Linnaeus, Sp. Pl. 2: 686. 1753 • Hemp-leaved marshmallow [I]

Althaea kotschyi Boissier; *A. narbonnensis* Pourret ex Cavanilles

Herbs perennial, (0.5–)1–2 m. **Stems** erect, clumped, branched distally, canescent, hairs stellate. **Leaves:** stipules mostly persistent, narrowly linear, simple or 3-fid, 2–8 mm; petiole as long as blade in proximal leaves, greatly reduced in distal leaves; blades of proximal leaves deltate-ovate, 3–5-palmately lobed or divided ± to base, 3–12 × 2–16 cm, lobes oblong to lanceolate, middle lobe longest, margins deeply dentate to serrate, surfaces stellate-hairy, blades of upper leaves similar but narrower, 3–5-palmately lobed or divided, 1–4.5 × 0.5–6 cm, base cuneate, margins serrate, apex acuminate. **Inflorescences** solitary flowers, usually long-pedunculate (1–15 cm) or peduncle 2- or 3-flowered apically; bracts 2, linear, 2 mm. **Pedicels** 0.5–16 cm; involucellar bractlets 7–9, erect, lanceolate, 3–7 mm, slightly accrescent, ½+ as long as calyx, stellate-tomentose. **Flowers:** calyx 5–7(–10) mm, lobes ovate, 2 times as long as tube, margins entire, apex obtuse, acuminate, or apiculate, stellate-tomentose; petals pink to pink-purple, 10–23 × 5–12 mm, 2 times as long as calyx, apex obtuse or notched; staminal column 0.5–6 mm, glabrous or sparsely clavate-hairy; anthers in upper ½, purplish to purplish brown; stigmas 12–16. **Fruits** somewhat longer than calyx and not concealed, 7 mm diam.; mericarps 12–16, brown, unwinged, reniform, 2.5–4 × 2–3.5 mm, striate with ribs radiating from rugose central area, back herringbone-veined, glabrous. **Seeds** brown, reniform, 3 mm, glabrous. $2n = 84$.

Flowering summer. Disturbed, usually moist areas; 0–200 m; introduced; D.C.; Europe; Asia (nw Iran, s Russia, Turkey, Turkistan); widespread.

Althaea cannabina is occasionally cultivated as an ornamental and rarely escapes or naturalizes.

3. Althaea hirsuta Linnaeus, Sp. Pl. 2: 687. 1753

• Rough marshmallow [I]

Herbs annual, 0.1–0.6 m. **Stems** erect to decumbent, simple or branched at base, coarsely hairy, hairs simple, rigid, stellate, pustulate-based. **Leaves:** stipules persistent, lanceolate, simple, 5–8 mm, hirsute; petiole as long as blade in proximal leaves, much reduced in distal leaves; blade reniform, those of proximal leaves shallowly 3–5-lobed, those of distal leaves deeply, palmately 3–5-lobed or divided, 1–4 × 1.5–3.5(–4) cm, lobes broad, blunt to acute, margins crenate to pinnatifid, surfaces hispid-hairy. **Inflorescences** solitary flowers, long-pedicellate, infrequently racemose-congested distally. **Pedicels/peduncles** 0.5–8 mm, 8 cm proximally, reduced distally; involucellar bractlets 7 or 8, spreading-erect, lanceolate, 4–13 × 1–4 mm, ½+ as long as calyx, slightly accrescent, rigid in fruit, hispid. **Flowers:** calyx 15 mm, lobes erect, lanceolate-acuminate, 5–15 × 1–3 mm, 2 times as long as tube, pustular-hispid; petals pinkish lilac, fading bluish, 12–20 mm, 1.5 times as long as calyx, apex entire or slightly notched; staminal column 0.5–1 mm, glabrous, sometimes sparsely glandular; anthers in upper ½, pale pink to almost white (yellow); stigmas 10–15. **Fruits** concealed by erect, accrescent calyx, 8–10 mm diam.; mericarps 10–15, brown, unwinged, reniform, with strong, radiating ridges on lateral face, 3 mm, margins rounded, surface glabrous, obscurely ridged abaxially. **Seeds** brown, reniform-round, 1–1.3 × 0.7–1.4(–2.5) mm, minutely rugulose, glabrous. $2n = 42$.

Flowering spring–summer. Disturbed areas; 0–100 m; introduced; Ont.; N.Y., Pa.; Europe; w, sw Asia.

There are very few recent North American collections of *Althaea hirsuta*. There is an old specimen of this plant from British Columbia dated 1924, but the species has not been collected in the province since then. Current molecular data suggest that it may belong within *Malva*.

The name *Althaea hispida* Moench, which pertains here, is superfluous and illegitimate.

4. Althaea officinalis Linnaeus, Sp. Pl. 2: 686. 1753

• Marshmallow, guimauve officinale [F] [I]

Althaea sublobata Stokes; *A. taurinensis* de Candolle; *Malva officinalis* (Linnaeus) Schimper & Spenner

Herbs perennial, to 1.5 m. **Stems** erect, clustered, branched distally or unbranched, softly stellate-tomentose. **Leaves:** stipules somewhat persistent, usually caducous, linear-lanceolate, subulate, sometimes 2-fid or dentate, (2–)5–8 mm, densely stellate-hairy; petioles 1–6 cm, reduced on distal leaves, usually shorter than blade; blades of proximal leaves ovate or obscurely 3-lobed less than ½ to midrib, distal leaves deltate-ovate to ovate, (2–)4–10 × 2–7 cm, base truncate to cuneate, lobes acuminate or broadly acute to obtuse, middle lobe larger than others, deeply plicate, margins irregularly dentate to crenate-serrate, surfaces softly stellate-tomentose, ribs very prominent abaxially. **Inflorescences** solitary flowers or 2–4-flowered fascicles in leaf axil, sometimes aggregated apically into terminal false racemes. **Pedicels/peduncles** 0.5–4 cm; involucellar bractlets 8–12, erect, linear-lanceolate, ½ length of calyx, 6 mm, lobes 2–6 × 1–2 mm, stellate-tomentose. **Flowers:** calyx 8–10 mm, lobes narrowly ovate-acuminate, 6 mm, 2 times as long as tube, stellate-velutinous; petals usually pale pink, rarely white, cuneate-obovate, 7.5–15 × 6–13 mm, 2–3 times as long as calyx, apex obtuse or notched; staminal column 3–5 mm, glabrous or sparsely papillose-hairy; anthers in upper ½, dark purple; style 15–20-branched. **Fruits** partially concealed by incurved, somewhat accrescent calyx lobes, 7–9 mm diam.; mericarps 15–20, brown, unwinged, orbiculate-reniform, 1.5–2.5 × 2–3 mm, rugose, lateral surface smooth, membranous, dorsal surface stellate-tomentose with medial furrow. **Seeds** brown, reniform-round, 1.5–2 × 1–1.5 mm, glabrous. $2n = 42$.

Flowering summer–fall. Wet, disturbed areas, along streams, brackish sand, coastal marshes; 0–200 m; introduced; N.B., Ont., Que.; Ark., Conn., Del., Ky., Md., Mass., Mich., Nebr., N.J., N.Y., N.Dak., Ohio, Pa., Va., Wis.; Eurasia; n Africa.

Althaea officinalis is occasionally cultivated for ornament, food (especially for the mucilaginous root sap once used with sugar to make marshmallows), and as a medicine; it occasionally escapes. There are few recent North American collections.

16. ANODA   Cavanilles, Diss. 1: 38, plate 10, fig. 3; plate 11, figs. 1, 2. 1785

* [Ceylonese vernacular name for a species of *Abutilon*]

Paul A. Fryxell†

Steven R. Hill

Cavanillea Medikus; *Sidanoda* (A. Gray) Wooton & Standley

Herbs or subshrubs, annual. **Stems** erect to decumbent, hispid or stellate-hairy to glabrescent. **Leaves:** stipules deciduous, inconspicuous, usually linear; blade usually linear, lanceolate, oblong, or ovate to triangular, sometimes lobed, base truncate, cordate, or cuneate, margins dentate to entire. **Inflorescences** axillary solitary flowers or terminal racemes or panicles; involucel absent. **Flowers:** calyx accrescent or not, not inflated, ribbed or not, base rounded, lobes ovate to triangular, apex acute or acuminate; corolla yellow, lavender, or purplish, rarely white; staminal column included; style 5–19-branched; stigmas usually abruptly capitate. **Fruits** schizocarps, erect, not inflated, oblate, not indurate, hairy; mericarps 5–19, 1-celled, with or without spur at dorsal angle, lateral walls usually disintegrating at maturity, irregularly dehiscent. **Seeds** 1 per mericarp, sometimes enclosed in persistent reticulate endocarp. $x = 15$.

Species 23 (7 in the flora): North America, Mexico, West Indies, Central America, South America, Australia.

Anoda is predominantly Mexican in both distribution and maximum diversity, the South American occurrences being predominantly of the weedy *A. cristata*.

SELECTED REFERENCES Bates, D. M. 1987. Chromosome numbers and evolution in *Anoda* and *Periptera* (Malvaceae). Aliso 11: 523–531. Fryxell, P. A. 1987. Revision of the genus *Anoda* (Malvaceae). Aliso 11: 485–522.

1. Petals usually lavender to purplish, rarely white.
   2. Plants frequently decumbent; petals 8–26(–30) mm, manifestly exceeding calyx, lavender or purplish, rarely white; adaxial leaf surface with appressed, simple hairs 1 mm; mericarps with dorsal spurs 1.5–4 mm . . . . 3. *Anoda cristata*
   2. Plants erect; petals 4–7 mm, barely exceeding calyx, blue-purple; adaxial leaf surface minutely stellate-hairy (rarely hairs simple); mericarps with dorsal spurs absent or to 1 mm.
      3. Mericarps 10 or 11, without dorsal spurs; endocarp present; midstem leaves with 3 narrowly linear lobes; staminal columns usually glabrous . . . . .6. *Anoda reticulata*
      3. Mericarps 6–8, with evident, small, dorsal spurs; endocarp incompletely developed or absent; midstem leaves ovate to hastate to triangular; staminal columns hairy
      . . . . .7. *Anoda thurberi*
1. Petals pale to bright yellow.
   4. Leaf blade surfaces with simple, appressed hairs adaxially; petals bright yellow; fruits hirsute, mericarps 10–12, dorsally spurred . . . . 4. *Anoda lanceolata*
   4. Leaf blade surfaces minutely tomentose; petals pale yellow, sometimes fading with reddish blush; fruits minutely or densely hairy, mericarps 5–13, dorsally rounded or spurred.
      5. Mericarps 10–13, dorsal spur 1–2 mm; endocarp present; petals 6–8 mm, not fading reddish . . . . .2. *Anoda crenatiflora*
      5. Mericarps 5–8, dorsally rounded or with spur to 0.5 mm; endocarp absent; petals 10 mm, often fading reddish.
         6. Calyces 5–7 mm; mericarps 5; fruits 6 mm diam.; leaf blades membranous, concolorous, broadly ovate, apex acuminate . . . . 1. *Anoda abutiloides*
         6. Calyces 3–5 mm; mericarps 5–8; fruits 4–5 mm diam.; leaf blades coriaceous, discolorous, often narrowly oblong or linear, apex acute . . . . 5. *Anoda pentaschista*

1. Anoda abutiloides A. Gray, Proc. Amer. Acad. Arts 22: 300. 1887 [F]

Anoda caudatifolia (B. L. Robinson & Greenman) B. L. Robinson & Greenman; *A. urophylla* L. Riley

Subshrubs, branching principally in inflorescence, 1 m. **Stems** erect, with simple hairs 0.5–1 mm and shorter glandular hairs. **Leaves:** petiole subequal to blade or shorter upward, with simple hairs 0.5–1 mm and shorter glandular hairs; blade concolorous, broadly ovate, gradually reduced and narrower upward, to 12 cm, membranous, base cordate, margins dentate, apex acuminate, surfaces minutely tomentose. **Inflorescences** usually panicles. **Pedicels** 1–5 cm. **Flowers:** calyx 5–7 mm, lobes with dark midrib, apex ± acuminate, tomentose; petals pale yellow, drying reddish, 10 mm, prominently bearded on claw; staminal column with recurved hairs; style 5-branched; stigmas glabrous. **Schizocarps** 6 mm diam., minutely hairy; mericarps 5, dorsally rounded. **Seeds** without enclosing endocarp. $2n = 30$.

Flowering spring–fall. Dry, open shrublands; 1000–1500 m; Ariz.; Mexico (Chihuahua, Jalisco, Sinaloa, Sonora).

Anoda abutiloides is found in Pima and Santa Cruz counties.

2. Anoda crenatiflora Ortega, Nov. Pl. Descr. Dec. 8: 96. 1798 • Pintapán del monte

Anoda parviflora Cavanilles; *Sida crenatiflora* (Ortega) Persoon

Herbs, to 1 m. **Stems** erect, stellate-hairy, hairs 0.1–0.3 mm. **Leaves:** petiole subequal to blade proximally, shorter distally, with stellate hairs 0.1–0.3 mm; blade concolorous, ovate to hastate, narrowly so upward, 3–9 cm, membranous, base truncate, margins coarsely crenate-dentate, apex acute, surfaces minutely and obscurely stellate-hairy, hairs simple, bifurcate adaxially. **Inflorescences** racemes or panicles. **Pedicels** 2–7 cm. **Flowers:** calyx 3–7 mm, accrescent to 6–8 mm, lobes without dark midrib, apex acute, densely tomentose; petals pale yellow, not fading reddish, 6–8 mm, ciliate on claw; staminal column glabrous or with few apical hairs; style 10–13-branched; stigmas glabrous. **Schizocarps** 7–9 mm diam., densely hairy; mericarps 10–13, with dorsal spur 1–2 mm. **Seeds** enclosed in endocarp. $2n = 60$.

Flowering fall. Dry, open shrublands, deciduous forests; 800–2300 m; Ariz., Tex.; Mexico (Chihuahua, Coahuila).

Anoda crenatiflora is known from five counties in Texas, primarily in the Big Bend region and Cameron County. In Arizona it has been reported only from Santa Cruz County.

3. Anoda cristata (Linnaeus) Schlechtendal, Linnaea 11: 210. 1837 • Violeta [W]

Sida cristata Linnaeus, Sp. Pl. 2: 685. 1753; *Anoda arizonica* A. Gray; *A. hastata* Cavanilles; *A. triangularis* (Willdenow) de Candolle

Herbs, to 1 m, usually much shorter (quite variable). **Stems** suberect to decumbent, with patent or retrorse, simple hairs, hairs 1 mm. **Leaves:** petiole ½ to equaling blade, hispid; blade concolorous, often with purple blotch along midvein, ovate, triangular, hastate, or sometimes palmately lobed, mostly 3–9 cm, membranous, base cordate, wide-rounded, or truncate, margins crenate to subentire, apex acute, surfaces sparsely hairy, hairs mostly simple, appressed, 1 mm. **Inflorescences** solitary flowers. **Pedicels** 4–12 cm, often exceeding leaf. **Flowers:** calyx 5–10 mm, accrescent to 12–20 mm, lobes without dark midrib, apex acute, hispid; petals purplish or lavender, rarely white, drying purplish, sometimes bluish, 8–26(–30) mm; staminal column hairy; style 10–19-branched; stigmas glabrous. **Schizocarps** 8–11 mm diam. (excluding spines), densely hispid; mericarps 10–19, with dorsal spur 1.5–4 mm. **Seeds** with or without enclosing endocarp. $2n = 30, 60, 90$.

Flowering summer–fall. Usually weedy in disturbed areas, fence rows, agricultural fields; 0–2300 m; Ont.; Ariz., Calif., Fla., Ga., Ill., Iowa, Kans., Ky., La., Mass., Miss., Mo., N.Mex., N.C., Okla., Pa., S.C., Tex., Va.; Mexico; West Indies; Central America; South America; introduced in Australia.

Anoda cristata appears to be increasing its range in North America and elsewhere. It is quite variable, with a decumbent to suberect habit and with varied flower sizes. The weedy form has generally small flowers while those to the south in Mexico have showy larger flowers even used in floral arrangements.

4. Anoda lanceolata Hooker & Arnott, Bot. Beechey Voy., 411. 1840

Anoda wrightii A. Gray

Herbs or subshrubs, branching, 0.5–1 m. **Stems** erect, minutely, roughly hairy, hairs to 0.5 mm. **Leaves:** petiole shorter than blade, minutely, roughly hairy, hairs to 0.5 mm; blade somewhat discolorous, ovate-triangular to lanceolate, 3–7(–12) cm, membranous, base truncate to cuneate, margins obscurely crenate to subentire, apex acute, surfaces hairy, hairs minute, stellate abaxially, simple, appressed adaxially. **Inflorescences** solitary flowers or panicles. **Pedicels** to 4(–6) cm. **Flowers:** calyx 6(–9) mm, accrescent to 9 mm, lobes without dark midrib, apex acute, minutely stellate-hairy; petals bright yellow, 9–16 mm; staminal column stellate-hairy; style 10–12-branched; stigmas glabrous. **Schizocarps** 9 mm diam., densely stellate-hairy, hairs 0.5–1 mm; mericarps 10–12, with dorsal spur 1–1.5 mm. **Seeds** enclosed in endocarp.

Flowering summer–fall. Arid habitats, disturbed sites, sometimes open, sometimes shady; 500–1000 m; Ariz., N.Mex., Tex.; Mexico (Baja California, Chihuahua, Coahuila, San Luis Potosí, Sinaloa).

Anoda lanceolata is found primarily in the western trans-Pecos region of Texas, with outliers in Sierra County, New Mexico, and Cochise County, Arizona.

5. Anoda pentaschista A. Gray, Smithsonian Contr. Knowl. 5(6): 22. 1853

Anoda pentaschista var. *obtusior* B. L. Robinson; *Sidanoda pentaschista* (A. Gray) Wooton & Standley

Subshrubs, widely branching, 1–2 m. **Stems** erect, often purplish, minutely stellate-hairy, hairs to 0.1 mm, glabrescent. **Leaves:** petiole commonly ¼ length of blade (longer on juvenile leaves), minutely stellate-hairy, hairs to 0.1 mm; blade markedly discolorous, highly variable in form and size, often narrowly oblong or linear, usually 1.5+ cm, coriaceous, base sometimes hastate, margins usually entire, apex acute, surfaces ± tomentulose. **Inflorescences** panicles. **Pedicels** 1.5–3.5 cm. **Flowers:** calyx 3–5 mm, lobes with dark midribs, apex acute, minutely hairy; petals pale yellow, sometimes fading reddish, 10 mm; staminal column apically scabrid, otherwise glabrous; style 5–8-branched; stigmas glabrous. **Schizocarps** 4–5 mm diam., minutely hairy; mericarps 5–8, with dorsal spur to 0.5 mm. **Seeds** without enclosing endocarp. *2n* = 30.

Flowering late summer–mid winter. Disturbed sites, roadsides, agricultural fields; 100–600(–1500) m; Ariz., Calif., N.Mex., Tex.; Mexico.

Anoda pentaschista occurs widely in Mexico mostly at lower elevations from Baja California south to Oaxaca. It has been found in the southern coastal bend of Texas south to Cameron County; in Hidalgo County, New Mexico; in Cochise, Maricopa, and Pima counties in Arizona; and as an introduction in Imperial County, California.

6. Anoda reticulata S. Watson, Proc. Amer. Acad. Arts 17: 368. 1882

Herbs, 1 m. **Stems** erect, scabridulous and with some glandular hairs, hairs to 0.2 mm. **Leaves:** petiole equaling or exceeding blade, shorter distally, scabridulous and with some glandular hairs, hairs to 0.2 mm; blade concolorous, with purplish blotch along midrib, typically 3-lobed, lobes narrowly linear, reduced distally to simple linear leaves, 3–6 cm, membranous, base cordate to truncate, margins remotely serrate to subentire, apex subacute, surfaces minutely stellate-hairy. **Inflorescences** solitary flowers or racemes. **Pedicels** to 8.5 cm. **Flowers:** calyx 5 mm, lobes with strong midrib, apex acute, densely hairy; petals bluish purple, 5–6 mm, glabrous; staminal column glabrous or nearly so; style 10- or 11-branched; stigmas glabrous. **Schizocarps** 6–7 mm diam., stellate-hairy; mericarps 10 or 11, dorsally rounded. **Seeds** enclosed in endocarp.

Flowering late summer. Desert habitats; 500–1000 m; Ariz.; Mexico (Sinaloa, Sonora).

Anoda reticulata has been found only in Pima and Santa Cruz counties.

7. Anoda thurberi A. Gray, Proc. Amer. Acad. Arts 22: 299. 1887

Herbs, 1 m. **Stems** erect, with minute stellate and glandular hairs, hairs to 0.2 mm. **Leaves:** petiole ½ to equaling blade, with minute stellate and glandular hairs (hairs to 0.2 mm); blade concolorous, sometimes with purple blotch along midvein, ovate-cordate to hastately 3-lobed at base (distally unlobed) or narrowly triangular, 5–8 cm (often smaller), membranous, base cordate to truncate, margins subentire, apex acute, surfaces minutely hairy, hairs usually stellate. **Inflorescences** usually racemes or panicles. **Pedicels** 1–3 cm. **Flowers:** calyx 3.5–6 mm, accrescent to 6–8 mm, lobes without dark midrib, apex acute, stellate-hairy; petals bluish lavender (darker at base), 4–7 mm; staminal column stellate-hairy; style 6–8-branched; stigmas glabrous. **Schizocarps** 6–8 mm diam., minutely hairy; mericarps 6–8, with dorsal spur to 1 mm. **Seeds** with endocarp absent or incomplete. $2n = 26, 28$.

Flowering late summer–early fall. Deciduous forests, dry, open shrublands; 900–2300 m; Ariz., N.Mex.; Mexico.

Anoda thurberi occurs only in Cochise County, Arizona, and in New Mexico west of the Chiricahua Mountains in low subsaline valley areas. In Mexico it is widespread from Sonora south to Oaxaca.

17. BASTARDIA Kunth in A. von Humboldt et al., Nov. Gen. Sp. 5(fol.): 197; 5(qto.): 254; plate 472. 1822 • [For Toussaint Bastard, 1784–1846, French botanist]

Paul A. Fryxell†

Steven R. Hill

Subshrubs or shrubs. Stems erect, not cespitose, with simple, stellate, or glandular hairs, never lepidote, viscid. **Leaves:** stipules persistent, subulate; blade ovate, unlobed or weakly lobulate, not maplelike, base cordate, margins serrate to subentire, surfaces stellate- and glandular-hairy. **Inflorescences** axillary, solitary flowers or terminal panicles; involucel absent. **Flowers:** calyx not accrescent, not inflated, divided nearly to base, shorter than mature fruits and not completely enclosing them, lobes unribbed, lanceolate; corolla yellow; staminal column included; style 6–8-branched; stigmas capitate. **Fruits** schizocarps but dehiscence imperfect, thus functionally capsular, erect, not inflated, oblate, papery, stellate-hairy; mericarps 6–8, 1-celled, without dorsal spur or endoglossum, lateral walls persistent, apex rounded to apiculate. **Seeds** 1 per mericarp, usually minutely hairy. $x = 7$.

Species 3 (1 in the flora): Texas, Mexico, West Indies, Central America, South America (Colombia, Ecuador, Peru, Venezuela).

1. **Bastardia viscosa** (Linnaeus) Kunth in A. von Humboldt et al., Nov. Gen. Sp. 5(fol.): 199; 5(qto.): 256. 1822 [F]

Sida viscosa Linnaeus, Syst. Nat. ed. 10, 2: 1145. 1759; *Bastardia guayaquilensis* Turczaninow; *B. parvifolia* Kunth; *S. bastardia* de Candolle; *S. foetida* Cavanilles

Plants 0.5–1.5 m. **Stems** often malodorous, sometimes also with simple hairs 2 mm. **Leaf blades** 4–8 cm (often smaller), apex acute to acuminate. **Inflorescences** usually leafy panicles. **Pedicels** slender. **Flowers:** calyx 4–5 mm, viscid; corolla usually 6+ mm, petals 4–5 mm. **Schizocarps** 5–6 mm diam.; mericarps muticous. **Seeds** 2 mm. $2n = 28$.

Flowering year-round. Shrublands, thickets, disturbed vegetation, pastures, roadsides; 0–500 m; Tex.; e Mexico; West Indies; Central America; South America (Colombia, Ecuador, Peru, Venezuela).

Bastardia viscosa is known only from the lower Rio Grande valley.

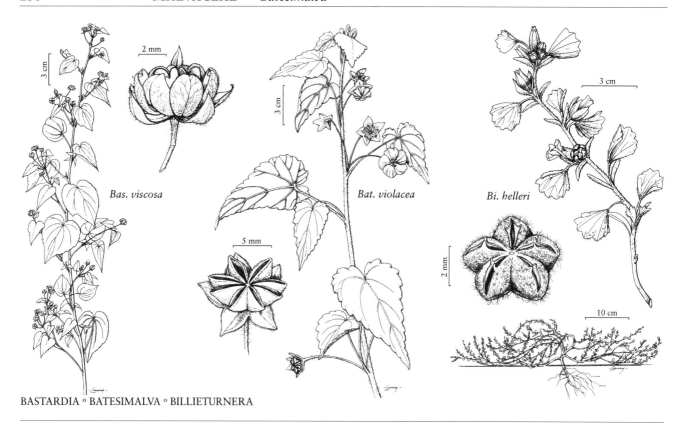

Bas. viscosa

Bat. violacea

Bi. helleri

BASTARDIA ∘ BATESIMALVA ∘ BILLIETURNERA

18. BATESIMALVA   Fryxell, Bol. Soc. Bot. México 35: 25, fig. 1. 1975   •   Bates's mallow [For David M. Bates, b. 1935, American botanist, and Latin *malva*, mallow]   C

Paul A. Fryxell†

Steven R. Hill

Shrubs [herbs]. Stems usually erect, hairy, not viscid. **Leaves:** stipules persistent, filiform, sometimes absent; blade ovate to ovate-lanceolate, not dissected or parted, base cordate, margins coarsely crenate. **Inflorescences** axillary, solitary flowers or 2–4-flowered clusters; involucel absent. **Flowers:** calyx not accrescent, not inflated, ca. ½-divided, shorter than mature fruit, lobes unribbed, lanceolate to ovate; corolla blue-violet [bluish lavender, white, or yellow]; staminal column included; style 8–10[–16]-branched; stigmas capitate. **Fruits** schizocarps, erect or semipendent, inflated, disciform, prominently lobed, papery, minutely tomentose; mericarps 8–10[–16], 2-celled, without dorsal spur, apex rounded, proximal cell indehiscent, enclosing 1 seed (each seed covered by endoglossum), distal cell dehiscent, empty (by ovule abortion), unwinged. **Seeds** 1 per mericarp, subglabrous to sparsely hairy. $x = 16$.

Species 5 (1 in the flora): Texas, n Mexico, n South America (w Venezuela).

1. **Batesimalva violacea** (Rose) Fryxell, Bol. Soc. Bot. México 35: 26. 1975 [C] [F]

Gaya violacea Rose, Contr. U.S. Natl. Herb. 12: 286. 1909

Plants 1.5–2 m. **Stems** stellate-tomentose and with simple hairs 1–1.5 mm. **Leaves:** stipules to 0.5 mm; petiole ¾ to as long as blade; blade discolorous, 6–10 cm, apex acute or acuminate, surfaces densely, velvety pubescent abaxially, sparsely hairy adaxially. **Pedicels** slender, elongate, 2–6.5 cm. **Flowers:** calyx 6–8 mm,

tomentulose and with long, simple hairs; petals 6–8 mm; staminal column 2–3 mm, glabrous; anthers yellow. **Schizocarps** 10–12 mm diam.; mericarps with proximal cell partially covered by endoglossum 1 mm. **Seeds** 2.5 mm. $2n = 32$.

Flowering fall. Dry, deciduous forests and shrublands; of conservation concern; 600–800 m; Tex.; Mexico (Coahuila, Nuevo León).

In the flora area, *Batesimalva violacea* is known from only a single station in Big Bend National Park.

19. **BILLIETURNERA** Fryxell, Sida 9: 195. 1982 • [For Billie Lee Turner, b. 1925, American botanist]

Paul A. Fryxell†

Steven R. Hill

Sida Linnaeus sect. *Icanifolia* Clement, Contr. Gray Herb. 180: 60. 1957

Subshrubs. Stems ascending or procumbent, softly hairy, not viscid. **Leaves:** stipules persistent, broadly oblanceolate; blade broadly oblanceolate, unlobed, not dissected or parted, base cuneate, margins dentate to subentire. **Inflorescences** axillary, solitary flowers; involucel absent. **Flowers:** calyx not accrescent, not inflated, not completely enclosing fruits, deeply divided, lobes unribbed, narrowly triangular; corolla pale yellow; staminal column included; style 5-branched; stigmas capitate. **Fruits** schizocarps, erect, not inflated, variable in shape, papery, stellate-hairy, apically dehiscent; mericarps 5, 1-celled, with abaxial keel and apical spine 1–1.5 mm. **Seeds** 1 per mericarp, minutely and obscurely pubescent. $x = 8$.

Species 1: Texas, ne Mexico.

Although originally included in *Sida* (because the mericarps are one-seeded), *Billieturnera* is more closely related to *Abutilon* than it is to *Sida*.

1. **Billieturnera helleri** (Rose ex A. Heller) Fryxell, Sida 9: 197. 1982 • Copper sida [F]

Sida helleri Rose ex A. Heller, Contr. Herb. Franklin Marshall Coll. 1: 66. 1895; *S. grayana* Clement

Plants to 1 dm. **Leaves:** stipules subequal to petioles; petiole 2–6 mm; blade 0.5–1.5 cm, ± as wide as long, softly stellate-hairy. **Flowers** subsessile; calyx 5–7 mm, lobes stellate-hairy; petals subequal to calyx, claw hairy, glabrous otherwise; staminal column hairy, anthers 20; style pallid. **Schizocarps** included in calyces, 4.5 mm diam. $2n = 16$.

Flowering year-round. Heavy saline soil in exposed sites; 0–200 m; Tex.; Mexico (Nuevo León, Tamaulipas).

In the flora area, *Billieturnera helleri* is found in scattered locations from Lampasas County south to Cameron County, but most frequently in the Rio Grande valley.

20. CALLIRHOË Nuttall, J. Acad. Nat. Sci. Philadelphia 2: 181. 1821 • Poppy mallow, wine cup, wild hollyhock [Derivation uncertain; possibly Greek *kallos*, beautiful, and *rhoias*, corn poppy, alluding to resemblance]

Laurence J. Dorr

Herbs, annual, perennial, or sometimes biennial, hairy, hairs stellate, 4-rayed, and/or simple, or plants glabrous and glaucous. **Stems** erect, ascending, or decumbent. **Leaves:** stipules persistent, caducous, or tardily deciduous, ovate, linear-lanceolate to subulate, auriculate, or rhombic-ovate; blade often pedate, suborbiculate, cordate, ovate, triangular, or hastate, palmately cleft or entire and crenate, base truncate, cordate, or sagittate to hastate, margins entire or dissected. **Inflorescences** racemose or paniculate or appearing corymbose or subumbellate; involucellar bractlets absent or persistent, (1–)3, distinct. **Flowers** bisexual or functionally pistillate; calyx not accrescent, not inflated, lobes with prominent midrib, deltate, ovate-lanceolate, or lanceolate, apex acute to acuminate, often attenuate; corolla cup-shaped, reddish purple, mauve, red, pink, or white; staminal column exserted, anthers borne equally along column, not divided into concentric series; ovary 10–28-carpellate; ovules 1 per carpel; styles 10–28-branched; stigmas introrsely decurrent, filiform. **Fruits** schizocarps, erect, not inflated, oblate or depressed-discoid, indurate, reticulate and rugose, strigose or glabrous, indehiscent or dehiscent (annual species only); mericarps 10–28, 2-celled, prominently obtusely beaked or not, drying tan or brown, distal locule sterile, proximal 1-seeded. **Seeds** 1 per locule, reniform or reniform-pyriform (annual species only), glabrous. $x = 14, 15$.

Species 9 (9 in the flora): United States, n Mexico.

Several species of *Callirhoë* are gynodioecious; populations of *C. alcaeoides*, *C. involucrata*, and *C. leiocarpa* have individuals with either bisexual or functionally pistillate (that is, male-sterile) flowers. In these species the functionally pistillate flowers can be recognized by their reduced number of anther sacs, failure of these anther sacs to dehisce, stigmatic lobes often conspicuous at early anthesis, reduced petal size, and in *C. alcaeoides* shorter calyx lobe length. A few populations of *C. pedata* in Arkansas exhibit a corolla size dimorphism suggesting that this species too may be gynodioecious. Several species of *Callirhoë* are cultivated and may escape. All taxa of this genus occur within the flora area except *C. involucrata* var. *tenuissima* Palmer ex Baker f., which is wholly Mexican.

SELECTED REFERENCES Bates, D. M., L. J. Dorr, and O. J. Blanchard. 1989. Chromosome numbers in *Callirhoë* (Malvaceae). Brittonia 41: 143–151. Dorr, L. J. 1990. A revision of the North American genus *Callirhoë* (Malvaceae). Mem. New York Bot. Gard. 56: 1–75.

1. Involucellar bractlets (1–)3.
    2. Calyx lobes divergent in bud, not forming point . . . . 5. *Callirhoë involucrata*
    2. Calyx lobes valvate in bud, forming apiculate or acuminate point.
        3. Involucellar bractlets spatulate or obovate; mericarps dehiscent; leaf blades triangular or ovate-lanceolate, unlobed or shallowly 3- or 5-lobed . . . . 9. *Callirhoë triangulata*
        3. Involucellar bractlets linear, lanceolate, or ovate; mericarps indehiscent; leaf blades cordate, ovate, suborbiculate, hastate, or triangular, 3-, 5-, or 7-lobed.
            4. Stems 1(–6), stiffly erect, densely hairy (hairs stellate, 6–8-rayed); mericarps hairy. . . . . 8. *Callirhoë scabriuscula*
            4. Stems 1–10, weakly erect, ascending, or decumbent, glabrate or hairy (hairs stellate, mostly 4-rayed); mericarps sparsely hairy.

5. Leaf blade lobes broad, oblong or obovate; involucellar bractlets lanceolate or ovate, 8–22 × 1–4 mm; stems hairy (hairs stellate, 4-rayed and often simple, spreading or retrorse).. 6. *Callirhoë bushii*
5. Leaf blade lobes narrowly lanceolate, linear, linear-falcate, or lanceolate-falcate; involucellar bractlets narrowly linear, 2–10.5 × 0.1–0.7 mm; stems hairy (hairs stellate, 4-rayed or simple) or sometimes glabrate......... ....7. *Callirhoë papaver* (in part)

1. Involucellar bractlets 0.
 6. Annuals (biennials); stipules auriculate; mericarp beaks subtended by 3-lobed collars ....4. *Callirhoë leiocarpa*
 6. Perennials; stipules linear-lanceolate, lanceolate to ovate or subulate; mericarp beaks subtended by 2-lobed, weakly developed collars or collars absent.
  7. Mericarps hairy, beaks prominent, protruding beyond seed-containing portions, forming distal ¼–⅓ of each mericarp.
   8. Inflorescences racemose; flowers (in population samples) usually bisexual, rarely functionally pistillate; petals usually reddish purple, rarely white or pink ....2. *Callirhoë pedata* (in part)
   8. Inflorescences racemose, racemes often appearing corymbose or subumbellate; flowers (in population samples) bisexual or functionally pistillate; petals white, pink, or mauve ....3. *Callirhoë alcaeoides*
  7. Mericarps glabrous or sparingly hairy, beaks not prominent, not or only slightly elevated beyond seed-containing portions, forming less than distal ¼ of each mericarp.
   9. Stipules caducous; inflorescences paniculate; leaf blades (3–)5–10-lobed .... 1. *Callirhoë digitata*
   9. Stipules persistent; inflorescences racemose; leaf blades 3–5-lobed.
    10. Stems glabrous; leaf blades with simple hairs abaxially ....2. *Callirhoë pedata* (in part)
    10. Stems hairy (hairs stellate, 4-rayed or simple) or sometimes glabrate; leaf blades hairy (hairs 4-rayed and simple) abaxially...........7. *Callirhoë papaver* (in part)

1. **Callirhoë digitata** Nuttall, J. Acad. Nat. Sci. Philadelphia 2: 181. 1821 • Fringed or finger poppy mallow, finger wine cup E

Plants perennial. **Stems** 1–4, erect, 5–20 dm, appearing glabrous, but often sparingly hairy, hairs simple, glaucous. **Leaves:** stipules caducous, subulate, 6–8 mm; petiole 12–30 cm; blade suborbiculate, cordate, or ovate, (3–)5–10-lobed, 3.5–15 × 6–20 cm, surfaces strigose, lobes linear to lanceolate. **Inflorescences** paniculate; involucellar bractlets absent. **Flowers** bisexual; calyx lobes valvate in bud, forming apiculate or acuminate point; petals reddish purple with white basal spot, 1.8–3.1 cm. **Schizocarps** 8–8.5 mm diam.; mericarps 10–20, 3.5–4.5 × 2–3 mm, glabrous, indehiscent; beaks not prominent, 0.5–1.7 mm; collars absent or very weakly developed. $2n = 28$.

Flowering spring–late summer. Limestone and dolomitic glades, bald knobs, barrens, rocky prairies, open, cherty woodlands; 200–500 m; Ark., Kans., Mo., Okla.

Callirhoë digitata occurs in the Ozark Plateaus, Ouachita Mountains, and adjacent Cherokee Plains. A collection made north of the Missouri River along a railroad right-of-way is clearly adventive.

2. **Callirhoë pedata** (Nuttall ex Hooker) A. Gray, Mem. Amer. Acad. Arts, n. s. 4: 17. 1849 E

Nuttallia pedata Nuttall ex Hooker, Exot. Fl. 3: 172, plate 172. 1827; *Callirhoë digitata* Nuttall var. *stipulata* Waterfall

Plants perennial. **Stems** (1)2–5 (–15), erect to weakly erect, 1.5–9 dm, essentially glabrous, but often with trace of simple or 4-rayed stellate hairs, glaucous. **Leaves:** stipules persistent, linear-lanceolate to subulate, 4–12.5(–15) mm; petiole 3.5–16 cm; blade cordate, suborbiculate, or ovate, crenate or 3–5-lobed, 2.5–8 (–16) × 2.5–8(–14) cm, surfaces sparsely hairy with simple hairs abaxially, glabrate adaxially, lobes oblanceolate to obtrullate. **Inflorescences** racemose; involucellar bractlets absent. **Flowers** bisexual, rarely functionally pistillate; calyx lobes valvate in bud, forming apiculate or acuminate point; petals reddish purple without white basal spot, rarely white or intergrading shades of pink,

1.6–3.2 cm. **Schizocarps** 6–7.5 mm diam.; mericarps 10–16, 2.5–3 × 2.5–3 mm, glabrous or with simple, appressed hairs, indehiscent; beaks prominent or not, 0.7–1.2 mm; collars absent or very weakly developed. $2n = 28$.

Flowering spring–summer. Open oak or oak-pine woods, mesquite woodlands, margins of woods, prairies, roadsides; 100–500 m; Ala., Ark., Ga., Ill., Okla., Tex.

Callirhoë pedata is variable with respect to indument, leaf size and shape, and mericarp shape; it is perhaps most closely related to *C. alcaeoides*, with which it intergrades in Oklahoma. In population samples, flowers of *C. pedata* are usually bisexual, and rarely functionally pistillate. *Callirhoë pedata* frequently is confused with *C. digitata*. *Callirhoë pedata* is introduced in Alabama, Georgia, and Illinois.

3. **Callirhoë alcaeoides** (Michaux) A. Gray, Mem. Amer. Acad. Arts, n. s. 4: 18. 1849 • Light or pale or plains poppy mallow, plains wine cup   E W

Sida alcaeoides Michaux, Fl. Bor.-Amer. 2: 44. 1803

Plants perennial. **Stems** 2–8 (–28), weakly erect or ascending, 1.5–8.5 dm, hairy, hairs 4-rayed, stellate. **Leaves:** stipules persistent or tardily deciduous, lanceolate to ovate, 5–12 mm; petiole 2–20(–30) cm; blade triangular, cordate, or ovate, unlobed or shallowly to deeply 5–7-lobed, 4–13(–16) × 3–10 cm, surfaces hairy, hairs both 4-rayed, stellate, and simple abaxially, glabrate or with simple hairs adaxially, lobes, when well developed, lanceolate or linear-falcate. **Inflorescences** racemose, often appearing corymbose or subumbellate; involucellar bractlets absent. **Flowers** bisexual or functionally pistillate; calyx lobes valvate in bud, forming apiculate or acuminate point; petals ± evenly white, pink, or mauve, 1.5–2.5(–3) cm (male sterile 0.7–1.7 cm). **Schizocarps** 7–10 mm diam.; mericarps 12–16, 4–6 × 2.5–3.5 mm, hairy, hairs dense, simple, appressed, indehiscent; beaks prominent, 1.5–2.6 mm; collars well developed, 2-lobed. $2n = 28$.

Flowering spring–summer. Plains, prairies, roadsides, waste places; 0–1000(–1500) m; Ala., Ark., Colo., Idaho, Ill., Ind., Iowa, Kans., Ky., Mo., Nebr., N.Mex., Okla., S.Dak., Tenn., Tex.

In population samples, flowers of *Callirhoë alcaeoides* are bisexual or functionally pistillate. This species is unique among the gynodioecious taxa of the genus in having dimorphisms of both petal and calyx lobe size. It is introduced in Idaho, Illinois, Indiana, central and eastern Missouri, New Mexico, and southeastern Texas. It probably is extirpated from Alabama and Kentucky.

4. **Callirhoë leiocarpa** R. F. Martin, J. Wash. Acad. Sci. 28: 108. 1938 • Annual or tall wine cup, tall poppy mallow   E F

Plants annual, sometimes biennial. **Stems** 1–7(–10), erect to weakly erect, 0.5–10(–12) dm, glabrous or sparingly hairy, hairs 4-rayed, stellate, somewhat glaucous. **Leaves:** stipules persistent, strongly auriculate, 4–8(–12) mm; petiole 0.7–8 (–12) cm; blade suborbiculate, reniform-cordate, or ovate, shallowly to deeply 3–7-lobed, 1–7(–9) × 1–9.5 cm, surfaces sparingly hairy, hairs 4-rayed, stellate, and simple abaxially, mostly simple adaxially, lobes oblanceolate. **Inflorescences** racemose; involucellar bractlets absent. **Flowers** bisexual or functionally pistillate; calyx lobes valvate in bud, forming apiculate or acuminate point; petals reddish purple with white basal spot, 1.5–2.3 (–3.5) cm (male sterile 1.1–1.6 cm). **Schizocarps** 5.5–7 mm diam.; mericarps 10–14, 2.8–4.5 × 1.5–2.7 mm, glabrous, dehiscent; beaks prominent, 1–2 mm; collars well developed, 3-lobed. $2n = 28$.

Flowering (late winter–)spring–mid summer, sporadically later. Prairies, mesquite-juniper woodlands, borders of woods and thickets; 0–1000 m; Kans., Okla., Tex.

Callirhoë leiocarpa has been confused with *C. pedata* by some authors. It is cultivated occasionally.

5. **Callirhoë involucrata** (Torrey & A. Gray) A. Gray, Mem. Amer. Acad. Arts, n. s. 4: 15. 1849   F W

Malva involucrata Torrey & A. Gray, Fl. N. Amer. 1: 226. 1838

Plants perennial. **Stems** 4–11 (–21), decumbent to weakly erect, 0.5–8 dm, hairy, hairs stellate, 4-rayed, and simple. **Leaves:** stipules persistent, ovate to ovate-lanceolate, somewhat auriculate, 2.5–15(–23) mm; petiole 0.7–13(–23.5) cm; blade suborbiculate to ovate, 3–5-lobed, 1–8.5 × 1–12 cm, surfaces hairy, hairs stellate and simple abaxially, mostly simple adaxially, lobes oblong, obovate, or inversely trowel-shaped. **Inflorescences** racemose; involucellar bractlets 3, linear to ovate, 4.5–17.5 × 0.5–3.5 mm. **Flowers** bisexual or functionally pistillate; calyx lobes distinct and divergent in bud, not forming point; petals reddish purple with white basal spot, white, or mauve with white margins. **Schizocarps** 7–11 mm diam.; mericarps 10–28, 2–5.5 × 1.7–3.8 mm, glabrous or hairy, indehiscent; beaks not prominent, 0.7–2.1 mm; collars weakly to well developed, 2-lobed.

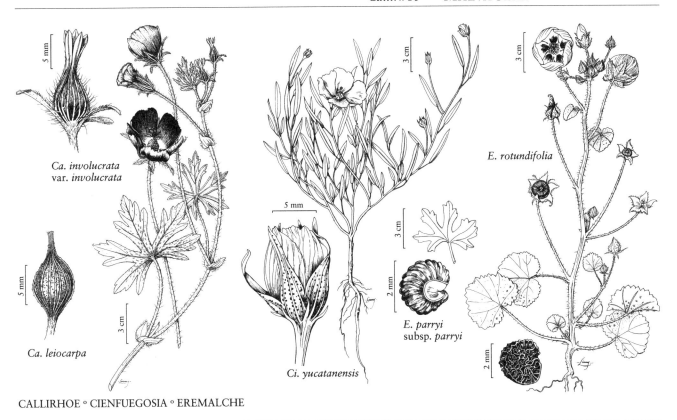

Ca. involucrata
var. involucrata

Ca. leiocarpa

Ci. yucatanensis

E. rotundifolia

E. parryi
subsp. *parryi*

CALLIRHOE ∘ CIENFUEGOSIA ∘ EREMALCHE

Varieties 3 (2 in the flora): United States, n Mexico.

The varieties of *Callirhoë involucrata*, including var. *tenuissima* Palmer ex Baker f., which is known only from Coahuila and Nuevo León, Mexico, are weakly differentiated. Both of the flora varieties are cultivated.

1. Sinuses between lobes of cauline leaves extending to within 5–15 mm of petiole; stipules 5–15(–23) × (3.5–)5.5–10(–15) mm; involucellar bractlets 6–17.5 × 1.5–3.5 mm; petals reddish purple with white basal spot, rarely entirely white; mericarps hairy . . . . 5a. *Callirhoë involucrata* var. *involucrata*
1. Sinuses between lobes of cauline leaves extending to within 2–5 mm of petiole; stipules 2.5–11.5 × 1.5–7(–9) mm; involucellar bractlets 4.5–10 (–13.5) × 0.5–2 mm; petals reddish purple with white basal spot, entirely white, or mauve with white margins; mericarps glabrous or hairy . . . . 5b. *Callirhoë involucrata* var. *lineariloba*

5a. Callirhoë involucrata (Torrey & A. Gray) A. Gray var. **involucrata** • Purple or low poppy mallow, buffalo rose E F

Callirhoë involucrata var. *novomexicana* Baker f.

Stems decumbent, 0.7–7.5 dm. **Leaves:** stipules 5–15(–23) × (3.5–)5.5–10(–15) mm; blade (1–)2–8.5 × 1.5–9.5 cm, sinuses between lobes of cauline leaves extending to within 5–15 mm of petiole. **Involucellar bractlets** 6–17.5 × 1.5–3.5 mm. **Petals** usually reddish purple with white basal spot, rarely white, (1.5–)1.9–3.2 cm (male sterile 0.8–2 cm). **Mericarps** hairy, hairs simple. $2n$ = 30, 60.

Flowering spring–early fall. Plains, prairies, waste places, along roads; 200–1500 m; Ariz., Ark., Colo., Ill., Ind., Iowa, Kans., Minn., Mo., Nebr., N.Mex., Okla., Oreg., Pa., Tex., Va., Wyo.

Variety *involucrata* is introduced in Arizona, central Arkansas, Illinois, Indiana, eastern Iowa, eastern Missouri, Oregon, Pennsylvania, and Virginia.

5b. Callirhoë involucrata (Torrey & A. Gray)
    A. Gray var. **lineariloba** (Torrey & A. Gray)
    A. Gray, Proc. Acad. Nat. Sci. Philadelphia 14: 161.
    1862 • Cowboy rose

Malva involucrata Torrey &
A. Gray var. *lineariloba* Torrey &
A. Gray, Fl. N. Amer. 1: 226.
1838; *Callirhoë geranioides* Small;
C. involucrata var. *parviflora*
Hochreutiner; *C. palmata* Buckley;
C. sidalceoides Standley

Stems decumbent to weakly
erect, 0.5–8 dm. **Leaves:** stipules
2.5–11.5 × 1.5–7(–9) mm; blade 1–8 × 1–9(–12) cm,
sinuses between lobes of cauline leaves extending to
within 2–5 mm of petiole. **Involucellar bractlets** 4.5–10
(–13.5) × 0.5–2 mm. **Petals** reddish purple with white
basal spot, white, or mauve with white margins, 1.4–3.5
cm (male sterile 1–2 cm). **Mericarps** glabrous or hairy,
hairs simple. *2n* = 28, 30, 60.

Flowering late winter–summer, sporadically later.
Pine, oak, and oak-hickory woods, prairies, roadsides,
other disturbed areas; 0–1200 m; Colo., Fla., N.Mex.,
Okla., Tex.; Mexico (Coahuila).

Variety *lineariloba* is introduced in Florida.

6. Callirhoë bushii Fernald, Rhodora 11: 51. 1909
    • Bush's poppy mallow   E

Callirhoë involucrata (Torrey &
A. Gray) A. Gray var. *bushii*
(Fernald) R. F. Martin; *C. papaver*
(Cavanilles) A. Gray var. *bushii*
(Fernald) Waterfall

Plants perennial. **Stems** 1–9,
weakly erect, ascending, or
decumbent, 4.8–14 dm, hairy,
hairs 4-rayed, stellate, and
often simple, spreading or retrorse, sometimes glabrate.
Leaves: stipules persistent, ovate, somewhat auriculate,
8–16(–21) × 3.5–10(–13) mm; petiole 2–27(–37) cm;
blade suborbiculate to ovate, (3–)5–7-lobed, 4–14
(–19) × (2.3–)5–15 cm, surfaces hairy, hairs stellate and
simple, lobes broad, oblong or obovate. **Inflorescences**
racemose; involucellar bractlets 3, lanceolate or ovate,
8–22 × 1–4 mm. **Flowers** bisexual; calyx lobes valvate in
bud, forming apiculate or acuminate point; petals red or
pale red without white basal spot, 2–3.2 cm. **Schizocarps**
8.5–11.5 mm diam.; mericarps 15–20, 4–4.6 × 2–3.5
mm, sparsely hairy, indehiscent; beaks not prominent,
0.7–2 mm; collars well developed, 2-lobed. *2n* = 56.

Flowering late spring–summer(–early fall). Rocky
woods, limestone glades, glade margins, meadows,
disturbed, open areas; 200–500 m; Ark., Iowa, Kans.,
Mo., Okla.

Callirhoë bushii is found in the Ozark Plateaus,
Ouachita Mountains, and Cherokee Plains. Adventive
populations have also been found north of the Missouri
River in Iowa and Missouri.

Callirhoë bushii is in the Center for Plant
Conservation's National Collection of Endangered
Plants.

7. Callirhoë papaver (Cavanilles) A. Gray, Mem.
    Amer. Acad. Arts, n. s. 4: 17. 1849 • Woods poppy
    mallow   E

Malva papaver Cavanilles, Diss.
2: 64, plate 15, fig. 3. 1786

Plants perennial. **Stems** 2–4
(–10), weakly erect, ascending,
or decumbent, 3–10 dm,
glabrate or hairy, hairs 4-rayed,
stellate, or scattered, simple.
Leaves: stipules persistent,
oblong, ovate, or rhombic-ovate,
4.3–10(–12) mm; petiole 2–25(–36) cm; blade hastate,
cordate, triangular, or ovate, 3- or 5(–7)-lobed, 3–11
× 3.5–13 cm, surfaces hairy, hairs 4-rayed and simple,
lobes narrowly lanceolate, linear, linear-falcate, or
lanceolate-falcate. **Inflorescences** racemose; involucellar
bractlets 3, rarely absent, narrowly linear, 2–10.5 ×
0.1–1.7 mm. **Flowers** bisexual; calyx lobes valvate in
bud, forming apiculate or acuminate point; petals reddish
purple without white basal spot, 2.2–4 cm. **Schizocarps**
7.7–11.2 mm diam.; mericarps 12–20, 2.8–4.2 × 2–3.5
mm, glabrous or sparsely hairy, indehiscent; beaks not
prominent, 0.7–1.7 mm; collars scarcely differentiated.
2n = 56, 112.

Flowering spring–summer. Pine, oak, and pine-
oak woods, margins of woods, dry prairies, old fields;
0–200 m; Ala., Ark., Fla., Ga., La., Miss., Tex.

Callirhoë papaver is known from the Gulf Coastal
Plain. It is local and uncommon in Alabama, northern
Florida, Georgia, and Mississippi, and more common
west of the Mississippi River in Louisiana and eastern
Texas.

8. **Callirhoë scabriuscula** B. L. Robinson in A. Gray et al., Syn. Fl. N. Amer. 1(1,2): 302. 1897 • Texas poppy mallow C E

Plants perennial. **Stems** 1(–6), stiffly erect, 3.2–10 dm, densely hairy, hairs 6–8-rayed, stellate. **Leaves:** stipules persistent, linear-lanceolate, 5.8–8.5 mm; petiole 1.5–10 cm; blade suborbiculate, 3- or 5-lobed, 4–6.5 × 3.8–7 cm, surfaces stellate-hairy, lobes oblong to oblanceolate or linear-oblanceolate. **Inflorescences** racemose; involucellar bractlets (1–)3, linear, 5.5–10 × 0.7–1.5 mm. **Flowers** bisexual; calyx lobes valvate in bud, forming apiculate or acuminate point; petals reddish purple with deep-red basal spot, 3–3.7(–4) cm. **Schizocarps** 7.8–12 mm diam.; mericarps 12–20, 4.2–5 × 3.2–4 mm, hairy, indehiscent; beaks not prominent, 0.7–2 mm; collars absent. $2n = 30$.

Flowering spring. Quaternary (Holocene) wind-blown sand deposits; of conservation concern; 600 m; Tex.

Callirhoë scabriuscula is found only in west-central Texas along the upper Colorado River where it has adapted to a rare edaphic niche, relict Quaternary sand dunes. Plants produce taproots up to one meter long. The U.S. Fish and Wildlife Service lists this species as endangered.

Callirhoë scabriuscula is in the Center for Plant Conservation's National Collection of Endangered Plants.

SELECTED REFERENCE Dorr, L. J. 1994. Plants in peril, 21. *Callirhoë scabriuscula*. Kew Magazine 11: 146–151.

9. **Callirhoë triangulata** (Leavenworth) A. Gray, Mem. Amer. Acad. Arts, n. s. 4: 16. 1849 • Clustered or purple poppy mallow, sand poppy, triangular-leaf mallow E

Malva triangulata Leavenworth, Amer. J. Sci. Arts 7: 62. 1824

Plants perennial. **Stems** 1–11 (–35), decumbent, ascending, or weakly erect, 4.5–5 dm, densely hairy, hairs stellate, 4(–6)-rayed. **Leaves:** stipules persistent, ovate to lanceolate, 3.7–7.2 mm; petiole 0.5–29(–34.5) cm; blade triangular or ovate-lanceolate, unlobed or shallowly 3- or 5-lobed, (3–)6–15(–20) × 3.3–10.5(–14) cm, surfaces stellate-hairy. **Inflorescences** paniculate; involucellar bractlets 3, spatulate or obovate, 3–8 × 1.5–3 mm. **Flowers** bisexual; calyx lobes valvate in bud, forming apiculate or acuminate point; petals red (each with white basal spot), 1.6–3.2(–3.5) cm. **Schizocarps** 6–7 mm diam.; mericarps 10–13, 2–4 × 2–3 mm, hairy, hairs simple and 2-rayed, dehiscent; beaks not prominent, to 0.5 mm or absent; collars absent. $2n = 30$.

Flowering spring–summer. Sand prairies, sand hills, sandy alluvium near streams; 100–300 m; Ala., Fla., Ga., Ill., Ind., Iowa, Miss., Mo., N.C., S.C., Wis.

Callirhoë triangulata is found in the upper Mississippi River drainage, especially along the Illinois, Kankakee, Mississippi, Wabash, and Wisconsin rivers. It is also scattered and uncommon in the Gulf and Atlantic coastal plains.

21. CIENFUEGOSIA Cavanilles, Diss. 2[app.]: [vi]. 1786 • [For Bernardo Cienfuegos, ca. 1580–ca. 1640, Spanish botanist]

Paul A. Fryxell†

Steven R. Hill

Herbs or subshrubs [shrubs], perennial. **Stems** decumbent or ascending to erect, glabrous or hairy, not viscid. **Leaves:** stipules persistent, subulate or lanceolate [foliaceous]; blade narrowly oblong-lanceolate or ovate to elliptic, unlobed [lobed], margins entire or serrate. **Inflorescences** axillary, solitary [sympodial] flowers; involucellar bractlets present [sometimes absent], persistent, 6–10, distinct. **Flowers:** calyx persistent, not accrescent, not inflated, not spathaceous, splitting, gland-dotted, lobes ribbed or unribbed, lanceolate; corolla campanulate or rotate, yellow, with or without dark center; staminal column included; ovary 3–5-carpellate; style usually connate to apex; stigmas connate, 3–5-lobed, clavate [capitate]. **Fruits** capsules, erect, not inflated, ovoid, papery, glabrous or hairy. **Seeds** 2–5 per locule, hairy. $x = 10$.

Species 26 (2 in the flora): sc, se United States, Mexico, Central America, West Indies, South America (to Argentina), sw Asia (Arabian Peninsula), Africa.

SELECTED REFERENCES Fryxell, P. A. 1969. The genus *Cienfuegosia* Cav. Ann. Missouri Bot. Gard. 56: 179–250. Karpovickas, A. 2003. Las especies Austroamericanas del género *Cienfuegosia* Cav. (Malvaceae-Gossypieae). Bonplandia (Corrientes) 12: 5–47.

1. Leaf blades ovate to elliptic, length ca. 1.5 times width, margins coarsely serrate; corolla 1.5–3.5 cm, campanulate, usually dark red proximally; involucellar bracts 6–14 mm, lanceolate to spatulate; stigmas 4 or 5, dark red; capsules 8–14 mm, 4- or 5-locular, glabrous on internal suture . . . . 1. *Cienfuegosia drummondii*
1. Leaf blades narrowly oblong-lanceolate, length 4–10 times width, margins entire; corolla 1–2 cm, rotate, not dark proximally; involucellar bracts 0.5–2 mm, subulate; stigmas 3, pallid; capsules 6–8 mm, 3-locular, copiously ciliate on internal suture . . . . 2. *Cienfuegosia yucatanensis*

1. **Cienfuegosia drummondii** (A. Gray) Lewton, Bull. Torrey Bot. Club 37: 475. 1910 • Yellow fugosia, sulphur mallow

Fugosia drummondii A. Gray, Smithsonian Contr. Knowl. 3(5): 23. 1852

Plants from woody rootstock. **Stems** decumbent to ascending, sparsely puberulent to glabrate. **Leaves:** stipules subulate or lanceolate, 1–8 mm; petiole to ½ length of blade, glabrate (or puberulent distally); blade ovate to elliptic, 5–10 cm, ca. 1.5 times as long as wide, margins coarsely serrate, apex acute, surfaces glabrate. **Pedicels** 2–11 cm, without involucellar nectaries at apex; involucellar bractlets 8–10, lanceolate to spatulate, 6–14 mm. **Flowers:** calyx 10–20 mm, inconspicuously punctate, minutely puberulent to glabrate; corolla campanulate, usually with dark red center, 1.5–3.5 cm; staminal column pallid or dark red, apically 5-dentate; style pallid; stigmas 4 or 5, dark red. **Capsules** obovoid-obtuse, 4- or 5-locular, 8–14 mm, glabrous. **Seeds** 3–4 mm, inconspicuously hairy, hairs tightly appressed, usually tan or brownish. $2n = 20$.

Flowering winter–spring. Open areas, usually in heavy, often saline soil; 30–100 m; Tex.; South America (Argentina, Brazil, Paraguay).

In the flora area, *Cienfuegosia drummondii* is known only from the coastal areas of southern Texas.

2. **Cienfuegosia yucatanensis** Millspaugh, Publ. Field Columbian Mus., Bot. Ser. 2: 74. 1900 [F]

Plants without woody rootstock. **Stems** erect, glabrous. **Leaves:** stipules subulate, 1–2 mm; petiole ¼–½ length of blade, glabrous; blade narrowly oblong-lanceolate, usually 2–4 cm, 4–10 times as long as wide (juvenile leaves sometimes wider), margins entire, apex acute, surfaces glabrous. **Pedicels** 1–5 cm, surmounted by 3 involucellar nectaries or nectaries absent; involucellar bractlets 6–9, subulate, 0.5–2 mm. **Flowers:** calyx 8–12 mm, punctate, glabrous; corolla rotate, 1–2 cm; staminal column pallid, apically 5-dentate; style pallid; stigmas 3, pallid. **Capsules** ovoid, 3-locular, 6–8 mm, glabrous externally, prominently ciliate internally. **Seeds** 2–3 mm, densely hairy, hairs brownish. $2n = 20$.

Flowering spring–summer. Open, arid sites; 0–100 m; Fla.; Mexico (Yucatán); West Indies (Bahamas, Cuba).

In the flora area, *Cienfuegosia yucatanensis* is known from Monroe County in the Florida Keys; it has sometimes been misidentified as *C. heterophylla* (Ventenat) Garcke.

22. EREMALCHE Greene, Leafl. Bot. Observ. Crit. 1: 208. 1906 • Desert mallow [Greek *eremia*, desert, and *alkea*, mallow, alluding to habitat]

David M. Bates

Malvastrum A. Gray subsect. *Pedunculosa* A. Gray in A. Gray et al., Syn. Fl. N. Amer. 1(1,2): 308. 1897

Herbs, annual, sometimes gynodioecious. **Stems** prostrate to erect, sparsely to densely hairy. **Leaves:** stipules caducous, subulate; blade reniform-round, unlobed or palmately 3-, 5-, [7-]cleft, base truncate to cordate, margins usually toothed, sometimes entire. **Inflorescences** axillary, solitary flowers in lower axils, congested and ± corymbose at branch tips; involucellar bractlets persistent, 3, distinct. **Flowers:** calyx not accrescent, not inflated, lobed to beyond middle, lobes not ribbed, acuminate; corolla campanulate, white to rose, lavender, or mauve, drying purple or blue; staminal column included; ovary 9–36-carpellate; ovules 1 per carpel; styles 9–36-branched (branches equal in number to carpels); stigmas terminal, capitate. **Fruits** schizocarps, oblique, not inflated, depressed-discoid, edge corrugate or reticulate, glabrous; mericarps 9–36, drying black, 1-celled, without spur, ± spheric, unarmed, lateral walls ± disintegrated, indehiscent. **Seeds** ascending, 1 per mericarp, black, obovoid-reniform, glabrous. *x* = 10.

Species 3 (3 in the flora): sw United States, nw Mexico.

1. Leaf blades unlobed; petals magenta-spotted at base; mericarps 2.8–3.5 mm . . . . . .1. *Eremalche rotundifolia*
1. Leaves lobed or parted; petals not spotted; mericarps 1.4–1.8 mm.
    2. Flowers bisexual; petals 4–5.5 mm, ± equaling calyx; plants prostrate to decumbent
    . . . . .2. *Eremalche exilis*
    2. Flowers bisexual or pistillate; petals 5.5–20(–25) mm, equaling or exceeding calyx;
    plants erect, main stem unbranched or with ascending branches from base. . . . . 3. *Eremalche parryi*

1. Eremalche rotundifolia (A. Gray) Greene, Leafl. Bot. Observ. Crit. 1: 208. 1906 • Desert five-spot F

Malvastrum rotundifolium A. Gray, Proc. Amer. Acad. Arts. 7: 333. 1868; *Sphaeralcea rotundifolia* (A. Gray) Jepson

Plants erect; main stem spotted or flushed purple-maroon, unbranched or branched from base, 8–30(–60) cm, sparsely or densely hairy, hairs mostly simple, bristly, to 3 mm. **Leaf blades** unlobed, 1.5–6 cm, margins crenate-dentate. **Inflorescences** usually exceeding leaves. **Pedicels** 1.5–8 cm; involucellar bractlets filiform, 6–10 mm. **Flowers:** calyx 9–14 mm, lobes 5.5–11 × 3.5–7 mm; petals rose-pink to lilac, magenta-spotted at base, (15–)20–30 mm, exceeding calyx. **Mericarps** 25–36, black, waferlike edges acute, reticulate, 2.8–3.5 mm. *2n* = 20.

Flowering mid winter–late spring. Dry desert scrub; –50–1200 m; Ariz., Calif., Nev., Utah; Mexico (Baja California, Sonora).

Eremalche rotundifolia is morphologically and perhaps generically distinct from the other species of *Eremalche*; it occurs widely in the Mojave and Sonoran deserts and in Death Valley.

2. Eremalche exilis (A. Gray) Greene, Leafl. Bot. Observ. Crit. 1: 208. 1906

Malvastrum exile A. Gray in J. C. Ives, Rep. Colorado R. 4: 8. 1861; *Sphaeralcea exilis* (A. Gray) Jepson

Plants prostrate to decumbent; stems sometimes flushed with purple-maroon, branched or unbranched, 2–40 cm, obscurely

stellate-hairy. **Leaf blades** 3- or 5-cleft or -lobed to ± ½ to base, 1–2.5 cm wide, divisions entire or 3-toothed at tip. **Inflorescences** seldom exceeding leaves, sometimes basal. **Pedicels** 0.2–1 cm, longer in fruit; involucellar bractlets threadlike to linear, 3.3–5.5(–7) mm. **Flowers:** calyx 4–5.5(–7) mm, lobes 3–5 × 1.5–2.5 mm; petals white or pale rose-lavender, 4–5.5 mm, equaling calyx. **Mericarps** 9–13, brownish to blackish, ± wedge-shaped, cushionlike, 1.4–1.8 mm, margins rounded, radially corrugated. *2n* = 20, 40.

Flowering late winter–spring. Sandy or rocky desert soil; 100–1500 m; Ariz., Calif., Nev., Utah; Mexico (Baja California, Sonora).

Eremalche exilis occurs in the Mojave and Sonoran deserts.

3. Eremalche parryi (Greene) Greene, Leafl. Bot. Observ. Crit. 1: 208. 1906 E F

Malvastrum parryi Greene, Fl. Francisc., 108. 1891; *Sphaeralcea parryi* (Greene) Jepson

Plants erect; main stem unbranched or with ascending branches from base, 2–50 cm, densely hairy distally, hairs stellate, 1–3-armed, arms to 2 mm. **Leaf blades** 3-cleft to beyond middle, usually 2–5 cm wide, divisions toothed to subcleft distally. **Inflorescences** usually exceeding leaves. **Pedicels** mostly 1–8 cm, longer in fruit; involucellar bractlets linear, (2.5–)4–10(–15) mm. **Flowers** bisexual or pistillate; calyx 4.5–13 mm, lobes 3.2–11 × 1.5–4 mm; petals white or pale to deep mauve, 5.5–20(–25) mm, equaling or exceeding calyx. **Mericarps** 9–22, brownish to blackish, ± wedge-shaped, cushionlike, 1.5–1.8 mm, margins rounded, radially corrugated.

Subspecies 2 (2 in the flora): California.

1. Flowers bisexual; petals mauve; calyx 10–14 mm, lobes 2.5–4 mm wide . . . . 3a. *Eremalche parryi* subsp. *parryi*
1. Flowers bisexual or pistillate; petals white or mauve, (proportionally smaller and carpels more numerous in female flowers); calyx 4.5–9(–10) mm, lobes 1.5–3(–3.5) mm wide . . . . 3b. *Eremalche parryi* subsp. *kernensis*

3a. Eremalche parryi (Greene) Greene subsp. **parryi**
⊞E ⊞F

Pedicels usually 2–8 cm in flower; involucellar bractlets 7–10(–15) mm. **Flowers** bisexual; calyx 10–14 mm, lobes 8–11 × 2.5–4 mm; petals mauve, 15–25 mm, exceeding calyx. **Mericarps** 14–22. **2***n* = 20.

Flowering late winter–spring. Valleys, foothills; 30–1500 m; Calif.

Subspecies *parryi* occurs in the South Coast Ranges through the northern portions of the Transverse Ranges to the southern foothills of the Sierra Nevada.

3b. Eremalche parryi (Greene) Greene subsp. **kernensis** (C. B. Wolf) D. M. Bates, Phytologia 72: 51. 1992
⊞C ⊞E

Eremalche kernensis C. B. Wolf, Occas. Pap. Rancho Santa Ana Bot. Gard. 2: 66, fig. 18. 1938; *Malvastrum kernense* (C. B. Wolf) Munz

Pedicels usually 1–5 cm in flower; involucellar bractlets (3–)4–7(–10) mm in bisexual flowers, (2.5–)4–6(–8) mm in pistillate flowers. **Flowers** bisexual or pistillate; petals white or mauve. **Bisexual flowers:** calyx 5–9(–10) mm, lobes 3.5–7(–8) × 1.7–3(–3.5) mm; petals 8–20(–25) mm, exceeding calyx. **Pistillate flowers:** calyx 4.5–7.5 (–10) mm, lobes 3.2–6.5(–8) × 1.5–2.5(–3.5) mm; petals 5.5–13 mm, ± equaling to slightly exceeding calyx. **Mericarps** 9–13 in bisexual flowers, 13–19 in pistillate flowers. **2***n* = 20.

Flowering late winter–spring. Eroded hillsides, alkali flats; of conservation concern; 60–1200 m; Calif.

Subspecies *kernensis* occurs in the southernmost inner South Coast Ranges and the southern San Joaquin Valley in Kern and San Luis Obispo counties.

SELECTED REFERENCE Bates, D. M. 1992. Gynodioecy, endangerment and status of *Eremalche kernensis* (Malvaceae). Phytologia 72: 48–54.

23. FRYXELLIA D. M. Bates, Brittonia 26: 95, fig. 1A. 1974 • [For Paul Arnold Fryxell, 1927–2011, American student of Malvaceae] ⊡

David M. Bates

Herbs, perennial, stellate-hairy. **Stems** cespitose. **Leaves:** stipules persistent, filiform to subulate; petiole subequal to or somewhat shorter than blade; blade ovate to oblong-lanceolate, unlobed, base subcordate to obtuse, margins toothed, apex acute or obtuse, without foliar nectaries. **Inflorescences** axillary, solitary flowers, ± equaling leaves; involucellar bractlets absent. **Flowers:** calyx accrescent, dehiscent with fruit, inflated, winged, 5-lobed to middle, lobes broadly triangular, midrib raised, veins several, parallel; corolla ± campanulate, deep orange, drying yellowish; staminal column included, anther-bearing at apex; filaments terminal and subterminal; ovary 12-carpellate; ovules 1 in each lower carpel cell, pendulous; styles 12-branched, (branches equal in number to carpels); stigmas capitate. **Fruits** schizocarps, not inflated, discoid, moderately indurate, stellate-hairy; mericarps 12, 1-celled, divided internally by endoglossum; distal cell reduced, barren, smooth-walled, hairy, each valve armed with suberect dorsal spur; proximal cell enclosing pendulous seed, reticulate-fenestrate, indehiscent, joined to columella by single median basal-dorsal vein. **Seeds** 1 per mericarp, puberulent about raphe. *x* = 8.

Species 1: Texas, n Mexico.

SELECTED REFERENCE Fryxell, P. A. and J. Valdés R. 1991. Observations of *Fryxellia pygmaea* (Malvaceae). Sida 14: 399–404.

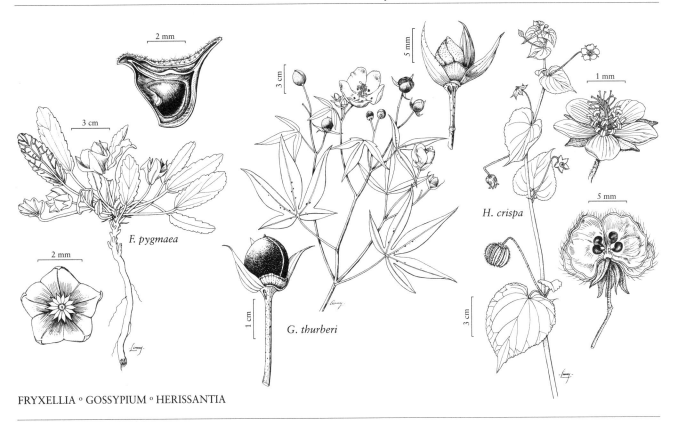

F. pygmaea

G. thurberi

H. crispa

FRYXELLIA ° GOSSYPIUM ° HERISSANTIA

1. Fryxellia pygmaea (Correll) D. M. Bates, Brittonia 26: 95. 1974 [C] [F]

Anoda pygmaea Correll, Wrightia 4: 75. 1968

Plants 10–15 cm diam., cespitose; growing points 1+, stems branched below soil surface. **Leaf blades** to 4 × 2.4 cm. **Pedicels** not articulated, 10–20 mm. **Flowers:** calyx subrotate, to 12 mm, to 17 mm in fruit, lobes to 7 × 8 mm (9 × 10 mm in fruit), papery, folded and slightly keeled along each commissural vein; corolla to 15 mm, exceeding calyx; staminal column to 4 mm. **Schizocarps** 9–10 mm diam., closely invested by calyx; mericarps pale brown, to 3 × 5 mm exclusive of dorsal spurs, spurs to 2 mm. **Seeds** 2 mm. $2n = 16$.

Flowering apparently summer. Dry, open slopes; of conservation concern; 1300 m; Tex.; Mexico (Coahuila).

The distribution of *Fryxellia pygmaea* in the flora area is not known. The type collection (the only collection known from the flora area) was made by Captain John Pope in 1854. It was taken on his survey route from Preston, Grayson County, on the Red River of Texas, southwest to the Pecos River and then nearly west to the valley of the Rio Grande near El Paso and Doña Ana, New Mexico.

Fryxellia pygmaea is in the Center for Plant Conservation's National Collection of Endangered Plants.

24. GOSSYPIUM Linnaeus, Sp. Pl. 2: 693. 1753; Gen. Pl. ed. 5, 309. 1754 • Cotton

[Greek *gossypion*, cotton, or Arabic *goz* or *gothn*, a silky or soft substance]

Paul A. Fryxell†

Steven R. Hill

Erioxylum Rose & Standley; *Ingenhouzia* de Candolle; *Selera* Ulbrich; *Thurberia* A. Gray

Shrubs [trees], hairy or glabrate, not viscid. **Stems** erect [or procumbent]. **Leaves:** stipules persistent, subulate or linear to falcate; blade ovate, unlobed, shallowly lobed, or deeply parted, base subcordate or cordate, margins entire, surfaces glabrous or stellate-hairy, often with abaxial foliar nectaries. **Inflorescences** axillary, solitary flowers or flowers sympodially arranged; involucellar bractlets deciduous or persistent, 3, distinct. **Flowers:** calyx not accrescent, not inflated, lobes sometimes unribbed, ovate or triangular, glabrous or hairy; corolla cream or yellow [rose], sometimes fading rose, with or without dark spot at center; staminal column included; ovary 3–5-carpellate; style unbranched; stigmas decurrent-clavate. **Fruits** capsules, erect, not inflated, ovoid or subglobose to oblong, leathery, usually glabrous. **Seeds** [2–]24, densely comose to glabrate or glabrous. $x = 13$.

Species ca. 50 (3 in the flora): s, sc United States, w Mexico, South America, Asia (Middle East), Afric, Australia; introduced nearly worldwide.

SELECTED REFERENCE   Fryxell, P. A. 1992. A revised taxonomic interpretation of *Gossypium* L. (Malvaceae). Rheedea 2: 108–165.

1. Involucellar bractlets broadly cordate-ovate, margins laciniate; petals 2–5 cm; capsules 2–4 cm; seeds comose (cottony); leaf blades shallowly lobed, lobes broadly ovate. . . . 2. *Gossypium hirsutum*
1. Involucellar bractlets ligulate, margins entire or apically toothed; petals 1.5–4.5 cm; capsules 1–1.5 cm; seeds glabrous, glabrate, or strigose (not cottony); leaf blades unlobed or lobes narrowly lanceolate.
    2. Involucellar bractlets deciduous before anthesis, margins entire; petals 2.5–4.5 cm, yellow; seeds 8 mm, strigose (hairs brownish, tightly appressed); leaf blades unlobed; petioles terete . . . . .1. *Gossypium armourianum*
    2. Involucellar bractlets persistent, margins entire or apically toothed; petals 1.5–2.5 cm, cream; seeds 3–4 mm, glabrous or glabrate; leaf blades 3–5-lobed, lobes narrowly lanceolate (4+ times as long as wide); petioles quadrangular. . . . . .3. *Gossypium thurberi*

1. Gossypium armourianum Kearney, J. Wash. Acad. Sci. 23: 558. 1933 • Wild cotton [1]

Plants 1 m, compact. **Stems** terete, glabrate. **Leaves:** stipules subulate, 1–3 mm; petiole terete, ½ to as long as blade; blade unlobed, ovate, 1.5–3 cm, coriaceous, base cordate, apex acute or subacuminate, surfaces glabrate. **Inflorescences** solitary flowers. **Pedicels** 2–6.5 cm, with 3 prominent nectaries apically; involucellar bractlets deciduous before anthesis, ligulate, 0.5–1 cm, margins entire. **Flowers:** calyx 5–7 mm, apex subtruncate or 5-toothed; petals yellow with red spot at base, 2.5–4.5 cm; staminal column 12–14 mm, glabrous; style somewhat shorter than petals; stigmas 3–5. **Capsules** 3- or 4-locular, subglobose or ovoid, 1.5 cm, with sunken glands, externally glabrous, internally ciliate. **Seeds** 8 mm, strigose, with tightly appressed brownish hairs. $2n = 26$.

Flowering spring–summer. Open, arid habitats; 100–200 m; introduced; Calif.; Mexico (Baja California).

Gossypium armourianum is native in restricted areas in Baja California, but naturalized near Palm Springs from an introduction in the 1930s.

2. Gossypium hirsutum Linnaeus, Sp. Pl. ed. 2, 2: 975. 1763 • Upland cotton, algodón [I] [W]

Gossypium hopi Lewton; *G. latifolium* Murray; *G. mexicanum* Todaro; *G. punctatum* Schumacher & Thonning; *G. religiosum* Linnaeus

Plants 1–2 m, usually widely branching. Stems terete, stellate-hairy. Leaves: stipules subulate to falcate, 5–15(–20) mm; petiole terete, ½ to as long as blade; blade shallowly 3–5-lobed, lobes broadly ovate, 4–10 cm, membranous, base cordate, apex acute to acuminate, surfaces glabrous or hairy. Inflorescences: flowers usually sympodial. Pedicels 2–4 cm, with 3-merous involucellar nectaries; involucellar bractlets persistent, foliaceous (enclosing bud), broadly cordate-ovate, 2–4.5 cm, margins laciniate. Flowers: calyx 5–6 mm (excluding teeth, if present), apex truncate or 5-toothed; petals cream, with or without red spot, 2–5 cm; staminal column 15 mm, glabrous; style somewhat exceeding androecium; stigmas 3–5. Capsules 3–5-locular, broadly ovoid or subglobose, 2–4 cm, smooth, glabrous. Seeds 8–10 mm, comose, hairs (cotton) usually white. 2*n* = 52.

Flowering year-round. Littoral vegetation, cultivated fields; 0–20 m; introduced; Ala., Fla., Ga., Ill., La., Mass., Miss., Mo., N.Mex., N.C., Okla., S.C., Tex., Va.; Central America; introduced nearly worldwide.

Gossypium hirsutum is part of the littoral vegetation in the Florida Keys and around the coasts of the Gulf of Mexico and the Caribbean Sea. The species is found also as the dominant agricultural crop of the Cotton Belt, from California to the Carolinas; it is cultivated worldwide in suitable climates. The species may be found also as an escape, or from cotton mulch used in gardens or from waste around areas of cotton agriculture (fields, gin yards, roadsides, and other places).

3. Gossypium thurberi Todaro, Relaz. Cult. Coton., 120. 1877 • Algodoncillo, wild cotton [F]

Thurberia thespesioides A. Gray, Pl. Nov. Thurb., 308. 1854, not *Gossypium thespesioides* (R. Brown ex Bentham) F. Mueller 1875

Plants 2 m, freely branching. Stems 5-angular when young, glabrate. Leaves: stipules linear, 5–10 mm; petiole quadrangular, ½–¾ as long as blade; blade deeply 3–5-lobed, lobes narrowly lanceolate (4+ times as long as wide), 5–15 cm, membranous, base subcordate, apex long-acuminate, surfaces glabrate. Inflorescences: flowers sympodial. Pedicels 1–3 cm, with 3 prominent nectaries; involucellar bractlets persistent, ligulate, 0.8–1.2 cm, margins entire or apically few-toothed. Flowers: calyx 3 mm, apex truncate; petals cream, with red spot at base, sometimes spot absent, 1.5–2.5 cm; staminal column 9 mm, glabrous; style slightly exceeding androecium; stigmas 3–5. Capsules 3-locular, subglobose to oblong, 1–1.5 cm, punctate, externally glabrous, internally ciliate. Seeds 3–4 mm, glabrous or glabrate. 2*n* = 26.

Flowering late summer. Open, arid habitats, rocky hillsides, banks of seasonal streams; 800–2400 m; Ariz.; Mexico (Chihuahua, Sonora).

Gossypium thurberi occurs in central to southern Arizona, from Yavapai to Cochise counties.

25. HERISSANTIA Medikus, Vorles. Churpfälz. Phys.-Ökon. Ges. 4: 244. 1788 • [For Louis Antoine Prospere Herissant, 1745–1769, French physician, naturalist, and poet]

John La Duke

Herbs, perennial, or shrubs, soft-tomentose, not viscid. Stems prostrate to erect. Leaves petiolate or sessile; stipules persistent, inconspicuous, linear to subulate; blade ovate to triangular, unlobed, base cordate, margins crenate to dentate. Inflorescences axillary, solitary flowers; involucel absent. Flowers: calyx not accrescent, not inflated, lobes not ribbed, ovate to ovate-lanceolate; corolla subrotate, yellow, orange, or white; staminal column included; ovules 1–3 per carpel; styles 10–14-branched; stigmas capitate. Fruits schizocarps, reflexed, inflated, spheric to subspheric, papery, setose; mericarps 10–12[–14], 1-celled, without dorsal spur, apex rounded, dehiscent dorsally. Seeds 2–6 per mericarp, spheric, glabrous or setose. *x* = 7.

Species 5 (1 in the flora): s United States, Mexico, West Indies, Central America, South America, Asia, Australia.

1. Herissantia crispa (Linnaeus) Brizicky, J. Arnold
   Arbor. 49: 279. 1968 [F]

Sida crispa Linnaeus, Sp. Pl.
2: 685. 1753; *Abutilon crispum*
(Linnaeus) Medikus; *A. sessifolium*
C. Presl; *Bogenhardia crispa*
(Linnaeus) Kearney; *Gayoides
crispa* (Linnaeus) Small; *Sida
amplexicaulis* Lamarck; *S. imberbis*
de Candolle; *S. retrofracta*
de Candolle; *S. sessilis* Vellozo

Plants sometimes sprawling, 1–6 dm. **Leaves** petiolate
to 52 mm, sessile distally; blade ovate to sometimes
elongate-triangular, 1–8 cm, apex acuminate. **Flowers:**
calyx rotate in flower to reflexed in fruit, 3–7 mm;
petals 6–11 mm. **Schizocarps** on jointed pedicel near
fruit; mericarps 10–12, to 10 mm, walls thin, papery.
$2n = 14$.

Flowering Mar–Jan. Forests, grasslands, roadsides,
weedy or disturbed sites; 0–2000 m; Ariz., Calif., Fla.,
N.Mex., Tex.; Mexico; West Indies; Central America;
South America; introduced in Asia, Australia.

Given the weedy nature of *Herissantia crispa*, it would
not be surprising to encounter it in other warm regions
adjacent to the currently documented distribution.

26. HIBISCUS Linnaeus, Sp. Pl. 2: 693. 1753; Gen. Pl. ed. 5, 310. 1754, name conserved
   • Rose-mallow [Greek *hibiscus* or *ibiscum*, alluding to cohabitation with *Ibis*, stork, in
   marshes]

Orland J. Blanchard Jr.

Herbs, annual or perennial, **subshrubs, shrubs, or trees,** glabrous or hairy. **Stems** erect,
ascending, or arching, unarmed or prickly. **Leaves:** stipules persistent or caducous, not enlarged
and protecting flower and leaves in bud, filiform to lanceolate or narrowly triangular; blade
lanceolate to ovate or elliptic or orbiculate, unlobed or lobed, base cordate, rounded, truncate,
or cuneate, margins entire or variously serrate, dentate, or crenate. **Inflorescences** axillary,
sometimes adnate to subtending petiole (*H. moscheutos*), flowers solitary or clustered; involucel
present, bractlets usually persistent, rarely deciduous (*H. mutabilis*), 6–16, rarely minute or
absent (*H. denudatus*), distinct, undivided, 2-fid or appendaged (*H. acetosella, H. aculeatus,
H. furcellatus, H. radiatus*). **Flowers** usually lasting for a day; calyx persistent, sometimes
accrescent, inflated or not, not spathaceous, primary veins usually ± straight, rarely zigzag
(*H. trionum*, sometimes *H. dasycalyx*), lobes triangular to trullate, sometimes strongly 3-ribbed
(*H. acetosella, H. aculeatus, H. furcellatus, H. radiatus*), sometimes with nectary on midrib;
corolla radial or weakly bilaterally symmetric, narrowly funnelform to broadly campanulate
or rotate or petals sometimes recurved (*H. clypeatus*), white, cream, yellow, orange, pink, red,
lavender, purple, or blue, usually red, purple, or brown basally; staminal column included or
exserted (*H. coccineus, H. poeppigii, H. schizopetalus*); ovary 5-carpellate; ovules 8–60 per
carpel; styles 5-branched from or beyond orifice of staminal column; stigmas capitate to discoid
or wedge-shaped (*H. striatus*). **Fruits** capsules, 5-valved, ovoid or spheroid, apex usually
apiculate, acute, or acuminate, sometimes rounded or depressed or impressed, glabrous or hairy.
Seeds to 60 per locule, reniform-ovoid, reniform-globose, or subglobose, sometimes angulate,
sometimes laterally compressed or impressed, papillose or not, glabrous or hairy. $x = 10, 11,$
[12], 14, [15–17], 18, 19, [20], 26, [27], and probably higher.

Species ca. 350 (21 in the flora): North America, Mexico, West Indies, Central America,
South America, Eurasia, Africa, Indian Ocean Islands, Pacific Islands, Australia.

Taxonomy of *Hibiscus* is in flux. Molecular studies place members of some other genera in
Hibiscus as traditionally circumscribed, indicating paraphyly (B. E. Pfeil and M. D. Crisp 2005).
Neither of the two solutions to this problem—vastly expanding the circumscription of *Hibiscus*
to include the nested genera, or breaking *Hibiscus* into smaller genera—has been attempted
here.

The 21 species of *Hibiscus* in the flora area are scattered among nine sections. The sections are not formally treated here since to do so would make the treatment unnecessarily cumbersome, and also because the morphological limits of some of the sections are poorly known on a worldwide basis. The sectional assignments of the species in the flora area are as follows: sect. *Hibiscus* (species 1), sect. *Bombicella* de Candolle (species 2–6), sect. *Furcaria* de Candolle (species 7–10), sect. *Lilibiscus* Hochreutiner (species 11 and 12), sect. *Muenchhusia* (Heister ex Fabricius) O. J. Blanchard (species 13–17), sect. *Venusti* Ulbrich (species 18), sect. *Striati* O. J. Blanchard (species 19), sect. *Trionum* de Candolle (species 20), and sect. *Clypeati* O. J. Blanchard (species 21).

Most *Hibiscus* species in the flora area are native; seven were introduced into North America either for horticultural purposes (*H. acetosella, H. mutabilis, H. rosa-sinensis, H. schizopetalus, H. syriacus,* and *H. trionum*) or as fiber crops (*H. radiatus*). Experimental crosses among some native species (*H. coccineus, H. grandiflorus, H. laevis,* and *H. moscheutos*) have been the basis of selections of hardy *Hibiscus* available in the horticultural trade. A popular cultivated form in the South is the result of an experimental cross involving a complex interspecific hybrid combining *H. coccineus, H. laevis,* and *H. moscheutos* as the staminate parent and *H. mutabilis* as the pistillate parent (H. F. Winters 1970).

North American taxa in the speciose sect. *Bombicella* share a base chromosome number of $x = 11$ (chromosome numbers are unknown for *Hibiscus coulteri*); those in sect. *Furcaria,* $x = 18$; and those in sect. *Muenchhusia,* $x = 19$. Base numbers of the two sections represented in the flora area by one native species each are 10 and 26 and cannot be readily reconciled between themselves or with the others. The remaining four sections together include five species, all of which are non-natives and, except for *H. trionum,* have long histories in cultivation and multiple different chromosome numbers that are not easily evaluated. This overall chromosomal diversity supports the molecular evidence for a polyphyletic *Hibiscus.*

Extensive experimental hybridization work in sect. *Furcaria,* and the study of chromosome pairing relationships in these hybrids, indicate that the native species *Hibiscus aculeatus* and *H. furcellatus,* both tetraploids, share the genomic makeup GGPP, while the introduced species *H. acetosella* and *H. radiatus,* also tetraploids, share AABB (F. D. Wilson 1994).

A naturally occurring hybrid between *Hibiscus coulteri* and *H. denudatus* (*Hibiscus* ×*sabei* Weckesser) has been documented from western Texas (W. Weckesser 2011).

Some desert species of *Hibiscus* sometimes produce cleistogamous flowers. Foliar nectaries are present abaxially near the base of the midvein in some species.

Three species of *Hibiscus* are of doubtful status in the flora area. *Hibiscus cannabinus* Linnaeus and *H. sabdariffa* Linnaeus have both been cultivated in the southern states as fiber crops, and the latter also for its fleshy calyces, which are used to flavor beverages. It is doubtful that either has become established in the flora area. *Hibiscus bifurcatus* Cavanilles is widely distributed from the West Indies and Guatemala to South America; it was documented with two collections from Dade County in Florida by R. Woodbury in 1949 and 1950, but since Woodbury was involved at the time in establishing what later became the John C. Gifford Arboretum in Coral Gables south of Miami, there is a good chance that the source was a plant in cultivation.

SELECTED REFERENCES Bates, D. M. 1965. Notes on the cultivated Malvaceae, 1. *Hibiscus.* Baileya 13: 56–130. Blanchard, O. J. 1976. A Revision of Species Segregated from *Hibiscus* Sect. *Trionum* (Medicus) de Candolle Sensu Lato (Malvaceae). Ph.D. dissertation. Cornell University. Fryxell, P. A. 1980. A Revision of the American Species of *Hibiscus* Section *Bombicella* (Malvaceae). Washington. [U.S.D.A. Techn. Bull. 1624.] Hochreutiner, B. P. G. 1900. Révision du genre *Hibiscus.* Annuaire Conserv. Jard. Bot. Genève 4: 23–190. Kearney, T. H. 1955b. A tentative key to the North American species of *Hibiscus* L. Leafl. W. Bot. 7: 274–284. Menzel, M. Y., P. A. Fryxell, and F. D. Wilson. 1983. Relationships among New World species of *Hibiscus* section *Furcaria* (Malvaceae). Brittonia 35: 204–221.

1. Leaf blades glabrous or glabrate abaxially, sometimes with few hairs or prickles on veins.
   2. Calyx lobes conspicuously 3-ribbed, 2 marginal, 1 medial.
      3. Calyx lobes with nectary on medial rib; herbage usually dark red . . . . 9. *Hibiscus acetosella*
      3. Calyx lobes without nectary; herbage not dark red. . . . . 10. *Hibiscus radiatus*
   2. Calyx lobes not conspicuously 3-ribbed.
      4. Plants annual, to 0.6(–1) m; petals mostly 1.5–3(–4) cm; calyx: primary veins zigzag; leaf blades 3(–5)-parted, segments pinnately lobed, lobes apically rounded or truncate . . . . 20. *Hibiscus trionum* (in part)
      4. Plants perennial, usually 1–5 m; petals 3.5–10.5 cm; calyx: primary veins not zigzag (except sometimes in *H. dasycalyx*, then petals 4–7 cm); leaf blades unlobed or, if parted, with segments not *both* pinnately lobed *and* lobes apically rounded or truncate.
         5. Leaf blades usually unlobed (rarely lobed in *H. rosa-sinensis*).
            6. Petals pinnatifid-laciniate, strongly reflexed; involucellar bractlets 0.06–0.18 cm. . . . . 12. *Hibiscus schizopetalus*
            6. Petals not pinnatifid-laciniate, spreading or slightly reflexed; involucellar bractlets 0.3–3 cm.
               7. Calyces conspicuously enlarging in fruit; widespread native herbaceous perennials of open wetlands . . . . 16. *Hibiscus laevis* (in part)
               7. Calyces not or little enlarging in fruit; woody plants of cultivation, occasionally escaping and then into nonwetland habitats.
                  8. Staminal column bearing filaments nearly throughout; capsules minutely densely stellate-hairy; temperate North America . . . . 1. *Hibiscus syriacus* (in part)
                  8. Staminal column bearing filaments in distal ½; capsules seldom produced, glabrous; s California, s Florida . . . . .11. *Hibiscus rosa-sinensis*
         5. Leaf blades 3-lobed, sometimes hastately, or palmately 3–5-lobed.
            9. Leaf blades 3–5-lobed; petals bright red. . . . . 15. *Hibiscus coccineus*
            9. Leaf blades 3-lobed; petals pink, lavender, blue, white, creamy white, or cream, reddish basally.
               10. Calyces divided ½ length, little enlarging in fruit; seeds flattened, reniform-ovoid, glabrous laterally, hairy dorsally; cultivated shrubs escaping into nonwetland habitats. . . . . 1. *Hibiscus syriacus* (in part)
               10. Calyces divided ⅓–½ length, conspicuously enlarging in fruit; seeds reniform-globose, hairy throughout; native herbs of open wetlands.
                  11. Calyces and capsules glabrous; e, c United States . . . . . . 16. *Hibiscus laevis* (in part)
                  11. Calyces and capsules hairy; Texas. . . . . .17. *Hibiscus dasycalyx*
1. Leaf blades variously hairy abaxially: stellate-hairy, scabrous, scabridulous, tomentose, stellate-tomentose, tomentulose, or velvety.
   12. Involucellar bractlets: apex 2-fid or appendaged; calyx lobes with nectaries on medial ribs; petals maroon, red, or purple-brown basally.
      13. Stems and leaf blades scabrous; petals pale yellow to white . . . . 7. *Hibiscus aculeatus*
      13. Stems and leaf blades stellate-tomentulose; petals pink. . . . . 8. *Hibiscus furcellatus*
   12. Involucellar bractlets (often absent in *H. denudatus*): apex not 2-fid or appendaged; calyx lobes without nectaries; petals with or without maroon, red, or purple-brown basally.
      14. Plants annual; calyces: primary veins zigzag; leaf blades 3(–5)-parted, segments pinnately lobed, lobes apically rounded or truncate . . . . 20. *Hibiscus trionum* (in part)
      14. Plants perennial: subshrubs, shrubs, trees, or herbaceous perennials; calyces: primary veins not zigzag; leaf blades unlobed or, if parted, with segments not *both* pinnately lobed *and* lobes apically rounded or truncate.
         15. Petals abaxially tomentose, somewhat revolute, dull red or dull orange; seeds glabrous; s Texas . . . . 21. *Hibiscus clypeatus*
         15. Petals not abaxially tomentose (sometimes sparsely hairy abaxially where exposed in bud), bright red, yellow, cream, pink, maroon, mauve, pale purple, or white, often with pink, red, or maroon lines or spot basally; seeds verrucose-papillose or hairy; s Florida and sw United States including Texas.

[16. Shifted to left margin.—Ed.]

16. Petals bright red throughout.
    17. Flowers horizontal or ascending, corolla broadly campanulate to rotate; involucellar bractlets 1.2–2(–2.4) cm; s Texas. . . . . 6. *Hibiscus martianus*
    17. Flowers nodding or pendulous, corolla narrowly funnelform; involucellar bractlets 0.5–0.9 cm; s Florida. . . . . 2. *Hibiscus poeppigii*
16. Petals yellow, cream, pink, mauve, pale purple, or white, usually with contrasting spots or lines near base.
    18. Pedicels nearly always much longer than subtending leaves.
        19. Stellate hairs of young stems dense, appressed, 4-armed, aligned with stem axis; seeds silky-hairy ± throughout . . . . 3. *Hibiscus coulteri*
        19. Stellate hairs of young stems sparse, few- to many-armed, 2- or 3-dimensionally radiate; seeds glabrous laterally . . . . 5. *Hibiscus biseptus*
    18. Pedicels shorter than subtending leaves (or else subtending leaves much reduced or absent in subcorymbose inflorescences of some *H. mutabilis*).
        20. Leaves mostly 1.2–3 cm; petals 1.3–3 cm; desert Southwest . . . . 4. *Hibiscus denudatus*
        20. Leaves mostly 4–30 cm; collectively widespread but almost never in desert Southwest.
            21. Stems at least remotely prickly; free portion of stamen filaments 1–1.5 mm; stigmas wedge-shaped; seeds tomentulose . . . . 19. *Hibiscus striatus*
            21. Stems unarmed; free portion of stamen filaments 2–9 mm; stigmas capitate or discoid; seeds verrucose-papillose or with long straight hairs dorsally and dorsolaterally.
                22. Subshrubs, shrubs, or trees; capsules: apex impressed; seeds with long straight hairs dorsally and dorsolaterally; plants of cultivation, escaping rarely (Alabama, Louisiana), and then into nonwetland habitats . . . . . 18. *Hibiscus mutabilis*
                22. Perennial herbs; capsules: apex apiculate; seeds verrucose-papillose; widespread (Ontario, e, sc United States and California) species of open wetlands.
                    23. Staminal columns: length ½ petals; filaments: free portions not secund; pedicels of later-produced flowers often adnate to petioles . . . . 13. *Hibiscus moscheutos*
                    23. Staminal columns: length ⅔ petals; filaments: free portions secund; pedicels never adnate to petioles . . . . 14. *Hibiscus grandiflorus*

1. **Hibiscus syriacus** Linnaeus, Sp. Pl. 2: 695. 1753

   • Rose-of-Sharon, shrub althaea, Korean hibiscus ☐

Shrubs to 4 m. **Stems:** twigs ± glabrous except for line of minute, curved hairs running length of internode. **Leaves:** stipules linear to filiform, 2.5–12 mm; petiole mostly ¼–½ blade, densely hairy adaxially; blade broadly rhombic-ovate, usually 3-lobed, sometimes unlobed in inflorescence, 3.5–9.5 × 2.5–8.5 cm, base cuneate or rounded, margins coarsely crenate-serrate, apex acute to short-acuminate, surfaces glabrate, midvein with nectary abaxially. **Inflorescences** solitary flowers or few-flowered clusters in axils of distal leaves. **Pedicels** jointed at or near base, to 1.5 cm, densely, minutely stellate-hairy; involucellar bractlets 7 or 8, linear or narrowly oblanceolate, 0.9–2.2 cm, unequal in length, margins not ciliate, densely stellate-hairy throughout. **Flowers** horizontal or ascending, sometimes double; calyx lobed ½ length, broadly campanulate, 1.6–2 cm, lobes triangular, apices acute or short-acuminate, minutely and densely stellate-hairy throughout, nectaries absent; corolla broadly funnelform, petals pink, lavender, blue, or white, usually dark red basally, obliquely obovate, 3.5–7.5 × 2.5–5.5 cm, margins entire or repand, sometimes undulate, finely hairy where exposed in bud; staminal column white, 2.5–3.5 cm, bearing filaments nearly throughout, free portion of filaments not secund, mostly 1.5–3 mm; pollen cream; styles white, 3–8 mm, branches sometimes of unequal lengths; stigmas white. **Capsules** greenish tan, ovoid, 1.5–2.5 cm, apex apiculate, minutely, densely stellate-hairy. **Seeds** reddish brown, reniform-ovoid, laterally flattened, 4–5 mm, glabrous laterally, long-hairy dorsally, hairs straight, reddish orange. $2n = 40, 80, 90$ (all cultivars).

Flowering mid–late summer. Old home sites, roadsides, disturbed areas; 0–300 m; introduced; Ala., Conn., Del., D.C., Ga., Ill., Kans., Ky., La., Miss., Mo., Nebr., N.Y., N.C., Pa., S.C., Tenn., Utah, Va., W.Va.; Asia (China); introduced also in Mexico, Bermuda,

Central America, w South America, s Eurasia, Pacific Islands (New Zealand), Australia.

Hibiscus syriacus is not the biblical rose-of-Sharon, which is identified as *Lilium candidum* or *Tulipa agenensis* subsp. *boissieri* (D. J. Mabberley 2008).

Hibiscus syriacus is widely cultivated as an ornamental and sometimes persists or escapes. It has been reported from southern Canada, Indiana, Ohio, and Texas; I have not seen specimens from those areas. Double and semidouble forms of *H. syriacus* are in cultivation.

2. **Hibiscus poeppigii** (Sprengel) Garcke, Jahresber. Naturwiss. Vereins Halle 2: 133. 1850 • Poeppig's rose-mallow

Achania poeppigii Sprengel, Syst. Veg. 3: 100. 1826

Subshrubs to 1.8 m, herbage with appressed-stellate pubescence throughout. **Stems** additionally with line of fine, curved, simple hairs extending from node to node and decurrent from leaf base; older twigs brown or greenish brown to gray. **Leaves:** stipules narrowly triangular, 2–4.5 mm; petiole mostly ¼–¾ blade, with line of fine, curved hairs adaxially; blade broadly ovate to transversely broadly ovate, usually 3-lobed or 3-fid, mostly 1.2–4.5 × 1.2–4.3 cm, base cordate to truncate, rarely broadly cuneate, margins irregularly crenate to serrate, apex obtuse to broadly acute, apex of lateral lobes obtuse or broadly acute, surfaces scabridulous, hairs appressed-stellate, nectary present abaxially at base of midvein. **Inflorescences** solitary flowers in axils of distal leaves. **Pedicels** jointed beyond middle, 2–5 cm, exceeding petioles and sometimes blades; involucellar bractlets 9–11, linear to narrowly oblanceolate or subulate-linear, 0.5–0.9 cm, margins not or indistinctly ciliate. **Flowers** nodding or pendulous; calyx divided ± ½ length, narrowly campanulate, 0.7–1.2 cm, exceeding involucel, lobes narrowly triangular-ovate, apices acute or short-acuminate, nectaries absent; corolla narrowly funnelform, petals bright red [pink], broadly oblanceolate, ± convolute, 1.5–2.6[–3.5] × 0.4–1.2 cm, margins ± entire, sparingly hairy abaxially where exposed in bud; staminal column exserted, straight, bright red, 1.8–3.4 cm, bearing filaments on distal ½, free portion of filaments not secund, 2–3 mm; pollen dark orange; styles red, 1–3.5 mm; stigmas red. **Capsules** dull brown, ± globose, 0.9–1.1 cm, equaling or surpassing calyces, apex apiculate, coarsely stellate-hairy. **Seeds** brown, angulately reniform-ovoid, 2–2.6 mm, pale silky-hairy ± throughout. 2*n* = 22 (Jamaica).

Flowering winter–summer. Open thickets and hammocks, limestone-derived soil; 0–10 m; Fla.; e, se Mexico; West Indies (Cuba, Jamaica); Central America (Guatemala).

Hibiscus poeppigii has been called *H. pilosus* (Swartz) Fawcett & Rendle (for example, J. K. Small 1933; R. W. Long and O. Lakela 1971), a name that pertains to a species of *Malvaviscus*. In the flora area, *H. poeppigii* is confined to Miami-Dade and Monroe counties; it has been erroneously cited from mid-peninsular Florida on the basis of a mislabeled specimen. The floral characteristics suggest hummingbird pollination.

3. **Hibiscus coulteri** Harvey ex A. Gray, Smithsonian Contr. Knowl. 3(5): 23. 1852
   • Desert rose-mallow F

Subshrubs, to 2 m, herbage with appressed-stellate and simple hairs throughout. **Stems:** stellate hairs of younger stems dense, appressed, 4-armed, arms approximate in pairs, aligned with stem axis, lines of fine, curved hairs absent or obscured; older growth grayish, rough, glabrescent. **Leaves:** stipules linear-subulate, 3–10 mm; petiole from less than ½ as long as to nearly equaling blade, sparingly hairy above with fine curved hairs in addition to normal hairs; blade broadly ovate to broadly transversely ovate (often folded in pressed specimens), usually 3-lobed or 3-parted, sometimes unlobed, mostly 1–3.5 × 1–4 cm, base broadly cuneate to truncate or rounded, margins coarsely, irregularly serrate or dentate, apex acute to truncate, lobes obovate to oblanceolate, toothed primarily in distal ½, sometimes themselves shallowly pinnately lobed, surfaces scabridulous, hairs appressed-stellate, obscure nectary present abaxially on midvein near base. **Inflorescences** solitary flowers in axils of distal leaves. **Pedicels** jointed below apices, to 17 cm, usually much exceeding subtending leaves; involucellar bractlets 8–14, linear-subulate, 1–2 cm, margins ciliate. **Flowers** erect or ascending; calyx divided ¾+ length, funnelform, 1.4–2.2 cm, equaling or slightly exceeding involucel, lobes narrowly lanceolate-triangular, margins ciliate, apices attenuate, nectaries absent; corolla rotate, petals yellow to cream, usually with dark to obscure maroon lines basally, asymmetrically obovate to broadly obovate, 1.6–4 × 1–3.5 cm, margins ± entire, sparingly hairy abaxially where exposed in bud; staminal column straight, yellow or cream, 0.6–1.4 cm, bearing filaments throughout, free portion of filaments not secund, 1–3 mm; pollen yellow-orange; styles cream, 1.5–5 mm; stigmas maroon or cream. **Capsules** pale olivaceous gray with darker median stripe on each valve, ovoid or ellipsoid, 0.7–1.7 cm, to ⅔ calyces, apex rounded, hairy near apex or glabrous throughout. **Seeds** dark brown, angulately reniform-ovoid, 2.4–3 mm, silky-hairy ± throughout.

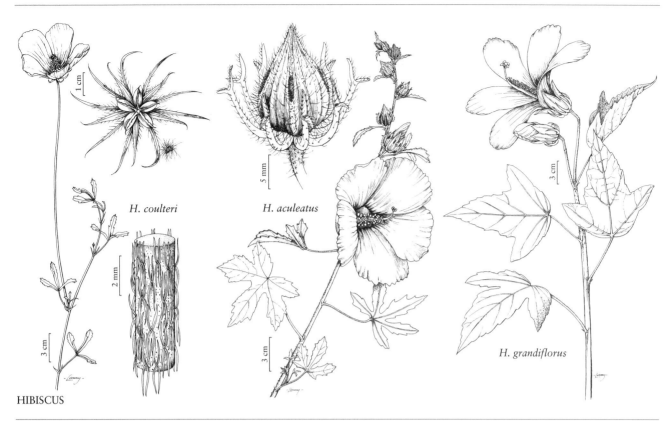

H. coulteri   *H. aculeatus*   *H. grandiflorus*

HIBISCUS

Flowering nearly year-round. Desert, rocky slopes; 600–1600 m; Ariz., N.Mex., Tex.; Mexico (Baja California, Chihuahua, Coahuila, Durango, Nuevo León, Sonora, Zacatecas).

Hibiscus coulteri has been recorded in the southern half of Arizona; in Otero County, New Mexico; and in the region west of the Pecos River in Texas.

A naturally occurring hybrid between *Hibiscus coulteri* and *H. denudatus* (*Hibiscus ×sabei* Weckesser) has recently been documented from western Texas (W. Weckesser 2011).

4. **Hibiscus denudatus** Bentham, Bot. Voy. Sulphur, 7, plate 3. 1844 • Paleface rose-mallow

Subshrubs, to 0.8 m, herbage densely ± yellowish stellate-tomentose throughout. **Stems:** older twigs yellowish brown, glabrescent, line of fine hairs obscured by other pubescence or absent. **Leaves:** stipules narrowly triangular to subulate, 1–3 mm; petiole to ½ as long as blade, adaxial fine, curved hairs absent or obscured by other pubescence; blade oblong or broadly to narrowly ovate (often folded in pressed specimens), usually unlobed, rarely shallowly 3-angulate-lobate, mostly 1.2–3 × 1–2.5 cm, base cuneate to rounded or truncate, margins irregularly and coarsely dentate or dentate-crenate, apex broadly acute to truncate, surfaces stellate-tomentose, ± obscured nectary present abaxially on midvein near base. **Inflorescences** solitary flowers in axils of distal leaves. **Pedicels** obscurely joined below apex, 0.5–2 cm, equaling or exceeding subtending petioles; involucellar bractlets 7–10, often absent, narrowly triangular to subulate, to 0.6 cm, margins not ciliate. **Flowers** ascending to erect; calyx divided ⅔–⅘ length, funnelform, 0.7–1.6 cm, lobes narrowly triangular-ovate, apices acuminate, nectaries absent; corolla rotate, petals pale purple or pink to nearly white, with maroon spot or lines basally, asymmetrically obovate, 1.3–3 × 0.9–2 cm, margins ± entire, finely hairy abaxially where exposed in bud; staminal column straight, pink, 0.4–1.1 cm, bearing filaments throughout, free portion of filaments not secund, 1–3 mm; pollen dark orange; styles pale pink, 1.5–4 mm; stigmas pale pink to maroon. **Capsules** dull yellow-green with darker medial stripe on each valve, drying straw-colored, ovoid to subglobose, 0.6–1 cm, shorter than calyces, apex apiculate, glabrous or with minute hairs near apex and on sutures. **Seeds** dark brown, angularly reniform-ovoid, 2.2–3 mm, whitish, silky-hairy ± throughout, hairs 3–4 mm. $2n = 22$ (Mexico: Durango).

Flowering year-round. Desert, often rocky; 30–1800 m; Ariz., Calif., Nev., N.Mex., Tex.; Mexico (Baja California, Baja California Sur, Chihuahua, Coahuila, Durango, Sonora).

Hibiscus denudatus has been recorded from Imperial, Riverside, San Bernardino, and San Diego counties in southern California; from southernmost Nevada (southern Clark County); from the southern half and near the Grand Canyon in Arizona; from southwestern New Mexico; and from the region west of the Pecos River in Texas.

As noted in the previous species, a naturally occurring hybrid between *Hibiscus denudatus* and *H. coulteri* (*Hibiscus* ×*sabei* Weckesser) has recently been documented from western Texas (W. Weckesser 2011).

5. Hibiscus biseptus S. Watson, Proc. Amer. Acad. Arts 21: 418. 1886 • Arizona rose-mallow

Subshrubs, to 1 m, herbage sparsely simple- and stellate-hairy throughout, few- to many-armed stellate hairs 2- or 3-dimensionally radiate. **Stems** also with a line of simple, fine, curved hairs decurrent from leaf base and extending from node to node. **Leaves:** stipules narrowly triangular, (4–)5–10 mm; petiole ½ to ± equaling blade, with fine curved hairs adaxially; blade transversely to broadly ovate, 3(–5)-lobed or -parted, mostly 2–8.5 × 3–9 cm, base broadly cuneate to truncate or rounded or cordate, margins irregularly and sometimes doubly serrate or crenate-serrate, apex acute to short-acuminate, lobes narrowly elliptic to lanceolate, margins entire basally, coarsely serrate distally, surfaces scabridulous abaxially, less so adaxially, nectary present abaxially on midvein near base. **Inflorescences** solitary flowers in axils of distal leaves. **Pedicels** obscurely jointed below apices, to 17.5 cm, usually exceeding subtending leaves, sometimes much elongated; involucellar bractlets 8–12, linear-subulate, 1.5–2.5 cm, margins ciliate. **Flowers** horizontal to ascending; calyx neither accrescent nor inflated, it and involucel in fruit tinged basally with pink or purple, divided ⁴⁄₅+ length, broadly campanulate, (1.4–)2–2.8 cm, equaling or usually exceeding involucel, lobes narrowly lanceolate-triangular, margins ciliate, apices attenuate, nectaries absent; corolla rotate, petals cream [white], with maroon lines or spot basally, these sometimes faint or absent, asymmetrically obovate to broadly obovate, 3–4.5 × 1.8–2.6 cm, margins ± entire, sparingly hairy abaxially where exposed in bud; staminal column straight, pale yellow or maroon, 1.1–1.5 cm, bearing filaments throughout, free portion of filaments not secund, 1.5–2.5 mm; pollen yellow to orange-red; styles white to maroon, 1–2 mm; stigmas white to maroon. **Capsules** dull yellow-green with darker medial stripe on each valve, ovoid to subglobose, 1–1.5 cm, much shorter than calyces, apex apiculate, glabrous. **Seeds** dark brown,

angulately reniform-ovoid, strongly depressed laterally, 2.5–3.5 mm, silky-hairy dorsally, glabrous laterally, with slightly raised pale yellow zone near hilum. $2n = 22$ (Mexico: Sinaloa).

Flowering Aug–Oct. Desert scrub and grasslands; 800–2300 m; Ariz.; Mexico (Baja California, Chihuahua, Jalisco, Nayarit, Sinaloa, Sonora).

Hibiscus biseptus is known from the southern Arizona counties of Maricopa, Pima, Pinal, and Santa Cruz.

6. Hibiscus martianus Zuccarini, Linnaea 24: 193. 1851 • Heartleaf rose-mallow

Hibiscus cardiophyllus A. Gray

Subshrubs, to 1.5 m, herbage stellate-tomentose throughout, lines of fine curved hairs absent or obscured. **Stems:** older twigs gray, glabrescent. **Leaves:** stipules linear-lanceolate, (2–)3–7(–10) mm; petiole subequal to blade, adaxial fine curved hairs absent or obscured; blade markedly discolorous, broadly ovate, unlobed or shallowly 3-angulate-lobate, 2–8 × 2–8.5 cm, base cordate, margins irregularly dentate or dentate-serrate, apex broadly acute to rounded, surfaces densely tomentose abaxially, less so adaxially, inconspicuous nectary abaxially on midvein near base. **Inflorescences** solitary flowers in axils of distal leaves. **Pedicels** jointed below apices, 4–10 cm, exceeding subtending petioles, elongating in fruit; involucellar bractlets (7)8–10, narrowly rhombic-elliptic, 1.2–2(–2.4) cm, enlarging in fruit, margins not or inconspicuously ciliate. **Flowers** horizontal or ascending; calyx rotate to campanulate, lobed nearly to base, 1.4–2.4(–2.7) cm, larger in fruit, lobes narrowly triangular-ovate, apices acute to short-acuminate, nectaries absent; corolla rotate to campanulate, petals bright red, asymmetrically obovate, 1.5–3 × 0.9–2.1 cm, margins ± entire, sometimes undulate, finely hairy abaxially where exposed in bud; staminal column somewhat declinate, bright red, 0.9–1.4 cm, bearing filaments throughout, free portion of filaments secund, 1–5 mm; pollen yellow-orange; styles red, 2–7 mm; stigmas red. **Capsules** yellowish brown, ovoid, 1.2–2 cm, apex apiculate, glabrous or with minute hairs near apex and on sutures. **Seeds** dark brown, angulately reniform-ovoid, 3–4 mm, stellate-hairy throughout. $2n = 22$ (Mexico: Nuevo León).

Flowering year-round. Dry, often rocky thorn-scrub and open woodlands; 10–800 m; Tex.; Mexico (Chihuahua, Coahuila, Hidalgo, Nuevo León, Puebla, San Luis Potosí, Tamaulipas).

In Texas, *Hibiscus martianus* occurs from the Big Bend region to the southernmost Gulf Coast, mostly in counties bordering or near the Rio Grande.

7. **Hibiscus aculeatus** Walter, Fl. Carol., 177. 1788

• Comfort-root, pineland hibiscus [E] [F]

Herbs, perennial, or subshrubs, to 0.5–1(–2) m, herbage scabrous throughout, hairs scattered, stellate, stout, pustular-based. **Stems** additionally with line of fine, curved hairs. **Leaves:** stipules linear or linear-filiform, 2–6 mm; petiole of lower leaf ⅔ to equaling blade, much shorter in inflorescence, with fine, curved hairs adaxially in addition to normal pubescence; blade broadly to transversely ovate, 3–5-fid, sometimes lobed, often unlobed in inflorescence, 3.5–9.5 × 4.5–13.5 cm, narrower in inflorescence, base cordate to cuneate, lobes obovate to oblanceolate, margins coarsely and irregularly crenate-serrate, apex acute to short-acuminate, surfaces scabrous, slitlike nectary present abaxially at or near base of midvein. **Inflorescences** solitary flowers in axils of distal leaves, or appearing racemose by reduction of subtending leaves. **Pedicels** inconspicuously jointed at base, to 1.5 cm; involucellar bractlets 8–11, linear-subulate, 1–1.6 cm, margins not or inconspicuously ciliate, apex 2-fid, bristly-hairy. **Flowers** horizontal or declinate; calyx divided ⅔ length, campanulate, 1.6–2.8 cm, lobes triangular, with 3 prominent, often reddish ribs, 1 medial, 2 marginal, medial with prominent nectary, apices acute or acuminate, veins and sometimes spaces between them with conspicuous simple or few-armed, stellate hairs; corolla funnelform, petals pale yellow to white, dark red basally, obovate, 5–8 × 2.5–5 cm, margins ± entire to crenate-dentate, sometimes undulate, minutely hairy abaxially where exposed in bud; staminal column straight, dark red, 2–4.5 cm, bearing filaments ± throughout, free portion of filaments not secund, 1.5–2.5 mm; pollen dark red; styles dark red, pink, or white, 9–22 mm; stigmas dark red, pink, or white. **Capsules** medium brown to stramineous, ovoid, 1.2–2 cm, apex acute to acuminate, variously antrorsely hispid, more minutely stellate-hairy. **Seeds** reddish brown to dark brown, sometimes with raised, pale concentric lines, angulately reniform-ovoid, 3.3–4.5 mm, sparingly to moderately papillose-verrucose. $2n = 72$.

Flowering Jun–Sep. Pine savannas, flatwoods, swales, roadside ditches; 0–100 m; Ala., Fla., Ga., La., Miss., N.C., S.C., Tex.

Hibiscus aculeatus ranges northeast to Carteret County in North Carolina, south to Lake County in central Florida, and west to Hardin County in eastern Texas. Within these limits its distribution is confined entirely to the coastal plain.

8. **Hibiscus furcellatus** Lamarck in J. Lamarck et al., Encycl. 3: 358. 1789

Herbs, perennial, or subshrubs, to 2(–4) m, herbage densely stellate-tomentulose throughout. **Stems** sometimes also with longer, stiff, simple hairs, fine, curved hairs absent or obscured. **Leaves:** stipules linear, 3–10 mm; petiole ⅔ to equaling blade, shorter in inflorescence, fine, curved hairs absent or obscured adaxially; blade somewhat discolorous, broadly to transversely ovate, unlobed or shallowly 3(–5)-lobed, rarely proximalmost deeply 5–7-lobed, mostly 6.5–11 × 6–12 cm, base cordate, often deeply so, margins unevenly serrate or crenate-serrate, apex broadly acute to short-acuminate, lobes broadly triangular, surfaces stellate-tomentulose, slitlike nectary present abaxially at or near base of midvein. **Inflorescences** solitary flowers in axils of distal leaves, sometimes appearing racemose by reduction of subtending leaves. **Pedicels** jointed at bases, to 2.5 cm, shorter than subtending petioles; involucellar bractlets 9–12, sometimes wide-spreading, terete, 0.8–1.6 cm, margins minutely pubescent, setose, apex 2-fid or appendaged, sometimes obscurely so. **Flowers** horizontal or declinate; calyx divided ½–⅔ length, campanulate, 1.5–2.4 cm, enlarging in fruit, lobes triangular, with 3 prominent ribs, 2 marginal, 1 medial, medial bearing conspicuous nectary, apices acute or acuminate, variously invested with both minute, stellate hairs and hispid with much larger, simple or stellate, pustular-based hairs, latter often largely confined to veins; corolla narrowly funnelform, petals pink, maroon at base, obliquely obovate, 5.5–9.5 × 2.5–4.5 cm, margins entire to repand or crenate, finely hairy abaxially where exposed in bud; staminal column straight, maroon, 3–4.5 cm, bearing filaments ± throughout, free portion of filaments not secund, 0.5–1.5 mm; pollen maroon; styles dark maroon, 1–3 mm; stigmas dark maroon. **Capsules** brown, ovoid, 2–2.5 cm, apex acute and apiculate or acuminate, surface obscured by pale yellowish, simple, dense, antrorsely appressed hairs. **Seeds** olivaceous brown to reddish or purplish brown, angulately reniform-ovoid, 2.8–3.8 mm, glabrous. $2n = 72$.

Flowering year-round. Freshwater marshes, pine flatwoods, sand pine scrub, fill, canal margins, waste areas; 0–50 m; Fla.; s Mexico; West Indies; Central America (Belize, Guatemala, Honduras, Nicaragua, Panama); n, c South America; Pacific Islands (Hawaii).

Hibiscus furcellatus is found primarily in counties along the central and southern parts of Florida's Atlantic coast, although there are a few inland records as well.

9. **Hibiscus acetosella** Welwitsch ex Hiern, Cat. Afr. Pl. 1: 73. 1896 • African rose-mallow, red-leaf or cranberry hibiscus, false roselle <u>I</u>

Subshrubs, 1–2(–4) m, herbage usually dark red throughout, glabrous, rarely sparsely hairy. **Stems** with line of fine, curved hairs. **Leaves:** stipules linear-lanceolate, (8–)10–15 mm; petiole ½ to ± equaling blade, with fine, curved hairs adaxially; blade usually dark red, broadly to transversely ovate, usually deeply 3–5-lobed, 4–10 × 3.5–10 cm, base broadly cuneate to truncate, margins crenate or crenate-serrate, apex acute to acuminate, lobes narrowly elliptic or narrowly obovate, surfaces glabrate, prominent slitlike nectary present abaxially on midvein near base. **Inflorescences** solitary flowers in axils of distal leaves, sometimes together appearing racemose by reduction of subtending leaves. **Pedicels** jointed near middle, to 1.2 cm; involucellar bractlets 8–10, terete, 0.6–1.6 cm, margins setose, apices 2-fid or appendaged. **Flowers** horizontal; calyx divided nearly ⅔ length, funnelform-campanulate, 1.2–2 cm, lobes triangular, with 3 prominent ribs, 2 marginal, 1 medial, medial bearing nectary, apices acuminate, veins setose with pustular-based, simple hairs; corolla funnelform-rotate, petals cream, yellow, or dull pink to dull red with veins usually darker pink, maroon basally, asymmetrically obovate, 3–5.5 × 2.5–4.5 cm, margins repand, finely hairy abaxially where exposed in bud; staminal column straight, maroon, 1.5–2.5 cm, bearing filaments nearly throughout, free portion of filaments not secund, 1.5–2.5 mm; pollen yellow; styles maroon, to 1 mm; stigmas maroon. **Capsules** reddish brown, ovoid, 1.6–2.5 cm, apex acute or short-acuminate, weakly antrorsely hispid with simple, scattered, loose hairs. **Seeds** olivaceous brown, angulately reniform-ovoid, 3.5–4 mm, papillose-scaly, scales pectinate. $2n = 72$.

Flowering mostly fall–winter. Roadsides, disturbed areas; 0–20 m; introduced; Fla.; Africa; introduced also in Mexico, West Indies, Central America, South America, Asia (Indonesia).

Hibiscus acetosella is cultivated as an ornamental and as a salad plant and occasionally escapes. It apparently originated in Africa, perhaps as an amphidiploidized hybrid between *H. asper* Hooker f. and *H. surattensis* Linnaeus (M. Y. Menzel 1986) and may no longer exist there or anywhere else truly in the wild (F. D. Wilson 1994).

10. **Hibiscus radiatus** Cavanilles, Diss. 3: 150, plate 54, fig. 2. 1787 • Monarch rose-mallow <u>I</u>

Hibiscus cannabinus Linnaeus var. *unidens* (Lindley) Hochreutiner

Herbs or subshrubs, erect or decumbent, to 1.5(–2) m, herbage usually glabrous, sometimes with prickles or simple hairs. **Stems** also with line of fine, curved hairs. **Leaves:** stipules linear-lanceolate, 10–16 mm, ciliate; petiole mostly ⅔ to equaling blade, with fine, curved hairs adaxially, sometimes sparingly prickly; blade broadly to transversely triangular-ovate, mostly 5-fid, 4.5–15 × 6–17 cm, base truncate or broadly and shallowly cordate, margins serrate, apex acuminate, segments lanceolate, surfaces glabrous but veins sometimes with retrorse prickles abaxially, nectary sometimes present abaxially on midvein near base. **Inflorescences** solitary flowers in axils of distal leaves. **Pedicels** jointed near middle, to 1.5 cm, prickly distal to joint; involucellar bractlets 8–10, flattened or canaliculate, 1–1.8 cm, margins setose, apex simple, 2-fid, or appendaged, spreading in flower to reflexed in fruit. **Flowers** horizontal or ascending; calyx divided ⅔+ length, cylindric-campanulate, 1.5–2.5 cm, accrescent, lobes narrowly triangular, with 3 prominent, setose ribs, 2 marginal, 1 medial, often darker and with stinging bristly hairs, apices acuminate-attenuate, nectary absent; corolla rotate, petals usually dark red to rose-purple, rarely yellow, with dark purple basally, asymmetrically obovate, 3.5–7 × 3–5.5 cm, margins repand, finely hairy abaxially where exposed in bud; staminal column straight, maroon, 2.4–3.5 cm, bearing filaments nearly throughout, free portion of filaments not secund, mostly 1.5–2.5 mm; pollen dull yellow; styles maroon, 1.5–3 mm; stigmas dark maroon. **Capsules** pinkish brown, ovoid, 1.8–2.5 cm, apex acute, apiculate, antrorsely hispid, hairs readily deciduous, simple. **Seeds** dark olivaceous, with fine, raised, concentric lines, angulately reniform-ovoid, 4.2–4.8 mm, moderately verrucose-lepidote, scales striate-fimbriate. $2n = 72$ (Trinidad, where cultivated).

Flowering fall. Disturbed sites; 0–80 m; introduced; Fla., Tex.; Asia (India); introduced also in Mexico, West Indies, Central America, South America, elsewhere in s Asia, Africa, Australia.

Hibiscus radiatus may have originated in India as a garden hybrid between *H. cannabinus* Linnaeus and *H. surattensis* Linnaeus and may not exist anywhere as a native (F. D. Wilson 1994; Wilson and M. Y. Menzel 1964). Other sources (for example, J. van Borssum Waalkes 1966) state that it is native to southern and southeastern Asia. It was originally grown in North America for its potential as a fiber crop.

Hibiscus cannabinus in the sense of J. K. Small (1933) pertains here.

11. **Hibiscus rosa-sinensis** Linnaeus, Sp. Pl. 2: 694. 1753 • Chinese hibiscus, shoe-black plant [1]

Shrubs or trees, 1–3(–5) m. **Stems:** new growth finely and sparingly stellate- or simple-hairy. **Leaves:** stipules linear to lanceolate, 8–16 mm; petiole usually to ⅓ blade, adaxial groove hairy with minute, ± sinuous hairs, sometimes villous; blade ovate, unlobed or only very rarely lobed, 5–12 × 3–8.5 cm, base rounded to cuneate, margins coarsely serrate in distal ⅔–¾, apex acute to short-acuminate, surfaces glabrate, nectary present abaxially on midvein near base. **Inflorescences** solitary flowers, in axils of distal leaves. **Pedicels** jointed closer to flower, 4–9.5 cm, sparsely stellate-pilose or ± glabrous; involucellar bractlets 6–8, narrowly lanceolate to narrowly triangular, 0.3–2.2 cm, width varying in same flower, margins not ciliate, surface and margins puberulent or glabrate. **Flowers** showy, horizontal or declinate, sometimes double; calyx divided ½–¾ length, narrowly campanulate, 2–3 cm, lobes triangular or narrowly so, apices acute to acuminate, often minutely, sparingly hairy, nectaries absent; corolla funnelform to rotate or petals slightly reflexed, petals usually red, sometimes pink, white, or yellow (or other colors in horticultural forms), usually darker at base, broadly to narrowly obovate, 6–10.5 × 4–6.5 cm, margins entire or crenate, often undulate, finely hairy abaxially mostly where exposed in bud; staminal column straight or moderately curved, usually red, often pink or white, 6.5–11.5 cm, bearing filaments in distal ½; free portion of filaments not secund, 3–9 mm, pollen yellow; styles red, pink, or white, 6–15 mm; stigmas usually reddish, sometimes golden yellow. **Capsules** seldom produced, brown, ovoid, 2.5–3 cm, apex rounded or beaked, glabrous. **Seeds** (rarely produced in cultivation), dark brown to black, reniform, 5 mm, minutely pubescent. $2n$ = 36, 46, 54, 63, 68, 72, 77, 84, 90, 92, 96, 112, 132, 144, 147, 150, 160, 165, 168, 180, 225 (all cultivars).

Flowering year-round. Disturbed sites; 0–50 m; introduced; Calif., Fla.; introduced also in Mexico, West Indies, Bermuda, Central America, South America, s Asia, Africa, Indian Ocean Islands, Pacific Islands, Australia.

Hibiscus rosa-sinensis is widely cultivated in the Tropics and subtropics, often as a hedge plant; it naturalizes sparingly, probably throughout its cultivated range. It is not known to exist anywhere as a native occurrence; it has been thought to have originated in China (C. Linnaeus 1753), Africa (J. van Borssum Waalkes 1966; A. C. Smith 1979–1996, vol. 2), or the New World Tropics (L. van der Pijl 1937; G. F. Carter 1954; see also H. D. V. Prendergast 1982). As broadly circumscribed here, *H. rosa-sinensis* includes a wealth of hybrids and other selections and, as its range of chromosome numbers suggests, it has had a complicated history in cultivation (F. Singh and T. N. Khoshoo 1970), which probably includes hybridization with the African *H. schizopetalus* and the Hawaiian *H. arnottianus* A. Gray, *H. kokio* Hillebrand, and *H. waimeae* A. Heller (E. V. Wilcox and V. S. Holt 1913). It usually fails to set seed and is generally propagated by cuttings.

12. **Hibiscus schizopetalus** (Dyer) Hooker f., Bot. Mag. 106: plate 6524. 1880 • Fringed rose-mallow or hibiscus, Chinese lantern [1]

Hibiscus rosa-sinensis Linnaeus var. *schizopetalus* Dyer, Gard. Chron., n. s. 11: 568. 1879

Shrubs or trees, to 3(–5) m. **Stems:** new growth essentially glabrous, lines of curved hairs absent. **Leaves:** stipules narrowly triangular, 1–2.5 mm; petiole to ⅓ blade, adaxial groove hairy with minute, ± sinuous hairs; blade lanceolate-ovate to ovate, unlobed, 3.5–10.5 × 1.5–4 cm, base rounded to cuneate, margins coarsely serrate in distal ⅔–¾, apex acute to short-acuminate, ± pinnately veined, surfaces glabrate, nectary present abaxially on midvein near base. **Inflorescences** solitary flowers in axils of distal leaves. **Pedicels** jointed at middle or distally, 7.5–15 cm; involucellar bractlets 6–8, triangular, 0.06–0.18 cm, margins not ciliate. **Flowers** pendulous; calyx divided ⅛–½ length, often 3-lobed, tubular to narrowly funnelform, (1–)1.4–2 cm, lobes broadly triangular, apices acute to obtuse, glabrate, neither accrescent nor inflated, nectaries absent; petals strongly recurved, rose-pink to red, darker on veins, broadly to narrowly obovate, deeply and irregularly pinnatifid-laciniate, 4–6.5 × 1.5–3.5 cm, glabrous; staminal column straight or curved apically, pendulous, pink to red, 5.5–9 cm, bearing filaments in distal ⅓–½, free portion of filaments not secund, 4.5–7.5 mm; pollen yellow; styles pink to red, 7–15 mm; stigmas pink to red. **Capsules** brown, oblong-cylindric, 3.5–4 cm, glabrous or puberulent. **Seeds** brown, angulately reniform-ovoid, 2–3 mm, smooth, glabrous or puberulent. $2n$ = 34, 40, 42, 45 (all cultivars).

Flowering year-round. Disturbed sites; 0–50 m; introduced; Fla.; e Africa; introduced also in Mexico, West Indies, Central America, South America, s Asia, elsewhere in Africa, Pacific Islands, Australia.

Apparently native only in Kenya, Tanzania, and perhaps Mozambique, *Hibiscus schizopetalus* is widely cultivated in the Tropics and occasionally escapes. The occurrence in many *H. rosa-sinensis* cultivars of semipendulous, long-pedicelled flowers with variously crenate, undulate petals suggests the involvement of *H. schizopetalus*. Hybrids between *H. schizopetalus* and *H. rosa-sinensis* can be called *H.* ×*archeri* W. Watson. Typification of *H. schizopetalus* was discussed by M. Cheek (1989).

13. **Hibiscus moscheutos** Linnaeus, Sp. Pl. 2: 693. 1753
• Common rose-mallow

Herbs, perennial, to 2.5 m. **Stems** glabrous or variously hairy, but without line of curved hairs. **Leaves:** stipules subulate, 1–4 mm; petiole ¼–¾ blade, glabrate or finely hairy; blade narrowly to broadly lanceolate to triangular-ovate or orbiculate, 3-lobed or unlobed, 8–20 × 3–13 cm, base cordate to cuneate, margins crenate to dentate or serrate, apex acute to acuminate, surfaces variously hairy, sometimes glabrous adaxially, nectary absent. **Inflorescences** solitary flowers in axils of distal leaves, pedicels of later-produced flowers often adnate to subtending petioles. **Pedicels** variously jointed sub-basally to distally, 2–15 cm, ½–1½ as long as petiole, glabrate or finely hairy; involucellar bractlets [8–]10–14 [or 15], linear-lanceolate, 0.5–4.5(–5) cm, margins ciliate or not, hairy. **Flowers** ± horizontal to slightly declinate; calyx divided ½–⅔ length, broadly campanulate, 1.5–4 cm, somewhat larger in fruit, lobes triangular or triangular-ovate, apices acute to subcaudate, surfaces hairy, nectaries absent; corolla funnelform to broadly so, petals pink or white, sometimes with red spot basally, narrowly to broadly, obliquely obovate, 4–12 × 3.5–6.5 cm, margins repand, sometimes undulate, minutely hairy abaxially mostly where exposed in bud; staminal column straight, white or cream, 1.2–5 cm, to ½ as long as petals, bearing filaments nearly throughout, free portion of filaments not secund, 2–8 mm; pollen yellow; styles white, 10–40 mm; stigmas creamy white to yellow. **Capsules** dark brown, ovoid to subglobose, 1.4–3.5 cm, apex apiculate, glabrous or hairy. **Seeds** brown, reniform-globose, 2.5–3 mm, verrucose-papillose. $2n = 38$.

Subspecies 2 (2 in the flora): North America, n Mexico; introduced in Europe (sw France, n Italy, nw Portugal), Asia (Georgia).

1. Capsules glabrous; involucellar bractlets usually not ciliate; leaf blade surfaces usually glabrous adaxially; mostly e of Mississippi River . . . . 13a. *Hibiscus moscheutos* subsp. *moscheutos*
1. Capsules hairy; involucellar bractlets usually ciliate; leaf blade surfaces usually hairy adaxially; mostly w of Mississippi River, Florida . . . . 13b. *Hibiscus moscheutos* subsp. *lasiocarpos*

13a. **Hibiscus moscheutos** Linnaeus subsp. **moscheutos** [E]

Hibiscus incanus J. C. Wendland; *H. moscheutos* subsp. *incanus* (J. C. Wendland) H. E. Ahles; *H. moscheutos* subsp. *palustris* (Linnaeus) R. T. Clausen; *H. oculiroseus* Britton; *H. palustris* Linnaeus

Leaf blades narrowly lanceolate to orbiculate, sometimes shallowly 3-lobed, usually glabrous adaxially. **Involucellar bractlets** 0.5–3.5(–4) cm, usually not ciliate. **Petals** white or pink, with or without red spot basally. **Capsules** glabrous. $2n = 38$.

Flowering mid Jun–Aug. Brackish and freshwater marshes, floodplain pools, beaver ponds, roadside ditches, farm ponds; 0–400 m; Ont.; Ala., Conn., Del., D.C., Fla., Ga., Ind., Kans., Ky., La., Md., Mass., Mich., Miss., N.H., N.J., N.Y., N.C., Ohio, Okla., Pa., R.I., S.C., Tenn., Tex., Va., W.Va., Wis.; introduced in Europe (sw France, n Italy, nw Portugal); Asia (Georgia).

Hibiscus incanus usually has been treated as having hairy capsules (for example, D. M. Bates 1965; G. E. Crow and C. B. Hellquist 2000; R. K. Godfrey and J. W. Wooten 1981; B. P. G. Hochreutiner 1900; T. H. Kearney 1955b; J. K. Small 1933), which places it with subsp. *lasiocarpos*; the original description by Wendland and later illustration of *H. incanus* show that it lacks hairs and belongs here (O. J. Blanchard 2008).

Subspecies *moscheutos* is widespread and variable in eastern North America. In the northeastern part of its range, it corresponds to *Hibiscus palustris*, synonymized here. That form also extends westward across the Ontario and Erie plains to the vicinity of the south end of Lake Michigan. Distinctly different plants that agree with *H. moscheutos* in the narrow sense occur in southern Indiana and Ohio and southward. Along the eastern seaboard, where *Hibiscus* occupies the coastal marshes of nearly every county from New Hampshire to northeastern Florida, the distinction between the two taxa breaks down. General north-to-south clinal variation occurs, with local exceptions, in features that in the Midwest readily separate the two, including leaf width, leaf lobing, capsule shape and size, and frequency of adnation of petiole and pedicel. Petal color in these strikingly large-flowered plants is mostly pink throughout in the north, and white with a red base in the south; in Maryland and Delaware, there are populations in which all four possible color combinations occur (O. J. Blanchard 1976). *Moscheutos*-like flowers with white petals and red bases occur in New York, and pink-petaled flowers are found in North Carolina. These and similar observations by others have led to reducing the two to subspecific rank (R. T. Clausen 1949b) or fully merging them, as has been done in most recent floras; M. L. Fernald (1942c) made a case for maintaining them as separate species. See Blanchard for further discussion.

Shinnecock Indians used subsp. *moscheutos* as a urinary aid (D. E. Moerman 1998).

13b. Hibiscus moscheutos Linnaeus subsp. **lasiocarpos** (Cavanilles) O. J. Blanchard, Novon 18: 4. 2008 • Woolly rose-mallow

Hibiscus lasiocarpos Cavanilles, Diss. 3: 159, plate 70, fig. 1. 1787; *H. californicus* Kellogg; *H. lasiocarpos* var. *occidentalis* (Torrey) A. Gray; *H. leucophyllus* Shiller; *H. moscheutos* var. *lasiocarpos* (Cavanilles) B. L. Turner; *H. moscheutos* var. *occidentalis* Torrey

Leaf blades lanceolate to broadly triangular-ovate, seldom lobed, usually hairy adaxially. **Involucellar bractlets** 1.5–4.5(–5) cm, usually ciliate. **Petals** usually white, sometimes pink, with red spot basally. **Capsules** hairy. $2n = 38$.

Flowering mid Jun–Aug. Freshwater marshes, edges of ponds and streams, roadside ditches, farm ponds; 0–400 m; Ala., Ark., Calif., Fla., Ill., Ind., Kans., Ky., La., Miss., Mo., N.Mex., Okla., Tenn., Tex.; Mexico (Chihuahua).

Subspecies *lasiocarpos* has been reported also from Utah; the relevant specimen at BRY was cultivated, grown from seeds sent from Hawaii (S. L. Welsh, pers. comm.).

Some authors have commented recently on the distinctness of subsp. *lasiocarpos* (O. J. Blanchard 2008; B. L. Turner 2008; S. R. Hill 2009). The first two are in agreement that *Hibiscus lasiocarpos* should be recognized at infraspecific rank; Hill has maintained it as a separate species and, at the same time, has recognized the California populations as *H. lasiocarpos* var. *occidentalis*.

R. L. Small (2004) used chloroplast and nuclear DNA sequences to construct a phylogeny of *Hibiscus* sect. *Muenchhusia*, which consists of *H. moscheutos* and the next four species of the present treatment. He found support for the distinctness of the latter four, but four other analyzed taxa (*H. incanus*, *H. lasiocarpos*, *H. moscheutos*, and *H. palustris*), all of which together comprise *H. moscheutos* in the broad sense used here, produced a distinct but mostly unresolved clade. This evidence favors the view that *H. lasiocarpos* is better treated at infraspecific rank.

S. R. Hill (2009, 2012) has made a case for maintaining the California populations as a distinct variety. The morphological and phenological features of the variety,

including its often orbiculate leaves, presence of rhizomes resulting in vegetative propagation, and conspicuous beaked flower buds have in the past supported its treatment as a distinct species, *Hibiscus californicus*. In the context of the present treatment, this taxon would be called *H. moscheutos* var. *occidentalis* Torrey.

14. Hibiscus grandiflorus Michaux, Fl. Bor.-Amer. 2: 46. 1803 • Swamp rose-mallow F

Hibiscus urbanii Helwig

Herbs, perennial, to 3 m. **Stems** glabrous or rarely stellate-hairy on younger parts, without line of minute, curved hairs. **Leaves:** stipules linear-subulate, 2–4 mm; petiole ½–¾ blade, glabrate or finely hairy; blade broadly ovate to transversely, broadly elliptic, 3-lobed or 3-angulate-lobate (maplelike), [4–]7–30 × 8–29 cm, base cordate to truncate, margins crenate to crenate-dentate, often coarsely so, apex broadly acute to long-acuminate, lateral lobe apices acuminate to broadly rounded, surfaces velvety stellate-hairy or sometimes scabridulous adaxially, nectary absent. **Inflorescences** solitary flowers in axils of distal leaves. **Pedicels** jointed at middle or distally, 2–10 cm, length ½–1 times petioles, glabrate or finely hairy; involucellar bractlets 9–13, linear-subulate, 1.3–2.7 cm, margins not ciliate, velvety-hairy. **Flowers** horizontal or ascending; calyx divided to ± middle, broadly campanulate, 2.9–6 cm, larger in fruit, lobes triangular, apices acute to subcaudate, velvety-hairy, nectaries absent; corolla broadly to narrowly funnelform, petals pale pink to white, red basally, narrowly obovate, usually not conspicuously overlapping, 8.5–14 × 4–8.5 cm, apical margins repand, finely hairy abaxially where exposed in bud; staminal column straight, pink to white, 6.2–9.5 cm, length ⅔ petals, bearing filaments throughout its length, free portion of filaments secund, 3–9 mm; pollen yellow; styles white, 7–17 mm; stigmas yellow. **Capsules** light to dark brown, ovoid to subglobose, 2.2–3.5 cm, apex apiculate, hispid with simple, yellowish-brown to reddish-brown hairs. **Seeds** brown to reddish brown, reniform-globose, 2.8–3.1 mm, verrucose-papillose. $2n = 38$.

Flowering (May–)Jun–Aug(–Sep). Freshwater and brackish marshes; 0–20 m; Ala., Fla., Ga., La., Miss., S.C., Tex.; West Indies (w Cuba).

The flowers of *Hibiscus grandiflorus* first open in the evening, emit a pleasant fragrance, and are pollinated by sphingid moths (O. J. Blanchard 1976). It is relatively common only in Florida. *Hibiscus grandiflorus* is sometimes cultivated and has been found to be hardy as far north as Illinois (S. R. Hill, pers. comm.).

15. Hibiscus coccineus Walter, Fl. Carol., 177. 1788
• Scarlet rose-mallow E

Hibiscus semilobatus Chapman

Herbs, perennial, to 3(–3.5) m, herbage glabrous throughout. **Stems** glaucous. **Leaves:** stipules caducous, linear-subulate, 1–3 mm; petiole ⅓ to equaling blade; blade orbiculate to transversely elliptic, deeply palmately 3–5-lobed, 10–19 × 13–25 cm, base cordate, segments linear-lanceolate, margins remotely, unevenly serrate, apices acuminate, surfaces glabrous, nectary absent. **Inflorescences** solitary flowers in axils of distal leaves. **Pedicels** jointed distally, 3–14 cm, ⅓–1¾ petioles; involucellar bractlets 9–15, linear-subulate, 2.5–4 cm, not ciliate. **Flowers** showy, horizontal or ascending; calyx divided ¾ length, rotate, 3.5–6 cm, larger in and longer than fruit, lobes narrowly triangular, apices acute to acuminate, nectaries absent; corolla rotate, petals not overlapping, bright red, narrowly spatulate-obovate, (6–)7.5–10 × 2.5–5.5 cm, minutely hairy abaxially where exposed in bud; staminal column straight, exserted, red, pink to white basally, 6.5–7 cm, bearing filaments in distal ⅓, free portion of filaments not secund, 4–8 mm; pollen dull yellow to dull red; styles red, 5–9 mm; stigmas red. **Capsules** brown, ovoid to globose, 2.8–3.5 cm, apex acute, apiculate, glabrous. **Seeds** brown, reniform-globose, 2.5–3.8 mm, hairy, hairs often in lines, brownish to reddish. $2n = 38$.

Flowering late May–early Aug. Riparian and other freshwater marshes, ditches, swamps; 0–40 m; Ala., Ark., Fla., Ga., La., Miss., N.C., Tex., Va.

Except for the Florida records and two very old ones from Georgia, the distribution of *Hibiscus coccineus* given here is based on relatively recent reports and almost certainly represents escapes from cultivation rather than a natural distribution.

A rare, white-flowered form is known from southern Florida and is now in the horticultural trade. Petal color in *Hibiscus coccineus* has been shown to be under the control of a simple diallelic locus in which red is completely dominant over white (L. A. Gettys 2012).

In 1871, A. W. Chapman found plants of *Hibiscus coccineus* in eastern Florida that bore distinctive, shallowly-lobed leaves, and his specimens form the basis for *H. semilobatus*. No extant populations of this variant have been rediscovered.

16. Hibiscus laevis Allioni, Auct. Syn. Meth. Stirp. Taurin., 31. 1773 • Halberd-leaved or smooth rose-mallow, sweating-weed, military hibiscus E

Hibiscus coccineus Walter var. *virginicus* Hochreutiner; *H. militaris* Cavanilles

Herbs, perennial, to 2.5 m, herbage glabrous or nearly so throughout. **Stems** often glaucous. **Leaves:** stipules caducous, linear-subulate, 2–10 mm; petiole ½ to somewhat exceeding blade; blade narrowly to broadly ovate or triangular- or lanceolate-ovate, usually hastately 3 (–5)-lobed, sometimes unlobed, 6–18 × 3–16 cm, base cordate to truncate, lobes, especially middle one, ovate to triangular, to 3 times as long as wide, margins crenate-serrate to serrate or serrate-dentate, apex acuminate to long-acuminate, surfaces glabrous, nectary absent. **Inflorescences** solitary flowers in axils of distal leaves. **Pedicels** jointed distally, 1–10 cm, ⅓ to slightly exceeding subtending petioles; involucellar bractlets (8)9–15(16), linear-subulate, 1–3 cm, margins not ciliate. **Flowers** horizontal; calyx divided ⅓–½ length, broadly cylindric-campanulate, 2.5–3 cm, conspicuously larger in fruit, lobes broadly triangular, apices acute, surfaces glabrous, nectaries absent; corolla broadly funnelform, petals pink to white, red basally, obovate, 5–8 × 2–5 cm, apical margins entire to repand, finely hairy abaxially where exposed in bud; staminal column straight, pale pink to white, 2.5–4 cm, ca. ½ as long as petals, bearing filaments nearly throughout, free portion of filaments not secund, 2–4 mm; pollen pale pink to white; styles pale pink to white, 5–12 mm; stigmas pink. **Capsules** brown, ovoid, 1.8–3 cm, apex truncate, apiculate, glabrous. **Seeds** reddish brown to brown, reniform-globose, 3–5 mm, hairy, hairs reddish. $2n = 38$.

Flowering Jun–Oct. Edges of freshwater lakes, larger, slow-moving streams, floodplain pools, wet roadside ditches, artificial ponds; 0–400 m; Ont.; Ala., Ark., Fla., Ga., Ill., Ind., Iowa, Kans., Ky., La., Md., Mich., Minn., Miss., Mo., Nebr., N.C., Ohio, Okla., Pa., S.C., Tenn., Tex., Va., W.Va., Wis.

B. P. G. Hochreutiner (1900) inexplicably merged *Hibiscus laevis* with the very different *H. coccineus*.

Hibiscus laevis has been recorded from Pelee Island in southernmost Ontario, but the 1904 collection is thought to have represented a short-lived population, as the species apparently has not since been found there (R. L. Stuckey 1968b). A report from the Bronx, New York (R. DeCandido 1991), well northeast of its closest previously known occurrence, is likely of an introduction or an escape.

There are reports of recent northward spread of *Hibiscus laevis* along larger streams (C. C. Deam 1940; M. L. Roberts and R. L. Stuckey 1992; Stuckey 1968b; F. H. Utech 1970). R. B. Kaul (pers. comm.) reported similar upriver increases in Nebraska on the Elkhorn, Missouri, and Platt rivers. *Hibiscus laevis* sometimes forms natural hybrids with *H. moscheutos* subsp. *moscheutos*, usually in man-made habitats (O. J. Blanchard 1976).

17. Hibiscus dasycalyx S. F. Blake & Shiller, J. Wash. Acad. Sci. 48: 277, fig. 1. 1958 • Neches River rose-mallow C E

Herbs, perennial, to 2.5 m. **Stems** glabrous. **Leaves:** stipules linear-subulate, 1.5–5 mm; petiole ½–¾ blade, glabrous; blade narrowly to broadly triangular-ovate, deeply hastately 3-lobed, 5–12 × 1–14 cm, base cordate to truncate, lobes linear-lanceolate, 3+ times as long as wide, margins coarsely and remotely serrate, apices long-acuminate, surfaces glabrous, nectary absent. **Inflorescences** solitary flowers in axils of distal leaves. **Pedicels** jointed medially to distally, 1–3 cm, ⅓–⅔ times subtending petioles, minutely hairy distal to joint; involucellar bractlets 8–10, linear-subulate, 1.8–2.2 cm, margins inconspicuously ciliate, simple-hairy. **Flowers** ± horizontal; calyx divided ⅓ length, broadly cylindric-campanulate, 1.5–3 cm, larger in fruit, lobes very broadly triangular and subtrullate, apices apiculate, hairy, hairs simple, 1+ mm, underlain by shorter stellate ones, veins sometimes zigzag, nectaries absent; corolla broadly funnelform, petals creamy white, deep red basally, obovate, 4.5–7 × 3–5.5 cm, apical margins repand, sometimes undulate, sometimes minutely hairy abaxially where exposed in bud; staminal column straight, pale pink to white, 2.5–3 cm, length ½ petals, bearing filaments nearly throughout, free portions of filaments not secund, 1–2 mm; pollen pale pink to purple; styles cream, 4–10 mm; stigmas cream. **Capsules** brown, ovoid, 1.6–2.8 cm, apex rounded-truncate, apiculate, hairy. **Seeds** reddish brown to brown, reniform-globose, 3.5 mm, hairy, hairs reddish brown. $2n = 38$.

Flowering mid Jun–mid Aug. Open marshy habitats, seasonally wet alluvial soil, edges of ponds; of conservation concern; 50–100 m; Tex.

R. A. Klips (1995) suggested that *Hibiscus dasycalyx* might better be treated as a subspecies or variety of *H. laevis*; R. L. Small (2004) found molecular evidence that it should be maintained as a species. It is known only from Cherokee, Harrison, Houston, and Trinity counties in eastern Texas.

Hibiscus dasycalyx is in the Center for Plant Conservation's National Collection of Endangered Plants.

18. Hibiscus mutabilis Linnaeus, Sp. Pl. 2: 694. 1753 • Confederate- or changeable-rose, Dixie rose-mallow I

Subshrubs, shrubs, or trees, to 8 m. **Stems:** new growth usually somewhat floccose, hairs stellate and simple or glandular, line of fine, curved hairs absent or obscured. **Leaves:** stipules subulate to narrowly triangular, 3–14 mm; petiole below inflorescence ½–1½ blade, fine, curved hairs absent or obscured adaxially; blade broadly to transversely ovate, 3–7-lobed, 6–20 × 6.5–22.5 cm below inflorescence, base deeply to shallowly cordate, lobes broadly triangular, margins subentire to coarsely crenate-dentate, apex broadly obtuse to long-acuminate, surfaces variably stellate-hairy, sometimes minutely so, nectary absent. **Inflorescences** solitary flowers, in axils of distal leaves or appearing subcorymbose by reduction of internodes and subtending leaves. **Pedicels** obscurely to conspicuously jointed distally, 2–12 cm, stellate- and glandular-hairy; involucellar bractlets (6)7–11(12), often deciduous after anthesis, linear-subulate to narrowly triangular, 0.8–2.2 cm, margins not ciliate, stellate- and glandular-hairy. **Flowers** horizontal or ascending, sometimes double; calyx divided ½–⅔ length, campanulate, 1.6–4 cm, lobes triangular, apices acute to acuminate, stellate- and glandular-hairy, nectaries absent; corolla broadly campanulate to rotate, petals usually opening white, changing to deep pink over course of day, rarely pink and unchanging, often with deep pink spot near base, ± obovate, (2.5–)4–7.5 × 1.2–7.5 cm, apical margins repand, usually undulate, finely hairy abaxially where exposed in bud; staminal column straight, white to pink, 1.4–2.6 cm, bearing filaments nearly throughout, free portions of filaments not secund, 2.5–5 mm; pollen cream to yellowish orange; styles white to pink, 3–16 mm; stigmas pink or yellow, rarely white. **Capsules** brown, broadly ovoid or globose, 1–2.5 cm, apex impressed, strigose. **Seeds** olivaceous to reddish brown or dark brown, reniform-ovoid, 2–2.8 mm, laterally glabrous, dorsally and dorsolaterally bearing long, straight, light brown to white, simple or 2- or 3-armed hairs. $2n = 84, 88, 92, 100, 110, 120$ (all cultivars).

Flowering mainly Jul–Oct. Old home sites, disturbed places; 40–700 m; introduced; Ala., La.; Asia (China); introduced also in Mexico, West Indies, Bermuda, Central America, South America, s Asia, Africa, Pacific Islands, Australia.

Hibiscus mutabilis is cultivated worldwide in warmer climates, including the southernmost part of the flora area, and it occasionally escapes. It has been reported as a volunteer in an avocado grove in southern California; I have seen no specimens from there. It is noteworthy for having flowers that usually open white or pale pink in the morning and change to deep pink by the evening. Double-flowered forms are common in cultivation.

19. Hibiscus striatus Cavanilles, Diss. 3: 146, plate 54, fig. 1. 1787 • Striped rose-mallow

Subspecies 3 (1 in the flora): Texas, Mexico, West Indies, Central America (Honduras), South America.

19a. Hibiscus striatus Cavanilles subsp. **lambertianus** (Kunth) O. J. Blanchard, J. Arnold Arbor. 63: 267. 1982

Hibiscus lambertianus Kunth in A. von Humboldt et al., Nov. Gen. Sp. 5(fol.): 226; 5(qto.): 291; plate 478. 1822; *H. cubensis* A. Richard

Subshrubs or shrubs, to 4 m, herbage stellate-tomentose throughout. **Stems** sparingly armed with ± stout prickles. **Leaves:** stipules linear-subulate, 3–14 mm; petiole ¼–¾ as long as blade, often prickly; blade lanceolate to narrowly ovate, unlobed [lobed], 8–17.5 × 3–8 cm, base truncate to rounded, margins crenate-serrate, apex acute, surfaces stellate-tomentose, nectary absent. **Inflorescences** solitary flowers in axils of distal leaves. **Pedicels** not jointed, abscising at base, 1.5–7 cm; involucellar bractlets 10–15, linear-subulate, 1–2.5 cm, margins ± ciliate. **Flowers** showy, horizontal or ascending; calyx divided ½ length, campanulate, 1.5–3.5 cm, somewhat larger in fruit, lobes triangular, apices acute to acuminate, densely stellate-tomentose, sparingly to moderately setose on veins, nectaries absent; corolla broadly funnelform, petals pale mauve, red at base, obovate, 4–12 × 2–6 cm, apical margins repand, sometimes undulate, sparingly coarsely hairy abaxially where exposed in bud; staminal column straight, white or pink, 1.5–4 cm, bearing filaments in ± discrete tiers, free portion of filaments not secund, 1–1.5 mm; pollen pale purple; styles white, wedge-shaped, 2–18 mm; stigmas white. **Capsules** dark brown, ovoid to ellipsoid, 1.5–3.5 cm, apex acute to depressed, hispid, hairs yellowish to whitish. **Seeds** dark brown, reniform-ovoid, 2.5–3.5 mm, tomentulose, hairs yellowish to reddish. $2n = 52$.

Flowering Jun–Oct. Marshes, edges of lakes, usually coastal; 0–70 m; Tex.; se Mexico; West Indies (Cuba, Jamaica); Central America (Honduras); South America (Bolivia, Brazil, Colombia, Peru, Suriname, Venezuela).

Subspecies *lambertianus* is occasionally cultivated as an ornamental.

20. Hibiscus trionum Linnaeus, Sp. Pl. 2: 697. 1753 • Flower-of-an-hour, bladder ketmia, Venice mallow, ketmie trilobée I W

Herbs, annual, to 0.6(–1) m, herbage throughout with mixture of coarse, simple or few-armed stellate hairs and fine, many-armed stellate hairs. **Stems** also with line of fine, curved hairs extending from node to node. **Leaves:** stipules narrowly triangular, 2–5(–8) mm; petiole ½ to ± equaling blade, with fine, curved hairs adaxially; blade ovate to broadly, transversely ovate, all but proximalmost leaves 3(–5)-parted, mostly 2.5–6.5 × 3–7 cm, base cuneate, segments ± rhomboidal, pinnately lobed to varying degrees, lobes triangular, entire, apices rounded or truncate, surfaces glabrous or sparsely hairy, nectary absent. **Inflorescences** solitary flowers in axils of distal leaves. **Pedicels** jointed distally, 1–4 cm, usually equaling or exceeding petioles; involucellar bractlets 10–15, linear, 1–2.5 cm, margins ± ciliate. **Flowers** lasting a few hours, ascending or erect; calyx divided ½ length, campanulate, 0.8–1.8 cm, accrescent and conspicuously inflating to enclose fruit, becoming scarious, primary veins zigzag, lobes broadly trullate, apices acute, nectaries absent; corolla rotate, petals yellow or cream with conspicuous purple-brown spot basally, purplish abaxially where exposed in bud, obovate to broadly obovate, 1.5–3(–4) × 1–2(–3) cm, apical margins repand, finely hairy abaxially where exposed in bud; staminal column dark red to purple, 0.4–0.7(–1.1) cm, abruptly expanded below to form cup over ovary, bearing filaments in distal ½, free portion of filaments not secund, 2–3(–5) mm; pollen yellow-orange; styles usually white to cream, yellowish, or maroon, 2–3(–6) mm; stigmas dark red to purple. **Capsules** dark brown-black, ellipsoid to ovoid, 1.2–1.5 cm, apex obtuse to broadly acute, hairy. **Seeds** dark gray-brown, reniform-ovoid, somewhat depressed laterally, 2–2.5 mm, sparingly and minutely papillose. $2n = 28, 56$ (both cultivars).

Flowering Jul–Nov. Cultivated and waste places; 10–1800 m; introduced; Man., N.B., N.S., Ont., P.E.I., Que., Sask.; Ala., Ariz., Ark., Calif., Colo., Conn., Del., D.C., Fla., Ga., Idaho, Ill., Ind., Iowa, Kans., Ky., La., Maine, Md., Mass., Mich., Minn., Miss., Mo., Mont., Nebr., N.H., N.J., N.Mex., N.Y., N.C., N.Dak., Ohio, Okla., Oreg., Pa., R.I., S.C., S.Dak., Tenn., Tex., Utah, Vt., Va., Wash., W.Va., Wis., Wyo.; Eurasia; Africa; Pacific Islands (New Zealand); Australia; introduced also in West Indies (Jamaica).

Hibiscus trionum is occasionally cultivated. It is native in Africa, Eurasia, and perhaps also in parts of New Zealand and Australia, where both diploids and tetraploids occur (B. G. Murray et al. 2008). Most wild plants in North America have low, widely spreading branches; some cultivated forms are strictly erect and much taller than the weedy forms.

Three species have been segregated from *Hibiscus trionum* in Australia and New Zealand (L. A. Craven et al. 2011), and the name *H. trionum* has now been lectotypified (D. Iamonico and L. Peruzzi 2014). A further splitting of the species in other parts of its range might consequently affect the name of plants in the flora area.

21. Hibiscus clypeatus Linnaeus, Syst. Nat. ed. 10, 2: 1149. 1759 • Congo mahoe

Subspecies 3 (1 in the flora): Texas, ne, se Mexico, West Indies, Central America (Belize, Guatemala).

21a. Hibiscus clypeatus Linnaeus subsp. **clypeatus**

Shrubs or trees, 2–5 m. **Stems:** new growth ± stellate- and usually simple-hairy, line of fine, curved hairs absent or obscured. **Leaves:** stipules subulate, 3–18 mm; petiole of principal leaves ½ to equaling blade, finely hairy; blade ovate to orbiculate, angulate-lobate (maple- or grapelike) or less frequently unlobed, 10–22 × 9–23 cm, base deeply and narrowly cordate to truncate, margins remotely and shallowly denticulate or undulate, apex acute to acuminate, surfaces stellate-tomentose, more densely so abaxially, nectary absent from base of midvein abaxially. **Inflorescences** solitary flowers in axils of distal leaves, or subcorymbose by reduction of internodes and subtending leaves. **Pedicels** jointed at base or not at all, 3–10 cm (often conspicuously elongated), subequal to or exceeding petioles, finely densely hairy; involucellar bractlets 7–10, linear-subulate to narrowly triangular, often randomly curved (sickle-shaped), 1–3 cm, margins not ciliate, densely short stellate-hairy. **Flowers** horizontal; calyx divided ⅓ length, campanulate, 3–4.5 cm, somewhat larger in fruit, more densely hairy basally, lobes triangular, apices short-acuminate, nectaries absent; corolla funnelform to narrowly campanulate, somewhat bilateral, petals dull red or dull orange, 2 upper little or not at all recurved, 3 lower recurved or revolute, 4–5.5 × 1–2 cm, margins apically entire or repand, tomentose throughout abaxially; staminal column declinate to recurved, pale yellowish to dull orange, 2.8–4.2 cm, bearing filaments in apical ⅓–½, free portion of filaments secund, 4–8(–10) mm; pollen yellow to orange; styles dull red to dull orange, 3–6 mm; stigmas dull red to dull orange. **Capsules** dull orange, ovoid to obovoid, 2.5–5 cm, apex apiculate, hispid, densely underlain by minute, coarse stellate hairs, hairs orangish. **Seeds** brown, mottled, subglobose, 3.5–4 mm, glabrous. $2n$ = 20 (Mexico: Veracruz).

Flowering Oct–Nov. Riparian woodlands, alluvial soil; 30–50 m; Tex.; ne, se Mexico; West Indies (Hispaniola, Jamaica, Puerto Rico, Saint Croix); Central America (Belize, Guatemala).

Subspecies *clypeatus* was collected for the first time in the flora area in 1994; it was not identified until 2009. It is known in the flora area only from near the Rio Grande in Hidalgo County and deserves state and federal protection. Hummingbird visits to *Hibiscus clypeatus* have been observed in Belize and Mexico (D. M. Bates, pers. comm.; J. C. Meerman 1993); bats are suspected as the primary agents for pollination (O. J. Blanchard 1976).

27. HORSFORDIA A. Gray, Proc. Amer. Acad. Arts 22: 296. 1887 • [For Frederick Hinsdale Horsford, 1855–1923, Vermont farmer and commercial seedsman, and probably also for Eben Norton Horsford, 1818–1893, chemist]

John La Duke

[Herbs] **shrubs**, hairy. **Stems** erect, stellate-tomentose or stellate-scabrous. **Leaves** petiolate to sessile distally; stipules deciduous, linear to subulate; blade ovate to triangular or lanceolate to cordate, unlobed, base cordate or truncate, margins entire, crenate, dentate, or denticulate, surfaces densely covered with short, stellate hairs, especially abaxially. **Inflorescences** axillary, solitary flowers or paniculate; involucel absent. **Flowers:** calyx not accrescent, not inflated, shorter than mature fruits, lobes not ribbed, lanceolate; corolla campanulate, yellow, orange, rose, pink, or lavender [white], often drying bluish, 2–3 times calyx; staminal column included, anthers at apex; ovules 1 per cell, 1–3 per carpel; styles 6–11-branched; stigmas capitate. **Fruits** schizocarps, sometimes on reflexed, jointed pedicel, spheric to subspheric, indurate; mericarps [6–]9 or 10[11], ± 2-celled, winged, flared at papery apex, lower cell rugose, stellate-hairy, sparsely hirsute, indehiscent, endoglossum absent, upper cell smooth, dehiscent apically. **Seeds** 1 in rugose portion, 1 or 2 apically, sometimes abortive, glabrous or minutely hairy. $x = 15$.

Species 4 (2 in the flora): sw United States, nw Mexico.

1. Petals rose, pink, or lavender, 12–21 mm; mericarp wings lanceolate; seeds minutely scabridulous . . . . .1. *Horsfordia alata*
1. Petals yellow or sometimes pale orange, 6–9 mm; mericarp wings ovate; seeds densely short-pubescent . . . . .2. *Horsfordia newberryi*

1. Horsfordia alata (S. Watson) A. Gray, Proc. Amer. Acad. Arts 22: 297. 1887 F

Sida alata S. Watson, Proc. Amer. Acad. Arts 20: 356. 1885; *Horsfordia palmeri* S. Watson

Shrubs, 1–2.5(–4) m, hairs dense, stellate, grayish yellow, rough. **Leaves** often folded, to 68 mm, petiolate to 8 cm to sessile distally; stipules 2–5 mm; blade ovate to triangular, 1–11.5 cm, base cordate, margins crenate or dentate to denticulate, apex acuminate. **Inflorescences** solitary flowers, rarely in groups of 3–5. **Flowers:** calyx campanulate, 3–10 mm; petals rose, pink, or pale lavender, 12–21 mm. **Schizocarps** 8–12 mm diam.; mericarps 9 or 10, 6–10 mm, sparsely stellate-hairy, lower cell reticulate, upper cell with 2 abortive ovules (rarely developing), wings lanceolate. **Seeds** 1 per mericarp, brown to blackish, 1.9 mm, minutely scabridulous. $2n = 30$.

Flowering Sep–Apr. Disturbed and undisturbed sites, deserts; 0–600 m; Ariz., Calif.; Mexico (Baja California, Baja California Sur, Sonora).

2. Horsfordia newberryi (S. Watson) A. Gray, Proc. Amer. Acad. Arts 22: 297. 1887 F

Abutilon newberryi S. Watson, Proc. Amer. Acad. Arts 11: 125. 1876

Shrubs, 2–3 m, hairs dense, stellate, yellow. **Leaves** not folded, petiolate 4–10 cm; stipules 2–5 mm; blade lanceolate to cordate, (3–)4–10 cm, base usually cordate or truncate, margins entire or denticulate, apex acute. **Inflorescences** solitary flowers or few-flowered panicles. **Flowers:** calyx campanulate, 5–6 mm; petals yellow, sometimes pale orange [white], 6–9 mm. **Schizocarps** 9–12 mm diam.; mericarps ca. 10, 6–9 mm, lower cell reticulate, upper cell 2-seeded, wings ovate. **Seeds** (2)3 per mericarp, brown to blackish, 2.2 mm, densely short-pubescent with whitish hairs. $2n = 30$.

Flowering Nov–Dec, Mar–Apr. Disturbed and undisturbed sites, deserts; 100–800 m; Ariz., Calif.; Mexico (Baja California, Baja California Sur, Sonora).

28. ILIAMNA Greene, Leafl. Bot. Observ. Crit. 1: 206. 1906 • Globe mallow, wild hollyhock [Derivation uncertain; perhaps after Lake Iliamna in Alaska] E

David M. Bates

Subshrubs or perennial herbs, often woody at base, glabrate to densely pubescent, hairs variously stellate to simple. **Stems** solitary to many, erect or ascending, rarely decumbent. **Leaves:** stipules ± deciduous, linear to subulate; blade ± orbiculate, (at least proximally) 3-, 5-, or 7-lobed, base truncate to cordate or cuneate, margins dentate or serrate, rarely nearly entire. **Inflorescences** axillary, solitary flowers or 2–5-flowered clusters, distally forming spikes, racemes, or panicles through reduction of leaves; involucellar bractlets persistent, 3, distinct. **Flowers:** calyx somewhat accrescent or not, not inflated, lobes not strongly ribbed, lanceolate to ovate or ovate-triangular; corolla campanulate to subrotate, whitish or pinkish to rose-purple, exceeding calyx; staminal column ± included; filaments terminal and subterminal; ovary (6–)10–15(16)-carpellate; ovules 2 or 3(4) per carpel; styles (6–)10–15(16)-branched, (branches equal in number to carpels);

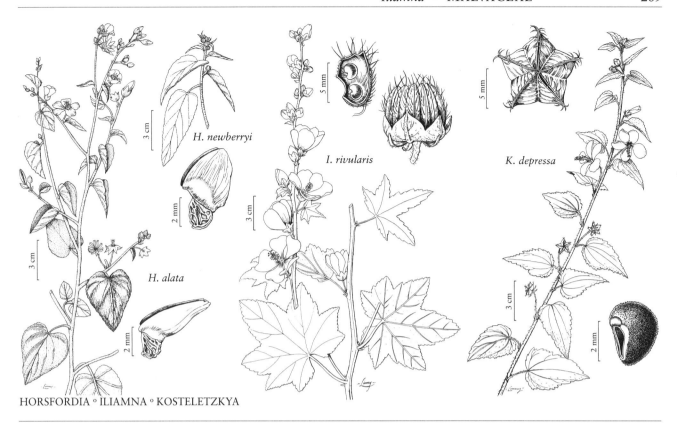

H. newberryi

I. rivularis

K. depressa

H. alata

HORSFORDIA ∘ ILIAMNA ∘ KOSTELETZKYA

stigmas terminal, obliquely capitate. **Fruits** schizocarps, erect, not inflated, subglobose, apically retuse, moderately indurate; mericarps (6–)10–15(16), drying black, 1-celled, oblong in lateral view, rounded at apex, thin-walled, smooth laterally, without dorsal spur, densely pubescent dorsally and apically with coarse, hirsute, velutinous simple hairs overlaying stellate hairs, sides smooth, glabrous, dehiscence loculicidal except ventral-basally where joined to columella by vascular bundles. **Seeds** 2 or 3(4) per mericarp, reniform to obovate-reniform, glabrate or puberulent marginally with simple, white to tawny hairs 1–2 mm. $x = 33$.

Species 8 (8 in the flora): North America.

Iliamna comprises one widespread species of the mountainous West (*I. rivularis*), and seven local species (two eastern and five western) differing principally but modestly in pubescence, leaf, involucellar bractlet, and calyx characters, and exhibiting varying degrees of intergradation. All, with the exception of *I. bakeri*, are species of relatively mesic, open woodlands, scrub, and meadows, often along stream banks. Each of the species apparently regenerates following disturbance or fire, sometimes in dense stands.

SELECTED REFERENCES Bodo Slotta, T. A., and D. M. Porter. 2006. Genetic variation within and between *Iliamna corei* and *I. remota* (Malvaceae): Implications for species delimitation. Bot. J. Linn. Soc. 151: 345–354. Wiggins, I. L. 1936. A resurrection and revision of the genus *Iliamna* Greene. Contr. Dudley Herb. 1: 213–229.

1. Involucellar bractlets 8–14 × 2–6 mm, ¾ to exceeding calyx length.
   2. Involucellar bractlets lanceolate-ovate, 8–12 × 2–3 mm, ¾ calyx length . . . . 4. *Iliamna grandiflora*
   2. Involucellar bractlets elliptic-lanceolate to -ovate, 10–14 × 4–6 mm, equaling or exceeding calyx length. . . . . 5. *Iliamna latibracteata*
1. Involucellar bractlets 3–10 × 1 mm, ⅓–⅔ calyx length.
   3. Calyces at anthesis 5–8(–11) mm; involucellar bractlets 3–6(–8) mm . . . . .1. *Iliamna rivularis*
   3. Calyces 9–20 mm; involucellar bractlets (5–)6–10 mm.
      4. Calyx lobes ovate to triangular-ovate, ± as wide as long, ± equaling tube in length.

5. Leaf blades 6–20 cm wide, (3-), 5-, or 7-lobed, terminal lobe triangular-ovate, sinuses broad ....2. *Iliamna remota*

5. Leaf blades 1.5–8 cm wide, shallowly and crenately 3-lobed or deeply 3- or 5-lobed, terminal lobe narrowly oblong, sinuses narrow .... 8. *Iliamna bakeri*

[4. Shifted to left margin.—Ed.]

4. Calyx lobes lanceolate to narrowly ovate, longer than wide, exceeding tube in length.

6. Calyx 9–12 mm; involucellar bractlets 6–8 mm .... 3. *Iliamna corei*

6. Calyx 15–20 mm; involucellar bractlets 5–10 mm.

7. Calyces and stems pubescent with simple and stellate hairs; petals deep rose-purple; plants 1–2 m.. 6. *Iliamna longisepala*

7. Calyces and stems obscurely pubescent with fine stellate hairs; petals whitish or pinkish; plants 0.5–0.7 m ....7. *Iliamna crandallii*

1. Iliamna rivularis (Douglas) Greene, Leafl. Bot. Observ. Crit. 1: 206. 1906 • Streambank globe-mallow or wild hollyhock E F

Malva rivularis Douglas in W. J. Hooker, Fl. Bor.-Amer. 1: 107. 1831; *Iliamna acerifolia* (Torrey & A. Gray) Greene; *I. rivularis* subsp. *diversa* (A. Nelson) Wiggins; *I. rivularis* var. *diversa* (A. Nelson) C. L. Hitchcock; *Phymosia acerifolia* (Torrey & A. Gray) Rydberg; *P. rivularis* (Douglas) Rydberg; *Sphaeralcea acerifolia* Torrey & A Gray; *S. rivularis* (Douglas) Torrey; *S. rivularis* var. *diversa* A. Nelson

Stems 0.5–2 m; herbage sparsely stellate-hairy. **Leaf blades** deeply (3-), 5-, or 7-lobed, 5–20 cm wide, lobes triangular-ovate, broadest at base, base truncate to cordate, margins coarsely dentate. **Inflorescences** clusters forming interrupted spikes to corymbose racemes distally; involucellar bractlets linear-subulate, 3–6(–8) × 1 mm, ⅓–⅔ as long as calyx. **Flowers:** calyx 5–8(–11) mm, lobes triangular-ovate to obtuse, 2.5–4(–5.5) mm, ± as wide as long, slightly exceeding tube; petals pinkish white to rose-purple, 1.8–2.5 cm. **Schizocarps** 10–12 mm diam.; mericarps 8–12, 6–10 mm. **Seeds** (2)3 or 4, dark brown, 2 mm, puberulent. $2n = 66$.

Flowering Jun–Aug. Stream banks, meadows, open woodlands, disturbed places, foothills to mountain slopes; 1400–2900 m; Alta., B.C.; Colo., Idaho, Mont., Nev., N.Mex., Oreg., Utah, Wash., Wyo.

Iliamna rivularis is variable in stature, pubescence, and leaf characters. Plants of Idaho and Wyoming with distally smaller (4–10 cm), three- to five-lobed blades, truncate bases, and inconspicuously toothed margins have been recognized as var. or subsp. *diversa*, but fall within the morphological range of the species as a whole.

Iliamna rivularis is in the Center for Plant Conservation's National Collection of Endangered Plants as *I. rivularis* var. *rivularis*.

2. Iliamna remota Greene, Leafl. Bot. Obs. Crit. 1: 206. 1906 • Kankakee mallow E

Phymosia remota (Greene) Britton; *Sphaeralcea remota* (Greene) Fernald

Stems 1–2.5 m; herbage densely to sparsely stellate-hairy. **Leaf blades** (3-), 5-, or 7-lobed, 6–20 cm wide, lobes broadly triangular-ovate, base truncate to cordate, margins crenate-dentate, sinuses broad, obtuse. **Inflorescences** clusters forming interrupted racemes distally; involucellar bractlets linear, 6–9 × 1 mm, ½–⅔ as long as calyx. **Flowers** fragrant; calyx 12–18 mm, lobes broadly ovate-acuminate, 6–8 mm, ± as wide as long, ± equaling tube; petals pale rose-purple, 2.5–3 cm. **Schizocarps** 15 mm diam.; mericarps 15, 10 mm. **Seeds** 2 or 3(4), dark brown, 3 mm, densely hairy. $2n = 66$.

Flowering Jun–Aug. Open woods and rocky slopes, riverbanks, shores and gravel bars of rivers, abandoned cultivated fields in sandy clay loam; 100–200 m; Ill., Ind., Va.

Iliamna remota is known from Langham Island opposite Altorf in the Kankakee River (about nine miles northwest of Kankakee, Illinois), where it persists as the only certain wild locality; it now may occasionally escape from cultivation or deliberate introduction, as apparently was the case of a naturalized population found in the 1940s in a swale about two miles east of New Paris, Indiana. It was apparently distributed along railroads by enthusiastic wildflower groups in the last century. *Iliamna remota* is listed as endangered by the state of Illinois.

3. **Iliamna corei** (Sherff) Sherff, Amer. J. Bot. 36: 503. 1949 • Core's wild hollyhock, Peter's mountain mallow C E

Iliamna remota Greene var. *corei* Sherff, Rhodora 48: 96, plate 1024. 1946

Stems 1–1.5 m; herbage stellate-hairy. **Leaf blades** 5- or 7-lobed, 5–10 cm wide, terminal lobe triangular, base ± truncate, margins serrate to nearly entire, sinuses narrow, acute. **Inflorescences** 2- or 3-flowered clusters, sometimes solitary flowers, forming interrupted spikes distally; involucellar bractlets filiform, 6–8 × 1 mm, ½–⅔ times as long as calyx. **Flowers** odorless; calyx 12 mm, lobes broadly lanceolate, long-acuminate, 6–8 mm, longer than wide, longer than tube; petals pale pink to deep rose, 2.5 cm. **Schizocarps** 12 mm diam.; mericarps 11–16, 10 mm. **Seeds** 2(3), dark brown, 2.5 mm, puberulent.

Flowering late Jun–Aug. Open, shrubby woodlands in pockets of soil among sandstone outcrops; of conservation concern; 700–800 m; Va.

Iliamna corei is known from a single locality on Peters Mountain at Narrows, Virginia. The species is considered critically imperiled globally. Whether it is distinct or not from *I. remota*, with which it is sometimes allied, has been problematic, but recent genetic studies support their recognition as distinct species (T. A. Bodo Slotta and D. M. Porter 2006).

Iliamna corei is in the Center for Plant Conservation's National Collection of Endangered Plants.

4. **Iliamna grandiflora** (Rydberg) Wiggins, Contr. Dudley Herb. 1: 223. 1936 • Large-flowered wild hollyhock E

Sphaeralcea grandiflora Rydberg, Bull. Torrey Bot. Club 31: 565. 1904; *Iliamna angulata* Greene; *Phymosia grandiflora* (Rydberg) Rydberg

Stems 1–2 m; herbage sparsely stellate-pubescent to glabrate. **Leaf blades** deeply 5- or 7-lobed, 6–13 cm wide, lobes lanceolate to triangular, middle lobe ± equaling to 2 times length of lateral lobes, base truncate to cordate, margins coarsely dentate. **Inflorescences** clusters forming interrupted spikes to racemes; involucellar bractlets lanceolate-ovate, 8–12 × 2–3 mm, ¾ as long as calyx. **Flowers:** calyx 11–18 mm, lobes triangular-ovate, 8–10 × 5–8 mm, broader than long, exceeding tube, densely villous-hirsute; petals pink, drying purplish, 2–3 cm. **Schizocarps** 12–15 mm diam.; mericarps 12, 10 mm. **Seeds** 2 or 3, dark brown, 2.5–3 mm, muricate, with few simple hairs.

Flowering Jun–Aug. Wet mountain meadows, open woodlands, stream courses; 2000–2700 m; Ariz., Colo., N.Mex., Utah.

5. **Iliamna latibracteata** Wiggins, Contr. Dudley Herb. 1: 225, plate 20. 1936 • California wild hollyhock E

Sphaeralcea rivularis (Douglas) Torrey var. *cismontana* Jepson

Stems 1–2 m; herbage harshly and moderately stellate-hairy. **Leaf blades** deeply (3-), 5-, or 7-lobed, 8–20 cm wide, lobes broadest at middle, base truncate to cordate, margins serrate or irregularly dentate. **Inflorescences** cymes, forming interrupted spikes or racemes; involucellar bractlets broadly elliptic-lanceolate to ovate, 10–14 × 4–6 mm, equaling or exceeding calyx. **Flowers:** calyx 8–10 mm, lobes triangular-ovate, 5–8 × 4–5 mm, slightly longer than broad, slightly longer than tube; petals rose-purple, 2–3 cm. **Schizocarps** 10–15 mm diam.; mericarps 10–14, 6–8 mm. **Seeds** 2 or 3, dark brown, 2 mm, puberulent.

Flowering Jun–Aug. Conifer forests, streamsides, sometimes shaded; 500–2000 m; Calif., Oreg.

Iliamna latibracteata is distinctive in having large involucral bracts, which more or less envelop the calyx and equal or surpass it in length. It occurs in the Klamath and Siskiyou mountains of Del Norte and Humboldt counties in California, north to Coos, Curry, Douglas, Jackson, and Josephine counties in Oregon, with an eastern outlying station reported from Mount Shasta, California.

6. **Iliamna longisepala** (Torrey) Wiggins, Contr. Dudley Herb. 1: 227. 1936 • Long-sepal globemallow, long-sepal wild hollyhock E

Sphaeralcea longisepala Torrey in C. Wilkes et al., U.S. Expl. Exped. 17: 255. 1874; *Phymosia longisepala* (Torrey) Rydberg

Stems 1–2 m, paniculately branched; herbage sparsely hispid, hairs simple, forked, and stellate. **Leaf blades** 5- or 7-lobed, 5–10 cm wide, lobes lanceolate to triangular, base truncate to cordate, margins with coarse rounded to pointed teeth. **Inflorescences** solitary flowers or few-flowered clusters forming open panicles; involucellar bractlets linear to linear-lanceolate, 5–10 × 1 mm, ⅓–½ calyx length. **Flowers:** calyx 15–20 mm, lobes lanceolate to ovate-lanceolate, 10–15 mm, longer than wide, exceeding tube, hirsute with few-rayed hairs 1–2 mm; petals deep rose-purple,

1.5–2.5 cm. **Schizocarps** depressed-globose, 8–10 mm diam.; mericarps 8–12, 6–8 mm. **Seeds** 2 or 3, dark brown, 2 mm, glabrous.

Flowering Jun–Aug. Gravelly streamsides and open hillsides, sage brush shrub-steppe to lower *Pinus ponderosa* zones; 100–1500 m; Wash.

Iliamna longisepala is distinctive in its long calyx and calyx lobes. The species is rare and limited to the eastern side of the Wenatchee Mountains in the arid transition zones over a total distance of about 120 kilometers in Chelan, Douglas, and Kittitas counties.

7. **Iliamna crandallii** (Rydberg) Wiggins, Contr. Dudley Herb. 1: 228. 1936 • Crandall's wild hollyhock E

Sphaeralcea crandallii Rydberg, Bull. Bot. Torrey Club 31: 564. 1904; *Phymosia crandallii* (Rydberg) Rydberg

Stems 0.5–0.7 m; herbage sparingly stellate to glabrate, hairs obscurely stellate. **Leaf blades** deeply 5- or 7-lobed, 5–10 cm wide, lobes narrowly triangular, central lobe 2–3 times longer than lateral lobes, base cordate to truncate, margins coarsely serrate. **Inflorescences** solitary flowers in distal leaf axils, appearing racemose; involucellar bractlets linear, 8–10 × 1 mm, ½–⅔ as long as calyx. **Flowers:** calyx 10–15 mm, lobes lanceolate to lanceolate-ovate, 9–12 mm, longer than wide, longer than tube, apex acuminate; petals whitish or pinkish, 2–2.5 cm. **Schizocarps** unknown. **Seeds** 2–4 per mericarp, glabrous.

Flowering Jul–Aug. Open forests, stream banks, mountain slopes; 2000–2300 m; Colo.

Iliamna crandallii was described from near Steamboat Springs, Routt County, and is restricted to that area. It has been reported from Rio Blanco County (according to a specimen at COLO). T. A. Bodo Slotta (2000) stated that she collected fresh material of the species at Buffalo Pass and Fish Creek Falls.

8. **Iliamna bakeri** (Jepson) Wiggins, Contr. Dudley Herb. 1: 228. 1936 • Baker's globemallow or wild hollyhock E

Sphaeralcea bakeri Jepson, Man. Fl. Pl. Calif., 635. 1925

Stems 0.3–1.2 m; herbage harshly stellate-hairy. **Leaf blades** shallowly and crenately 3-lobed or deeply 3- or 5-lobed with terminal lobe narrowly oblong and toothed apically, 1.5–8 cm wide, lobes triangular, base cuneate to truncate, margins irregularly serrate, sinuses narrow. **Inflorescences** solitary flowers or few-flowered clusters; involucellar bractlets linear to subulate, 5–8 × 1 mm, ca. ½–⅔ calyx length. **Flowers:** calyx 9–12 mm, lobes broadly ovate, 4–6 × 4–6 mm, ± as wide as long, ± equaling tube; petals rose-purple, 1.5–3 cm. **Schizocarps** depressed-globose, 10–15 mm diam.; mericarps ca. 10–12, 8–10 mm. **Seeds** 3 or 4, dark brown, 2 mm, finely stellate-puberulent.

Flowering Jun–Aug. Juniper woodlands, lava beds, conifer forests, chaparral, mountain slopes, rangeland, sometimes disturbed areas; 1000–2500 m; Calif., Oreg.

Iliamna bakeri is found in Modoc, Shasta, and Siskiyou counties in California, and Klamath County in Oregon.

29. KOSTELETZKYA C. Presl, Reliq. Haenk. 2: 130, plate 70. 1835, name conserved • [For Vincenz Franz Kosteletzky, 1801–1887, Czech botanist]

Orland J. Blanchard Jr.

Thorntonia Reichenbach, name rejected

Herbs, perennial [annual], **or subshrubs,** variously scabrous. **Stems** erect or ascending [scrambling or creeping]. **Leaves:** stipules persistent, linear-subulate or filiform; blade narrowly to transversely ovate [lanceolate], unlobed to palmately, often hastately or sagittately, 3–5-lobed, maplelike, base cordate to rounded or truncate, margins crenate, serrate, or nearly entire. **Inflorescences** axillary, solitary flowers or, by upper leaf reduction, forming open panicles [racemes, spikes, or pedunculate glomerules]. **Pedicels** jointed, at least in fruit; involucellar bractlets persistent, 6–10, distinct. **Flowers:** calyx persistent, somewhat accrescent, not inflated, not spathaceous, lobes veined, not strongly ribbed, triangular-ovate, often narrowly so, apex acute; corolla rotate or funnelform [convolute], pink or white [yellow, sometimes with red spot at base]; staminal column ½ length of to nearly equaling petals [much exceeding them]; ovary 5-carpellate; ovules 1 per carpel; styles 5-branched at or beyond orifice of staminal column; stigmas capitate. **Fruits** capsules, erect, not inflated, oblate, depressed, slightly impressed, 5-angled or -winged, 5-valved, 5-locular, apex apiculate, flattened or rounded, not indurate, not fleshy, hairy or minutely so, ± scabrid, often transversely rugose or striate, sections dehiscent. **Seeds** 1 per locule, reniform-ovoid [reniform-globose], glabrous or scabridulous, often with curved, concentric lines. *x* = 19.

Species 17 (2 in the flora): United States, Mexico, West Indies, Bermuda, Central America, n South America, Eurasia, tropical Africa, Pacific Islands (Philippines).

Kosteletzkya is characterized by carpels that are uni-ovulate and capsules that are more or less depressed and five-angled, often bristly on the sutures. At maturity, the capsule valves separate from the fruiting axis as well as from one another.

A specimen of *Kosteletzkya thurberi* A. Gray at MO is labeled "Arizona (Gadsden Purchase) A. Schott leg."; it is probably a duplicate of a collection by Schott from the Cocospera River near Cocospera in the Mexican state of Sonora in September 1855 (F, NY). *Kosteletzkya thurberi* is not known otherwise from north of Mexico.

1. Petals 5–14 mm, usually white, sometimes with pink blush, sometimes drying yellowish; calyx 3.2–6 mm; fruits variously hairy but with long, curved or hooked, simple hairs on sutures; seeds minutely hairy . . . . .1. *Kosteletzkya depressa*
1. Petals 15–45 mm, usually pink, rarely white; calyx 7–11 mm; fruits hairy throughout, sometimes minutely so; seeds glabrous. . . . . .2. *Kosteletzkya pentacarpos*

1. **Kosteletzkya depressa** (Linnaeus) O. J. Blanchard, P. A. Fryxell & D. M. Bates, Gentes Herbarum 11: 357. 1978 • Stinging mallow, white fen-rose [F]

Melochia depressa Linnaeus, Sp. Pl. 2: 674 (as 774). 1753; *Kosteletzkya pentasperma* Grisebach

Plants often multistemmed, from fibrous-thickened root crown, 1–2 m, herbage variously hispid or scabrid, hairs stellate or simple. **Stems** sparingly to freely branched. **Leaves:** stipules linear-subulate, 2.5–6 mm; petiole of lower leaves ¼–⅔ times blade; blade narrowly ovate to transversely ovate, sometimes palmately, hastately, or sagittately 3(–5)-lobed, 3–8 × 1.5–6.5 cm, margins irregularly serrate to crenate, apex broadly acute to acuminate. **Inflorescences** solitary flowers in axils of distal leaves or leafy, open panicles. **Pedicels** exceeding petioles, sometimes blades; involucellar bractlets 6 or 7, linear to subulate, 2–3 mm. **Flowers:** calyx divided for ⅔–⅘+ its length, campanulate to rotate, 3.2–6 mm; corolla rotate, petals usually white, usually with yellow base, sometimes with pink blush, sometimes drying yellowish, asymmetrically ovate to orbiculate, 5–14 × 5–10 mm; staminal column straight or declinate, yellow, 4–10 mm, bearing filaments mostly in distal ½–⅔; filaments 1.5 mm; anthers subsessile, yellow; pollen yellow; styles white or pink, 1.5–2 mm; stigmas pink. **Capsules** olivaceous to brown, 8–11 mm diam., valve margins angulate or subangulate in outline, variously short-hairy, with prominent curved or hooked, simple hairs on sutures. **Seeds** brown, 2.5–2.8 mm, minutely hairy. *2n* = 38 (Mexico, as *K. pentasperma*).

Flowering fall. Freshwater and brackish marshes near coast; 0–10 m; Fla., Tex.; Mexico; West Indies; Central America; South America (Colombia, Ecuador, Peru, Venezuela).

Kosteletzkya depressa is variable and barely enters the flora area. The southern Texas plants conform more to Mexican populations in characters of leaves, fruits, and seeds; in the same characters, the southern Florida plants show affinity with populations of the Greater Antilles.

An earlier chromosome count of *2n* = 34 for *Kosteletzkya depressa* (A. Skovsted 1941) has been shown to be incorrect (O. J. Blanchard 1974).

Plants of *Kosteletzkya depressa* are said to have stinging or irritating hairs.

2. Kosteletzkya pentacarpos (Linnaeus) Ledebour, Fl. Ross. 1: 437. 1842 (as pentacarpa) • Fen-rose, saltmarsh or seashore or seaside or Virginia saltmarsh mallow

Hibiscus pentacarpos Linnaeus, Sp. Pl. 2: 697. 1753; *Kosteletzkya althaeifolia* (Chapman) A. Gray ex S. Watson; *K. pentacarpos* var. *smilacifolia* (Chapman) S. N. Alexander; *K. smilacifolia* (Chapman) Chapman; *K. virginica* (Linnaeus) C. Presl ex A. Gray; *K. virginica* var. *althaeifolia* Chapman; *K. virginica* var. *aquilonia* Fernald

Plants to 3 m, herbage variously pubescent, mostly sparsely to densely stellate-hairy to glabrate. **Stems** sparingly to freely branched. **Leaves:** stipules 2.5–10 mm; petiole of lower leaves ⅓–⅔ times blade; blade narrowly ovate to transversely ovate, often palmately or hastately 3(–5)-lobed, 5.5–17.5 × 3.5–16 cm, margins coarsely to finely serrate or irregularly dentate or ± entire, apex acute to acuminate. **Inflorescences** solitary flowers in axils of upper leaves or axillary on lateral branches. **Pedicels** exceeding petiole but not blade; involucellar bractlets 8–10, linear to subulate, 4–10 mm. **Flowers:** calyx divided for ca. ¾ its length, campanulate to rotate, 7–11 mm; corolla rotate, becoming funnelform-campanulate as day progresses, petals usually pink, rarely white, yellow at base, asymmetrically obovate, 15–45 × 10–40 mm; staminal column ± declinate, yellow, 15–35 mm, bearing filaments nearly to base; filaments mostly 1–2.5 mm; anthers subsessile, yellow; pollen yellow; styles deep pink, 2.5–9 mm; stigmas deep pink. **Capsules** dark brown, 10–14 mm diam., minutely to coarsely hairy, valve margins strongly curved in outline. **Seeds** brown with paler concentric lines, 3–4 mm, glabrous. $2n = 38$ (as *K. virginica*).

Flowering late summer, sometimes winter–spring in Florida. Coastal brackish and fresh marshes; 0–20 m; Ala., Del., D.C., Fla., Ga., La., Md., Miss., N.J., N.Y., N.C., Pa., S.C., Tex., Va.; West Indies (w Cuba); Bermuda; Eurasia (e Azerbaijan, Balearic Islands, Corsica, w Georgia, n Iran, n Italy, sw Russia, s Spain).

Kosteletzkya pentacarpos has been known widely in the United States as *K. virginica*; the Eurasian *K. pentacarpos* is virtually indistinguishable from our northeastern plants (O. J. Blanchard 2008). The two basionyms were published simultaneously by Linnaeus; A. J. Cavanilles ([1785]–1790, diss. 3) was the first to unite them, and his choice of the epithet *pentacarpos* must be followed.

A chromosome number of $2n = 34$, at variance with that accepted here, has been reported for *Kosteletzkya pentacarpos* (N. Tornadore et al. 2000); it is probably an error (O. J. Blanchard 2012).

Kosteletzkya pentacarpos is variable, and this is reflected in the number of infraspecific names that have appeared. S. N. Alexander et al. (2012) examined the named variants and concluded that only a narrow-leaved extreme from southern Florida should be retained: *K. pentacarpos* var. *smilacifolia*. White-flowered plants are found occasionally.

30. KRAPOVICKASIA Fryxell, Brittonia 30: 456. 1978 • [For Antonio Krapovickas, b. 1921, Argentinian botanist] ⊡

Paul A. Fryxell†

Steven R. Hill

Physalastrum Monteiro, Anais Congr. Soc. Bot. Brasil. 20: 395. 1969 (not *Physaliastrum* Makino 1914 [Solanaceae]), based on *Sida* Linnaeus sect. *Physalodes* Grisebach ex K. Schumann in C. F. P. von Martius et al., Fl. Bras. 12(3): 280. 1891

Herbs, perennial. **Stems** procumbent [suberect], with fine stellate pubescence and long simple hairs, not viscid. **Leaves** alternate and distichous; stipules persistent, inconspicuous, subulate; blade ovate-oblong [lanceolate-ovate], unlobed, base cordate, margins crenate-dentate, surfaces tomentose. **Inflorescences** axillary, solitary flowers [fasciculate]; involucel absent. **Flowers:** calyx accrescent, inflated in fruit, brownish-membranous at maturity, completely enclosing fruit, lobes unribbed, ovate, stellate-hairy; corolla white or rose; staminal column included; ovary 5[–9]-carpellate; style 5[–9]-branched; stigmas capitate. **Fruits** schizocarps, erect, not inflated, subglobose, not indurate, fragile-walled, glabrous; mericarps 5, 1-celled, without dorsal spurs, indehiscent. **Seeds** 1 per mericarp, glabrous. $x = 8$.

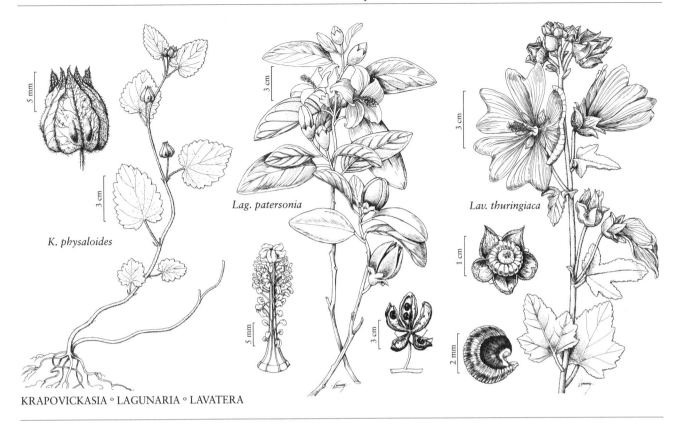

Lag. patersonia

Lav. thuringiaca

K. physaloides

KRAPOVICKASIA ° LAGUNARIA ° LAVATERA

Species 4 (1 in the flora): introduced, Texas; n Mexico, South America (Argentina, Bolivia, Brazil, Paraguay, Peru, Uruguay).

1. **Krapovickasia physaloides** (C. Presl) Fryxell, Brittonia 30: 457. 1978 F I

Sida physaloides C. Presl, Reliq. Haenk. 2: 105. 1835; *S. standleyi* Clement

Stems with simple hairs 1–1.5 mm and smaller, stellate hairs. **Leaves:** stipules 4–5 mm; petiole ¼–½ as long as blade; blade discolorous, 2.5–4.5 cm, apex obtuse or subacute. **Pedicels** slender, 0.5–1.5 cm. **Flowers:** calyx 7–8(–12) mm; petals subequal to calyx; staminal column glabrous, anthers 10. **Schizocarps** blackish; mericarps blackish, 2.5–3 mm, relatively fragile and unornamented. $2n = 16$.

Flowering spring–summer. Oak woodlands, grassy sites; 100 m; introduced; Tex.; Mexico (Nuevo León, San Luis Potosí, Tamaulipas).

Krapovickasia physaloides is known in the flora area from Karnes and San Saba counties, where it is currently thought to have been introduced.

31. LAGUNARIA (de Candolle) Reichenbach, Consp. Regn. Veg., 202. 1828/1829

• Norfolk Island hibiscus, cow itch tree [Genus *Laguna*, for Andrés de Laguna, 1499–1559, Spanish botanist and physician to Pope Julius III, and Latin *-aria*, similarity]   [I]

Steven R. Hill

Hibiscus Linnaeus sect. *Lagunaria* de Candolle in A. P. de Candolle and A. L. P. P. de Candolle, Prodr. 1: 454. 1824

Trees, evergreen. **Stems** erect; bark gray, smooth; twigs stellate-hairy and with peltate scales, glabrescent. **Leaves:** stipules deciduous, inconspicuous, filiform; blade ovate, unlobed, base rounded, margins entire. **Inflorescences** axillary, solitary flowers; involucel cuplike, bractlets persistent or deciduous, 3–5, often obscure or suppressed, connate basally. **Flowers:** calyx not accrescent, not inflated, lobes unribbed, wide-ovate, apex acute to rounded or irregularly lacerate; corolla wide-campanulate, pink to rose-pink, fading white; staminal column exserted; ovary 5-carpellate; ovules (3–)5–10 per locule; style not branched; stigmas club-shaped. **Fruits** capsules, erect to pendulous, ovate, 5-locular, somewhat indurate-chartaceous, brown stellate-hairy covered with fine spicules, filled with irritant hairs reminiscent of fiberglass. **Seeds** (3–)5–10 per locule, glabrous.

Species 1 or 2 (1 in the flora): introduced, California; Pacific Islands (Lord Howe Island, Norfolk Island), e Australia.

The pinnately veined leaves of *Lagunaria* are exceptional for the family. The Queensland, Australia, population has recently (2006) been reclassified as *L. queenslandica* Craven, but there is controversy about its distinctness.

1. Lagunaria patersonia (Andrews) G. Don, Gen. Hist. 1: 485. 1831 (as patersonii)   [F] [I]

Hibiscus patersonius Andrews, Bot. Repos. 4: plate 286. 1803

Plants 2–10(–15) m. **Leaves** short-petiolate; petiole 1 cm; blade olive green, pinnately-veined, 5–10 cm, somewhat thick and leathery, apex blunt, densely felty stellate-hairy abaxially, peltate scales also present. **Inflorescences** 1+ per branch, usually produced over a short time period; peduncle short, thick; involucellar bractlets enclosing flower in bud, wide at base. **Flowers** 3.5–8 cm diam.; calyx connate, lobes very short; petals showy, reflexed, waxy-textured; staminal column with anthers throughout; stigmas with 5 radiating blunt lobes. **Capsules** 4 × 2 cm before opening, apex rounded to acute, walls persistent, firm. **Seeds** bright orange when fresh, smooth.

Flowering late spring–summer. Riparian forests, estuaries; 0–10 m; introduced; Calif.; Pacific Islands (Lord Howe Island, Norfolk Island); e Australia.

Lagunaria patersonia is widely cultivated but apparently only recently naturalized in San Diego County; it is salt tolerant.

32. LAVATERA Linnaeus, Sp. Pl. 2: 690. 1753; Gen. Pl. ed. 5, 308. 1754 • Rose mallow

[For Lavater family, 17th-century physicians and naturalists of Zurich]   [I]

Steven R. Hill

Olbia Medikus; *Stegia* de Candolle

Herbs, subshrubs, or shrubs, perennial or annual [biennial], stellate-hairy or glabrate. **Stems** erect. **Leaves:** stipules early-deciduous, narrowly triangular, lanceolate, oblanceolate, or linear; blade lanceolate or ovate to orbiculate, narrower above, base rounded to wide-cuneate, unlobed or palmately 3–7-lobed, margins crenate or dentate or nearly entire. **Inflorescences** axillary solitary flowers, [fascicles], or terminal racemes; involucel present, bractlets persistent, 3, connate ca. ½ toward base, cupuliform. **Flowers:** calyx not accrescent, not inflated, lobes triangular to ovate, not ribbed; corolla broadly trumpet-shaped to nearly rotate, rose-pink, white, or purple, usually with darker purplish veins; staminal column included; ovary [9–]12–22[–40]-carpellate; ovules 1 per cell; style [9–]12–22[–40]-branched (same number as locules); stigmas introrsely decurrent, filiform. **Fruits** schizocarps, erect, not inflated, flattened-globose, discoid, somewhat indurate, persistent style base swollen, conic or disc-shaped; mericarps [9–]12–22[–40], drying tan or brown, 1-celled, elliptic in cross section, edges rounded, walls readily separating from seed, without dorsal spur, not beaked, sometimes slightly ridged or keeled, apex rounded, glabrous or hairy, indehiscent. **Seeds** 1 per mericarp, glabrous. $x = 14$.

Species ca. 12–25 (3 in the flora): introduced; s Europe (Mediterranean regions), sw Asia (to Kashmir), Africa (Ethiopia), Atlantic Islands (Canary Islands); introduced also widely.

Lavatera has traditionally been distinguished from *Malva* by its cuplike, basally fused involucellar bractlets. Based on molecular evidence, *Lavatera* and *Malva* are nearly indistinguishable, and while still accepted here based upon different criteria, *Lavatera* taxa eventually may be combined with *Malva*. *Lavatera arborea* Linnaeus, *L. assurgentiflora* Kellogg, and *L. cretica* Linnaeus of previous treatments have been transferred to *Malva*, and some epithet changes were necessary.

Lavatera cachemiriana Cambessèdes is commonly cultivated and may escape; it is similar to *L. thuringiaca* but has denser, long-stellate indument and the involucellar bractlets are two-thirds as long as the calyx, are broadly ovate with acute apices, fused at base into a cup about 1 cm long, barely appressed to the calyx, and are accrescent in fruit.

SELECTED REFERENCE   Ray, M. F. 1995. Systematics of *Lavatera* and *Malva* (Malvaceae, Malveae)—A new perspective. Pl. Syst. Evol. 198: 29–53.

1. Annual herbs, 0.5–1.4 m; petals rose-pink or white, bases usually overlapping and obscuring calyx surface . . . . 3. *Lavatera trimestris*
1. Perennial herbs or subshrubs, or shrubs, usually 1–2 m; petals pale purplish pink or darker purple-violet, bases not overlapping, calyx surface clearly visible between.
   2. Subshrubs or shrubs; stems ± woody at base; inflorescences solitary flowers, sometimes racemelike, subsessile, pedicels 0.2–0.7 cm; plants usually hispid throughout . . . . 1. *Lavatera olbia*
   2. Herbs; inflorescences solitary flowers, sometimes racemelike but not congested, pedicellate, pedicels 2–5 cm; plants usually puberulent to tomentose throughout . . . . 2. *Lavatera thuringiaca*

1. **Lavatera olbia** Linnaeus, Sp. Pl. 2: 690. 1753 • Tree lavatera or mallow [1]

Althaea olbia (Linnaeus) Kuntze; *Lavatera hispida* Desfontaines

Subshrubs or shrubs, usually 1–2 m. **Stems** woody at base, bristly-hairy, young stems tomentose, infrequently glabrescent. **Leaves** reduced distally; stipules narrowly lanceolate, 2–4 mm; petioles to 10 cm in proximal leaves, reduced distally to ¼ blade length; blade green, to 15 cm, surfaces densely stellate-tomentose, proximal blades broadly to narrowly ovate, 3–5-lobed, lobes lanceolate to ovate, margins crenate-dentate to nearly entire, apex acute to rounded, distal blades oblong-ovate to lanceolate, often slightly 3-lobed and subhastate. **Inflorescences** solitary flowers or congested racemelike distally, subsessile. **Pedicels** 0.2–0.7 cm, not elongating in fruit; involucellar bractlets connate halfway, subequal to calyx, lobes wide-ovate, apex acuminate. **Flowers** showy; calyx divided halfway, campanulate, ciliate, hispid-hairy, lobes abruptly acuminate, ± enclosing fruit; corolla 4–6 cm diam., petals pink to purple-violet, veins darker rose, 2–3 cm, bases not overlapping (calyx visible between), apex shallowly notched; staminal column glabrous; anthers on upper ¾ of stamen column, pink to yellowish; style 17–19-branched (same number as locules). **Mericarps** 17–19, tomentose or hispid, sides sometimes glabrate. $2n = 42$.

Flowering Jun–Aug(–Oct). Disturbed habitats; 0–100 m; introduced; Calif.; sw Europe.

Lavatera olbia is widely cultivated as a garden ornamental and along highway medians; it only rarely escapes or naturalizes, but a few apparent escapes have been found in Monterey, Orange, and San Francisco counties. *Lavatera olbia* and *L. thuringiaca* hybridize to produce the occasionally cultivated hybrid *Lavatera ×clementii* Cheek.

2. Lavatera thuringiaca Linnaeus, Sp. Pl. 2: 691. 1753

F I

Althaea ambigua (de Candolle) Alefeld; *A. vitifolia* (Schur) Borbás; *Malva thuringiaca* (Linnaeus) Visiani

Herbs, sometimes appearing subshrublike, perennial, usually 1.5–2 m. **Stems** short stellate-tomentose. **Leaves** reduced distally; stipules linear to oblanceolate, 6 mm; petioles equaling blades proximally, progressively reduced distally on stem to ¼ blade length; blade grayish, cordate-orbiculate, usually palmately 3–5-lobed, usually 7–9 cm, lobes subacute to obtuse, margins crenate, apex acute, surfaces short-hairy. **Inflorescences** solitary flowers or racemelike, usually not congested, pedicellate. **Pedicels** 2–5 cm, not elongating in fruit; involucellar bractlets connate in lower ⅓, lobes wide-ovate, usually shorter than calyx, apex obtuse to abruptly acute. **Flowers** showy; calyx divided halfway, campanulate, puberulent, not hispid, lobes narrowly acute to acuminate, ± enclosing fruit; corolla 3–7 cm diam., petals pale purplish pink, only occasionally with somewhat darker pink veins, 2–3.5 cm, bases not overlapping (calyx surface visible between), apex notched; staminal column glabrous; anthers on upper ¾ of stamen column, pale pink to white; style 20–22-branched (same number as locules). **Mericarps** 20–22, glabrous. **2*n*** = 42.

Flowering Jun–Oct. Disturbed habitats; 0–1800 m; introduced; Man., Sask.; Minn., Wyo.; c, e Europe; c Asia.

Lavatera thuringiaca is showy and occasionally cultivated as a garden ornamental. There are only a few records of the species as escaped or naturalized in North America. Old reports have indicated its presence in New Brunswick and Ontario based on escapes from cultivation that did not become established; there was also an apparently erroneous report from Quebec.

3. Lavatera trimestris Linnaeus, Sp. Pl. 2: 692. 1753

• Annual or rose mallow I

Althaea trimestris (Linnaeus) Kuntze; *Stegia trimestris* (Linnaeus) T. Luque & Devesa

Herbs, annual, usually 0.5–1.4 m. **Stems** sparsely hispid-hairy, hairs simple or few-rayed, glabrate in age. **Leaves** reduced distally; stipules narrowly triangular, 2–6 mm; petioles equaling blades proximally, 1–7(–13) cm, reduced distally ca. ½ blade length; blade green, orbiculate, 2.5–7(–12) cm, proximal leaves unlobed to shallowly 3–5-lobed or -angled, distal leaves shallowly palmately 3–7-lobed, lobes triangular, margins crenate or dentate (serrate), apex acuminate or acute to obtuse, surfaces sparsely bristly to glabrous. **Inflorescences** solitary flowers, clustered towards stem tips, pedicellate. **Pedicels** 2(–7) cm, elongating in fruit; involucellar bractlets connate halfway, subequal to or ¾ length of calyx, lobes wide-deltate, 0.6–1 cm, distinctly enlarging in fruit, apex acute. **Flowers** showy; calyx divided ca. ½ its length, campanulate, 1–1.5 cm, enlarging in fruit, ± enclosing fruit, sparsely short stellate-hairy, lobes triangular to narrowly ovate, apex acute to acuminate; corolla 3–7(–10) cm diam., petals rose-pink to white with darker rose veins, 2.5–3.5(–7) cm, bases strongly overlapping (obscuring calyx between), apex blunt to rounded; staminal column puberulent; anthers on upper ¾ of stamen column, pinkish; style ca. 12+-branched. **Mericarps** ca. 12+, 3.3–3.7 mm, ridged, covered and obscured by disclike expansion of central axis, glabrous, transverse-reticulately veined on back. **2*n*** = 14.

Flowering Jun–Oct. Disturbed habitats; 0–600 m; introduced; Calif., Conn., Mo., Pa., Tex., Vt.; s Europe; w Asia.

Lavatera trimestris, the type species of the genus, is occasionally cultivated as a garden ornamental; it rarely escapes or naturalizes. Most records are of plants that did not persist. It is easily distinguished from other commonly cultivated mallows by the disclike expansion of the central axis of the fruit above the mericarps.

33. MALACHRA Linnaeus, Syst. Nat. ed. 12, 2: 458. 1767; Mant. Pl. 1: 13. 1767

• [Ancient name, perhaps from Greek *malache*, mallow]

Paul A. Fryxell†

Steven R. Hill

Herbs or subshrubs, annual or perennial. **Stems** erect, puberulent or hispid, not viscid. **Leaves:** stipules persistent, filiform; blade broadly ovate, lyrate, or palmately 3–5-lobed, base rounded or truncate, margins crenate or serrate, surfaces usually hairy. **Inflorescences** axillary [or terminal], headlike racemes, subtended by prominently veined, sessile or subsessile, boat-shaped bracts; involucel absent [present]. **Flowers:** calyx not accrescent, not inflated, deeply divided, lobes unribbed or somewhat ribbed, lanceolate-ovate, hispid; corolla yellow, [white or lavender]; staminal column included; styles 10-branched; stigmas capitate. **Fruits** schizocarps, erect, not inflated, oblate, not indurate, glabrous or puberulent; mericarps 5, 1-celled, without dorsal spur, smooth, indehiscent. **Seeds** 1 per mericarp, glabrous. $x = 28$.

Species 8–10 (2 in the flora): sc United States, Mexico, West Indies, Central America, South America; introduced in Asia (including Malesia), Africa.

1. Peduncles usually exceeding corresponding petioles; floral bracts deeply cordate; herbage stellate-hairy; fruits glabrous . . . . 1. *Malachra capitata*
1. Peduncles shorter than corresponding petioles; floral bracts ovate; herbage hispid; fruits minutely puberulent to pubescent . . . . 2. *Malachra urens*

1. Malachra capitata (Linnaeus) Linnaeus, Syst. Nat. ed. 12, 2: 458. 1767 (as capitat) [F]

Sida capitata Linnaeus, Sp. Pl. 2: 685. 1753

Plants 0.4–1.5 m. **Stems** densely stellate-hairy, sometimes with longer, simple hairs 1.5 mm. **Leaves:** stipules 9–15 mm; petiole to ½ as long as blade; blade ovate, lyrate, or 3–5-lobed, 4–10 cm, lobes basally narrowed, apex acute or obtuse, surfaces stellate-hairy. **Racemes** pedunculate; peduncle usually exceeding corresponding petiole; boat-shaped floral bracts subsessile, with prominent, white markings at base, deeply cordate, margins entire, often hispid. **Flowers:** calyx 6–8 mm; petals 7–10 mm. **Mericarps** brownish, 3 mm, smooth with reticulate veins, glabrous. $2n = 56$.

Flowering year-round. Disturbed vegetation, thickets, roadsides, cultivated fields; 0–300 m; Fla., La., Tex.; Mexico; West Indies; Central America; South America; introduced in Asia (Malesia).

Malachra capitata is introduced in Malesia according to J. van Borssum Waalkes (1966), who synonymized *M. capitata* and *M. alceifolia* Jacquin. The Louisiana records are thought to represent adventive or introduced populations. Those in southern Florida and Texas are accepted as native populations.

2. Malachra urens Poiteau ex Ledebour & Alderstam, Diss. Bot. Pl. Doming., 22. 1805

Urena urens (Poiteau ex Ledebour & Alderstam) M. Gómez

Plants 0.5–1 m. **Stems** hispid with long, simple hairs and with longitudinal lines of minute hairs. **Leaves:** stipules 7–15 mm; petiole ½+ as long as blade; blade broadly ovate to weakly 3-lobed, 3–12 cm, apex acute to obtuse, surfaces hispid. **Racemes** short-pedunculate; peduncle shorter than corresponding petiole; boat-shaped floral bracts sessile, without markings, ovate, margins entire, hispid. **Flowers:** calyx 3–5 mm; petals 7–11 mm. **Mericarps** 3 mm, smooth with reticulate veins, minutely puberulent to pubescent.

Flowering year-round. Disturbed vegetation, savannas, secondary vegetation, often in poorly drained soil; 0–500 m; Fla.; West Indies (Cuba, Puerto Rico).

The name *Malachra fasciata* Jacquin is sometimes misapplied to *M. urens*, but the former is white-flowered rather than yellow-flowered, and differs in other characters, as noted by F. Areces B. and P. A. Fryxell (2007).

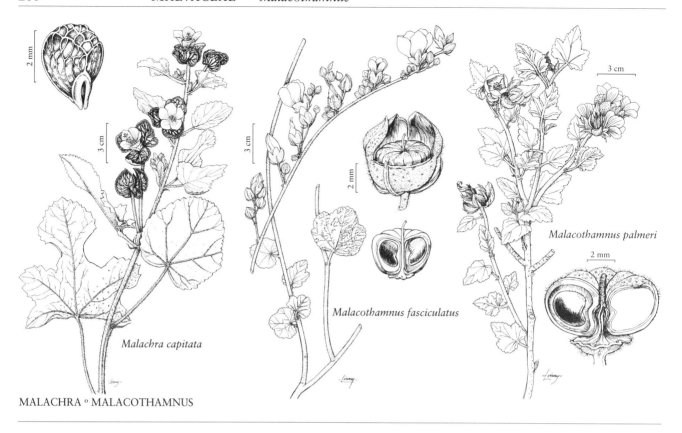

Malachra capitata

Malacothamnus fasciculatus

Malacothamnus palmeri

MALACHRA ° MALACOTHAMNUS

34. MALACOTHAMNUS   Greene, Leafl. Bot. Observ. Crit. 1: 208. 1906   •   Chaparral mallow, bushmallow [Greek *malakos*, soft, or *malache*, mallow, and *thamnos*, shrub, alluding to habit]

David M. Bates

Subshrubs or shrubs, sometimes root-suckering, glabrate to densely hairy, hairs various, simple or bifurcate, simple-glandular, and stellate, ca. 6–30-armed. **Stems** erect or ascending, ultimate branches strict to diffuse. **Leaves:** stipules ± persistent, filiform to subulate; blade ovate to round, rhombic, or reniform, unlobed or palmately 3-, 5-, or 7-lobed, base cordate to truncate or cuneate, margins usually toothed. **Inflorescences** usually axillary, rarely terminal, flowers solitary or in clusters in dense to open heads, spikes, racemes, or panicles; involucellar bractlets persistent, 3, distinct (basally connate in *M. aboriginum*). **Flowers:** calyx not accrescent, not inflated, lobes triangular to ovate, not ribbed; corolla exceeding calyx, campanulate to subrotate, usually rose to pale pink, pinkish mauve, or mauve, rarely white; staminal column ± included; filaments terminal and subterminal; ovary 7–14-carpellate; ovules 1 per cell; styles 7–14-branched, (branches equal in number to carpels); stigmas capitate. **Fruits** schizocarps, erect, not inflated, disclike, subglobose-obovate in lateral view, not indurate, fragile, apically minutely stellate-hairy; mericarps 7–14, drying tan, 1-celled, asymmetrically suborbicular to obovoid-reniform, smooth-walled, without dorsal spur, apex muticous, dehiscence loculicidal, walls falling away as 2 fragile valves. **Seeds** 1 per mericarp, ascending, brown or black, obovoid-reniform, usually papillate-stellate or minutely stellate-hairy or rarely glabrous. $x = 17$.

Species 11 (10 in the flora): California, nw Mexico.

Most species of *Malacothamnus* are locally common in early burn successions of chaparral and adjacent vegetation types in California and northern Baja California. *Malacothamnus foliosus* (S. Watson) Kearney is the only species restricted to Mexico; it is found in northern Baja

California. Major morphological variants in *Malacothamnus* have been recognized as species, as treated by T. H. Kearney (1951). Insofar as known, most taxa are interfertile and in some instances intergrade in areas of proximity. Interpopulational variation within most species in indument and inflorescence characters is high and of moderately complex patterning.

SELECTED REFERENCES Bates, D. M. 1963. The Genus *Malacothamnus*. Ph.D. dissertation. University of California, Los Angeles. Benesh, D. L. and W. J. Elisens. 1999. Morphological variation in *Malacothamnus fasciculatus* (Torrey & A. Gray) E. Greene (Malvaceae) and related species. Madroño 46: 142–152. Kearney, T. H. 1951. The genus *Malacothamnus* Greene (Malvaceae). Leafl. W. Bot. 6: 113–140.

1. Involucellar bractlets 1–9 mm wide.
    2. Inflorescences open-paniculate, flowers solitary or in loose clusters, not subtended by conspicuous bracts; indument white and dense (hairs simple, fine), sometimes sublepidote (hairs stellate, sessile) . . . . 8. *Malacothamnus abbottii*
    2. Inflorescences spicate, racemose, or headlike, flower clusters glomerate, subtended by conspicuous bracts; indument grayish, tawny, or white, dense to sparse, hairs stellate, sessile or stalked, sometimes mixed with glandular and/or armed hairs.
        3. Flower clusters in elongate, spicate or racemose inflorescences, involucellar bractlets distinct or basally connate . . . . .9. *Malacothamnus aboriginum*
        3. Flower clusters in headlike or spicate inflorescences, involucellar bractlets distinct . . . . .10. *Malacothamnus palmeri*
1. Involucellar bractlets 0.2–0.8(–1) mm wide.
    4. Calyx lobes: length 1–1.5 times width, 1(–2) times tube, triangular to ovate, apex acute or acuminate to short-mucronate; involucellar bractlets ½–⅔ times calyx.
        5. Calyces sparsely bristly- or stellate-hairy, stellate hairs overlain by 1–few-armed hairs, arms to 3 mm; flower clusters sessile, glomerate, inflorescences interrupted, spicate. . . . . 6. *Malacothamnus densiflorus* (in part)
        5. Calyces sparsely hairy to densely stellate-hairy, if sparsely, not bristly, hairs usually 10–30-armed, arms 0.2–1.5 mm; flower clusters sessile or pedunculate, glomerate or loose, inflorescences non-interrupted, cymose, spicate, racemose, or open-paniculate.
            6. Leaf blades 2–6(–11) cm, usually thin, surfaces: adaxial usually gray-green and sparsely hairy, basal lobes not overlapping; branches usually slender . . . . .1. *Malacothamnus fasciculatus*
            6. Leaf blades 5–11(–20) cm, thick, surfaces: adaxial tawny and densely stellate-hairy, basal lobes usually overlapping; branches usually stout . . . . .2. *Malacothamnus davidsonii*
    4. Calyx lobes: length 2–3 times width, 1–3 times tube, triangular, deltate-lanceolate, lanceolate, or ovate, apex acute to acuminate; involucellar bractlets (⅓–)½ times to exceeding calyx.
        7. Calyces angled and/or winged in bud, lobes lanceolate, ovate, or triangular, apex acuminate; indument usually tawny, sparse, moderately dense, or dense, hairs stellate or simple, overlain by coarser hairs.
            8. Flower clusters usually glomerate; inflorescences not basally leafy, interrupted, spicate; calyx indument sparse, hairs usually 1- or 2–few-armed, bristly . . . . 6. *Malacothamnus densiflorus* (in part)
            8. Flower clusters glomerate to open; inflorescences usually basally leafy, interrupted or not, spicate or racemose; calyx indument dense, stellate-hairy, not bristly. . . . . 7. *Malacothamnus marrubioides*
        7. Calyces not angled and/or winged in bud, lobes usually triangular or deltate-lanceolate, apex long-acute to ± acuminate; indument densely lanate, white-lanate, or villous, or sublepidote to tomentose.
            9. Petals white, fading lavender; leaf blades 3- or 5-lobed, surfaces: adaxial glabrate, dark green . . . . 3. *Malacothamnus clementinus*
            9. Petals pale pink to deep mauve; leaf blades unlobed or 3- or 5-lobed, surfaces: adaxial soft tomentose-white, pale ash green and velvety white or sparsely hairy and green.

[10. Shifted to left margin.—Ed.]

10. Inflorescences interrupted and spicate or narrow-paniculate to open-paniculate, flower clusters dense or loose, 6–10-flowered; branches stout or slender, not flexuous; leaf blades ovate, broadly ovate, suborbiculate, or reniform, 4–6(–11) cm . . . . 4. *Malacothamnus fremontii*

10. Inflorescences usually open-paniculate or subracemose, flower clusters loose, 3(–6)-flowered, sometimes flowers solitary; branches slender, flexuous; leaf blades ovate, ± rhombic, or suborbiculate, 2.5–4.5(–7) cm . . . . .5. *Malacothamnus jonesii*

1. **Malacothamnus fasciculatus** (Nuttall ex Torrey & A. Gray) Greene, Leafl. Bot. Observ. Crit. 1: 208. 1906 • Chaparral bushmallow F

Malva fasciculata Nuttall ex Torrey & A. Gray, Fl. N. Amer. 1: 225. 1838; *Malacothamnus arcuatus* (Greene) Greene; *M. fasciculatus* subsp. *catalinensis* (Eastwood) Thorne; *M. fasciculatus* var. *catalinensis* (Eastwood) Kearney; *M. fasciculatus* var. *laxiflorus* (A. Gray) Kearney; *M. fasciculatus* var. *nesioticus* (B. L. Robinson) Kearney; *M. fasciculatus* var. *nuttallii* (Abrams) Kearney; *M. hallii* (Eastwood) Kearney; *M. mendocinensis* (Eastwood) Kearney; *M. nesioticus* (B. L. Robinson) Abrams; *M. nuttallii* Abrams; *M. parishii* (Eastwood) Kearney; *Malvastrum arcuatum* (Greene) B. L. Robinson; *M. catalinense* Eastwood; *M. fasciculatum* (Nuttall ex Torrey & A. Gray) Greene; *M. fasciculatum* var. *laxiflorum* (A. Gray) Munz & I. M. Johnston; *M. hallii* Eastwood; *M. laxiflorum* (A. Gray) Davidson & Moxley; *M. mendocinense* Eastwood; *M. nesioticum* B. L. Robinson; *M. nesioticum* subsp. *nuttallii* (Abrams) Wiggins; *M. nuttallii* (Abrams) Davidson & Moxley; *M. parishii* Eastwood; *Sphaeralcea arcuata* (Greene) Arthur; *S. nesiotica* (B. L. Robinson) Jepson

Shrubs, 1–3(–5) m; branches usually slender; indument white to tawny, sparse to dense, stellate hairs sessile or stalked, 10–30-armed, arms 0.2–1.5 mm. **Leaf blades** ovate, broadly ovate, or round, unlobed or 3-, 5-, or 7-lobed, 2–6(–11) cm, usually thin, surfaces: adaxial gray-green and sparsely hairy, basal lobes not overlapping. **Inflorescences** cymose, spicate, racemose, or open-paniculate, flower clusters subsessile to pedunculate, usually many-flowered, glomerate to loose; involucellar bractlets subulate to linear, 1–6(–8) × to 1 mm, usually less than ½(–⅔) calyx length. **Flowers:** calyx campanulate, 4–9(–11) mm, lobes triangular to ovate, 1.8–6(–8) × 1.5–3(–4) mm, usually slightly longer than wide, slightly longer than tube, apex acute to short-acuminate, densely stellate-hairy, hairs usually 10–30-armed, arms 0.2–1.5 mm; petals pale pink to mauve, to 2 cm. **Mericarps** 2–3.5(–4.5) mm. $2n = 34$.

Flowering Apr–Jul. Coastal sage scrub, chaparral; 10–600(–2500) m; Calif.; Mexico (Baja California, Sonora).

Malacothamnus fasciculatus occurs principally in the Coast and Transverse ranges. It recently was found in natural Sonora Desert scrub in Maricopa County, Arizona (S. Hunkins 2012). Whether introduced or native is not known; its presence supports the nativity of collections made by Thurber in Sonora in the early 1850s. In southern California, some variants have been recognized infraspecifically: var. *laxiflorus*, with open-paniculate inflorescences, occurs essentially through the range of the species except in the coastal Peninsula Ranges; var. *nuttallii*, robust plants with paniculate inflorescences, often densely white-hairy adaxial leaf surfaces, and floral and carpel measurements in the upper ranges, occurs in the Transverse Ranges of Santa Barbara and Ventura counties; var. *nesioticus*, presumably with fastigiate inflorescence branches, but essentially the same as var. *nuttallii*, occurs on Santa Cruz Island; var. *catalinensis*, robust plants with spicate to narrowly paniculate inflorescences, occurs only on Santa Catalina Island. In northern California, populations generally have floral measurements in the lower range, although in the Santa Cruz Mountains plants sometimes have involucellar bractlets in the upper range. Otherwise, populations vary in patterns similar to those of southern California. *Malacothamnus fasciculatus* intergrades with *M. davidsonii* and *M. fremontii* in the Transverse Ranges and *M. densiflorus* in the Peninsula Ranges.

2. **Malacothamnus davidsonii** (B. L. Robinson) Greene, Leafl. Bot. Observ. Crit. 1: 208. 1906 • Davidson's bushmallow C E

Malvastrum davidsonii B. L. Robinson in A. Gray et al., Syn. Fl. N. Amer. 1(1,2): 312. 1897; *Sphaeralcea davidsonii* (B. L. Robinson) Jepson

Shrubs, to 3(–5) m, branches usually stout, indument white to tawny, dense (shaggy-tomentose), stellate hairs sessile or stalked, 10–30-armed, arms 0.2–1.5 mm. **Leaf blades** ± round, 3-, 5-, or 7-lobed, 5–11(–20) cm, thick, surfaces: adaxial tawny and densely stellate-hairy (velvety to scabrous-tomentose), basal lobes overlapping. **Inflorescences** spicate to racemose, flower clusters usually short-pedunculate, glomerate; involucellar bractlets subulate, 1.5–4 × to 1 mm, usually less than ½ calyx length. **Flowers:** calyx campanulate, 5–9 mm, lobes triangular to ovate, 3–5 × 1.8–3.2(–4) mm, slightly longer than wide, slightly longer than tube, apex acute to short-mucronate, densely stellate-hairy, hairs 10–30-armed; petals mauve, to 1.5 cm. **Mericarps** 2–3 mm. **2*n*** = 34.

Flowering Jun–Aug. Slopes, washes, chaparral; of conservation concern; 200–700 m; Calif.

Malacothamnus davidsonii occurs on the central Coast Ranges of San Luis Obispo and Monterey counties and Transverse Ranges in Los Angeles County. It intergrades with *M. fasciculatus* in Little Tujunga and Pacoima canyons in Los Angeles County.

3. **Malacothamnus clementinus** (Munz & I. M. Johnston) Kearney, Leafl. W. Bot. 6: 127. 1951
   · San Clemente Island bushmallow C E

Malvastrum clementinum Munz & I. M. Johnston, Bull. Torrey Bot. Club 51: 296. 1924; *Sphaeralcea orbiculata* (Greene) Jepson var. *clementina* (Munz & I. M. Johnston) Jepson

Subshrubs, 0.4–1 m, branches slender, indument white or grayish, sparse to dense, shaggy-tomentose, canescent, hairs mostly stellate, stalked, 10–30-armed. **Leaf blades** ± round, 3- or 5-lobed, to 5(–8) cm, thin to moderately thick, surfaces: adaxial dark green and glabrate, abaxial soft-tomentose, lobes triangular to rounded, basal sinus not overlapping. **Inflorescences** short-spicate, flower clusters subsessile, congested; involucellar bractlets filiform to linear, 3–9 × 0.5 mm, ¾ to exceeding calyx length. **Flowers:** calyx campanulate, 5–9 mm, lobes coherent, narrowly triangular, 3.5–6.5 × 1.5–2.3 mm, ca. 2 times as long as wide, 2–3 times tube length, apex long-acute to acuminate, densely villous; petals white, fading lavender, 1.5 cm. **Mericarps** 2–3 mm. **2*n*** = 34.

Flowering Mar–May. Coastal sage scrub; of conservation concern; 100–400 m; Calif.

Malacothamnus clementinus is known from San Clemente Island and is considered endangered. In leaf characters it is similar to *M. fasciculatus*; in indument and calyx characters, it approaches *M. fremontii*. The petals are often not overlapping, a feature rare in *Malacothamnus*. *Malacothamnus clementinus* tends to propagate mainly via rhizomes and rarely produces fertile seeds.

Malacothamnus clementinus is in the Center for Plant Conservation's National Collection of Endangered Plants.

4. **Malacothamnus fremontii** (Torrey ex A. Gray) Greene, Leafl. Bot. Observ. Crit. 1: 208. 1906
   · Fremont's bushmallow E

Malvastrum fremontii Torrey ex A. Gray, Mem. Amer. Acad. Arts, n. s. 4: 21. 1849; *Malacothamnus fremontii* subsp. *cercophorus* (B. L. Robinson) Munz; *M. helleri* (Eastwood) Kearney; *M. howellii* (Eastwood) Kearney; *M. orbiculatus* (Greene) Greene; *Malvastrum fremontii* var. *cercophorum* B. L. Robinson; *M. helleri* Eastwood; *M. howellii* Eastwood; *M. orbiculatum* Greene; *Sphaeralcea fremontii* (Torrey ex A. Gray) Jepson; *S. fremontii* var. *exfibulosa* Jepson; *S. orbiculata* (Greene) Jepson

Subshrubs or shrubs, 0.5–2(–3) m, branches stout or slender, indument white, sparsely to densely woolly, stellate hairs stalked, 20–30-armed, glandular hairs often abundant. **Leaf blades** ovate, broadly ovate, suborbiculate, or reniform, 3- or 5-lobed, 4–6(–11) cm, thin or thick, surfaces: adaxial white, densely velvety-tomentose or green, sparsely hairy, basal sinus open, not overlapping. **Inflorescences** interrupted, spicate or narrow- or, rarely, open-paniculate, flower clusters sessile or pedunculate, 6–10-flowered; involucellar bractlets subulate to filiform, 3–12(–15) × to 0.5 mm, ⅓–1⅓ times calyx length. **Flowers:** calyx campanulate, 5.5–13 mm, lobes usually narrowly triangular, 3–10 × 1–4 mm, ca. 2–3 times as long as wide, ± equaling to 3 times tube length, apex acute or short-acuminate, densely white-lanate; petals pale pink, to 1.8 cm. **Mericarps** 2.5–4 mm. **2*n*** = 34.

Flowering Apr–Jun. Chaparral, lower margins of pine woodlands; 50–2800 m; Calif.

Malacothamnus fremontii occurs about the Central Valley in the inner Coast Ranges, Sierra Nevada, and Transverse Ranges. In the northern Coast Ranges, plants with calyx measurements in the lower range have been recognized as *M. helleri* or *Sphaeralcea fremontii* var. *exfibulosa*; plants with a more southerly distribution and measurements in the upper range have been called *M. howellii* or *M. fremontii* subsp. *cercophorus*; these forms occur elsewhere in the range of *M. fremontii*. In the Transverse Ranges, plants are generally less woolly (*M. orbiculatus*) and may intergrade with *M. fasciculatus*.

5. **Malacothamnus jonesii** (Munz) Kearney, Leafl. W. Bot. 6: 135. 1951 • San Luis Obispo or Jones's bushmallow E

Malvastrum jonesii Munz, Bull. S. Calif. Acad. Sci. 24: 88. 1925; *Malacothamnus gracilis* (Eastwood) Kearney; *M. niveus* (Eastwood) Kearney; *Malvastrum fremontii* Torrey ex A. Gray var. *niveum* (Eastwood) McMinn; *M. gracile* Eastwood; *M. niveum* Eastwood

Shrubs, 1–2.5 m, branches slender, flexuous, indument white, usually velvety, not shaggy, stellate hairs stalked, sometimes sublepidote, ca. 10–30-armed, glandular hairs often abundant. **Leaf blades** ash green, ovate, ± rhombic, or suborbiculate, unlobed or 3- or 5-lobed, 2.5–4.5(–7) cm, thin to thick, surfaces: soft tomentose-white, adaxial pale ash green, white-velvety, basal sinus open, not overlapping. **Inflorescences** open-paniculate or subracemose, flower clusters pedunculate, loose, 3(–6)-flowered, or solitary flowers; involucellar bractlets awl-shaped to filiform, 2.5–7(–8) × 1 mm, 1/3–1/2(–2/3) calyx length. **Flowers:** calyx campanulate, not angled in bud, 5–9(–10) mm, lobes distinct in bud, narrowly triangular or deltate-lanceolate, 3–6(–7.5) × 1.5–3(–4) mm, length 2 times width, equaling to 3 times tube length, apex acute to short-acuminate, usually sublepidote to tomentose, hairs 10–30-armed, sometimes densely lanate; petals pale pink, to 1 cm. **Mericarps** 2.5–3.8 mm. $2n = 34$.

Flowering late Apr–Jun. Open chaparral, foothill woodlands; 200–900 m; Calif.

Malacothamnus jonesii is known from the central Coast Ranges.

6. **Malacothamnus densiflorus** (S. Watson) Greene, Leafl. Bot. Observ. Crit. 1: 208. 1906 • Dense-flowered or yellowstem bushmallow

Malvastrum densiflorum S. Watson, Proc. Amer. Acad. Arts 17: 368. 1882; *Malacothamnus densiflorus* var. *viscidus* (Abrams) Kearney; *Malvastrum densiflorum* var. *viscidum* (Abrams) Estes; *M. viscidum* Abrams; *Sphaeralcea densiflora* (S. Watson) Jepson; *S. densiflora* var. *viscida* (Abrams) Jepson

Shrubs, 1–2 m, branches slender, indument tawny, sparse to moderately dense, stellate hairs sessile or subsessile, usually overlain by 1- or 2-few-armed, bristly hairs to 3 mm, glandular hairs often abundant. **Leaf blades** ovate, broadly ovate, or round, unlobed or 3- or 5-lobed, 3–6 cm, often leathery, surfaces: abaxial sometimes glossy green, adaxial densely tawny-hairy, basal sinus open, not

overlapping. **Inflorescences** interrupted, spicate, flower clusters sessile, glomerate, 10+-flowered; involucellar bractlets awl-shaped to filiform, 5–15 × 1 mm, 2/3–1 1/2 times calyx length. **Flowers:** calyx angled in bud, 6–14 mm, lobes triangular to ovate, 4–11 × 2.3–3.5(–5) mm, slightly less to slightly more than 2 times as long as wide, equaling to 3 times tube length, apex acute to acuminate, sparsely bristly-hairy, hairs 1–few-armed, to 3 mm; petals pale pink, 1–2 cm. **Mericarps** 2.4–3.2 mm. $2n = 34$.

Flowering Mar–Jun(Sep–Oct). Chaparral, coastal sage scrub ecotone; (100–)300–1100 m; Calif.; Mexico (Baja California).

Malacothamnus densiflorus occurs in the southern Coast Ranges. Plants with relatively short involucellar bractlets and calyx measurements in the lower range have been recognized as var. *viscidus*; intergradation with more typical forms is complete. *Malacothamnus densiflorus* intergrades with *M. fasciculatus* at lower elevations of the coastal Peninsula Ranges in San Diego County.

7. **Malacothamnus marrubioides** (Durand & Hilgard) Greene, Leafl. Bot. Observ. Crit. 1: 208. 1906 • Pink-flowered bushmallow

Malvastrum marrubioides Durand & Hilgard, Pl. Heermann., 38. 1854; *M. gabrielense* Munz & I. M. Johnston

Subshrubs, 1–2 m, branches slender, indument tawny, moderately dense, often viscid, not shaggy, stellate hairs sessile or stalked, not bristly, few-many-armed, glandular hairs usually abundant. **Leaf blades** broadly ovate or suborbiculate, unlobed or 3- or 5-lobed, 3–6(–8) cm, surfaces: moderately to copiously hairy, hairs grayish to tawny, 10–30-armed, basal sinus open, not overlapping. **Inflorescences** usually short, interrupted, spicate to racemose, flower clusters sessile or short-pedunculate, glomerate to open, usually leafy; involucellar bractlets filiform, 5–13 × 1 mm, mostly 2/3 to equaling calyx length. **Flowers:** calyx angled or slightly winged in bud, 7–15 mm, lobes lanceolate, triangular, or ovate, 4.5–12 × 1.7–3(–4) mm, 2–3 times as long as wide, 2–3 times tube length, apex long-acuminate, densely stellate-hairy, hairs many-armed; petals pink, to 2 cm. **Mericarps** 2–3.2 mm. $2n = 34$.

Flowering Apr–Jun(–Aug). Chaparral, washes, hillsides; 400–1100 m; Calif.; Mexico (Baja California).

Malacothamnus marrubioides is reported from lower elevations of the Sierra Nevada, and is known otherwise from the Transverse Ranges of southern California and Coast Ranges of Baja California.

8. **Malacothamnus abbottii** (Eastwood) Kearney, Leafl. W. Bot. 6: 129. 1951 • Abbott's bushmallow C E

Malvastrum abbottii Eastwood, Leafl. W. Bot. 1: 215. 1936

Shrubs, to 1.5 m, branches slender, flexuous, indument complex: white, dense, sublepidote, hairs simple, fine, not shaggy, or stellate, sessile, many-armed. **Leaf blades** ovate, unlobed or 3-lobed, 3–6.5 cm, thin, surfaces: copiously white-stellate, basal sinus open, not overlapping. **Inflorescences** open-paniculate, flowers solitary or in pedunculate clusters, loose, not subtended by conspicuous bracts, flowers 3 or 4 per node; involucellar bractlets lanceolate, 5–8 × 1–1.5 mm, ½–¾ calyx length. **Flowers:** calyx slightly winged in bud, 9–11 mm, lobes ovate-acuminate, 6–7.5 × 2.5–3.5 mm, ca. 2 times as long as wide, 2–3 times tube length, apex acuminate, densely stellate; petals pale pink, 1.5–2 cm. **Mericarps** to 3 mm.

Flowering Oct. Stream banks, open chaparral; of conservation concern; 100–500 m; Calif.

Malacothamnus abbottii was once considered to be extinct. It is known from three localities in southern Monterey County.

9. **Malacothamnus aboriginum** (B. L. Robinson) Greene, Leafl. Bot. Observ. Crit. 1: 208. 1906 • Indian Valley bushmallow E

Malvastrum aboriginum B. L. Robinson in A. Gray et al., Syn. Fl. N. Amer. 1(1,2): 311. 1897; *Sphaeralcea aboriginum* (B. L. Robinson) Jepson

Subshrubs or shrubs, 2–3 m, branches ± stout, indument grayish to tawny, shaggy-tomentose, stellate hairs sessile or stalked, many-armed. **Leaf blades** ovate, broadly ovate, or round, 3- or 5-lobed, 4–7(–12) cm, thin or thick, surfaces: sparsely to densely grayish- to tawny-short-stellate-hairy, basal sinus open, not overlapping. **Inflorescences** interrupted, spicate or racemose, flower clusters sessile or subsessile, subtended by conspicuous bracts, usually densely flowered, usually 10 flowers per node; involucellar bractlets distinct or basally connate, subcordate, narrow-elliptic, or ± round, 6–15 × (1–)3–9 mm, ⅔ to exceeding calyx length. **Flowers:** calyx strongly plicate, angled or winged in bud, 8–17 mm, lobes subcordate, ovate, or ± round, 5–11 × 3.5–10 mm, slightly longer than broad, ca. 2 times tube length, apex abruptly acuminate, stellate-hairy, hairs many-armed; petals pale pink to rose, 1–2 cm. **Mericarps** 2.5–3.5 mm. $2n = 34$.

Flowering Apr–Jun, Aug–Oct. Chaparral; 100–800 (–1700) m; Calif.

Malacothamnus aboriginum occurs principally in the central inner Coast Ranges and in the Laguna Mountains in San Diego County. There, plants have involucellar bractlets about 1 mm wide but otherwise appear to be *M. aboriginum*; these have sometimes been assigned to *M. densiflorus*. The character of the often connate, wide involucellar bractlets is distinctive of *M. aboriginum*.

10. **Malacothamnus palmeri** (S. Watson) Greene, Leafl. Bot. Observ. Crit. 1: 208. 1906 • Palmer's or Santa Lucia bushmallow E F

Malvastrum palmeri S. Watson, Proc. Amer. Acad. Arts 12: 250. 1877; *Malacothamnus palmeri* var. *involucratus* (B. L. Robinson) Kearney; *M. palmeri* var. *lucianus* Kearney; *Malvastrum involucratum* B. L. Robinson; *M. palmeri* var. *involucratum* (B. L. Robinson) McMinn

Shrubs, 1.5–2.5 m, branches stout, indument white-canescent, grayish, or tawny, sparse to dense, stellate hairs sessile or stalked, glandular and 1–few-armed hairs sometimes also present, arms to 3 mm. **Leaf blades** ovate to broadly ovate, unlobed or 3- or 5-lobed, to 7–8 cm, thin or thick, surfaces: sparsely to densely grayish- to tawny-hairy, basal sinus open, not overlapping. **Inflorescences** terminal, headlike or, sometimes, short-spicate, flower clusters sessile, densely flowered, subtended by conspicuous bracts; stipules marginally connate, enclosing young flower clusters; involucellar bractlets distinct, sometimes basally adnate to calyx, linear to lanceolate, 8–16(–21) × 1–3(–5) mm, ⅔–1⅓ times calyx length. **Flowers:** calyx (8–)10–16(–20) mm, lobes ovate to deltate-lanceolate, 5–15 × 2.5–5.5 mm, ca. 2 times as long as wide, slightly exceeding to 4 times tube length, apex acuminate, grayish- to tawny-hairy, hairs stellate, arms to 3.5 mm and sometimes simple-glandular; petals rose-pink or pale pink to nearly white, 1.4–2(–3) cm. **Mericarps** 2.5–4(–5) mm. $2n = 34$.

Flowering Mar–Jul. Chaparral; 30–800 m; Calif.

Malacothamnus palmeri occurs in the central, outer Coast Ranges; its morphology is variable. Plants from Jolon and the Carmel Valley, Monterey County, have been called var. *involucratus*; they are recognized by cordate leaves, sparse indument (leaves adaxially glabrate), broad bracts to 20 mm wide subtending and partially enclosing the young flower clusters, relatively broad involucellar bractlets, calyx measurements in the lower range (to 1.5 cm), and nearly white petals 1–3 cm. They intergrade with the typical form and, in plants with short-spicate inflorescences, approach *M. aboriginum*. Plants from near Arroyo Seco, Monterey County, with simple and glandular hairs and generally darker rose petals have been named var. *lucianus*; they are otherwise similar to the typical variety.

35. MALVA Linnaeus, Sp. Pl. 2: 687. 1753; Gen. Pl. ed. 5, 308. 1754 • Mallow, mauve [Latin name derived from Greek *malacho*, to soften, alluding to emollient qualities of some species]

Steven R. Hill

Axolopha (de Candolle) Alefeld; *Bismalva* Medikus; *Saviniona* Webb & Berthelot

Herbs, annual, biennial, or perennial, **subshrubs, or shrubs,** glabrous or hairy, hairs stellate or simple. **Stems** erect, ascending, or trailing. **Leaves:** stipules persistent or deciduous, linear, lanceolate, triangular, or ovate to ± falcate; blade orbiculate or reniform, unlobed or palmately 3–7(–9)-lobed or divided, base cordate to truncate, margins crenate to dentate. **Inflorescences** usually axillary, flowers usually in fascicles, sometimes solitary, sometimes terminal racemes; involucel present, bractlets persistent, 3, distinct or basally connate. **Flowers:** calyx accrescent or not, not inflated, lobes reticulate-veined or not, not prominently ribbed, ovate or triangular; corolla rotate to campanulate, white, pink, lavender, rose, or dark purple, veins darker or not; staminal column included; ovary 6–15(–20)-carpellate; ovules 1 per carpel; style 6–15(–20)-branched (same number as locules); stigmas introrsely decurrent (on inside surface of style branches), filiform. **Fruits** schizocarps, erect (sometimes on sigmoid or recurved pedicel), not inflated, oblate-discoid, usually depressed in center, around broad axis, without persistent swollen style base, ± indurate, glabrous or hairy; mericarps 6–15(–20), drying tan or brown, 1-celled, wedge-shaped (triangular in cross section), oblong to reniform, beak or cusp absent, sides thin and papery or thicker, margins usually edged, apex rounded, indehiscent. **Seeds** 1 per mericarp, adherent to mericarp wall, usually not readily separated from it, reniform-rounded, notched, glabrous. $x = 21$.

Species 30–40 (11 in the flora): North America, Mexico, Eurasia, n Africa (especially Mediterranean region); introduced nearly worldwide.

Some species of *Malva* are weedy; five or six in the flora area generally occur in cultivation as ornamentals or as vegetables and occasionally escape. Some species previously treated within *Lavatera* (see M. F. Ray 1995, 1998) are here included in *Malva* based upon molecular evidence. Traditionally, *Lavatera* and *Malva* were separated by the presence of partially connate relatively wide involucellar bractlets in the former and distinct generally narrow bractlets in the latter. The annual species of *Althaea* (sect. *Hirsutae* Iljin ex Olyanitskaya & Tzvelev) may also belong within *Malva*, but have been kept separate here. Nomenclature in *Althaea*, *Lavatera*, and *Malva* is still in flux, and a satisfactory classification is not yet available. Intergeneric hybrids among some species of all three genera suggest a close relationship.

SELECTED REFERENCE Ray, M. F. 1995. Systematics of *Lavatera* and *Malva* (Malvaceae, Malveae)—A new perspective. Pl. Syst. Evol. 198: 29–53.

1. Shrubs, 1–4 m; petals 25–45 mm; calyces 12–15 mm; mericarps 6–10 . . . . .3. *Malva assurgentiflora*
1. Herbs or subshrubs, 0.2–3 m; petals 3–35(–45) mm (if subshrubs, petals to 20 mm); calyces 3–10(–15) mm; mericarps 6–15(–20).
  2. Distal leaf blades deeply (3–)5–7-lobed; petals 20–35 mm.
    3. Hairs stellate-canescent; involucellar bractlets ovate or ovate-deltate to obovate; mericarps 18–20, glabrous or sparsely hairy. . . . . 1. *Malva alcea*
    3. Hairs usually simple, sometimes stellate; involucellar bractlets linear to narrowly oblanceolate or elliptic; mericarps 11–15, densely hirsute apically . . . . 4. *Malva moschata*
  2. Distal leaf blades unlobed or shallowly lobed; petals 3–30(–45) mm.

[4. Shifted to left margin.—Ed.]

4. Petals (12–)16–30(–45) mm, length 2½–3(–4) times calyx; involucellar bractlets distinct, sometimes adnate to calyx in basal 1 mm . . . . 10. *Malva sylvestris*
4. Petals 3–15(–20) mm, length 1–2(–5) times calyx; involucellar bractlets distinct or partly connate and/or adnate to calyx (if petals longer than 14 mm, involucellar bractlets connate).
   5. Involucellar bractlets connate in proximal ⅓–½, adnate to calyx, ovate-deltate, ovate, or round; petals 10–20 mm; stems usually erect, rarely prostrate, 1–3 m.
      6. Involucel longer than calyx; leaf blade surfaces densely soft-stellate-hairy; petals rose to lavender with 5 darker veins; stem base usually woody. . . . . 2. *Malva arborea*
      6. Involucel shorter than calyx; leaf blade surfaces sparsely stellate-hairy; petals pale pink to white, usually with 3 darker veins; stem base not woody . . . . 8. *Malva pseudolavatera*
   5. Involucellar bractlets distinct, basally adnate to calyx (in *M. nicaeensis*) or not, filiform, linear, lanceolate, oblong-lanceolate, ovate, or obovate; petals 3–15 mm; stems erect or ascending to decumbent, procumbent, prostrate, or trailing, 0.2–0.8(–2.5) m.
      7. Involucellar bractlets basally adnate to calyx, broadly lanceolate to ovate or obovate; petals drying bluish, usually with darker veins . . . . 6. *Malva nicaeensis*
      7. Involucellar bractlets not adnate to calyx, filiform, linear, lanceolate, or oblong-lanceolate; petals drying pinkish or whitish, or faded, veins not much darker.
         8. Involucellar bractlets filiform to linear; calyces accrescent, lobes spreading outward exposing mericarps; petals 3–4.5(–5) mm, white to pale lilac; mericarp margins narrowly winged, toothed. . . . . .7. *Malva parviflora*
         8. Involucellar bractlets linear, oblanceolate, or lanceolate; calyces not accrescent, or, if so, lobes usually enclosing mericarps; petals (3–)5–13 mm, pale lilac, pink, pinkish, purplish, to nearly white or whitish; mericarp margins not winged, sometimes toothed.
           9. Stems erect; plants 0.5–2.5 m; leaf blades 3–10(–25) cm; pedicels stout and rigid in fruit . . . . 11. *Malva verticillata*
           9. Stems prostrate or trailing to ascending; plants usually 0.2–0.6 m; leaf blades 1–3.5(–6) cm; pedicels slender and flexible in fruit.
              10. Petals 6–13 mm, length 2 times calyx; mericarps hairy, smooth to slightly roughened or reticulate . . . . .5. *Malva neglecta*
              10. Petals 3–6 mm, length subequal to or slightly exceeding calyx; mericarps hairy or glabrate, strongly rugose-reticulate . . . . 9. *Malva pusilla*

1. **Malva alcea** Linnaeus, Sp. Pl. 2: 689. 1753
  • Hollyhock or Vervain mallow, mauve alcée

Malva alcea var. *fastigiata* (Cavanilles) K. Koch

Herbs, perennial, 0.3–1.3 m, usually stellate-canescent. **Stems** erect, sparsely hirsute proximally, stellate-hairy distally, hairs often pustulose. **Leaves:** stipules deciduous, lanceolate, slightly falcate, 5(–10) × 1–2.5 mm, ciliate; petioles of lower leaves 1.5–2.5 times as long as blade, reduced distally to ½ blade length, stellate-hairy; blade 2–8 × 2–8 cm, base cordate to somewhat truncate, those most distal sometimes wide-cuneate, surfaces stellate-hairy, proximal leaf blades cordate-orbiculate, margins crenate to dentate, shallowly lobed, apex rounded, distal leaf blades deeply (3–)5-lobed, lobe margins obtusely dentate or pinnatifid, apex narrowly acute. **Inflorescences** axillary, flowers solitary or distal flowers in racemes. **Pedicels** conspicuously jointed distally, 1.4–2 cm, not much longer in fruit; involucellar bractlets distinct, not adnate to calyx, ovate or ovate-deltate to obovate, narrowed to base, 5–8(–12) × 2.5(–5) mm, shorter than calyx, margins entire, surfaces stellate-hairy or glabrate. **Flowers:** calyx 9–12(–15) mm, lobes enclosing mericarps, stellate-hairy; petals usually bright pink, rarely white, 20–35 mm, length 2.5–3 times calyx; staminal column 9–10 mm, sparsely stellate-hairy; style 18–20-branched; stigmas 18–20. **Schizocarps** 4–8 mm diam.; mericarps 18–20, black, 2.4–2.8 mm, apical surface and margins rounded, smooth or faintly ridged, glabrous or sparsely hairy. **Seeds** brown, 2.5 mm. $2n = 84$.

Flowering Jun–Aug(–Sep). Disturbed areas, roadsides, old farm sites; 0–400 m; introduced; N.B., N.S., Ont., Que., Sask.; Conn., Idaho, Ind., Maine, Mass., Mich., N.H., N.J., N.Y., Ohio, Pa., R.I., Vt., Wash., Wis.; Europe; w Asia.

Malva alcea is found in most of Europe, but is rare in the Mediterranean region, and barely extends into Turkey in western Asia. The leaf shape, indument, and shape and size of the petals are variable, the most extreme forms having deeply 2-fid petals and deeply divided distal leaves with narrow, almost simple lobes. It occasionally hybridizes with *M. sylvestris* (*Malva* ×*egarensis* Cadevall) and *M. moschata* (*Malva* ×*intermedia* Boreau).

Malva alcea is sparingly naturalized in North America, primarily in New England and around the Great Lakes into eastern Canada; it is sometimes cultivated as an ornamental and naturalizes locally.

2. **Malva arborea**   (Linnaeus) Webb & Berthelot, Hist. Nat. Îles Canaries 3(2,1): 30. 1836 • Tree mallow   [I]

Lavatera arborea Linnaeus, Sp. Pl. 2: 690. 1753

Herbs, biennial or perennial, **or subshrubs,** 1–3 m, stellate-tomentose. **Stems** erect, base usually woody. **Leaves:** stipules deciduous, ovate, 4–5 × 2–3 mm, papery, apex acute to obtuse, sparsely stellate-hairy and ciliate; petiole longer than blade; blade rounded, shallowly and unequally 5–7(–9)-lobed (lobes obtuse), 5–20 × 5–20 cm, base cordate, margins crenate, apex obtuse to rounded, surfaces densely soft stellate-hairy especially abaxially. **Inflorescences** axillary, flowers in fascicles. **Pedicels** jointed distally, 0.5–1 cm, not much longer in fruit; involucellar bractlets connate in proximal ⅓, adnate to calyx, lobes broadly ovate to round, 8 × 5–6 mm, longer than calyx, margins entire, apex acute or obtuse, surfaces stellate-hairy. **Flowers:** calyx 3–4 mm, not much enlarged in fruit, densely stellate-canescent; petals rose to lavender with 5 darker veins, dark purple basally, 15–20 mm, length 4–5 times calyx, apex emarginate; staminal column 8–10 mm, glabrous proximally, stellate-hairy distally; anthers purplish; style (6–)8(or 9)-branched; stigmas (6–)8(9). **Schizocarps** 8–10 mm diam.; mericarps (6–)8(9), 4–5 mm, margins sharp-angled, apical surface and sides ridged, surfaces glabrous or hairy. **Seeds** dark brown, 3 mm. $2n$ = 36, 40, 42, 44.

Flowering Apr–May(–Sep). Disturbed areas, coastal bluffs, dunes; 0–200 m; introduced; Calif., Oreg.; Europe; introduced also in Mexico (Baja California), Africa (Libya), Atlantic Islands (Canary Islands), Pacific Islands (New Zealand), Australia.

Malva arborea is infrequently cultivated as a garden ornamental. It is traditionally placed in *Lavatera* and has three prominent, spreading, rounded, earlike involucellar bractlets and inconspicuous sepals.

3. **Malva assurgentiflora**   (Kellogg) M. F. Ray, Novon 8: 290. 1998 • Island mallow, malva rosa   [E][F][W]

Lavatera assurgentiflora Kellogg, Proc. Calif. Acad. Sci. 1: 14. 1854; *L. assurgentiflora* subsp. *glabra* Philbrick; *Saviniona assurgentiflora* (Kellogg) Greene; *S. clementina* Greene; *S. reticulata* Greene

Shrubs, 1–4 m, stellate-hairy to glabrate. **Stems** erect to decumbent, base woody. **Leaves:** stipules early-deciduous, lanceolate to ovate, 2–4 × 0.6–1.5 mm, minutely stellate-puberulent; petiole as long as or longer than blade; blade 5–15 × 5–15 cm, base cordate, surfaces glabrous or stellate-puberulent, upper leaf blades 5–7-lobed ca. ½ to base (maplelike) with deep sinus, lobes triangular-ovate, margins dentate or coarsely crenate, apex acute to obtuse. **Inflorescences** axillary, 1- or 2-flowered. **Pedicels** usually S-shaped, slender, 2–4 cm, to 6 cm in fruit; involucellar bractlets connate to ¼ length, base adnate to calyx, lobes lanceolate to wide-lanceolate, 6–8 × 3–4 mm, shorter than calyx, margins entire, surfaces canescent-puberulent. **Flowers:** calyx 12–15 mm, slightly enlarging in fruit, stellate-puberulent; petals reflexed or not with age, pale rose, rose, or purplish, sometimes white at base, veins darker, broadly obovate, 25–45 mm, length 3 times calyx; staminal column 15–20 mm, glabrous or sparsely hairy; free filaments 1–2 mm; anthers on distal ½; style 6–10-branched (same number as locules), purplish; stigmas 6–10 (same number as locules), purplish. **Schizocarps** 12–16 mm diam.; mericarps 6–10, 6–7 mm, apical face and margins sharp-edged, surfaces smooth to faintly ribbed, glabrous or puberulent on apical surface. **Seeds** dark brown, 4 mm, nearly as thick as long, notch slight. $2n$ = ca. 40.

Flowering Feb–Jun and Sep–Oct, sporadically year-round. Coastal bluffs, disturbed areas; 0–400 m; Calif.; introduced in Mexico, Central America (Guatemala), South America (Bolivia, Chile, Ecuador, Peru), Pacific Islands (New Zealand), Australia.

Malva assurgentiflora, traditionally placed in *Lavatera*, has long been cultivated as an ornamental or windbreak in California and is native only on the Channel Islands. It has become naturalized on the mainland as well as in Mexico and sparingly elsewhere. The shrubby habit, large flowers with dark-veined petals, and thick, hemispheric fruits make it distinctive; it is our only native species of *Malva*. The petals are often recurved with age, and the corky mericarps float and are tolerant of salt water. Further study may indicate that there are two distinct subspecies, as suggested by R. N. Philbrick (1980).

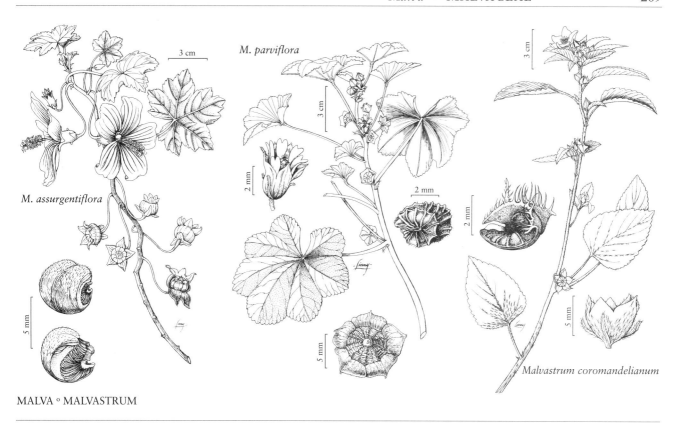

M. parviflora

M. assurgentiflora

Malvastrum coromandelianum

MALVA ° MALVASTRUM

4. Malva moschata Linnaeus, Sp. Pl. 2: 690. 1753

• Musk-mallow, mauve musquée ⊡ ⊡

Herbs, perennial, 0.3–1.3 m, hairs usually spreading, simple, sometimes stellate-hairy distally. **Stems** erect to ascending, sparsely hirsute proximally, stellate-hairy distally. **Leaves:** stipules persistent, linear to narrowly oblong-lanceolate, 3–8 × 2–3 mm; petioles of proximal leaf blades 3 times as long as blade, reduced to ½ blade length distally, mid-stem petioles 2 times as long as blade, hairs simple; distal blades usually round to reniform, deeply 5–7-lobed, lobes acutely 2-pinnatifid, 2–6 × 5–6 cm, base deeply cordate, margins irregularly toothed, apex rounded, obtuse, or acute, surfaces glabrous or sparsely hairy, hairs simple or stellate. **Inflorescences** axillary, flowers solitary or in fascicles, often appearing short-racemose or subumbellate terminally, long-stalked. **Pedicels** (0.5–)0.8–2.5 cm, to 10–35 cm in fruit, hairs simple; involucellar bractlets distinct, not adnate to calyx, linear to narrowly oblanceolate or elliptic, 5 × 1–1.5 mm, to 7–8 mm in fruit, length ½ calyx, margins entire, surfaces glabrous or sparsely hirsute and long-ciliate. **Flowers:** calyx reticulate-veined, 6–8 mm, to 15 mm in fruit, outer surface hairy, hairs both simple and stellate; petals bright pink to pale purple or white, 20–35 mm, length 2.5–3 times calyx; staminal column 7–8(–10) mm, glabrate; style 11–15-branched; stigmas 11–15. **Schizocarps** 9–11 mm diam.; mericarps 11–15, black, 1.5–2 mm, apical face and margins rounded, sides thin and papery, smooth, surfaces densely hirsute at least apically. **Seeds** 1.2–1.5 mm. $2n = 42$.

Flowering May–Oct. Disturbed areas, roadsides; 0–1300 m; introduced; B.C., Man., N.B., Nfld. and Labr. (Nfld.), N.S., Ont., P.E.I., Que.; Conn., Del., D.C., Idaho, Ill., Ind., Ky., Maine, Md., Mass., Mich., Mo., Mont., N.H., N.J., N.Y., N.C., Ohio, Oreg., Pa., R.I., Tenn., Vt., Va., Wash., W.Va., Wis., Wyo.; Europe; sw Asia (Turkey); n Africa; introduced also in South America (Chile), Australia.

Malva moschata is native from Spain to the British Isles, Poland, southern Russia, and Turkey. It has become naturalized in North America, especially in temperate northern and coastal areas. It is widely cultivated as an ornamental and frequently escapes. It occasionally hybridizes with *M. sylvestris* (*Malva* ×*inodora* Ponert) and *M. alcea* (*Malva* ×*intermedia* Boreau). It is similar to *M. alcea*, from which it can be distinguished by its narrower involucellar bractlets and densely hirsute mericarps.

5. **Malva neglecta** Wallroth in C. F. Hornschuch, Syll. Pl. Nov. 1: 140. 1824 • Common mallow, cheeses, mauve négligée [I] [W]

Herbs, annual, biennial, or perennial, 0.2–0.6 m, trailing stems sometimes to 1 m, usually sparsely stellate-hairy and with simple hairs. Stems usually prostrate to ascending, sometimes trailing, sparsely stellate-hairy with simple hairs persistent on older stems. Leaves: stipules persistent, narrowly triangular, 3–6 × 2.5 mm, papery; petiole usually 2–5 times as long as blade, gradually reduced distally; blade reniform to orbiculate-cordate, unlobed or very shallowly 5–7-lobed, 1.5–3.5(–6) × 1–4(–5) cm, base cordate, margins crenate-dentate, apex obtuse or rounded, surfaces glabrous or sparsely stellate-hairy. Inflorescences axillary, 2–6-flowered fascicles, long-stalked. Pedicels 1–5 cm, usually 10+ mm in fruit, several times longer than calyx, slender and flexible in fruit; involucellar bractlets distinct, not adnate to calyx, linear to narrowly oblong-lanceolate, 3–5(–6) × 1 mm, shorter than calyx, margins entire, surfaces sparsely stellate-puberulent and short-ciliate. Flowers: calyx 4–7 mm, slightly accrescent, to 8 mm in fruit, lobes enclosing mericarps, not veined, triangular-ovate, stellate-hairy, ciliate; petals pale lilac to whitish, drying pinkish or whitish, or faded, veins not darker, (6–)9–13 mm, length 2 times calyx, apex notched; staminal column 4–4.5 mm, retrorsely stellate-puberulent; style 12–15-branched; stigmas 12–15 (same number as locules), purple. Schizocarps 6 mm diam.; mericarps 12–15, 1.5–2 mm, apical face and margins rounded-angled, not winged or toothed, sides thin and papery, smooth to slightly roughened or reticulate, surfaces puberulent apically. Seeds 1–1.5 mm. $2n = 42$.

Flowering Apr–Oct. Disturbed areas, vacant lots, farm yards; 0–2700 m; introduced; Alta., B.C., Man., N.B., Nfld. and Labr. (Nfld.), N.S., N.W.T., Ont., P.E.I., Que., Sask.; Ala., Alaska, Ariz., Ark., Calif., Colo., Conn., Del., D.C., Ga., Idaho, Ill., Ind., Iowa, Kans., Ky., La., Maine, Md., Mass., Mich., Minn., Mo., Mont., Nebr., Nev., N.H., N.J., N.Mex., N.Y., N.C., N.Dak., Ohio, Okla., Oreg., Pa., R.I., S.C., S.Dak., Tenn., Tex., Utah, Vt., Va., Wash., W.Va., Wis., Wyo.; Europe; Asia; introduced also in Mexico (Chihuahua), West Indies (Dominican Republic), Central America (Panama), South America (Argentina, Brazil), s Asia (India, Pakistan), Africa, Pacific Islands (New Zealand), Australia.

Malva neglecta is the most commonly found mallow in most of North America. It has been introduced essentially worldwide in temperate areas and is usually considered a weed. In some older treatments, it was included within *M. rotundifolia* Linnaeus, a name rejected because of its inconsistent use for this species as well as for *M. pusilla* and other species. The immature, mucilaginous fruits are sometimes eaten; they have the appearance and texture of an old-fashioned wheel of cheese, hence one of the common names.

6. **Malva nicaeensis** Allioni, Fl. Pedem. 2: 40. 1785 • Bull mallow [I] [W]

Herbs, annual or biennial, 0.2–0.6 m, sparsely to densely hairy, hairs both simple and stellate. Stems procumbent or trailing to ascending, villous-hirsute. Leaves: stipules persistent, ovate to broadly ovate, 4–6 × 3–5 mm; petiole 2–5 times as long as blade; blade semicircular or reniform, 2–4 × 2–4 cm, sometimes to 12 cm in young plants, base cordate to nearly truncate, distalmost leaves sometimes wide-cuneate, margins crenate or dentate, undulate, or with 5–7 shallow lobes, apex obtuse, rounded, or acute, surfaces sparsely hairy, hairs simple. Inflorescences axillary, flowers solitary or in 2–6-flowered fascicles. Pedicels slender, subequal to calyx, 0.3–0.7 cm, increasing 2–5 times in fruit; involucellar bractlets distinct, basally adnate to calyx 1–1.5 mm, broadly lanceolate to ovate or obovate, 4–5 × 1.5 mm, subequal to or slightly longer than calyx, shorter in fruit, margins entire, ciliate. Flowers: calyx 4–6 mm, to 6–12 mm in fruit, lobes enclosing fruit, hirsute, hairs mostly unbranched, some forked or stellate; petals pink to lavender, drying bluish, usually with darker veins, 5–15 mm, subequal to or length slightly less than 2 times calyx, glabrous or nearly so; staminal column 2–2.5(–3) mm, densely, retrorsely puberulent, hairs simple; style 7–10-branched; stigmas 7–10. Schizocarps 6–7 mm diam.; mericarps 7–10, 3 mm, thick, as wide as long, margins sharp-angled but not winged, conspicuously, deeply reticulate-pitted apically, surfaces densely hirsute or glabrous. Seeds dark brown, 2–2.5 mm. $2n = 42$.

Flowering year-round. Disturbed sites; 0–400 (–1200) m; introduced; Ala., Calif.; s Europe; w Asia; n Africa; introduced also in Mexico, South America (Argentina, Chile), Atlantic Islands (Macaronesia), Pacific Islands (New Zealand), Australia.

Malva nicaeensis has been collected as a waif in Massachusetts and New Jersey and has been reported in British Columbia and Montana; vouchers have not been found. One vouchered collection has been reported from Mobile County, Alabama. In the flora area, it is found most commonly in the Mediterranean climate of California. It may not be established elsewhere within our range. In some older treatments it was identified as or included within *M. rotundifolia*, a name rejected because of its inconsistent use for this as well as for *M. pusilla* and other species. It is similar to *M. sylvestris*, except for its decumbent habit and smaller flowers.

7. Malva parviflora Linnaeus, Demonstr. Pl., 18. 1753

• Small-flowered mallow, mauve parviflore F I W

Herbs, annual, 0.2–0.8 m. **Stems** usually erect or ascending, rarely decumbent, wide-branched, glabrous or sparsely stellate-hairy distally. **Leaves:** stipules persistent, broadly lanceolate, 4–5 × 2–3 mm; petiole 2–3(–4) times as long as blade; blade suborbiculate-cordate or reniform, mostly shallowly 5–7-lobed or angled, 2–8(–10) × 2–8(–10) cm, base cordate (to nearly truncate), lobes deltate or rounded, margins evenly crenate, apex rounded to broadly acute, surfaces glabrous or hairy, especially at base, hairs simple and stellate. **Inflorescences** axillary, flowers solitary or in 2–4-flowered fascicles. **Pedicels** 0.2–0.4 cm, usually to 1 cm in fruit, shorter than calyx; involucellar bractlets distinct, not adnate to calyx, linear to filiform, (1–)2–3 × 0.3 mm, shorter than calyx, margins entire, surfaces glabrous or slightly ciliate. **Flowers:** calyx 3–4.5 mm, to 7–8 mm in fruit, glabrous or stellate-hairy, lobes wide-spreading outward in fruit, orbiculate-deltate, reticulate-veined, apex often abruptly acuminate, short-ciliate or not, scarious in fruit; petals white to pale lilac, drying pinkish or whitish, or faded, veins not darker, 3–4.5(–5) mm, subequal to or only slightly longer than calyx, glabrous; staminal column 1.5 mm, glabrous; style 10- or 11-branched; stigmas 10 or 11. **Schizocarps** 6–7 mm diam.; mericarps 10 or 11, 2–2.5 mm, apical face strongly reticulate-wrinkled, sides appearing strongly, radially ribbed, margins sharp-edged, toothed, narrowly winged, surface glabrous or hairy. **Seeds** 1.5–2 mm. $2n = 42$.

Flowering year-round. Disturbed, usually dry, warm sites; 0–2500 m; introduced; Alta., B.C., Man., N.B., Ont., P.E.I., Que., Sask.; Ariz., Calif., Colo., Fla., Ga., Idaho, Iowa, Kans., La., Md., Mass., Mo., Mont., Nebr., Nev., N.J., N.Mex., N.Y., N.Dak., Okla., Oreg., S.C., S.Dak., Tex., Utah, Wash., Wyo.; Eurasia (possibly as far east as India); n Africa; introduced also in Mexico, West Indies, Central America, South America, elsewhere in Africa, Pacific Islands, Australia.

Malva parviflora is native in southwestern Europe and the Mediterranean region to India; it is commonly introduced in many parts of the world. It is distinguished from similar species by its short petals (often equaling the calyx), the lack of darker lines on the petals, and the wide-spreading calyx lobes in fruit. The sharp-edged or winged mericarp with a conspicuously reticulate-pitted surface is likewise distinctive.

Malva parviflora is more heat-tolerant than most *Malva* species. It is especially common as a weed from California to Texas. Northern records should be checked because some may be based upon waifs and others may be based on misidentifications. In some older floras, *M. parviflora* was confused with *M. rotundifolia*, a name rejected because of its inconsistent use for this as well as for *M. pusilla* and other species. It is sometimes cultivated as a forage crop in semi-arid regions.

8. Malva pseudolavatera Webb & Berthelot, Hist. Nat. Îles Canaries 3(2,1): 29. 1836 (as pseudo lavatera)

• Cretan or smaller tree mallow I

Lavatera cretica Linnaeus, Sp. Pl. 2: 691. 1753, not *Malva cretica* Cavanilles 1786

Herbs, annual, biennial, or perennial, 1–3 m, sparsely stellate-hairy. **Stems** usually erect, rarely prostrate, base not woody. **Leaves:** stipules deciduous, broadly lanceolate to ovate, 4–4.5 × 2 mm, papery, sparsely puberulent, ciliate, hairs simple; petioles of proximal leaves to 3.5 times as long as blade, usually equaling blade on distal portion of stems, stellate-hairy to glabrate; blade suborbiculate, 4–10 × 4–10 cm, base cordate, surfaces sparsely stellate-hairy, proximal leaves often unlobed, distal blades 5-lobed, suborbiculate to ovate, base truncate to somewhat cordate, margins crenate, apex rounded to acute. **Inflorescences** axillary, loose, 3–6-flowered fascicles (condensed cymes). **Pedicels** 0.5–0.7 cm, to 0.6–1.2 cm in fruit, hairy, hairs stellate and/or simple; involucellar bractlets connate in proximal ⅓–½, adnate to calyx, forming a cup, 5–6 mm, to 7 mm in fruit, lobes ovate-deltate to round, 3–4 × 3–4 mm, equaling to shorter than calyx, margins entire, apex obtuse, surfaces stellate-hairy. **Flowers:** calyx (4–)7–8 mm, to 11 mm in fruit, densely soft stellate-hairy; petals pale pink to white, sometimes drying bluish, usually with 3 darker veins, 10–16 mm, length 2½–4 times calyx; staminal column 5 mm, densely stellate-hairy; style 7–10-branched; stigmas 7–10 (same number as locules). **Schizocarps** 9–12 mm diam., style base often expanded and disclike; mericarps 7–10, 4 × 4 mm, margins rounded, smooth, usually cross-ridged, surfaces usually glabrous. **Seeds** brown, 2.5–2.7 mm. $2n = 44, 112$.

Flowering mostly Apr–Jun. Disturbed areas, coastal bluffs, dunes; 0–800 m; introduced; Calif.; Europe; w Asia; n Africa; Atlantic Islands (Canary Islands); introduced also in South America (Ecuador), s Africa, Pacific Islands (New Zealand), Australia.

Malva pseudolavatera is naturalized in California along the coast and apparently is spreading. The name *M. multiflora* (Cavanilles) Soldano, Banfi & Golasso (based on *Malope multiflora* Cavanilles) has been proposed as the correct name.

9. **Malva pusilla** Smith in J. E. Smith et al., Engl. Bot. 4: plate 241. 1795 • Dwarf mallow, mauve à feuilles rondes [I]

Herbs, usually annual, rarely biennial or perennial, 0.2–0.6 m, trailing stems sometimes to 1 m, sparsely hairy, hairs usually simple and stellate. **Stems** trailing to ascending, hairs usually both simple and stellate. **Leaves:** stipules persistent, narrowly to ovate-triangular, 3–5(–6) × 2–3 mm; petiole on proximal and midstem leaves to 3 times as long as blade, usually 2 times longer on distal leaves, sparsely hairy, hairs simple and stellate; blade cordate to broadly reniform, unlobed or broadly and very shallowly 5-lobed, 1–5 × 1–5 cm, base cordate, margins finely crenate or dentate, apex rounded, surfaces glabrous or sparsely stellate-hairy. **Inflorescences** axillary, flowers solitary or in 2–10-flowered fascicles (compact cymes), obviously stalked. **Pedicels** 0.4–0.5 cm, to 1–2.4 cm in fruit, slender and flexible, sparsely hairy, hairs simple and stellate; involucellar bractlets distinct, not adnate to calyx, linear to lanceolate, 3–4 × 0.5 mm, shorter than calyx, margins entire, surfaces sparsely ciliate and stellate-puberulent. **Flowers:** calyx 3–5 mm, 5–6 mm in fruit, lobes incompletely enclosing mericarps, mostly hispid, hairs both simple and stellate, conspicuously long-ciliate, lobes not obviously veined; petals pale pink or nearly white, drying pinkish or whitish, or faded, veins not darker, 3–6 mm, subequal to slightly longer than calyx; staminal column 1–1.5 mm, glabrous; style 10- or 11-branched; stigmas 10 or 11. **Schizocarps** 5–7 mm diam.; mericarps 10 or 11, 1.5–2 mm, margins sharply angled, toothed, not winged, lateral faces radially veined, apical surface strongly rugose-reticulate ridged, surface glabrate or sparsely stellate-puberulent. **Seeds** 1–1.5 mm. $2n = 42$.

Flowering May–Oct. Disturbed sites; 100–1000 m; introduced; Alta., B.C., Man., N.B., N.S., Ont., P.E.I., Que., Sask.; Ill., Iowa, Kans., Mass., Minn., Mo., Mont., Nebr., N.Dak., Oreg., Pa., S.Dak., Va., Wis., Wyo.; Europe.

Malva pusilla appears to be more cold-tolerant than *M. neglecta* and is often confused with it and tends to replace it in the north. In most older treatments, it was included within *M. rotundifolia*, a name rejected because of its inconsistent use for this as well as for *M. neglecta* and other species. Reports of *M. pusilla* in older floras of California and some southern states probably are based on misidentifications. It is found only occasionally, but the species is probably more widely distributed in the flora area than the records indicate.

10. **Malva sylvestris** Linnaeus, Sp. Pl. 2: 689. 1753 • High or garden mallow, mauve des bois [I] [W]

Malva mauritiana Linnaeus; *M. sylvestris* var. *mauritiana* (Linnaeus) Boissier

Herbs, annual, biennial, or perennial, 0.5–1.5(–3) m, hairy to glabrate, hairs simple and stellate. **Stems** erect or ascending, glabrous or sparsely hairy, hairs both simple and stellate. **Leaves:** stipules persistent, lanceolate to ovate-triangular, 3–5(–8) × 3 mm; petiole ½–1½ times as long as blade, pubescent in adaxial groove, otherwise glabrous; blade reniform to suborbiculate-cordate, unlobed or shallowly 3–7-lobed, sinuses to ½ to base, (2–)5–10(–14) × (2–)5–10(–14) cm, base cordate to ± truncate, lobes semicircular to oblong, margins crenate, apex rounded to wide-acute, surfaces glabrous or sparsely hairy, hairs simple or stellate. **Inflorescences** axillary, flowers solitary or in 2–4-flowered fascicles, long-stalked. **Pedicels** 1–2.5 cm, 1–4.5 cm in fruit, much longer than calyx, glabrous or sparsely stellate-hairy; involucellar bractlets distinct, sometimes adnate to calyx in basal 1 mm, oblong-lanceolate to elliptic or narrowly obovate, reticulate-veined, (3–)4–5(–7) × 2.5–4 mm, shorter than calyx, margins entire, surfaces glabrous, sparsely ciliate. **Flowers:** calyx reticulate-veined, 5–6 mm, to 10 mm in fruit, lobes incompletely enclosing mericarps, stellate-puberulent; petals pink to purple or reddish purple with darker veins, usually drying blue, (12–)16–30(–45) mm, length 2½–3(–4) times calyx; staminal column 5 mm, minutely, retrorsely stellate-puberulent; style 10–12-branched; stigmas 10–12. **Schizocarps** 7 mm diam.; mericarps 10–12, 2–2.5 mm, margins sharp, not winged, sides thin and papery, with radiating veins, surface strongly to obscurely reticulate-wrinkled, usually glabrous, sometimes sparsely hairy. **Seeds** 1.5–2.2 mm. $2n = 42$.

Flowering mostly Apr–Oct. Disturbed areas, roadsides, farm yards, pastures; 0–1000 m; introduced; Alta., B.C., Man., N.B., N.S., Ont., Que., Sask.; Calif., Conn., Del., D.C., Idaho, Ill., Ind., Iowa, Kans., Ky., La., Maine, Md., Mass., Mich., Minn., Mo., Mont., Nebr., N.H., N.J., N.Y., N.C., N.Dak., Ohio, Oreg., Pa., R.I., S.C., S.Dak., Tenn., Tex., Utah, Vt., Va., Wash., W.Va., Wis., Wyo.; Europe; temperate Asia; n Africa; introduced also in Mexico, Central America (Guatemala, Honduras), South America (Argentina, Brazil, Colombia, Ecuador, Peru, Venezuela), Pacific Islands (New Zealand), Australia.

Malva sylvestris is native throughout Europe except in the extreme north and into temperate Asia and northern Africa and is widely cultivated for food and ornament. It is the most commonly cultivated *Malva* in most of the United States. It is not very tolerant of hot, arid conditions. It is variable in habit, indument, leaf shape, and corolla size and color; most variants originated as selections/cultivars. The flowers and fruits indicate its close relationship with *M. nicaeensis*; the upright habit and much larger flowers allow an easy distinction.

11. Malva verticillata Linnaeus, Sp. Pl. 2: 689. 1753
• Whorled or clustered or Chinese or curled mallow, mauve verticillée [I]

Malva crispa (Linnaeus) Linnaeus; *M. verticillata* var. *chinensis* (Miller) S. Y. Hu; *M. verticillata* var. *crispa* Linnaeus

Herbs, annual, 0.5–2.5 m, glabrous or hairy, hairs usually stellate. **Stems** erect, usually stellate-hairy. **Leaves:** stipules persistent, ovate-triangular, slightly falcate, 4–7 × (2–)4–5 mm, usually papery; petiole shorter than to equaling or longer than blade, surfaces often glabrate abaxially, stellate-hairy adaxially; blade round to broadly reniform, unlobed or broadly, shallowly 5–7-lobed, 3–10(–25) × (2–)5–10(–25) cm, base cordate or sometimes ± truncate, lobes rounded, margins finely dentate to doubly crenate, often strongly and irregularly crisped-undulate (ruffled) to plane, apex rounded to acute, surfaces sparsely to moderately stellate-hairy. **Inflorescences** axillary, flowers solitary or in 2–8-flowered fascicles (compact cymes), subsessile. **Pedicels** 0.2–0.3 cm, to 1–1.4(–4) cm in fruit, usually shorter than calyx, stellate-puberulent, rather stout and rigid in fruit; involucellar bractlets distinct, not adnate to calyx, linear to lanceolate, reticulate-veined, 2–3 × 0.5 mm, to 6 × 1 mm in fruit, shorter than calyx, thin and translucent, margins entire, sparsely stellate-puberulent and ciliate. **Flowers:** calyx reticulate-veined, 4–6 mm, to 10 mm in fruit, lobes enclosing mericarps, papery, stellate-puberulent; petals pale lilac to whitish with lavender-pink tips, drying pinkish or whitish, or faded, veins not darker, 5–8 mm, subequal to or longer than calyx; staminal column 2 mm, usually sparsely hairy or glabrous, hairs minute; style 8–11-branched; stigmas 8–11 (same number as locules). **Schizocarps** 7–9 mm diam.; mericarps 8–11, 2.5–3 mm, glabrous, apical face smooth to obscurely reticulate at margins, not toothed or winged, lateral faces radially ribbed, very thin over seed. **Seeds** 2–2.5 mm. $2n = 84, 112$.

Flowering (May–)Jul–Oct. Disturbed areas, old gardens, roadsides; 0–3000 m; introduced; Alta., B.C., Man., N.B., N.S., Ont., P.E.I., Que., Sask.; Ariz., Calif., Colo., Conn., Del., D.C., Ill., Iowa, Maine, Md., Mass., Mich., Minn., Mo., Nebr., N.H., N.J., N.Mex., N.Y., N.Dak., Ohio, Pa., R.I., Utah, Vt., Wash., W.Va., Wis., Wyo.; Eurasia; n Africa; introduced also in South America (Peru) and in temperate regions worldwide.

Malva verticillata is commonly grown as a vegetable and medicinal herb. Variety *verticillata* is apparently native in eastern Asia, particularly China, and has a long history of cultivation there for use as a cooked vegetable and salad plant, and in traditional medicine. It differs from var. *crispa* Linnaeus in its flatter leaf blades and generally shorter stems. It has only rarely been found naturalized in North America; its garden use is increasing and it is becoming more frequently planted by eastern Asian immigrants. Variety *crispa* (with conspicuously undulate/ruffled/crisped leaves) is also grown as a vegetable and the leaves are sometimes used in salads. It is more commonly found as an escape or naturalized plant than the typical variety and it is sometimes treated as a species; it appears to be a selection derived from var. *verticillata* and is unknown in the wild. Variety *crispa* is more widely cultivated and naturalized in Europe and is naturalized in Asia.

36. MALVASTRUM A. Gray, Mem. Amer. Acad. Arts, n. s. 4: 21. 1849, name conserved

• False mallow [Genus *Malva* and Latin *-astrum*, incomplete resemblance]

Steven R. Hill

Sidopsis Rydberg

Herbs, annual or perennial, **or subshrubs,** hairy, hairs closely appressed or tufted, sometimes pustular-based, (2–)4–10[–12]-rayed, stellate, sometimes bilateral, infrequently sublepidote or simple. **Stems** erect or ascending to decumbent. **Leaves:** stipules persistent [deciduous], lanceolate to linear [wide-ovate], usually subfalcate or falcate; blade wide-ovate to lanceolate, unlobed or sometimes obscurely 3-lobed, base rounded, slightly cordate, nearly truncate, to cuneate, margins crenate-dentate to dentate-serrate or denticulate. **Inflorescences** axillary solitary flowers, terminal racemes or spikes in distal ½ of plant; involucel present, bractlets persistent, 3, distinct, free or adnate basally to calyx. **Flowers:** calyx not inflated (slightly so in *M. hispidum*), somewhat accrescent, lobes 3–5-ribbed, deltate to narrowly triangular; corolla campanulate to wide-spreading, yellow to yellow-orange; staminal column included; ovary (5–)8–18-carpellate; ovules 1 per cell; style 5–18-branched (equal in number to locules); stigmas capitate. **Fruits** schizocarps, erect, not inflated, oblate-discoid, usually depressed in center, somewhat indurate at maturity; mericarps (5–)8–18, drying tan or brown, without dorsal spurs or with 1–3 apical (dorsal) spurs (mucros or cusps) 0.1–2.3 mm, sparsely to densely hairy, rarely glabrous, indehiscent or rarely dehiscent (in *M. hispidum*). **Seeds** 1 per mericarp, glabrous. *x* = 6.

Species 15 (6 in the flora): United States, Mexico, West Indies, Central America, South America, ne Australia.

Two or three species of *Malvastrum* have been widely introduced worldwide in tropical and warm-temperate regions; none is usually cultivated; several are considered to be quite weedy. At least one species (*M. coromandelianum*) has some medicinal use.

SELECTED REFERENCE　Hill, S. R. 1982. A monograph of the genus *Malvastrum* A. Gray (Malvaceae: Malveae). Rhodora 84: 1–83, 159–264, 317–409.

1. Mericarps unornamented or with 1 small apical mucro/cusp to 0.2 mm.
　2. Mericarps dehiscent, separating into 2 valves, edges of fruit rounded, mucro/cusp absent; inflorescences axillary solitary flowers; leaf blades lanceolate to linear-lanceolate, at least 3 times longer than wide; temperate c United States . . . . 6. *Malvastrum hispidum*
　2. Mericarps indehiscent, not separating into valves, edges of fruit angled, minute mucro/cusp usually present; inflorescences dense, terminal spikes, sometimes axillary solitary flowers below spikes; leaf blades ovate-lanceolate to ovate, sometimes narrower toward branch tips, usually 1–2 times longer than wide; warm-temperate to subtropical areas.
　　3. Leaves and stems with usually dense, 5–12-rayed, often tufted, stellate hairs; inflorescences dense terminal spikes 3–10 cm . . . . 1. *Malvastrum americanum*
　　3. Leaves and stems with sparse, 3–6-rayed, appressed, bilateral and stellate hairs; inflorescences solitary flowers, axillary or congested or loose terminal spikes 1–2 cm. . . . . 4. *Malvastrum corchorifolium* (in part)
1. Mericarps with 1–3 apical cusps 0.1–2.3 mm.
　4. Mericarps 2-cusped, cusps conspicuous, distal (pointing away from fruit axis); stems erect, branched usually in distal ½; stem hairs not lepidote, 4–8-rayed, radially symmetric; filament tubes puberulent . . . . 3. *Malvastrum bicuspidatum*
　4. Mericarps 3-cusped, cusps conspicuous or minute, 1 halfway between proximal and distal ends of top surface, 2 distal; stems erect or ascending to decumbent, branched in proximal ½; stem hairs sublepidote and 6–10-rayed, or appressed and 2–4-rayed, stellate hairs sometimes mixed with simple hairs, 4-rayed hairs distinctly bilateral; filament tubes puberulent or glabrous.
　　5. Mericarp mucros/cusps 0.1–0.4 mm; 2-fid floral bracts usually present; filament tubes sparsely puberulent . . . . 4. *Malvastrum corchorifolium* (in part)
　　5. Mericarp mucros/cusps 0.5–2.3 mm; 2-fid floral bracts absent; filament tubes glabrous.
　　　6. Stem hairs sublepidote (stellate-lepidote), 6–10-rayed; mericarps with 1 prominent medial-apical cusp 1.5–2.3 mm and 2 contiguous, flattened, obtuse cusps 1 mm at distal margins . . . . 2. *Malvastrum aurantiacum*
　　　6. Stem hairs not sublepidote, 2–4-rayed, bilateral hairs; mericarps with 1 prominent medial-apical cusp 1–2 mm and 2 divergent distal-apical cusps 0.3–1 mm. . . . . 5. *Malvastrum coromandelianum*

1. **Malvastrum americanum** (Linnaeus) Torrey in W. H. Emory, Rep. U.S. Mex. Bound. 2(1): 38. 1859 • American false mallow

Malva americana Linnaeus, Sp. Pl. 2: 687. 1753; *M. spicata* Linnaeus; *Malvastrum spicatum* (Linnaeus) A. Gray

Herbs, perennial, **or subshrubs,** (0.5–)1–2 m, often bushy-branched in distal ½. **Stems** erect, canescent, hairs tufted (not appressed), 6–8-rayed, infrequently glabrate. **Leaves:** stipules persistent, lanceolate, subfalcate, 3–5 × 1 mm, apex acuminate; petioles 35–80 mm on proximal leaves, reduced to 10–15 mm on distal leaves and usually on xerophytes; blade wide-ovate to ovate-lanceolate, very shallowly 3-lobed in distal ½ or unlobed (in most plants in the flora area), varying from 5–12 × 4–10 cm on proximal leaves to 2–4 × 1.5–3 cm on distal leaves, usually 1–2 times longer than wide, 2 times longer than petiole of proximal leaves to 3–5 times longer on distal leaves, base slightly cordate or rounded to truncate or cuneate, margins dentate to denticulate, apex acute, surfaces stellate-hairy, hairs (5)6–12-rayed. **Inflorescences:** first 1 or 2 flowers solitary, axillary, remainder in dense terminal spikes 3–10 cm, these terminating each branch; floral bracts 2-fid, 4–5 × 2 mm. **Pedicels** 0.1–3 mm, not lengthening in fruit; involucellar bractlets adnate basally to calyx for 1.5–2 mm, lanceolate, subfalcate, 5–7 × 0.8–1.5 mm, equaling to barely exceeding calyx lobes, apex acute to acuminate. **Flowers:** calyx connate ¼–⅓ its length, broadly campanulate, 5–6 mm, to 6–10 mm in fruit, surface densely hirsute, hairs scattered, appressed, apically directed, 1–1.5 mm, mixed with minute, closely appressed, 5–8-rayed, stellate hairs; corolla wide-spreading, orange-yellow, 12–17 mm diam., petals obovate, shortly asymmetrically lobed, 6–10 × 4–6 mm, exceeding calyx by 2–3 mm; staminal column 2–3 mm, stellate-puberulent; style (9)10–15 (–18)-branched. **Schizocarps** 4–6 mm diam.; mericarps tardily shed from calyx, (9)10–15(–18), 1.5–3 × 1.5–2 × 0.8 mm, margins angled, sides radially ribbed, narrowly notched, with 1 minute, proximal-apical mucro to 0.1 mm, minutely hirsute, hairs ascending, restricted to top, simple, 0.1–0.5 mm. **Seeds** 1.5 mm. 2*n* = 24.

Flowering nearly year-round when sufficiently wet and warm; frost-sensitive. Open, usually secondary and disturbed habitats, near coast; 0–100 m; Fla., Tex.; Mexico; West Indies; Central America; South America (to Argentina); introduced in Asia (China), Africa (Cape Verde Island), Pacific Islands (Indonesia), Australia.

Plants of *Malvastrum americanum* from within the flora area tend to be shorter, and to have smaller, narrower unlobed leaves than those of the wet Tropics. The calyces and mericarps can attach easily to clothing and fur. The species is more cold-sensitive than *M. coromandelianum* and is not as widespread or weedy.

2. **Malvastrum aurantiacum** (Scheele) Walpers, Ann. Bot. Syst. 2: 153. 1851 • Texas false mallow [E]

Malva aurantiaca Scheele, Linnaea 21: 469. 1848; *Malvastrum wrightii* A. Gray

Subshrubs, 0.4–1 m, sparsely branched in proximal ½. **Stems** usually procumbent and ascending, hairs appressed, (6–)8–10-rayed, sublepidote. **Leaves:** stipules persistent, lanceolate, subfalcate, 4–6 × 1–1.5 mm, apex acute; petiole 10–25 mm; blade wide-ovate to ovate, unlobed or rarely with 2 obscure lateral lobes halfway from base, (1.8–)3–4(–5.5) × (1–)1.5–3(–4.5) cm, usually slightly (ca. 1.2 times) longer than wide, 2 times longer than petiole, base cordate to rounded or truncate, margins crenate-dentate, apex rounded, surfaces hairy, hairs scattered, appressed, (6–)8–10-rayed, sublepidote. **Inflorescences** axillary solitary flowers; floral bracts absent. **Pedicels** 3–5(–10) mm in flower, to 2 mm in fruit; involucellar bractlets basally adnate to calyx for 2–3 mm, deltate-cordiform in distal 4–6 mm, abruptly narrowed, 8–10 × 3–5 mm, shorter than calyx lobes, apex acute. **Flowers:** calyx connate ⅓–½ its length, broadly campanulate, lobes slightly auriculate at base, 11–12 mm, to 13–17 mm in fruit, surface hairy, hairs scattered, appressed, stellate-lepidote; corolla wide-spreading, golden yellow to pale orange-yellow, 30 mm diam., petals obovate, conspicuously asymmetrically lobed, 12–16 × 10–15 mm, exceeding calyx by 7–8 mm; staminal column 4 mm, glabrous; style 12–16-branched. **Schizocarps** 9–10 mm diam.; mericarps readily shed from calyx, 12–16, 4–6 × 4–6 × 1–1.5 mm, widely notched, with 1 conspicuous, distally-directed, medial-apical cusp 1.5–2.3 mm and 2 distal-apical, contiguous, flattened, obtuse cusps 1 mm, surface sparsely hairy, hairs both erect and appressed, rigid, simple on apex and minute, stellate on dorsal ⅓, sides smooth. **Seeds** 2–3 mm. 2*n* = 36.

Flowering spring–summer, into fall when sufficiently wet and warm. Restricted to heavy clay soil of river floodplains in prairies, especially in the Texas Coastal Bend; 0–300 m; Tex.

Malvastrum aurantiacum is easily distinguished from the other species of the genus in the flora area by its wide, auriculate involucellar bractlets and its raspberry-red fruits when fresh, these later drying brown.

3. **Malvastrum bicuspidatum** (S. Watson) Rose, Contr. U.S. Natl. Herb. 12: 286. 1909 • Sonoran Desert false mallow

Malvastrum tricuspidatum A. Gray var. *bicuspidatum* S. Watson, Proc. Amer. Acad. Arts 21: 417. 1886

Herbs, perennial, **or subshrubs,** (0.4–)1–1.5 m, sparsely horizontally branched, usually with 1 main stem and usually unbranched secondary branches in distal ½. **Stems** erect, canescent, hairs radially symmetric, appressed, not lepidote, with both larger (4)5-rayed hairs and minute 4–8-rayed stellate hairs, sometimes shed in age. **Leaves:** stipules persistent, narrowly lanceolate, subfalcate, 5–6 × 1 mm, apex acuminate; petiole 10–25(–70) mm; blade wide-ovate to ovate-lanceolate to nearly lanceolate on distal stem, unlobed, usually 3–5 × 1.5–3 cm, 1.3–1.7 times longer than wide, 2.5–3.5 times longer than petiole, base usually rounded to wide-cuneate, margins dentate to sharply dentate, apex acute to acuminate, surfaces stellate-hairy, hairs 4–8-rayed, adaxial surface also with simple hairs. **Inflorescences** axillary solitary flowers at first, later elongated terminal racemes, flowers sometimes appearing clustered on reduced racemes, these terminating each branch; floral bracts 2-fid, 4 × 2 mm. **Pedicels** 0.5–2 mm, increasing by 0.5–1 mm in fruit; involucellar bractlets adnate basally to calyx for 1–2 mm, narrowly lanceolate to lanceolate, subfalcate, 4–6 × 0.5–1 mm, ½ length of calyx lobes, apex acute. **Flowers:** calyx connate ¼–½ its length, rotate, 6–8 mm, to 7–9 mm in fruit, surface hairy, hairs sparse, tufted, stellate, 2–6-rayed, and few, simple, marginal hairs mixed with minute, dense, stellate, 5–8-rayed hairs; corolla wide-spreading, yellow to orange-yellow, 15–18 mm diam., petals obovate, slightly asymmetrically lobed, 6–8 × 7 mm, exceeding calyx by 3–4 mm; staminal column 2–3 mm, stellate-puberulent; style 9–12-branched. **Schizocarps** 5.5–8 mm diam.; mericarps tardily shed from calyx, 9–12, 3–4 × 1.8–2 × 1 mm, conspicuously notched, without proximal cusp or with small mucro to 0.1 mm and with 2 conspicuous divergent distal-apical cusps (0.5–)1–1.5 mm, dorsal surface hirsute, hairs simple, rigid, 0.1–1 mm, mixed with simple and stellate, smaller hairs, lateral faces ± glabrous, conspicuously ribbed and thickened. **Seeds** 1.5–2 mm. $2n = 24$.

Flowering early–mid spring after winter rains and again in late summer–fall after rains. On slopes in arid to semi-arid regions, primarily in dry shrublands in the Sonoran Desert; 30–1300 m; Ariz.; Mexico (Baja California Sur, Chihuahua, Durango, Sinaloa, Sonora).

Subspecies *bicuspidatum* is the northernmost representative of *Malvastrum bicuspidatum*, in sect. *Tomentosum* S. R. Hill, a primarily Central and South American group of five similar species found mainly in arid regions locally from Arizona south to Paraguay and Brazil; the other three subspecies, subsp. *campanulatum* S. R. Hill, subsp. *oaxacanum* S. R. Hill, and subsp. *tumidum* S. R. Hill, are found in central and southern Mexico and differ by their generally shorter cusps, thinner-walled, non-ribbed and even somewhat leathery mericarps, and usually by more congested racemes.

4. **Malvastrum corchorifolium** (Desrousseaux) Britton ex Small, Fl. Miami, 200. 1913 • Caribbean false mallow

Malva corchorifolia Desrousseaux in J. Lamarck et al., Encycl. 3: 755 [as 743]. 1792, based on *M. scoparia* Jacquin, Collectanea 1: 59. 1787, not L'Héritier 1786; *Malvastrum rugelii* S. Watson

Herbs, annual or perennial, suffruticose in age, 0.6–1.5 m, sparsely branched in proximal ½, usually with 1 main stem. **Stems** erect, hairs scattered, appressed, distinctly bilateral, 4-rayed, not sublepidote, swollen-based, or similar 3–5-rayed stellate hairs 0.8–1.5 mm. **Leaves:** stipules persistent, lanceolate to narrowly lanceolate, subfalcate, 3–5 × 0.8–1.2 mm, apex acuminate; petiole 15–35(–70) mm; blade ovate, unlobed, (1.6–)3–5(–8) × (0.8–)2.5–4(–7.5) cm, usually 1–2 times longer than wide, 1.8–2.2 times longer than petiole, base broadly cuneate to rounded, margins dentate, apex acute to rounded, surfaces sparsely stellate-hairy, hairs 4 or 5(6)-rayed, and few, minute, marginal, simple hairs. **Inflorescences** axillary, solitary flowers at first, later congested or loose terminal spikes 1–2 cm, these in distal leaf axils or terminating each branch; floral bracts usually 2-fid, 3–6 × 1 mm, or flowers subtended by leaf and stipules. **Pedicels** 0.5–2 mm; involucellar bractlets basally adnate to calyx for 0.5–1 mm, lanceolate, subfalcate, 4–6 × 0.8–1 mm, ± equaling calyx lobes, apex acuminate. **Flowers:** calyx connate for ¼–⅓ its length, broadly campanulate, 5–6 mm, to 7–11 mm in fruit, surface moderately hairy, hairs simple, 2–4-rayed, mixed with scattered, 4–6-rayed, stellate, minute hairs; corolla campanulate to wide-spreading, yellow to pale yellow-orange, 12–17 mm diam., petals obovate, asymmetrically lobed, 6–7 × 3–4 mm, exceeding calyx by 2 mm; staminal column 2–2.5 mm, sparsely stellate-puberulent; style (9–)11–13(–16)-branched. **Schizocarps** 4–7 mm diam.; mericarps tardily shed from calyx, (9–)11–13(–16), 2.5–3 × 2–2.5 × 1.1 mm, margins angled, sides radially ribbed, narrowly-notched, with 3 minute, apical cusps 0.1–0.4 mm, 1 at proximal-apical surface, 2 at distal-apical surface, moderately hairy on dorsal ⅓, hairs erect, minute, simple or 2-rayed, and erect, simple, rigid hairs

0.5–1 mm, minutely hirsute with ascending, simple hairs 0.1–0.5 mm mixed with minute, simple or 2- or 3-rayed, stellate hairs. **Seeds** 1.5–1.7 mm. $2n = 48$.

Flowering nearly year-round when sufficiently wet and warm. Open, usually coastal, calcareous soil; 0–20 m; Ala., Fla.; Mexico; West Indies; Central America; introduced in Africa (Ghana).

Malvastrum corchorifolium is most likely a stabilized allopolyploid hybrid between *M. americanum* and *M. coromandelianum*, and it appears to have originated and stabilized at least twice, in eastern Central America (Honduras or Nicaragua) and somewhere in the West Indies (the two groups are morphologically distinguishable). Plants in the flora area have typical West Indian morphology, as do those from Ghana, the latter possibly introduced during the time of the slave trade. The Alabama record was historic only; the specimen was collected on ship ballast brought in from Jamaica.

5. **Malvastrum coromandelianum** (Linnaeus) Garcke, Bonplandia (Hanover) 5: 295. 1857 • Common false mallow [F]

Malva coromandeliana Linnaeus, Sp. Pl. 2: 687. 1753; *Malvastrum lindheimerianum* (Scheele) Walpers

Herbs, annual or perennial, 0.2–0.6(–1) m, with 1 main stem, freely branching in proximal ½. **Stems** erect or decumbent, hairs scattered, appressed, bilateral, (2–)4-rayed, swollen-based, not sublepidote, hairs 1–3 mm. **Leaves:** stipules persistent, lanceolate, subfalcate to falcate, 3–6 × 0.5–1 mm, apex acuminate; petiole 10–20(–40) mm; blade ovate to ± lanceolate, unlobed, (1.7–)3–4(–6.5) × (0.6–)1.5–3 (–5.5) cm, 1.1–2.8 times longer than wide, 2.5–4.5 times longer than petiole, not greatly reduced on stem distally, base truncate to broadly-rounded to often wide-cuneate, margins dentate to serrate, apex acute, surfaces sparsely hairy, hairs bilateral, 2–4-rayed, stellate or with simple hairs on adaxial surface. **Inflorescences** axillary, solitary flowers, flowers sometimes congested towards branch tips; floral bracts absent. **Pedicels** 1–2 mm, to 3–5 mm in fruit; involucellar bractlets basally adnate to calyx for 0.5–1 mm, lanceolate, subfalcate, 4–6 × 0.6–1 mm, shorter than calyx lobes, apex acute. **Flowers:** calyx connate ½ its length, broadly campanulate, 5–7 mm, to 8–11 mm in fruit, surface sparsely hairy, hairs primarily bilateral, appressed, (3)4-rayed, stellate; corolla wide-spreading, golden yellow to orange-yellow, 19–21 mm diam., petals obovate, asymmetrically 2-lobed, 8–10 × 6–7 mm, exceeding calyx by 2–3 mm; staminal column 2–2.5 mm, glabrous; style (9)10–14(15)-branched. **Schizocarps** 6–7 mm

diam.; mericarps very readily shed from calyx, (9)10–14 (15), 3.5–4 × 3–3.5 × 1–2 mm, narrowly notched, with 3 conspicuous, apical cusps 0.5–2 mm, 1 at medial-apical surface 1–2 mm, 2 at distal-apical surface 0.3–1 mm, these divergent, moderately hairy dorsally, hairs erect, simple, 0.6–1 mm, firm, mixed with few, minute, 2–5-rayed, stellate or simple hairs, lateral sides ± glabrous, radially ribbed. **Seeds** 1.7–1.9 mm. $2n = 24$.

Flowering spring–frost at northern limit as an annual (cold-sensitive), nearly year-round when sufficiently wet and warm as a perennial. River floodplains and banks, disturbed areas, often in alkaline soil; 0–100 m; Fla., La., Tex.; Mexico; West Indies; Central America; South America (to Argentina); introduced worldwide from Tropics and subtropics to warm temperate zones.

Malvastrum coromandelianum is a widespread weed and the most common species in the genus; it is apparently native from Texas to Argentina. The introduced and widespread form has simple hairs on the adaxial surface of the leaf, while the native form has stellate hairs on that surface. Both forms are found in Texas. The species historically has been introduced in ballast in Alabama, Massachusetts, New Jersey, and Pennsylvania, but did not persist.

Subspecies *coromandelianum* occurs in the flora area and is a widespread weed in tropical and warm-temperate areas worldwide; the other two subspecies occur only in South America and on the Galapagos Islands.

6. **Malvastrum hispidum** (Pursh) Hochreutiner, Annuaire Conserv. Jard. Bot. Genève 20: 129. 1917 • Hispid false mallow [E]

Sida hispida Pursh, Fl. Amer. Sept. 2: 452. 1813; *Malvastrum angustum* A. Gray; *Sidopsis hispida* (Pursh) Rydberg; *Sphaeralcea angusta* (A. Gray) Fernald

Herbs, annual, (0.05–)0.1–0.5 (–0.8) m, bushy-branched. **Stems** erect, sparsely or moderately hairy, hairs appressed, 4(–6)-rayed, normally bilateral and swollen-based. **Leaves:** stipules persistent, linear, 3–5 × 0.3 mm, apex acuminate; petiole 5–10 mm; blade lanceolate to linear-lanceolate, unlobed, 2.5–3.5 × 0.5–1 cm, not much reduced distally, at least 3 times longer than wide, 3–4 times longer than petiole, base rounded to cuneate, margins sparsely serrate-dentate, apex acute, surfaces stellate-hairy, hairs scattered, appressed, (2–)4–6(–8)-rayed, reduced to simple hairs at margins. **Inflorescences** axillary solitary flowers in distal ½ of plant; floral bracts absent. **Pedicels** 2–4(–10) mm, to 5–8(–17) mm in fruit; involucellar bractlets free from or basally adnate to calyx for 0.5–1 mm, narrowly lanceolate to linear, subfalcate, 3–7 × 0.3–0.6 mm,

½ times length of calyx lobes, apex acuminate. **Flowers:** calyx connate ⅖–½ its length, broadly campanulate, distinctly plicate-angular due to prolonged reniform-auriculate lobe bases, 5–7 mm, accrescent to 8–10 mm, somewhat inflated in fruit, surface hairy, hairs both 4–6-rayed, stellate and simple to 2-rayed marginal; corolla wide-spreading, yellow to pale yellow-orange, 5 mm diam., petals asymmetrically obovate, shallowly lobed, 3–4.5 × 2–3.5 mm, subequal to or exceeding calyx by 0.5–1 mm; staminal column 1–1.6 mm, glabrous; style 5- or 6-branched. **Schizocarps** 6–8 mm diam.; mericarps readily shed from calyx, 5 or 6, 2.6–3.5 × 2.6–3.5 × 1.5 mm, dehiscent, separating into 2 valves, short-notched, mucro absent, hairy, hairs both minute, simple and longer erect, simple, to 1 mm on proximal-apical surface, sides rounded, smooth and unribbed. **Seeds** 2–2.5 mm. $2n = 36$.

Flowering summer; frost-sensitive. Open limestone and dolomite outcrops, cedar glades, prairies, shallow soil; 50–300 m; Ala., Ark., Ill., Iowa, Kans., Ky., Mo., Okla., Tenn., Va.

Malvastrum hispidum is local in distribution. The seeds are very thick-walled and persist in the seed bank for many years, but the plants survive only in areas with little other vegetation and will disappear when there is much competition from other species. It appears to benefit from occasional fires. While it is clearly related to the other species of the genus, its dehiscent mericarps, narrow leaves, distribution, and strictly annual duration easily distinguish it. Open flowers are only infrequently seen; most appear to be cleistogamous. Its origins and exact relationships are somewhat obscure.

Collections of the species from Iowa are historic; no recent ones are known from there. It is endangered in Illinois and Virginia. There are very few recent collections from Kentucky. Its status in Arkansas, Kansas, and Oklahoma is uncertain. In Alabama it is quite rare and occurs only in a few cedar glades near the Tennessee border. The species appears to be most secure in Missouri and Tennessee.

37. **MALVAVISCUS** Fabricius, Enum., 155. 1759 • Turk's-cap [Latin, *malva*, mallow, and *viscidus*, sticky, alluding to sap]

Meghan G. Mendenhall

Paul A. Fryxell|

Shrubs [trees], in *M. arboreus* var. *drummondii* forming dense clones propagated by root proliferation. **Stems** erect, hairy or glabrous, not viscid. **Leaves:** stipules deciduous, linear or subulate; blade elliptic to broadly ovate, unlobed or 3–5-lobed, base rounded to cordate, margins serrate or crenate to subentire. **Inflorescences** axillary solitary flowers, rarely apical 2–5-flowered cymes; involucel present, bractlets persistent, usually 8 or 9, distinct. **Flowers:** calyx not accrescent, not inflated, lobes unribbed, lanceolate or deltate; corolla tubular, usually red [yellow]; staminal column exserted [included]; styles 10-branched, 2 per carpel; stigmas capitate. **Fruits** schizocarpic berries, erect [pendent], not inflated, oblate, fleshy, glabrous; mericarps 5, 1-celled, without dorsal spur, indehiscent. **Seeds** 1 per mericarp, ellipsoid wedge-shaped, glabrous. $x = 14$.

Species 5 (1 in the flora): sc, se United States, Mexico, West Indies, Central America, South America.

Malvaviscus penduliflorus de Candolle is found in gardens throughout the Old and New World Tropics and subtropics. This species appears to be a sterile hybrid; it never fruits and is not known in the wild.

Malvaviscus is readily distinguished by its auriculate petals and red, edible, baccate fruits. This fruit type is very rare in the Malvoideae. The genera thought to be most closely related to *Malvaviscus* are *Anotea* (de Candolle) Ulbrich and *Pavonia*. Of the two, only *Pavonia*, with dry fruits, occurs within the flora area. Morphological variation within *Malvaviscus* has led to proposals of more than 50 specific names. Much of the variation is due to the existence of populational forms and their intergradation. Modern treatments recognize from one to a dozen species.

Only *Malvaviscus arboreus* var. *drummondii* is common in the flora area; two other taxa may also be encountered. Variety *drummondii* has uniformly three-lobed, broad, cordate leaves and is the only taxon in *Malvaviscus* to form clones through root proliferation and to occur in a fully temperate climate. *Malvaviscus arboreus* var. *arboreus* and *M. penduliflorus* de Candolle have leaves that vary from unlobed to shallowly lobed and are usually much longer than wide and have rounded bases. In the flora area, var. *arboreus* (including *M. arboreus* var. *mexicanus*) occurs rarely in the Rio Grande valley of southern Texas and is also infrequently cultivated from Texas to Florida. *Malvaviscus penduliflorus* is frequently cultivated in the Rio Grande plains of Texas and eastward to Florida and throughout the Tropics; it is not known to occur in the wild. Although plants may persist after cultivation, this species does not truly escape domestication in the flora area or elsewhere. Plants of *M. penduliflorus* are more robust than those of *M. arboreus* and have spectacular, pendulous flowers up to 7 cm long. Unlike other species of *Malvaviscus*, the stamens are not fully exserted although the styles and stigmas generally are. Individuals are not known to set fruit and are propagated by cuttings. *Malvaviscus penduliflorus* may be found where there is evidence of a homestead, such as remnants of a building foundation.

SELECTED REFERENCES Schery, R. W. 1942. Monograph of *Malvaviscus*. Ann. Missouri Bot. Gard. 29: 183–244. Turner, B. L. and M. G. Mendenhall. 1993. Revision of *Malvaviscus* (Malvaceae). Ann. Missouri Bot. Gard. 80: 439–457.

1. **Malvaviscus arboreus** Cavanilles, Diss. 3: 131. 1787 [F]

Hibiscus malvaviscus Linnaeus, Sp. Pl. 2: 694. 1753

Plants 0.5–3[–10] m. **Leaf blades** elliptic to broadly ovate, 4–20[–25] × 3–12 cm, surfaces sparsely to densely hairy with simple and stellate trichomes. **Involucellar bractlets** linear-spatulate. **Flowers** ascending or erect; calyx persistent, lobes connate for ½–⅔ their lengths, 8–15 mm, glabrous or hirsute; petals imbricate at anthesis, asymmetrically obovate-cuneate and auriculate toward base, 1.5–4(–5) cm; staminal column with 5 apical teeth; stigmas exserted. **Fruits** berrylike, usually red, not winged, 8–13[–16] × 10–17 mm, fleshy, edible; mericarps broadly ellipsoid wedge-shaped, smooth, glabrous. **Seeds** reniform, ¾ as wide as long. $2n = 56$.

Varieties 2 (2 in the flora): sc, se United States, Mexico, West Indies, Central America, n South America (Colombia).

1. Leaf blades usually unlobed, longer than wide, apex acute, base rounded (to subcordate); stems stellate-hairy to glabrate (glabrous) . . . . 1a. *Malvaviscus arboreus* var. *arboreus*
1. Leaf blades deeply 3-lobed, at least as long as wide, apex obtuse (acute), base usually strongly cordate; stems glabrous proximally, densely and minutely tomentose distally . . . . 1b. *Malvaviscus arboreus* var. *drummondii*

1a. **Malvaviscus arboreus** Cavanilles var. **arboreus** [F] [I]

Malvaviscus arboreus var. *mexicanus* Schlechtendal; *M. grandiflorus* Kunth

Plants not clone-forming, 1–3[–10] m, subglabrous to densely hairy. **Stems** stellate-hairy to glabrate (glabrous). **Leaf blades** usually unlobed, sometimes shallowly 3–5-lobed, elliptic to ovate, 4–20[–25] × 3–12 cm, often longer than wide, base rounded (to subcordate), margins serrate to subentire, apex acute. **Petals** 1.5–4(–5) cm. $2n$ = ca. 28, 56.

Flowering summer–fall. Palm forests and thorn scrub, sometimes cultivated; 0–100 m; introduced; Tex.; Mexico; West Indies; Central America; n South America (Colombia).

Variety *arboreus* is infrequently cultivated in the Rio Grande Valley and eastward to Florida. It does not appear to be native within the flora area, and the Texas record was likely an escape. Individuals may escape and persist.

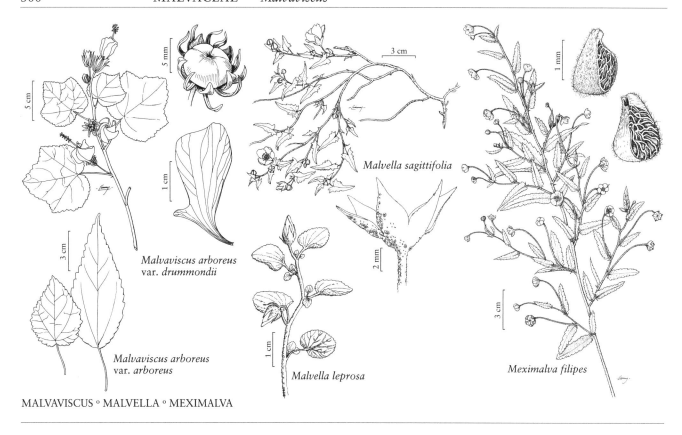

Malvaviscus arboreus
var. *drummondii*

Malvaviscus arboreus
var. *arboreus*

Malvella sagittifolia

Malvella leprosa

Meximalva filipes

MALVAVISCUS ∘ MALVELLA ∘ MEXIMALVA

1b. Malvaviscus arboreus Cavanilles var. **drummondii**
(Torrey & A. Gray) Schery, Ann. Missouri Bot.
Gard. 29: 215. 1942 • Texas mallow, Turk's cap,
Drummond waxmallow [F]

Malvaviscus drummondii Torrey &
A. Gray, Fl. N. Amer. 1: 230.
1838; *Pavonia drummondii* Torrey
& A. Gray

Plants clone-forming, 0.5–1.5(–3)
m, softly hairy. **Stems** glabrous
proximally, densely and minutely
tomentose distally. **Leaf blades**
deeply 3-lobed, broadly ovate,
4–9 × 4–12 cm, at least as long as wide, base usually
strongly cordate, margins crenate-dentate, apex obtuse
(acute). **Petals** 1.5–3.5 cm. **2***n* = 28.

Flowering summer–fall. Limestone slopes and
ledges, wooded arroyos, along streams in shaded areas;
0 200 m; Ala., Ark., Fla., Ga., La., Miss., Tex.; Mexico
(Coahuila, Tamaulipas).

It is questionable whether var. *drummondii* is native
to states east of Texas. The remarkably uniform
populations that grow in central Texas suggest that
the variety is native there. The taxon apparently was,
early on, taken into cultivation in coastal southeastern
United States, where it has escaped and persisted.
Variety *drummondii* is also grown in California, North
Carolina, and South Carolina but has not naturalized.

38. MALVELLA Jaubert & Spach, Ill. Pl. Orient. 5: 47, plate 444. 1855 • Alkali mallow

[Generic name *Malva* and Latin *-ella*, diminutive]

Paul A. Fryxell†

Steven R. Hill

Disella Greene

Herbs, perennial. **Stems** prostrate, with stellate and lepidote pubescence, not viscid. **Leaves** alternate and distichous; stipules persistent, subulate, 3–5 mm; blade asymmetric basally, [± flabellate], reniform, or ovate to triangular, not dissected or parted, base asymmetric, margins dentate, serrate, or entire, surfaces stellate or lepidote. **Inflorescences** axillary solitary flowers; involucel present or absent, bractlets 0–3, persistent or caducous, distinct. **Flowers:** calyx not accrescent, not inflated, lobes unribbed, ovate or cordate, stellate or lepidote; corolla whitish or pale yellow, sometimes with pink flush or fading pale rose; staminal column included; style 7–10-branched; stigmas capitate. **Fruits** schizocarps, erect, not inflated, oblate, not indurate, minutely hairy; mericarps 7–10, 1-celled, dorsally rounded, essentially unornamented, lateral walls persistent, essentially indehiscent. **Seeds** 1 per mericarp, glabrous. *x* = probably 16.

Species 4 (3 in the flora): United States, Mexico, South America (Peru to Uruguay), Europe, Asia (Asia Minor).

1. Leaf blades wider than long, ± reniform, apex obtuse or subacute, indument predominantly stellate; involucellar bractlets usually 3, sometimes 0; calyx lobes ovate, bases not overlapping . . . . .2. *Malvella leprosa*
1. Leaf blades longer than wide, ovate to triangular, apex acute, indument ± silvery-lepidote; involucellar bractlets usually 0, rarely present; calyx lobes ± cordate, bases plicate-overlapping.
    2. Leaf blades ± triangular, length 1–2(–3) times width, margins irregularly dentate, surfaces often with at least some stellate hairs mixed with silvery scales . . . . 1. *Malvella lepidota*
    2. Leaf blades narrowly triangular, length 3–5(–6) times width, margins hastate-toothed at base, surfaces silvery-lepidote . . . . 3. *Malvella sagittifolia*

1. Malvella lepidota (A. Gray) Fryxell, SouthW. Naturalist 19: 101. 1974

Sida lepidota A. Gray, Smithsonian Contr. Knowl. 3(5): 18. 1852; *Disella lepidota* (A. Gray) Greene; *S. lepidota* var. *depauperata* A. Gray

Stems prostrate, trailing, densely hairy with mixture of intergrading stellate and silvery-lepidote hairs. **Leaves:** petiole ½–2 times as long as blade; blade ± triangular to somewhat ovate, mostly 1–2 cm, usually 1–2(–3) times longer than wide, base truncate or cuneate, margins irregularly dentate, apex acute, surfaces densely hairy, hairs predominantly stellate abaxially, predominantly silvery-lepidote adaxially. **Pedicels** long, subequal to subtending leaves; involucellar bractlets usually 0, sometimes 3, filiform. **Flowers:** calyx 6–8 mm, silvery-lepidote, lobes cordate-ovate, bases plicate, apex acuminate; petals whitish or pale yellow, sometimes fading rose, asymmetric, 10–15 mm; stamens pallid, glabrous, staminal column antheriferous at apex; style ca. 7-branched, pallid, glabrous. **Schizocarps** 5–6 mm diam.

Flowering year-round. Heavy, saline soil on mud flats, lake shores; to 1500 m; Ariz., Nev., N.Mex., Tex.; Mexico (Chihuahua, Coahuila, Durango, San Luis Potosí, Sonora, Zacatecas).

Malvella lepidota is somewhat intermediate between the other two species in our range, having both stellate hairs and lepidote scales, and sometimes having an involucel. It is less common and less weedy than *M. leprosa*.

2. **Malvella leprosa**   (Ortega) Krapovickas, Bonplandia (Corrientes) 3: 59. 1970 • Dollar-weed, oreja de ratón, scurfy sida   F   W

Malva leprosa Ortega, Nov. Pl. Descr. Dec. 8: 95. 1798; *Disella hederacea* (Douglas) Greene; *M. hederacea* Douglas; *M. sulphurea* Gillies ex Hooker & Arnott; *Malvastrum sulphureum* (Gillies ex Hooker & Arnott) Grisebach; *Sida hederacea* (Douglas) Torrey ex A. Gray; *S. leprosa* (Ortega) K. Schumann; *S. sulphurea* (Gillies ex Hooker & Arnott) A. Gray

Stems prostrate, indument mixture of stellate and appressed, sublepidote hairs. **Leaves:** petiole ½–1 times as long as blade; blade ± reniform, 1–3.5 cm, wider than long, base obliquely truncate, margins serrate, apex obtuse or subacute, surfaces densely hairy, hairs appressed, sublepidote and stellate. **Pedicels** long, subequal to subtending petiole; involucellar bractlets 3, inconspicuous, filiform, or 0. **Flowers:** calyx 8–10 mm, with mixture of stellate and appressed sublepidote hairs, lobes ovate, bases not plicate-overlapping, apex acuminate; petals pale yellow, sometimes with rose flush on fading, asymmetric, 12–15 mm; stamens pallid, glabrous, staminal column antheriferous at apex; style 7–10-branched, pallid, glabrous. **Schizocarps** 7 mm diam. *2n* = 22, 32.

Flowering year-round in warmer areas. Heavy, saline soil; 800–1500 m; Ariz., Calif., Colo., Idaho, Kans., Nev., N.Mex., Okla., Oreg., Tex., Utah, Wash.; Mexico (Baja California, Chihuahua, Coahuila, Durango, San Luis Potosí, Sinaloa, Sonora); South America (Argentina, Chile, Peru, Uruguay).

Malvella leprosa is possibly introduced in Colorado. The species is considered to be a noxious weed in Arizona and California and increasing as saline soils increase.

3. **Malvella sagittifolia**   (A. Gray) Fryxell, SouthW. Naturalist 19: 102. 1974   F

Sida lepidota A. Gray var. *sagittifolia* A. Gray, Smithsonian Contr. Knowl. 3(5): 18. 1852 (as sagittaefolia); *Disella sagittifolia* (A. Gray) Greene; *S. sagittifolia* (A. Gray) Rydberg

Stems prostrate, trailing, invested with silvery-white, lepidote scales. **Leaves:** petiole to ⅓ as long as blade; blade narrowly triangular, 1.5–3.5 cm, usually 3–5(–6) times as long as wide, base truncate, margins entire with 2–4 hastate teeth at base, apex acute, surfaces with sparse, silvery-lepidote scales. **Pedicels** long, usually shorter than subtending leaves, involucellar bractlets 0. **Flowers:** calyx 7–9 mm, silvery-lepidote, lobes cordate-ovate, bases plicate-overlapping, apex acuminate; petals whitish or pale yellow, sometimes fading rose, asymmetric, 15 mm; stamens pallid, glabrous, staminal column antheriferous at apex; style 7- or 8-branched, pallid, glabrous. **Schizocarps** 5–6 mm diam.

Flowering year-round in warmer areas. Heavy, saline soil, mud flats, along lake shores; 30–1500 m; Ariz., Colo., N.Mex., Tex.; Mexico (Chihuahua, Coahuila, Durango, San Luis Potosí, Sonora, Tamaulipas).

Malvella sagittifolia is known from Otero County, Colorado; Dona Ana County, New Mexico; in scattered locations in Texas but primarily in the Big Bend region; and in much of central and southern Arizona. The comparatively long leaves with hastate teeth, the lack of any stellate hairs, and the consistent lack of an involucel distinguishes this species.

39. **MEXIMALVA**   Fryxell, Bol. Soc. Bot. México 35: 30, fig. 2. 1975 • [Country name Mexico and Latin *malva*, mallow]

Paul A. Fryxell†

Steven R. Hill

Subshrubs. Stems erect, stellate-tomentose. **Leaves** distichous and alternate; stipules persistent, linear; blade narrowly ovate or oblong-lanceolate, unlobed, base cordate, margins serrate, surfaces stellate-hairy. **Inflorescences** axillary, solitary flowers; involucel absent. **Flowers:** calyx not accrescent, not inflated, lobes unribbed, wide-ovate; corolla rotate, dark purple [lavender or purple]; staminal column included; style 7- or 8-branched; stigmas capitate. **Fruits** schizocarps, erect, sometimes nodding, not inflated, oblate, indurate, puberulent; mericarps 7 or 8, 1-celled, without dorsal spur, lateral walls persistent, prominently reticulate, partially dehiscent apically between 2 empty, doubled-walled "horns." **Seeds** 1 per mericarp, glabrous (subglabrous) or with a few stellate hairs. $x = 8$.

Species 2 (1 in the flora): Texas, n, c Mexico.

The second species, *Meximalva venusta* (Schlechtendal) Fryxell, is known only from Mexico.

1. **Meximalva filipes** (A. Gray) Fryxell, Bol. Soc. Bot. México 35: 32. 1976 F

Sida filipes A. Gray, Boston J. Nat. Hist. 6: 164. 1850

Plants 1–1.5 m. **Stems** distichously branched, stellate-hairy, hairs ferruginous, 1 mm diam. **Leaves:** stipules equaling or exceeding petioles; petiole 1–3(–5) mm; blade discolorous, 2–5(–8) cm, apex acute, surfaces densely stellate-hairy abaxially, sparsely hairy adaxially, hairs simple. **Pedicels** capillary, often equaling or exceeding subtending leaves. **Flowers:** buds nodding; calyx 3–4 mm, rounded basally, surfaces ferruginous stellate-hairy; petals 4–6 mm, glabrous throughout (including claw); staminal column purplish, 2–2.5 mm, apically antheriferous; anthers yellowish; styles purple, glabrous. **Schizocarps** 5 mm diam.; mericarps 2.5 mm. **Seeds** 1.7 mm, subglabrous. $2n = 16$.

Flowering spring–summer. Arid shrublands; 50–500 m; Tex.; Mexico (Coahuila, Nuevo León, San Luis Potosí, Tamaulipas).

Meximalva filipes is very reminiscent of the widespread *Sidastrum paniculatum*, differing from that primarily by its solitary, axillary flowers each subtended by a reduced leaf and its non-reflexed petals. In the flora area, it is known from scattered areas of central, southern, and southwestern Texas.

40. **MODIOLA** Moench, Methodus, 619. 1794 • [Latin *modiolus*, wheel hub, alluding to fruit shape] I

Steven R. Hill

Abutilodes Kuntze; *Diadesma* Rafinesque; *Modanthos* Alefeld

Herbs, perennial, usually glabrate. **Stems** procumbent, not viscid, sometimes with few simple hairs on herbage. **Leaves:** stipules persistent, ovate; blade orbiculate, usually palmately 5–7-parted or -lobed, early-season and proximal leaves more shallowly lobed, base shallowly to deeply cordate, margins dentate. **Inflorescences** axillary, solitary flowers; involucel present, bractlets persistent, 3, distinct. **Flowers:** calyx slightly accrescent, not inflated, lobes not ribbed, narrowly triangular; corolla wide-campanulate, salmon-orange, often with darker center, drying brick red; staminal column included; ovary 16–22-carpellate; ovules 1 per cell; style 16–22-branched; stigmas capitate. **Fruits** schizocarps, erect, not inflated, flattened disc indented in center, not indurate, setose; mericarps 16–22, 2-celled, dorsally setose, laterally glabrous, with 2 apical spurs/spines, chamber divided by endoglossum or partial septum, lower cell strongly rugose, indehiscent, upper cell smoother, dehiscent. **Seeds** 1 per cell, rounded, with proximal notch, sparsely, minutely hairy. $x = 9$.

Species 1: introduced; Mexico, West Indies, Central America, South America (to n Argentina), Europe.

Modiola is undoubtedly adventive over much of its range and possibly native only in northern Argentina and the Paraná basin of South America.

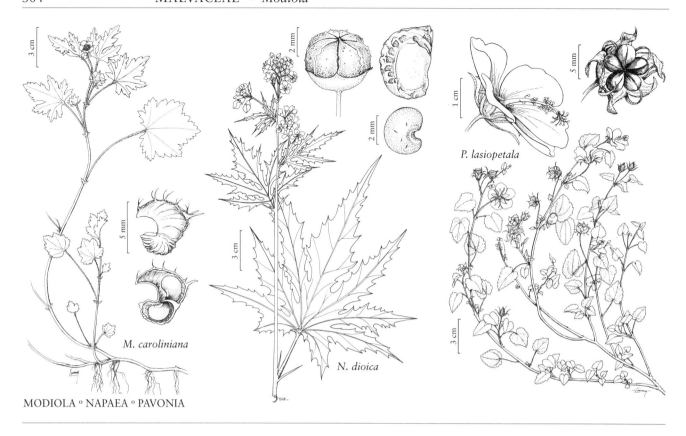

M. caroliniana

N. dioica

P. lasiopetala

MODIOLA ∘ NAPAEA ∘ PAVONIA

1. Modiola caroliniana (Linnaeus) G. Don, Gen. Hist. 1: 466. 1831 • Carolina bristlemallow [F] [I] [W]

Malva caroliniana Linnaeus, Sp. Pl. 2: 688. 1753; *Modiola prostrata* (Cavanilles) A. St.-Hilaire; *M. reptans* A. St.-Hilaire; *M. urticifolia* (Kunth) G. Don

Stems: flowering apices often ascending, branched, usually 0.2–0.5 m, often rooting at nodes. **Leaves:** stipules 3–4 × 1.5–3 mm; petiole length 1–2 times blade; blade 1.5–4 × 1.5–4 cm. **Pedicels** usually shorter than subtending petioles, hairy; involucellar bractlets lanceolate, 4–5 mm. **Flowers:** calyx 5–7 mm, hairy, hairs simple, 1–2 mm; corolla erect, 6–8 mm; staminal column yellowish; anthers crowded at apex; stigmas equaling number of locules. **Mericarps** drying black, 5–6 mm, apical spines 1.5–3 mm. **Seeds** 1.5 mm. *2n* = 18.

Flowering Mar–Nov. Disturbed, usually moist habitats, shores of ponds and reservoirs, low sandy areas, lawns, roadsides; 0–400 m; introduced; Ala., Ariz., Ark., Calif., Del., Fla., Ga., Ky., La., Mass., Miss., N.C., Okla., Oreg., Pa., S.C., Tenn., Tex., Va.; South America; introduced also in Mexico, Central America, Pacific Islands (Hawaii).

Modiola caroliniana is somewhat weedy but not a serious agricultural weed. It has been reported in Delaware, Massachusetts, New Jersey, and Pennsylvania as a waif but doubtfully persists that far north. It is well established in the southeastern United States and is rather common as a lawn weed in some locations and as a garden weed in California. It probably came from southern South America in wool or cotton. Its closest relative, *Modiolastrum* K. Schumann, is known from southern South America.

41. NAPAEA Linnaeus, Sp. Pl. 2: 686. 1753; Gen. Pl. ed. 5, 307. 1754 • [Greek *napaea*, wood nymph, alluding to woodland habitat) E

John W. Thieret†

Herbs, perennial. **Stems** erect, unbranched below inflorescence, simple- and/or stellate-hairy. **Leaves:** stipules persistent, broadly lanceolate; blade ± reniform to orbiculate, palmately lobed, base truncate to cordate, margins coarsely toothed. **Inflorescences** terminal panicles, bracteate; involucel absent. **Flowers** unisexual, staminate and pistillate on different plants (plants dioecious); calyx not accrescent, not inflated, lobes unribbed, triangular; corolla white; staminal column ± included; style 6–10-branched; stigmas introrsely decurrent, linear or filiform. **Fruits** schizocarpic, erect, not inflated; mericarps 6–10, reniform, 1-celled, apex apiculate (sometimes minutely so), indehiscent or tardily dehiscent. **Seeds** 1 per mericarp, glabrous. $x = 15$.

Species 1: nc United States.

H. H. Iltis and S. Kawano (1964) suggested that *Napaea* may have originated as an allopolyploid.

1. **Napaea dioica** Linnaeus, Sp. Pl. 2: 686. 1753
   • Glade mallow E F

Plants subscapose, 1–2.2(–3) m, from stout taproot. **Leaves** at base of plant largest and with longest petioles, progressively smaller and shorter-petioled distally, passing into bracts of inflorescence; petiole to 1.2 m (at base of plant); blade deeply 5–9(–11)-lobed, 5–50 × 7–65 cm, mostly somewhat wider than long, lobe apices acute or, especially on distal leaves, acuminate, surfaces simple- or stellate-hairy (especially abaxially). **Flowers:** calyx lobes 1.5–2 mm, shorter than tube; petals 5–10 mm; staminate flowers with 16–20 stamens, carpels 0; pistillate flowers with column of abortive stamens.

Mericarps 4–5 mm, rugose abaxially, especially near margins, and often laterally when mature, glabrous. **Seeds** 3–4 mm. $2n = 30$.

Flowering early–mid summer(–early fall). Stream banks, flood plains, meadows and thickets, roadsides, alkaline soil; 200–400 m; Ill., Ind., Iowa, Minn., Ohio, Wis.

Napaea dioica is the only dioecious member of Malvoideae native in the Western Hemisphere. Reports of glade mallow in areas outside its natural range derive from confusion between it and *Sida hermaphrodita* (for example, Virginia; H. H. Iltis 1963) or from garden escapes (for example, Vermont; J. P. Brown 1941).

The degree, distribution, and type of indument are variable. Forma *stellata* Fassett includes plants with mostly stellate hairs.

Napaea dioica is in the Center for Plant Conservation's National Collection of Endangered Plants.

42. PAVONIA Cavanilles, Diss. 2[app.]: [v]. 1786, name conserved • [For José Antonio Pavón, 1754–1844, Spanish physician and botanist]

Paul A. Fryxell†

Steven R. Hill

Lebretonia Schrank; *Malache* B. Vogel, name rejected

[Herbs] subshrubs or shrubs, [perennial]. **Stems** usually erect [prostrate], often stellate-hairy, sometimes glabrate, not [sometimes] viscid. **Leaves:** stipules usually persistent (early-deciduous in *P. paludicola*), subulate to filiform; blade usually symmetric, ovate-triangular to hastate-oblong or ovate, not [sometimes] dissected or parted, base truncate to cordate, margins dentate or crenate to subentire. **Inflorescences** terminal racemes [panicles, capitula], or axillary solitary flowers; involucel present, bractlets persistent, [4]5–8[–18+], distinct. **Flowers:** calyx persistent, not splitting symmetrically, not spathaceous, not or scarcely accrescent, not inflated [somewhat inflated], lobes ribbed or unribbed, usually lanceolate-ovate; corolla rotate to tubular, lavender to pink or yellow [white, purple], petal bases auriculate or not; staminal column usually included [exserted]; style 10-branched, (2 per carpel); stigmas 10, capitate. **Fruits** schizocarps, usually erect, not inflated, not angled, often oblate, dry, [sometimes] indurate, minutely hairy or glabrous; mericarps 5, 1-celled, unornamented [sometimes winged], with spines, or otherwise ornamented, usually indehiscent. **Seeds** 1 per mericarp, glabrous or hairy. $x = 14$.

Species ca. 250 (4 in the flora): s United States, Mexico, West Indies, Central America, South America, s Asia, Africa.

Of the many species of *Pavonia*, more than 160 are South American; the genus is also well represented in Africa. The South American *P. multiflora* A. St.-Hilaire has been introduced to California gardens as an ornamental shrub; it is self-sterile and not known to have become naturalized.

SELECTED REFERENCE Fryxell, P. A. 1999. *Pavonia*. In: Organization for Flora Neotropica. 1968+. Flora Neotropica. 109+ nos. New York. No. 76.

1. Leaf blades broadly ovate, margins obscurely dentate to subentire; inflorescences terminal racemes; petals pale yellow or yellow-green, 12–18 mm; stream banks and brackish estuaries. . . . . .3. *Pavonia paludicola*
1. Leaf blades ovate to hastate, margins crenate or dentate; inflorescences axillary solitary flowers; petals yellow or lavender to pink, 15–25 mm; habitats away from coast.
   2. Petals yellow; mericarps 3-spined, spines 6–7 mm, retrorsely barbed . . . . .4. *Pavonia spinifex*
   2. Petals lavender to pink; mericarps without spines.
      3. Leaf blades ovate-triangular to hastate-oblong, petioles to ⅓ length of blades; calyces 6–8 mm; fruits 6 mm diam.. . . . . 1. *Pavonia hastata*
      3. Leaf blades ovate, petioles ½–1 times length of blades; calyces 9–12 mm; fruits 8–9 mm diam.. . . . . 2. *Pavonia lasiopetala*

1. Pavonia hastata Cavanilles, Diss. 3: 138, plate 47, fig. 2. 1787 • Pale or pink pavonia [I]

Malva lecontei Buckley; *Pavonia jonesii* Feay ex Alph. Wood; *P. lecontei* (Buckley) Torrey & A. Gray

Subshrubs, 0.5 m. **Stems** stellate-hairy, hairs 0.1 mm. **Leaves:** stipules subulate, 2 mm; petiole to ⅓ length of blade; blade discolorous, ovate-triangular to hastate-oblong, to 7 cm, base cordate, margins coarsely crenate, apex acute or subobtuse, surfaces minutely stellate-hairy, more densely so abaxially. **Inflorescences** axillary solitary flowers. **Pedicels** 1.7–2.3 cm, exceeding subtending petiole; involucellar bractlets 5, alternate with calyx lobes, ovate, 4–6 × 2–3 mm, shorter than calyx. **Flowers:** calyx 6–8 mm, stellate-hairy, lobes 3-veined; corolla rotate, petals pale pink with dark maroon base, usually with darker reddish veins, not auriculate, 15–22 mm; staminal column with 5 apical teeth, glabrous; stigmas included, glabrous; flowers sometimes cleistogamous, smaller, petals shorter than calyx, stamens 5. **Schizocarps** pale green, maturing dark brown to ± black, 6 mm diam., puberulent; mericarps without spines, dorsally keeled, 4 mm, reticulate-costate. **Seeds** minutely puberulent. $2n = 56$.

Flowering spring–fall. Sandy soil on coastal plains; 100 m; introduced; Fla., Ga., Tex.; South America (Argentina, Brazil, Paraguay, Uruguay); introduced also in Mexico, Australia.

Pavonia hastata is often cultivated and has escaped in limited areas. The flowers appear to be seasonally cleistogamous, and these are usually not present when chasmogamous flowers predominate.

2. Pavonia lasiopetala Scheele, Linnaea 21: 470. 1848 • Rose or Wright's pavonia, Texas rock rose [F]

Pavonia wrightii A. Gray

Shrubs, 0.5–1 m. **Stems** densely to sparsely stellate-hairy, hairs to 0.5 mm. **Leaves:** stipules subulate, 2–5 mm; petiole ½–1 times length of blade; blade slightly discolorous, ovate, 2–5 cm, slightly longer than wide, base cordate, margins coarsely dentate, apex acute, surfaces stellate-hairy. **Inflorescences** axillary solitary flowers. **Pedicels** 2–5 cm, usually subequal to petiole; involucellar bractlets 5, alternate with calyx lobes, usually linear-lanceolate, 1–2 mm wide, shorter than to subequal to calyx, hirsute. **Flowers:** calyx 9–12 mm, hirsute, lobes prominently 3–5-veined; corolla rotate, petals lavender to pink, not auriculate, 15–25 mm; staminal column usually declinate resulting in somewhat bilateral flower, with 5 apical teeth, glabrous; stigmas included, usually villous. **Schizocarps** pallid, 8–9 mm diam., subglabrous; mericarps pale brown, without spines, obscurely carinate dorsally, otherwise smoothly rounded, 3.5–4 mm. **Seeds** tufted on hilum.

Flowering summer–fall. Open shrublands; 500–1000 m; Tex.; Mexico (Coahuila, Nuevo León).

Pavonia lasiopetala has become a popular cultivated plant in Texas, where it is also native. It is also used as a component in seed mixtures for ranges and pastures.

3. **Pavonia paludicola** Nicolson ex Fryxell in R. A. Howard, Fl. Less. Antill. 5: 241. 1989 • Mangrove mallow

Malache scabra B. Vogel in C. J. Trew, Pl. Select. 9: 50, plate 90. 1772, not *Pavonia scabra* C. Presl 1835

Shrubs, 1–4 m, sometimes supported on other vegetation. **Stems** minutely stellate-hairy. **Leaves:** stipules early-deciduous, obscure; petiole to ½ length of blade; blade concolorous, broadly ovate, 6–18 cm, base subcordate, margins obscurely dentate to subentire, apex acuminate, sparsely stellate-hairy. **Inflorescences** terminal racemes. **Pedicels** 1–4 cm; involucellar bractlets ca. 8, lanceolate, 8.8–10 × 2.5–4 mm, subglabrous. **Flowers:** calyx 8–11 mm, minutely stellate-hairy, lobes not prominently veined; corolla tubular, petals pale yellow or yellow-green, not auriculate, 12–18 mm; staminal column with 5 apical teeth; stigmas included. **Schizocarps** pale green, maturing brown, 10–13 mm diam., woody, glabrous; mericarps narrowed basally, 7–9 mm, smooth, usually 3-pointed apically. **Seeds** not tufted on hilum.

Flowering year-round. Stream banks, brackish estuaries; 0 m; Fla.; West Indies; Central America; n South America.

Pavonia paludicola has a circum-Caribbean distribution, extending northward to the Bahamas and southern Florida. The species is endangered in Florida, and is known in Collier, Miami-Dade, and Monroe counties. It is often found with mangroves. The mericarps are distributed by water.

4. **Pavonia spinifex** (Linnaeus) Cavanilles, Diss. 3: 133. 1787 • Gingerbush

Hibiscus spinifex Linnaeus, Syst. Nat. ed. 10, 2: 1149. 1759; *Malache spinifex* (Linnaeus) Kuntze

Subshrubs, 1–2 m. **Stems** minutely hairy to glabrate, hairs minute, recurved, often in well-defined longitudinal rows. **Leaves:** stipules subulate to filiform, 5–10 mm; petiole to ⅓ length of blade; blade concolorous, ovate, 4–12 cm, base truncate to subcordate, margins irregularly dentate, apex acute, surfaces sparsely hairy to glabrate. **Inflorescences** axillary solitary flowers. **Pedicels** 1–3 cm, somewhat longer than petiole; involucellar bractlets 5–7, not alternate with calyx lobes, ligulate or slightly spatulate, 10–12 × 1 mm, subequal to calyx, ciliate. **Flowers:** calyx 8–11 mm, ciliate, hairs 1–2 mm; petals yellow, auriculate, 20–25 mm; staminal column with 5 apical teeth; stigmas exserted. **Schizocarps** pallid, 8–10 mm diam., rugose; mericarps 3-spined, spines 6–7 mm, retrorsely barbed, central spine erect, lateral spines divergent. **Seeds** not tufted on hilum.

Flowering probably year-round. Forests, shrublands, savannahs; 0–100 m; Fla.; West Indies; introduced in Bermuda.

Pavonia spinifex has been reported from elsewhere in the flora area; such reports generally are based on misidentifications of other species. The species may be native in Florida, but the fruits easily attach to clothing and fur and the species has been introduced widely; its native range is not known. In Florida, it is found in seven counties north of Lake Okeechobee, from Duval to Brevard and Highlands counties. It appears to be most frequent on shell middens on the coast north of Cape Canaveral.

43. PSEUDABUTILON   R. E. Fries, Kongl. Svenska Vetensk. Acad. Handl., n. s. 43: 96, plate 5, fig. 2; plate 7, figs. 19–27. 1908 • [Greek *pseud-*, false, and genus *Abutilon*]

Brian R. Keener

Subshrubs or shrubs. Stems erect or rarely decumbent, setose or stellate-hairy, not viscid. **Leaves:** stipules persistent, linear to lanceolate; blade ovate to broadly ovate, unlobed or sometimes shallowly 3-lobed, not dissected, base cordate, margins crenate to serrate. **Inflorescences** axillary umbels or solitary flowers, sometimes aggregated in terminal inflorescences; involucel absent. **Flowers:** calyx accrescent, subequal to mature fruit, not inflated, lobes unribbed, triangular to ovate, basally rounded; corolla rotate or shallowly campanulate, yellow; staminal column included; ovary 5–8[–10]-carpellate; ovules (2)3 per cell; styles 5–8[–10]-branched; stigmas capitate. **Fruits** schizocarps, erect to weakly ascending, not inflated, oblate, slightly indurate, hairy; mericarps 5–8[–10], 1-celled sometimes with endoglossum which partially divides chamber into upper and lower portions (this may be interpreted as 2 cells), apex rounded, with divergent apical distal spines, lateral walls usually disintegrating at maturity, dehiscent. **Seeds** (2 or)3 per mericarp, turbinate, papillate, glabrous. *x* = 8.

Species 18 (2 in the flora): sc, sw United States, Mexico, West Indies, Central America, South America.

SELECTED REFERENCE   Fryxell, P. A. 1997b. A revision and redefinition of *Pseudabutilon* (Malvaceae). Contr. Univ. Michigan Herb. 21: 175–195.

1. Stems sparsely setose, hairs spreading, glabrescent, sometimes also with few stellate hairs
     . . . . .1. *Pseudabutilon thurberi*
1. Stems densely stellate-hairy, hairs tufted, sometimes also with few simple hairs. . . . .
     . . . . .2. *Pseudabutilon umbellatum*

1. Pseudabutilon thurberi (A. Gray) Fryxell, Contr. Univ. Michigan Herb. 21: 189. 1997   C F

Abutilon thurberi A. Gray, Pl. Nov. Thurb., 307. 1854

Subshrubs, erect, sometimes decumbent, 1 m, sometimes branching near base and in inflorescence. **Stems** sparsely setose, hairs spreading, glabrescent, rarely also with few appressed, stellate hairs. **Leaves:** stipules linear, to 12 × 1 mm, setose; petiole ½–1 times blade length, setose; blade ovate, unlobed, to 10 × 6 cm, gradually reduced and narrower upward, apex acuminate, surfaces strigose and stellate-hairy. **Inflorescences** solitary or paired flowers. **Pedicels** 0.5–2.5 cm, setose throughout, distally glandular-hairy, sometimes distally stellate-hairy also. **Flowers:** calyx 4–7 mm, lobes ovate, apex acuminate, setose, glandular and stellate-hairy; petals 4–6 mm, glabrous; staminal column 2–3 mm, glabrous to setose; styles 5-branched. **Schizocarps** straw colored at maturity, 5–6 mm diam., stellate-hairy; mericarps 5, spur 2–3 mm. **Seeds** 2–3 mm. *2n* = 30.

Flowering late summer–fall. Dry, open shrublands; of conservation concern; 800–1100 m; Ariz.; Mexico (Sonora).

Pseudabutilon thurberi is known only from Pima County.

A taxon from south of our range, *Pseudabutilon sonorae* Wiggins was placed in synonymy here by P. A. Fryxell (1997b). However, while the two taxa are very similar morphologically, the type illustration of *P. sonorae* clearly indicates the presence of an endoglossum dividing the carpel into two chambers. This character was not observed in specimens of *P. thurberi*, which raises uncertainty about the placement in synonymy by Fryxell.

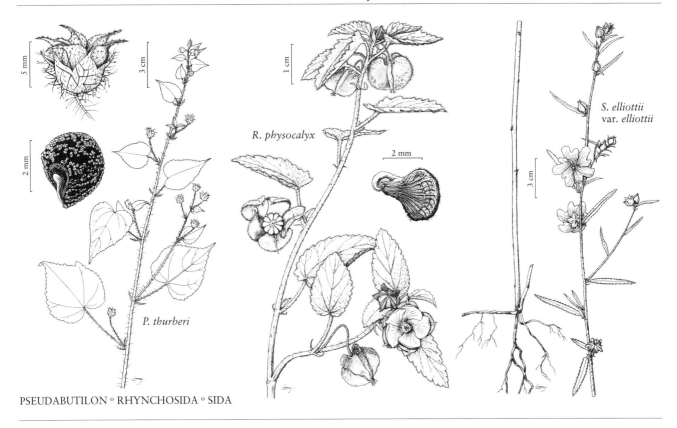

R. physocalyx

S. elliottii
var. elliottii

P. thurberi

PSEUDABUTILON ° RHYNCHOSIDA ° SIDA

2. **Pseudabutilon umbellatum** (Linnaeus) Fryxell, Contr. Univ. Michigan Herb. 21: 190. 1997

Sida umbellata Linnaeus, Syst. Nat. ed. 10, 2: 1145. 1759; *Abutilon umbellatum* (Linnaeus) Sweet

Shrubs, erect, 1–2 m, sometimes branching in inflorescence. **Stems** densely stellate-hairy, hairs tufted, sometimes also with few appressed, simple hairs. **Leaves:** stipules linear to lanceolate, to 10 × 1 mm, stellate-hairy; petiole ½–1 times blade length, stellate-hairy; blade broadly ovate, sometimes shallowly 3-lobed apically, to 12 × 6 cm, gradually reduced and narrower upward, apex acute or acuminate, surfaces with bifurcate hairs abaxially, stellate-hairy adaxially. **Inflorescences** 2–5-flowered umbels, sometimes appearing compound terminally; peduncle 1–5 cm, stellate-hairy. **Pedicels** 1–2 cm, stellate-hairy. **Flowers:** calyx 6–8 mm, lobes triangular, apex acute, stellate-hairy and sometimes hirsute with simple hairs; petals 6–8 mm, glabrous; staminal column 3–4 mm, setose and stellate-hairy; styles 6–8-branched. **Schizocarps** grayish brown at maturity, 5–9 mm diam., stellate-hairy; mericarps 6–8, spur 2–4 mm. **Seeds** 2–3 mm. $2n = 30$.

Flowering year-round. Dry shrublands, roadside ditches, fencerows; 300–800 m; Tex.; Mexico; West Indies; Central America; South America.

In the flora area, *Pseudabutilon umbellatum* is known only from the lower Rio Grande of southern Texas.

44. RHYNCHOSIDA Fryxell, Brittonia 30: 458, fig. 4. 1978 • [Greek *rhynchos*, beak, and generic name *Sida*, alluding to beaked mericarps]

Paul A. Fryxell†

Steven R. Hill

Herbs [shrubs], perennial. **Stems** trailing to ascending, stellate-hairy, not viscid. **Leaves:** stipules persistent, linear; blade oblong-ovate to oblong-lanceolate [orbiculate], unlobed, base shallowly cordate, margins crenate or serrate. **Inflorescences** axillary solitary flowers; involucel absent. **Flowers:** calyx accrescent, green-membranous and inflated in fruit, enclosing it, lobes overlapping, unribbed, cordate, stellate-pubescent; corollas yellow, [with] without dark spot at base; staminal column included; style 8–10[–14]-branched; stigmas capitate. **Fruits** schizocarps, pendulous, somewhat inflated, oblate, indurate, glabrous; mericarps 8–10[–14], 1-celled, each with long, horizontal, obtuse rostrum, lateral walls reticulate, indehiscent. **Seeds** 1 per mericarp, puberulent. $x = 8$.

Species 2 (1 in the flora): sc, sw North America, Mexico, South America (Argentina, Bolivia, Brazil, Uruguay).

Rhynchosida kearneyi Fryxell is endemic to Bolivia.

1. **Rhynchosida physocalyx** (A. Gray) Fryxell, Brittonia 30: 458. 1978 • Beaked sida, buffpetal F W

Sida physocalyx A. Gray, Boston J. Nat. Hist. 6: 163. 1850

Plants with large taproot. **Stems** with hairs scattered. **Leaves:** stipules 3–4 mm; petiole ½–¾ as long as blade; blade 2–5 cm, apex obtuse to acute, surfaces coarsely hairy, hairs stellate abaxially, simple adaxially. **Pedicels** slender, 1–2 cm. **Flowers:** calyx green-membranous, lobes apiculate; petals yellow [red basally], 5–8 mm, subequal to calyx; staminal column 3–4 mm, glabrous. **Schizocarps** 8–9 mm diam.; mericarp blackish. **Seeds** 2 mm. $2n = 16$.

Flowering spring–late fall. Deciduous forests, shrublands, roadsides, fencerows, disturbed habitats, sometimes in pastures and lawns; 0–1500 m; Ariz., N.Mex., Okla., Tex.; Mexico; South America (Argentina, Bolivia, Brazil, Uruguay).

While this distinctive species, with its *Physalis*-like inflated fruits, is thought to be native over a wide area and disjunct between southern South America and Mexico and the United States, it is most commonly found in disturbed habitats as are so many other herbaceous mallows.

45. SIDA Linnaeus, Sp. Pl. 2: 683. 1753; Gen. Pl. ed. 5, 306. 1754 • Fanpetals, wireweed [Greek *side*, name used by Theophrastos for plants now called *Nymphaea alba* Linnaeus]

Paul A. Fryxell†

Steven R. Hill

Dictyocarpus Wight; *Malvinda* Boehmer; *Pseudomalachra* (K. Schumann) Monteiro

Herbs, annual or perennial, **subshrubs, or shrubs. Stems** erect, ascending, or reclining to procumbent, glabrous or hairy, sometimes viscid (*S. glabra*). **Leaves** spirally arranged (distichous in *S. planicaulis* and *S. ulmifolia*), petiolate or subsessile; stipules persistent, usually linear to lanceolate or falcate; blade usually unlobed (lobed with maplelike leaves in *S. hermaphrodita*), base cuneate, cordate, subcordate, truncate, or rounded, margins crenate, dentate, serrate, or entire. **Inflorescences** axillary solitary (sometimes paired or clustered) flowers or terminal panicles or corymbs, including fascicles, subumbels, and glomerules; involucel absent. **Flowers:** calyx not accrescent, not inflated, not completely enclosing fruit, often plicate in bud, usually ½ divided, often 10-ribbed at base (unribbed in *S. hermaphrodita*) or angulate, lobes acute or acuminate to triangular or ovate; corolla white, cream, yellow, yellow-orange, salmon-pink, red-orange, or reddish [purplish], sometimes with dark red center; staminal column included; style 5–14-branched; stigmas capitate. **Fruits** schizocarps, erect, not inflated, oblate to conic, indurate, glabrous or hairy; mericarps 5–14, 1-celled, without endoglossum, without dorsal spurs, usually with 2 apical spines, ± well developed or muticous, reticulate, glabrous or hairy, lateral walls usually persistent, indehiscent below with well-differentiated dorsal wall, indehiscent or partially dehiscent apically. **Seeds** 1 per mericarp, glabrous. $x = 7, 8$.

Species ca. 150 (19 in the flora): North America, Mexico, West Indies, Central America, South America, Asia, Africa, Australia; warm-temperate and tropical areas.

In the flora area, *Sida linifolia* Cavanilles, flax-leaved sida, is known from a single collection (Alabama, Mobile, introduced from West Indies on ballast, Sep 1886, *Mohr s.n.*, F) and treated here as a waif; it is distinguished from other sidas in North America by its entire leaf margins. *Sida cordata* (Burman f.) Borssum Waalkes has been reported in Maryland (Baltimore City); it is a generally prostrate herb with cordate leaves and filiform pedicels that are nearly the same length as the leaves; no vouchers have been found; if it was present, it can be regarded as a waif. Reports of *S. aggregata* K. Presl, a variable and rather common Neotropical species, have not been verified; no vouchers have been located. *Sida acuta* Burman f. and *S. carpinifolia* Linnaeus f. are names often used for ballast specimens of plants found in temperate seaports that have not persisted.

Most sidas have apical spines on the fruits that adhere to fur, wool, and clothing, and therefore it may be difficult to pinpoint their native ranges versus the areas to which they have been introduced. Some are considered to be pan-tropical roadside weeds.

SELECTED REFERENCES Fryxell, P. A. 1985. Sidus sidarum V. The North and Central American species of *Sida*. Sida 11: 62–91. Kearney, T. H. 1954. A tentative key to the North American species of *Sida* L. Leafl. W. Bot. 7: 138–150. Krapovickas, A. 2003b. *Sida* sección *Distichifolia* (Monteiro) Krapov. comb. nov., stat. nov. (Malvaceae–Malveae). Bonplandia (Corrientes) 12: 83–121. Krapovickas, A. 2007. Las especies de *Sida* secc. *Malachroideae* del Cono Sur de Sudamérica. Bonplandia (Corrientes) 16: 209–225. Siedo, S. J. 1999. A taxonomic treatment of *Sida* sect. *Ellipticifolia* (Malvaceae). Lundellia 2: 100–127.

1. Leaf blades palmately 5–7-lobed, maplelike, to 24 cm; calyces not ribbed or angled . . . 7. *Sida hermaphrodita*
1. Leaf blades unlobed, 1–9 cm; calyces ribbed or angled.
   2. Stems procumbent.
      3. Petals white; leaves distributed evenly along stems, blade margins crenate to base; stems usually with 1–2 mm simple hairs in addition to multirayed stellate hairs; mericarps slightly rugose. . . . . 1. *Sida abutilifolia*
      3. Petals salmon-pink, red-orange, or yellowish; leaves crowded distally on stems, blade margins dentate only at apex; stems with appressed, normally 4-rayed stellate hairs; mericarps prominently muricate . . . . .3. *Sida ciliaris*
   2. Stems erect, sometimes ascending to reclining but not procumbent.
      4. Styles 5-branched; mericarps 5; leaf blades cordate or subcordate at base.
         5. Stems and petioles minutely stellate-hairy, hairs to 0.5 mm; petioles usually with small spinelike tubercle on stem just below its attachment . . . . 16. *Sida spinosa*
         5. Stems and petioles glandular-viscid and/or with simple hairs 1–3 mm; petioles without spinelike tubercle just below attachment to stem.
            6. Stems usually glandular-viscid; petals white or yellow-orange without darker base, not fading rose-pink; calyces usually glandular, not setose, lobes triangular, acute to short-acuminate (not beaked in bud) . . . . 6. *Sida glabra*
            6. Stems not glandular-viscid; petals yellowish to orange or salmon usually with dark orange or reddish base, fading rose-pink; calyces setose, not glandular, lobes trullate, attenuate-aristate (beaked in bud). . . . . .19. *Sida urens*
      4. Styles 7–14-branched; mericarps 7–14; leaf blades usually cuneate or truncate at base, sometimes rounded or subcordate to cordate.
         7. Leaves and branches distichous.
            8. Mericarps 7 or 8, spines 2 mm; inflorescences usually axillary glomerules, sometimes flowers solitary or paired; calyces 5–6 mm; staminal columns glabrous. . . . . 12. *Sida planicaulis*
            8. Mericarps 8–12, spines 0.5–1 mm; inflorescences axillary, flowers solitary or paired; calyces 6–8 mm; staminal columns hairy, sometimes glabrous . . . . .18. *Sida ulmifolia*
         7. Leaves spirally arranged.

[9. Shifted to left margin.—Ed.]

9. Leaf blades ± ovate or ovate-oblong with broadly cordate base, infrequently ovate-lanceolate to lanceolate; petioles (5–)10–25 mm.
    10. Leaf blades velvety-tomentose, 1–2 times longer than wide; calyces prominently ribbed, densely stellate-tomentose; inflorescences axillary, usually corymbs or panicles, sometimes solitary flowers; fruits 6–7 mm diam.; mericarp spines to 2 mm; Alabama, Florida, Texas . . . . 4. *Sida cordifolia*
    10. Leaf blades not velvety, usually 5+ times as long as wide; calyces obscurely ribbed, stellate-hairy and with long simple hairs; inflorescences axillary solitary flowers; fruits 5–6 mm diam.; mericarp spines to 1 mm; Texas . . . . 17. *Sida tragiifolia*
9. Leaf blades rhombic to subrhombic or elliptic to oblong, sometimes oblong-lanceolate, lanceolate-elliptic, round, lanceolate, or linear; petioles 2–10(–40) mm.
    11. Leaf blade margins usually entire basally, distally dentate or serrate.
        12. Pedicels 0.4–1.2 cm; leaf blades lanceolate-elliptic to round, 1.5–5 cm . . . . 2. *Sida antillensis*
        12. Pedicels 0.5–4(–16) cm; leaf blades rhombic, subrhombic, or elliptic, 2.5–9 cm.
            13. Petals yellow; stem hairs to 0.1 mm; pedicels (1–)3–4 cm . . . . .13. *Sida rhombifolia*
            13. Petals cream or pale yellow with reddish spot at base; stem hairs to 0.5 mm; pedicels to 2 cm . . . . 15. *Sida santaremensis*
    11. Leaf blade margins crenulate-serrate to dentate to base.
        14. Pedicels 8–12(–16) cm . . . . 10. *Sida longipes*
        14. Pedicels 0.5–6 cm.
            15. Leaf blades narrowly oblong-lanceolate or elliptic to linear, 4–20 times as long as wide.
                16. Plants usually little-branched from base; flowers little, if at all, apically congested; petals yellow-orange . . . . 5. *Sida elliottii*
                16. Plants freely branching from base; flowers apically congested; petals yellow-orange to reddish, sometimes drying lavender . . . . .11. *Sida neomexicana*
            15. Leaf blades narrowly lanceolate or elliptic to subrhombic, 2.5–10 times as long as wide.
                17. Pedicels 2–6 cm, sometimes equaling subtending leaf; mericarps 8–10. . . . 8. *Sida lindheimeri*
                17. Pedicels 0.5–3 cm, shorter than subtending leaf; mericarps 9–12.
                    18. Pedicels unarticulated; stipules subequal to corresponding petioles . . . . . 9. *Sida littoralis*
                    18. Pedicels articulated; stipules 2 times length of corresponding petioles . . . . 14. *Sida rubromarginata*

1. **Sida abutilifolia** Miller, Gard. Dict. ed. 8, Sida no. 12. 1768 (as abutifolia) • Creeping sida, hierba del buen dia, axocatzin

Sida diffusa Kunth; *S. filicaulis* Torrey & A. Gray; *S. procumbens* Swartz; *S. supina* L'Héritier

Herbs, perennial, 0.3–0.6(–1) m. **Stems** procumbent, stellate-hairy, hairs multirayed, usually also with simple 1–2 mm hairs. **Leaves** distributed evenly along stems; stipules inconspicuous, free from petiole, subulate, 1.5–3 mm; petiole 5–15 mm, ½ to equaling or exceeding blade, often with simple 1–2 mm hairs; blade ovate to oblong, to 1.5+ cm, 1.5–3 times longer than wide, base cordate, margins crenate to base, apex obtuse to acute, surfaces hairy. **Inflorescences** axillary solitary flowers. **Pedicels** slender, 1–2.5 cm, 2–5 times as long as calyx. **Flowers:** calyx angulate, 4–5(–7) mm, hirsute, lobes ovate-acuminate; petals white, 5–6 (–10) mm; staminal column puberulent; style 5-branched. **Schizocarps** conic, 4 mm diam., hairy; mericarps 5, 2–3 mm, basal portion slightly rugose, apex spined, spines 0.1–0.5 mm, antrorsely hairy. $2n = 14$.

Flowering year-round. Open, arid areas, disturbed habitats; 0–2000 m; Ariz., Fla., N.Mex., Okla., Tex.; Mexico; West Indies; Central America (Costa Rica, Guatemala); South America (Brazil, Colombia, Ecuador, French Guiana, Guyana, Peru, Venezuela).

Sida abutilifolia is apparently native from the southern United States to northern South America. Within the flora area, the procumbent-prostrate even mat-forming habit with freely branched, long, flexible stems is quite distinctive.

2. Sida antillensis Urban, Symb. Antill. 5: 418. 1908

• West Indian sida

Subshrubs, 0.5(–1) m. **Stems** erect, glabrescent. **Leaves:** stipules free from petiole, 1-veined, linear, 3–6 mm, subequal to petiole; petiole 2–5 mm, ca. ¼ length of blade, glabrescent; blade lanceolate-elliptic to round, 1.5–5 cm, somewhat longer than wide, base truncate, margins dentate apically, entire basally, apex acute or obtuse, surfaces minutely hairy. **Inflorescences** axillary solitary subsessile flowers. **Pedicels** not slender, 0.4–1.2 cm, usually shorter than calyx. **Flowers:** calyx ribbed, 5–8 mm, glabrescent, lobes acute or acuminate; petals yellow, 6–8 mm; style 8–10-branched. **Schizocarps** subconic, 5–6 mm diam., glabrous; mericarps 8–10, laterally reticulate, apex spined, spines to 1.5 mm, puberulent.

Flowering year-round. Roadsides, disturbed sites, often in sandy areas; 0–50 m; Fla.; West Indies; Central America.

Sida antillensis has been found in Collier, Hendry, Lee, Miami-Dade, and Monroe counties, where it is generally said to be introduced. It is widespread in the West Indies and could be native to parts of coastal Florida.

3. Sida ciliaris Linnaeus, Syst. Nat. ed. 10, 2: 1145. 1759

• Bracted or fringed or salmon sida, huinar

Malvastrum linearifolium Buckley; *Sida anomala* A. Saint-Hilaire; *S. ciliaris* var. *anomala* (A. Saint-Hilaire) K. Schumann; *S. ciliaris* var. *mexicana* (Moricand) Shinners; *S. involucrata* A. Richard

Herbs, probably perennial, 0.1–0.3 m. **Stems** procumbent, branched from base, with appressed, stellate, usually 4-rayed hairs. **Leaves** usually crowded at stem apex; stipules partially adnate to petiole, 1-veined, linear to oblanceolate, 4–12 mm, usually longer than petiole; petiole 2–10 mm, ¼–½ length of blade, with appressed stellate hairs; blade narrowly elliptic, 1–2 cm, usually 2–3 times longer than wide, base truncate to subcordate, margins dentate apically, entire basally, apex acute or obtuse, surfaces stellate-hairy abaxially, glabrous adaxially. **Inflorescences** terminal, subsessile, usually 1–10-flowered, flowers crowded at branch apices because of shortening of internodes, obscurely solitary, axillary. **Pedicels** adnate to petiole of leaflike bract, 0.1–0.4 cm, shorter than calyx. **Flowers:** calyx obscurely angulate, 4–6 mm, hirsute,

lobes ovate; petals usually salmon-pink, red-orange, sometimes yellowish, 5–11 mm; staminal column hairy; style 5–8-branched. **Schizocarps** conic, 5–6 mm diam., subglabrous; mericarps 5–8, prominently muricate, otherwise glabrous. $2n = 16$.

Flowering year-round. Roadsides, pastures, disturbed habitats, usually in open areas; 0–100 m; Fla., Tex.; Mexico; West Indies; South America.

Sida ciliaris is found in Broward and Miami-Dade counties and the Florida Keys and in central and southern Texas. The stems can be procumbent but not distinctly mat-forming, and they are often ascending, not flexible, and tufted. The flowers are sometimes described as being salmon-colored; that feature, the congested terminal leaves and flowers, and the adnate stipules are quite distinctive.

4. Sida cordifolia Linnaeus, Sp. Pl. 2: 684. 1753

• Great-leaved sida I W

Sida althaeifolia Swartz; *S. pellita* Kunth

Subshrubs or shrubs, to 1.5 m. **Stems** erect, stellate-tomentose. **Leaves:** stipules free from petiole, 1-veined, linear, 5–8 mm, shorter than petiole; petiole 10–25 mm, to ½ length of blade, stellate-tomentose; blade broadly cordate to ovate-lanceolate, to 6 cm, reduced distally, 1–2 times longer than wide, base cordate, margins dentate to base, apex acute, surfaces softly velvety-tomentose. **Inflorescences** axillary, usually subsessile, crowded panicles or corymbs, sometimes solitary flowers. **Pedicels** 0.2–0.4 cm, enlarging slightly in fruit, shorter than calyx. **Flowers:** calyx prominently ribbed, 6–7 mm, densely stellate-tomentose, lobes ovate; petals yellow-orange, often with darker reddish base, 8–11 mm; staminal column hairy; style 8–14-branched. **Schizocarps** oblate-conic, 6–7 mm diam., apically hairy; mericarps 8–14, 4–5 mm, dorsally smooth, apex spined, spines to 2 mm, retrorsely barbed (variably developed, rarely suppressed). $2n = 28$.

Flowering year-round. Disturbed sites, savannas, open shrublands, pinelands; 0–300 m; introduced; Ala., Fla., Tex.; Asia; introduced also in Mexico, West Indies, Central America, South America, Africa, Australia.

A velvety-tomentose herb sometimes used in herbal medicines, *Sida cordifolia* is believed to have originated in India, but has been widely spread in warmer regions globally. In many areas it is considered to be an invasive weed. There is considerable variation in the flower color patterns; the velvety-tomentose indument and retrorsely barbed, relatively large or conspicuous spines can help in identification.

5. Sida elliottii Torrey & A. Gray, Fl. N. Amer. 1: 231. 1838 • Elliott's sida F

Sida gracilis Elliott, Sketch Bot. S. Carolina 2: 159. 1822, not Richard 1792

Herbs or subshrubs, perennial, 0.5–1 m, usually little-branched from base. **Stems** erect, puberulent to glabrate. **Leaves:** stipules free from petiole, 1-veined, subulate, 5–8 mm, subequal to petiole; petiole 2–9 mm, much shorter than blade, minutely puberulent to glabrate; blade narrowly linear to narrowly elliptic, 2–8 cm, 4–20 times longer than wide, base truncate, margins dentate to base, apex acute, surfaces minutely hairy abaxially, glabrous adaxially. **Inflorescences** axillary solitary flowers, scattered along stem, little, if at all, apically congested. **Pedicels** slender, 0.5–4 cm, longer than calyx. **Flowers:** calyx ribbed, 6–9 mm, basally hairy, lobes ovate; petals yellow-orange, 12–15 mm; staminal column sparsely and minutely hairy; style 8–11-branched. **Schizocarps** oblate, 6–8 mm diam., sparsely hairy apically, otherwise glabrous; mericarps 8–11, 3 mm, laterally reticulate, apex blunt to spinose, spines 0.2–1 mm.

Varieties 2 (2 in the flora): sc, se United States, s, e Mexico, Central America (Guatemala).

1. Plants 0.5–1 m; mericarp spines to 1 mm . . . . . .
. . . . 5a. *Sida elliottii* var. *elliottii*
1. Plants to 0.5 m; mericarp spines 0.2–0.7 mm . . .
. . . . 5b. *Sida elliottii* var. *parviflora*

5a. Sida elliottii Torrey & A. Gray var. **elliottii** E F

Sida inflexa Fernald; *S. leptophylla* Small

Herbs or subshrubs, 0.5–1 m. **Leaf blades** 3–8 cm, with or without purplish margins. **Pedicels** 0.5–2.5 cm. **Flowers:** calyx 6–10 mm, minutely stellate-hairy and hirsute at base and along ribs. **Mericarps** blunt to spinose apically, spines to 1 mm.

Flowering spring–fall. Disturbed sites, stream banks, grasslands, open, shrubby areas, preferring sandy soil; 0–300 m; Ala., Ark., Fla., Ga., La., Miss., Mo., N.C., S.C., Tenn., Va.

Variety *elliottii* occurs generally north of 29°N latitude.

5b. Sida elliottii Torrey & A. Gray var. **parviflora** Chapman, Fl. South. U.S. ed. 3, 48. 1897

Herbs, to 0.5 m. **Leaf blades** 2–6 cm, margins usually purplish. **Pedicels** 0.5–4 cm. **Flowers:** calyx 6–9 mm, stellate-hairy basally. **Mericarps** spinose apically, spines 0.2–0.7 mm.

Flowering spring–fall. Sandy soil to rocky limestone slopes; 0–100 m; Fla., Tex.; Mexico; Central America (Guatemala).

Variety *parviflora* occurs generally south of 29°N latitude.

6. Sida glabra Miller, Gard. Dict. ed. 8, Sida no. 14. 1768 • Sticky sida I

Sida glutinosa Commerson ex Cavanilles; *S. viscidula* Blume

Subshrubs, 0.4–1.2 m. **Stems** erect to often reclining, glandular-puberulent, viscid and with simple 1–2 mm hairs. **Leaves:** stipules free from petiole, 1-veined, subulate, 1–3 mm, shorter than petiole; petiole 8–30 mm, ¼–½ times length of blade, glandular-puberulent and with simple 1–2 mm hairs; blade ovate, 3–6 cm, 1.5–2 times longer than wide, base cordate, margins serrate-crenate or dentate to base, apex acute, surfaces sparsely stellate-hairy. **Inflorescences** axillary solitary flowers and 2–4-flowered fascicles, sometimes appearing paniculate. **Pedicels** jointed 2–5 mm below calyx, slender, 1–2 cm, longer than calyx. **Flowers:** calyx angulate, 4–5 mm, not beaked in bud, with both stellate and glandular hairs, lobes triangular, acute to short-acuminate; petals white or yellow-orange, 5–6 mm; staminal column hairy; style 5-branched. **Schizocarps** subconic, 4–5 mm diam., puberulent; mericarps 5, 2.5 mm, dorsally reticulate, apex spined, spines 1–2 mm, minutely antrorsely hairy. $2n = 16$.

Flowering spring–fall. Disturbed areas, often in shade, urban weed; 100 m; introduced; Fla.; Mexico; West Indies; Central America; n South America.

Sida glabra is apparently a casual introduction to Miami-Dade County. The later homonym *S. glabra* Nuttall from 1834 is a synonym of *S. rhombifolia*.

7. Sida hermaphrodita (Linnaeus) Rusby, Mem. Torrey Bot. Club 5: 223. 1894 • Virginia mallow E

Napaea hermaphrodita Linnaeus, Sp. Pl. 2: 686. 1753

Herbs, perennial, 1–2.5(–5) m. **Stems** erect, minutely stellate-hairy when young, soon glabrate. **Leaves:** stipules free from petiole, linear-lanceolate, 3–4 mm, shorter than petiole; petiole to 0.9 mm, shorter than blade, glabrous; blade palmately 5–7-lobed, maplelike, to 24 cm, ± as long as wide, smaller upward, base cordate, margins serrate, apex long-acuminate, surfaces glabrous. **Inflorescences** axillary, subumbellate, 2–10-flowered pedunculate corymbs, forming terminal panicles. **Flowers:** calyx dark-pigmented basally, unribbed, not angulate, 4–5 mm, minutely stellate-hairy, lobes wide-triangular; petals white, 8–10 mm; staminal column hairy; style 8-branched. **Schizocarps** subconic, 6–8 mm diam., minutely stellate-hairy; mericarps 8, not reticulate, apex beaked. $2n = 28$.

Flowering late summer. Along streams, roadsides, railroad embankments, disturbed sites; 50–200 m; Ont.; D.C., Ind., Ky., Md., Mich., Ohio, Pa., Va., W.Va.

Some occurrences of *Sida hermaphrodita* may be the result of escapes from cultivation. It is generally rare except locally common along the Kanawha and Ohio rivers in Ohio and West Virginia (D. M. Spooner et al. 1985); it has been extirpated from Tennessee. Reports from Massachusetts, New Jersey, and New York refer to garden escapes. The species may or may not be native in Michigan.

8. Sida lindheimeri Engelmann & A. Gray, Boston J. Nat. Hist. 5: 213. 1845 • Lindheimer's sida

Sida elliottii Torrey & A. Gray var. *texana* Torrey & A. Gray; *S. texana* (Torrey & A. Gray) Small

Herbs or subshrubs, perennial, 1 m. **Stems** erect, minutely and sparsely stellate-hairy. **Leaves:** stipules free from petiole, 1-veined, linear, 4–8 mm, ½–1 times length of corresponding petiole; petiole 6–17 mm, to ¼ length of blade, obscurely hairy; blade narrowly lanceolate or elliptic, 2.5–7 cm, 6–10 times longer than wide, base truncate, margins dentate to base, apex acute, surfaces obscurely hairy abaxially, glabrate adaxially. **Inflorescences** axillary solitary flowers. **Pedicels** slender, 2–4(–6) cm, often equaling subtending leaf, much longer than calyx. **Flowers:** calyx ribbed, 7–10 mm, obscurely stellate-hairy, lobes triangular; petals yellow, 12–17 mm; staminal column sparsely hairy; style 8–10-branched.

Schizocarps oblate, 8–9 mm diam., apically hairy; mericarps 8–10, laterally reticulate, apex spined, spines 1 mm. $2n = 28$.

Flowering spring–fall. Open, sandy shrublands and woodlands; 10–300 m; La., Tex.; Mexico (Tamaulipas).

Specimens from Florida identified as *Sida lindheimeri* are generally misidentified individuals of *S. elliottii*. *Sida lindheimeri* is widespread and occasionally common in south-central and southernmost Texas in approximately 40 counties, and it has been reported from Cameron and East Feliciana parishes in Louisiana.

9. Sida littoralis Siedo, Phytoneuron 2014-75: 1. 2014 • Florida sida E

Sida floridana Siedo, Lundellia 4: 69, figs. 1, 2. 2001, not Gandoger 1924

Herbs, perennial, 1–2 m. **Stems** erect, minutely stellate-hairy to glabrescent. **Leaves:** stipules free from pedicel, filiform to linear, often curved or twisted, 4–9 mm, subequal to corresponding petiole; petiole 4–7 mm, ca. ¹⁄₁₀ length of blade, sparsely stellate-hairy; blade lanceolate-elliptic to subrhombic, 5–9 cm, 4–7 times longer than wide, base cuneate, margins crenulate-serrate to base, apex acute, surfaces sparsely stellate-hairy abaxially, glabrescent adaxially. **Inflorescences** axillary solitary flowers, often congested apically. **Pedicels** unarticulated, 0.5–1 cm, shorter than subtending leaf, shorter than or subequal to calyx. **Flowers:** calyx ribbed, 7–9 mm, glabrous, lobes trullate; petals yellow, 15 mm; staminal column hairy; style 9–12-branched. **Schizocarps** oblate, 6–7 mm diam., apically hairy; mericarps 9–12, 5 mm, laterally reticulate, apex spined, spines 1.5–2 mm.

Flowering spring–summer. Sand and gravel substrates in forests; 0–30 m; Fla.

Sida littoralis is found on Captiva and La Costa islands, Lee County.

10. Sida longipes A. Gray, Smithsonian Contr. Knowl. 3(5): 19. 1852 • Long-stalked sida

Herbs or subshrubs, perennial, to 0.5 m. **Stems** ascending to erect, with stellate 0.1 mm hairs and glandular hairs. **Leaves:** stipules free from petiole, 1-veined, subulate, 3–4 mm, ½–1 times length of petiole; petiole 6.3–40 mm, ¼–½ length of blade, minutely stellate-hairy; blade narrowly linear, 2.5–8 cm, 6–20 times longer than wide, base truncate, margins dentate to base, apex subacute, surfaces stellate-hairy. **Inflorescences** axillary

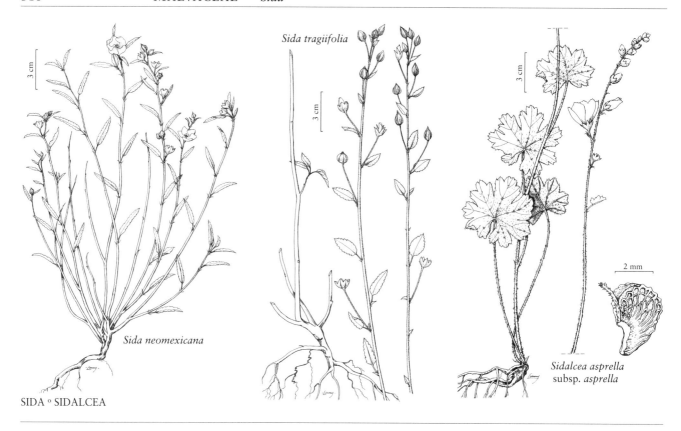

Sida tragiifolia

Sida neomexicana

Sidalcea asprella subsp. *asprella*

SIDA ○ SIDALCEA

solitary flowers. **Pedicels** slender, 8–12(–16) cm, usually 2+ times length of subtending leaves, much longer than calyx. **Flowers:** calyx obscurely 10-ribbed basally, 6–8 mm, stellate-hairy, lobes triangular; petals pale orange, 12 mm; staminal column minutely hairy; style 8–10-branched. **Schizocarps** subconic, 5–7 mm diam., minutely glandular-puberulent to subglabrous; mericarps 8–10, 3.5 mm, strongly reticulate laterally, apex muticous.

Flowering spring–fall. Arid shrublands, often on limestone; 1000–1800 m; Tex.; Mexico (Coahuila).

Sida longipes is known from at least six counties in the Big Bend region of southwestern Texas.

11. Sida neomexicana A. Gray, Proc. Amer. Acad. Arts 22: 296. 1887 (as neo-mexicana) • New Mexico sida F

Herbs or subshrubs, perennial, usually less than 0.5 m, freely branching from base. **Stems** erect, stellate-puberulent. **Leaves:** stipules free from petiole, 1-veined, linear to falcate, 5–7 mm, ½–1 times length of petiole; petiole 2–10 mm, to ¼ length of blade, obscurely puberulent; blade narrowly oblong-lanceolate, 2–4.5 cm, 6–15 times longer than wide, base truncate, margins dentate to base, apex acute, surfaces sparsely hairy abaxially, glabrous adaxially. **Inflorescences** axillary, solitary flowers, usually congested apically. **Pedicels** usually to 1 cm, subequal to calyx, much shorter than subtending leaf. **Flowers:** calyx ribbed, 6–7 mm, minutely hairy, lobes triangular; petals yellow-orange to reddish, sometimes drying lavender, 10–12 mm; style 10–12-branched. **Schizocarps** oblate, 6–7 mm diam., apically puberulent; mericarps 10–12, 3 mm, obscurely reticulate laterally, apex muticous. $2n$ = ca. 14.

Flowering late summer. Open, arid habitats; 400–2400 m; Ariz., N.Mex., Tex.; Mexico (Chihuahua, Coahuila, Durango).

Sida neomexicana resembles *S. ciliaris* in several respects, especially in the congested terminal inflorescences resulting from shorter internodes.

12. Sida planicaulis Cavanilles, Diss. 1: 24, plate 3, fig. 11. 1785 • Flatstem sida, Brazilian wire-weed I

Malvastrum carpinifolium (Medikus) A. Gray; *Malvinda carpinifolia* Medikus; *Sida acuta* Burman f. subsp. *carpinifolia* (Medikus) Borssum Waalkes; *S. acuta* var. *carpinifolia* (Medikus) K. Schumann; *S. betulina* Cavanilles 1803; *S. bracteolata* de Candolle; *S. carpinifolia* Linnaeus f. var. *antillana* Millspaugh; *S. carpinifolia* var. *betulina* de Candolle

Herbs or subshrubs, perennial, branches distichous, planar, 0.3–1 m. **Stems** erect, with simple 1–2 mm hairs, sometimes also minutely stellate-hairy. **Leaves** distichous; stipules free from petiole, subfalcate, 4–10 × 0.5–1 mm or less, often exceeding petiole, margins ciliate; petiole 5–6 mm, $\frac{1}{15}$–$\frac{1}{5}$ blade length, hirsute; blade elliptic-lanceolate, 2.5–9 × 1–4 cm, 2 times longer than wide, base rounded, margins short-serrate at least distally, entire basally, apex acute, surfaces glabrous or with minute scattered stellate hairs and simple appressed antrorse hairs. **Inflorescences** axillary congested glomerules, sometimes solitary or paired flowers. **Pedicels** 0.3–0.5 cm, subequal to or shorter than subtending petiole, shorter than to subequal to calyx. **Flowers:** calyx ribbed, 5–6 mm, often ciliate, with scattered minute stellate hairs, lobes triangular; petals yellow, 8–10 mm; staminal column glabrous; style 7- or 8-branched. **Schizocarps** subconic, 6–7 mm diam., glabrous or axially minutely puberulent; mericarps 7 or 8, 3 mm, smooth dorsally, laterally somewhat reticulate, apex spined, spines 2 mm, apically minutely puberulent. $2n = 28$.

Flowering year-round. Disturbed sites; 10–100 m; introduced; Fla.; South America (Brazil); introduced also in Indian Ocean Islands (Mauritius), Pacific Islands (Hawaii, Tubuai Islands).

Sida planicaulis is weedy and thought to have originated in Brazil, where it is quite common.

Sida planicaulis was first reported from south-central Florida (Glades, Highlands, Okeechobee, Osceola, and Polk counties) by K. R. DeLaney (2010) as new to North America; it has been reported (as *S. acuta*) also from New Jersey and Pennsylvania on ballast as a waif.

13. **Sida rhombifolia** Linnaeus, Sp. Pl. 2: 684. 1753 • Arrowleaf sida, axocatzín, Cuba jute, escobilla, huinar [I] [W]

Sida hondensis Kunth;
S. rhomboidea Roxburgh

Subshrubs, 1 m. **Stems** erect, stellate-puberulent, hairs to 0.1 mm. **Leaves:** stipules free from petiole, 1-veined, subulate, 5–6 mm, subequal to petiole; petiole 5–7 mm, $\frac{1}{10}$–$\frac{1}{4}$ length of blade, stellate-puberulent; blade ± rhombic, 2.5–9 cm, smaller distally, 2–3(–4) times longer than wide, base usually cuneate, sometimes somewhat truncate to subcordate, margins serrate distally, entire basally, apex acute to subobtuse,

surfaces stellate-puberulent or glabrescent adaxially. **Inflorescences** axillary solitary flowers. **Pedicels** slender, (1–)3–4 cm, 4–6 times length of calyx, much shorter than to ± equaling subtending leaf, at least distalmost. **Flowers:** calyx ribbed, 5–6 mm, puberulent, lobes ovate; petals yellow, 7–9 mm; staminal column hairy; style 10–14-branched. **Schizocarps** subconic, 4–5 mm diam., glabrous; mericarps 10–14, 3–4 mm, laterally reticulate, apex muticous to spined, sometimes 1-spined through failure of dehiscence, glabrous. $2n = 14, 28$.

Flowering year-round in warmer localities. Disturbed sites, roadsides, pastures, urban areas; 0–1500 m; introduced; Ala., Ark., Fla., Ga., Kans., La., Miss., N.C., Okla., S.C., Tex., Va.; s Asia (China); introduced also in Mexico, West Indies, Central America, South America, tropical Africa, Pacific Islands (Philippines, Polynesia), Australia.

Sida rhombifolia is found occasionally on ballast in New Jersey and Pennsylvania. It is a common weed in warm-temperate, subtropical, and tropical areas; its genetic diversity seems to indicate that it was introduced from the Old World. The species has been cultivated for medicinal and cordage use.

14. **Sida rubromarginata** Nash, Bull. Torrey Bot. Club 23: 102. 1896 (as rubro-marginata) • Redleaf sida [E]

Subshrubs, 1 m. **Stems** erect, subglabrous. **Leaves:** stipules free from petiole, linear, 6–11 mm, 2 times length of corresponding petiole; petiole 3–7 mm, $\frac{1}{10}$ length of blade, subglabrous; blade lanceolate-elliptic to subrhombic, 3–9 cm, 2.5–5 times longer than wide, base truncate, margins crenate or serrate to base, apex usually acute, surfaces glabrous. **Inflorescences** axillary solitary or paired flowers. **Pedicels** articulated, 0.5–2(–3) cm, usually 2–3 times calyx length, $\frac{1}{4}$–$\frac{1}{2}$ length of subtending leaf. **Flowers:** calyx ribbed, 7–10 mm, glabrous, lobes ovate; petals yellow, 15 mm; staminal column hairy; style 10–12-branched. **Schizocarps** subconic, 6–7 mm diam., minutely hairy apically; mericarps 10–12, 5 mm, laterally reticulate, apex spined, spines 1.5 mm.

Flowering summer. Disturbed sites, waste places; 0–30 m; Fla.

Sida rubromarginata has been reported from Hillsborough, Leon, and Sarasota counties.

15. **Sida santaremensis** Monteiro, Monogr. Malv. Bras. 1: 44, plate 8. 1936 • Brazilian sida, moth fanpetals

Subshrubs, to 1 m. **Stems** erect, sparsely stellate-hairy, hairs to 0.5 mm. **Leaves:** stipules free from petiole, 1-veined, linear, 7 mm, subequal to petiole; petiole 3–10 mm, ⅕ times length of blade, densely stellate-hairy distally; blade broadly elliptic to subrhombic, to 5.5 cm, 2–3.5 times longer than wide, smaller and narrower upward, base truncate to rounded, margins dentate almost to base, apex acute, surfaces evenly stellate-hairy, densely so abaxially. **Inflorescences** usually axillary solitary flowers. **Pedicels** slender, to 2 cm, usually 3+ times longer than calyx, 2 times as long as subtending petiole. **Flowers:** calyx ribbed, 6–7 mm, stellate-hairy, lobes triangular; petals cream or pale yellow with reddish spot at base, 10 mm; staminal column glabrous; style 11-branched. **Schizocarps** oblate-conic, 5–6 mm diam., subglabrous apically; mericarps 11, 4–5 mm, dorsal wall somewhat sunken, lateral walls smooth to obscurely reticulate, apex short-apiculate, with few antrorse hairs. $2n = 14$.

Flowering spring–fall. Sandy, disturbed areas; 0–100 m; introduced; Fla.; South America (Argentina, Bolivia, Brazil).

Sida santaremensis is a relatively recent introduction around Tampa (P. A. Fryxell et al. 1984). It often is infected with the sida golden mosaic virus.

16. **Sida spinosa** Linnaeus, Sp. Pl. 2: 683. 1753 • False or Indian or prickly mallow, prickly sida

Sida alba Linnaeus; *S. alnifolia* Linnaeus; *S. angustifolia* Miller; *S. heterocarpa* Engelmann

Subshrubs or herbs, annual or perennial, 0.2–1 m, rarely taller. **Stems** erect, minutely stellate-hairy, hairs to 0.5 mm. **Leaves:** stipules free from petiole, 1-veined, subulate, 3–6 mm, ½ as long as petiole; petiole 5–15 mm, usually ¼–½ length of blade, sometimes shorter, minutely stellate-hairy, hairs to 0.5 mm, usually with small spinelike tubercle on stem just below its attachment; blade ovate, lanceolate, or narrowly oblong, 2–6 cm, smaller apically, 2–5 times longer than wide, base subcordate, margins crenate-serrate to base, apex usually acute, surfaces stellate-tomentulose abaxially, glabrate adaxially. **Inflorescences**

axillary solitary or 2–4 clustered flowers. **Pedicels** 0.5–1 cm, subequal to calyx and subtending petiole. **Flowers:** calyx angulate, 5–7 mm, minutely tomentose, lobes triangular; petals yellow, rarely white, 5 mm; staminal column glabrous; style 5-branched. **Schizocarps** subconic, 4–5 mm diam., hairy; mericarps 5, 3–4 mm, somewhat rugose, apex spined, spines 1 mm, antrorsely hairy. $2n = 14, 28$.

Flowering year-round in warmer areas, summer elsewhere. Roadsides, pastures, disturbed ground; 0–1500 m; Ont.; Ala., Ariz., Ark., D.C., Fla., Ga., Ill., Ind., Iowa, Kans., Ky., La., Md., Mass., Mich., Miss., Mo., Nebr., N.J., N.C., Ohio, Okla., Pa., S.C., Tenn., Tex., Va., W.Va.; Mexico; West Indies; Central America; South America.

A small spur sometimes is present on the abaxial side of the petiole at the juncture with the stem, to which the specific epithet refers. It is not a spine and occasionally is absent.

17. **Sida tragiifolia** A. Gray, Boston J. Nat. Hist. 6: 164. 1850 (as tragiaefolia) • Noseburn-leaf sida, earleaf fanpetals

Herbs or subshrubs, perennial, 0.5 m. **Stems** erect, minutely stellate-hairy, hairs 0.3–0.4 mm, sometimes also with simple 2–2.5 mm hairs. **Leaves:** stipules free from petiole, 1-veined, subulate, 6–7 mm, ½ length of petiole; petiole (5–)10–20 mm, ⅓–½ length of blade, minutely hairy; blade ovate-oblong to narrowly lanceolate, 1.5–4 × 0.5–2 cm, 5+ times longer than wide, base truncate to subcordate, margins dentate-serrate to base, apex acute to obtuse, surfaces stellate-hairy. **Inflorescences** axillary solitary flowers. **Pedicels** articulated 1–2 mm below flowers, 0.5–2 cm, 1–2 times length of calyx, exceeding subtending petiole. **Flowers:** calyx obscurely ribbed, 6.5–9 mm, stellate-hairy and with long simple hairs, lobes triangular, midrib conspicuous; petals yellow, fading rose, 10 mm; staminal column hairy; style 8-branched. **Schizocarps** oblate, 5–6 mm diam., apically hairy; mericarps 8+, 3 mm, strongly reticulate laterally, apically dehiscent, apex 2-spined or not, spines to 1 mm, sometimes suppressed, apex hispid.

Flowering spring–fall. Arid shrublands; 500–1500 m; Ariz., Tex.; Mexico (Coahuila, Nuevo León, Tamaulipas).

Within the flora area, *Sida tragiifolia* is known in Graham, Pima, and Santa Cruz counties, Arizona, and in Brewster, Cameron, Hidalgo, and Presidio counties, Texas.

18. **Sida ulmifolia** Miller, Gard. Dict. ed. 8, Sida no. 1. 1768 • Broom weed, escobilla, southern sida, common wire-weed

Sida acuta Burman f. var. *intermedia* S. Y. Hu; *S. balbisiana* de Candolle; *S. brachypetala* de Candolle; *S. carpinifolia* Miller; *S. carpinifolia* Linnaeus f. var. *balbisiana* (de Candolle) Millspaugh; *S. carpinifolia* Linnaeus f. var. *brevicuspidata* Grisebach

Herbs or subshrubs, perennial, branches distichous, 1 m. **Stems** erect, minutely stellate-hairy. **Leaves** distichous; stipules free from petiole, 1–3(–5)-veined, broadly falcate, 6–12 mm, often exceeding petiole; petiole (1–)4–5(–8) mm, ca. $^1/_{10}$ blade length, obscurely stellate-hairy; blade lanceolate to ovate, 3–9 cm, 2–4 times longer than wide, base cuneate to rounded, margins serrate at least distally, apex acute, surfaces hirsute to glabrate. **Inflorescences** axillary solitary or paired flowers, sometimes more and subumbellate. **Pedicels** jointed near base, 0.2–0.5(–0.8) cm, subequal to calyx and subtending petiole. **Flowers:** calyx ribbed, 6–8 mm, often ciliate, lobes triangular; petals yellow, 7–10(–12) mm; staminal column glabrous or hairy; style 7–12-branched. **Schizocarps** subconic, 5–7 mm, glabrous; mericarps 7–12, 3–4 mm, laterally reticulate, apex spined, spines to 0.5 mm. $2n = 28$.

Flowering year-round. Disturbed sites, principally coastal; 0–50 m; Fla.; Mexico; West Indies; Central America; South America (Bolivia, Brazil, Colombia, Guyana, Venezuela); s Asia; Africa; Pacific Islands; Australia.

Sida ulmifolia is pantropical and weedy but thought to have originated in Central America. In previous floras it and *S. planicaulis* have been treated as *S. acuta* Burman f., but that is a different species from those from Brazil, Guatemala, Mozambique, Sri Lanka, southeastern Asia, several Pacific islands, and Australia. Under different names, *S. ulmifolia* has been reported also from New Jersey and Pennsylvania on ballast.

19. **Sida urens** Linnaeus, Syst. Nat. ed. 10: 1145. 1759 • Bristly sida, tropical fanpetals [I]

Sida verticillata Cavanilles

Herbs or subshrubs, perennial, often scandent, 0.5–1.5 m. **Stems** erect or reclining, with simple 1.5–3 mm hairs mixed with shorter stellate hairs, rarely only stellate-hairy. **Leaves:** stipules free from petiole, 1-veined, subulate, 2–5 mm; petiole 10–30 mm, $^1/_4$–$^1/_2$ (to nearly equaling) blade length, pubescence like stem; blade ovate to triangular, 4–9 cm, 1.5–2 times longer than wide, base cordate, margins crenate-serrate or coarsely serrate to base, apex acuminate or attenuate, surfaces sparsely pubescent, abaxial surface stellate-pubescent, adaxial surface stellate-pubescent or with simple, often antrorsely-oriented hairs. **Inflorescences** axillary, dense, subsessile, 3–8-flowered glomerules or in pedunculate, to 3 cm "heads." **Pedicels** 0.5–2(–12) mm, shorter than calyx and subtending petioles. **Flowers:** calyx ribbed, 5–8 mm, setose, lobes trullate, attenuate-aristate, beaked in bud, margins dark green; petals yellowish to orange or salmon, fading rose-pink, often with dark orange or reddish base, 5–7(–9) mm, exceeding calyx; staminal column hairy; style 5-branched. **Schizocarps** subconic, 3.5–4 mm, glabrous or nearly so; mericarps 5, 3 × 1.5 mm, laterally faintly striate to smooth, apex muticous. $2n = 32$.

Flowering year-round. Disturbed sites; 0–10 m; introduced; Fla.; Mexico; West Indies; Central America; South America; introduced also in Africa, Indian Ocean Islands (Madagascar), Pacific Islands (Hawaii).

Sida urens was found only recently (2008) in Broward County. The species is easily distinguished by its long-acuminate beaked flower buds, setose calyx, cordate-acuminate leaves, and tendency to have long, reclining stems. It is rather common in tropical regions.

46. SIDALCEA A. Gray, Mem. Amer. Acad. Arts, n. s. 4: 18. 1849 • Checkerbloom, checker mallow [Generic names *Sida* and *Alcea*, alluding to resemblances]

Steven R. Hill

Hesperalcea Greene

Herbs, annual or perennial, **or subshrubs,** sometimes glaucous, usually hairy, sometimes glabrate, hairs stellate or simple or both, with taproot, clustered fleshy roots, caudex, or adventitious roots, sometimes with shallow, elongated or compact rhizomes often termed rootstocks. **Stems** erect or ascending, reclining to decumbent at base, these often rooting, stolonlike stems sometimes present. **Leaves** evenly spaced on stem or crowded near base (sometimes rosettelike); stipules usually persistent (sometimes deciduous in *S. candida, S. celata, S. gigantea, S. hartwegii, S. hendersonii, S. hirsuta, S. robusta, S. setosa*; persistent or deciduous in *S. cusickii, S. nelsoniana, S. oregana*; basal early-deciduous in *S. diploscypha*), linear or lanceolate to ovate; blade ovate to orbiculate or reniform, unlobed or palmately lobed or parted, usually variable even on one individual, base cordate, rounded, or truncate to wide-cuneate, margins crenate or dentate; leaves often of 3 types—long-petioled basal leaves that are scarcely lobed or merely crenate, mid-stem leaves that are palmately 3–7(–9)-lobed, and upper short-petiolate to subsessile cauline leaves that are palmately divided nearly to base and often have narrow, entire or dentate lobes. **Inflorescences** usually terminal, spiciform, capitate, or racemose, sometimes axillary solitary flowers; involucre absent, bracts involucrelike in *S. diploscypha* and *S. stipularis*, bracts at pedicel base 2, usually stipulelike, often connate into single structure and appearing to be a single bract with 2-fid apex, sometimes apex undivided, rarely multidivided in *S. diploscypha*; involucellar bractlets present or absent, usually 0, sometimes 1–3, usually persistent, distinct. **Flowers** bisexual or unisexual and pistillate, rarely staminate (in *S. cusickii, S. malachroides, S. nelsoniana*), plants bisexual, gynodioecious, or trioecious, bisexual flowers protandrous; calyx often somewhat accrescent, not inflated, lobes ribbed or unribbed, ovate or triangular; corolla cuplike or funnelform, usually pale to dark pink, rose-pink, magenta, or purplish to lavender, sometimes white, veins often paler, infrequently with paler or darker patch at petal base; staminal column included, divided into concentric inner and outer series (tubes) of connate filaments; anthers distal on filament tube; ovary (4)5–10-carpellate; ovules 1 per cell; style (4)5–10-branched; stigmas linear, introrsely decurrent. **Fruits** schizocarps, erect, not inflated, oblate-discoid, indurate (but lateral walls thin), glabrous or pubescent; mericarps (4)5–10, drying tan or brown, 1-celled, indehiscent, sometimes beaked, cuspidate, or mucronate. **Seeds** 1 per mericarp, glabrous. $x = 10$.

Species 31 (31 in the flora): w North America, n Mexico.

Some species of *Sidalcea* are cultivated for ornament. The common name checkerbloom comes from the pattern of veins on the petals of some species.

Some species of *Sidalcea* are similar and difficult to distinguish; details of the plant habit and fruits are generally necessary for identification. Most *Sidalcea* species, especially those in the *S. oregana* group and *S. malviflora* group, show great plasticity. The presence of gynodioecy is widespread in *Sidalcea*, and pistillate and bisexual flowers often have different sizes within a population (the pistillate being smaller). As with most other mallows that have schizocarps, the mericarp features usually are not evident until full dry maturity. The inflorescences are generally rather congested initially and later may elongate; both alternatives in the key should be considered. Inflorescence descriptions in the key are for plants in full flower; staminal column measurements are for bisexual flowers only.

There is some difficulty in applying the term rhizome in *Sidalcea*, although it has widespread traditional use in keys to the species; most structures called rhizomes are at or near the soil surface and may be decumbent rooting stems; rhizomes are best developed in *S. calycosa* subsp. *rhizomata, S. gigantea*, and *S. hirtipes*.

Molecular studies have suggested multiple evolutionary hypotheses including: identification of the basal species, separate derivations of the annual species, and five groups of closely related species with recent evolutionary radiation that are still in need of further study.

The rank subspecies is widely used in this treatment in accordance with C. L. Hitchcock (1957) and S. R. Hill (2012b). Varieties were distinguished within some of the subspecies in past treatments. See Hitchcock for more information on the published varieties.

SELECTED REFERENCES Andreasen, K. and B. G. Baldwin. 2001. Unequal evolutionary rates between annual and perennial lineages of checker mallows (*Sidalcea*, Malvaceae): Evidence from 18S-26S rDNA internal and external transcribed spacers. Molec. Biol. Evol. 18: 936–944. Andreasen, K. and B. G. Baldwin. 2003. Reexamination of relationships, habital evolution, and phylogeography of checker mallows (*Sidalcea*, Malvaceae) based on molecular phylogenetic data. Amer. J. Bot. 90: 436–444. Andreasen, K. and B. G. Baldwin. 2003b. Nuclear ribosomal DNA sequence polymorphism and hybridization in checker mallows (*Sidalcea*, Malvaceae). Molec. Phylogen. Evol. 29: 563–581. Halse, R. R., B. A. Rottink, and R. Mishaga. 1989. Studies in *Sidalcea* taxonomy. NorthW. Sci. 63: 154–161. Hill, S. R. 1993. *Sidalcea*. In: J. C. Hickman, ed. 1993. The Jepson Manual. Higher Plants of California. Berkeley, Los Angeles, and London. Pp. 755–760. Hill, S. R. 2012b. *Sidalcea*. In: B. G. Baldwin et al., eds. 2012. The Jepson Manual: Vascular Plants of California, ed. 2. Berkeley. Pp. 887–896. Hitchcock, C. L. 1957. A Study of the Perennial Species of *Sidalcea*. Seattle. [Univ. Wash. Publ. Biol. 18.] Roush, E. M. F. 1931. A monograph of the genus *Sidalcea*. Ann. Missouri Bot. Gard. 18: 117–244.

1. Herbs, annual, except *S. calycosa* subsp. *rhizomata*.
    2. Cauline leaf blades: lobes of mid and distal usually apically 2–5-toothed, linear to obovate; stems bristly-hairy, hairs erect; mericarps usually glabrous, mucro absent.
        3. Stipules at mid and distal nodes divided into 2–5 filiform or linear segments, subequal to or longer than calyx, becoming involucrelike; distal cauline leaves: blades 5–7-lobed, lobes linear, sometimes 3-toothed, mid tooth longer than laterals, or lobe margins entire; calyces seldom glandular; mericarps without bristles . . . . 8. *Sidalcea diploscypha*
        3. Stipules and mid and distal nodes linear-filiform, rarely divided, usually much shorter than calyx, not involucrelike; distal cauline leaves: blades 3-lobed, lobes obovate, apically 2–5-toothed, teeth subequal; calyces usually glandular; mericarps usually with 1–5 minute proximal apical bristles . . . . 17. *Sidalcea keckii*
    2. Cauline leaf blades: lobes of mid and distal entire, not apically 2–5-toothed, lobes linear to linear-elliptic or oblanceolate-obtuse; stems usually glabrate at least proximally; mericarps glabrous or sparsely puberulent, mucro present.
        4. Stamens: outer filaments distally distinct, anthers not attached to connate portion of filaments; mericarps reticulate-veined on back, deeply pitted especially on top, glabrous or glandular-puberulent; calyces 8–10(–12) mm . . . . 12. *Sidalcea hartwegii*
        4. Stamens: outer filaments connate to apex, staminal column funnel-like with continuous rim to which the unstalked anthers attach; mericarps glabrous or puberulent, longitudinally grooved or reticulate-veined; calyces 4–12 mm.
            5. Stems glabrous or sparsely stellate-hairy; plants annual or perennial; bracts ovate to wide-elliptic, 2–5 mm wide; mericarps glabrous, longitudinally grooved; calyces stellate-puberulent and strigose-bristly . . . . 2. *Sidalcea calycosa*
            5. Stems usually densely bristly-hairy distally, rarely glabrescent; plants annual; bracts linear, to 2 mm wide; mericarps puberulent, reticulate-veined, pitted; calyces prominently tawny-hirsute and densely stellate-canescent. . . . . 15. *Sidalcea hirsuta*
1. Herbs, perennial, or subshrubs.
    6. Mericarps: mucro absent; involucellar bractlets (0)1–3; leaves evenly arrayed on stem throughout season, blades all similar in shape.
        7. Involucellar bractlets (0)1 or 2; petals white or purple-tinged; leaf blades lobed, maplelike; plants (0.4–)0.8–1.5(–2) m. . . . . 18. *Sidalcea malachroides*
        7. Involucellar bractlets (2)3; petals usually pale pink to pink-lavender, rarely white; leaves unlobed or lobed, but not maplelike; plants 0.1–0.5(–0.8) m.
            8. Inflorescences usually open or spiciform, sometimes dense early; bracts: proximalmost not involucrelike; stipule width sometimes, but not obviously, exceeding stem diam.; leaf blades orbiculate to flabelliform, unlobed or lobed, base truncate or cordate; widely scattered, local in California but not in Sierra Nevada, also in sw Oregon . . . . 14. *Sidalcea hickmanii*
            8. Inflorescences ± dense, capitate; bracts: proximalmost involucrelike; stipule width exceeding stem diam.; leaf blades ovate to elliptic, unlobed, base cordate; local in n Sierra Nevada, Nevada County, California . . . . 30. *Sidalcea stipularis*
    6. Mericarps: mucro present; involucellar bractlets absent; leaves not evenly arrayed, often proximally crowded at least early in season or not (*S. gigantea*), sometimes rosettelike, proximal and distal blades usually dissimilar.

[9. Shifted to left margin.—Ed.]

9. Plants (0.8–)2(–2.5) m, usually in colonies, rhizomes (6–)10 mm diam., with reflexed, appressed bristle hairs; stems erect proximally, hollow, densely bristly, bristles reflexed; high Cascades and high Sierra Nevada, California. . . . . 10. *Sidalcea gigantea*
9. Plants (0.1–)0.5–1(–2) m, seldom in colonies, without rhizomes or rhizomes usually 3–10 mm diam., hairs usually erect, retrorse or appressed, simple or stellate or glabrescent, sometimes glabrous; stems erect, sprawling, or ascending to decumbent, proximally usually solid, seldom hollow, hairy, hairs sometimes bristlelike; widespread.
10. Inflorescences dense, subcapitate or spiciform, not 1-sided; calyces usually overlapping others in flower, sometimes in fruit.
11. Mericarps smooth, slightly reticulate-veined, or pitted (sometimes slightly wrinkled); inflorescences continuous (flowers sometimes clustered but not regularly interrupted), usually dense in flower, sometimes elongating in fruit; rhizomes present or not.
12. Petals white to pale pinkish, drying yellowish, 10–20 mm; rhizomes present; herbage glabrous, hirsute, puberulent, or stellate-hairy . . . . .4. *Sidalcea candida*
12. Petals pale pink, pinkish rose, pink, or pinkish lavender, dark rose-pink, or magenta, not drying yellowish, (5–)10–15(–23) mm; rhizomes present or not; herbage hairy.
13. Rhizomes absent; inflorescences unbranched or branched, flowers opening and closing sequentially from base to apex, usually 3–10 open at same time; stems usually solid; calyx lobes not strongly veined, usually green; widespread in w North America . . . . .23. *Sidalcea oregana* (in part)
13. Rhizomes present; inflorescences branched, most flowers usually open at same time; stems usually hollow proximally; calyx lobes strongly veined, usually purplish; inland w Oregon and immediate coast from Oregon to Alaska.
14. Mericarps 3 mm, apical margins ± rounded, mucro 0.5–1 mm; calyces usually stellate-puberulent, sometimes glabrate; inland . . . . .7. *Sidalcea cusickii*
14. Mericarps 4 mm, apical margins sharp-edged, not winged, mucro 0.8–1.3 mm; calyces stellate-hairy or proximally glabrous; coastal marshes. . . . . .13. *Sidalcea hendersonii*
11. Mericarps usually reticulate-veined, sometimes pitted; inflorescences dense in flower, elongated or interrupted in fruit; rhizomes or rooting stems sometimes present.
15. Plants without rhizomes, stems not proximally rooting; stems single or multiple, clustered, not rooting; petals not notably whitish- or pale-veined.
16. Calyces 3.5–10 mm in flower and fruit, usually uniformly stellate-puberulent, sometimes also with bristles; inflorescences spicate, 10–30 cm . . . . .23. *Sidalcea oregana* (in part)
16. Calyces 5–9 mm, to 10 mm in fruit, stellate-hairy and bristly; inflorescences subcapitate or spicate, 3–7(–10) cm. . . . . .28. *Sidalcea setosa*
15. Plants with rhizomes or stems proximally rooting; stems usually scattered, not clustered; petals usually whitish- or pale-veined.
17. Inflorescences interrupted, rachis exposed between flower clusters; inland mountains in Kern and Tulare counties, California . . . . 25. *Sidalcea ranunculacea*
17. Inflorescences not interrupted, rachis not exposed between flower clusters; coastal bluffs or mountains, California, Oregon, Washington.
18. Stems erect; inflorescences usually long-pedunculate; plants usually in colonies, with cordlike rhizomes; stems coarsely bristly; calyces 11–16 mm in fruit; nw Oregon, sw Washington . . . . 16. *Sidalcea hirtipes*
18. Stems decumbent to suberect; inflorescences not long-pedunculate; plants in colonies or not, stem bases usually rooting or with compact rhizomes but without cordlike rhizomes; stems often soft-bristly; calyces 8–11(–13) mm in fruit; California, sw Oregon . . . . 19. *Sidalcea malviflora* (in part)

[10. Shifted to left margin.—Ed.]

10. Inflorescences usually ± open, sometimes dense, sometimes 1-sided; calyces usually not conspicuously overlapping others in flower or in fruit.
    19. Mericarps usually glabrous or glandular-puberulent; stems erect, proximally stellate-hairy, distally glabrous, glaucous; inflorescences 30–40(–45) cm; Butte County, California . . . . .27. *Sidalcea robusta*
    19. Mericarps glabrous or puberulent; stems erect, decumbent, sprawling, prostrate, or ascending, glabrous or hairy, glaucous; inflorescences 2–20(–45) cm; not limited to Butte County, California.
        20. Mericarps usually not, sometimes slightly, reticulate-veined or pitted, usually glabrous, rarely sparsely glandular-puberulent; without rhizomes.
            21. Basal leaves deeply incised, lobes again incised, segments linear to oblong-elliptic; cauline leaves 1–3; stems usually unbranched (plants nearly scapose); inflorescences much elongated in fruit; San Bernardino County, California . . .24. *Sidalcea pedata*
            21. Basal leaves absent or unlobed or shallowly lobed; cauline leaves usually 3+; stems often branched (plants usually not appearing to be scapose); inflorescences usually somewhat elongated in fruit; not restricted to San Bernardino County, California.
                22. Roots fleshy; plants usually with nonwoody taproot and fleshy roots; calyces sparsely hirsute, hairs pustulose and sometimes stellate, surface not obscured . . . . .22. *Sidalcea neomexicana* (in part)
                22. Roots not fleshy; plants with fibrous or woody caudex or taproot; calyces densely stellate-puberulent and sometimes with nonpustulose bristles, surface sometimes obscured . . . . .23. *Sidalcea oregana* (in part)
        20. Mericarps usually slightly or coarsely reticulate-veined or pitted, usually densely to sparsely stellate-puberulent or glandular, rarely glabrous; with rhizomes or not.
            23. Roots fleshy; rhizomes and caudices absent; stems clustered; inflorescences usually 20+-flowered; alkaline flats, desert seeps, mountain meadows.
                24. Leaves mostly basal; blades of basal leaves glaucous, 5–7-lobed, deeply incised; calyces stellate-puberulent, without pustulose bristles; e of Sierra Nevada, Inyo County, California. . . . . 6. *Sidalcea covillei*
                24. Leaves basal and cauline; blades of basal leaves glaucous or not, unlobed or 5(–9)-lobed, shallowly incised; calyces stellate-puberulent and with pustulose bristles; widespread . . . . .22. *Sidalcea neomexicana* (in part)
            23. Roots not fleshy; rhizomes and caudices present or absent; stems single or clustered or scattered; inflorescences 2–30+-flowered; usually not alkaline flats.
                25. Inflorescences usually 30+-flowered, 5–20(–30) in *S. virgata*, loosely or densely spiciform, usually branched, usually not 1-sided; petals of bisexual flowers 5–15(–30) mm; stems usually erect, sometimes proximally decumbent.
                    26. Stems erect, decumbent-ascending, or trailing; plants 0.3–0.6(–0.8) m, with rhizomes and taproot; stems proximally hairy, hairs usually soft, tangled, and stellate; inflorescences: proximalmost 1 or 2 flowers usually leafy-bracted . . . . 31. *Sidalcea virgata*
                    26. Stems erect; plants (0.3–)1–2 m, with rhizomes or not; stems proximally hairy, hairs simple or stellate, not soft or tangled; inflorescences: proximalmost 1 or 2 flowers not leafy-bracted.
                        27. Plants 0.5–2 m; petals nearly white to pale pink or pale lavender, usually not overlapping; calyces 8–10 mm in fruit . . . . .3. *Sidalcea campestris*
                        27. Plants 0.3–1.5 m; petals pink, pinkish purple, pink-purple, purplish rose, or magenta, usually overlapping, sometimes not in *S. nelsoniana*; calyces 3.5–10 mm in fruit.

28. Calyces 4–6 mm, lobes usually purple-tinged; plants with rhizomes; mericarps finely reticulate to faintly rugose or nearly smooth; stems glabrous or sparsely hirtellous; primarily Willamette Valley of Oregon, and sw Washington . . . . . 21. *Sidalcea nelsoniana*

28. Calyces 3.5–10 mm, lobes green; plants without rhizomes; mericarps usually slightly reticulate-veined or pitted; stems usually stellate-hairy, sometimes sparsely so; widespread, usually s, e and ne of Willamette Valley . . . . .23. *Sidalcea oregana* (in part)

[25. Shifted to left margin.—Ed.]

25. Inflorescences usually 2–10(–30)-flowered, usually racemose, sometimes spiciform, branched or not, 1-sided or not; petals of bisexual flowers (5–)10–33 mm; stems erect, ascending to decumbent, sprawling, or nearly prostrate.

29. Plants glaucous, without rhizomes; stems not freely rooting, proximally glabrous or hairy, hairs stellate; calyces uniformly stellate-puberulent; frequently at elevations above 2000 m (900–3000 m).

30. Stems usually sprawling or decumbent to ascending, rarely erect; basal leaves 9 or fewer or deciduous, blades usually 5(–7)-lobed, lobes dentate or entire; California, n, c high Sierra Nevada, nw to the Cascades and Klamath ranges and e to Washoe County, Nevada . . . . .11. *Sidalcea glaucescens*

30. Stems usually erect or ascending; basal leaves 10+, persistent, blades usually (5–)7–9-lobed, lobes pinnately or ternately lobed; California, high Sierra Nevada, e to Nevada . . . . .20. *Sidalcea multifida*

29. Plants seldom glaucous, with or without rhizomes; stems freely rooting or not, proximally hairy, hairs simple and stellate; calyces stellate-puberulent, usually bristly; usually at elevations below 2000 m (0–2700 m).

31. Stems ascending-decumbent or creeping, freely rooting, usually proximally hairy, rarely glabrate, hairs simple, 2–3 mm, distally ± stellate-puberulent; leaf blade surfaces hairy, hairs simple; mericarps densely stellate-puberulent apically, on back, and on mucro . . . . 26. *Sidalcea reptans*

31. Stems usually erect to ascending, sometimes prostrate or sprawling to decumbent and rooting or not, proximally hairy, hairs simple, stellate, or mixed; leaf blade surfaces hairy, hairs stellate or not; mericarps sparsely stellate-puberulent or glandular-puberulent apically, on back, and/or on mucro.

32. Leaves mostly basal, cauline usually to 5 and distally smaller, plants sometimes appearing scapose; inflorescences 10+-flowered, (15–)30–45 cm; mericarps slightly to moderately reticulate-veined and pitted . . . . .29. *Sidalcea sparsifolia*

32. Leaves mostly cauline, sometimes clustered at base, distally smaller or not; inflorescences 2–10(–30)-flowered, 2–30(–40) cm; mericarps usually strongly reticulate-veined, roughened, pitted or honeycomb-pitted.

33. Calyces usually both stellate-hairy and bristly-hairy, if bristles absent, some stellate hairs larger; plants not glaucous; leaf blades unlobed or lobed; usually coastal, sometimes inland, seldom on serpentine . . . . 19. *Sidalcea malviflora* (in part)

33. Calyces uniformly stellate-puberulent; plants sometimes glaucous; leaf blades lobed; usually inland, mountains, often on serpentine.

34. Rhizomes usually present, usually 2–4 mm diam., freely rooting; inflorescences ascending; stems sometimes proximally decumbent or prostrate.

35. Stems proximally usually hairy, sometimes glabrate, hairs coarse, stellate; stems not especially brittle; n high Sierra Nevada, Cascade Ranges, high North Coast Ranges, Klamath Ranges, California to sw Oregon . . . . .1. *Sidalcea asprella* (in part)

35. Stems proximally hairy or glabrate, hairs soft, simple and stellate; stems distally brittle; Klamath Ranges, nw California, sw Oregon . . . . 9. *Sidalcea elegans*

34. Rhizomes absent or present, 4–6 mm diam., freely rooting or not; inflorescences usually erect; stems usually erect.

[36. Shifted to left margin.—Ed.]

36. Plants with or without caudex, usually with rooting rootstocks or rhizomes; stems sometimes sprawling, supported by other plants, proximally hairy, hairs stellate; leaf blades usually lobed, all similar in shape, lobe margins usually crenate; c, n High Sierra Nevada, Cascades, California, sw Oregon . . . . .1. *Sidalcea asprella* (in part)

36. Plants with caudex and rootstocks, not freely rooting, without rhizomes; stems usually erect, free-standing, proximally hairy, hairs reflexed, stiff, simple, sometimes also coarse, 2- or 3-rayed, stellate; leaf blades lobed, proximals shallowly incised, lobe margins crenate-dentate, distals deeply incised, lobe margins entire or 1–5-toothed; n inner North Coast Ranges, California . . . . 5. *Sidalcea celata*

1. **Sidalcea asprella** Greene, Bull. Calif. Acad. Sci. 1: 78. 1885 E F

Sidalcea malviflora (de Candolle) A. Gray subsp. *asprella* (Greene) C. L. Hitchcock

Herbs, perennial, 0.1–1(–1.2) m, infrequently ± glaucous, with caudex or not, usually with freely-rooting fibrous rootstocks or rhizomes (5–) 10–30 cm × 2–4 mm, matted or not. **Stems** usually single, erect and sometimes supported by adjacent plants (sprawling), base prostrate or decumbent-ascending to erect, often rooting, solid, not brittle, sometimes ± glaucous distally, proximally stellate-hairy, glabrate, hairs minute or larger and coarse (never simple only), usually 4-rayed, 0.5–1 mm. **Leaves** basal and/or cauline, similar in size and shape; stipules linear to lanceolate, 2–3 × 1.1 mm; petiole (1–)5–10(–15) cm, longest on proximal leaves, 1–4 times longer on proximal leaves to ½ times to as long as blade on distal leaves; blade usually shallowly to deeply palmately 3–7-lobed usually halfway to base, proximal and distal cauline blades rounded to reniform, 2–3 × 2–5 cm, usually wider than long, base cordate to truncate, margins crenate, apex blunt or rounded, lobes narrowest at base, margins usually apically coarsely toothed, rarely entire, surfaces stellate-puberulent. **Inflorescences** ascending or erect, often spiciform, open, calyces not overlapping in flower or fruit, unbranched or branched, 2–15(–30)-flowered, elongate in both flower and fruit, usually 1-sided, 6–11(–30) cm; bracts leaflike to linear, usually 2-fid, (2–)3–5(–15) mm. **Pedicels** 2–5 (–10) mm; involucellar bractlets absent. **Flowers** bisexual or unisexual and pistillate, plants gynodioecious; calyx 5–12 mm, uniformly densely stellate-puberulent; petals pink to pale purple, pale-veined, (5–)10–28 mm, pistillate flowers darker, 5–15 mm; staminal column 4–5 mm, stellate-puberulent; anthers white; stigmas (6)7 or 8. **Schizocarps** 6–8 mm diam.; mericarps (6)7 or 8, 3–4 mm, usually glandular-puberulent to stellate-puberulent, sometimes glabrous, roughened, strongly reticulate-veined, sides and back pitted, mucro 0.5–1 mm. **Seeds** 1.5–2.8 mm.

Subspecies 2 (2 in the flora): w United States.

Sidalcea asprella is variable and occurs from the central Sierra Nevada to southwestern Oregon. Typical plants in the central Sierra Nevada have weak, elongated stems that are often supported by neighboring vegetation; they lack simple recurved hairs at the stem base and may have either elongated rhizomes or a caudex. It has been confused with *S. celata*, *S. elegans*, *S. gigantea*, and *S. glaucescens*; formerly it was included within *S. malviflora*; molecular study has shown that it is different from *S. malviflora*. It belongs to a group including *S. celata*, *S. elegans*, *S. gigantea*, and *S. hirtipes* (K. Andreasen and B. G. Baldwin 2003).

1. Plants (0.3–)0.5–1(–1.2) m, with caudex or usually compact rootstocks or rhizomes to 10(–30) cm × 4 mm; leaves mostly cauline; inflorescences 8–15 (–30)-flowered, erect; stems erect, sometimes weak and supported by other vegetation, sometimes proximally decumbent . . . . 1a. *Sidalcea asprella* subsp. *asprella*

1. Plants 0.1–0.3(–0.4) m, with rhizomes 5–20 cm × 2(–3) mm; leaves mostly basal; inflorescences usually 2–10(–19)-flowered, ascending; stems decumbent-ascending to erect, sometimes proximally prostrate . . . . 1b. *Sidalcea asprella* subsp. *nana*

1a. **Sidalcea asprella** Greene subsp. **asprella** • Harsh checkerbloom E F

Plants (0.3–)0.5–1(–1.2) m, with caudex or usually compact rootstocks or rhizomes to 10 (–30) cm × 4 mm. **Stems** erect, sometimes supported by other vegetation, sometimes proximally decumbent and often rooting to erect, roughly minutely- or long-stellate-hairy, hairs spreading, usually on swollen bases, minutely hairy to glabrate distally. **Leaves** mostly cauline, evenly distributed, not concentrated at stem base, gradually reduced distally; petiole (1–)5–10(–15) cm, those of proximal leaves 4 times as long as blade, reduced distally to ½ times to as long as blade; blades: proximal usually lobed, distal palmately (5–)7-lobed, (3–)5–12(–15) cm wide, base with wide sinus to truncate, margins crenate-serrate, lobe tips oblong, usually 3-toothed (sometimes entire), distalmost 3–5-parted; surfaces stellate-puberulent, sometimes with some simple hairs. **Inflorescences** erect, unbranched or branched, 8–15 (–30)-flowered, sometimes 1-sided; proximalmost bract usually leaflike, 15 × 12 mm, distal bracts linear, 3 mm, shorter than pedicel. **Pedicels** 2–5(–10) mm. **Flowers:** calyx 5–7 mm, to 7–10 mm in fruit; petals pink to pale purple, pistillate usually (5–)10–15 mm, bisexual (15–)20–25 mm; stigmas (6)7 or 8. **Schizocarps** 6–8 mm diam.; mericarps (6)7 or 8, 3–4 mm; prominently reticulate-veined, pitted, margins and back rugose-pitted but less so on back, sometimes glabrous, mucro 0.5–0.8 mm. **Seeds** 1.5 mm. $2n$ = 20, 40, 60.

Flowering (Apr–)May–Jun(–Jul). Open woodlands, wet meadow margins, foothill woodlands, conifer forests, sometimes serpentine; 200–1000(–1800) m; Calif., Oreg.

Subspecies *asprella* has been confused with subsp. *nana*, *Sidalcea celata*, *S. elegans*, and *S. glaucescens*, and it previously has been placed within *S. malviflora*. Taller individuals can be confused with *S. celata* and *S. gigantea*; the adaxial leaf surfaces of *S. celata* generally have simple or two-branched hairs and those of the others have stellate hairs. *Sidalcea gigantea* always has long, retrorse bristle hairs at the stem base, and well-developed rhizomes. *Sidalcea celata* usually has dense, short, retrorse simple hairs at the stem base and rarely some stellate hairs; subsp. *asprella* always has stellate hairs at the stem base, but they can be less dense and short-scabrous or longer and softer. All three taxa occur in Shasta County, California, and some specimens from there may not be easily determined to species; young plants of *S. gigantea* may be easily mistaken as subsp. *asprella* before the stems and rhizomes are fully developed, and the fruits essentially match in both species. Subspecies *asprella* appears to be replaced by subsp. *nana* in northern California and Oregon, and it is especially difficult to be certain of identifications in that area.

1b. Sidalcea asprella Greene subsp. **nana** (Jepson) S. R. Hill, Madroño 56: 105. 2009 • Dwarf harsh checkerbloom E

Sidalcea reptans Greene var. *nana* Jepson, Fl. Calif. 2: 489. 1936; *S. malviflora* (de Candolle) A. Gray subsp. *nana* (Jepson) C. L. Hitchcock

Plants 0.1–0.3(–0.4) m, with caudex or not, with rhizomes freely rooting, 5–20 cm × 2(–3) mm. **Stems** decumbent-ascending to erect, sometimes proximally prostrate, rooting, not brittle, densely, harshly stellate-puberulent to glabrate, hairs 0.1–0.5(–0.7) mm, usually less dense distally. **Leaves** mostly basal, cauline 1–3+, gradually reduced distally; petioles of proximal leaves 5–10 cm, 1–4 times as long as blade, those of distal leaves ½ times to as long as blade; blades: basal usually palmately 7-lobed, or deeply crenate, 2.2–2.7(–4) × 2–2.3(–4) cm, base cordate or sinus wide to narrow, apex rounded, cauline deeply 3–7-lobed, 1.5–3 cm, lobes usually apically 3-toothed or distalmost entire, surfaces hairy, more densely stellate-puberulent adaxially, hairs usually 2–4-rayed (simple), bristly. **Inflorescences** ascending, usually unbranched, subscapose, usually 1-sided, pistillate usually 9–14-flowered, bisexual 2–9(–19)-flowered; bracts linear to lanceolate, stipulelike, 2–2.5(–4) mm, shorter than to as long as pedicel. **Pedicels** 3–4 mm. **Flowers:** calyx 5–7(–12) mm; petals pink, pistillate 9–11 mm, bisexual 15–20(–28) mm; stigmas 7 or 8. **Schizocarps** 7–8 mm diam.; mericarps 7 or 8, 4 mm, reticulate-rugose-veined, sides and back pitted, mucro 1 mm, with few minute bristles. **Seeds** 2.8 mm.

Flowering Jun–Aug(–Sep). Open woodlands, grassy margins, yellow pine-Douglas fir forests, usually serpentine; 400–1900 m; Calif., Oreg.

Subspecies *nana* has been confused with subsp. *asprella*, *Sidalcea celata*, *S. elegans*, and *S. glaucescens* and has been placed within *S. malviflora*. The type specimen (*Jepson 14061*) has small individuals with only two or three flowers and long rhizomes, and it does superficially resemble *S. reptans*, as Jepson suggested, but more robust individuals have more flowers and clearly show vegetative and reproductive similarity to subsp. *asprella*.

Subspecies *nana* occurs from California in the northern Sierra Nevada to southwestern Oregon.

2. Sidalcea calycosa M. E. Jones, Amer. Naturalist 17: 875. 1883 (as Sidalcia) E

Herbs, annual, rarely perennial, (0.2–)0.3–0.5(–0.9) m, not glaucous, with taproot or rhizomelike stolons to 35 cm. **Stems** single, erect or prostrate, usually somewhat reddened, unbranched or branched, solid, hollow in older robust specimens, glabrous or sparsely stellate-hairy. **Leaves** basal and cauline; stipules green or purple, lanceolate to ovate, 2–5(–18) × 1.5–2(–5) mm; petiole 2–5(–30) cm, longest on proximalmost leaves, ½–5 times as long as blade; basal blades usually orbiculate, unlobed or slightly palmately lobed, 1–2(–10) × 2–5(–10) cm, wider than long, margins crenate, shiny; cauline blades rounded, deeply 5–9(–11)-lobed, 1–2 × 2–4 cm, lobes linear-elliptic to oblanceolate-obtuse, margins entire, ciliate, surfaces slightly hirsute. **Inflorescences** erect, spicate, dense, calyces sometimes overlapping, unbranched or branched, distal stem sometimes leafless, many-flowered, not 1-sided, 2–10 cm, elongating at maturity; bracts green or purplish, ovate to wide-elliptic, usually 2-fid, sometimes undivided, 2–6(–12) × 2–5 mm. **Pedicels** (2–)4–5 mm; involucellar bractlets absent. **Flowers** bisexual or infrequently unisexual and pistillate, plants gynodioecious; calyx often purple tinted or scarious, 4–12 mm, silky strigose-bristly at base and on veins, stellate-puberulent; petals usually pale purple or pink, rarely white, base pale or white, (9–)10–25 mm, pistillate shortest; stamens: outer filaments connate to apex, tube funnel-like, with continuous rim to which unstalked anthers attach; staminal column 4–9 mm, hairy in proximal ½; anthers white to pale purple; stigmas 5–9. **Schizocarps** 5–9 mm diam.; mericarps 5–9, often purple tinted, 2.5–4.5 mm, glabrous, not especially roughened, sides reticulate-veined, back deeply longitudinally grooved, mucro often appressed, 0.5 mm. **Seeds** 2–3.5 mm.

Subspecies 2 (2 in the flora): California.

Sidalcea calycosa is variable and found locally in marshes and vernal pools in northern California. The annual, short-taprooted, subsp. *calycosa* varies in height and branching and is consistent in other vegetative and reproductive features. The perennial subsp. *rhizomata* has elongated, amphibious stolons somewhat like those of the fern *Marsilea* and is found in marshes near the coast. It and subsp. *calycosa* have the same fragile nature and similar morphology, and their distinctive fruits are essentially indistinguishable, having the only multifurrowed or striate surfaces in *Sidalcea*. Because *S. calycosa* is not at all fibrous and tough like the other perennials, it is keyed with the annuals.

1. Plants annual, with taproot, stem base rarely decumbent, rooting; bracts 2–6 mm, not obscuring calyx, glabrous or ciliate; calyces 4–7 mm; petals (9–)10–25 mm; relatively widespread in n, c California. . . . . 2a. *Sidalcea calycosa* subsp. *calycosa*
1. Plants perennial, with rhizomelike stolons rooting at nodes; bracts 8–12 mm, mostly obscuring calyx, silky-hairy; calyces 6–12 mm; petals 20–25 mm; restricted to n, c coastal California. . . . . .2b. *Sidalcea calycosa* subsp. *rhizomata*

2a. Sidalcea calycosa M. E. Jones subsp. **calycosa**

• Vernal-pool or annual or hog-walla checkerbloom E

Sidalcea sulcata Curran ex Greene

Plants annual, (0.2–)0.3–0.6 (–0.9) m, with taproot, sometimes fleshy, usually glabrous. **Stems** usually erect, rarely proximally decumbent, rooting. **Leaves** mostly cauline, basal leaves usually persistent, 1–3; stipules lanceolate to ovate, 2–5 mm; petiole of proximalmost leaves often 4–5 times as long as blade, that of distal leaves ½ times to as long

as blade; blades: basal unlobed, 2–5 cm wide, margins crenate, distal as in species. **Inflorescences** single or multiple from distal nodes; bracts ovate to wide-elliptic or narrower, sometimes 2-fid, 2–6 mm, shorter than or equaling pedicel, not obscuring calyx, glabrous or ciliate. **Pedicels** 4–5 mm. **Flowers:** calyx 4–7 mm; petals (9–)10–25 mm. **Schizocarps** 5–9 mm diam.; mericarps 2.5 mm. **Seeds** 2 mm.

Flowering Mar–Jun(–Aug). Wet places, especially vernal pools, hog wallows, swales, foothill woodlands and chaparral openings; 0–1200 m; Calif.

Subspecies *calycosa* is typically a spring-flowering, annual, taprooted plant of the vernal pool regions of northern and central California and can be locally common. It is variable in size dependent on local growing conditions; dwarfed or nearly leafless plants are often found.

2b. Sidalcea calycosa M. E. Jones subsp. **rhizomata**

(Jepson) Munz ex S. R. Hill, Madroño 56: 106. 2009

• Point Reyes checkerbloom C E

Sidalcea rhizomata Jepson, Man. Fl. Pl. Calif., 629. 1925

Plants perennial, 0.3–0.6 m, with long, narrow, rhizomelike stolons, rooting at nodes. **Stems:** flowering erect-ascending, vegetative prostrate, proximally rooting. **Leaves** basal and cauline, scattered on rhizomes, denser on flower stems; stipules ovate, 12–18 mm; petiole of proximal and rhizome leaves 3–4 times as long as blade, that of distal leaves ½ times to as long as blade; blade 2.5–10 cm wide, unlobed with margins shallowly crenate on proximal and vegetative stems to deeply lobed with margins entire on flowering stems. **Inflorescences:** bracts ovate to wide-elliptic, simple or usually 2-fid, 8–12 mm, 3–4 times longer than pedicel, mostly obscuring calyx, silky-hairy. **Pedicels** 2–4 mm. **Flowers:** calyx often reddish-purple tinted, 6–12 mm; petals 20–25 mm. **Schizocarps** 8–9 mm diam.; mericarps 4.5 mm. **Seeds** 3–3.5 mm.

Flowering May–Jul(–Sep). Freshwater and salt marshes near coast; of conservation concern; 0–30 m; Calif.

Subspecies *rhizomata* is known from only a few marshes near the coast in Marin, Mendocino, and Sonoma counties; it is easily distinguished by its relatively large, connate, ciliate bracts and prostrate fleshy rhizomes with scattered leaves and relatively large stipules.

3. Sidalcea campestris Greene, Bull. Calif. Acad. Sci. 1: 76. 1885 • Meadow checkerbloom E

Sidalcea asplenifolia Greene; *S. sylvestris* A. Nelson

Herbs, perennial, 0.5–2 m, sometimes glaucous in age, with thick, rather woody taproot and short rhizomes usually to 10 cm × 10 mm. **Stems** single or clustered, erect, base often decumbent-ascending, solid (proximally hollow on older stems), proximally densely bristly-hirsute, hairs simple or forked, 0.5–1 mm, sometimes mixed with minute, stellate hairs, sometimes glabrous and glaucous in age, hairs usually more appressed, simple, stellate, or sparse distally. **Leaves** mostly cauline; stipules lanceolate, 4–11 × 1–1.5 mm; petioles of proximal leaves 15–20 cm, 2–3 times as long as blades, distal reduced to 0.5–10 cm, ½ times to as long as blades; blade 10–15 × 10–15 cm, smaller distally, surfaces scabrid-hairy, hairs dense, simple or stellate, stiff, stellate hairs mostly on adaxial surfaces, proximal blades orbiculate, shallowly to deeply palmately 7–9-lobed, 5–15 × 5–15 cm, base cordate, margins coarsely crenate-serrate, lobes apically 2–5-toothed; distal cauline leaves variable, mid blades usually palmately divided nearly to base into 5–7 cuneate lobes, 15 × 15 cm, lobes deeply cut to laciniately dissected, distal blades divided into (3–)5–7 linear, marginally subentire segments, surfaces glabrescent or with few hairs on abaxial surface veins, ciliate. **Inflorescences** erect, open, spiciform, calyces not conspicuously overlapping except sometimes in bud, often branched from distal leaves, 15+-flowered, elongate, not 1-sided, 10–35 cm, proximal flowers spaced several cm apart, not leafy-bracted; bracts linear, distal undivided or 2-fid, proximal separate to base, 4–10 mm, usually equaling or longer than pedicels. **Pedicels** 3–6(–20) mm; involucellar bractlets absent. **Flowers** bisexual or unisexual and pistillate, plants gynodioecious; calyx 5–9 mm, pistillate 5–7 mm, bisexual 6–9 mm, 8–10 mm in fruit, uniformly, densely stellate-hairy or with coarser, longer, stellate hairs to 2 mm; petals usually not overlapping, nearly white to pale pink or pale lavender, pistillate 9–12 mm, bisexual 13–25 mm; staminal column 5–7 mm, hairy; anthers white to pale pink; stigmas (6)7 or 8. **Schizocarps** 7–8 mm diam.; mericarps (6)7 or 8, 3.5 mm, roughened, sides prominently reticulate-rugose and pitted, back less so and glandular-puberulent, mucro 0.5–1 mm. **Seeds** 2.5 mm. $2n = 60$.

Flowering May–Aug. Open shrublands, meadows, hedgerows, prairies; 40–200 m; Oreg.

Sidalcea campestris is one of the taller species of *Sidalcea* and can be distinguished also by its range, hirsute indument, long inflorescences with spaced, long-stalked flowers, narrow very pale petals, and deeply incised pinnatifid stem leaves. It has been confused with *S. hirtipes*, *S. nelsoniana*, and *S. oregana*; it differs especially in characters of rhizomes, inflorescences, and leaves. It is a candidate for listing as threatened or endangered in Oregon. Specimens from British Columbia and Washington identified as *S. campestris* are either *S. campestris* escaped from cultivation or *S. hendersonii*. It is known as a native only from the Willamette Valley area (Multnomah and Washington to Benton and Linn counties). Apparently, it was introduced near Seattle, Washington; it may not persist there.

4. Sidalcea candida A. Gray, Mem. Amer. Acad. Arts, n. s. 4: 24. 1849 • White checkerbloom E

Sidalcea candida var. *glabrata* C. L. Hitchcock; *S. candida* var. *tincta* Cockerell

Herbs, perennial, 0.3–1 m, ± glaucous, with rhizomes wide-spreading, compact to elongate, 5 mm diam. **Stems** single, clustered in older plants, erect, sometimes branched distally, solid, glabrous or moderately hirsute, hairs retrorse, simple proximally, becoming ± finely stellate-hairy distally. **Leaves** basal and mostly cauline, not evenly arrayed; stipules deciduous, ovate, 10 × 2–2.5 mm; petiole (4–)6–18 cm, usually ½ times to as long as blade; blade orbiculate, (4–)8–20 × (4–)8–20 cm, surfaces glabrous or sparsely hairy with erect or appressed simple hairs most commonly on veins abaxially, somewhat ciliate; basal unlobed or shallowly palmately 5–7-lobed, 4–15 × 4–15 cm, margins coarsely crenate, distal successively more deeply incised ½+ to base, 5–20 cm wide, lobes 2- or 3-dentate at apex, distalmost divided to base, 3–5(–7)-lobed, lobes 4–10 cm wide, margins entire. **Inflorescences** erect, ± spiciform, dense distally, calyces overlapping, branched or unbranched, 20+-flowered, not 1-sided, 7–10 cm; bracts elliptic to ovate, deeply 2-fid, 5–7 mm, shorter than or subequal to calyx, longer than pedicel. **Pedicels** 2–7 mm (10–20 mm in proximalmost flower); involucellar bractlets absent. **Flowers** bisexual; calyx 7–9 mm, somewhat accrescent, stellate-hairy, ciliate with simple to stellate hairs, puberulent (sometimes short-glandular); petals often not overlapping, white to pale pinkish, drying yellowish, 10–20 mm; staminal column 6–9 mm, hairy; anthers bluish pink; stigmas 6–9. **Schizocarps** 5–7 mm diam.; mericarps 6–9, 3–5.5 mm, sides smooth or slightly reticulate-veined, back less so, not pitted, top minutely hairy, mucro 0.5–0.8 mm. **Seeds** 2 mm. $2n = 20$.

Flowering (May–)Jun–Aug(–Sep). Moist stream banks and meadows, mountains; (1400–)2000–3200 m; Colo., Nev., N.Mex., Utah, Wyo.

Sidalcea candida is one of the more eastern species of *Sidalcea* (along with *S. neomexicana*) and is also one of the more easily distinguished because of its relatively large, crowded flowers with white petals, deep, wide-spreading rhizomes, pinkish-blue anthers, and nearly glabrous stems and leaves. It has become a popular garden plant. Hitchcock recognized two varieties based on calyx indument: var. *candida* with the calyx uniformly hairy, and var. *glabrata* with the calyx much more hairy at the base than on the lobes. Cockerell distinguished plants with somewhat pink petals as var. *tincta*. These differences do not appear to be taxonomically significant, and the varieties lack geographic coherence.

5. **Sidalcea celata** (Jepson) S. R. Hill, Madroño 56: 106. 2009 • Redding checkerbloom E

Sidalcea malviflora (de Candolle) A. Gray var. *celata* Jepson, Fl. Calif. 2: 493. 1936 (as malvaeflora); *S. malviflora* subsp. *celata* (Jepson) C. L. Hitchcock

Herbs, perennial, 0.4–0.8(–1) m, sometimes glaucous, with caudex and thick, woody rootstocks, not freely rooting, without rhizomes. **Stems** mostly single, usually erect, sometimes branched distally, solid (sometimes hollow in older, robust plants), sometimes glaucous, proximally densely bristly-hairy, hairs reflexed, stiff, simple, sometimes also coarse, 2- or 3-rayed, stellate, 1–1.6 (–2.5) mm, pustulate, sparser distally, with minute, stellate hairs in inflorescence. **Leaves** mostly basal, some cauline; stipules usually early-deciduous, linear-lanceolate, 3–7 × 1 mm; proximal petioles 15–18 cm, 3–4 times as long as blade, distal usually to ½ times to as long as blade; blades: basal rounded, usually palmately 7-lobed, shallowly incised, 4–8 × 4–8 cm, lobes 1.5–2 cm wide, margins crenate-dentate, 3–7-toothed, apices obtuse; distal 3–10, palmately 5-lobed, deeply incised, 2–6 × 2–6 cm, lobes narrow, linear, 2–3.5 × 2–4 mm, margins entire or 1–5-toothed, surfaces: abaxial stellate-hairy, hairs 6-rayed, 1.5 mm, adaxial hairy, hairs simple, 1.5 mm, distalmost glabrate adaxially and sometimes simple-hairy. **Inflorescences** erect, open, calyces not conspicuously overlapping except sometimes in bud, unbranched or branched, (5–)10–12(–23)-flowered, elongate, sometimes 1-sided, 10–40 cm; bracts lanceolate, usually undivided, 3–4 mm, shorter than pedicels and calyx. **Pedicels** (3–)5–10(–15) mm; involucellar bractlets absent. **Flowers** usually bisexual, infrequently unisexual and pistillate, plants gynodioecious; calyx (7–)9–10 mm, to 12–14 mm in fruit, densely stellate-puberulent; petals pale pinkish lavender, often pale-veined especially when dry, pistillate 10–20 mm, bisexual 20–25(–31) mm; staminal column 7–8(–10) mm, sparsely puberulent; anthers white; stigmas (6)7(8). **Schizocarps** 6–8 mm diam.; mericarps (6)7(8), 3–4 mm, minutely glandular-puberulent, glabrescent, margins sometimes sharp-edged, roughened, strongly reticulate-veined, sides and back deeply pitted, honeycomblike, median line on back but not furrowed, mucro 1 mm. **Seeds** 2.5 mm. $2n = 60$.

Flowering May–Jun(–Aug). Open oak woodlands, wet sites, sometimes on serpentine; 100–400(–1600) m; Calif.

Sidalcea celata has been recognized as a local species endemic to Shasta and Tehama counties (especially near Redding) in the northern inner North Coast Ranges; it was previously included within *S. malviflora*. Hitchcock's concept of *S. celata* as a subspecies was broader and included plants from additional counties that do not match the type very well; most have been re-identified as *S. asprella* or *S. oregana*. *Sidalcea celata* has been confused with *S. asprella*, *S. gigantea*, and *S. robusta* and appears to be variable. Its narrow distribution, preference for serpentine, typical lack of elongated rhizomes, basally retrorsely hirsute stems, presence of basal leaves, and relatively large bisexual flowers and fruits are distinctive.

6. **Sidalcea covillei** Greene, Cybele Columb. 1: 35. 1914 • Owens Valley checkerbloom E

Sidalcea neomexicana A. Gray var. *covillei* (Greene) Roush

Herbs, perennial, 0.2–0.6 m, often glaucous, with fleshy, simple to clustered roots, without caudex or rhizomes. **Stems** several, clustered, erect, solid, often glaucous proximally, base sparsely, finely to coarsely stellate-hairy or hispid, hairs smaller distally. **Leaves** 2–5 per stem, mostly basal; stipules linear-lanceolate, 3–6 × 1 mm; petiole (4–)5–10 cm, reduced on cauline leaves, proximal 1–4 times as long as blade, distal ½ times to as long as blade; blade fleshy, glaucous, rather densely stellate-hairy, proximalmost usually shallowly to deeply, ternately, palmately 5–7-lobed, 1.5–3 × 1.5–4 cm, lobes obovate, margins crenate-dentate, distal deeply 3–7-lobed, lobes linear, distalmost 2–4 cm wide. **Inflorescences** erect, open, calyces not conspicuously overlapping except sometimes in bud, branched or unbranched, nearly scapose, often 20+-flowered, slender, elongate, 1-sided or not, 6–30 cm; bracts inconspicuous, linear, 2-fid, 2–4 mm, shorter than calyx and pedicels. **Pedicels** 2–8(–10) mm; involucellar bractlets absent. **Flowers** bisexual; calyx 5–8 mm, uniformly, densely stellate-puberulent or few with longer rays; petals pale pink-lavender, veins paler, 10–15 mm; staminal column 4–5 mm, hairy; anthers white; stigmas 5 or 6. **Schizocarps** 5 mm diam.; mericarps 5 or 6, 2.5 mm, sparsely glandular-puberulent, roughened, back reticulate-veined, sides strongly so, not pitted, mucro 0.1–0.3 mm. **Seeds** 1 mm. $2n = 20$.

Flowering (Apr–)May–Jun. Alkaline flats, springs, meadows; 1100–1400 m; Calif.

Sidalcea covillei is one of two species of *Sidalcea* (along with *S. neomexicana*) with fleshy roots and adapted to alkaline conditions on flats. Its range (Owens Valley in Inyo County) and specialized habitat have made it vulnerable to any lowering of the water table and to grazing; it is listed as endangered in California. Most individuals of it were destroyed by construction of the Haiwee Reservoir. Once thought to have been extirpated, it was subsequently rediscovered.

7. **Sidalcea cusickii** Piper, Proc. Biol. Soc. Wash. 29: 99. 1916 • Cusick's checkerbloom E

Sidalcea cusickii subsp. *purpurea* C. L. Hitchcock; *S. oregana* (Nuttall ex Torrey & A. Gray) A. Gray var. *cusickii* (Piper) Roush

Herbs, perennial, 0.4–1.8 m, not glaucous, with thick taproot and freely-rooting rhizomes usually 3–10 mm diam. **Stems** usually several, clustered, erect, often purplish, often thick and hollow proximally, usually proximally ± scabrous, hairs stellate and distally stellate-hairy, sometimes glabrous. **Leaves** mostly cauline; stipules persistent or deciduous, lanceolate, 6–8 × 1–1.5 mm; proximal petioles 25–30 cm, 4–5 times as long as blade, reduced to 7–13 cm on midstem leaves and to 2–3 cm on distalmost leaves, these to ½ times as long as blades; basal blades wide-ovate to orbiculate, shallowly to deeply, palmately 5–7(–9)-lobed, 6–13 × 6–13 cm, lobe margins toothed, teeth rounded, sometimes *Ribes*-like, base cordate, apex rounded, surfaces: abaxial densely stellate-pubescent, adaxial glabrous or sparsely stellate-pubescent on veins, cauline blades round, divided nearly to base, 5–9-lobed, lobes again deeply incised to subentire, 10–20 × 10–20 cm, distalmost 5–7-cleft to base. **Inflorescences** erect, spiciform, dense, calyces usually conspicuously overlapping in flower and sometimes in fruit, usually 20–30-branched per primary stem, 20+-flowered, elongate, not 1-sided, most flowers usually open at same time, branches relatively short, each spike unit 6 cm, longer in fruit; bracts ovate-lanceolate to subulate, undivided, 1–6(–10) mm, subequal or longer than pedicels, much shorter than calyx. **Pedicels** 1–2(–5) mm; involucellar bractlets absent. **Flowers** bisexual or unisexual and pistillate, pistillate plants more frequent than bisexual ones, some flowers may be staminate, plants gynodioecious; calyx urceolate with swollen base especially in young fruit, pistillate 6–8 mm, bisexual 6–10 mm, strongly reticulate-veined, usually finely stellate-puberulent, sometimes glabrate, lobes usually purple tinted; petals 7 or 8, pale pink, pinkish rose, or rose-purple, usually neither pale-veined nor white at base, pistillate 8–12(–14) mm, staminate or bisexual (10–)11–19(–23) mm; staminal column 5–6 mm, hairy; anthers white; stigmas 7 or 8. **Schizocarps** 6 mm diam.; mericarps 7 or 8, 3 mm, apical margins ± rounded, sides smooth or slightly reticulate-veined, not pitted, back essentially smooth, very sparsely glandular-puberulent near tip, mucro 0.5–1 mm. **Seeds** 2 mm. $2n = 20$.

Flowering Jun–Aug. Moist to wet, mostly black, adobe soil, lowland and mountain meadows, often with *Juncus* and *Camassia*; 100–200(–500), 1000–1400 m; Oreg.

Sidalcea cusickii is showy and distinctive. It has been confused with *S. campestris*, *S. hendersonii*, *S. nelsoniana*, *S. oregana* (subsp. *spicata*), *S. setosa*, and *S. virgata*. Molecular data support a close relationship with *S. hendersonii* and *S. virgata* (K. Andreasen and B. G. Baldwin 2003b). Hitchcock distinguished subsp. *cusickii* and subsp. *purpurea*, the latter with purple, ciliate calyx lobes and stem proximally glabrous and thought to be more closely related to *S. hendersonii* and *S. nelsoniana*. The two variations overlap in range and characteristics and are not recognized here. There appear to be two groups of populations at different elevation ranges with little overlap. Overall, *S. cusickii* can be distinguished by its range and by its spikelike, many-flowered, thick, compounded racemes in which essentially all of the flowers are open at the same time. It is also unusual in the preponderance of pistillate-flowered individuals in a given population and the presence of what appear to be truly staminate, rather than bisexual, flowers on some individuals. It is found in the Willamette and Umpqua valley region in Douglas, Josephine, and Lane counties.

8. **Sidalcea diploscypha** (Torrey & A. Gray) A. Gray, Mem. Amer. Acad. Arts, n. s. 4: 19. 1849 • Fringed checkerbloom E F

Sida diploscypha Torrey & A. Gray, Fl. N. Amer. 1: 234. 1838; *Sidalcea diploscypha* var. *minor* A. Gray; *S. secundiflora* Greene

Herbs, annual, 0.2–0.7 m, not glaucous, with taproot. **Stems** single, erect, usually branched distally, solid, both short-stellate-puberulent and long soft bristly-hairy, hairs erect. **Leaves** basal, early-deciduous, and cauline; mid to distal stem stipules divided into 2–5 filiform or linear segments, involucrelike, 10+ × 1 mm; petiole (4–)6–20(–50) cm, usually ½ times to as long as blade; basal leaf blades orbiculate, unlobed, 1–2.5 × 1–2.5 cm, base cordate, margins crenate, apex rounded; cauline leaf blades orbiculate, palmately 5–7-lobed, (1–)2–6 × (1–)2–6 cm, lobes linear distally, sometimes 3-toothed or -lobed, then

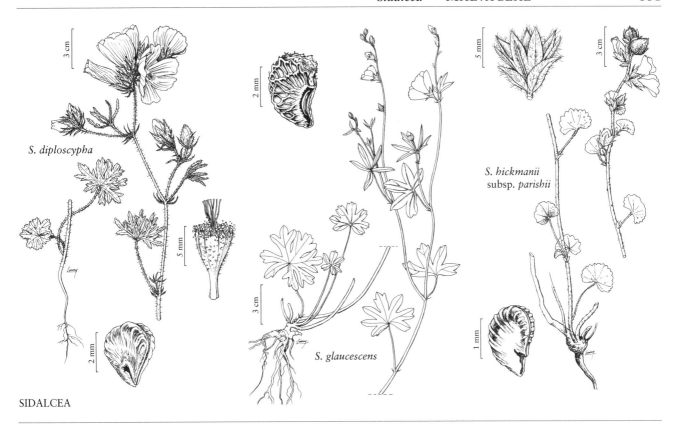

S. *diploscypha*

S. *glaucescens*

S. *hickmanii*
subsp. *parishii*

SIDALCEA

midtooth or lobe much longer than laterals, margins entire, surfaces bristly-puberulent. **Inflorescences** erect, dense, calyces overlapping, occasionally short-branched, clusters to 10-flowered, subumbellate to elongate in age, not 1-sided; bracts linear or filiform, palmately 2–7-lobed, 8–12 mm, lobes linear, usually becoming involucrelike, 1–2.5 cm, subequal to or longer than calyx. **Pedicels** 1–3 mm, (short branches may easily be mistaken for pedicels); involucellar bractlets absent. **Flowers** bisexual; calyx 8–12 mm, not much accrescent, lobes often with narrow purple line or spot at lobe base inside, outer surface bristly-hairy and stellate-puberulent, seldom densely glandular, multicellular hairs usually few or absent; petals dark pink to deep purple, veins often paler, darker patch sometimes at base, 20–35 mm; staminal column 4–6 mm, hairy; anthers sessile on rim, white; stigmas 5 or 6. **Schizocarps** 6–7 mm diam.; mericarps 5 or 6, sometimes pinkish when fresh, 2.5 mm, glabrous, back minutely hairy, back and sides reticulate-veined, back with prominent midvein, not pitted, mucro absent. **Seeds** 2 mm. $2n = 20$.

Flowering Apr–May(–Jun). Grasslands, open woodlands, valleys, near vernal pools, usually on serpentine; 0–900 m; Calif., Oreg.

Sidalcea diploscypha is widespread in central and northern California and occurs also in Douglas County, Oregon, where it is apparently introduced. Young plants, even in flower, may resemble *S. keckii*, and transitional plants are known; *S. diploscypha* generally differs from *S. keckii* by its longer divided bracts, usually entire lobes

on its distal stem leaves, simple bristles on the calyx, bristles absent at the standard mucro position on the mericarp, relatively few glandular and multicellular hairs, and generally clustered flowers and fruits. Plants in Colusa, Napa, Solano, and Yolo counties, California, are sometimes hard to distinguish from *S. keckii*, and vice versa. Some plants in Butte and Lake counties, California, also show some transitional features; none have yet been assigned to *S. keckii*.

9. **Sidalcea elegans** Greene, Cybele Columb. 1: 35. 1914
• Del Norte checkerbloom [E]

Sidalcea malviflora (de Candolle) A. Gray subsp. *elegans* (Greene) C. L. Hitchcock

Herbs, perennial, 0.2–0.6(–0.8) m, sometimes ± glaucous, with woody taproot or caudex and rhizomes, freely rooting, 20–30 cm × 2–4 mm, often mat-forming, forming clones often 1–8 m². **Stems** scattered, erect to ascending, base decumbent to erect, usually rooting freely, solid, proximally hairy or glabrate, hairs soft, simple and stellate, distally brittle, easily broken, glabrous-glaucous. **Leaves** mostly basal or cauline 3 or 4 on proximal ⅓ of stem, much reduced; stipules wide-lanceolate to ovate, 3–4 × 1–1.5 mm; petioles of proximal leaves 6–12 cm, 2–4 times as long as blades in basal leaves, those of cauline leaves greatly reduced to ½ times or as long as blades; blade rounded to reniform, palmately (3–)5–7-lobed, usually (1–)2–5(–10) × (1–)2–5(–10) cm, apex rounded, surfaces: abaxial harshly stellate-hairy, adaxial usually simple-haired, basal blades shallowly incised, lobes with 3 deep crenations, cauline blades usually deeply 3–5(–7)-lobed nearly to base, lobe margins dentate or entire. **Inflorescences** ascending, open, calyces not conspicuously overlapping except sometimes in bud, unbranched or few-branched, loosely (3–)5–10(–20)-flowered, not greatly elongated, 1-sided, 10–20 cm; bracts narrowly elliptic, shallowly 2-fid, proximal bracts divided to base and often with leaf remnant between, 3–4 mm, usually shorter than to equaling pedicels. **Pedicels** 3–4(–10) mm; involucellar bractlets absent. **Flowers** bisexual or unisexual and pistillate, plants gynodioecious; calyx 7–10 mm, slightly enlarged in fruit, uniformly, coarsely stellate-puberulent (some rays sometimes longer than others); petals dark pink, pale-veined at least when dry, pistillate 10–15(–20) mm, bisexual 20–25(–33) mm; staminal column 4–5 mm, sparsely hairy; anthers white to pale pink; stigmas 6 or 7(8). **Schizocarps** 6–7 mm diam.; mericarps 6 or 7(8), 3–3.5 mm, back not ribbed, roughened, sides and back strongly reticulate-veined-rugose and pitted (honeycomblike), top minutely glandular-puberulent, mucro 0.8–1 mm. **Seeds** 2 mm. $2n = 40, 60$.

Flowering (May–)Jun–Jul(–Sep). Open, dry woodlands, usually on serpentine; 100–200(–900) m; Calif., Oreg.

Sidalcea elegans has been included within *S. malviflora*; it is easily distinguished by its relatively long, slender, shallow rhizomes; open, one-sided inflorescences; and thin, brittle stems. It resembles *S. glaucescens* in its leaves and inflorescence and is easily distinguished by its rhizomes and soft, simple hairs at the stem base. It has been confused with *S. asprella*, from which it is distinguished by its simple, flexible hairs at the stem base and by its more developed and elongated rhizomes and variable leaves. *Sidalcea elegans* occurs in northwestern California and southwestern Oregon.

10. **Sidalcea gigantea** G. L. Clifton, R. E. Buck & S. R. Hill, Madroño 56: 285, figs. 1–3. 2010 • Giant checkerbloom E

Herbs, perennial, (0.8–)2(–2.5) m, not glaucous except on stems, usually in colonies, with rhizomes to 40–60 cm × (6–)10 mm, glabrescent with reflexed-appressed bristle hairs 2.5 mm. **Stems** single, scattered, erect, usually purple tinted proximally, hollow especially toward base, pithy distally, 10–14 mm diam. just distal to base, often glaucous, proximally densely retrorsely bristly-hairy and stellate-hairy, hairs pustulate, 1.5–2.5 mm, distally sometimes glabrous. **Leaves** mostly cauline, basal usually absent; stipules deciduous, with pink band on stem at base, narrowly lanceolate, (3.5–)5(–8) × 0.7(–1.5) mm; petioles of proximal leaves 6–8 cm, those of midstem leaves 10–14 cm, 1–2 times length of blade, apex curved with swollen portion or pulvinus 5–6 × 1.8 mm; blades of proximalmost cauline leaves rounded, shallowly, palmately 4- or 5-lobed, 0.5–0.7 × 0.5–0.7 cm, lobe margins irregularly, sparsely dentate, apices rounded, mid-cauline blades 5–7-lobed, 6.5–12 × 10–13 cm, wider than long, gradually reduced distally, lobes straplike, divided ¾ to base, lobes 5.5 × 1.5–1.8 cm, margins coarsely dentate, surfaces sparsely, minutely hispid and stellate-puberulent, distal leaves deeply 5-lobed, otherwise similar in shape, leaves immediately below inflorescence greatly reduced, irregularly 2- or 3-lobed. **Inflorescences** erect, spiciform, open, calyces usually not overlapping, usually branched, branches 14–18 cm, each branch 10–20 flowered, not 1-sided, elongate, 5–20 cm; bracts 1, usually undivided, narrowly lanceolate, often 2-fid, 2.5 mm, equaling or slightly longer than pedicel, much shorter than calyx. **Pedicels** 2–3(–5) mm; involucellar bractlets absent. **Flowers** bisexual or unisexual and pistillate, plants gynodioecious; calyx 5–6 mm, to 8 mm in fruit, uniformly, densely stellate-puberulent, surface obscured; petals pale pink, pale-veined, pistillate 7–9 mm, bisexual (10–)14–20(–25) mm; staminal column 6–8 mm, stellate-puberulent; anthers white to cream or pale yellow; stigmas (6)7 or 8. **Schizocarps** 6–7 mm diam.; mericarps (6)7 or 8, sides 3 mm, thick, back and margins thick, rounded, reticulate-veined, pitted, back with prominent groove, top sparsely glandular-stellate-puberulent, mucro 1 mm. **Seeds** 1.5 mm.

Flowering Jul–Sep. Moist to wet, forested slopes, seeps, stream margins, meadows, coniferous forests; (600–)900–1700 m; Calif.

Sidalcea gigantea is likely the tallest *Sidalcea* species; it can be distinguished also by its range (high Cascades and the northern high Sierra Nevada), retrorse hirsute stem bases, thick, hollow stems, and massive, hirsute rhizome systems. Most large patches appear to be clonal and generally produce either bisexual or unisexual, pistillate stems. *Sidalcea gigantea* is closely related to, and has been confused with, both *S. asprella* and *S. celata*. Its leaves are most like those of *S. asprella* in that they are usually similar in shape throughout the stem; the tall stems and massive rhizome systems distinguish it from *S. asprella*, as do its occurrence at higher elevations and its later flowering time.

11. Sidalcea glaucescens Greene, Bull. Calif. Acad. Sci. 1: 77. 1885 • Waxy checkerbloom E F

Sidalcea montana Congdon

Herbs, perennial, 0.2–0.5 (–0.7) m, glaucous, with taproot and caudex, without rhizomes. **Stems** usually few to many, clustered, sprawling or decumbent to ascending, rarely erect, not rooting, solid, glaucous, proximally usually stellate-puberulent, sometimes glabrous, distally glabrous. **Leaves** basal and cauline, basal leaves 9 or fewer or deciduous; stipules lanceolate, (2–)3–5(–6) × 0.5–1.5 mm; petioles of basal and proximal cauline leaves 6–14 cm, 3–4 times as long as blades, reduced distally to ½ times to as long as blades; blade reniform-orbiculate, palmately 5(–7)-lobed, deeply incised, 2–6(–8) × 2–6(–8) cm, glaucous, surfaces glabrous or minutely stellate-puberulent, lobes shallowly dentate, more deeply divided on distal leaves, margins entire, distalmost sometimes linear, unlobed. **Inflorescences** ascending, open, calyces not conspicuously overlapping except sometimes in bud, usually unbranched, 3–10 (–20)-flowered, elongate, 1-sided, 8–20 cm, axis curved between flowers, sometimes zigzag in appearance; bracts linear to lanceolate, distinct or connate and 2-fid, 5 mm, proximal divided to base, distal often undivided, shorter than to equaling pedicels. **Pedicels** 2–3(–10) mm; involucellar bractlets absent. **Flowers** bisexual or unisexual and pistillate, plants gynodioecious; calyx 5–10 mm, enlarging in fruit, hairy, hairs scattered, minute, stellate and sometimes capitate, glandular; petals pink to pink-purple, pale-veined at least when dry, pistillate (7–)9–12 mm, bisexual 15–20(–25) mm; staminal column 4–7 mm, stellate-hairy; anthers pale yellow or pinkish to white; stigmas 6–8. **Schizocarps** 5–7 mm diam.; mericarps 6–8, 3–3.5 mm, roughened, sides reticulate-veined and deeply pitted, back reticulate-veined and glandular-puberulent, mucro 0.3–1 mm. **Seeds** 2 mm. $2n = 40$.

Flowering (May–)Jun–Aug(–Sep). Dry, grassy meadows, open, usually red fir, juniper, or ponderosa pine forests, often serpentine; (900–)1500–3000 m; Calif., Nev.

Sidalcea glaucescens is a relatively common, generally low-growing plant of relatively high elevations; it occurs from the central and northern Sierra Nevada to the southern Cascade and Klamath ranges and to north of Reno, Nevada. It usually can be distinguished by its highly glaucous, waxy stems and leaves, 3–5-lobed, entire-margined leaves, and basal leaves that wither by the time fruit is mature; additionally, proximal flowers are spaced several centimeters apart and leafy-bracted, and the inflorescence axis is curved between flowers.

It has been confused with *S. asprella*, *S. elegans*, and *S. multifida*, to which it appears to be closely related. It can generally be distinguished from *S. multifida* by its 5(–7)-lobed leaves, the lobes shallowly incised or entire, its nonpersisting, fewer basal leaves, and its more-procumbent habit. *Sidalcea elegans* and *S. virgata* in southwestern Oregon also have been confused with *S. glaucescens*.

12. Sidalcea hartwegii A. Gray, Mem. Amer. Acad. Arts, n. s. 4: 20. 1849 (as hartwegi) • Hartweg's checkerbloom E

Sidalcea hartwegii var. *tenella* (Greene) Baker f.; *S. tenella* Greene

Herbs, annual, 0.1–0.4(–0.6) m, not glaucous, with taproot. **Stems** single, erect, unbranched or distally branched, often zigzag, solid, proximally usually glabrous or sparsely stellate-puberulent. **Leaves** cauline; stipules deciduous or inconspicuous, sometimes purplish, subulate, 1–1.5(–3) × 0.5–1 mm; petiole 0.5–1.5(–3) cm, usually ½ times to as long as blade; blade rounded, proximal blades early-deciduous, shallowly lobed, 0.7–2.5 × 0.7–2.5 cm, distal blades deeply palmately (3–)5–7-lobed, 2–4(–6) × 2–4(–6) cm, lobes linear, usually unlobed, rarely 2- or 3-lobed, lobe margins entire, surfaces: abaxial sparsely stellate-puberulent, adaxial glabrate. **Inflorescences** erect, calyces overlapping, unbranched or branched, usually (2–)4–6(–10)-flowered, not elongate, not 1-sided, 4–10 cm, not much longer in fruit; bracts lanceolate, usually shallowly 2-fid, 2–3 mm, shorter than pedicels. **Pedicels** 2–3 mm; involucellar bractlets absent. **Flowers** bisexual or unisexual and pistillate, plants gynodioecious; calyx 8–10(–12) mm, not much enlarging in fruit, stellate-canescent, sparsely ciliate; petals pink to rose-purple or white, pale-veined, often whitened at base, 18–20(–25) mm; stamens: outer filaments incompletely connate, distally distinct, anthers not attached to connate portion of filaments; staminal column 6–7 mm, hairy; anthers white, stalked, aborted in pistillate flowers; stigmas 6 or 7. **Schizocarps** 5–7 mm diam.; mericarps 6 or 7, 2.5–4 mm, sides smooth, margins rugose, back reticulate-veined, deeply pitted especially on top, glabrous or glandular-puberulent, mucro 0.5–0.8(–1) mm. **Seeds** 1.5–2 mm.

Flowering (Mar–)Apr–Jun. Dry to moist, grassy hillsides, foothill woodlands, vernal pools, often on serpentine; 30–800(–1000) m; Calif.

Sidalcea hartwegii is widespread in California from Tulare to Shasta counties, a range similar to those of the other vernal-pool annuals. It is most easily recognized by its stamen column, on which the anthers are borne on free portions of filaments, unlike in the other annual species and more typical of the perennial species. This helps to support the suggestion that the annual species were separately derived and not part of a single lineage. *Sidalcea hartwegii* often grows with *S. calycosa*, from which it can be distinguished also by its rugose rather than deeply longitudinally furrowed dorsal mericarp surfaces.

13. Sidalcea hendersonii S. Watson, Proc. Amer. Acad. Arts 23: 262. 1888 (as hendersoni) • Henderson's checkerbloom E

Herbs, perennial, 0.5–1.5 m, not glaucous, sparsely hairy to nearly glabrous throughout, with woody taproot and short, thick, ascending rootstock, usually also with compact rhizomes to 10 mm diam. **Stems** usually several, clustered, erect, usually purplish tinted especially near base, mostly unbranched, sometimes proximally hollow, base glabrous or sparsely hairy, hairs minute, simple or stellate, more densely scabrous distally. **Leaves** basal and cauline, ± fleshy; stipules deciduous, often purplish, lanceolate to linear, 6–11 × 1–3 mm; petiole often purplish tinted, those of basal leaves to 30 cm, proximalmost 3 times as long as blades, mid cauline 1–2 times as long as blades, distalmost to ½ times as long as blades and leaves subsessile; basal blades orbiculate, unlobed or shallowly 5-lobed, 7–15 × 7–15 cm, margins crenate or dentate, ciliate, surfaces glabrous or sparsely hirsute, hairs sometimes forked, not stellate, mid cauline palmately 7–9-lobed, incised ½+ length, 7–12(–17) × 7–12(–17) cm, lobes cuneate, margins coarsely dentate to pinnatifid, apex toothed, surfaces: adaxial sparsely hairy or only on veins, ciliate, distal cauline leaves 3–5-parted nearly to base, lobes narrowly lanceolate to linear, margins coarsely dentate-laciniate. **Inflorescences** erect, spiciform, dense, especially when young, sometimes elongate, calyces usually conspicuously overlapping in flower and sometimes in fruit, usually branched, 20-flowered, usually several to many open at same time, not 1-sided, 5–20 cm; bracts purplish, lanceolate to linear, undivided, rarely 2-fid, 4–5 mm, longer than pedicels, shorter than calyx. **Pedicels** 1–4 mm, to 7–8 mm in fruit; involucellar bractlets absent. **Flowers** bisexual or often unisexual and pistillate, plants gynodioecious; calyx somewhat urceolate, 8–9 mm, to 10–12 mm in fruit, lobes strongly reticulate-veined, sparsely stellate-hairy or proximally glabrous, lobe tips usually purple, rarely green, margins ciliate; petals bright pinkish lavender, drying deep purple, veins not paler, bases not white, pistillate usually 8–10 mm, bisexual (10–)15–20 mm; staminal column 6–7 mm, hairy; anthers white; stigmas (5–)7 or 8(9). **Schizocarps** 8–9 mm diam.; mericarps (5–)7 or 8(9), 4 mm, smooth,

glabrous (rarely sparsely glandular-puberulent), thin-walled, apical margins sharp-edged, not winged, back sometimes slightly wrinkled, mucro 0.8–1.3 mm. **Seeds** 3 mm. 2*n* = 20.

Flowering Jun–Aug. Coastal marshes, sandy to boggy tidal areas, upper beach meadows; 0–50 m; B.C.; Alaska, Oreg., Wash.

Sidalcea hendersonii is showy and distinctive; it has been confused with *S. oregana* subsp. *spicata* and with *S. cusickii*, to which it is similar and apparently closely related. Like *S. cusickii*, *S. hendersonii* generally has purplish tinted vegetative parts, compact inflorescences with flowers that are mostly open at the same time, and ciliate calyx lobes. *Sidalcea hendersonii* is generally considered vulnerable; the number of extant populations has declined in recent decades, especially in Oregon, where it is close to extirpation. It can be distinguished by its marshy seacoast habitat, its often hollow stems, its tendency to be glabrous, its dense, compound inflorescences, and its relatively large mericarps and mucro. It is the northernmost species of *Sidalcea* and is the one better adapted to brackish or saline marshes. *Sidalcea hendersonii* occurs from Douglas County, Oregon, to British Columbia, and a single specimen is known from Alaska.

14. Sidalcea hickmanii Greene, Pittonia 1: 139. 1887 (as hickmani) C E F

Herbs, perennial, 0.1–0.8 m, not glaucous, with thick, woody taproot or caudex, without rhizomes. **Stems** several to many (ca. 3–20+), clustered, erect to ascending, branched or unbranched, solid, usually densely stellate-canescent. **Leaves** cauline, evenly arrayed on stem, usually similar in size, shape; stipules linear-lanceolate to ovate, 2–9 × 1–3 mm, widest above base, width sometimes exceeding stem diam.; petiole 0.6–3(–9) cm, ½–3 times as long as blade, apex often with pulvinus; blade orbiculate or reniform to flabelliform, unlobed and margins coarsely crenate to shallowly or deeply lobed, 1–7 × 1–7 cm, usually wider than long, base truncate or cordate, apex rounded, surfaces stellate-hairy. **Inflorescences** erect, infrequently ascending, usually spiciform, dense or open, calyces overlapping or not, branched or unbranched, 2–20+-flowered, proximal flowers scattered, usually more congested distally, not notably elongate in flower, not 1-sided, (1.5–)3–25 cm, usually longer in fruit; bracts linear to ovate-lanceolate or oblong, undivided, 2-fid, or divided, 2–8(–12) mm, not involucrelike, distal entire to 2-fid, stipulelike, proximalmost not involucrelike, divided to base, much shorter than to nearly equaling calyx. **Pedicels** 1–4(–5) mm; involucellar bractlets (2)3, 2–10 mm, shorter to slightly longer than calyx. **Flowers** usually bisexual, infrequently unisexual and pistillate; calyx 4–12 mm, densely to sparsely stellate-puberulent to long-bristly; petals usually pale pink to pink-lavender, rarely white, veins not conspicuously whitened, 5–17 mm; staminal column 4–7 mm, hairy; anthers white to pale pinkish or pale yellow; stigmas (4–)6 or 7(–10). **Schizocarps** 4–7 mm diam.; mericarps usually (4–)6 or 7(–10), (1.5–)2–2.5 mm, glabrous, sides usually smooth, thin, margins and back usually lightly reticulate-veined, transversely corrugated, back usually with medial, raised line, not pitted, mucro absent. **Seeds** 1–2 mm.

Subspecies 7 (7 in the flora): w United States.

Sidalcea hickmanii is found in isolated populations from southern California to southwestern Oregon and appears to have a relict distribution. K. Andreasen and B. G. Baldwin (2001, 2003) suggested that it is basal within *Sidalcea*. It is distinctive in having three (normally two in subsp. *petraea*) involucellar bractlets attached to the calyx, no mucro on the mericarps, and leaves that are almost the same size and shape throughout the stem. Each subspecies apparently represents a distinct relictual colony; the sexuality of these is not well known because of the paucity of specimens. As in many sidalceas, this species in particular appears to be fire-dependent.

1. Leaf blades usually lobed, incised ± to base; California.
    2. Bracts (7–)10–12 mm, equaling or shorter than calyx; c San Luis Obispo County . . . . 14b. *Sidalcea hickmanii* subsp. *anomala*
    2. Bracts 5.5–7 mm, shorter than calyx; Napa County. . . . 14c. *Sidalcea hickmanii* subsp. *napensis*
1. Leaf blades unlobed, incised to ½ length; California, Oregon.
    3. Involucellar bractlets 2(3); petals white to pale pink; flowers bisexual or pistillate; sw Oregon . . . . . 14e. *Sidalcea hickmanii* subsp. *petraea*
    3. Involucellar bractlets 3; petals pink, pale pink, pinkish lavender, or pale lavender; flowers bisexual; California.
        4. Bracts broadly lanceolate, 5–7(–10) × 2.5–4 mm, slightly shorter than calyx; involucellar bractlets equaling or slightly shorter than calyx; leaf blades: distal unlobed or incised to ¼ length; Santa Barbara and San Bernardino counties. . . . . 14d. *Sidalcea hickmanii* subsp. *parishii*
        4. Bracts linear, lanceolate, or oblong, 2–7 × 0.5–2 mm, shorter than calyx; involucellar bractlets shorter than calyx; leaf blades: distal usually unlobed, sometimes incised to ¼ length; Lake, Marin, or Monterey counties.

[5. Shifted to left margin.—Ed.]
5. Plants 0.4–0.8 m; stems brick red, greenish, or grayish; calyces stellate-puberulent, hairs longest at margins; involucellar bractlets 2–7 mm; largest leaf blades deeply cordate, 2.5–7 cm wide; inflorescences dense; Monterey County. . . . . 14a. *Sidalcea hickmanii* subsp. *hickmanii*
5. Plants 0.1–0.4 m; stems greenish or reddish; calyces stellate-puberulent, hair lengths uniform; involucellar bractlets 2–4 mm; largest leaf blades truncate to wide-cuneate, 0.7–2.7 cm wide; inflorescences open; Lake or Marin counties.
    6. Bracts of distal flowers 1, cupped; leaf blades 0.6–1.5 × 0.7–2.2 cm; stems distally stellate-hairy, hairs appressed, 0.2–0.5 mm; plants 0.1–0.4 m; inflorescences not spiciform, to 10-flowered; calyces 4–5.5 mm; n Lake County . . . . . . 14f. *Sidalcea hickmanii* subsp. *pillsburiensis*
    6. Bracts of distal flowers usually 2, flat or cupped; leaf blades (1–)2–4 × (1–)2.7 cm; stems distally hairy, hairs tufted, 0.5–1.2 mm; plants (0.2–)0.3(–0.4) m; inflorescences spiciform in age, 10+-flowered; calyces 6–7 mm; Marin County . . . . 14g. *Sidalcea hickmanii* subsp. *viridis*

14a. Sidalcea hickmanii Greene subsp. **hickmanii**

• Hickman's checkerbloom [C][E]

Plants 0.4–0.8 m, with taproot. **Stems** erect, proximally often brick red, otherwise greenish to grayish, canescent, hairs coarse, stellate. **Leaves:** stipules linear-lanceolate, 7–9 × 1.5 mm; petiole 2–3(–7) cm, ½–3 times as long as blade, shortest on distalmost leaves; blade usually rounded to reniform with narrow sinus, or deeply cordate especially on largest proximal leaves, to flabelliform with wide-cuneate base in distal leaves, unlobed, rarely incised to ¼ length, 2.5–7 × 2.5–7 cm, margins coarsely crenate, grayish canescent. **Inflorescences** dense, usually branched, 10–20 cm; bracts distinct, linear to wide-lanceolate, often somewhat falcate, 2–5(–7) × 0.5–2 mm, much longer than pedicels, much shorter than calyx. **Pedicels** 1–3 mm, usually obscured by bracts; involucellar bractlets 3, 2–7 mm, shorter than calyx. **Flowers** bisexual; calyx 8–10 mm, stellate-puberulent, hairs longest at margins; petals pale lavender, sometimes lightly pale-veined, 16–17 mm; staminal column 6–7 mm, hairy; anthers white to pale yellow; stigmas usually 6. **Schizocarps** 5–7 mm diam.; mericarps usually 6, (1.5–)2–2.5 mm, smooth except for dorsal-medial line and several transverse corrugations especially at margins. **Seeds** 1.5–2 mm.

Flowering May–Jul. Chaparral; of conservation concern; 300–1200 m; Calif.

Subspecies *hickmanii* is recognized by its relatively large, usually unlobed leaves and linear bracts, relatively robust size, and dense inflorescences, as well as its restriction to the Santa Lucia Mountains, outer South Coast Ranges, Monterey County.

14b. Sidalcea hickmanii Greene subsp. **anomala** C. L. Hitchcock, Perenn. Sp. Sidalcea, 79. 1957

• Cuesta Pass checkerbloom C E

Plants often mound-forming, 0.3–0.6 m, with taproot. **Stems** usually crowded in older plants, erect to ascending, usually reddish, grayish stellate-hairy, hairs coarse, longest 1.5 mm. **Leaves:** stipules wide-lanceolate to ovate, 6–8 × 2.5–3 mm; petiole 3–6 cm, 1–3 times as long as blade; blades on proximal portion of stem rounded, those distally with wide sinus, deeply, palmately (3–)5–7-lobed, incised ± to base, 2–6 × 2–6 cm, narrowed to base, margins crenate-dentate, surfaces coarsely stellate-hairy. **Inflorescences** open, usually branched, 10–20 cm; bracts distinct, ovate-lanceolate to ovate, (7–)10–12 mm, 3–4 times longer than pedicels, equaling or shorter than calyx. **Pedicels** 1–4 mm, obscured by bracts; involucellar bractlets 3, 8–10 mm, ± equaling calyx. **Flowers** bisexual or pistillate, plants gynodioecious; calyx 10–12 mm, densely long-bristly; petals pinkish lavender, veins often paler, 12–15 mm; staminal column 6–7 mm, hairy; anthers white to pale pinkish; stigmas usually 6. **Schizocarps** 5–6 mm diam.; mericarps usually 6, (1.5–)2–2.5 mm, smooth except for dorsal-medial line and several transverse corrugations especially at margins. **Seeds** 1.5–2 mm.

Flowering Apr–May(–Jun). Closed-cone conifer forests and chaparral, usually serpentine; of conservation concern; 600–800 m; Calif.

Subspecies *anomala* is known from the vicinity of Cuesta Pass, outer South Coast Ranges, San Luis Obispo County, and is distinctive in its hemispheric-mounding habit, deeply palmately lobed leaves, and relatively large bracts that equal the calyx in length. It is listed as endangered in California and is occasionally cultivated.

Subspecies *anomala* is in the Center for Plant Conservation's National Collection of Endangered Plants.

14c. Sidalcea hickmanii Greene subsp. **napensis** S. R. Hill, J. Bot. Res. Inst. Texas 2: 783, fig. 1. 2008

• Napa checkerbloom C E

Plants 0.2–0.3(–0.4) m, with erect caudex and elongated taproot. **Stems** erect to ascending, usually maroon, glabrous or stellate-canescent, hairs scattered, tufted, stellate and simple. **Leaves:** stipules often purplish, ovate, (4–)5–6 × 2–3 mm; petiole 0.6–0.9 cm on distalmost leaves, 1–3.6 cm on mid-stem leaves, to 6 cm on proximal leaves, ½ times to longer than blade; blade rounded, 1–1.7(–2) × 1.5–2.5(–3.4) cm, surfaces stellate-hairy, proximalmost rarely unlobed and margins deeply crenate; mid-stem blades deeply palmately 3–5(–7)-lobed, incised ± to base, middle lobe largest, lobes narrowed to base, tips deeply few-toothed. **Inflorescences** open, usually unbranched, 5–25 cm; bracts paired, stipulelike, distinct, ovate, 5.5–7 × 2.3 mm, longer than pedicels, shorter than calyx. **Pedicels** 2 mm, obscured by bracts; involucellar bractlets 3, 5–6(–8) mm, shorter or slightly longer than calyx. **Flowers** bisexual; calyx 7–10 mm, stellate-puberulent; petals pale pink, 9–11 (–15) mm; staminal column 5 mm, minutely hairy; anthers white; stigmas 6 or 7. **Schizocarps** 4–5 mm diam.; mericarps 6 or 7, 1.5–2 mm, lateral walls smooth; back and margins corrugated with 2–5 wavy ridges. **Seeds** 1.5 mm.

Flowering Apr–Jun. Chamise chaparral, rocky, rhyolitic, volcanic soil; of conservation concern; 400–500 m; Calif.

Subspecies *napensis* is known only from a population in the inner North Coast Ranges, Napa County, and is threatened by agricultural development. It has deeply incised leaves like subsp. *anomala* and generally fewer and shorter stems, fewer leaves, and shorter bracts.

14d. Sidalcea hickmanii Greene subsp. **parishii** (B. L. Robinson) C. L. Hitchcock, Perenn. Sp. Sidalcea, 79. 1957 • Parish's checkerbloom C E F

Sidalcea hickmanii var. *parishii* B. L. Robinson in A. Gray et al., Syn. Fl. N. Amer. 1(1,2): 307. 1897 (as hickmani); *S. parishii* (B. L. Robinson) Davidson & Moxley

Plants 0.4–0.8 m, with caudex and taproot. **Stems** erect, gray-green, sometimes brick reddish proximally, coarsely stellate-hairy, hairs grayish, to 2 mm diam. **Leaves:** stipules lanceolate, 5–6 × 2 mm; petiole 2–3(–7) cm, ½–3 times as long as blade, shortest on distalmost leaves; blade rounded to truncate or wide-cordate, unlobed or very shallowly lobed, margins crenate, proximal 3–7(–10) × 3–7(–10) cm, distal usually flabelliform, unlobed or very shallowly incised to ¼ length, 2–3 × 1.8–2.5 cm, surfaces stellate-hairy, hairs usually overlapping. **Inflorescences** erect or ascending, often dense and spiciform when young, more open and elongating in age, unbranched or few-branched, 2–10 cm; bracts paired, distinct, broadly lanceolate, 5–7(–10) × 2.5–4 mm, much longer than pedicels, slightly shorter than calyx. **Pedicels** 1–2 mm, obscured by bracts; involucellar bractlets 3, 5–7 mm, equaling or slightly shorter than calyx. **Flowers** bisexual; calyx 8–10 mm, stellate-hairy, often sparsely so, sometimes with coarser, longer, stellate hairs; petals pale pink to pinkish lavender, 11–15 mm; staminal column 6–7 mm, hairy; anthers white to pale yellow; stigmas 6 or 7(–10). **Schizocarps** 5–7 mm diam.; mericarps 6 or 7(–10), (1.5–)2–2.5 mm; back and margins with 9 corrugations. **Seeds** 1.5–2 mm. $2n = 20$.

Flowering Jun–Aug. Chaparral, cis/montane woodlands, open conifer forests; of conservation concern; 1000–2200 m; Calif.

Subspecies *parishii* is the most variable of the subspecies and is found in at least two widely distant, isolated populations, each with metapopulations. It most closely resembles subsp. *hickmanii*; it can be distinguished from that by its wider, longer bracts, its more deeply crenate leaves, and its range. It is threatened by grazing, urbanization, fire-suppression, and road maintenance, and is listed as rare in California. Subspecies *parishii* is found in the outer South Coast Ranges, western Transverse Ranges, and San Bernardino Mountains in Santa Barbara and San Bernardino counties.

14e. Sidalcea hickmanii Greene subsp. petraea

S. R. Hill & Halse, Phytoneuron 2014-113: 2, figs. 1–4. 2014 • Neil Rock checkerbloom C E

Plants (0.3–)0.5(–0.9) m, with caudex and taproot. **Stems** erect, green, often tinted pale brick-red proximally, finely stellate-pubescent (some parts ciliate, and minute simple hairs sometimes present), hairs 0.1–1 mm diam., 2(–5)-rayed. **Leaves:** stipules falcate, ovate-lanceolate, 5–6 × 2–2.5 mm; proximal petioles (3.5–)5–6.5(–9) cm, ½–3 times as long as blade, shortest (0.5–)1–2(–3.2) cm on distalmost leaves; blade usually flabelliform or reniform, often wider than long, rounded at apex, base wide-cuneate to wide-cordate, unlobed, margins coarsely crenate-dentate; proximal (2–)3–4.5 × (2.3–)3–5.5 cm, surfaces stellate-hairy, hairs 2–3(4)-rayed, hairs usually not overlapping, sparser adaxially. **Inflorescences** erect or ascending, often 8 racemes present per major stem, each terminating a stem branch, flowers apically

congested and not or only slightly elongating in age, unbranched or few-branched, (1.5–)3.5–5(–7.5) cm (excluding peduncle if present); bracts paired and distinct to base, or 2-fid or undivided distally, narrowly lanceolate to oblanceolate, 5–6(–8) × 1–1.5 mm, (1–)2 times longer than pedicels, slightly shorter than calyx. **Pedicels** 1.5–2 mm, usually not obscured by bracts; involucellar bractlets 2(3), linear to narrowly lanceolate, 4–5 × 0.2–1 mm, slightly shorter than calyx. **Flowers** bisexual or unisexual and pistillate, plants gynodioecious; calyx 5–6 mm, finely stellate-hairy, often sparsely so; petals white to very pale pink, pistillate 6 mm, bisexual 10–11 mm; staminal column 5–6 mm, minutely hairy; anthers white; stigmas 6–8. **Schizocarps** 5–7 mm diam.; mericarps 6–8, 2 mm, back and margins corrugated-ridged and with medial ridge. **Seeds** 1.5–2 mm.

Flowering (May–)Jun(–Jul). Rocky sandstone summits, pine-oak-*Ceanothus*-*Cercocarpus* *Toxicodendron* association; of conservation concern; 800–900 m; Oreg.

Subspecies *petraea* is the most isolated of the subspecies and is known only from a single population on the summit of Neil Rock, Jackson County, Oregon. The white petals and predominantly two instead of three involucellar bractlets are distinctive, and its gynodioecious habit is also quite unusual for the species overall. It is likely threatened by fire suppression and possibly by recreational activity and/or mining. Subspecies *petraea* grows on sandstone with significant deposits of cinnabar rather than the serpentine characteristic of most of the other subspecies.

14f. Sidalcea hickmanii Greene subsp. pillsburiensis

S. R. Hill, J. Bot. Res. Inst. Texas 2: 786, figs. 2–4. 2008 • Lake Pillsbury checkerbloom C E

Plants greenish, 0.1–0.4 m, with small, erect, branched caudex and taproot. **Stems** erect, green, stellate-hairy, hairs appressed, moderately dense, uniform, 0.2–0.5 mm. **Leaves:** stipules wide-lanceolate, 3 × 1.3 mm; petioles of midstem leaves 0.6–2 cm, ½ times to as long as or longer than blade; blade broadly flabelliform, usually truncate, sometimes wide-cuneate, unlobed, 0.6–1.5 × 0.7–2.2 cm, margins coarsely crenate, surfaces tufted-stellate-hairy. **Inflorescences** open, not spiciform, unbranched, to 10 cm, to 10-flowered, proximal flowers solitary, axillary; bracts 2 in proximal flowers, 1 in distal flowers, distinct, oblong, narrow, cupped, 3 × 1.2 mm, narrowest at base, longer than pedicels, much shorter than calyx. **Pedicels** 2 mm, usually not obscured by bracts; involucellar bractlets 3, oblong, 3 mm, ½ as long as calyx. **Flowers** bisexual; calyx 4–5 mm, to 5.5 mm in fruit, densely stellate-puberulent, hairs 0.1–0.6 mm diam.; petals pale pink, slightly pale-veined, 8–10 mm; staminal column 5 mm, minutely puberulent; anthers white; stigmas (4–)6 or 7. **Schizocarps** 4–5 mm diam.; mericarps usually (4–)6 or 7, 1.5–2 mm, side walls smooth, back and margins with 2–5 corrugations. **Seeds** 1–1.5 mm.

Flowering (Jun–)Jul–Aug. Chaparral, ephemeral drainages; of conservation concern; 700–800 m; Calif.

Subspecies *pillsburiensis* is known from a single population near Lake Pillsbury, North Coast Ranges, Lake County. It appears to be a dwarf most closely related to subsp. *viridis* and differing in its lower stature, undivided (non two-fid) bracts, fewer flowers, and less conspicuous, appressed indument.

14g. Sidalcea hickmanii Greene subsp. viridis

C. L. Hitchcock, Perenn. Sp. Sidalcea, 78. 1957 • Marin checkerbloom C E

Plants usually greenish, (0.2–)0.3(–0.4) m, with taproot and caudex. **Stems** erect, usually reddish, sparsely stellate-hairy, distally with hairs tufted, not appressed, 0.5–1.2 mm. **Leaves:** stipules lanceolate, 2–4 × 1–1.5 mm; petiole 1–3 cm, to ½–2½ times as long as blade; blade reniform to flabelliform, rounded, usually truncate or with wide sinus, unlobed or very shallowly incised, (1–)2–4 × 1–2.7 cm, margins coarsely crenate, surfaces stellate-hairy. **Inflorescences** open, spiciform in age, branched or unbranched, 10+-flowered, 10–20 cm, flowers evenly spaced, proximalmost often solitary, axillary; bracts (1)2, distinct, flat or cupped, distalmost undivided or 2-fid, proximal distinct to base, lanceolate to narrowly oblong, 2–5 × 0.5–1 mm, shorter or longer than pedicels, shorter than calyx. **Pedicels** 1–3(–5) mm, usually not obscured by bracts; involucellar bractlets 3, (2–)3–4 mm, shorter than calyx. **Flowers** bisexual, probably unisexual and pistillate also, plants gynodioecious; calyx 6–7 mm, densely stellate-puberulent, hair length uniform; petals pink or pale lavender, (5–)11–14 mm, pistillate smaller; staminal column 4–5 mm, hairy; anthers white; stigmas (4–)7. **Schizocarps** 5–6 mm diam.; mericarps usually (4–)7, 1.5–2 mm, side walls smooth, back and margins with 2–5 corrugations. **Seeds** 1–1.5 mm.

Flowering May–Jun. Dry ridges, with serpentine or serpentinite; of conservation concern; 300–600 m; Calif.

Subspecies *viridis* is similar to subsp. *hickmanii* but is generally smaller in stature, more greenish than grayish green, and has smaller leaves and a more open inflorescence. It is most similar to the dwarf subsp. *pillsburiensis*. Plants from Mendocino and Sonoma counties reported as this subspecies are usually

Sidalcea malviflora subsp. *rostrata*, confused with this because of their generally unlobed leaves; *S. malviflora* has no involucellar bractlets, has mucronate fruits, and is generally coastal. Subspecies *viridis* is known only from Carson Ridge, north Central Coast Range, Marin County.

15. Sidalcea hirsuta A. Gray, Smithsonian Contr. Knowl. 3(5): 16. 1852 • Hairy checkerbloom E

Herbs, annual, (0.1–)0.3–0.8 m, not glaucous, with taproot. **Stems** single, erect, usually branched distally with erect branches, solid, not glaucous, proximally glabrate, distally usually softly, densely bristly-hirsute, rarely glabrescent. **Leaves** cauline; stipules inconspicuous or deciduous, purplish, lanceolate to subulate, 3–12 × 1–2 mm; petiole 2–7 cm, longest on proximal leaves and gradually reduced distally, proximalmost to 3 times as long as blade, reduced distally to ½ times or as long as blade; blades: proximalmost early-deciduous, orbiculate, unlobed, 1–2.5 × 1–2.5 cm, base cordate, margins crenate, surfaces ± bristly, distal deeply palmately 5–7(–9)-lobed to base, 3–8 × 3–8 cm, lobes linear, margins entire, apex acute, surfaces: abaxially younger blades hirsute, older glabrous except on veins. **Inflorescences** erect, spiciform, dense, calyces usually overlapping, ca. 20–30-flowered, proximalmost 1 or 2 flowers in leaf axils, not elongate, not 1-sided, 2–5 cm, to 20 cm in fruit; bracts inconspicuous or deciduous, often purplish, linear, 4–8 × to 2 mm, slightly longer than pedicels, usually 2-fid, sometimes undivided. **Pedicels** 2–3 mm; involucellar bractlets absent. **Flowers** bisexual, less often unisexual and pistillate and plants gynodioecious; calyx 8–10 mm, to 10–13 mm in fruit, prominently tawny-hirsute and densely stellate-canescent; petals pale pink to dark rose-pink or rose-purple, often with paler veins, 13–25 mm; stamens: filaments connate to apex, funnel-like, with rim to which unstalked anthers attach; staminal column 6–7 mm, hairy; anthers white; stigmas 5 or 6. **Schizocarps** 8–9 mm diam.; mericarps 5 or 6, 3–4 mm, back and sides reticulate-veined and pitted, wrinkled, ± stellate-puberulent, mucro 1 mm. **Seeds** 1.5–2 mm.

Flowering Apr–May(–Jun). Vernally wet places: pools, ditches, grasslands; 20–1000 m; Calif.

Sidalcea hirsuta is widespread but local in central and northern California, at least from Merced to southern Shasta counties and is sometimes locally common. The dense, terminal, spiciform inflorescences combined with the relatively small bracts and distally hirsute stems are distinctive; the lack of stalked anthers also helps to distinguish it from *S. hartwegii*.

16. Sidalcea hirtipes C. L. Hitchcock, Perenn. Sp.
Sidalcea, 42. 1957 • Bristly-stemmed checkerbloom
C E

Herbs, perennial, usually in colonies, 0.7–1.3(–1.8) m, not glaucous, with thick, rather woody taproot and coarse, elongate (cordlike) rhizomes 20–100 × 5 mm. **Stems** several, scattered, erect, arising from rhizome apices, usually slightly hollow, densely, harshly bristly-hirsute, hairs stiff, pustular, simple, forked, or stellate, often 2–2.5 mm. **Leaves** basal and cauline; stipules linear-subulate, 6–8 × 1–1.5 mm; petioles of proximal leaves 20–30 cm, 3 times as long as blades, reduced distally to ½ times as long as blades; blades: basal and proximalmost orbiculate to reniform, shallowly 5–9-lobed, 10–15 × 10–15 cm, thick, base narrowly cordate, lobe margins coarsely crenate-dentate, apex rounded, surfaces coarsely hirsute, hairs stiff; distal orbiculate, deeply palmately 5–7-lobed, incised ± to base, lobes sometimes lobed again, base cuneate, apex acute, deeply 2- or 3-toothed, surfaces long-hirsute or with 2–4-rayed, stellate hairs abaxially. **Inflorescences** erect, spiciform to subcapitate, dense, calyces usually conspicuously overlapping in flower and sometimes in fruit, proximals usually long-pedunculate, unbranched or branched, 20+-flowered, 10+ flowers usually open on spike at same time, not interrupted, not 1-sided, usually to 8 cm, usually not elongate but sometimes slightly elongated in some populations and to 20 cm in fruit; bracts paired or single, linear, distal undivided, proximal distinct to base, 6 mm, mostly slightly longer than pedicels. **Pedicels** 1–3(–5) mm; involucellar bractlets absent. **Flowers** bisexual or unisexual and pistillate, plants gynodioecious; calyx often purple tinted, 9–11 mm, to 11–16 mm in fruit, margins ciliate, hairs 1–2 mm, surfaces finely stellate-hairy at base and with coarser, longer, simple and stellate hairs apically; petals usually pale pink to rose-lavender, rarely white, slightly or not pale-veined, (9–)10–21 mm, pistillate often 9–14 mm; staminal column 5–7(–10) mm, hairy; anthers white; stigmas 5–10. **Schizocarps** 7–8 mm diam.; mericarps 5–10, 3.5–4 mm, glabrous or sparsely stellate-puberulent, roughened, prominently reticulate-veined, sides rugose and pitted, back less so, mucro 0.6–0.8 mm. **Seeds** 2.5–3.5 mm. $2n = 60$.

Flowering (Apr–)May–Jul(–Aug). Prairie remnants, coastal bluffs, open shrublands, fencerows, meadows, usually mesic, basaltic soil; of conservation concern; 0–200(–1200) m; Oreg., Wash.

Sidalcea hirtipes is uncommon and known from Clatsop, Lincoln, and Tillamook counties in Oregon and Clark, Lewis, and Wahkiakum counties in Washington. Its elevation and habitat vary, and it seems as much at home on steep coastal cliffs as in more inland, historic prairies and mountain meadows. Populations can appear to be large because of the long-rhizomatous and clonal nature of the plants; they are few and local. It is threatened by grazing, loss of habitat, fire suppression, road construction and maintenance, and changes in hydrology. It is a candidate for listing in Oregon and has been listed as endangered in Washington. *Sidalcea hirtipes* is characterized by its coarse indument of bristle hairs, its generally compact spikelike inflorescences, its relatively few, large, erect, hirsute leaves, and, especially, its extensive, coarse rhizomes. The inflorescences in some populations are elongated in fruit; its range, hirsute indument, and thick leaves along with coarse rhizomes help to distinguish it from other species. Stem internode length varies depending on habitat, as in many other *Sidalcea*. Molecular data suggest a relationship among *S. hirtipes* and *S. asprella*, *S. celata*, and *S. gigantea* (K. Andreasen and B. G. Baldwin 2003).

17. Sidalcea keckii Wiggins, Contr. Dudley Herb.
3: 56, plate 13, figs. 2–6. 1940 • Keck's
checkerbloom C E

Herbs, annual, 0.1–0.4 m, not glaucous, with taproot. **Stems** single, erect, sometimes branched distally, solid, short-stellate-puberulent and long, soft bristly-hairy, distally with multicellular trichomes intermixed, hairs erect. **Leaves** cauline, 3–5 per stem; stipules linear-filiform, undivided or rarely few-divided in robust plants, 3–5 × 1 mm; petiole 2–4.5(–6) cm, usually 1–2 times as long as blade; blades: proximalmost orbiculate, unlobed, 1–2.5 × 1–2.5 cm, base cordate with narrow sinus, margins crenate, apex rounded; other proximals rounded, unlobed or shallowly palmately 7–9-lobed, 2.5–4.8(–6) × 2.5–4.8(–6) cm, margins coarsely crenate, sinus wide to narrow, surfaces stellate-hairy; distals gradually reduced, orbiculate, more deeply 3-lobed, 1.5–2.5 × 1.5–2.5 cm, lobes obovate, apically widened to nearly truncate and narrowed to base, margins entire, apex 2–5-toothed, teeth subequal. **Inflorescences** erect to ascending, usually open, calyces not overlapping except possibly in bud, unbranched or infrequently branched, 5–12-flowered per branch, not especially elongate, usually 1-sided, 5–10 cm, to 5–15 cm in fruit, ½ times plant height; bracts linear-filiform, undivided or rarely few-divided in robust plants, (3–)7–11 mm, longer than pedicels, usually much shorter than calyx, not involucrelike, densely stellate-hairy and pilose-hirsute. **Pedicels** 2–4 mm, to 5–6 mm in fruit; involucellar bractlets absent. **Flowers** bisexual; calyx 8–10 mm, to 11–14 mm in fruit, lobe base within with conspicuous, purplish spot 1–2 mm wide, hairy, hairs glandular and non-glandular, multicellular as in inflorescence; petals dark pink, without pale veins, with or without reddish basal spot, 10–22(–26) mm; filaments connate to apex of tube; staminal column 3–5 mm, hairy, hairs relatively long, simple; anthers white; stigmas 4 or 5. **Schizocarps** 5–7 mm diam.; mericarps 4 or 5, usually tinted pink when fresh, 3–4 mm, usually glabrous, back reticulate-veined, pitted, with prominent midvein, mucro absent but with 1–5 minute bristles in its place. **Seeds** 1.5–2 mm.

Flowering Apr–May(–Jun). Grassy slopes, tolerant of, not restricted to, serpentine; of conservation concern; 70–700 m; Calif.

Sidalcea keckii, federally listed as endangered, was once thought to have been extirpated (S. R. Hill 1993); it was rediscovered in 1992. It appears to have occurred historically in at least seven counties; extant populations are thought to be very few. It is found in the southern inner North Coast Ranges in Colusa, Napa, Solano, and Yolo counties, and in the Sierra Nevada foothills in Fresno, Merced, and Tulare counties. It is closely related and similar to *S. diploscypha* and is often mistaken for that species; differences between the two are discussed under 8. *S. diploscypha*. Some plants of *S. keckii* in Colusa, Solano, and Yolo counties have divided bracts like those of *S. diploscypha*. *Sidalcea keckii* is vulnerable to agricultural and residential land development.

18. Sidalcea malachroides (Hooker & Arnott) A. Gray, Proc. Amer. Acad. Arts 7: 332. 1868 • Maple-leaved checkerbloom E

Malva malachroides Hooker & Arnott, Bot. Beechey Voy., 326. 1838; *Hesperalcea malachroides* (Hooker & Arnott) Greene; *Sidalcea vitifolia* A. Gray

Herbs, perennial, **or subshrubs,** (0.4–)0.8–1.5(–2) m, not glaucous, with thick, rather woody caudex or taproot, without rhizomes. **Stems** clustered, erect, solid, bristly-hirsute, hairs simple, forked, and stellate. **Leaves** cauline, evenly arrayed on stem, similar in size and shape; stipules linear-lanceolate, 5–15 × 0.5 mm; petioles 4–7 cm on distal leaves, ½ times to as long as blades, longer on proximal leaves; blade maplelike, usually palmately (3–)5–7-lobed, shallowly incised, 4–10(–20) × 4–10(–20) cm, base cordate, lobe margins coarsely to finely crenate, apex acute, surfaces rough-hairy, hairs stiff, simple, forked, or stellate. **Inflorescences** erect, spiciform or capitate, dense, calyces overlapping, often much-branched, paniclelike, usually 20+-flowered, not elongate, not 1-sided, 3–5(–7) cm; bracts sometimes purple tinted, linear-lanceolate, undivided (proximal deeply divided), 10 mm, longer than flower buds and

pedicels, proximal ± equaling calyx. **Pedicels** 1–3 mm; involucellar bractlets (0–)1–2, 5–8 mm, shorter than calyx. **Flowers** bisexual, staminate, pistillate, or mixed, plants trioecious; calyx often purplish tinted, 6–9(–12) mm, accrescent, densely bristly-hairy, hairs stellate, both coarse and fine; petals white or purple-tinged, 7–15 mm, pistillate 6–7 mm, bisexual or staminate 10–12 mm; staminal column 5–7 mm, hairy; anthers white to pale purplish or pale yellowish; stigmas 5–9. **Schizocarps** 5–6 mm diam.; mericarps 5–9, 2.5 mm, glabrous or sparsely stellate-hairy, margins rounded, back ridged, sides smooth or with slight corrugations near margins, not pitted, mucro absent. **Seeds** 1–1.5 mm. $2n = 20$.

Flowering May–Jul(–Aug). Woodlands, redwood forests, moist clearings near coast; 20–700 m; Calif., Oreg.

Sidalcea malachroides has long been considered to be the basal or so-called most ancient extant species of *Sidalcea* based on its morphology (E. M. F. Roush 1931; C. L. Hitchcock 1957). Molecular data support this conclusion (K. Andreasen and B. G. Baldwin 2001, 2003). These robust plants are distinguished by maplelike leaves that vary little in size and shape from base to apex of the stem, by relatively numerous, relatively small flowers with white or pale pink petals in dense, spiciform clusters on branched inflorescences, and by the coastal habitat. Formerly, it occurred in widely scattered sites from Monterey County, California, to Curry County, Oregon; fewer populations are extant; it has sometimes been cultivated. It is usually found in clearings and disturbed areas; it is threatened by logging and associated road usage, development, and non-native plant competition.

19. Sidalcea malviflora (de Candolle) A. Gray, Smithsonian Contr. Knowl. 3(5): 16. 1852 (as malvaeflora) • California checkerbloom E

Sida malviflora de Candolle in A. P. de Candolle and A. L. P. P. de Candolle, Prodr. 1: 474. 1824 (as malvaeflora); *Nuttallia malviflora* (de Candolle) Fischer & Trautvetter

Herbs, perennial, colonial or not, 0.2–0.6(–1.1) m, not glaucous, with woody caudex, usually with woody taproot, without rhizomes or rhizomes not cordlike, 3–10 mm diam. **Stems** clustered or scattered, erect to ascending, or decumbent to suberect, sometimes rooting, unbranched or branched, solid, usually densely to sparsely hirsute, stellate-hairy, or glabrescent, hairs 1–2 mm, distal stem usually more sparsely hairy to glabrate. **Leaves** basal and/or cauline; stipules green or purplish, linear-lanceolate or wide-lanceolate to oblong or ovate, 3–8(–12) × 1–3.5(–5) mm, often hairy to ciliate; petioles of proximalmost leaves (5–)6–15(–30) cm, 3–9 times as long as blades, much reduced distally to ½–2 times as long as blade; blades: proximalmost usually orbiculate, sometimes reniform, unlobed or shallowly lobed, (1–)4–15 × (1–)4–12 cm, base cordate, margins crenate, apex rounded, surfaces hairy, hairs stiff, either or both simple and stellate or forked, lobes cuneate-obovate, margins dentate; mid stem unlobed or palmately 5–9-lobed, margins coarsely crenate-serrate, lobes toothed or lobed to dissected; distalmost smaller, sometimes subsessile. **Inflorescences** erect or somewhat ascending, dense, subcapitate or spicate especially when young to elongated and open, calyces overlapping or not, not long-pedunculate, usually unbranched, 2–21-flowered, proximal flowers usually in leaf axils, elongate or not, 1-sided or not, (2–)5–20(–30) cm; bracts (2–)4–6(–9) mm, shorter or longer than pedicels, proximalmost bracts often leaflike, usually divided to base, distal bracts linear or lanceolate to oblong or ovate, undivided or 2-fid, ciliate-puberulent. **Pedicels** 1–6(–30) mm; involucellar bractlets absent. **Flowers** bisexual or unisexual and pistillate, plants gynodioecious; calyx 6.5–8 mm, to 8–11(–13) mm in fruit, sparsely or densely stellate-hairy and bristly with some forked hairs or simple bristles to 2 mm (these sometimes pustulate), if bristles absent, some stellate hairs larger, lobes green or purplish, margins ciliate; petals usually light or bright pink to lavender or dark rose-purple, rarely white, pale-veined, 7–20(–30) mm, pistillate 7–11(–30) mm, bisexual 10–25(–30) mm; staminal column 4–8 mm, hairy; anthers white to pale pink or pale yellow; stigmas (6)7 or 8(9). **Schizocarps** 4–8 mm diam.; mericarps (6)7 or 8(9), 2.5–4 mm, usually minutely hirtellous, stellate-puberulent, or glandular, rarely glabrous, roughened, sides ± honeycomb-pitted and reticulate-veined, back less so, mucro 0.3–1 mm. **Seeds** 1.5–3 mm. $2n$ = 20, 40, 60.

Subspecies 7 (7 in the flora): w United States.

Sidalcea malviflora includes local variants and intermediates. It was first described as a coastal taxon; later researchers subsequently included interior populations that shared a few characteristics. Molecular study has shown that the majority of the interior plants are not very closely related to the coastal forms; the species is again considered to be primarily coastal. Two inland taxa remain, subspp. *californica* and *dolosa*.

Sidalcea malviflora can generally be distinguished by its coastal distribution, its decumbent-based stems, its relatively short pedicels, its relatively large and showy petals that usually have conspicuous whitish veins, especially when dry, and its generally prominently reticulate-pitted and usually glandular-puberulent mericarps that have a mucro. The leaves can be extremely variable, from unlobed to highly dissected; the indument tends to be harsh to the touch.

1. Inflorescences usually dense, flowers obviously overlapping, fruits either congested or spaced and not overlapping on elongated axis.
2. Rhizomes freely rooting; leaves at midstem usually lobed; stems softly bristly-hairy proximally, stellate-hairy distally, usually not densely . . . . 19e. *Sidalcea malviflora* subsp. *patula*
2. Rhizomes not freely rooting; leaves at midstem unlobed; stems densely, softly bristly-hairy or stellate-hairy . . . . .19g. *Sidalcea malviflora* subsp. *rostrata* (in part)
1. Inflorescences usually open, elongate, flowers not obviously overlapping except in bud, axis elongated in flower, fruits not congested.
3. Leaves: blades, except proximalmost, deeply lobed, lobes ternate or dissected, segments linear (*Geranium*-like); stem base usually decumbent, often rooting. . . . . .19d. *Sidalcea malviflora* subsp. *laciniata*
3. Leaves: blades, especially at midstem, unlobed or lobes not ternate or dissected, segments linear; stem base decumbent, rooting or not.
4. Leaves: mid and distal blades usually unlobed, similar in shape; calyces and stipules sometimes purple tinted; coastal.
5. Leaf blades: basal 1–2(–2.5) cm wide; calyces and stipules purplish; stems glabrous, short stellate-hairy, or sparsely bristly. . . . . .19f. *Sidalcea malviflora* subsp. *purpurea*
5. Leaf blades: basal 2–6 cm wide; calyces and stipules usually green, not purple tinted (except sometimes at base of stipule only); stems softly bristly-hairy or stellate-hairy . . . . . 19g. *Sidalcea malviflora* subsp. *rostrata* (in part)
4. Leaves: mid and distal blades obviously lobed, basal and distal leaves dissimilar in shape; calyces and stipules green (not purple tinted); coastal or inland.
6. Stem base hairy, hairs simple, soft, 2 mm; mericarps slightly reticulate-veined, wrinkled; rhizomes usually 10+ cm, freely rooting, sometimes matted; calyces sparsely stellate-puberulent and hirsute-bristly; San Bernardino Mountains. . . . . 19c. *Sidalcea malviflora* subsp. *dolosa*
6. Stem base hairy, hairs stellate and/or simple, to 2 mm; mericarps strongly reticulate-veined, wrinkled or pitted; rhizomes usually to 10 cm, not freely rooting, not matted; calyces usually densely simple- or stellate-hairy; s outer South Coast and w Transverse ranges from Monterey to Los Angeles counties.

[7. Shifted to left margin.—Ed.]

7. Stems usually stellate-hairy and/or spreading-bristly to glabrate; calyces densely stellate-puberulent and coarsely bristly; petals: bisexual 15–29 mm; coastal and insular . . . . .19a. *Sidalcea malviflora* subsp. *malviflora*

7. Stems densely, softly velvety stellate-hairy, hairs 1 mm; calyces densely stellate-puberulent; petals: bisexual 10–30 mm; inland . . . . 19b. *Sidalcea malviflora* subsp. *californica*

19a. Sidalcea malviflora (de Candolle) A. Gray subsp. **malviflora** E

Sida delphiniifolia Nuttall ex Torrey & A. Gray; *Sidalcea delphiniifolia* (Nuttall ex Torrey & A. Gray) A. Gray; *S. delphiniifolia* var. *humilis* (A. Gray) Greene; *S. humilis* A. Gray

Plants 0.3–0.6 m, with fibrous-woody taproot and widespreading rootstocks or rhizomes to 10 cm, not freely rooting, not matted. **Stems** erect or ascending, base decumbent, usually rooting, usually stellate-hairy and/or spreading-bristly to glabrate, hairs to 2 mm. **Leaves** basal and cauline; stipules lanceolate, 3–4 × 1.1 mm; petioles 8–15 cm on proximal leaves, reduced distally, proximal 3–4 times as long as blade, distal ½–2 times as long as blade; blade 2–6 cm wide, often fleshy; basal blades orbiculate to reniform, shallowly, palmately 7–9-lobed, margins coarsely crenate, surfaces usually appressed-hirsute or stellate-hairy with few-rayed hairs; mid and distal similar to basal or more deeply incised, 5–7(–9)-lobed ¾–⅘ times their length, lobes usually entire or shallowly toothed. **Inflorescences** open, 5–10-flowered, 5–20 cm, flowers evenly spaced 0.5–1 cm, especially in age, several often open at same time, axis longer in fruit; proximal bracts leaflike, often divided to base, distal bracts ovate to lanceolate, 2-fid or undivided, 3–6 mm, usually shorter or longer than pedicels. **Pedicels** 3–5 (–10) mm, longer on proximalmost flowers. **Flowers:** calyx sometimes purplish, 8–12 mm, densely stellate-puberulent and sparsely, coarsely bristly, bristles often on swollen pad, hairs at base shorter, denser, marginal hairs longest; petals pink to rose or lavender-purple, pale-veined, pistillate 7–11 mm, bisexual 15–29 mm; staminal column 5–7 mm; anthers white; stigmas (7)8 or 9. **Schizocarps** 5–7 mm diam.; mericarps (7)8 or 9, 3.5–4 mm, sparsely glandular-puberulent, strongly reticulate-veined, wrinkled and pitted on sides, mucro 0.5 mm. **Seeds** 2–3 mm. $2n$ = 40, 60.

Flowering (Feb–)Mar–Jun(–Jul). Coastal prairies, bluffs, hillsides, scrub, open forests; 0–500 m; Calif.

Subspecies *malviflora* is coastal and found also on the Channel Islands. It has deeply incised leaf blades, the lobes generally entire or not laciniate or ternate, with indument of scattered, coarse, long, simple and stellate hairs. It can be distinguished from subsp. *rostrata*, which has essentially unlobed leaves, and from subsp. *laciniata* by its lesser degree of division of the leaf lobes. It is far less frequent than formerly, having been extirpated in the Los Angeles area, where it was common. Plants from Monterey to San Francisco best match the original illustrations of subsp. *malviflora*, and the more typical form appears to persist on the Channel Islands. By distinguishing subsp. *laciniata* and subsp. *rostrata*, the typical subspecies is more clearly defined and more fully matches the original description and type.

19b. Sidalcea malviflora (de Candolle) A. Gray subsp. **californica** (Nuttall ex Torrey & A. Gray) C. L. Hitchcock, Perenn. Sp. Sidalcea, 32. 1957 (as malvaeflora) • Chaparral checkerbloom E

Sida californica Nuttall ex Torrey & A. Gray, Fl. N. Amer. 1: 233. 1838; *Sidalcea californica* (Nuttall ex Torrey & A. Gray) A. Gray; *S. malviflora* var. *californica* (Nuttall ex Torrey & A. Gray) Jepson

Plants grayish, 0.4–0.8(–1) m, with thick, rather woody taproot and usually non-matted, long-spreading rootstocks or rhizomes to 10 cm. **Stems** erect, base usually decumbent, rooting or not, densely, softly velvety stellate-hairy, hairs 1 mm. **Leaves** mostly cauline, evenly distributed along stem; stipules linear-lanceolate, 4–10 × 1–3 mm; petioles of basal leaves 15–20 cm, reduced distally to ½ times as long as blade; blade orbiculate, especially at midstem and proximally, to more cordate in distal lobed leaves, 3–8 cm wide, surfaces uniformly, densely, softly stellate-hairy (some hairs forked on adaxial surface), proximalmost usually very shallowly lobed, margins coarsely crenate, distal usually more deeply incised, 7-lobed, lobes cuneate-oblong, apically toothed. **Inflorescences** open, 10+ -flowered, 10–20 cm, flowers evenly spaced 0.5–1 cm, especially in age, axis elongated in fruit; proximalmost bracts leaflike, often divided to base, distal linear to oblong, undivided or 2-fid, 3–6 mm, usually equaling pedicels. **Pedicels** 3–6(–20) mm, proximal longest. **Flowers:** calyx 9–13 mm, uniformly, densely, coarsely stellate-puberulent, hairs to 1 mm, margins often long-ciliate; petals pale pinkish rose, often pale-veined, 10–30 mm, pistillate darker and shorter; staminal column 5–6 mm; anthers white; stigmas (6)7–9. **Schizocarps** 6–7 mm diam.; mericarps (6)7–9, 3–3.5 mm, glandular-puberulent, rugose, prominently, coarsely reticulate-veined-pitted and wrinkled, mucro 0.5 mm. **Seeds** 2–2.5 mm.

Flowering Mar–May(–Jun). Coastal scrub, oak savannas, chaparral; 20–800(–1800) m; Calif.

Subspecies *californica* is similar to and intergrades with subsp. *laciniata* and with *Sidalcea sparsifolia*; it is best recognized by its long-petioled, very shallowly incised basal leaves and shallowly to more deeply incised seven-lobed distal leaves, all of which are densely velvety stellate-pubescent. DNA evidence suggests that it may not belong within *S. malviflora* (K. Andreasen and B. G. Baldwin 2001, 2003). Subspecies *californica* is found in the southern outer South Coast and the western Transverse ranges from Monterey to Los Angeles counties.

19c. Sidalcea malviflora (de Candolle) A. Gray subsp. **dolosa** C. L. Hitchcock, Perenn. Sp. Sidalcea, 36. 1957 (as malvaeflora) • Bear Valley checkerbloom E

Plants 0.2–0.6(–1) m, with rhizomes wide-spreading, slender, usually 10+ cm, freely rooting, sometimes matted. **Stems** erect, base decumbent, rooting, sometimes reddish tinted distally, usually long-hirsute, sometimes glabrous, hairs simple, 2 mm, slender, soft. **Leaves** basal and cauline; proximalmost stipules wide-lanceolate to ovate, 4–5 × 2–3 mm; proximal petiole 5–20(–30) cm, 3–4 times as long as blade, greatly reduced distally to ½–1 times as long as blade; blade (2–)4–8(–12) cm wide, basal cordate-rounded, unlobed or shallowly lobed, margins coarsely crenate, distal deeply palmately 5–9-lobed, lobes usually 3-toothed to 3-lobed, surfaces coarsely, sparsely stellate-hairy abaxially, hirsute and stellate-puberulent adaxially, distalmost not much reduced, palmately 5–7-divided nearly to base, lobe margins entire. **Inflorescences** open and elongate or dense when young, 5–14(–21)-flowered, 15–20 cm, calyces overlapping or not, several flowers often open at same time; bracts linear, divided nearly to base, (2–)5–7(–9) mm, distal undivided or 2-fid, usually equaling or longer than pedicels, much shorter than calyx. **Pedicels** 1–4(–6) mm. **Flowers:** calyx 6.5–10 (–12) mm, sparsely stellate-puberulent and hirsute-bristly on margins, longest hairs simple; petals rose-pink to purplish, pale-veined, usually with pale to white base, 20 mm, pistillate darker and smaller; staminal column 5–6 mm; anthers white to pale pink; stigmas (6)7–9. **Schizocarps** 5–6 mm diam.; mericarps (6)7–9, 3 mm, back glabrous or sparingly glandular-puberulent, slightly reticulate-veined and wrinkled, ridged, back less so, mucro 1 mm. **Seeds** 2 mm.

Flowering Jun–Jul(–Aug). Open pine forests, grassy areas; 1500–2300(–2700) m; Calif.

According to DNA evidence, subsp. *dolosa* may not belong within *Sidalcea malviflora*. Its range matches that of *S. pedata*, and it also appears to be rare. It is found in the San Bernardino Mountains. It somewhat resembles *S. neomexicana*, but rhizomelike stems are present, the fruit is rougher, and the stem is more leafy; it strongly resembles and may be a close relative of *S. reptans*. It is threatened by development.

19d. Sidalcea malviflora (de Candolle) A. Gray subsp. **laciniata** C. L. Hitchcock, Perenn. Sp. Sidalcea, 29. 1957 (as malvaeflora) • Geranium-leaved checkerbloom E

Sidalcea malviflora var. *sancta* C. L. Hitchcock; *S. scabra* Greene

Plants 0.2–0.8(–1.1) m, with fibrous-woody taproot and often short rootstocks or compact rhizomes. **Stems** erect to ascending, base decumbent, often rooting, short-hirsute and/or stellate-hairy or glabrous proximally, usually more sparsely, softly stellate-hairy distally. **Leaves** basal and cauline, proximal clustered, distal well-spaced; stipules ovate to lanceolate, 5–12 × 1.5–3(–5) mm; petioles of basal leaves 15–20(–25) cm, proximalmost often 8–9 times as long as blade, much reduced distally to ½ times as long as blade; blade 2–7 cm wide, surfaces usually sparsely stellate-hairy or hirsute, more densely so abaxially; basal rounded to reniform, usually shallowly 7-lobed, margins crenate, distal reniform, deeply, ternately lobed and dissected, lobes wide-linear, 1–2 mm wide, total segments usually 13+. **Inflorescences** dense, becoming open and elongate, usually unbranched, often 20+-flowered, 15+ cm in age, proximal flowers and fruits spaced 1+ cm apart; proximal bracts leaflike, dissected to base, shorter than pedicels, distal bracts linear to ovate, undivided or 2-fid, (2–)4–5 mm, longer than pedicels. **Pedicels** 5–15(–30) mm on proximalmost flowers, distal pedicels 2–5 mm. **Flowers:** calyx 7–11 mm, sparsely stellate-puberulent and bristly, marginal and vein hairs longest; petals usually light pink to dark rose, rarely white, usually conspicuously pale-veined especially when dried, pistillate darker, 9–15 mm, bisexual (12–)15–25 mm; staminal column 6–7 mm; anthers white to pale yellow; stigmas (6)7 or 8(9). **Schizocarps** 5–7 mm diam.; mericarps (6)7 or 8(9), 2.5–3.5 mm, glandular-puberulent, usually prominently reticulate-veined-pitted, mucro 0.5 mm. **Seeds** 2–3 mm. $2n = 20, 40$.

Flowering (Feb–)Mar–Jun. Grasslands, open foothill woodlands, sometimes serpentine; 0–700 m; Calif.

Subspecies *laciniata* is the most commonly found component of *Sidalcea malviflora*. It can be distinguished from most of the other subspecies by its deeply incised leaves with deeply incised lobes. This difference appears to be the only one between it and subsp. *malviflora*, and they intergrade. In the southern part of its range, subsp. *laciniata* appears to intergrade with subsp. *californica* and with *S. sparsifolia*. C. L. Hitchcock (1957) distinguished the southern plants as var. *sancta*, with $2n = 20$, stating that both varieties pass imperceptibly into one another. Plants in Monterey and San Luis Obispo counties are difficult to determine. Subspecies *laciniata* is found in the southern North Coast Ranges and central western California, from Lake to San Luis Obispo counties; it is most common in the San Francisco Bay area on foothills.

19e. **Sidalcea malviflora** (de Candolle) A. Gray subsp. **patula** C. L. Hitchcock, Perenn. Sp. Sidalcea, 21. 1957 (as malvaeflora) • Torch-flowered or spreading checkerbloom C E

Plants (0.2–)0.5–0.6(–0.9) m, usually with elongated, freely-rooting, sometimes matted rhizomes 2–5 mm diam. **Stems** erect or ascending, base decumbent, freely rooting, softly bristly-hairy proximally, stellate-hairy distally, usually not densely so. **Leaves** mostly basal, some cauline; stipules sometimes purplish, lanceolate to ovate, 5–6(–9) × 1.5–3 mm; petioles of basal leaves 7–30 cm, 5–8 times as long as blades, gradually reduced distally to 1–2 times as long as blades, distalmost petioles ½–1 times as long as blade; blade 3–12 cm wide, basal unlobed and margins crenate to shallowly 7–9-lobed and margins coarsely dentate-crenate, distal more deeply incised to digitate, distalmost divided nearly to base, 5–7-lobed, lobes dentate to laciniate, surfaces densely stellate-hairy abaxially, usually hirsute adaxially, hairs simple or forked. **Inflorescences** dense, spicate to capitate, usually unbranched or with reduced lateral branches, 10–20+-flowered, 2–6 cm, to 15–20(–30) cm in fruit, fruits congested or spaced, not overlapping on elongated axis, flowers obviously overlapping, stiffly erect, usually 5+ open at same time, proximalmost 1–3 flowers usually in leaf axils and separated from spike above, flowers in pistillate plants smaller and more separated; bracts lanceolate, 5–7 × 2 mm, shorter or longer than pedicels, distal undivided or 2-fid, proximal separated to base. **Pedicels** usually 2–3(–10) mm. **Flowers:** calyx 8–13 mm, hairy, hairs small, stellate and with longer, usually forked bristles; petals bright pinkish to rose-pink, drying dark purple, pale-veined at least when dry, pistillate 9–11(–15) mm, bisexual 15–25 mm; staminal column 5–7 mm; anthers white; stigmas (6)7 or 8. **Schizocarps** 6–8 mm diam.; mericarps (6)7 or 8, 3.5–4 mm, glandular-puberulent, prominently reticulate-veined, rugose, and pitted, mucro 0.3–0.5(–1) mm. **Seeds** 2.5 mm. $2n = 40$.

Flowering May–Jul. Open coastal forests, grassy areas, sometimes serpentine; of conservation concern; 0–700 m; Calif., Oreg.

Subspecies *patula* is rare and is distinguished by its dense, relatively short, spiciform inflorescences that elongate in fruit, its usually rough indument, its tendency to retain basal leaves, and its relatively long, freely rooting rhizomes that sometimes form mats. Some collections show a transition with subsp. *rostrata*, which usually has unlobed stem leaves, a denser tomentum, and shorter rhizomes, and with *Sidalcea virgata*, which is less hairy, has shorter rootstocks, and is more open-flowered. Molecular data support the Malviflora clade including subsp. *patula* but not including *S. virgata* (K. Andreasen, pers. comm.). Subspecies *patula* is a candidate for threatened or endangered listing in Oregon; fewer than 20 populations of it are known. It is known from coastal Del Norte and Humboldt counties in California and Curry County in Oregon.

19f. **Sidalcea malviflora** (de Candolle) A. Gray subsp. **purpurea** C. L. Hitchcock, Perenn. Sp. Sidalcea, 23. 1957 (as malvaeflora) • Purple checkerbloom C E

Plants purple tinted, 0.2–0.6 m, with thick, rather woody taproot, roots often ± fleshy, with compact, rhizomelike stem bases. **Stems** decumbent or prostrate to ascending, base often rooting, glabrous or short stellate-hairy or sparsely bristly. **Leaves** mostly cauline; stipules purplish, oblong to ovate-lanceolate or ovate, 3–8 × 1–2.5 mm; proximalmost petiole 5–8(–15) cm, 3–4 times as long as blade, reduced distally to ½ times as long as blade, distalmost leaves subsessile; cauline blades 1.5–3(–4) cm wide, proximalmost reniform, unlobed, 1–2(–2.5) cm wide, subcordate to cuneate-based distally, margins usually coarsely crenate, surfaces sparsely short-hirsute to subglabrate, distal usually unlobed, rarely very shallowly 7–9-lobed. **Inflorescences** open, branched or not, calyces not overlapping except sometimes in bud, not congested in fruit, usually 2–10-flowered, 7–15 cm, elongated in age, 2–5 flowers usually open at same time, proximal flowers in axils of reduced leaves; bracts lanceolate to ovate, usually 2-fid or deeply 2-lobed, distal usually 2-fid, proximal separated to base and often with leaf remnant between, 3–4 mm, usually equaling to much shorter than pedicels, much shorter than calyx. **Pedicels** sometimes recurved, (2–)3–10(–30) mm. **Flowers:** calyx purplish, 7–11 mm, sparsely stellate-puberulent with some coarser bristles; petals bright pinkish rose, pale-veined, pistillate 8–11 mm, bisexual 11–19(–25) mm; staminal column 4–5 mm; anthers white; stigmas 6–8. **Schizocarps** 4–5 mm diam.;

mericarps 6–8, 3–3.5 mm, sparsely glandular-puberulent on top and back, prominently reticulate-veined-pitted, mucro 0.5 mm. **Seeds** 1.5–2 mm. $2n = 20$.

Flowering (Mar–)Apr–May(–Jul). Open coastal forests, coastal prairies, meadows; of conservation concern; 0–30 m; Calif.

Subspecies *purpurea* is one of the more easily recognized taxa in *Sidalcea* because of the purplish tint to vegetative parts, the delicate, dwarfed to prostrate habit, the limited indument, and the relatively few and small leaves and flowers. Its range is the north-central California coast in Marin, southern Mendocino, San Mateo, and northern Sonoma counties. There are some transitions to subsp. *rostrata* which is overall more robust and very bristly-hairy with larger, more numerous, congested flowers on shorter, thicker pedicels. Subspecies *purpurea* has been confused with *Sidalcea hickmanii*, which is more inland and has two or three involucellar bractlets attached to the base of the calyx.

19g. Sidalcea malviflora (de Candolle) A. Gray subsp. **rostrata** (Eastwood) Wiggins in L. Abrams and R. S. Ferris, Ill. Fl. Pacific States 3: 105. 1951 • Sea-cliff checkerbloom E

Sidalcea rostrata Eastwood, Bull. Torrey Bot. Club 29: 80. 1902

Plants (0.2–)0.3–0.5(–0.6) m, with short, thick, rather woody taproot or caudex, without rooting rhizomes. **Stems** erect or procumbent, base decumbent, usually not rooting, usually coarsely, densely, softly bristly-tomentose or long stellate-hairy, basal hairs 1–2 mm. **Leaves** basal and cauline; stipules proximal sometimes purplish, distal green, wide-lanceolate to ovate, 7–8 × 2–4 mm; petioles of basal leaves 6–15 cm, 2 times as long as blade, reduced distally, much less than ½ times as long as blade, distalmost blades subsessile; blade often reniform, unlobed, 2–6 cm wide, margins crenate, sometimes deeply so, surfaces densely, softly or coarsely bristly-hairy. **Inflorescences** spicate to subcapitate, usually dense, unbranched, often 10+-flowered, often 4–7 cm, flowers mostly open at same time and obviously overlapping, rarely not overlapping, proximalmost in axils of reduced leaves, fruits usually spaced and not overlapping on a somewhat elongated axis; bracts cuneate 2-fid, proximal separated to base, 5 mm, usually equaling or longer than pedicels, margins sometimes toothed. **Pedicels** 1–2 mm. **Flowers:** calyx green (without purplish tint), 7–11 mm, densely stellate-puberulent and coarsely bristly, bristles often on swollen pad, hairs at base shorter, denser, marginal hairs longer; petals pink to rose, usually pale-veined, (10–)23–24 mm; staminal column 8 mm; anthers white; stigmas (6)7

or 8. **Schizocarps** 5–7 mm diam.; mericarps (6)7 or 8, 3.5–4 mm, sparsely glandular-puberulent, prominently reticulate-veined, rugose, and pitted, mucro 0.3–0.5(–1) mm. **Seeds** 2–3 mm.

Flowering (Feb–)Apr–May(–Jun). Open, coastal bluffs; 0–200 m; Calif.

Subspecies *rostrata* intergrades with subspp. *patula* and *purpurea*; it has been confused with *Sidalcea hickmanii* because of its unlobed and similar cauline leaves; it lacks involucellar bractlets. The unlobed and similarly shaped leaves, congested flowers, and dense, soft-bristly hairs together are distinctive. It has been included within subsp. *malviflora* but is very different, the unlobed leaves and shaggy hairs being the most noteworthy differences. Subspecies *rostrata* is known from the central and southern north coast to the north-central coast in Marin, southern Mendocino, San Mateo, and Sonoma counties.

20. Sidalcea multifida Greene, Cybele Columb. 1: 34. 1914 • Cut-leaf checkerbloom E

Herbs, perennial, (0.1–)0.2–0.4 (–0.6) m, pale-glaucous, with thick, rather woody taproot and simple or branched caudex, without rhizomes. **Stems** clustered, usually erect or ascending, sometimes proximally decumbent or prostrate, not rooting, solid, sparsely to densely appressed stellate-hairy. **Leaves** mostly basal, persistent, to 10+; stipules linear-lanceolate to elliptic, 5–6 × 2 mm on proximal stem, 4–5 × 0.5 mm on distal stem; petiole 5.5–16 cm, basal 3–5 times longer than blades, reduced distally to ½ times as long as blades; blades: basal reniform-orbiculate, palmately (5–)7–9-lobed, deeply incised, (1.5–)2.5–4(–6) × (1.5–)2.5–4(–6) cm, base cordate, margins entire, apex rounded to acute, lobes linear to oblong, again deeply pinnately or ternately 3–5-lobed; cauline ternately lobed, lobes linear to filiform, narrowest on distalmost leaves, somewhat fleshy, surfaces hairy, hairs appressed, stellate; distalmost leaf lobes unlobed. **Inflorescences** erect, open, calyces not conspicuously overlapping except sometimes in bud, usually unbranched, usually 3–9(–15)-flowered, elongate, 1-sided, to 7–25 cm; bracts usually paired, linear or lanceolate to narrowly ovate, proximal divided to base, distalmost 2-fid or simple, 5 mm, shorter to longer than pedicels. **Pedicels** 3–8(–10) mm; involucellar bractlets absent. **Flowers** usually bisexual or, infrequently, unisexual and pistillate, plants gynodioecious; calyx (6–)7–10 mm, slightly enlarged in fruit, uniformly minutely stellate-puberulent; petals pink to rose, pale-veined at least when dry, 9–20(–25) mm, pistillate shortest; staminal column 3.3–7.5 mm, sparsely stellate-hairy; anthers pale yellowish to white; stigmas 6 or 7. **Schizocarps** 5–7 mm diam.; mericarps 6 or 7, 3.5–4.3 mm, roughened, back finely glandular-puberulent, sides and back coarsely reticulate-veined, pitted, mucro 0.5–1 mm. **Seeds** 1.5–2 mm. $2n = 20$.

Flowering May–Jul(–Sep).   Dry places, sagebrush scrub, pinyon-juniper or pine forests; (1200–) 2000–2500(–2900) m; Calif., Nev.

Sidalcea multifida is generally a low-growing plant of high elevations, and usually can be distinguished by its highly glaucous, waxy, generally erect stems and leaves, and the generally basal and persistent seven- to nine-lobed leaf blades with pinnate or ternate lobes. It has been confused with *S. glaucescens*, to which it appears to be closely related and of which it may be found to be a variant or subspecies; as in *S. glaucescens*, the narrow inflorescence is often slightly curved between flowers. *Sidalcea multifida* can generally be distinguished from *S. glaucescens* by its seven- to nine-lobed leaf blades with more finely divided, ternate lobes, persisting basal leaves, range, and more erect habit. It occurs in Alpine, Mono, and Tulare counties in California, and from Lyon to Washoe counties in Nevada.

21. **Sidalcea nelsoniana** Piper, Proc. Biol. Soc. Wash. 32: 41. 1919 • Nelson's checkerbloom C E

Herbs, perennial, 0.4–1 m, often glaucous, with thick, woody taproot and lateral rhizomes to 500 mm. **Stems** clustered, erect, base decumbent-ascending, solid or ± hollow in age, usually ± glaucous distally, glabrous or sparsely hirtellous, hairs short, appressed, simple. **Leaves** basal and cauline; stipules sometimes deciduous, lanceolate to ovate, (4–)7–9(–14) × (1.2–)3–4 mm; petioles of proximal leaves 18–37 cm, 3–5 times as long as blades, distal 5–20 cm, to ½–1 times as long as blades; blades: basal reniform to rounded, unlobed and marginally deeply crenate or very shallowly palmately 5–7-lobed, 6–20 × 6–20 cm, base cordate, apex rounded; cauline deeply 5–9-lobed, lobes linear-elliptic to oblong-elliptic, margins usually entire, sometimes toothed, surfaces finely, sparsely hairy, hairs mostly simple. **Inflorescences** erect, spiciform, proximally open, distally dense, otherwise calyces not conspicuously overlapping except sometimes in bud, branched, 20+-flowered, proximalmost flowers spaced usually to 1 cm apart, not leafy-bracted, elongate, not 1-sided, 10–20 cm; bracts single, purple, linear, undivided, 4–6 mm, usually longer than pedicels and shorter than calyx. **Pedicels** (2–)3–4(–10) mm; involucellar bractlets absent. **Flowers** bisexual, often unisexual and pistillate, some apparently staminate, plants gynodioecious or trioecious; calyx 4–6 mm, to 4–7(–9) mm in fruit, subglabrous to minutely puberulent,

margins minutely ciliate, lobes usually purple-tinged, greener in shaded plants; petals overlapping or not, pinkish purple to purplish rose, usually pale-veined, 5–16 mm, pistillate darker, 5–8 mm, bisexual 10–16 mm; staminal column 5–7 mm, hairy; anthers white; stigmas (6)7 or 8. **Schizocarps** 4–5 mm diam.; mericarps (6)7 or 8, 2 mm, roughened, sparsely glandular-puberulent apically, sides finely reticulate to faintly rugose, sometimes also on back, to nearly smooth, not pitted, mucro 0.5–1 mm. **Seeds** 1.5 mm. $2n = 20$.

Flowering (May–)Jun–Jul(–Sep).   Open fields, meadows, fencerows, remnant prairies; of conservation concern; (40–)100–600(–1300) m; Oreg., Wash.

Sidalcea nelsoniana is uncommon and persists generally in degraded habitats. It resembles *S. campestris* and *S. virgata*, with similar ranges, but is generally less hairy and has a smaller, usually purplish calyx. It has been listed as endangered in Washington and as threatened in Oregon; the number of populations and individuals has greatly declined as a result of land development. It is found in the Willamette Valley area from Multnomah and Washington to Benton and Linn counties in Oregon, and in Cowlitz and Lewis counties in Washington.

Sidalcea nelsoniana is in the Center for Plant Conservation's National Collection of Endangered Plants.

22. **Sidalcea neomexicana** A. Gray, Mem. Amer. Acad. Arts, n. s. 4: 23. 1849 (as neo-mexicana) • New Mexico or salt-spring checkerbloom

Sidalcea confinis Greene; *S. crenulata* A. Nelson; *S. neomexicana* subsp. *crenulata* (A. Nelson) C. L. Hitchcock; *S. neomexicana* subsp. *diehlii* (M. E. Jones) C. L. Hitchcock; *S. neomexicana* var. *diehlii* M. E. Jones; *S. neomexicana* var. *parviflora* (Greene) Roush; *S. neomexicana* subsp. *thurberi* (B. L. Robinson) C. L. Hitchcock; *S. nitrophila* Parish; *S. parviflora* Greene; *S. parviflora* var. *thurberi* B. L. Robinson

Herbs, perennial, 0.2–0.8(–1.2) m, glaucous or not, with thick, fleshy, tuberous or fibrous taproot and other roots clustered, fleshy, without caudex and rhizomes. **Stems** 1–several, clustered, erect or ascending from slightly decumbent base, unbranched or branched, solid, infrequently hollow in robust plants, sometimes glaucous, proximally usually coarsely, densely to moderately bristly-hirsute, hairs often pustulose to sparsely stellate-hairy, sometimes glabrous, distally usually minutely puberulent or glabrous. **Leaves** basal and cauline, cauline 3+; stipules linear to wide-lanceolate, 4–8 × 1–2 mm; petioles of proximalmost leaves 10–25 cm, reduced distally to shorter than blade, to 5 times blade length; blade fleshy, margins often short-ciliate, surfaces sparsely hairy, less so adaxially, hairs simple-hirsute to stellate, appressed, basal orbiculate, unlobed, margins crenate, or shallowly 5–7(–9)-lobed, (1.5–)2–6(–8) × (1.5–) 2–6(–8) cm, base cordate, apex rounded, lobes with margins crenate to dentate, apex acute, distal cauline highly reduced or not, deeply, palmately (3–)5–7(–9)-lobed, smaller, ultimate divisions linear, margins entire. **Inflorescences** erect to ascending, open or dense, calyces not conspicuously overlapping except sometimes in bud, unbranched or branched, 20+-flowered, proximal flowers spaced 1+ cm apart, elongate, sometimes 1-sided, 10–25 cm, elongating in fruit; bracts linear to lanceolate, undivided or deeply 2-fid, 4–10 mm, usually equaling or longer than pedicels. **Pedicels** 5–8(–40) mm, equaling to much longer than calyx in fruit; involucellar bractlets absent. **Flowers** bisexual or, less frequently, unisexual and pistillate, plants gynodioecious; calyx 5–8(–10) mm, often accrescent, sparsely hairy, hairs simple, to 1 mm, pustulose, sometimes small, stellate, infrequently glandular, surface not obscured; petals pale pinkish rose or pale lavender-purple, bases paler, rarely white, veins usually pale, (6–)10–20(–25) mm, pistillate 8–12 mm, bisexual 18–20 mm; staminal column 5–6 mm, hispid-puberulent; anthers white; stigmas (7)8 or 9. **Schizocarps** 5 mm diam.; mericarps (7)8 or 9, 2–3 mm, ± glabrous, sides thin, smooth to slightly reticulate-veined, not pitted, mucro 0.5–0.8(–1) mm. **Seeds** 1.5 mm. $2n$ = 20.

Flowering (Mar–)Apr–Aug. Desert alkaline springs, moist mountain meadows, wet ditches, marshes; 10–2800 m; Ariz., Calif., Colo., Idaho, Nev., N.Mex., Oreg., Tex., Utah, Wyo.; Mexico (Chihuahua, Coahuila, Durango).

Sidalcea neomexicana is variable among and within populations. It is similar to *S. covillei* and *S. sparsifolia*, the former possibly derived from it. *Sidalcea neomexicana* usually can be distinguished by its fleshy roots; hirsute stems; slender pedicels (especially in fruit); pustulate, coarse calyx hairs; and relatively smooth mericarp surfaces. Some plants (in California and Mexico) are well adapted to hot desert springs; overall, *S. neomexicana* appears to have roots and a rootstock adapted to marshy conditions. It ranges farther south than any other *Sidalcea*. E. M. F. Roush (1931) recognized no subspecific taxa; C. L. Hitchcock (1957) accepted four geographically and morphologically defined subspecies, three of which (*crenulata*, *neomexicana*, *thurberi*) occur north of Mexico.

23. **Sidalcea oregana** (Nuttall ex Torrey & A. Gray) A. Gray, Mem. Amer. Acad. Arts, n. s. 4: 20. 1849 [E] [F]

Sida oregana Nuttall ex Torrey & A. Gray, Fl. N. Amer. 1: 234. 1838

Herbs, perennial, (0.3–)0.4–1.5 m, glaucous or not, with short, thick, rather woody taproot and branching caudex, without rhizomes or rhizomelike rootstocks (subsp. *valida* sometimes with rhizomes). **Stems** single or clustered, erect, rarely rooting at base, unbranched or distally branched, proximally usually solid, sometimes hollow in age, base glabrous, coarsely stellate-hairy to long-bristly, or glabrate, hairs usually becoming appressed, simple or stellate. **Leaves** basal and cauline, basal sometimes deciduous, cauline 3+; stipules usually deciduous, linear to lanceolate, 4–6 (–14) × 0.5–1(–2) mm; petioles of basal and proximal leaves (5–)7–10(–35) cm, 3–5 times as long as blades, reduced distally, distalmost leaves sometimes subsessile; blade cordate or reniform-orbicular, 3–10(–15) × 3–10 (–15) cm, base cordate, apex rounded, lobe apex often acute, surfaces glabrous or sparsely hairy, hairs minute, simple, forked, or stellate, proximal usually shallowly palmately 5–7(–9)-lobed, sometimes unlobed with margins crenate; midstem more deeply (3–)5–9-lobed, lobes again palmately or pinnately lobed, distalmost unlobed or 3(–5)-lobed, segments unlobed or deeply lobed, narrow. **Inflorescences** erect, usually spiciform, sometimes subcapitate, congested in bud, dense, calyces sometimes conspicuously overlapping in flower and sometimes in fruit, to open and elongate, few-branched or unbranched, 10–20+-flowered, flowers opening and closing sequentially from base to apex, sometimes 3–10 open on same day, not leafy-bracted, not 1-sided, (1.5–) 10–30 cm, elongating in flower or fruit; bracts linear to linear-lanceolate, undivided to 2-fid, proximal sometimes divided to base, 4–6(–7) mm, sometimes exceeding flower buds, usually equaling or longer than pedicels, shorter than calyx. **Pedicels** 1–3(–10) mm; involucellar bractlets absent. **Flowers** bisexual or unisexual and pistillate, plants gynodioecious; calyx usually green, 3.5–10 mm, usually lightly reticulate-veined, glabrous or densely, uniformly stellate-puberulent or bristly, surface often obscured; petals usually overlapping, pink or pink-lavender to dark rose-pink or magenta, not notably pale-veined, pistillate 5–10 mm, bisexual 8–15(–20) mm; staminal column 4–6(–9) mm, hairy; anthers white to pinkish; stigmas 6–9. **Schizocarps** 4–7 mm diam.; mericarps 6–9, 2–3 mm, sparsely glandular-puberulent, sometimes glabrous, not stellate-hairy, back and margins rounded, smooth or slightly reticulate-veined or pitted, infrequently prominently roughened at least on margins and/or back, mucro 0.1–0.7 mm. **Seeds** 1.5–2.5 mm. $2n$ = 20, 40, 60.

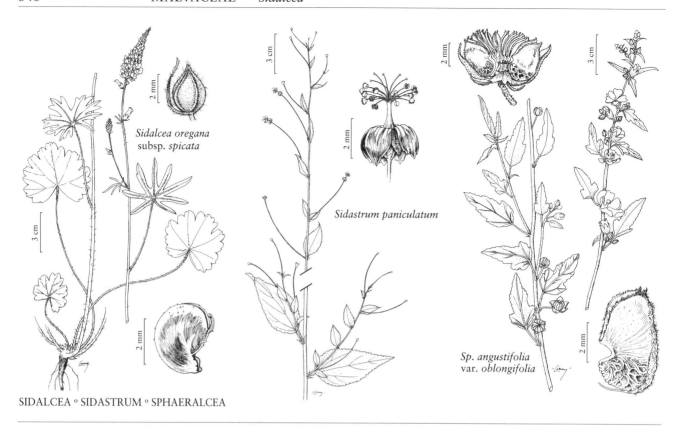

Sidalcea oregana subsp. *spicata*

Sidastrum paniculatum

Sp. angustifolia var. *oblongifolia*

SIDALCEA ∘ SIDASTRUM ∘ SPHAERALCEA

Subspecies 5 (5 in the flora): w North America.

Sidalcea oregana is variable, and parts of it have been treated as distinct species, subspecies, varieties, or extremes of a continuum. The plants are generally characterized by their strictly erect, leafless inflorescences that are congested in bud, their variable, sparsely hairy, lobed leaves that are both basal and cauline, and their usual lack of rhizomes. They often have been characterized and distinguished from the *S. malviflora* group by smooth mericarps; this feature depends upon the subspecies and is not true of all plants of *S. oregana*. These are usually mountain plants; some grow at lower elevations toward the northern parts of the range.

1. Primary peduncles (distal to distalmost leaves) longer than mature inflorescence; inflorescences dense, short-spiciform or subcapitate, 1.5–5 cm in age; calyces usually stellate-puberulent with sparse or no bristles; usually wetlands; California, Nevada.
    2. Stems to 5 mm diam. near base, base stellate-bristly or hirsute, bristle hairs 2 mm; inflorescences 1.5–2.5 cm; 1100–2300 m. . . .
    . . . . .23c. *Sidalcea oregana* subsp. *hydrophila*
    2. Stems 5–10 mm diam. near base, base usually bristly-hirsute, sometimes with minute stellate hairs, bristle hairs 1–1.5 mm; inflorescences 2–5 cm; 100–200 m. . . . .
    . . . . .23e. *Sidalcea oregana* subsp. *valida*

1. Primary peduncles usually equaling or shorter than mature inflorescence; inflorescences open or dense, spiciform, 3–30 cm in age; calyces stellate-puberulent, with or without longer bristle hairs 0.5–2.5 mm; meadows, prairies, streamsides, not characteristically wetlands; British Columbia to California, Montana, Nevada, Utah, Wyoming.
    3. Calyces to 8–13 mm in fruit, bristly-tomentose, some hairs 1.5–2.5 mm; Humboldt County, California . . . . . 23b. *Sidalcea oregana* subsp. *eximia*
    3. Calyces to 3.5–7(–10) mm in fruit, short-stellate-hairy, sometimes also with bristles but not bristly-tomentose; British Columbia to California, Montana, Nevada, Utah, Wyoming.
        4. Stem bases glabrous or stellate-hairy, hairs usually to 1 mm, sometimes bristly with appressed hairs, infrequently spreading; calyces usually uniformly stellate-puberulent; bracts usually shorter than young flower buds; rachis apex usually blunt or rounded . . . .
        . . . . .23a. *Sidalcea oregana* subsp. *oregana*
        4. Stem bases usually softly bristly-hirsute with hairs 1–2 mm or stellate-hairy, sometimes glabrous; calyces usually densely stellate-hairy, sometimes bristly; bracts usually longer than young flower buds; rachis apex usually acute. . . . .
        . . . . .23d. *Sidalcea oregana* subsp. *spicata*

23a. Sidalcea oregana (Nuttall ex Torrey & A. Gray) A. Gray subsp. **oregana** • Oregon checkerbloom [E]

Sidalcea maxima M. Peck; *S. nervata* A. Nelson; *S. oregana* var. *calva* C. L. Hitchcock; *S. oregana* var. *maxima* (M. Peck) C. L. Hitchcock; *S. oregana* var. *nevadensis* C. L. Hitchcock; *S. oregana* var. *procera* C. L. Hitchcock

Plants (0.3–)0.8–1.5 m, clumps often 20–30 cm diam., with fibrous-woody taproot or caudex. **Stems** 3–20, branched or not, usually solid, sometimes ± hollow on older, robust plants, base glabrous or stellate-hairy, hairs to 1 mm, sometimes bristly with appressed, infrequently spreading, hairs, stem distally less hairy, sometimes glabrous or glaucous. **Leaves:** stipules sometimes purple, linear, 4–6 × 0.5–1 mm; petioles 10–20(–35) cm on basal leaves, proximalmost 3–5 times as long as blades, greatly reduced distally to ½ times blade length; blade 5–15 × 5–15 cm, surfaces usually minutely stellate-hairy abaxially, basal orbiculate, unlobed and margins merely crenate to shallowly to deeply palmately 7–9-lobed, 2–8(–10) × 2–8(–10) cm, lobes deeply 2–5-toothed at apex, cauline distally variable, usually palmately 5–9-lobed, divided ± to base, usually largest at midstem, 6–10 × 7–12 cm, reduced distally, lobes cuneate, margins entire, deeply dentate, to laciniately dissected, distalmost divided into 3–5 linear segments with subentire margins, 2–4(–6) × 2–6(–8) cm, minutely hairy. **Inflorescences** spiciform, open, calyces not conspicuously overlapping except sometimes in bud, branched or not, usually 10+-flowered per branch, 10–30 cm, flowers spaced, not especially overlapping but, if overlapping, rachis apex blunt or rounded, axis elongating especially in fruit; primary peduncles (above distalmost leaf) usually equaling or shorter than mature inflorescence; bracts linear, distal simple and undivided to 2-fid, proximal divided to base, 4–5 mm, usually shorter than young flower buds, usually equaling or longer than pedicels, shorter than calyx. **Pedicels** 2–5(–10) mm, 3 mm in flower, 5 mm in fruit. **Flowers** bisexual or, infrequently, unisexual and pistillate, plants gynodioecious; calyx 5–8 mm, to 10 mm in fruit, usually densely, uniformly stellate-puberulent, marginal hairs sometimes longer, sometimes nearly glabrous; petals bright pink to rose-pink, often pale at base, 9–20 mm, pistillate 5–10 mm, bisexual 10–15 mm; staminal column 4–6 mm, hairy; anthers white; stigmas (6)7 or 8. **Schizocarps** 6–7 mm diam.; mericarps (6)7 or 8, 2.5–3 mm, usually glabrous, sometimes nearly smooth, back usually rugose-roughened, sides reticulate-veined, sometimes rugose-roughened, mucro 0.3–0.7 mm. **Seeds** 2 mm. 2*n* = 20, 40, 60.

Flowering (May–)Jun–Aug(–Sep). Moist meadows; 500–2500 m; B.C.; Calif., Idaho, Mont., Nev., Oreg., Utah, Wash., Wyo.

Subspecies *oregana* is variable; it intergrades with subsp. *spicata* and *Sidalcea setosa*. C. L. Hitchcock (1957) accepted five varieties within the typical subspecies; morphological intergrades exist. A case can be made for recognition of var. *calva* C. L. Hitchcock, which has been listed as endangered both federally and in Washington, where it is endemic. These plants are generally robust, sparsely appressed-hairy with four-rayed hairs, the leaves are generally nearly glabrous and fleshy-textured, and the calyx lobes are subglabrous and ciliate. Found in the Wenatchee Mountains, an area of high endemism, var. *calva* does not appear to be much different from other, nearly glabrous populations elsewhere; it appears to be the only variety with a chromosome count of 2*n* = 60. This treatment does not accept both subspecies and varieties within *Sidalcea*; therefore, it has been placed here into synonymy with the wide-ranging, variable typical subspecies. Subspecies *oregana* can generally be distinguished from the other subspecies by its more open inflorescences that are elongated in fruit, its multistemmed clumps as much as 30 cm in diameter, its generally stellate-hairy to glabrescent stem bases, its generally uniformly stellate-hairy calyces, its somewhat reticulate-roughened mericarps, and its bracts that are generally equal to or shorter than the young flower buds. Subspecies *oregana* appears to be the source of commonly sold cultivars. It has been listed as sensitive in Montana and as rare in British Columbia.

23b. Sidalcea oregana (Nuttall ex Torrey & A. Gray) A. Gray subsp. **eximia** (Greene) C. L. Hitchcock, Perenn. Sp. Sidalcea, 66. 1957 • Humboldt checkerbloom [C][E]

Sidalcea eximia Greene, Cybele Columb. 1: 34. 1914

Plants 0.9–1.5 m, with short, thick, rather woody taproot and branching caudex. **Stems** sometimes rooting at base, usually branched distally, often hollow but firm, base conspicuously densely bristly, hairs simple with swollen base, 1–2.5 mm. **Leaves:** stipules lanceolate, 10–14 × 2 mm; petioles of basal and proximal leaves (5–)10–20(–30) cm, 3–4+ times as long as blade, reduced distally and distalmost leaves nearly sessile; blade reniform-orbiculate, cordate, unlobed and margins merely crenate or shallowly 5–7(–9)-lobed, 3–9.5(–12) × 3–13 cm, blades distally all deeply 5–7-lobed essentially to base, hairs simple on both sides, 1–1.5 mm, distalmost blades usually 5-parted with irregularly-toothed lobes. **Inflorescences** spiciform, dense, calyces usually overlapping, branched, 20+-flowered, 3–8 cm, little elongating in fruit, flowers and fruits overlapping; primary peduncles (above distalmost leaf) usually equaling or shorter than mature inflorescence; bracts linear-lanceolate, undivided, 7 mm, usually longer than young flower buds and conspicuous then, much longer than pedicels, shorter than to nearly equaling calyx. **Pedicels** 1–2 mm. **Flowers** bisexual; calyx 8–10 mm, to 13 mm in fruit, stellate-puberulent, conspicuously bristly-tomentose especially on base, hairs 1.5–2.5 mm; petals pink, 15–16 mm; staminal column 9 mm, hirtellous-hairy; anthers white; stigmas 7–9. **Schizocarps** 5–6 mm diam.; mericarps 7–9, 2.5 mm, sparsely puberulent with a few bristles, smooth or nearly so, margins rounded, mucro 0.5–0.7 mm. **Seeds** 2 mm.

Flowering Jun–Aug(–Sep). Meadow openings, prairies; of conservation concern; 0–1200 m; Calif.

Subspecies *eximia* is uncommon and somewhat obscure. It was recognized by Hitchcock because of its notably accrescent and long-hirsute calyces and its robust habit and narrow range. It clearly intergrades with subsp. *spicata* and also strongly resembles *Sidalcea setosa*. There are no recent collections of subsp. *eximia* and only a limited number of older ones; it may be extirpated. It is known from the northern coast and North Coast Ranges in Humboldt County.

23c. Sidalcea oregana (Nuttall ex Torrey & A. Gray) A. Gray subsp. **hydrophila** (A. Heller) C. L. Hitchcock, Perenn. Sp. Sidalcea, 67. 1957 • Meadow checkerbloom [C] [E]

Sidalcea hydrophila A. Heller, Muhlenbergia 1: 107. 1904

Plants 0.3–0.9 m, with thick, woody taproot and caudex, sometimes forming small patches or colonies. **Stems** decumbent-based, often rooting, branched or not, solid or hollow near base in age, usually to 5 mm diam. near base, base stellate-bristly or hirsute, bristle hairs 2 mm. **Leaves:** stipules linear or lanceolate, 4–6(–9) × 0.5–1 mm; petioles of basal leaves (5–)10–20(–30) cm, 3–4+ times as long as blade, reduced distally and distalmost leaves nearly sessile; blade reniform-orbiculate, cordate, proximalmost unlobed and margins merely crenate or shallowly 5–7(–9)-lobed, 3–10(–12) × 3–13 cm, blades progressively more deeply lobed distally, lobes ternate or pinnately lobed, distalmost blades 3–5-lobed to base, lobes narrowly lanceolate, margins often entire, surfaces glabrous or sparsely appressed-hairy, hairs simple and/or minute-stellate. **Inflorescences** short-spiciform or subcapitate, dense, calyces overlapping, branched, to 20-flowered, 1.5–2.5 cm, flowers and fruits overlapping; primary peduncles longer than mature inflorescence; bracts linear, undivided, 4–6 mm, sometimes longer than young flower buds, much longer than pedicels, shorter than calyx. **Pedicels** 1–2 mm. **Flowers** bisexual, sometimes also unisexual and pistillate, plants gynodioecious; calyx 6–8 mm, stellate-puberulent, not bristly; petals bright pink, 8–12 mm; staminal column 4–6 mm, hairy; anthers white; stigmas 7–9. **Schizocarps** 4–5 mm diam.; mericarps 7–9, 2–2.5 mm, usually glabrous or sparsely glandular-puberulent on back, smooth, mucro 0.3 mm. **Seeds** 1.5–2 mm. $2n = 20$.

Flowering Jul–Sep. Wet soil of stream banks, seeps, meadows; of conservation concern; 1100–2300 m; Calif., Nev.

Subspecies *hydrophila* is generally recognized by its few-flowered, compact, terminal, headlike spikes in branched inflorescences; its long bristles on the proximal stems; and its high wet mountain-meadow habitat. It intergrades somewhat with subsp. *spicata*. Similar plants, possibly referable to subsp. *hydrophila*, have been found near Spooner Lake in Douglas County, Nevada. Subspecies *valida* also has compact spicate inflorescences, but it can be distinguished by its longer spikes, greater flower number, thicker stems with stellate hairs, and marshy habitat at lower elevations. Subspecies *hydrophila* occurs in the high and inner North Coast Ranges in Colusa, Glenn, Lake, Mendocino, and Napa counties, California, and possibly in Douglas County, Nevada.

23d. Sidalcea oregana (Nuttall ex Torrey & A. Gray) A. Gray subsp. **spicata** (Regel) C. L. Hitchcock, Perenn. Sp. Sidalcea, 64. 1957 • Spiked or spicate checkerbloom [E] [F]

Callirhoë spicata Regel, Gartenflora 21: 291, plate 737, figs. 3, 4. 1872; *Sidalcea oregana* var. *spicata* (Regel) Jepson; *S. spicata* Greene; *S. spicata* var. *tonsa* M. Peck

Plants 0.3–0.8(–0.9) m, with thick taproot, woody, branching crown. **Stems** branched or not, solid, base usually softly bristly-hirsute, hairs 1–2 mm, or stellate-hairy, sometimes glabrous. **Leaves:** stipules linear or lanceolate, 4–6(–9) × 0.5–1 mm; petioles of basal leaves (5–)10–20(–30) cm, 3–4+ times as long as blade, reduced distally and distalmost leaves nearly sessile; blade reniform-orbiculate, cordate, unlobed and margins merely crenate or shallowly 5–7(–9)-lobed, progressively more deeply lobed distally, 3–10(–12) × 3–13 cm, lobes ternate or pinnately lobed, distalmost blades 3–5-lobed to base, lobes narrowly lanceolate, margins often entire, surfaces glabrous or sparsely appressed-hairy with simple hairs and/or minute, stellate hairs. **Inflorescences** spiciform, rachis apex usually acute in bud, usually dense, calyces overlapping, unbranched or branched, 20+-flowered, 10–15 cm, overlapping in flower and fruit (less often fruits spaced); primary peduncles usually equaling or shorter than mature inflorescence; bracts linear, undivided, 4–6 mm, usually longer than young flower buds and conspicuous, much longer than pedicels, shorter than to nearly equaling calyx. **Pedicels** 1–2(–3) mm. **Flowers** bisexual or unisexual and pistillate, plants gynodioecious; calyx (3.5–)5–6(–7) mm, not accrescent, usually densely stellate-puberulent, sometimes bristly with hairs 0.5–1 (–2.5) mm; petals pink to rose-pink or magenta, pistillate 6–10 mm, bisexual 10–15(–20) mm; staminal column 4–6 mm, hairy; anthers white to pinkish; stigmas 7–9. **Schizocarps** 4–7 mm diam.; mericarps 7–9, 2.5–3 mm, back moderately glandular-puberulent, usually smooth, margins rounded, sides sometimes lightly reticulate-veined, mucro 0.1–0.5 mm. **Seeds** 2 mm. $2n = 20, 40$.

Flowering (Jun–)Jul–Aug(–Sep). Moist, open meadows, streamsides, subalpine forests, yellow pine slopes; 1100–3000 m; Calif., Nev., Oreg.

Subspecies *spicata* has frequently been recognized as a distinct species, but it clearly fits within the *Sidalcea oregana* complex. The extremes are readily recognizable, but the variation often makes for confusion with other subspecies, as well as with *S. setosa*. Similar species and subspecies have either been confused with it or included within it, including subspp. *eximia*, *hydrophila*, and *valida*, and even *S. nelsoniana*. Subspecies *spicata* is distinguished primarily by its elongated bud bracts, dense spicate inflorescences, essentially smooth mericarps, relatively small calyx, spreading-hirsute stem base, and lack of rhizomes. It is more restricted in range and elevation than subsp. *oregana* and occurs in the mountains from Mono County, California, north to Jefferson and Union counties, Oregon, and to Douglas and Washoe counties, Nevada.

23e. Sidalcea oregana (Nuttall ex Torrey & A. Gray) A. Gray subsp. **valida** (Greene) C. L. Hitchcock, Perenn. Sp. Sidalcea, 67. 1957 • Kenwood Marsh checkerbloom C E

Sidalcea valida Greene, Pittonia 3: 157. 1897; *S. spicata* Greene subsp. *valida* (Greene) Wiggins

Plants 0.9–1.2(–1.5) m, with thick, ± fleshy roots or rootstock. **Stems** decumbent-based, base often rooting, branched or not, solid, often hollow in age, 5–10 mm diam. near base, base usually bristly-hirsute, sometimes with minute stellate hairs, bristle hairs 1–1.5 mm, becoming glabrate distally. **Leaves:** stipules linear or lanceolate, 4–6(–9) ×

0.5–1 mm; petioles of basal leaves (5–)10–20(–30) cm, 3–4+ times as long as blade, reduced distally and distalmost leaves nearly sessile; blade 3–10(–12) × 3–13 cm, basal orbiculate, cordate, unlobed and margins merely crenate or shallowly 5–7-lobed, progressively more deeply lobed distally, lobes ternate or pinnately lobed, distalmost blades 3–5-lobed to base with subentire linear segments. **Inflorescences** short-spiciform, dense, calyces overlapping, branched, 20+-flowered, 2–5 cm, axis sometimes slightly elongating in fruit; primary peduncles longer than mature inflorescence; bracts linear, undivided, 4–6 mm, sometimes longer than young flower buds, much longer than pedicels, shorter than calyx. **Pedicels** 1–2 mm. **Flowers** bisexual or unisexual and pistillate, plants gynodioecious; calyx 7.5 mm, accrescent, finely stellate-puberulent, sparsely, if at all, bristly; petals dark rose-pink to pink-lavender, 10–15 mm; staminal column 4–6 mm, hairy; anthers white; stigmas (6)7 or 8. **Schizocarps** 5–6 mm diam.; mericarps (6)7 or 8, 3 mm, sparsely glandular-puberulent, smooth, mucro 0.5 mm. **Seeds** 1.5–2.5 mm. $2n = 20$.

Flowering Jun–Sep. Marshes; of conservation concern; 100–200 m; Calif.

Subspecies *valida* is known from a small area and is threatened by grazing and marsh alteration. It is found in freshwater marshes near Kenwood, southern outer North Coast Ranges in Sonoma County. It has been listed as endangered in California and also federally. It is similar to subsp. *hydrophila* except in overall size, spike length, number of flowers, and indument patterns as well as range and habitat. The somewhat fleshy roots are a reflection of its marshy habitat; it is not closely related to the other fleshy-rooted *Sidalcea* species.

24. Sidalcea pedata A. Gray, Proc. Amer. Acad. Arts 22: 288. 1887 • Bird-foot checkerbloom C E

Sidalcea spicata Greene var. *pedata* (A. Gray) Jepson

Herbs, perennial, much of plant often tinted reddish purple, 0.2–0.4 m, not glaucous, with fleshy, simple to branched taproot, without rhizomelike rootstocks. **Stems** clustered, erect to slightly ascending, base erect to decumbent-ascending, usually unbranched, (plants nearly scapose), solid, long-bristly with hairs often 2 mm, sometimes also stellate-hairy near base, usually becoming finely hispid distally. **Leaves** mostly basal, cauline 1–3; stipules lanceolate, 3–5 × 1 mm; petiole 6–9 cm, basal 2 times blade length, cauline ½ times blade length; blade usually orbiculate, 2–5(–6) × 2–5(–6) cm, base cordate, apex rounded, basal deeply dissected into 3–7 primary lobes, each less deeply incised and somewhat ternate, ultimate segments linear to oblong-elliptic, margins entire, distal blades repeatedly dissected with linear segments, surfaces densely hirtellous and stellate-hairy abaxially, less hairy with mostly simple hairs adaxially. **Inflorescences** erect or ascending, spiciform, initially dense, later open, calyces not conspicuously overlapping except sometimes in fruit, usually unbranched, usually 20+-flowered, proximal flowers remotely, evenly spaced, elongated in fruit, often 1-sided if ascending and not 1-sided when strictly erect, 15–30(–40) cm, elongating with fruits spaced, axis often wine-red; bracts linear, usually undivided, 2.5–4 mm, usually ± equaling or longer than pedicels. **Pedicels** 1–2(–3) mm; involucellar bractlets absent. **Flowers** bisexual or unisexual and pistillate, similar in size, plants gynodioecious; calyx usually wine-red, 4–5(–7) mm, usually not accrescent, finely stellate-puberulent, marginal hairs longer and often simple; petals dark rose-pink, sometimes pale-veined, 7–10(–12) mm; staminal column 3–5 mm, sparsely hairy; anthers white; stigmas 5–8. **Schizocarps** 5 mm diam.; mericarps 5–8, somewhat inflated, 2.5 mm, glabrous, smooth, back lightly grooved, not reticulate-veined or pitted, mucro 0.2–0.3 mm. **Seeds** 2 mm. $2n = 20$.

Flowering May–Aug. Moist meadows, open woodlands; of conservation concern; (1500–) 1600–2500 m; Calif.

Sidalcea pedata is known from Bear Valley and Bluff Lake in San Bernardino County. It is generally easily distinguished by its relatively small flowers; its stature; its wine-red inflorescence axis, calyx, and buds; its fleshy taproot and lack of rhizomes; and its dense basal cluster of palmately dissected leaves. It is threatened by development, vehicles, and grazing, and has been listed as endangered in California and also federally.

25. **Sidalcea ranunculacea** Greene, Leafl. Bot. Observ. Crit. 1: 75. 1904 • Big Tree checkerbloom [E]

Sidalcea interrupta Greene; *S. reptans* Greene var. *ranunculacea* (Greene) Jepson; *S. spicata* Greene var. *ranunculacea* (Greene) Roush

Herbs, perennial, 0.2–0.5 m, not glaucous or stems sometimes glaucous, with taproot from branched crown, usually wide-spreading and clonal from elongated, horizontal, freely-rooting rhizomes or elongated horizontal stem bases, 2–5 mm diam. **Stems** usually scattered, ascending to erect, solid, sometimes glaucous, proximally usually long bristly-hirsute, hairs simple or few stellate, rarely glabrous, distally stellate-hairy. **Leaves** mostly cauline (on reproductive stems), basal present mostly when stems young; stipules lanceolate to ovate, (3–)4–5 × 0.8–3 mm; proximal petioles 7–12 cm, 3–4 times blade length, distal usually ½–1 times blade length; blade often grayish green,

reniform or reniform-orbiculate, 2.5–6 × 2.5–6 cm, base wide-cordate, margins ciliate, apex rounded overall (lobes can be acute), surfaces softly stellate-hairy to silky villous-hirsute; basal shallowly, palmately 5-lobed, lobe margins coarsely crenate or dentate; distal more deeply 5 (–7)-lobed, distalmost 3–5-lobed, lobes entire or sparsely toothed. **Inflorescences** erect, spiciform to subcapitate, calyces generally conspicuously overlapping in flower and sometimes in fruit, branched or unbranched, 6–8-flowered, flowers crowded distally, usually (especially in more robust or older plants) in interrupted clusters with rachis exposed between clusters, sometimes with single cluster, not 1-sided, 3–8 cm, elongating in fruit; bracts linear, unlobed or infrequently 2-fid, 2–5 mm, subequal to or longer than pedicels, much shorter than calyx. **Pedicels** 1–3(–5) mm; involucellar bractlets absent. **Flowers** bisexual or unisexual and pistillate, plants gynodioecious; calyx 5–9 mm, stellate-puberulent and softly long-hirsute with marginal bristles 1.5 mm; petals magenta-pink, drying dark purple, sometimes pale-veined, 5–15 mm, pistillate darker, 5 mm; staminal column 3–6 mm, hairy; anthers white; stigmas 6–8. **Schizocarps** 4–5 mm diam.; mericarps 6–8, 2.5 mm, glabrous or sparsely stellate-puberulent, sides slightly reticulate-veined, back rougher, pitted, mucro 0.5 mm. **Seeds** 1.5–2 mm. $2n = 20$.

Flowering Jun–Aug. Moist meadows, stream banks; 1800–2800(–3100) m; Calif.

Sidalcea ranunculacea is uncommon and generally found in the vicinity of *Sequoiadendron*, hence the common name, in the Greenhorn Mountains of Kern and Tulare counties. The interrupted inflorescences, with as much as 5–15 cm of exposed rachis between flower clusters in well-developed individuals, are distinctive, as are the slender rhizomes and relatively small flowers. In some respects it resembles *S. reptans*, which is more widely ranging in similar habitats and also has long bristles at the base of its stems and slender rhizomes.

26. **Sidalcea reptans** Greene, Pittonia 3: 159. 1897 • Sierra checkerbloom [E]

Sidalcea favosa Congdon; *S. spicata* Greene var. *reptans* (Greene) Jepson

Herbs, perennial, 0.2–0.5 m, not glaucous, with woody taproot and rootstocks, without well-developed rhizomes, but horizontal stems rhizomelike. **Stems** scattered, ascending-decumbent or creeping, freely rooting, solid, flowering stems ascending to erect, proximal and midstem densely spreading-long-hirsute, bristles 2–3 mm, often on swollen pads, rarely subglabrous, base without stellate hairs, distally somewhat stellate-puberulent. **Leaves** basal and cauline; stipules on prostrate stems wide-lanceolate to ovate, 5–7 × 2–5 mm, elsewhere 3–6 × 1.3–2 mm; petiole 3–8(–15) cm, longest on basal leaves, proximalmost 3 times blade length, reduced distally; blades: proximalmost suborbiculate, unlobed or very shallowly lobed, 2.5–5 × 2.5–5 cm, base wide-cordate, margins deeply crenate, apex rounded, surfaces usually bristly, hairs long, simple, to subglabrous, not stellate-hairy; distal rather similar to proximalmost or more deeply palmately 3–5(–7)-lobed or toothed, lobes simple or 3-dentate. **Inflorescences** erect, open or some flowers overlapping, calyces not conspicuously overlapping except sometimes in bud, unbranched, usually 2–10-flowered, flowers 1+ cm apart, not much elongated, usually 1-sided, (5–)10–20 cm; bracts linear or narrowly oblong-lanceolate, 2-fid, usually inrolled, proximal separated to base, 2–4 mm, usually equaling or slightly longer than pedicels, much shorter than calyx. **Pedicels** 2–5 mm; involucellar bractlets absent. **Flowers** bisexual; calyx (6.5–)8–10 mm, somewhat accrescent, stellate-puberulent, marginal hairs longer, sometimes few glandular, multicellular, lobes sometimes dull purplish at tips; petals dark pink to pale lavender-pink, usually pale-veined, 12–20 mm; staminal column 6–7 mm, minutely hirtellous; anthers white; stigmas (6–)8–10. **Schizocarps** 6 mm diam.; mericarps (6–)8–10, 3 mm, roughened, top, back, mucro densely stellate-puberulent, deeply reticulate-veined, pitted, mucro 1 mm. **Seeds** 1.5–2 mm. $2n = 20$.

Flowering Jun–Aug. Moist meadows, dry places in pine forests; 1100–2500 m; Calif.

Sidalcea reptans is generally distinguished by its long-creeping, freely rooting stems with ascending ends that are long bristly-hairy proximally, its wide stipules on the proximal stems, its proximal leaves that are unlobed, crenate, long bristly-hairy, and on relatively long petioles, and its mericarps that are densely stellate-puberulent on their top and back and on the relatively long mucro. It occurs at relatively high elevations for *Sidalcea* and is found mainly in the central high Sierra Nevada.

27. **Sidalcea robusta** A. Heller ex Roush, Ann. Missouri Bot. Gard. 18: 205. 1931 • Butte County checkerbloom C E

Sidalcea asprella Greene var. *robusta* (A. Heller ex Roush) Jepson

Herbs, perennial, (0.5–)0.8–1.2 (–1.8) m, glaucous, with caudex and usually well-developed rhizomes to 0.5 mm diam. **Stems** often single, usually scattered, erect, usually unbranched, solid or somewhat hollow in age, proximally densely, finely stellate-hairy, hairs spreading, distally glaucous, glabrous. **Leaves** cauline, mostly on proximal ⅓ of stem;

stipules deciduous, inconspicuous, lanceolate, 2–3 × 1.1 mm; proximal petioles 10–15 cm, 2–3 times longer than blades, gradually reduced distally; blade 3–8 × 3–8 cm, basal rounded to reniform, cordate, shallowly 5–7-lobed and margins coarsely crenate, distal progressively more deeply incised, lobes narrowed to base, deeply toothed to ternate with linear segments, distalmost 3-lobed, surfaces glaucous, sparsely stellate-hairy abaxially, bristly-hirsute adaxially. **Inflorescences** erect, open, calyces not conspicuously overlapping except sometimes in bud, usually unbranched, loosely 10+-flowered, flowers 1+ cm apart, elongate, sometimes 1-sided, 30–40(–45) cm; bracts inconspicuous, lanceolate to linear, distal unlobed, 2–4 mm, proximal divided ± to base, 4–6 mm, usually equaling or shorter than pedicels. **Pedicels** 2–5 mm; involucellar bractlets absent. **Flowers** usually bisexual, sometimes pistillate, plants gynodioecious; calyx 10–15 mm, uniformly, densely stellate-puberulent; petals: bisexual pale pink, often drying yellowish, pale-veined or not, base pale pink to white, (15–)20–35 mm, pistillate usually darker purple, base white, 5(–10) mm; staminal column 6–8 mm, hairy; anthers white; stigmas (6)7 or 8. **Schizocarps** 6–8 mm diam.; mericarps (6)7 or 8, 3–3.5 mm, usually glabrous or very sparsely glandular-puberulent, distinctly narrowly wing-margined dorsally, sides lightly reticulate-veined, pitted, back less so, mucro 0.3–0.5 mm. **Seeds** 2–2.5 mm. $2n = 20$.

Flowering Apr–May(–Jun). Dry banks in chaparral at ecotone with foothill woodlands, often basaltic soil, with *Quercus douglasii*; of conservation concern; 100–400(–1300) m; Calif.

Sidalcea robusta is one of the taller species of *Sidalcea* and can be distinguished also by its relatively long inflorescences with widely-spaced, showy flowers, its limited range, and its notable winged mericarps. Rare and threatened by development, it is known from Butte County in the southern Cascade Range foothills and the northern Sierra Nevada foothills.

28. **Sidalcea setosa** C. L. Hitchcock, Perenn. Sp. Sidalcea, 53. 1957 • Bristly-fruited or Edgewood checkerbloom E

Sidalcea setosa subsp. *querceta* C. L. Hitchcock

Herbs, perennial, 0.5–1(–1.5) m, not glaucous, with thick, fibrous taproot or caudex and short, thick rootstocks that are not rhizomelike. **Stems** 1–3, clustered, erect, solid, proximally hairy, hairs a mix of soft, appressed, stellate and/or longer, bristlelike, 2 mm, distally sparsely puberulent. **Leaves** basal and cauline; stipules deciduous, lanceolate, 4–7(–15) × 1–2 mm; petioles of proximal leaves 10–20 cm, 3–4 times as long as blades, reduced distally to ½ times blade length; blades: proximal orbiculate, shallowly or deeply 5–9-lobed, 5–10 × 5–10 cm, base wide-cordate to ± truncate, lobe margins coarsely crenate and dentate, apex 2–5-toothed, mid stem largest, 5–7(–9)-lobed, usually incised ± to base, 10–25 × 10–25 cm, lobe margins coarsely dentate to laciniate or entire, surfaces scabrid-hairy, hairs simple or forked, stiff; distalmost 5–7-lobed, lobes linear, subentire, margins ciliate, surfaces glabrescent or with few hairs on abaxial veins. **Inflorescences** erect, spiciform or subcapitate, dense, calyces usually conspicuously overlapping in flower and sometimes in fruit, branched, ca. 10-flowered, flowers/calyces overlapping, not 1-sided, 3–7(–10) cm, elongated in fruit; bracts lanceolate-elliptic, undivided, 3–8 mm, longer than pedicels, often longer than young flower buds, ± equaling calyx. **Pedicels** 1–2 mm, to 4 mm in fruit; involucellar bractlets absent. **Flowers** bisexual or unisexual and pistillate, plants gynodioecious; calyx 5–9 mm, to 10 mm in fruit, margins ciliate, minutely stellate-puberulent and bristly with longer, coarser, usually pustulate hairs usually on pads, bristle hairs 1–2 mm; petals pink to pinkish lavender, not notably whitish- or pale-veined, (5–)8–30 mm, pistillate 5–15 mm, bisexual to 20–30 mm; staminal column 5–7 mm, hairy; anthers white to pale pink; stigmas (6)7 or 8. **Schizocarps** 6–7 mm diam.; mericarps (6)7 or 8, 2.5 mm, sparsely glandular-puberulent, margins usually sharp-edged but not winged, sides coarsely reticulate-veined to nearly smooth, back lightly reticulate-veined, usually prominently roughened at least on margins and/or back, pitted, mucro 0.5 mm. **Seeds** 1.8 mm. $2n = 40, 60$.

Flowering Jun–Jul. Meadows, rocky hillsides, roadsides; 300–2300 m; Calif., Oreg.

Sidalcea setosa is distinguished by its branched inflorescence with each branch tipped by a dense spike, and by its conspicuously accrescent, membranous, bristly calyx. It intergrades with, and has been confused with, *S. oregana* subspp. *oregana* and *spicata*, and some authors have considered it to be doubtfully distinct from them. It is similar also to *S. oregana* subsp. *eximia*, and some plants show similarities to *S. asprella*. When mature, the fruit separates as a whole from the plant and can adhere to clothing or fur, an apparent adaptation for dispersal. It appears to be best developed and most frequent near Grants Pass, in the Klamath Range of Oregon.

29. Sidalcea sparsifolia (C. L. Hitchcock) S. R. Hill, Madroño 56: 107. 2009 • Southern checkerbloom

Sidalcea malviflora (de Candolle) A. Gray subsp. *sparsifolia* C. L. Hitchcock, Perenn. Sp. Sidalcea, 32. 1957 (as malvaeflora); *S. malviflora* var. *hirsuta* C. L. Hitchcock; *S. malviflora* var. *stellata* C. L. Hitchcock; *S. malviflora* var. *uliginosa* C. L. Hitchcock

Herbs, perennial, 0.2–0.8 m, not glaucous, with thick, fibrous caudex or taproot and short rootstocks, rhizomes developed or not, 3–5 mm diam. **Stems** 1–20, clustered, erect to ascending or decumbent, usually not rooting, solid, proximally hairy, hairs simple, stellate, or mixed, 0.5–1.5 mm, distally usually stellate-hairy or glabrous. **Leaves** mostly basal, cauline to 5, (plant sometimes scapose); stipules linear to lanceolate, 2–8 × 1–1.5 mm; petioles of basal leaves 5–10(–25) cm, often 5 times blade length, distally reduced to ½ times blade length; blade orbiculate to reniform, 2–6(–8) × 2–6(–8) cm, base cordate, apex rounded, surfaces stellate-hairy abaxially, moderately minutely stellate-hairy to appressed-hairy adaxially, basal leaves usually shallowly 7-lobed, sometimes unlobed and margins crenate, cauline smaller, palmately 3–5-lobed ± to base, lobe margins entire or incised. **Inflorescences** usually ascending, ± open, calyces not conspicuously overlapping except sometimes in bud, branched or unbranched, 10+-flowered, most flowers spaced 1+ cm apart, elongated, extending well above leaves, narrow, often 1-sided, (15–)30–45 cm; bracts lanceolate, mostly 2-fid, proximal bracts divided ± to base, distal sometimes undivided. **Pedicels** 2–8 (–15) mm; involucellar bractlets absent. **Flowers** bisexual or unisexual and pistillate, plants gynodioecious; calyx 6–10 mm, densely to sparsely stellate-puberulent, hairs usually longer, coarser, 2–4-rayed especially at margins and over veins; petals pinkish rose to pinkish lavender or magenta, pale-veined or not, base often white, 10–25 mm, pistillate darker, 6–15 mm, bisexual 12–25(–30) mm; staminal column 5–6 mm, puberulent; anthers white; stigmas (6)7–9. **Schizocarps** 5–6 mm diam.; mericarps (6)7–9, 2.5–3 mm, roughened, glandular-puberulent, lightly to moderately reticulate-veined, pitted, mucro 0.5–1 mm. **Seeds** 2.5 mm. $2n = 20, 40$.

Flowering Mar–Jun(–Sep). Moist, often grassy, open places, often on mesas, chaparral, pine-oak openings; 10–2200 m; Calif.; Mexico (Baja California).

Sidalcea sparsifolia intergrades with *S. malviflora* subsp. *californica* and *S. neomexicana*. None of the varieties described within *S. malviflora* subsp. *sparsifolia* is recognized here, but one or more of them may merit distinction. Along with *S. neomexicana*, *S. sparsifolia* is one of the southern species of *Sidalcea* and one of the more tolerant of hot, dry, desert conditions. *Sidalcea sparsifolia* is the most common of the southern California species, occurring in the Sierra Nevada foothills and South Coast Ranges. It has been confused with *S. neomexicana* and with subspecies of *S. malviflora*. It can be distinguished from the former by its nonpustulate hairs, fibrous crown, and nonfleshy roots, and from the latter by its inland habitat, larger, ascending inflorescences, and tendency to have mostly basal leaves.

30. **Sidalcea stipularis** J. T. Howell & G. H. True, Four Seasons 4(4): 20, fig. 16. 1974 • Scadden Flat checkerbloom C E

Herbs, perennial, 0.3–0.7 m, not glaucous, with slender, elongated rhizomes to 1 cm diam. **Stems** single or in dense patches, erect, solid, hairy, hairs simple, spreading, bristly. **Leaves** cauline, evenly arrayed on stem; stipules asymmetric-ovate and auriculate, 10–20 × 10–20 mm, width exceeding stem diam.; petioles of proximal leaves 7–10 cm, gradually reduced distally to 2 cm, 1–1½ times blade length proximally to ½ blade length distally; blade ovate to elliptic, unlobed, usually (4–)7–8 × (2.5–)4–5 cm, reduced distally, base cordate, margins crenate-serrate, apex acute to rounded, surfaces glabrous abaxially, sparsely hirsute adaxially. **Inflorescences** erect, capitate, ± dense, calyces overlapping, unbranched, 2–10-flowered, not elongate, not 1-sided, 3–5 cm; proximalmost bracts involucrelike, similar to stipules in size and shape, narrowed to linear distally, usually divided to base, 10–20 mm, longer than pedicels, equaling or slightly shorter than calyx. **Pedicels** 1–2 mm; involucellar bractlets 3, 8–12 mm, equaling or longer than calyx. **Flowers** bisexual; calyx 8 mm, sparsely to densely bristly, sometimes also stellate-hairy; petals pink, usually pale-veined, 15 mm; staminal column 4–6 mm, hairy; anthers white; stigmas 7 or 8. **Schizocarps** 4–5 mm diam.; mericarps 7 or 8, 2 mm, glabrous, smooth, back with medial line, mucro absent. **Seeds** 1.9 mm.

Flowering Jun–Aug. Marshes; of conservation concern; 700 m; Calif.

Sidalcea stipularis is the most distinctive and easily recognized species of *Sidalcea*; it is also among the rarest. It is listed as endangered in California. The ovate unlobed leaves are found in no other *Sidalcea* species. The relatively large stipules, bracts, and involucellar bractlets are also unique in *Sidalcea*, making the inflorescence appear involucrate. Studies by K. Andreasen and B. G. Baldwin (2001, 2003) suggested that it is one of the basal species within *Sidalcea*, and it has probably been long isolated. *Sidalcea stipularis* is known from the northern Sierra Nevada foothills in Nevada County.

Sidalcea stipularis is in the Center for Plant Conservation's National Collection of Endangered Plants.

31. **Sidalcea virgata** Howell, Fl. N.W. Amer. 1: 101. 1897 • Virgate or rose checkerbloom E

Sidalcea malviflora (de Candolle) A. Gray subsp. *virgata* (Howell) C. L. Hitchcock; *S. malviflora* var. *virgata* (Howell) Dimling

Herbs, perennial, 0.3–0.6(–0.8) m, not glaucous, with woody taproot and compact rhizomes to 1 cm diam., these sometimes longer and freely rooting. **Stems** scattered or clustered, erect, decumbent-ascending, or reclining, freely rooting proximally, solid, proximally densely stellate-hairy, hairs long, soft, tangled, often 1.5 mm, distally hirsute to subglabrous, hairs smaller and appressed. **Leaves** basal and cauline; stipules sometimes purplish, linear-lanceolate, 4–5(–10) × 0.5–1.5 mm; petioles 10–20 cm on basal leaves, 3 times blade length, gradually reduced distally to ½ times blade length; blades: proximalmost orbiculate or semi-orbiculate to cordate, shallowly 5–7-lobed, 2–10(–15) × 2–10(–15) cm, base cordate, apex rounded, lobe margins coarsely dentate, surfaces densely stellate-hairy abaxially, hairs often simple and appressed adaxially; mid cauline deeply, palmately 5–7(–9)-lobed, lobes oblong, margins coarsely dentate; distal cauline smaller, deeply divided ± to base, lobe margins often entire. **Inflorescences** erect or ascending, often ± spiciform, usually open, sometimes dense, calyces usually not conspicuously overlapping except sometimes in bud, unbranched or rarely 1–3-branched, 5–20(–30)-flowered, proximalmost 1 or 2 flowers usually leafy-bracted, spaced 1+ cm, elongate, slender and virgate, often 1-sided, 20–25 cm; bracts often purplish, linear to oblanceolate, usually 2-fid, 3–6 mm, usually equaling pedicels. **Pedicels** (2–)3–8(–15) mm; involucellar bractlets absent. **Flowers** bisexual or unisexual and pistillate, plants gynodioecious; calyx 6–12 mm, densely, finely stellate-hairy, without longer hairs, lobes green or purple-tinged; petals pink or pinkish lavender to magenta, usually drying purple, usually pale-veined, pistillate 9–10 mm, bisexual 15–28(–30) mm; staminal column 6–8 mm, hairy; anthers white; stigmas 6 or 7(8). **Schizocarps** 6–7 mm diam.; mericarps 6 or 7(8), 3–3.5(–4) mm, roughened, back glandular-puberulent to finely stellate, prominently reticulate-veined, pitted, mucro 0.5 mm. **Seeds** 1.5–2 mm. $2n = 20, 40$.

Flowering May–Jun(–Aug). Dry hillsides, open shrublands, meadows, forest margins; 70–500(–800) m; Oreg., Wash.

Sidalcea virgata was included as a subspecies within *S. malviflora* by C. L. Hitchcock (1957). It does have some resemblance to *S. malviflora* subsp. *patula*; the inflorescence is generally much more open, the rhizomes are not as long, the stems are narrower and less hairy, and it tends to occur farther inland. It has been confused also with *S. asprella* and *S. elegans*, and the three appear to be closely related. *Sidalcea virgata* is somewhat difficult to define because it overlaps with other taxa in most of its characters, yet it has been generally accepted as distinct. Its range is well delineated, but it is not always easily distinguished from sympatric species, especially in fruit. Its proximalmost flowers consistently being in the axils of well-developed leaves may be its most useful identification feature (especially in herbarium specimens). Hitchcock noted that it does not occur south of Oregon and considered it to be more geographically than morphologically distinct. It has been listed as endangered in Washington (as *S. malviflora* subsp. *virgata*); its single occurrence there needs more investigation. *Sidalcea virgata* is found in the Willamette Valley area and in Josephine to Yamhill counties, Oregon, and, possibly, in Thurston County, Washington.

47. SIDASTRUM   Baker f., J. Bot. 30: 137. 1892 • [Generic name *Sida* and Latin *-astrum*, resembling]

Paul A. Fryxell†

Steven R. Hill

Shrubs [subshrubs]. Stems erect, ± stellate-hairy, not viscid. **Leaves** spirally arranged, petiolate or subsessile; stipules persistent, subulate; blade ovate or lanceolate [elliptic], not dissected or parted, base truncate or subcordate [rounded], margins dentate [serrate or crenate], surfaces stellate-hairy. **Inflorescences** terminal panicles [solitary flowers or racemes]; involucel absent. **Flowers:** calyx not accrescent, not inflated, not completely enclosing fruit, lobes cupuliform, unribbed; corolla reflexed [rotate], purple [white, yellow, or orange]; staminal column ± included; style 5[–10]-branched; stigmas capitate. **Fruits** schizocarps, erect, not inflated, oblate to conic, not indurate, often stellate-hairy; mericarps 5[–10], 1-celled, without dorsal spurs or endoglossum, smooth or weakly reticulate, lateral walls firm, not evanescent, essentially indehiscent. **Seeds** 1 per mericarp, glabrous [subglabrous, sparsely hairy]. $x = 16$.

Species 7 (1 in the flora): Texas, Mexico, West Indies, Central America, South America, Africa; not in saline habitats.

1. Sidastrum paniculatum (Linnaeus) Fryxell, Brittonia 30: 453. 1978 • Cadillo liso F

Sida paniculata Linnaeus, Syst. Nat. ed. 10, 2: 1145. 1759; *S. atrosanguinea* Jacquin; *S. capillaris* Cavanilles; *S. floribunda* Kunth

Plants 1–3 m. **Stems** roughly stellate-hairy, hairs somewhat ferruginous. **Leaves:** stipules 6–7 mm; petiole: proximal ⅓ length of blade, distal subsessile; blade slightly discolorous, 3–13 × 1–7 cm (reduced distally), apex acute. **Panicles** 100+-flowered. **Pedicels** capillary, 1–3.5 cm, often subtended by 3 stipuliform bracts. **Flowers:** calyx 2–3 mm, stellate-hairy; petals 3 mm, glabrous; staminal column purple, subequal to corolla, hairy; style exserted, pallid. **Schizocarps** 4–5 mm diam., puberulent; mericarps 2.5 mm. $2n = 32$.

Flowering year-round. Open shrublands, coastal thickets; 0–300 m; Tex.; e, s Mexico; West Indies; Central America; South America; Africa.

48. SPHAERALCEA   A. St.-Hilaire in A. St.-Hilaire et al., Fl. Bras. Merid. 1(fol.): 163. 1825

  • Globemallow [Greek *sphaera*, sphere, and *alkea*, mallow, alluding to arrangement of mericarps in a spherical head]

John La Duke

Herbs, annual or perennial. **Stems** erect or decumbent to ascending, stellate-canescent to stellate-silvery-lepidote, rarely glabrous. **Leaves** petiolate or sessile; stipules persistent or deciduous, linear; blade linear, lanceolate, orbiculate, or ovate to triangular, trullate, or cordate, unlobed or pedately dissected, base tapered, cuneate, or truncate to cordate, margins entire or crenate to serrate. **Inflorescences** terminal, racemes or panicles, or flowers sometimes fascicled or in axillary cymose racemes; involucel absent or present, involucellar bractlets present, persistent or deciduous (*S. coccinea*), 2 or 3, distinct. **Flowers:** calyx not accrescent, not inflated, lobes connate for ½ length, not ribbed, linear to lanceolate; corolla usually orange, red-orange, or red, sometimes lavender, purple, pink, or white, campanulate; staminal column included; styles 7–30-branched; stigmas capitate. **Fruits** schizocarps, erect, not inflated, cylindric, ovoid, flattened-spheric, flattened-spheric-conic, hemispheric, widely conic, truncate-conic, spheric, truncate-ovoid, ellipsoid, short-urceolate, helmet-shaped, or truncate-spheric, indurate, hairy; mericarps [7–]9–30, drying tan to brown, 1- or 2-celled, without dorsal spur, distally extended or not, usually smooth, apex cuspidate, truncate, or acute, with or without seeds, proximally usually rugose-reticulate and indehiscent, distally ± dehiscent, often remaining attached to fruit axis after maturity by threadlike extension of dorsal vein. **Seeds** 1 or 2(3) per mericarp, glabrous or slightly pubescent. $x = 5$.

Species ca. 50 (26 in the flora): w, c North America, Mexico, South America (Argentina, Bolivia, Brazil, Chile, Paraguay).

Sphaeralcea is often collected and somewhat difficult to identify. The species boundaries are not always sharp because there is frequent hybridization, polyploidy, and morphological variation in response to environmental conditions, particularly rainfall. Plants growing in Arizona are particularly difficult to identify.

SELECTED REFERENCE   Kearney, T. H. 1935. The North American species of *Sphaeralcea* subgenus *Eusphaeralcea*. Univ. Calif. Publ. Bot. 19: 1–128.

1. Plants annual, biennial, or perennial; mericarps: dehiscent part 10–75% of height, indehiscent part ± equaling to usually wider than dehiscent part.
   2. Leaf blade surfaces silvery stellate-lepidote . . . . 15.   *Sphaeralcea leptophylla*
   2. Leaf blade surfaces stellate-pubescent or stellate-canescent.
       3. Plants annual; mericarps 1.5–2 mm; leaf blades gray-green pubescent, unlobed or 3–5-lobed . . . . 5.   *Sphaeralcea coulteri*
       3. Plants usually perennial, sometimes annual or biennial; mericarps 2–5.5 mm; leaf blades green, pale green, yellow, yellow-green, or gray, light gray, or gray-green pubescent, unlobed, or if green or gray to green pubescent, leaves unlobed, 3–5-lobed, or pedately divided.
           4. Leaf blades trullate or triangular, not deeply divided . . . . 19.   *Sphaeralcea orcuttii*
           4. Leaf blades ovate, cordate-ovate, triangular, deltate, spatulate, or cordate, pedately divided or 3–5-lobed.
               5. Stems (1.5–)3–10 dm.
                   6. Sepals 5–13 mm; petals 15–20(–25) mm . . . . 10.   *Sphaeralcea gierischii*
                   6. Sepals 4.5–8.5 mm; petals 8–15 mm . . . . 17.   *Sphaeralcea moorei*
               5. Stems 1–3(–5) dm.

7. Leaf blades pedately divided, lobes relatively narrow . . . . 21. *Sphaeralcea pedatifida*
7. Leaf blades usually 3–5-lobed, sometimes pedately divided with relatively broad lobes.
    8. Leaf blades 3–5-lobed or pedately divided, green to gray-green, stellate-canescent; bractlets deciduous. . . . . 4. *Sphaeralcea coccinea*
    8. Leaf blades 3–5-lobed, yellow-green, sparsely stellate-canescent, bractlets persistent . . . . 24. *Sphaeralcea psoraloides*
1. Plants perennial; mericarps: dehiscent part 50–85% of height, indehiscent part usually not notably wider than dehiscent part.
    9. Inflorescences usually racemose, sometimes paniculate.
        10. Leaf blades 3-lobed to pedately divided; inflorescences racemose . . . . 6. *Sphaeralcea digitata*
        10. Leaf blades unlobed, 3-lobed, or pedately divided; inflorescences racemose or paniculate.
            11. Stems decumbent; leaf blades abaxially white to silvery, adaxially green; coastal Texas . . . . 16. *Sphaeralcea lindheimeri*
            11. Stems usually ascending to erect, sometimes decumbent; leaf blades green, yellow-green, or gray-green; w Texas and sw United States.
                12. Leaf blades ovate to deltate, unlobed or 3-lobed.
                    13. Stems 5–10 dm; widely distributed . . . . 1. *Sphaeralcea ambigua* (in part)
                    13. Stems 2–2.5 dm; restricted to Beaver and Millard counties, Utah, or Nye County, Nevada. . . . . 3. *Sphaeralcea caespitosa*
                12. Leaf blades oblong-ovate to ovate-lanceolate and unlobed, or widely ovate or triangular to orbiculate and pedately divided.
                    14. Leaf blades unlobed; sepals forming beak in bud. . . . . 12. *Sphaeralcea hastulata* (in part)
                    14. Leaf blades usually pedately divided (all but proximalmost); sepals not forming beak in bud . . . . 26. *Sphaeralcea wrightii*
    9. Inflorescences paniculate.
        15. Inflorescences open.
            16. Leaf blades unlobed or weakly 3-lobed; stems yellow- to yellow-green canescent or gray- to white-canescent; anthers yellow, gray, or purple; petals red-orange, apricot, lavender, pink, or white . . . . 1. *Sphaeralcea ambigua* (in part)
            16. Leaf blades 3-lobed or 3- or 5-parted; stems gray- to white-tomentose; anthers usually purple; petals red-orange . . . . 14. *Sphaeralcea laxa*
        15. Inflorescences crowded, at least distally.
            17. Leaf blades linear to lanceolate or broadly trullate, unlobed to hastately lobed to angulate; inflorescences usually with leaves throughout . . . . 2. *Sphaeralcea angustifolia*
            17. Leaf blades deltate, elongate-deltate, oblong-ovate, ovate-lanceolate, subhastate, lanceolate, ovate-triangular, ovate, triangular, or cordate, if lobed, with or without elongate central lobe, or pedately divided; inflorescences usually without leaves throughout, sometimes blades pedately divided and inflorescences leafy to tip.
                18. Leaf blades unlobed, elongate-deltate, oblong-ovate, ovate-lanceolate, lanceolate, triangular, ovate, or deltate, or weakly 3–5-lobed.
                    19. Stems yellow to yellow-green, rubbery; leaf blades deltate or elongate-deltate, sometimes lobed, center lobe sometimes elongate, lobes relatively broad; sepal tips not forming beak in bud . . . . 13. *Sphaeralcea incana*
                    19. Stems green, gray-green, or white, not rubbery; leaf blades rounded, ovate, oblong-ovate, ovate-lanceolate, lanceolate, triangular, unlobed or lobed; sepal tips sometimes forming beak in bud.
                        20. Leaf blades oblong-ovate to ovate-lanceolate; stems 1–3(–5) dm; sepal tips forming beak in bud . . . . 12. *Sphaeralcea hastulata* (in part)
                        20. Leaf blades ovate to triangular, unlobed or weakly lobed; stems 1.5–10 dm; sepal tips not forming beak in bud.
                            21. Stems green to gray-green canescent; leaf margins coarsely toothed; primarily Nevada, Utah, and north . . . . 18. *Sphaeralcea munroana*
                            21. Stems white- to yellow-canescent; leaf margins entire or crenate to serrate; primarily Nevada, Utah, and south . . . . 20. *Sphaeralcea parvifolia*

[18. Shifted to left margin.—Ed.]
18. Leaf blades usually strongly 3–5-lobed or pedately divided.
    22. Leaf blades 3(–5)-lobed; stems (2–)4–30 dm.
        23. Stems (2–)4–7(–10) dm, usually sparsely canescent or hirsute; petals red-orange
        . . . . 8. *Sphaeralcea fendleri* (in part)
        23. Stems 10–30 dm, glabrous or hairy; petals red-orange, red, pink, white, lavender,
           purple, or rose-purple.
           24. Leaf blades ovate-triangular to lanceolate, 3-lobed; inflorescence tip not leafy;
               stem glabrous or coarse-canescent . . . . 7. *Sphaeralcea emoryi*
           24. Leaf blades deltate, lanceolate, ovate, triangular, or cordate, subhastate to
               3-lobed; inflorescence tip leafy; stem indument soft.
               25. Mericarps 12–14, 3.5–5.5 mm. . . . . .22. *Sphaeralcea polychroma*
               25. Mericarps 10, usually 3 mm . . . . 23. *Sphaeralcea procera*
    22. Leaf blades usually deeply lobed or pedately divided, sometimes highly dissected; stems
        1–10 dm.
        26. Stems (2–)4–10 dm.
           27. Leaf blade surfaces hirsute, sparsely pubescent, or densely soft-pubescent,
               lobed, appearing green to gray-green . . . . 8. *Sphaeralcea fendleri* (in part)
           27. Leaf blade surfaces stellate-pubescent, commonly lobed or pedately divided,
               appearing yellow-green to gray-green to green . . . . 11. *Sphaeralcea grossulariifolia*
        26. Stems 1–3(–6) dm.
           28. Stems green to gray-green; bractlets and leaf margins not red, reddish, or
               purple, usually green or tan . . . . 12. *Sphaeralcea hastulata* (in part)
           28. Stems gray-green to purple (red-purple basally in *S. fumariensis*); bractlets and
             sometimes leaf margins red, reddish, or purple.
             29. Leaf blades densely stellate-pubescent; anthers yellow; mericarps 10–14,
               3–4 mm . . . . 9. *Sphaeralcea fumariensis*
             29. Leaf blades sparsely stellate-pubescent; anthers usually red to purple,
               sometimes yellow; mericarps 13, 4.5–6 mm. . . . . 25. *Sphaeralcea rusbyi*

1. **Sphaeralcea ambigua** A. Gray, Proc. Amer. Acad. Arts 22: 292. 1887 • Apricot mallow

Plants perennial. **Stems** erect, yellow-green or gray to white, 5–10 dm, yellow-green- or gray-to white-canescent. **Leaf blades** green or yellow-green, deltate with rounded lobes, unlobed to weakly 3-lobed, 1.5–5 cm, wrinkled or markedly rugose or not, base cuneate, truncate, or cordate, margins crenate, wavy, surfaces stellate-pubescent. **Inflorescences** paniculate, open, flowers clustered or solitary, inflorescence tip not leafy; involucellar bractlets green to tan. **Flowers:** sepals 6–13 mm; petals red-orange, apricot, lavender, pink, or white, 7–23 mm; anthers yellow, gray, or purple. **Schizocarps** cylindric; mericarps 9–13, 4–6.5 × 2–3.5 mm, chartaceous, nonreticulate dehiscent part 50–75% of height, with or without cusp, indehiscent part not notably wider than dehiscent part. **Seeds** 2 per mericarp, brown, glabrous or pubescent.

Varieties 4 (3 in the flora): w United States, nw Mexico.

Sphaeralcea ambigua var. *versicolor* (Kearney) Kearney is known from Baja California.

1. Petals lavender, pink, or white. . . . .
    . . . . 1b. *Sphaeralcea ambigua* var. *rosacea*
1. Petals apricot or red-orange.
    2. Leaf blades not markedly rugose . . . .
        . . . . 1a. *Sphaeralcea ambigua* var. *ambigua*
    2. Leaf blades markedly rugose . . . .
        . . . . .1c. *Sphaeralcea ambigua* var. *rugosa*

1a. Sphaeralcea ambigua A. Gray var. **ambigua**

Sphaeralcea ambigua var. *keckii* Munz; *S. ambigua* subsp. *monticola* Kearney; *S. ambigua* var. *monticola* (Kearney) Kearney; *S. macdougalii* Rose & Standley

Leaf blades wrinkled, not markedly rugose. **Flowers:** petals apricot or red-orange; anthers yellow. $2n = 10$.

Flowering spring–summer. Dry washes and rocky areas; 100–2300 m; Ariz., Calif., Nev., Utah; Mexico (Baja California, Sonora).

1b. Sphaeralcea ambigua A. Gray var. **rosacea** (Munz & I. M. Johnston) Kearney, J. Wash. Acad. Sci. 29: 486. 1939

Sphaeralcea rosacea Munz & I. M. Johnston, Bull. Torrey Bot. Club 49: 353. 1923; *S. purpurea* Parish ex Jepson

Leaf blades wrinkled, not markedly rugose. **Flowers:** petals lavender, pink, or white; anthers purple to gray.

Flowering spring–summer. Dry washes and rocky areas; 100–2100 m; Ariz., Calif., Nev.; Mexico (Baja California, Sonora).

1c. Sphaeralcea ambigua A. Gray var. **rugosa** (Kearney) Kearney, J. Wash. Acad. Sci. 29: 486. 1939 [E]

Sphaeralcea ambigua subsp. *rugosa* Kearney, Univ. Calif. Publ. Bot. 19: 49. 1935

Leaf blades markedly rugose. **Flowers:** petals red-orange; anthers yellow. $2n = 20$.

Flowering spring–summer. Dry washes and rocky areas; 10–2600 m; Ariz., Calif., Nev.

2. Sphaeralcea angustifolia (Cavanilles) G. Don, Gen. Hist. 1: 465. 1831 • Copper or narrow-leaved globemallow [F]

Malva angustifolia Cavanilles, Diss. 2: 64, plate 20, fig. 3. 1786

Plants perennial. **Stems** erect, green to gray-green, (3–)6–20 dm, grayish stellate-hairy. **Leaf blades** green to gray-green, linear to linear-lanceolate or lanceolate to broadly trullate, unlobed to hastately lobed to angulate, if lobed, center lobe elongate, not deeply dissected, (1.5–)3.5–10(–15) cm, not rugose, base cuneate, margins finely dentate to crenate, surfaces stellate-pubescent. **Inflorescences** paniculate, crowded, flowers clustered, tip usually leafy with distinct mature subtending leaves throughout; involucellar bractlets green to tan. **Flowers:** sepals 5–9 mm; petals mauve, lavender, red, red-orange, pink, or white, 8–15 mm; anthers yellow or purple. **Schizocarps** ovoid or truncate-conic; mericarps 9–15, 3.5–7 × 1.5–2 mm, chartaceous, nonreticulate dehiscent part 65–85% of height, muticous to cuspidate, indehiscent part not wider than dehiscent part. **Seeds** 2 per mericarp, brown to black, pubescent.

Varieties 2 (2 in the flora): w, c United States, Mexico.

1. Leaf blades linear-lanceolate to lanceolate, with or without hastate to angulate lobes . . . . 2a. *Sphaeralcea angustifolia* var. *angustifolia*
1. Leaf blades broadly linear to broadly trullate, with broad lobes . . . . 2b. *Sphaeralcea angustifolia* var. *oblongifolia*

2a. Sphaeralcea angustifolia (Cavanilles) G. Don var. **angustifolia**

Malva longifolia Sessé & Mociño; *Sphaeralcea angustifolia* subsp. *cuspidata* (A. Gray) Kearney; *S. angustifolia* var. *cuspidata* A. Gray; *S. angustifolia* var. *violacea* Burtt Davy; *S. cuspidata* (A. Gray) Britton; *S. emoryi* Torrey ex A. Gray var. *nevadensis* (Kearney) Kearney; *S. stellata* Torrey & A. Gray; *Sphaeroma angustifolium* (Cavanilles) Schlechtendal

Leaf blades linear-lanceolate to lanceolate, with or without hastate or angulate lobes. $2n = 10, 20, 30$.

Flowering spring–fall. Arroyos, disturbed roadsides; 900–2200 m; Ariz., Calif., Colo., Nebr., Nev., N.Mex., Okla., Tex., Utah; Mexico.

2b. Sphaeralcea angustifolia (Cavanilles) G. Don var. **oblongifolia** (A. Gray) Shinners, Sida 1: 384. 1964 [F]

Sphaeralcea incana Torrey ex A. Gray var. *oblongifolia* A. Gray, Smithsonian Contr. Knowl. 5(6): 21. 1853; *S. fendleri* A. Gray subsp. *elongata* Kearney; *S. fendleri* var. *elongata* (Kearney) Kearney; *S. fendleri* var. *perpallida* (Cockerell) Cockerell; *S. incana* subsp. *cuneata* Kearney; *S. incana* var. *cuneata* (Kearney) Kearney; *S. lobata* Wooton; *S. lobata* var. *perpallida* Cockerell

Leaf blades broadly linear to broadly trullate, with broad lobes.

Flowering spring–fall. Arroyos, disturbed roadsides; 1200–2400 m; Ariz., Calif., Colo., Kans., N.Mex., Tex.; Mexico (Chihuahua).

3. Sphaeralcea caespitosa M. E. Jones, Contr. W. Bot. 12: 4. 1908 • Tufted globemallow C E

Plants perennial. **Stems** 1–3(–5), erect, gray-green, 2–2.5 dm, white-gray canescent. **Leaf blades** gray-green or green, ovate, unlobed or 3-lobed, 5–8 cm, wrinkled, not rugose, base cuneate, margins coarsely toothed, surfaces stellate-pubescent. **Inflorescences** racemose, open, 3–6-flowered, tip not leafy; involucellar bractlets tan. **Flowers:** sepals 11–15 mm; petals red-orange, 15–22 mm; anthers yellow. **Schizocarps** flattened-spheric; mericarps 13, 3–6 × 2–4.5 mm, chartaceous, nonreticulate dehiscent part 60% of height, tip rounded, indehiscent part not wider than dehiscent part. **Seeds** 1 or 2 per mericarp, brown or black, minutely pubescent.

Varieties 2 (2 in the flora): w United States.

1. Leaf blades thick, gray-green, densely pubescent; staminal column 3.5–5 mm; Beaver and Millard counties, Utah . . . . .3a. *Sphaeralcea caespitosa* var. *caespitosa*
1. Leaf blades thin, green, sparsely pubescent; staminal column 6–9 mm; Nye County, Nevada . . . . .3b. *Sphaeralcea caespitosa* var. *williamsiae*

3a. Sphaeralcea caespitosa M. E. Jones var. **caespitosa** C E

Leaf blades gray-green, thick, densely pubescent. **Staminal columns** 3.5–5 mm.

Flowering spring–summer. Shallow soil, calcareous soil; of conservation concern; 1400–2000 m; Utah.

Variety *caespitosa* is known from Beaver and Millard counties.

3b. Sphaeralcea caespitosa M. E. Jones var. **williamsiae** N. H. Holmgren, Sida 20: 49, figs. 1E–H. 2002 C E

Leaf blades green, thin, sparsely pubescent. **Staminal columns** 6–9 mm.

Flowering spring–summer. Shallow soil or alluvium; of conservation concern; 1400–1600 m; Nev.

Variety *williamsiae* is known from Nye County.

4. Sphaeralcea coccinea (Nuttall) Rydberg, Bull. Torrey Bot. Club 40: 58. 1913 • Scarlet or common globemallow F

Malva coccinea Nuttall, Cat. Pl. Upper Louisiana, no. 51. 1813; *Malvastrum coccineum* (Nuttall) A. Gray; *Sida coccinea* (Nuttall) de Candolle

Plants perennial, rhizomatous. **Stems** 3–6, ascending or decumbent, light green to grayish, 1–3(–5) dm, stellate-canescent. **Leaf blades** green to gray-green, broadly to elongate-deltate, 3–5-lobed or pedately divided with relatively broad lobes, midlobe and often primary side-divisions pinnately few-cleft to parted, spatulate to narrowly spatulate, 1–6 cm, not rugose, base cuneate, margins entire, surfaces stellate-canescent. **Inflorescences** racemose or paniculate, crowded, few–many-flowered, tip not leafy; involucellar bractlets deciduous, green to gray-green. **Flowers:** sepals 5–10 mm; petals red-orange, 5–20 mm; anthers yellow. **Schizocarps** flattened spheric-conic; mericarps 10–14, 3–3.5 × 2.5–3 mm, thick, coriaceous, nonreticulate dehiscent part 10–35% of height, muticous, indehiscent part usually wider than dehiscent part. **Seeds** 1 per mericarp, gray to black, ± glabrous.

Varieties 2 (2 in the flora): w, c North America, n Mexico.

Sphaeralcea coccinea is variable; it is one of the first sphaeralceas to bloom and is commonly found on roadsides.

SELECTED REFERENCE La Duke, J. C. and D. K. Northington. 1978. The systematics of *Sphaeralcea coccinea* (Nutt.) Rydberg. SouthW. Naturalist 23: 651–660.

1. Leaf blades: midlobe ± equaling secondary lobes . . . . 4a. *Sphaeralcea coccinea* var. *coccinea*
1. Leaf blades: midlobe longer than secondary lobes . . . . 4b. *Sphaeralcea coccinea* var. *elata*

4a. Sphaeralcea coccinea (Nuttall) Rydberg var. **coccinea** F

Malvastrum coccineum (Nuttall) A. Gary var. *dissectum* (Nuttall ex Torrey & A. Gray); *M. cockerellii* A. Nelson; *M. dissectum* (Nuttall ex Torrey & A. Gray) Cockerell var. *cockerellii* (A. Nelson) A. Nelson; *Sida dissecta* Nuttall ex Torrey & A. Gray; *Sphaeralcea coccinea* subsp. *dissecta* (Nuttall ex Torrey & A. Gray) Kearney; *S. coccinea* var. *dissecta* (Nuttall ex Torrey & A. Gray) Kearney

Leaf blades: midlobe ± equaling secondary lobes. **Mericarps:** dehiscent part 10–20% of height. $2n = 10$.

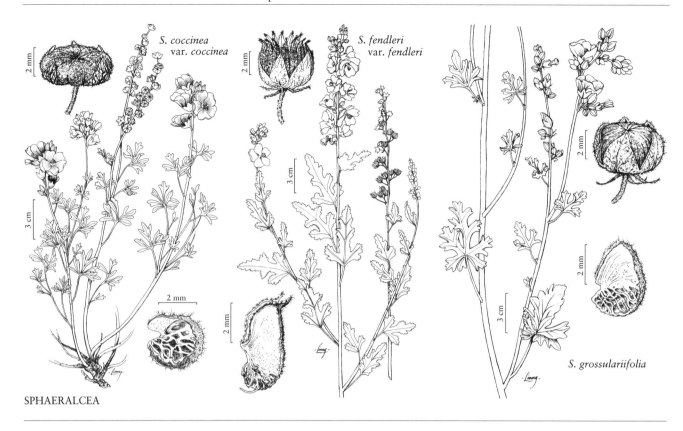

SPHAERALCEA

Flowering spring(–fall). Dry plains, prairies, disturbed roadsides; 300–2500 m; Alta., B.C., Man., Sask.; Ariz., Colo., Idaho, Iowa, Kans., Mont., Nebr., N.Mex., N.Dak., Okla., S.Dak., Tex., Utah, Wyo.; Mexico (Chihuahua).

4b. Sphaeralcea coccinea (Nuttall) Rydberg var. **elata** (Baker f.) Kearney, J. Wash. Acad. Sci. 29: 486. 1939 [E]

Malvastrum coccineum (Nuttall) A. Gray var. *elatum* Baker f., J. Bot. 29: 171. 1891; *M. elatum* (Baker f.) A. Nelson; *M. micranthum* Wooton & Standley; *Sphaeralcea coccinea* subsp. *elata* (Baker f.) Kearney; *S. elata* (Baker f.) Rydberg

Leaf blades: midlobe longer than secondary lobes. **Mericarps:** dehiscent part 10–35% of height. $2n = 20$.

Flowering spring. Rocky, sandy, or silty soil, limestone, gypsum, or igneous substrates, slopes, flats, or washes, desert scrub or shrublands; 300–1600 m; N.Mex., Tex.

5. Sphaeralcea coulteri (S. Watson) A. Gray, Proc. Amer. Acad. Arts 22: 291. 1887 • Annual globemallow

Malvastrum coulteri S. Watson, Proc. Amer. Acad. Arts 11: 125. 1876

Varieties 3 (1 in the flora): sw United States, Mexico.

5a. Sphaeralcea coulteri (Greene) Rydberg var. **coulteri**

Plants annual. **Stems** 4–6, sprawling [erect], green to gray-green, 1.5–15 dm, stellate-canescent, hairs few, slender, long, soft. **Leaf blades** gray-green, triangular or cordiform, rounded, unlobed or 3–5-lobed, 1.5–4.5 cm, sometimes wider than long, not rugose, base typically cordate, margins coarsely toothed, surfaces stellate-canescent. **Inflorescences** paniculate, open, flowers clustered, tip appearing leafy; involucellar bractlets green to tan. **Flowers:** sepals 5–7 mm; petals orange [salmon-orange], [6–]8–11[–20] mm; anthers yellow. **Schizocarps** ± hemispheric; mericarps 15, 1.5–2 × 2–2.5 mm, chartaceous, nonreticulate dehiscent part flat on top, 30% of height, projecting toward fruit axis, reticulations often blackish, indehiscent part not notably wider than dehiscent part. **Seeds** 1 per mericarp, brown or black, glabrous or pubescent. $2n = 10$.

Flowering winter–spring. Dry sandy places; 50–800 m; Ariz., Calif.; Mexico (Baja California, Sinaloa, Sonora).

6. Sphaeralcea digitata (Greene) Rydberg, Bull. Torrey Bot. Club 40: 58. 1913 • Juniper globemallow

Malvastrum digitatum Greene, Leafl. Bot. Observ. Crit. 1: 154. 1905; *Sphaeralcea digitata* var. *angustiloba* (A. Gray) Shinners; *S. digitata* subsp. *tenuipes* (Wooton & Standley) Kearney; *S. digitata* var. *tenuipes* Wooton & Standley) Kearney & Peebles; *S. pedata* Torrey ex A. Gray var. *angustiloba* A. Gray; *S. tenuipes* Wooton & Standley

Plants perennial. **Stems** decumbent to erect, green to gray-green, (1–)3–4(–5.5) dm, sparsely pubescent. **Leaf blades** green to gray-green, wide-triangular, 3-lobed or pedately divided, (0.5–)1.5–2.5(–4) cm, not rugose, base cuneate, margins entire, surfaces stellate-pubescent. **Inflorescences** racemose, flowers solitary or proximal nodes 2- or 3-flowered, tip ± leafless; involucellar bractlets green to tan. **Flowers:** sepals 3.5–8 mm; petals red-orange, 8–14 mm; anthers purple. **Schizocarps** cylindric to widely conic; mericarps 9–13, 3–5 × 2–2.5 mm, thickish, chartaceous, nonreticulate dehiscent part 50–60% of height, tip muticous to short-cuspidate, indehiscent part not wider than dehiscent part, sides finely, not highly reticulate. **Seeds** 1(2) per mericarp, brown or black, glabrous or pubescent. $2n = 10$.

Flowering spring–summer. Diverse substrates including gypsum, desert scrub; 1200–2200 m; Ariz., N.Mex., Tex., Utah; Mexico (Chihuahua, Sonora).

7. Sphaeralcea emoryi Torrey ex A. Gray, Mem. Amer. Acad. Arts, n. s. 4: 23. 1849 • Emory's globemallow

Sphaeralcea arida Rose; *S. emoryi* var. *arida* (Rose) Kearney; *S. emoryi* var. *variabilis* (Cockerell) Kearney; *S. fendleri* A. Gray var. *californica* Parish; *S. fendleri* var. *variablis* (Cockerell) Cockerell; *S. variablis* Cockerell

Plants perennial. **Stems** erect, gray to gray-green, to 21 dm, glabrous or coarse-canescent. **Leaf blades** green to gray-green, ovate-triangular to lanceolate, 3-lobed, 2.5–5.5 cm, not rugose, base cordate to truncate, margins crenate to serrate, surfaces stellate-pubescent. **Inflorescences**

paniculate, proximally open, distally crowded, flowers clustered, tip not leafy; involucellar bractlets green to tan, rarely red. **Flowers:** sepals 6–8 mm; petals red-orange, pink, or lavender, 10–12 mm; anthers yellow. **Schizocarps** truncate-conic; mericarps 10–16, 4.5–5 × 2.5 mm, chartaceous, nonreticulate dehiscent part 60% of height, tip acute, indehiscent part not wider than dehiscent part. **Seeds** 1 or 2 per mericarp, brown or black, pubescent. $2n = 20, 30, 50$.

Flowering spring–summer. Roadsides, disturbed areas; (–20–)200–1200 m; Ariz., Calif., Nev.; Mexico (Baja California, Chihuahua, Sinaloa, Sonora).

Sphaeralcea emoryi may intergrade with *S. angustifolia* and *S. laxa.*

8. Sphaeralcea fendleri A. Gray, Smithsonian Contr. Knowl. 3(5): 21. 1852 • Fendler's globemallow [F]

Plants perennial. **Stems** erect or ascending, usually gray to gray-green or green, sometimes purple to black, (2–)4–7(–10) dm, sparsely to densely canescent or hirsute. **Leaf blades** green or gray to gray-green, lanceolate to triangular, shallowly to deeply 3-lobed, (1.5–)3–7 cm, not rugose, base truncate to cuneate, margins crenate to dentate, surfaces hirsute or sparsely to densely stellate soft-pubescent. **Inflorescences** paniculate, narrow, usually crowded, tip leafy or not; involucellar bractlets green to tan. **Flowers:** sepals 4.5–6(–9) mm; petals red-orange or lavender, 8–13 mm; anthers yellow. **Schizocarps** cylindric to conic; mericarps 9–16, 4–5.5 × 2–2.5 mm, chartaceous, nonreticulate dehiscent part 70–80% of height, tip reflexed and cuspidate, indehiscent part not wider than dehiscent part, sides faintly, very finely reticulate, tips reflexed. **Seeds** 2 per mericarp, black, pubescent.

Varieties 3 (3 in the flora): sw United States, n Mexico.

1. Leaf blade surfaces distinctly hirsute. . . . .
. . . . 8c. *Sphaeralcea fendleri* var. *venusta*
1. Leaf blade surfaces soft-pubescent or sparsely pubescent.
  2. Leaf blade surfaces sparsely pubescent, appearing green; petals red-orange. . . . .
. . . . .8a. *Sphaeralcea fendleri* var. *fendleri*
  2. Leaf blade surfaces densely soft white-pubescent; petals usually lavender . . . .
. . . . 8b. *Sphaeralcea fendleri* var. *albescens*

8a. Sphaeralcea fendleri A. Gray var. **fendleri** [F]

Sphaeralcea fendleri subsp. *elongata* Kearney; *S. fendleri* var. *elongata* (Kearney) Kearney; *S. fendleri* subsp. *tripartita* (Wooton & Standley) Kearney; *S. fendleri* var. *tripartita* (Wooton & Standley) Kearney; *S. leiocarpa* Wooton & Standley; *S. tripartita* Wooton & Standley

Stems usually green, sometimes purple or black, sparsely pubescent. **Leaf blades** green to light green, surfaces sparsely pubescent. **Petals** red-orange. $2n$ = 10, 20.

Flowering summer–fall. Forested areas; 900–2500 m; Ariz., Colo., Kans., N.Mex.; Mexico (Chihuahua, Sonora).

8b. Sphaeralcea fendleri A. Gray var. **albescens** (Kearney) Kearney, J. Wash. Acad. Sci. 29: 486. 1939

Sphaeralcea fendleri subsp. *albescens* Kearney, Univ. Calif. Publ. Bot. 19: 62. 1935

Stems gray-green, stellate-pubescent. **Leaf blades** gray-green, surfaces densely soft white-pubescent. **Petals** usually lavender. $2n$ = 20.

Flowering summer–fall. Washes, disturbed sites; 1000–1100 m; Ariz.; Mexico (Coahuila, Sonora).

8c. Sphaeralcea fendleri A. Gray var. **venusta** (Kearney) Kearney, J. Wash. Acad. Sci. 29: 486. 1939

Sphaeralcea fendleri subsp. *venusta* Kearney, Univ. Calif. Publ. Bot. 19: 63. 1935

Stems gray to gray-green, hirsute. **Leaf blades** gray to gray-green, surfaces hirsute. **Petals** lavender to red-orange. $2n$= 20.

Flowering summer–fall. Dry roadsides, rocky terrain; 1300–2100 m; Ariz.; Mexico (Chihuahua, Sonora).

9. Sphaeralcea fumariensis (S. L. Welsh & N. D. Atwood) N. D. Atwood & S. L. Welsh, Novon 12: 160. 2002 • Smoky Mountain globemallow [E]

Sphaeralcea grossulariifolia (Hooker & Arnott) Rydberg var. *fumariensis* S. L. Welsh & N. D. Atwood, Rhodora 103: 82, fig. 4. 2001

Plants perennial. **Stems** ascending, gray-green, red-purple basally, 2–5.5 dm, pubescent. **Leaf blades** gray-green, cordate, pedately divided with usually 5 main lobes, 1–4.5 cm, not rugose, base cordate to truncate, margins entire, surfaces densely stellate-pubescent. **Inflorescences** paniculate, crowded, tip not leafy; involucellar bractlets reddish. **Flowers:** sepals 7.5–12 mm; petals red-orange, 12–17 mm; anthers yellow. **Schizocarps** hemispheric; mericarps 10–14, 3–4 × 2–3 mm, chartaceous, nonreticulate dehiscent part 50–65% of height, tip rounded, indehiscent part not wider than dehiscent part, sides reticulate. **Seeds** 1(2) per mericarp, brown to black, glabrous or stellate-pubescent.

Flowering May–Jun. Alluvium of mudstone, shale, and sandstone formations; 1300–1700 m; Utah.

Sphaeralcea fumariensis is similar to *S. moorei* but is notably gray-green pubescent in contrast to the less pubescent and green *S. moorei*. See 11. *S. grossulariifolia* for discussion. *Sphaeralcea fumariensis* is known from Kane County.

10. Sphaeralcea gierischii N. D. Atwood & S. L. Welsh, Novon 12: 161, fig. 1. 2002 • Gierisch's globemallow [C][E]

Plants perennial. **Stems** ascending, usually dark red-purple, 4–10 dm, glabrous or stellate-canescent. **Leaf blades** green, gray, or gray-green, ovate to cordate-ovate, 3- or 5-lobed, main lobe unlobed in 3-lobed leaves, cleft in 5-lobed leaves, lobes spatulate, 1.2–4 cm, not rugose, base truncate to cordate, margins entire, surfaces stellate-pubescent. **Inflorescences** paniculate, open, 2–few-flowered per node, tip not leafy; involucellar bractlets green. **Flowers:** sepals 5–13 mm; petals red-orange, 15–20(–25) mm; anthers purple or yellow. **Schizocarps** hemispheric; mericarps 10–15, 4.5–5.5 × 3 mm, chartaceous, nonreticulate dehiscent part 40–50% of height, tip subacute, indehiscent part wider than dehiscent part, sides prominently reticulate. **Seeds** 2 per mericarp, black, pubescent.

Flowering Apr–Jun. Gypsum soil; of conservation concern; 700–1200 m; Ariz., Utah.

Sphaeralcea gierischii is found in Mohave County, Arizona, and Washington County, Utah. Material of *S. gierischii* is limited and eventually may be found to be conspecific with *S. grossulariifolia*.

11. **Sphaeralcea grossulariifolia** (Hooker & Arnott) Rydberg, Bull. Torrey Bot. Club 40: 58. 1913 (as grossulariaefolia) • Gooseberry-leaf or currant-leaf globemallow E F

Sida grossulariifolia Hooker & Arnott, Bot. Beechey Voy., 326. 1838 (as grossulariaefolia); *Malvastrum coccineum* (Nuttall) A. Gray var. *grossulariifolium* (Hooker & Arnott) Torrey; *M. grossulariifolium* (Hooker & Arnott) A. Gray; *Sphaeralcea grossulariifolia* subsp. *pedata* (Torrey ex A. Gray) Kearney; *S. grossulariifolia* var. *pedata* (Torrey ex A. Gray) Kearney; *S. pedata* Torrey ex A. Gray

Plants perennial. **Stems** erect, green or purple, (2–)6–10 dm, white-canescent or glabrate. **Leaf blades** green to gray-green or yellow-green, broadly triangular, 3-lobed or pedately divided, tips rounded or acute, 1.7–3.5(–9) cm, not rugose, base cordate to truncate, margins dentate or crenate, surfaces stellate-pubescent. **Inflorescences** paniculate, crowded, flowers solitary or clustered, tip not leafy; involucellar bractlets green to tan. **Flowers:** sepals 5–8 mm; petals red-orange, 8–15 mm; anthers yellow. **Schizocarps** truncate-conic to spheric; mericarps 10–12, 2.5–3.5 × 2–2.5 mm, chartaceous, reticulate dehiscent part 60% of height, tip rounded-acute, indehiscent part not wider than dehiscent part. **Seeds** 1(2) per mericarp, gray or black, glabrous or pubescent. $2n$ = 20.

Flowering spring–summer. Dry, volcanic, rocky or sandy soil; 100–2000 m; Ariz., Idaho, Nev., N.Mex., Oreg., Utah, Wash.

Sphaeralcea grossulariifolia is variable and may include *S. fumariensis*, *S. gierischii*, and *S. moorei*. Material of the latter taxa is limited, and although select specimens can be attributed to each, the full morphological range of *S. grossulariifolia* can readily include these other taxa.

12. **Sphaeralcea hastulata** A. Gray, Smithsonian Contr. Knowl. 3(5): 17. 1852 • Spear globemallow

Sphaeralcea arenaria Wooton & Standley; *S. glabrescens* Wooton & Standley; *S. martii* Cockerell; *S. pumila* Wooton & Standley; *S. simulans* Wooton & Standley; *S. subhastata* J. M. Coulter; *S. subhastata* subsp. *connata* Kearney; *S. subhastata* var. *connata* (Kearney) Kearney; *S. subhastata* subsp. *martii* (Cockerell) Kearney; *S. subhastata* var. *martii* (Cockerell) Kearney; *S. subhastata* subsp. *pumila* (Wooton & Standley) Kearney; *S. subhastata* var. *pumila* (Wooton & Standley) Kearney; *S. subhastata* subsp. *thyrsoidea* Kearney; *S. subhastata* var. *thyrsoidea* (Kearney) Kearney

Plants perennial. **Stems** ascending to decumbent, green to gray-green, 1–3(–5) dm, canescent. **Leaf blades** green to gray-green, oblong-ovate to ovate-lanceolate, unlobed, lobed, or pedately divided, (1–)2–6 cm, not rugose, base cuneate to cordate, margins entire, crenate, or dentate, usually green or tan, surfaces stellate-pubescent. **Inflorescences** paniculate or racemose, open or crowded, flowers clustered or solitary, tip leafy or not; involucellar bractlets green to tan. **Flowers:** sepals 4–11 mm, tips forming beak in bud; petals red-orange, pink, or purple, 10–20 mm; anthers yellow or purple. **Schizocarps** widely conic; mericarps 10–30, 3–7 × 1.5–2.5 mm, thickish, chartaceous, nonreticulate dehiscent part 60–80% of height, tip obtuse to acute, usually muticous (to mucronate to cuspidate), cusp 1–2 mm, indehiscent part sometimes wider than dehiscent part. **Seeds** 1 or 2 per mericarp, brown or black, glabrous or pubescent. $2n$ = 10, 30.

Flowering early spring–fall. Plains, alkaline or gypsum areas; 800–2100 m; Ariz., N.Mex., Tex.; Mexico (Chihuahua, Coahuila, Durango, Nuevo León, Sonora, Tamaulipas).

T. H. Kearney (1935) recognized subspecies of *Sphaeralcea hastulata*; the subspecies are not clearly and consistently distinguishable and are herein combined.

13. **Sphaeralcea incana** Torrey ex A. Gray, Mem. Amer. Acad. Arts, n. s. 4: 23. 1849 • Gray globemallow

Sphaeralcea incana subsp. *cuneata* Kearney; *S. incana* var. *cuneata* (Kearney) Kearney

Plants perennial. **Stems** 3–5(–7), erect, yellow to yellow-green, (4–)6–18(–30) dm, rubbery when fresh, sometimes appearing slightly ridged or fasciated when dry, densely canescent. **Leaf blades** light green to yellow-green, deltate or elongate-deltate, unlobed or weakly 3-lobed, center lobe sometimes elongate, lobes relatively broad, 3–5(–7) cm, not rugose, rubbery, base cuneate or truncate to cordate, margins entire or crenulate to undulate, surfaces stellate-pubescent. **Inflorescences** paniculate, crowded, flowers long-interrupted, tip not leafy; involucellar bractlets green to tan. **Flowers:** sepals 3.5–6.5 mm, spheric in bud; petals red-orange to pale red, 10–17 mm; anthers yellow. **Schizocarps** hemispheric; mericarps 11–15, 4–5.5 × 2–2.5 mm, chartaceous, nonreticulate dehiscent part 60–75% of height, tip reflexed, indehiscent part not wider than dehiscent part, sides faintly to prominently and finely reticulate with thin-translucent areolae. **Seeds** 2 per mericarp, brown, glabrous or pubescent. $2n$ = 10, 20.

Flowering summer. Sandy areas; 1000–1900 m; Ariz., N.Mex.; Mexico (Chihuahua, Coahuila, Sonora).

Sphaeralcea incana is often seen in relatively small groups. Fresh stems tend to be very rubbery and flower buds notably rounded.

14. **Sphaeralcea laxa** Wooton & Standley, Bull. Torrey Bot. Club 36: 108. 1909 • Caliche globemallow

Sphaeralcea incana Torrey ex A. Gray var. *dissecta* A. Gray; *S. ribifolia* Wooton & Standley

Plants perennial. **Stems** ascending, (2–)4–7(–9) dm, densely gray- or white-tomentose. **Leaf blades** gray or white, broadly ovate-deltate, 3-lobed or 3- or 5-parted, (1–)1.5–5 cm, not rugose, base typically distinctly truncate (to cordate), margins crenate, surfaces stellate-pubescent. **Inflorescences** paniculate, open, few-flowered, individual flowers widely spaced, tip not leafy; involucellar bractlets red to purple. **Flowers:** sepals 7–11 mm; petals red-orange, 12–18 mm; anthers usually purple. **Schizocarps** truncate-ovoid; mericarps 11–16, 4–6 × 2–3 mm, thin-walled, papery, nonreticulate dehiscent part 55–65% of height, tip acute-cuspidate, cusp 1 mm, indehiscent part not wider than dehiscent part, sides prominently, finely reticulate with ± transparent areolae. **Seeds** 2 or 3 per mericarp, gray, pubescent. $2n$ = 10, 20.

Flowering early spring–late fall. Rock outcrops and caliche; 600–1900 m; Ariz., N.Mex., Tex.; Mexico (Sonora).

15. **Sphaeralcea leptophylla** (A. Gray) Rydberg, Bull. Torrey Bot. Club 40: 59. 1913

Malvastrum leptophyllum A. Gray, Smithsonian Contr. Knowl. 3(5): 17. 1852; *Malveopsis leptophylla* (A. Gray) Kuntze; *Sphaeralcea janeae* (S. L. Welsh) S. L. Welsh; *S. leptophylla* var. *janeae* S. L. Welsh

Plants perennial. **Stems** ascending, white to silvery, 1–4(–6) dm, silvery stellate-lepidote. **Leaf blades** white to silvery, linear to triangular, mostly unlobed, proximalmost blades 3-parted, rarely with nonfiliform lobes, 1–3.5 cm, not rugose, base truncate to cuneate, margins entire, surfaces silvery stellate-lepidote. **Inflorescences** racemose, lax, open, 3–12-flowered, tip not leafy; involucellar bractlets silvery-lepidote. **Flowers:** sepals 4.5–5.5(–7) mm; petals red-orange, 8–15 mm; anthers yellow. **Schizocarps** flattened-spheric to conic; mericarps 7–9, 2.5–3.5 × 2–3 mm, thick-coriaceous, nonreticulate dehiscent part 20–40% of height, tip muticous-mucronulate, indehiscent part usually wider than dehiscent part, sides coarsely reticulate. **Seeds** 1 per mericarp, brown, usually glabrous. $2n$ = 20.

Flowering spring–summer. Dry rocky areas; 1500–1800 m; Ariz., Colo., Nev., N.Mex., Tex., Utah; Mexico (Chihuahua, Sonora).

16. **Sphaeralcea lindheimeri** A. Gray, Boston J. Nat. Hist. 6: 162. 1850 • Woolly globemallow [E]

Plants perennial. **Stems** decumbent, whitish, 2–7 dm, soft-pubescent. **Leaf blades** abaxially white to silvery, adaxially green, deltate-ovate, unlobed or 3-lobed, 4 cm, not rugose, base cordate to truncate, margins broadly crenate, surfaces stellate-pubescent. **Inflorescences** racemose, crowded, few-flowered, tip not leafy; involucellar bractlets green to purple. **Flowers:** sepals 8–15 mm; petals red to red-pink, 15–25 mm; anthers yellow. **Schizocarps** hemispheric; mericarps 18, 4 mm, chartaceous, nonreticulate dehiscent part 60–70% of height, tip acute, indehiscent part not wider than dehiscent part. **Seeds** 2 or 3 per mericarp, brown to black, slightly pubescent.

Flowering fall–spring. Sandy soil, open thickets or roadsides; 0–300 m; Tex.

Sphaeralcea lindheimeri has unusually long and soft hairs and is found usually near the coast.

17. **Sphaeralcea moorei** (S. L. Welsh) N. D. Atwood & S. L. Welsh, Novon 12: 163. 2002 [E]

Sphaeralcea grossulariifolia (Hooker & Arnott) Rydberg var. *moorei* S. L. Welsh, Great Basin Naturalist 40: 35. 1980

Plants perennial. **Stems** erect, red-purple basally, 3–8 dm, stellate-canescent. **Leaf blades** green, ovate to deltate, 3- or 5-lobed, digitate, spatulate, secondary divisions of lobes frequent, basal lobes frequently not divided, 1–4(–7) cm, not rugose, base truncate or cuneate, margins entire, surfaces gray-green stellate-canescent. **Inflorescences** paniculate, open, 1–few-flowered, very few-leaved, tip not leafy; involucellar bractlets green to red-purple. **Flowers:** sepals 4.5–8.5 mm; petals red-orange, 8–15 mm; anthers yellow. **Schizocarps** hemispheric; mericarps 10(–14), 3–5 × 2 mm, chartaceous, nonreticulate dehiscent part 50% of height, tip rounded to acute, indehiscent part wider than dehiscent part, sides reticulate. **Seeds** 2 per mericarp, brown to black, glabrous or sparsely pubescent.

Flowering May–Jul. Sandy soil, riparian; 800–1900 m; Utah.

Sphaeralcea moorei intergrades with *S. grossulariifolia* and has similarities to *S. parvifolia*. See under 11. *S. grossulariifolia* for discussion.

18. **Sphaeralcea munroana** (Douglas ex Lindley) Spach, Hist. Nat. Vég. 14: 403. 1847 • Munro's globemallow [E]

Malva munroana Douglas ex Lindley, Edward's Bot. Reg. 16: plate 1306. 1830; *M. creeana* Graham; *Malvastrum munroanum* (Douglas ex Lindley) A. Gray; *Malveopsis munroana* (Douglas ex Lindley) Kuntze; *Nuttallia munroana* (Douglas ex Lindley) Nuttall; *Sphaeralcea munroana* subsp. *subrhomboidea* (Rydberg) Kearney; *S. munroana* var. *subrhomboidea* (Rydberg) Kearney; *S. subrhomboidea* Rydberg

Plants perennial. **Stems** erect, green to gray-green, 2.5–9 dm, gray-green canescent. **Leaf blades** green to gray-green, triangular, unlobed or weakly 5-lobed, to 4.5 cm, not rugose, base cuneate to truncate, margins coarsely toothed, surfaces stellate-pubescent. **Inflorescences** narrowly paniculate, crowded, few–many-flowered, flowers in clusters, tip not leafy; involucellar bractlets green to tan. **Flowers:** sepals 4–9 mm, tips forming weak beak in bud; petals red-orange, 11–14 mm; anthers yellow. **Schizocarps** spheric; mericarps 12,

3.5–4 × 2.5–3 mm, chartaceous, nonreticulate dehiscent part 55% of height, tip acute, with reflexed apical cusp, indehiscent part not wider than dehiscent part. **Seeds** 1 per mericarp, brown, slightly hairy.

Flowering spring–summer. Xeric plains and slopes; 100–2300 m; B.C.; Calif., Idaho, Mont., Nev., Oreg., Utah, Wash., Wyo.

Sphaeralcea munroana and *S. parvifolia* are difficult to distinguish and may be conspecific. *Sphaeralcea parvifolia* is more southern and smaller overall; *S. munroana* is more northern and more robust.

19. **Sphaeralcea orcuttii** Rose, Contr. U.S. Natl. Herb. 1: 289. 1893 • Carrizo Creek globemallow

Plants usually annual, sometimes biennial. **Stems** erect, yellow, 5–12 dm, stellate-canescent. **Leaf blades** yellow-green to pale green, trullate with straight or rounded appearance or triangular with subhastate lobes, apical lobes distinctly triangular, 2-lobed, 3–5 cm, not rugose, base tapered to truncate, margins entire or wavy, surfaces stellate-canescent. **Inflorescences** paniculate, crowded, many-flowered, flowers clustered, tip not leafy; involucellar bractlets green to tan. **Flowers:** sepals 4–7 mm; petals red-orange, 10–12 mm; anthers yellow. **Schizocarps** ± hemispheric; mericarps 12–17, 2.5–3 × 2–3 mm, chartaceous, nonreticulate dehiscent part 30–40% of height, tip rounded, indehiscent part wider than dehiscent part. **Seeds** 1 per mericarp, brown, glabrous or pubescent. $2n = 10$.

Flowering spring. Dry, sandy, ± alkaline places; –50–900 m; Ariz., Calif.; Mexico (Baja California, Baja California Sur, Sonora).

Sphaeralcea orcuttii has distinctive trullate leaf blades and an erect habit.

20. **Sphaeralcea parvifolia** A. Nelson, Proc. Biol. Soc. Wash. 17: 94. 1904 • Small-leaf globemallow [E]

Sphaeralcea arizonica A. Heller ex Rydberg; *S. marginata* York ex Rydberg

Plants perennial. **Stems** erect, green or gray-green, 1.5–4 (–10) dm, white- to yellow-canescent. **Leaf blades** gray to green, ovate, unlobed or weakly 3–5-lobed, 1–5.5 cm, not rugose, base cuneate to cordate, margins entire or crenate to serrate, surfaces stellate-pubescent. **Inflorescences** paniculate, crowded, flowers clustered with distinct internodes between clusters, tip not leafy; involucellar bractlets usually green to tan, sometimes red-purple. **Flowers:** sepals 6–9 mm, tip not forming distinct beak in bud; petals red-orange, 8–14 mm; anthers yellow. **Schizocarps** ellipsoid; mericarps 12, 3.5–5.5 × 1.5–3 mm, chartaceous, nonreticulate dehiscent part 60–70% of height, with or without apical cusp, indehiscent part not wider than dehiscent part. **Seeds** 1 per mericarp, gray or black, ± pubescent. $2n$ = 10, 20.

Flowering summer. Dry slopes; 1500–2100 m; Ariz., Calif., Colo., Nev., N.Mex., Utah.

21. **Sphaeralcea pedatifida** (A. Gray) A. Gray, Proc. Amer. Acad. Arts 22: 291. 1887 • Palmleaf globemallow

Malvastrum pedatifidum A. Gray, Boston J. Nat. Hist. 6: 160. 1850; *Sidalcea atacosa* Buckley

Plants perennial. **Stems** erect, green to light gray, 1.5–3 dm, stellate-pubescent. **Leaf blades** green to light gray, triangular, pedately divided, lobes relatively narrow, 1–2.8 cm, not rugose, base truncate to cordate, margins coarsely toothed, surfaces green to light gray stellate-canescent. **Inflorescences** racemose, open, few-flowered, tip not leafy; involucellar bractlets green to tan. **Flowers:** sepals 6–8 mm; petals red-orange, 8–12 mm; anthers yellow. **Schizocarps** hemispheric to flat-hemispheric; mericarps 15–20, 4 × 2–2.4 mm, chartaceous, nonreticulate dehiscent part 60–75% of height, tip acute, indehiscent part wider than dehiscent part. **Seeds** 1 per mericarp, brown to black, glabrous or pubescent. $2n$ = 10.

Flowering spring–fall. Sandy or rocky slopes; 50–200 m; Tex.; Mexico (Coahuila, Tamaulipas).

Sphaeralcea pedatifida has finely divided, thin leaves. It is known from western Texas.

22. **Sphaeralcea polychroma** La Duke, SouthW. Naturalist 30: 433, fig. 1. 1985 • Hot Springs globemallow

Plants perennial. **Stems** erect, white or yellow, to 10–20 dm, rubbery, densely soft stellate-pubescent. **Leaf blades** white or yellow, deltate to lanceolate, subhastate to 3-lobed, 4–7 cm, secondary lobes to 2.2 cm, not rugose, base cuneate, margins crenate to dentate, surfaces green- to white-canescent. **Inflorescences** paniculate, open, many-flowered, interrupted, tip leafy; involucellar bractlets green to tan. **Flowers:** sepals 6–7 mm; petals white, pink, lavender, purple, red-orange, or red, 10–13 mm; anthers yellow. **Schizocarps** short-urceolate; mericarps 12–14, (3.5–)4–5.5 × 2–3 mm, chartaceous,

nonreticulate dehiscent part 60% of height, with usually reflexed cusps to 2 mm, indehiscent part not wider than dehiscent part. **Seeds** 1 or 2 per mericarp, gray, pubescent. $2n$ =20.

Flowering summer. Desert lowlands; 1900 m; Ariz., N.Mex., Tex.; Mexico (Chihuahua).

Sphaeralcea polychroma is frequent in central New Mexico and western Texas. It is closely related to *S. procera*.

23. **Sphaeralcea procera** Ced. Porter, Bull. Torrey Bot. Club. 70: 531, figs. 1, 2. 1943 • Luna County globemallow [E]

Plants perennial. **Stems** erect, white, 30 dm, densely soft stellate-pubescent. **Leaf blades** white or yellow, lanceolate or ovate to triangular or cordate, 3-lobed, central lobe larger than laterals, 1–5 cm, rugose, coriaceous, base cuneate, margins irregularly dentate, wavy, surfaces stellate-pubescent. **Inflorescences** paniculate, narrow, crowded, multiflowered, interrupted, tip leafy; involucellar bractlets green. **Flowers:** sepals 5 mm; petals rose-purple, 10–13 mm; anthers yellow. **Schizocarps** helmet-shaped; mericarps usually 10, usually 3 × 2 mm, chartaceous, moderately reticulate dehiscent part 60% of height, tip acute, indehiscent part not wider than dehiscent part. **Seeds** 1 per mericarp, gray to black, pubescent.

Flowering fall. Sandy soil; 1400 m; N.Mex.

Sphaeralcea procera is known only from the type collection, from Luna County. It is closely related to *S. polychroma*, which is usually common when it is found. It is surprising that *S. procera* is known only from a single collection. Botanists, including me, have looked for it around the type locality without success. It is retained here pending evidence that it is conspecific with *S. polychroma* or extinct.

24. **Sphaeralcea psoraloides** S. L. Welsh, Great Basin Naturalist 40: 36. 1980 • Psoralea globemallow [C][E]

Plants perennial. **Stems** erect, yellow-green, 1.4–2(–3) dm, stellate-canescent. **Leaf blades** yellow-green, triangular to deltate, usually 3(5)-lobed, lobes unlobed or broadly oblanceolate, 1.3–3.5 cm, not rugose, base cuneate(-truncate), margins entire, surfaces sparsely stellate-canescent. **Inflorescences** racemose, open, flowers usually 1 per node, tip leafy; involucellar bractlets tan. **Flowers:** sepals 4.5–8 mm; petals red-orange, 10–17 mm;

anthers yellow. **Schizocarps** flattened-hemispheric; mericarps 9–13, 2–2.5 × 2 mm, chartaceous, nonreticulate dehiscent part 10–15% of height, tip acute, indehiscent part wider than dehiscent part, sides reticulate. **Seeds** 1 per mericarp, dark brown to black, glabrous or pubescent.

Flowering spring–summer. Clay or gravel soil; of conservation concern; 1200–1900 m; Utah.

Sphaeralcea psoraloides resembles *S. coccinea* but has markedly greenish, simply-lobed leaves; it occurs in Emery, Grand, and Wayne counties.

25. **Sphaeralcea rusbyi** A. Gray, Proc. Amer. Acad. Arts 22: 293. 1887 • Rusby's globemallow E

Plants perennial. **Stems** erect, gray-green to purple, 3–6 dm, sparsely pubescent. **Leaf blades** light green, sometimes with purple margins, triangular to broadly ovate, pedately divided or 3–5-lobed, 1.5–5 cm, not rugose, base truncate to cordate, margins entire, surfaces sparsely stellate-pubescent. **Inflorescences** paniculate, crowded, few-flowered per node, tip ± leafy; involucellar bractlets red to purple. **Flowers:** sepals sometimes red or purple, 6–15 mm, lobes equaling to 3 times tube length; petals red-orange, 10–20 mm; anthers usually red to purple, sometimes yellow. **Schizocarps** truncate-spheric; mericarps 13, 4.5–6 × 2–2.5 mm, chartaceous, nonreticulate dehiscent part 55–80% of height, tip acute, indehiscent part not wider than dehiscent part. **Seeds** 1 or 2 per mericarp, black-gray, glabrous or pubescent.

Varieties 2 (2 in the flora): sw United States.

1. Sepals 6–10 mm, lobes deltate; calyx length ± equaling fruit . . . . . 25a. *Sphaeralcea rusbyi* var. *rusbyi*
1. Sepals 13–15 mm, lobes elongate; calyx length 2 times fruit . . . . . . 25b. *Sphaeralcea rusbyi* var. *eremicola*

25a. **Sphaeralcea rusbyi** A. Gray var. **rusbyi** E

Sphaeralcea rusbyi subsp. *gilensis* Kearney; *S. rusbyi* var. *gilensis* (Kearney) Kearney

Sepals 6–10 mm, lobes deltate, calyx length ± equaling fruit. $2n = 10, 20$.

Flowering spring–fall. Slopes and forest openings; 200–2000 m; Ariz., Calif., Utah.

25b. **Sphaeralcea rusbyi** A. Gray var. **eremicola** (Jepson) Kearney, J. Wash. Acad. Sci. 29: 486. 1939 • Rusby's desert or panimint mallow C E

Sphaeralcea eremicola Jepson, Man. Fl. Pl. Calif., 635. 1925; *S. rusbyi* subsp. *eremicola* (Jepson) Kearney

Sepals 13–15 mm, lobes elongate, calyx length 2 times fruit.

Flowering May. Rocky limestone; of conservation concern; 1300–1500 m; Calif.

Variety *eremicola* is known from the Emigrant Canyon area of Inyo County.

26. **Sphaeralcea wrightii** A. Gray, Smithsonian Contr. Knowl. 5(6): 21. 1853 • Wright's globemallow

Plants perennial. **Stems** ascending to erect, gray-green, 2–5(–7.5) dm, canescent to tomentose. **Leaf blades** gray-green, widely ovate or triangular to orbiculate, all but proximalmost deeply pedately divided, (1–)2–4 cm, not rugose, base truncate to cordate, margins entire, surfaces stellate-pubescent. **Inflorescences** racemose or paniculate, narrow, open, few–many-flowered, tip not leafy; involucellar bractlets green to tan. **Flowers:** sepals 6–7 mm, not forming beak; petals lavender, red-orange, or pink, 10–13.5(–18) mm; anthers yellow or purple. **Schizocarps** hemispheric to truncate-conic; mericarps 12–15, 4–7 × 2.5–3 mm, chartaceous, nonreticulate dehiscent part 55–65% of height, with prominent ventral beak, tip subacute, cuspidate, indehiscent part not wider than dehiscent part, sides prominently reticulate. **Seeds** 2 per mericarp, black, pubescent. $2n = 20$.

Flowering spring. Dry rocky areas; 1200 m; Ariz., N.Mex., Tex.; Mexico (Chihuahua).

49. TALIPARITI Fryxell, Contr. Univ. Michigan Herb. 23: 231, figs. 1–5. 2001 • [Malayam *thaali*, shampoo, and *paruthi*, cotton, alluding to use and resemblance, respectively] I

Paul A. Fryxell†

Steven R. Hill

Hibiscus Linnaeus sect. *Azanzae* de Candolle

Trees. **Stems** spreading [erect], minutely hairy [to glabrescent], not viscid. **Leaves:** stipules usually early-deciduous leaving annular scars, prominent, oblong, enclosing terminal bud; blade broadly ovate [orbiculate, elliptic], not dissected or parted [3-lobed], base deeply cordate [rarely cuneate], margins usually entire (sometimes obscurely crenulate or denticulate), with nectaries near base abaxially. **Inflorescences** axillary, solitary flowers; involucel present, bractlets persistent, connate. **Flowers:** calyx persistent, splitting asymmetrically, not accrescent or inflated, lobes often ribbed, triangular to ovate, not gland-dotted; corolla campanulate, yellow, fading to orange or red, sometimes drying dark greenish; staminal column included; style 5-branched from or beyond orifice or staminal column; stigmas capitate. **Fruits** capsules, erect, not inflated, carpels 5, subglobose or ovoid, not indurate, not fleshy, densely hairy, dehiscent. **Seeds** 2–many per carpel, densely hairy or seemingly glabrous. $x = 40$–48.

Species 22 (1 in the flora): introduced, Florida; Mexico, West Indies, Central America, South America, Asia, Pacific Islands (especially New Guinea).

Two taxa are sometimes cultivated for ornament in southern Florida: *Talipariti elatum* (Swartz) Fryxell and *T. tiliaceum* var. *tiliaceum*, the latter sometimes naturalized. Species of *Talipariti* usually are estuarine and/or littoral.

1. **Talipariti tiliaceum** (Linnaeus) Fryxell, Contr. Univ. Michigan Herb. 23: 258. 2001 F I

Hibiscus tiliaceus Linnaeus, Sp. Pl. 2: 694. 1753; *Pariti tiliaceum* (Linnaeus) A. Saint-Hilaire

Trees 3–8 m. **Leaves:** petiole commonly subequal to blade, with pubescence like stem; stipules sessile, oblong-lanceolate, 15–40 × 8–14 mm; blade discolorous, palmately 7–9-veined, mostly 6–13 cm, coriaceous, apex short-acuminate, surfaces minutely and densely stellate-hairy or nearly glabrous abaxially. **Inflorescences:** flowers sometimes aggregated distally with reduced leaves; involucel cupuliform. **Pedicels** minutely stellate-hairy; involucellar bractlets connate. **Flowers:** calyx deeply divided, 1.5–2 cm, lobes not gland-dotted, densely stellate-hairy, with nectary on each midrib; staminal column pallid, apically 5-dentate, glabrous; style emergent from staminal column; stigmas 5. **Capsules** 5-locular, 1.5–2 × 1.5–2 cm, subequal to calyx, antrorsely hairy, hairs yellowish or brownish. **Seeds** 4 mm, minutely papillate.

Varieties 2 (2 in the flora): introduced, Florida; Mexico, West Indies, Central America, South America, Asia.

1. Corollas yellow with prominent red center; stipule scars nearly straight or slightly curved; epidermis of involucel and stipules visible through pubescence . . . . . . 1a. *Talipariti tiliaceum* var. *tiliaceum*
1. Corollas yellow without red center; stipule scars markedly curved; epidermis of involucel and stipules obscured by pubescence . . . .
. . . . 1b. *Talipariti tiliaceum* var. *pernambucense*

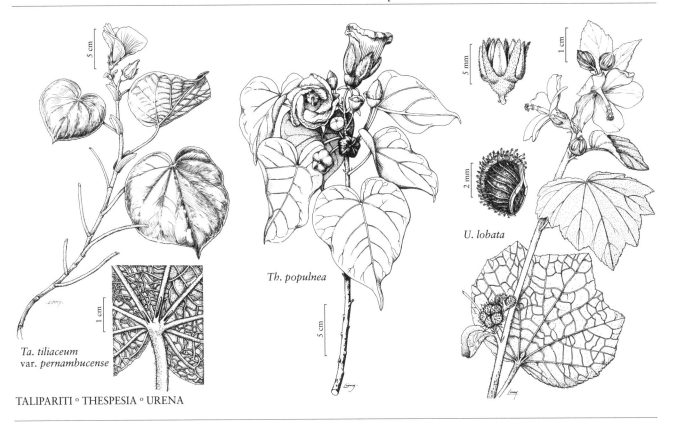

Th. populnea

U. lobata

Ta. tiliaceum
var. *pernambucense*

TALIPARITI ∘ THESPESIA ∘ URENA

1a. Talipariti tiliaceum (Linnaeus) var. **tiliaceum** [I]

Leaves: stipules leaving straight or slightly curved annular scars, epidermis visible through pubescence. **Inflorescences:** involucel to ¼ length of calyx, 8–10-dentate, teeth triangular, 2–3 mm, epidermis visible through pubescence. **Pedicels** (above articulation) 0.5–2 cm. **Flowers:** calyx lobes 12–15 × 5–7 mm; corolla yellow with prominent red center, 4–6.5 cm; staminal column 2–2.5 cm; filaments 1–2 mm.

Flowering year-round. Wet ground; 0–50 m; introduced; Fla.; Asia.

Variety *tiliaceum* is cultivated and escaped in southern Florida. It has been vouchered from Broward, Collier, Hendry, Lee, Manatee, Martin, Miami-Dade, Monroe, Palm Beach, and St. Lucie counties, but some specimens may have been from cultivated plants.

1b. Talipariti tiliaceum (Linnaeus) Fryxell var. **pernambucense** (Arruda) Fryxell, Contr. Univ. Michigan Herb. 23: 262. 2001 [F][I]

Hibiscus pernambucensis Arruda, Diss. Pl. Brazil, 44. 1810 (as pernambucencis); *H. tiliaceus* Linnaeus subsp. *pernambucensis* (Arruda) A. Castellanos; *H. tiliaceus* var. *pernambucensis* (Arruda) I. M. Johnston; *Paritium pernambucense* (Arruda) G. Don

Leaves: stipules leaving prominent, markedly curved annular scars, epidermis obscured by pubescence. **Inflorescences:** involucel to ½ length of calyx, 8–12-dentate, teeth lanceolate, 10 (–20) mm, epidermis obscured by pubescence. **Pedicels** (above articulation) 0.5–3 cm. **Flowers:** calyx lobes 12–15 × 6–8 mm; corolla yellow without red center, 4–6(–8) cm; staminal column 2.5–3 cm; filaments 1–3 mm.

Flowering year-round. Littoral vegetation and stream banks; 0 m; introduced; Fla.; Mexico; West Indies; Central America; South America.

Variety *pernambucense* is a New World native, but apparently is only introduced in the flora area, to Lee, Miami-Dade, and Monroe counties in southern Florida.

50. THESPESIA   Solander ex Corrêa, Ann. Mus. Natl. Hist. Nat. 9: 290, plate 25, fig. 1. 1807, name conserved • [Greek *thespesios*, divine, wondrous, or excellent, alluding to planting in sacred groves and use for carving religious sculpture] ⊡

Paul A. Fryxell†

Steven R. Hill

Trees [shrubs]. Stems erect, glabrous or hairy when young, usually glandular-punctate, not viscid. **Leaves:** stipules persistent or deciduous, lanceolate or falcate; blade ovate, unlobed [3-lobulate], base deeply cordate [shallowly cordate to ± truncate], margins entire, surfaces glabrate [hairy], with abaxial foliar nectaries. **Inflorescences** axillary solitary flowers, [sometimes aggregated apically]; involucel present, bractlets caducous, 3, distinct. **Flowers:** calyx not accrescent, not inflated, lobes truncate [to 5-lobed], not ribbed; corolla yellow [white or rose], with [without] maroon spot at base, usually fading pinkish orange; staminal column usually included; ovary 3–5-carpellate, style 3–5-branched; stigmas clavate. **Fruits** capsules, erect, somewhat inflated, oblate, coriaceous [ligneous], lepidote [glabrous or hairy], indehiscent [dehiscent]. **Seeds** 3–5 per locule, short-hairy [glabrous]. $x = 13$.

Species 17 (1 in the flora): introduced, Florida; Asia, Africa, Pacific Islands (Papua New Guinea), Australia; introduced also in Mexico, West Indies, n South America.

SELECTED REFERENCE   Fosberg, F. R. and M.-H. Sachet. 1972. *Thespesia populnea* (L.) Solander ex Corrêa and *Thespesia populneoides* (Roxburgh) Kosteletzky (Malvaceae). Smithsonian Contr. Bot. 7: 1–13.

1. **Thespesia populnea** (Linnaeus) Solander ex Correa, Ann. Mus. Natl. Hist. Nat. 9: 290. 1807 • Cork-tree, portia tree, seaside mahoe, Spanish cork, majagua Ⓕ Ⓘ

Hibiscus populneus Linnaeus, Sp. Pl. 2: 694. 1753

Trees 2–12 m. **Stems** lepidote to glabrate when young. **Leaves:** stipules 3–7 mm; petiole mostly ⅔–1 times length of blade; blade 6–13 cm, apex acute or acuminate, venation palmate, with nectariferous zone near base of midrib. **Inflorescences:** flowers large. **Pedicels** erect, stout, shorter than subtending petiole; involucellar bractlets irregularly inserted, ligulate. **Flowers:** calyx 8–10 mm, subglabrous, minutely lepidote; petals 4–6 cm, punctate; staminal column pallid, ca. ½ length of petals, apically 5-dentate, glabrous; style exceeding stamens; stigmas decurrent. **Capsules** (3–)5-locular, 3–3.5 cm diam. **Seeds** 8–9 mm. $2n = 24, 26$.

Flowering year-round.   Littoral vegetation; 0 m; introduced; Fla.; Asia; Africa; Pacific Islands (New Guinea); Australia; introduced also in Mexico (Veracruz), West Indies, n South America.

Thespesia populnea is sometimes grown as a shade tree, and has been found in coastal Florida from Brevard and Sarasota counties south to Monroe County, most commonly on the Florida Keys. The species is thought to be native to coastal areas of the Indian and Pacific oceans and widely introduced and naturalized in the New World. The capsules float and have distributed the seeds widely. It has been used widely for food, lumber, fiber, and medicine.

51. URENA Linnaeus, Sp. Pl. 2: 692. 1753; Gen. Pl. ed. 5, 309. 1754 • [Malayalam *ooren*, to loosen or soak, alluding to retting process to extract fibers from stem tissues of *U. lobata*] ☐

Steven R. Hill

Herbs, perennial, **subshrubs, or shrubs,** stellate-hairy or glabrate. **Stems** erect. **Leaves:** stipules early-deciduous, subulate; blade variable, sometimes ovate, [oblong or lanceolate], often 3–5-angled, -lobed, or -parted, base truncate to cordate, margins crenate [serrate], surfaces minutely stellate-hairy, with 1+ prominent, foliar nectaries (glands) on abaxial side at base of central vein. **Inflorescences** axillary solitary flowers or fascicles, [terminal racemes]; involucel present, bractlets persistent, 5, basally connate. **Pedicels** 1.5–7 mm. **Flowers:** calyx not accrescent, not inflated, lobes keeled, 1-veined, lanceolate, elliptic to ovate, usually acute, usually obscured by involucel; corolla shallowly campanulate to rotate, rose-pink, drying lavender; staminal column included; ovary 5-carpellate; ovules 1 per cell; style 10-branched, 2 per carpel; stigmas capitellate. **Fruits** schizocarps, erect, not inflated, oblate, unwinged, slightly indurate; mericarps 5, 1-celled, without dorsal spur, apex rounded, dorsal surface prominently glochidiate [smooth], otherwise hirtellous or glabrous, flat lateral surfaces striate, indehiscent. **Seeds** 1 per mericarp, obovoid or reniform, glabrous. $x = 7$.

Species 6 (1 in the flora): introduced; nearly worldwide in tropical and warm regions.

1. **Urena lobata** Linnaeus, Sp. Pl. 2: 692. 1753

• Common urena, caesarweed, bur-mallow, Congo jute

F I W

Urena trilobata Vellozo

Plants 1 m. **Stems** minutely stellate-hairy. **Leaves** reduced distally; stipules 2–4 mm; petioles equaling blades proximally, reduced distally to ¼ blade length; blade paler abaxially (slightly discolorous), broadly to narrowly ovate, angulate to shallowly lobate with ± acute sinuses, usually 3–9 cm, apex acute. **Inflorescences** subsessile or pedicels to 7 mm; involucel 5–6 mm, subequal to calyx, lobes alternate with calyx lobes. **Flowers:** calyx ½-divided, 5–9 mm, hairy, lobes ciliate; petals 15–20 mm, abaxially hairy; staminal column 5-dentate apically, glabrous; filaments nearly absent; anthers subsessile, usually pinkish; style reflexed, slender; stigmas 2 times number of locules. **Schizocarps** 5-lobed, 8 mm diam., lobes convex, stellate-hairy, glochidiate-spiny spines numerous. **Seeds** 3–5 mm. $2n = 28, 56$.

Flowering nearly year-round. Disturbed habitats; 0–200 m; introduced; Fla., La., S.C.; se Asia (India); Africa; Pacific Islands (Philippines); introduced also in Mexico, Central America, South America.

Urena lobata is a pantropical weed, probably native to the Old World Tropics. The species has been reported from East Baton Rouge and Saint Charles parishes in Louisiana and appears to be naturalized. In South Carolina it has been introduced to the southeastern part of the state in landscape plantings and is spreading to vacant lots and roadsides. It has also been collected relatively recently in Scott County, Tennessee, as a waif, and will likely be found elsewhere on occasion. *Urena lobata* is well established in peninsular Florida and common in some locations, and has been grown for its jutelike fiber in some regions.

The similar *Urena sinuata* Linnaeus [*U. lobata* var. *sinuata* (Linnaeus) Hochreutiner] is known in Mexico, the West Indies, and elsewhere in tropical America and may occasionally appear as a waif within the flora area. It was once collected in ballast at Mobile, Alabama. This species differs from *U. lobata* by its deeply, palmately five-parted leaves with broadly rounded sinuses, its smaller (4–5 mm) calyx, and its shorter (8–18 mm) petals. Transitions are found between *U. sinuata* and *U. lobata*, and they may not be specifically distinct. The fruits in both taxa readily adhere to clothing, and the plants have become widely distributed as weeds.

52. WISSADULA   Medikus, Malvenfam., 24. 1787   •   [Presumably Sinhalese *wissa*, poison, and *duvili*, dust or powder; common name *wissaduli* used for plants of *Centipeda minima* (Linnaeus) A. Braun & Ascherson and misapplied here]

Paul A. Fryxell†

Steven R. Hill

Subshrubs [perennial herbs]. **Stems** usually erect, hairy [glabrate], not viscid. **Leaves** distalmost sometimes subsessile; stipules usually persistent, filiform, subulate, or minute; blade broadly ovate to ovate-triangular [narrowly triangular], unlobed, base cordate, margins entire [crenate-dentate], surfaces usually stellate-hairy [sometimes glabrate]. **Inflorescences** terminal panicles or racemes; involucel absent. **Flowers:** calyx not accrescent, not inflated, shorter than mature fruits, lobes not ribbed, triangular; corolla usually yellowish, sometimes white, rotate; staminal column exserted; style 3–6-branched; stigmas capitate. **Fruits** schizocarps, erect, not inflated, obovoid, not indurate; mericarps 3–6, 2-celled, apex bulbous-apiculate, proximal cell indehiscent, distal cell dehiscent. **Seeds** (1–)3 per mericarp, lower cell 1-seeded, upper cell usually 2-seeded, hairy, proximal seed relatively more densely hairy. $x = 7$.

Species 25 (3 in the flora): sc United States, Mexico, West Indies, South America, s Asia, Africa.

SELECTED REFERENCE   Fries, R. E. 1908. Entwurf einer Monographie der Gattungen *Wissadula* und *Pseudabutilon*. Kongl. Svenska Vetensk. Acad. Handl., n. s. 43(4): 1–114.

1. Leaf blades 2.5–3.5 cm, apex acute to subobtuse; stipules minute; petals yellow, fading to orange . . . . .2. *Wissadula parvifolia*
1. Leaf blades 4–11 cm, apex acute or acuminate; stipules 4–12 mm; petals usually yellowish or white, sometimes with dark red basal spot.
    2. Leaf blades broadly ovate, base deeply cordate (except uppermost), margins curved; petals yellowish; stipules 7–12 mm . . . . 1. *Wissadula hernandioides*
    2. Leaf blades ovate-triangular, base shallowly cordate, margins straight; petals yellowish or white, sometimes with dark red basal spot; stipules 4–5 mm. . . . . .3. *Wissadula periplocifolia*

1. Wissadula hernandioides (L'Héritier) Garcke, Z. Naturwiss. 63: 122. 1890

Sida hernandioides L'Héritier, Stirp. Nov., 121, plate 58. 1789; *S. polyantha* Schlechtendal ex Link; *Wissadula mucronulata* A. Gray ex Torrey

Plants usually unbranched below inflorescence, to 2 m. **Stems** slender, minutely puberulent (and with stipitate stellate hairs, especially at distal ends of petioles). **Leaves** distalmost subsessile; stipules filiform, 7–12 mm; petiole 5–11 cm (reduced distally), subequal to blade in lower leaves, obscurely stellate-puberulent; blade discolorous, broadly ovate, 5–11 cm (reduced distally), base deeply cordate (except distalmost), margins curved, apex acuminate.

Inflorescences open panicles, essentially ebracteate. **Pedicels** 1–4 cm. **Flowers:** calyx ½-divided, 3 mm; petals yellowish, 5 mm. **Schizocarps** minutely puberulent; mericarps 3–5, 6–7 mm. **Seeds** 2.5 mm, sparsely hairy. $2n = 14$.

Flowering fall–early winter. Deciduous forests, disturbed vegetation; 10–100 m; Fla., La., Tex.; Mexico; West Indies; Central America; South America; Africa.

Wissadula hernandioides is very uncommon within the flora area, having been reported [as *W. amplissima* (Linnaeus) R. E. Fries, following Fries (1908)] from only Cameron County, Texas; Collier County, Florida; and Louisiana (unvouchered).

Wissadula hernandioides has been widely, but incorrectly, referred to *W. amplissima*. A. Krapovickas (1996) clarified the correct application of the latter name to a different Neotropical species.

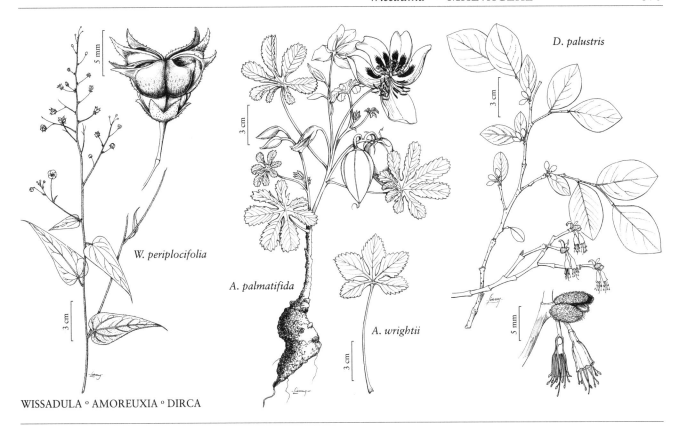

W. *periplocifolia*

A. *palmatifida*

A. *wrightii*

D. *palustris*

WISSADULA ∘ AMOREUXIA ∘ DIRCA

2. **Wissadula parvifolia** Fryxell, Lundellia 10: 3, fig. 2. 2007 C E

Plants branched, 1 m. **Stems** slender, minutely and obscurely stellate-hairy. **Leaves:** stipules minute; petiole 1–1.5(–2) cm, to ⅓ times blade length, minutely and obscurely stellate-hairy; blade discolorous, ovate, 2.5–3.5 cm, base ± cordate, margins ± straight, apex acute to subobtuse. **Inflorescences** open racemes or panicles, ebracteate. **Pedicels** 2–3 cm. **Flowers:** calyx ca. ½-divided, 3–5 mm; petals yellow, fading to orange, 5 mm. **Schizocarps** minutely and obscurely hairy; mericarps 5, 7–9 mm. **Seeds** 2.5 mm, minutely hairy.

Flowering summer. Roadsides, disturbed sites; of conservation concern; 100 m; Tex.

Wissadula parvifolia is found only in Hidalgo County.

3. **Wissadula periplocifolia** (Linnaeus) C. Presl ex Thwaites, Enum. Pl. Zeyl., 27. 1858 F

Sida periplocifolia Linnaeus, Sp. Pl. 2: 684. 1753; *Abutilon periplocifolium* (Linnaeus) Sweet

Plants widely branched, 1–2 m. **Stems** slender, stellate-hairy, hairs often stipitate (especially on younger growth). **Leaves** distalmost subsessile; stipules subulate, 4–5 mm; petiole 1–5 cm, reduced distally, ¼–½ length of blade; blade markedly discolorous, ovate-triangular, 4–11 cm (reduced distally), base shallowly cordate, margins straight, apex acute. **Inflorescences** open panicles, ± ebracteate. **Pedicels** 1.5–5 cm. **Flowers:** calyx ½-divided, 2.5–3 mm; petals yellowish or white, sometimes with dark red basal spot, 3–4 mm. **Schizocarps** minutely puberulent; mericarps 4 or 5, 5–6 mm. **Seeds** 2.5 mm, sparsely to densely hairy. $2n = 14$.

Flowering year-round. Shrublands, disturbed vegetation; 10 m; Tex.; Mexico; West Indies; Central America; South America; s Asia; Africa.

In the flora area, *Wissadula periplocifolia* has been found only in Cameron County.

COCHLOSPERMACEAE Planchon

• Yellowshow Family

Susannah B. Johnson-Fulton

Linda E. Watson

Herbs [shrubs, trees], perennial, suffrutescent. **Leaves** alternate, stipulate, petiolate, simple, venation palmate, [slightly to] deeply, palmately lobed [palmately compound]. **Inflorescences** terminal [axillary], cymose or racemose [paniculate]. **Flowers:** sepals 5, actinomorphic, imbricate, distinct; petals 5, zygomorphic [actinomorphic], distinct; stamens 50–100+, hypogynous, distinct; anthers 2-celled, straight, dehiscing via apical pores; pistil 1; ovary superior, 1–3-locular, carpels 3[–5], connate; placentation axile [parietal]; ovules 18–100+; styles simple. **Fruits** capsular, 3[–5]-valved. **Seeds** 18–100+, reniform or globose to ovoid [cochleate], glabrous or hairy; endosperm oily; embryo same shape as seed.

Genera 2, species 16 (1 genus, 3 species in the flora): sw, sc United States, Mexico, West Indies, Central America, South America, Asia, Africa, Australia.

Cochlospermaceae was established as a family, encompassing *Cochlospermum* and *Amoreuxia*, by J. É. Planchon (1847). It was included in the Parietales by H. G. A. Engler and K. Prantl (1887–1915) and by R. K. F. Pilger (1925). A. Cronquist (1981) placed *Cochlospermum* and *Amoreuxia* in Bixaceae in the Violales, suggesting a close relationship to Flacourtiaceae and Cistaceae. A. L. Takhtajan (1987, 1997) at first placed Cochlospermaceae in the Bixales, then later transferred it to the Cistales. The more recent placement within Malvales is supported by both morphological and molecular evidence (G. Dahlgren 1989b; W. S. Judd and S. R. Manchester 1997; P. F. Stevens, www.mobot.org/MOBOT/research/APweb/; S. D. Smith and D. A. Baum, tolweb.org/Malvales/21050/2003.03.20). There has been some discussion about the inclusion of *Cochlospermum* and *Amoreuxia* in Bixaceae. Takhtajan (1980) and Cronquist placed the two genera, along with *Bixa*, in Bixaceae; others, such as V. H. Heywood (1978) and H. H. Poppendieck (1980, 1981, 2003, 2003b) placed *Cochlospermum* and *Amoreuxia* in Cochlospermaceae. This segregation of *Cochlospermum* and *Amoreuxia* from *Bixa* has been done chiefly on the basis of their palmately lobed or compound versus simple, entire leaves, 3–5-valved versus 2-valved fruit, and oily versus starchy endosperm (Heywood). Molecular and morphological studies support the distinction of Cochlospermaceae (M. F. Fay et al. 1998; S. B. Johnson-Fulton 2014; R. C. Keating 1970, 1972; H. H. Poppendieck 1980; V. Savolainen et al. 2000b).

A thorough monograph of Cochlospermaceae was conducted by H. H. Poppendieck (1980), and the morphology and anatomy of Cochlospermaceae have been well studied by R. C. Keating (1968, 1970, 1972). Large tropical bees have been observed pollinating many Central and South American *Cochlospermum* species and flying foxes have been observed pollinating *C. religiosum* in India (P. Erancheri et al. 2013; D. W. Roubik et al. 1982; A. A. Snow and Roubik 1987).

Many species of Cochlospermaceae are ethnobotanically important. The starchy, tuberous roots as well as other parts of *Amoreuxia* species have been used for food by indigenous tribes in Mexico and southwestern United States (W. C. Hodgson 2001; D. Yetman and T. R. Van Devender 2002). The bark fiber of *Cochlospermum* has been used traditionally throughout its range to make cordage. The seed hairs of many species of *Cochlospermum* have been used as filling for pillows in South America. Some of the documented medicinal uses for certain species of *Cochlospermum* include treatments for jaundice, stomach problems, leprosy, and asthma (D. K. Abbiw 1990; J. F. Morton 1981; H. H. Poppendieck 1980; P. K. Warrier et al. 1993+, vol. 2). Certain species of *Cochlospermum* are grown as ornamentals in tropical and subtropical gardens throughout the world (N. P. Smith et al. 2004). *Cochlospermum religiosum* and *C. vitifolium* are cultivated in the United States; none has escaped.

SELECTED REFERENCES Johnson-Fulton, S. B. 2014. Systematics, Biogeography, and Ethnobotany of the Pantropical Family Cochlospermaceae (Malvales). Ph.D. dissertation. Miami University. Keating, R. C. 1968. Comparative morphology of Cochlospermaceae. I. Synopsis of the family and wood anatomy. Phytomorphology 18: 379–392. Keating, R. C. 1970. Comparative morphology of the Cochlospermaceae. II. Anatomy of the young vegetative shoot. Amer. J. Bot. 57: 889–898. Keating, R. C. 1972. The comparative morphology of the Cochlospermaceae. III. The flower and pollen. Ann. Missouri Bot. Gard. 59: 282–296. Poppendieck, H. H. 1980. A monograph of the Cochlospermaceae. Bot. Jahrb. Syst. 101: 191–265. Poppendieck, H. H. 1981. Cochlospermaceae. In: Organization for Flora Neotropica. 1968+. Flora Neotropica. 109+ nos. New York. No. 27.

1. AMOREUXIA de Candolle in A. P. de Candolle and A. L. P. P. de Candolle, Prodr. 2: 638. 1825 • Yellowshow [For Pierre-Joseph Amoreux, 1741–1824, French physician and naturalist]

Roots (xylopodia) thick, tuberous or woody. **Stems** to 60 cm, glabrous or hairy. **Leaves** long-petiolate; stipules inconspicuous, brown; blade suborbiculate to reniform, 5 or 7(9)-lobed [shallowly lobed]. **Inflorescences** usually cymose, rarely racemose, loosely 2–7-flowered, bracteate; peduncle 2.5–6 cm; bracts brown to roseate, linear. **Flowers** 3–8 cm diam.; sepals linear to oblong-ovate, apex acute to acuminate; petals yellow or pinkish to orange, unequal in width, proximal petal with no marks, lateral petals with red to purple mark at bases, distal petals with 2 conspicuous red to purple marks at bases; stamens in 2 dimorphic sets, filaments of proximal set longer with cream to dark red anthers, filaments of distal set shorter with yellow or yellow-cream anthers; anthers 2-pored; ovary usually pubescent; style usually surpassing stamens. **Capsules** pendulous, oblong-obovoid or ovoid to ellipsoid. **Seeds** glabrous or pilose, hairs sometimes spinelike. $x = 6$.

Species 4 (3 in the flora): sw United States, Mexico, West Indies (Curaçao), Central America, South America (Colombia, Peru).

The one species of *Amoreuxia* not in the flora area is *A. malvifolia* A. Gray, which occurs in the Mexican states of Chihuahua and Durango.

Most of the species of *Amoreuxia* have been important food plants for indigenous tribes of northern Mexico and southwestern United States (W. C. Hodgson 2001; H. H. Poppendieck 1980).

Amoreuxia is usually characterized as herbaceous, despite having slightly woody aerial parts and a woody subterranean organ. The woody subterranean organs, known as xylopodia, are made up of rootlike stems. The roots originate from below the xylopodium. They are similar in size and shape to thick carrots; they are brownish outside and creamy white inside.

The species of *Amoreuxia* are very similar. To separate them, leaf, fruit, and seed characters can generally be used reliably; flower characters appear to be relatively uniform across species (R. McVaugh 2001). According to H. H. Poppendieck (1980), it is probable that *Amoreuxia* is a tropical taxon that moved northward into Central America and southern North America in relatively recent times with a secondary center of diversification in northern Mexico. Poppendieck hypothesized that the scattered and apparently relictual distribution of *A. palmatifida* is evidence that it may be most closely related to the primitive species of the genus from which all the currently known species may have evolved. This hypothesis is not strongly supported by the interesting disjunct distributions of *A. wrightii* in Peru and in Curaçao and the molecular clock data which indicate an earlier divergence for *A. wrightii* (S. B. Johnson-Fulton 2014).

The flowers of *Amoreuxia* are very bright and easy to spot in the field; because they open early and usually close by late morning or early afternoon, it is best to search for this genus in the morning.

SELECTED REFERENCE   Sprague, T. A. 1922. A revision of *Amoreuxia*. Bull. Misc. Inform. Kew 1922: 96–105.

1. Leaf blades 5(7)-lobed, lobes subrhombic, apex rounded to acute, tooth of central lobe apex most distal; seeds glabrous . . . . 1. *Amoreuxia wrightii*
1. Leaf blades 7(9)-lobed, lobes obovate, spatulate, or linear, apex usually truncate to slightly obtuse, sometimes emarginate, 3–5 teeth of central lobe apex usually equally distal; seeds sparsely pilose or hispid, hairs sometimes spinelike.
  2. Capsules broadly ovoid, 2–4 cm; seeds reniform; ovaries densely short-haired lanate to tomentose or puberulent; stems and petioles subglabrous to sparsely lanulose-puberulent. . . . . .2. *Amoreuxia palmatifida*
  2. Capsules ellipsoid, 5–8 cm; seeds globose; ovaries densely silky-pilose; stems and petioles pilose (hairs wavy, slightly tangled) . . . . .3. *Amoreuxia gonzalezii*

1. **Amoreuxia wrightii** A. Gray, Smithsonian Contr. Knowl. 5(6): 26. 1853 • Wright's yellowshow [F]

Stems 15–50 cm, glabrous or sparsely lanulose-puberulent (mostly on young growth), glandular. **Leaves:** petiole 3–8 cm, subglabrous to sparsely lanulose-puberulent (mostly on young growth), glandular; blade 5(7)-lobed, 3–5 × 4–7 cm, lobes subrhombic, distal half of margins serrate, apex rounded to acute, tooth of central lobe apex most distal. **Peduncles** 2.5–5 cm. **Flowers** 3–7.5 cm diam.; sepals narrowly oblong-ovate, 17–20 × 5–8 mm, apex acute; petals pinkish to orange, distal petals with 2 conspicuous purplish marks at bases, lateral petals each with 1 purplish mark at base; distal set of filaments 5–10 mm with yellow anthers, proximal set of filaments 11–17 mm with usually dark red anthers; ovary usually short-haired-lanate to tomentose or puberulent (glandular). **Capsules** oblong-obovoid, 3–6 cm, sparsely pubescent (glandular). **Seeds** globose to ovoid, glabrous. $2n = 12$.

Flowering May–Jul(–Oct). Dry, rocky or gravelly hills, plains, deserts, grasslands, shrublands, along roads or railroads, limestone soil, silty flats; 0–400 m; Tex.; Mexico; West Indies (Curaçao); South America (Peru).

Amoreuxia wrightii is morphologically homogeneous throughout its disjunct range (H. H. Poppendieck 1981).

2. **Amoreuxia palmatifida** de Candolle in A. P. de Candolle and A. L. P. P. de Candolle, Prodr. 2: 638. 1825 • Mexican yellowshow, saya, saiya, Temaquí [F]

Stems 20–60 cm, subglabrous to sparsely lanulose-puberulent (mostly on young growth), glandular. **Leaves:** petiole 3–10 cm, sparsely lanulose-puberulent (mostly on young growth), glandular; blade 7(9)-lobed, 2–7.5 × 2.5–10(–14) cm, lobes obovate, spatulate, or linear, distal ½–¾ of margins serrate (crenate), apex usually truncate to slightly obtuse, sometimes emarginate, 3–5 teeth of central lobe apex usually equally distal. **Peduncles** 2.5–6 cm. **Flowers** 4–6.5 cm diam.; sepals linear to narrowly ovate, 15–20 × 3–5 mm, apex acute to acuminate; petals yellow to orange, distal petals each with 2 conspicuous red marks at base, lateral petals each with 1 red mark at base; distal set of filaments 6–12 mm with yellow (sometimes dark red-maroon) anthers, proximal set of filaments 13–19 mm with usually dark red-maroon anthers; ovary densely short-haired-lanate to tomentose or puberulent (glandular). **Capsules** broadly ovoid, 2–4 cm, sparsely pubescent. **Seeds** reniform, sparsely pilose or hispid, hairs sometimes spinelike.

Flowering Jun–Sep; fruiting Jul–Oct. Gravelly, sometimes loamy, soil, dry rocky slopes and mesas, deserts, grasslands, openings in oak forests or secondary deciduous forests, disturbed sites, forest clearings, roadsides; 0–1600 m; Ariz., N.Mex.; Mexico; Central America; South America (Colombia).

The center of diversity for *Amoreuxia palmatifida* is in Sonora and Sinaloa in northwestern Mexico. Seeds range from subglabrous to densely pilose and the leaves vary from longer (to 5.5 cm) with very long and narrow linear lobes to shorter (to 3 cm) with shorter spatulate lobes (H. H. Poppendieck 1981).

All parts of *Amoreuxia palmatifida*, especially the root, have been used as important food sources by indigenous tribes in northern Mexico and southwestern United States (W. C. Hodgson 2001). The roots, which have been described as tasting like potatoes or parsnips, have been eaten raw, boiled, toasted, roasted, baked, or preserved (Hodgson; H. H. Poppendieck 1981). In Baja California, the roots were used in soups, and baked or dried and ground into a flour to make tortillas. One Seri tribe inhabiting Tiburon Island strung thin slices of the root as necklaces (Hodgson). The roots are used medicinally in Sinaloa to lower fevers (Poppendieck). The flowers, young tender fruits, and seeds are eaten raw, and the fruits are also boiled. The dried mature seeds have been used as a coffee substitute, and the stems and leaves as a potherb (Hodgson; D. Yetman and T. R. Van Devender 2002).

3. **Amoreuxia gonzalezii** Sprague & L. Riley, Bull. Misc. Inform. Kew 1922: 102. 1922 • Santa Rita Mountain yellowshow, saya, saiya [C]

Stems 20–60 cm, pilose (hairs wavy, slightly tangled, mostly on young growth), sparsely glandular. **Leaves:** petiole 3–10 cm, sparsely pilose (hairs wavy, slightly tangled, mostly on young growth), glandular; blade 7(9)-lobed, 2.5–6 × 3.5–9 cm, lobes obovate to linear, distal half of margins serrate, apex usually truncate to slightly obtuse, sometimes emarginate, 3–5 teeth of central lobe apex usually equally distal. **Peduncles** 2.5–5 cm. **Flowers** 6–8 cm diam.; sepals narrowly ovate, 15–20 × 5–8 mm, apex acute; petals pale yellow to orange, lateral petals each with 1 red mark at base, distal petals each with 2 conspicuous red marks at base, proximal set of filaments 13–19 mm with usually dark red, sometimes cream, anthers, distal set of filaments 8–10 mm with yellow-cream anthers; ovary densely silky-pilose (hairs long). **Capsules** ellipsoid, 5–8 cm, sparsely pilose. **Seeds** globose, sparsely pilose, hairs sometimes spinelike.

Flowering Jul–Sep. Limestone outcrops/hills, granitic talus slopes, deserts, cliffs, upper Sonoran desert scrub/grassland transition zone; of conservation concern; 400–1400 m; Ariz.; Mexico (Sinaloa, Sonora).

Amoreuxia gonzalezii appears to be intermediate in morphology between *A. palmatifida* and *A. wrightii*. It has leaves with seven, sometimes nine, spatulate lobes and seed coats hairy, similar to those of *A. palmatifida*, and seeds globose, similar to *A. wrightii*. Molecular phylogenetic data indicate that *A. gonzalezii* and *A. wrightii* are more closely related to each other than either is to *A. palmatifida* (S. B. Johnson-Fulton 2014).

Amoreuxia gonzalezii is very rare. In the United States, fewer than five populations have been found, primarily in two locations. Over its entire range, it is known to occur in fewer than ten localities. Major threats are alien species and herbivory by cattle, which find it very palatable. As with *A. palmatifida*, all parts of *A. gonzalezii*, especially the root, have been used as a food source by indigenous tribes in southwestern United States and northern Mexico (W. C. Hodgson 2001; H. H. Poppendieck 1980).

Amoreuxia gonzalezii is in the Center for Plant Conservation's National Collection of Endangered Plants.

THYMELAEACEAE Jussieu

• Daphne Family

Lorin I. Nevling Jr.

Kerry Barringer

Herbs or shrubs, deciduous or evergreen. **Stems** flexible; wood usually poorly lignified; bark difficult to tear because of tough, meshed fibers. **Leaves** alternate, [opposite, whorled], estipulate, petiolate or sessile; blade pinnately veined, margins entire. **Inflorescences** terminal, extra-axillary, or axillary, usually racemose or cymose, capitate, fasciculate, [umbelliform], rarely solitary flowers. **Pedicels** present or absent. **Flowers** bisexual [unisexual], radially symmetric or slightly asymmetric, [3]4[5]-merous; hypanthium perigynous, usually tubular, campanulate, urceolate, or funnelform, rarely reduced; sepals usually imbricate, rarely valvate, white, green, yellow, or red; petals alternisepalous, usually absent, scalelike or connate into ring; disc usually present (minute or absent in *Thymelaea*), surrounding gynoecium; stamens usually diplostemonous; anthers introrse, 4-sporangiate; ovary 1-locular, 1(–5)-carpellate; ovules 1+, pendent, raphe ventral, bitegmic, arillate or carunculate, anatropous or hemianatropous, crassinucellate; style 1, rarely absent, terminal or lateral; stigma capitate, clavate, or punctiform. **Fruits** capsular or drupaceous, usually fleshy, rarely dry, indehiscent [dehiscent]. **Seeds** 1[–8], usually with endosperm, testate; cotyledons 2, embryo achlorophyllous.

Genera ca. 50, species ca. 515 (4 genera, 7 species in the flora): North America, ne Mexico, s, c Europe, e, se Asia, n Africa, Australia.

Classification of Thymelaeaceae within the dicotyledons is uncertain. It was placed in Myrtales, near Elaeagnaceae, because of its internal phloem and prominent hypanthium; that treatment is not supported by embryological and genetic data. Euphorbiaceae and Flacourtiaceae also have been suggested as possible relatives based on similarities in pollen and biochemistry. Other classifications place Thymelaeaceae in the Malvanae, either in its own order, Thymelaeales, or in Malvales.

All species contain tricyclic daphnane and tiglane diterpenes, esters of daphnetoxin, and coumarins, which are poisonous to mammals. The bark and fruit can be extremely irritating to mucous membranes even after drying and cause contact dermatitis in sensitive individuals.

A dozen species of *Daphne* are grown for their sweet-scented flowers and early bloom. Species of *Pimelea* Banks ex Gaertner and *Dais cotinifolia* Linnaeus also are cultivated as ornamentals in warm climates.

SELECTED REFERENCES Domke, F. W. 1934. Untersuchungen über die systematische und geographische Gliederung der Thymelaeaceen. Biblioth. Bot. 27: 1–151. Fuchs, A. 1938. Beiträge zur Embryologie der Thymelaeaceae. Oesterr. Bot. Z. 87: 1–41. Gilg, E. 1894. Thymelaeaceae. In: H. G. A. Engler and K. Prantl, eds. 1887–1915. Die natürlichen Pflanzenfamilien.... 254 fascs. Leipzig. Fasc. 106[III,6a], pp. 217–245. Nevling, L. I. Jr. 1962. Thymelaeaceae in the southeastern United States. J. Arnold Arbor. 43: 428–434.

1. Herbs; leaf blades linear to linear-lanceolate, 0.6–1.5 cm; flowers 2–3 mm; fruits capsular, indehiscent. . . . . 2. *Thymelaea*, p. 383
1. Shrubs; leaf blades ovate to obovate, oblong, elliptic, or lanceolate, 2–25 cm; flowers 1–18 mm; fruits drupaceous.
  2. Stems appearing jointed; leaves scattered along stem, not clustered distally; stamens exserted. . . . . 1. *Dirca*, p. 381
  2. Stems not jointed; leaves clustered distally; stamens included.
    3. Styles short or absent; fruits fleshy, yellow, red, or black, hypanthium not persistent . . . . 3. *Daphne*, p. 384
    3. Styles linear; fruits dry to leathery, brown, enclosed at base by persistent hypanthium . . . . .4. *Edgeworthia*, p. 385

1. DIRCA Linnaeus, Sp. Pl. 1: 358. 1753; Gen. Pl. ed. 5, 167. 1754 • Leatherwood [Greek mythological Dirce, wife of Lycus who was transformed by Dionysus into a fountain]

Aaron Floden

Lorin I. Nevling Jr.

Shrubs, deciduous, to 3 m. **Stems** erect, branched, appearing jointed, woody, sericeous or glabrous, buds hairy. **Leaves** scattered along stem, not clustered distally, petiolate; petiole forming hood over axillary bud; blade ovate, obovate, or elliptic-oblong, surfaces glabrous or sericeous, glabrescent. **Inflorescences** axillary on previous year's twigs, racemose or fasciculate, sessile, subsessile, or pedunculate; bracts 4, elliptic to obovate-oblong. **Flowers** appearing before or with leaves; hypanthium tubular to narrowly campanulate; calyx unlobed or (3)4-lobed; petals (absent in *D. palustris*), included, minute, forming ring at base of staminal filaments; stamens 8, exserted, inserted within calyx; style exserted, elongate; stigma punctiform. **Fruits** drupaceous, yellow, yellow-green, or green, fleshy, oil-rich, hypanthium not persistent.

Species 4 (3 in the flora): North America, ne Mexico.

Dirca mexicana G. L. Nesom & Mayfield is known from Tamaulipas, Mexico.

SELECTED REFERENCES Peterson, B. J., W. R. Graves, and J. Sharma. 2009. Color of pubescence on bud scales conflicts with keys for identifying species of *Dirca* (Thymelaeaceae). Rhodora 111: 126–130. Schrader, J. A. and W. R. Graves. 2004. Systematics of *Dirca* (Thymelaeaceae) based on ITS sequences and ISSR polymorphisms. Sida 21: 511–524. Schrader, J. A. and W. R. Graves. 2005. Seed germination of *Dirca* (leatherwood); pretreatments and interspecific comparisons. HortScience 40: 1838–1842. Vogelman, H. A. 1953. Comparison of *Dirca palustris* and *Dirca occidentalis* (Thymelaeaceae). Asa Gray Bull. 2: 77–82.

1. Inflorescences pedunculate, flowers pedicellate; calyx unlobed or 4- or 5-lobed, margins shallowly crenate, erose, or undulate; bracts brown- to tan-pubescent; leaf blades usually glabrous, rarely with persistent indument . . . . 1. *Dirca palustris*
1. Inflorescence epedunculate, flowers sessile or subsessile; calyx (3)4-lobed, margins entire or erose; bracts white-tan tomentose; leaf blades sericeous.

[2. Shifted to left margin.—Ed.]

2. Pedicels glabrous; apex of ovary and fruit glabrous; leaf margins with 6–9 cilia per mm
    . . . . 2. *Dirca occidentalis*
2. Pedicels sericeous; apex of ovary and fruit with trichomes; leaf margins with 11–18 cilia
    per mm . . . . .3. *Dirca decipiens*

1. Dirca palustris Linnaeus, Sp. Pl. 1: 358. 1753

• Eastern leatherwood, wicopy, moosewood, bois de plomb, dirca des marais [E] [F]

Shrubs to 3 m; branches turning gray, glabrous. **Leaves:** petiole 2–5 mm, glabrescent; blade obovate to ovate, 5–10 × 3.5–7 cm, base cuneate, apex acute or obtuse, surfaces usually glabrous, rarely with persistent indument. **Inflorescences** racemose, becoming pendent, (2)3(–6)-flowered, pedunculate, flowers pedicellate; bracts elliptic, apex obtuse to acute, brown- to tan-pubescent, occasionally hoary; peduncles 5–13 mm. **Pedicels** 2–10 mm, glabrous. **Flowers:** calyx yellow-green, unlobed or 4- or 5-lobed, lobes to 1 mm, margins shallowly crenate, erose, or undulate; ovary glabrous. **Drupes** green or yellow, drying red, ovoid, 6–12 mm, glabrous. $2n = 18, 36$.

Flowering spring (Mar–May); fruiting summer (Jul–Aug). Rich, moist to wet, lowland woods, calcareous rich slopes, along braided streams, swamps; 0–1500 m; N.B., N.S., Ont., Que.; Ala., Conn., Del., Fla., Ga., Ill., Ind., Iowa, Ky., La., Maine, Md., Mass., Mich., Minn., Mo., N.H., N.J., N.Y., N.C., Ohio, Okla., Pa., R.I., S.C., Tenn., Vt., Va., Wis.

P. H. Raven (pers. comm.) has reported that populations of *Dirca palustris* south of the last glacial advance are diploid and those north of it are tetraploid. It is suspected that pollen diameter (volume) may be a useful surrogate to predict ploidy level in *D. palustris*.

The bark of *Dirca palustris* is strong and pliable; it was once used in ropes and baskets. The plants may cause contact dermatitis. The fruits are mildly poisonous to mammals and have been used as purgatives and fish poisons. The plants, sometimes browsed by deer, are slow-growing in cultivation and prefer moist to wet soil. The leaves are sometimes parasitized by the miner *Leucanthiza dircella* Braun. Insect visitors to the flowers include species from five genera of generalist bees; the plants may be facultatively autogamous. Fruit dispersal may be by birds or small mammals.

The flowers appear in early spring before or as the leaves expand, after the flowers of red maple and before the canopy closes.

SELECTED REFERENCES Fernald, M. L. 1943g. The fruit of *Dirca palustris*. Rhodora 45: 117–119. Holm, T. 1921. *Dirca palustris* L. A morphological study. Amer. J. Sci., ser. 5, 2: 177–182. McVaugh, R. 1941. The fruit of the eastern leatherwood. Castanea 6: 83–86. Williams, C. E. 2004. Mating system and pollination biology of the spring-flowering shrub, *Dirca palustris*. Pl. Sp. Biol. 19: 101–106.

2. Dirca occidentalis A. Gray, Proc. Amer. Acad. Arts 8: 631. 1873 • Western leatherwood [E]

Shrubs to 2 m; branches turning dark brown to red-brown, sericeous. **Leaves:** petiole 3–6 mm, sericeous; blade obovate to broadly ovate, 2–7 × 1–3.5 cm, base rounded, margins ciliate with 6–9 cilia per mm, apex rounded, surfaces sericeous. **Inflorescences** fasciculate, nodding, (1–)3(–6)-flowered, flowers sessile or subsessile; bracts obovate-oblong, finely white-tan tomentose. **Pedicels** 0–1 mm, glabrous. **Flowers:** calyx clear yellow, distinctly (3)4-lobed, lobes 1–3 mm, margins entire; ovary glabrous. **Drupes** yellow-green, ovoid, 8–10 mm, glabrous. $2n = 36$.

Flowering winter (Nov–Mar); fruiting late spring–summer (May–Jun). Moist slopes in woods; 50–300 m; Calif.

Dirca occidentalis has been found in only six counties in the San Francisco Bay area. The plants prefer north- and northeast-facing slopes and appear to rely on fog for moisture. Leaves are parasitized by an unknown miner.

SELECTED REFERENCE McMinn, H. and B. Forderhase. 1935. Notes on western leatherwood, *Dirca occidentalis* Gray. Madroño 3: 117–120.

3. Dirca decipiens Floden, J. Bot. Res. Inst. Texas 3: 494, fig. 1. 2009 • Ozark leatherwood [E]

Shrubs to 3 m; branches turning gray, to gray-brown, sericeous. **Leaves:** petiole 2–4 mm, sericeous; blade obovate to elliptic-oblong, 7.5–9 × 4.5–6 cm, base rounded, margins ciliate with 11–18 cilia per mm, apex round, abaxial surface sericeous, adaxial surface glabrous except on primary veins near petiole. **Inflorescences** fasciculate, deflexed, (2)3(–6)-flowered; bracts obovate-oblong, finely white-tan tomentose, flowers sessile or subsessile. **Pedicels** 0–1 mm, sericeous. **Flowers:** calyx yellow-green to pale yellow, irregularly (3)4-lobed, lobes 1–3 mm, margins erose; ovary with apical trichomes. **Drupes** yellow-green, ovoid to pyriform, 8–9 mm, with 10–17 apical trichomes.

Flowering spring (Mar–Apr); fruiting late spring (Apr–Jun). Northeast- and northwest-facing bluffs above streams, calcareous bluffs; 100–500 m; Ark., Kans., Mo.

Dirca decipiens occurs typically in more xeric sites than *D. palustris* and is frequently associated with *Quercus* and *Juniperus*. It is currently known from only four populations.

Flowers of *Dirca decipiens* appear about one week after those of nearby populations of *D. palustris*. The plants are relatively quick growing in cultivation, and seed germination rates appear to be much higher than in other *Dirca* species. The leaves are parasitized, likely by *Leucanthiza dircella* Braun (Lepidoptera: Gracillariidae).

2. THYMELAEA Miller, Gard. Dict. Abr. ed. 4, vol. 3. 1754 ☐

Lorin I. Nevling Jr.

Kerry Barringer

Herbs, annual, to 6 m. **Stems** erect, branched, not jointed, slender, glabrous or sparsely pilose. **Leaves** scattered along new growth, not clustered distally, often appressed to stem when young, sessile or subsessile; blade linear to linear-lanceolate, surfaces glabrous. **Inflorescences** axillary, solitary flowers or cymose clusters, sessile [short pedicellate]; bracts 2, [linear] lanceolate, often with tuft of white hairs. **Flowers:** hypanthium tubular or urceolate; calyx 4-lobed, lobes spreading; petals absent; disc minute or absent; stamens 8, included or distal whorl partially exserted, adnate to hypanthium; style slightly exserted; stigma capitate. **Fruits** capsular, green, becoming black, hard, enclosed by persistent hypanthium.

Species 32 (1 in the flora): introduced; Europe, w Asia, n Africa.

SELECTED REFERENCES Pohl, R. W. 1955. *Thymelaea passerina*, new weed in the United States. Proc. Iowa Acad. Sci. 62: 152–154. Tan, K. 1980. Studies in the Thymelaeaceae: 2. A revision of the genus *Thymelaea*. Notes Roy. Bot. Gard. Edinburgh 38: 189–246.

1. Thymelaea passerina (Linnaeus) Cosson & Germain, Syn. Anal. Fl. Paris ed. 2, 360. 1859 • Passarine annuelle, spurge flax ☐ ☐

Stellera passerina Linnaeus, Sp. Pl. 1: 559. 1753

Stems green or yellow-green, turning red in fall. **Leaves** upright; petiole to 1 mm; blade 0.6–1.5 × 0.1–0.2 cm, stiff, herbaceous to coriaceous, apex acute. **Inflorescences** cymose, 1–7-flowered; bracts 1.5 cm, with tuft of white trichomes at base. **Flowers:** hypanthium green to yellow, tubular, becoming urceolate, 2–3 mm, appressed-hairy; calyx lobes minute; stamens: distal whorl inserted at throat; distal anthers subexserted; ovary 0.75 mm, apex hairy; style 0.7–0.8 mm, glabrous, becoming ± eccentric in fruit; stigma exserted. **Capsules** pyriform. $2n = 18$.

Flowering summer (late Jun–Aug); fruiting summer-fall (Jul–early Sep). Dry, disturbed ground, including prairies and old fields; 10–300 m; introduced; Ont., Que.; Ala., Ill., Iowa, Kans., Mich., Miss., Nebr., Ohio, Tex., Wash., Wis.; s Europe; Mediterranean regions.

Thymelaea passerina is sometimes accidentally introduced in fodder and spreads quickly. It is sometimes mistaken for one of the delicate species of *Polygala*.

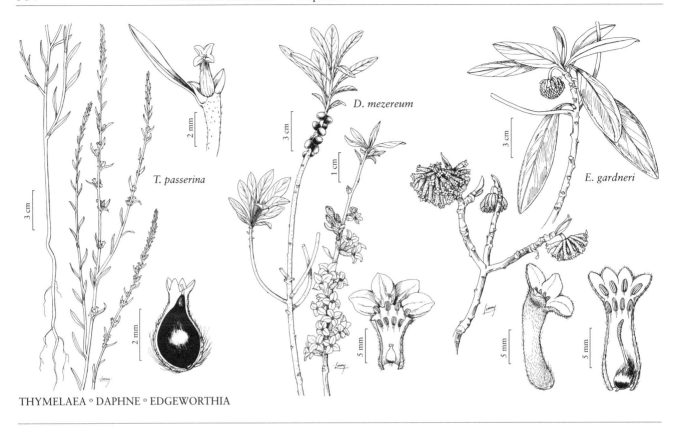

T. passerina

D. mezereum

E. gardneri

THYMELAEA ° DAPHNE ° EDGEWORTHIA

3. DAPHNE Linnaeus, Sp. Pl. 1: 356. 1753; Gen. Pl. ed. 5, 167. 1754 • [Greek, laurel]

1

Lorin I. Nevling Jr.

Kerry Barringer

Shrubs, deciduous or evergreen, to 1.5 m. **Stems** erect, procumbent, or prostrate, branched, not jointed, thick, glabrous or glabrescent. **Leaves** clustered distally, sessile or subsessile; blade obovate or oblong to lanceolate, surfaces glabrous or hairy. **Inflorescences** terminal or axillary, capitate, fasciculate, or racemose, flowers sessile [petiolate]; bracts 0. **Flowers:** hypanthium tubular to narrowly funnelform; calyx 4-lobed, lobes spreading or slightly reflexed; petals absent; stamens 8, usually included, in distal ½ of tube; style included, short or absent; stigma capitate. **Fruits** drupaceous, yellow, red, or black, fleshy, hypanthium not persistent.

Species ca. 100 (2 in the flora): introduced; s, c Europe, n Africa, Asia; temperate and subtropical, widely cultivated in temperate areas.

SELECTED REFERENCES Brickell C. D. and B. Mathew. 1976. *Daphne*: The Genus in the Wild and in Cultivation. Woking. Halda, J. J. 2001. The Genus *Daphne*. Dobré.

1. Leaves deciduous; branches glabrescent; leaf blades membranous, adaxial surface matte; hypanthia pink, red, or white; drupes red or yellow .... 1. *Daphne mezereum*
1. Leaves persistent; branches glabrous; leaf blades coriaceous, adaxial surface shiny; hypanthia yellow-green or yellow; drupes black .... 2. *Daphne laureola*

1. Daphne mezereum Linnaeus, Sp. Pl. 1: 356. 1753

• Mezereon, paradise-plant [F] [I]

Shrubs to 1 m; branches glabrescent. **Leaves** deciduous, sessile; blade oblong to lanceolate, 3–8 × 1–2.5 cm, membranous, adaxial surface matte. **Inflorescences** 2–4-flowered. **Flowers** fragrant, sessile, appearing before leaves, clustered in axils of previous year's growth; hypanthium pink, red, or white, tubular, 5–8 mm, sericeous; calyx lobes ovate or obtuse to rounded, 3–5 mm. **Drupes** red or yellow. $2n = 18$.

Flowering summer (Jun–Aug). Cool, damp woods, especially on limestone; 0–300 m; introduced; B.C., N.B., N.S., Ont., P.E.I., Que.; Alaska, Conn., Maine, Mass., Mich., Mont., N.H., N.Y., Ohio, R.I., Vt.; Europe; w, c Asia.

Daphne mezereum was used in the herbal pharmacopoeia as a purgative. It contains daphnin and is highly poisonous.

2. Daphne laureola Linnaeus, Sp. Pl. 1: 357. 1753

• Spurge-laurel, February daphne [I]

Shrubs to 1.5 m, branches glabrous. **Leaves** persistent, subsessile; blade obovate, oblong, or lanceolate, 3–8 × 1–1.5 cm, coriaceous, adaxial surface shiny. **Inflorescences** 2–10-flowered. **Flowers** sessile or short pedicellate, appearing after leaves have matured, clustered in axils on current year's growth; hypanthium yellow-green or yellow, tubular to narrowly funnelform, 3–8 mm, glabrous; calyx lobes ovate, 2.5–4 mm. **Drupes** black.

Flowering late winter–early spring (Feb–Apr). Oak and conifer woods; 0–300 m; introduced; B.C.; Oreg., Wash.; s Europe; sw Asia; n Africa; Mediterranean region.

Daphne laureola is locally invasive in the Pacific Northwest.

4. EDGEWORTHIA Meisner, Denkschr. Königl.-Baier. Ges. Regensburg 3: 280. 1841

• [For Michael Pakenham Edgeworth, 1812–1881, Irish botanist and British civil servant in Bengal] [I]

Lorin I. Nevling Jr.

Kerry Barringer

Shrubs, deciduous, to 3.5 m. **Stems** erect, ternately branched, not jointed, thick, sericeous, glabrescent. **Leaves** usually clustered distally, petiolate; blade lanceolate to oblong, surfaces glabrescent. **Inflorescences** axillary, capitate, often drooping, pedunculate; bracts absent. **Flowers:** hypanthium tubular, often curved; calyx 4-lobed, lobes spreading; petals absent; stamens 8, included, adnate to hypanthium; style included, linear; stigma clavate. **Fruits** drupaceous, brown, dry to leathery, enclosed at base by persistent hypanthium.

Species 4 (1 in the flora): introduced, Georgia; Asia; cultivated in warm, temperate regions.

SELECTED REFERENCE Duncan, W. H. and M. Mellinger. 1972. *Edgeworthia* (Thymelaeaceae) new to the Western Hemisphere. Rhodora 74: 436–439.

1. Edgeworthia gardneri (Wallich) Meisner, Denkschr. Königl.-Baier. Bot. Ges. Regensburg 3: 280. 1841

• Mitsumata, paperbush, Japanese paper plant [F] [I]

Daphne gardneri Wallich, Asiat. Res. 13: 388, plate 9. 1820; *Edgeworthia chrysantha* Lindley; *E. papyrifera* Siebold & Zuccarini

Branches yellow-brown when young. **Leaves:** petiole 5–7 mm; blade 8–25 × 2–5 cm, membranous, base cuneate, apex usually acute, rarely obtuse, surfaces sericeous. **Flowers** fragrant; hypanthium curved, 10–18 mm, outer surface silvery-sericeous or villous; calyx yellow or white, lobes distinct, 2–5 mm; anthers oblong; pistil villous; style 2–4 mm. **Drupes** ellipsoid to ovoid, 6–8 mm, dry. $2n = 36$.

Flowering spring (May–early Jun). Mesophytic forests; 400–500 m; introduced; Ga.; Asia (China).

Edgeworthia chrysantha is cultivated and escapes to moist woods (W. H. Duncan, pers. comm.). The pith is a source of pulp for fine Japanese papers.

CISTACEAE Jussieu

• Rockrose Family

John L. Strother

Herbs, annual or perennial, **subshrubs, or shrubs**, usually hairy. **Leaves** alternate, opposite, or whorled, usually estipulate, sometimes stipulate (*Tuberaria*), stipules caducous, petiolate or sessile; blade 1- or 3- [5-]veined from base, not lobed, sometimes scalelike, margins entire [crenate, serrate], sometimes revolute and/or undulate. **Inflorescences** usually corymbose, cymose, paniculate, racemose, thyrsiform, or umbellate, seldom solitary flowers. **Pedicels** present or absent; bracts present or absent. **Flowers** chasmogamous or cleistogamous; sepals persistent or tardily falling, 3–5; petals usually caducous [marcescent], usually 3–5, sometimes 0 in cleistogamous flowers, imbricate, distinct, crumpled in bud, green, dark red, pink, purple, red, white, or yellow; stamens (3–)5–150+; filaments distinct or basally connate; ovaries superior, 2-, 3-, 5-, or 6–12-carpellate; placentation parietal; styles 0 or 1; stigmas 1 or 3; ovules orthotropous [anatropous], bitegmic, crassinucellate. **Fruits** capsular, dehiscence loculicidal [septifragal]. **Seeds** (1–)3–800+ per capsule, often with thin outer integument.

Genera 8, species 170–180 (5 genera, 40 species in the flora): North America, Mexico, West Indies, Central America, South America, sw Europe, n Africa; mostly of temperate areas.

Affinities of Cistaceae are evidently with Malvales. Members of Cistaceae are widely cultivated, especially cultivars of hybrids and species of *Cistus*, *Crocanthemum*, *Halimium* (Dunal) Spach, and *Helianthemum* Miller.

Hairs on Cistaceae plants may be simple or stellate (comprising tight clusters or tufts of simple, unbranched hairs) and glandular or eglandular.

Two species of Cistaceae have been collected in the flora area as waifs. *Helianthemum nummularium* Miller is known from Colorado, Missouri, and Oregon; it differs from species of *Crocanthemum* by the combination of glabrous abaxial surfaces of sepals and stellate-tomentose ovaries. *Helianthemum salicifolium* (Linnaeus) Miller is known from New York; it differs from species of *Crocanthemum* by its opposite leaves and erect, curved pedicels.

SELECTED REFERENCES Arrington, J. M. 2004. Systematics of the Cistaceae. Ph.D. dissertation. Duke University. Arrington, J. M. and K. Kubitzki. 2003. Cistaceae. In: K. Kubitzki et al., eds. 1990+. The Families and Genera of Vascular Plants. 10+ vols. Berlin etc. Vol. 5, pp. 62–70. Guzmán, B. and P. Vargas. 2009. Historical biogeography and character evolution of Cistaceae (Malvales) based on analysis of plastid *rbc*L and *trn*L-*trn*F sequences. Organisms Diversity Evol. 9: 83–99.

1. CISTUS Linnaeus, Sp. Pl. 1: 523. 1753; Gen. Pl. ed. 5, 234. 1754 • Rockrose [Ancient Greek name for plants of the genus] ⊡

John L. Strother

Shrubs, 5–25 dm, hairy, hairs simple, sometimes clustered (stellate), glandular and eglandular. **Leaves** opposite, petiolate or sessile; blade 1- or 3-veined from base, margins sometimes revolute and/or undulate. **Inflorescences** solitary flowers or corymbiform, scorpioid [paniculiform or racemiform] cymes. **Pedicels** present or absent; bracts present or absent. **Flowers** chasmogamous; sepals tardily falling, 3 or 5; petals [0] (4)5, pink, purple, red, or white, sometimes drying yellowish; stamens [50–]100–150+; filaments distinct or basally connate; carpels 5 or 6–12; styles 0 or 1, to 0.5 or 1–3+ mm; stigmas 1, capitate or truncate. **Capsules** 5–12-valved. **Seeds** 10–800+ per capsule. *x* = 9.

Species 20 (5 in the flora): introduced, California; Europe, n Africa; Mediterranean areas.

Species of *Cistus* are grown as ornamentals in warm climates. Cultivars (some hybrids) are common in gardens and parks and may be encountered in the flora area. Bona fide records of non-cultivated cistuses from the flora area are from urban areas in coastal counties of California. Leaves of *C. incanus* and *C. ladanifer* are the source of a scented resin, labdanum or ladanum, which may be the myrrh of ancient references. *Cistus* is not the source of the opiate laudanum.

SELECTED REFERENCES Dansereau, P. M. 1939. Monographie du genre *Cistus* L. Boissiera 4: 1–90. Guzmán, B. and P. Vargas. 2005. Systematics, character evolution, and biogeography of *Cistus* L. (Cistaceae) based on ITS, *trn*L-*trn*F, and *mat*K sequences. Molec. Phylogen. Evol. 37: 644–660.

1. Petals pink, purple, or reddish; styles 1–3+ mm.................................1. *Cistus incanus*
1. Petals white, sometimes with red or yellow near bases; styles 0–0.5 mm.
  2. Leaves petiolate; blade 1-veined from base ....2. *Cistus salviifolius*
  2. Leaves sessile; blade 3-veined from base.
    3. Inflorescences solitary flowers; sepals 3; petals 30–50 mm; capsules 6–12-locular
       ....3. *Cistus ladanifer*
    3. Inflorescences usually cymes; sepals 5; petals 10–25 mm; capsules 5-locular.
      4. Leaf blades oblong-ovate to oblong-linear, (5–)10–20+ mm wide, margins not notably revolute, usually ciliate ....4. *Cistus psilosepalus*
      4. Leaf blades lanceolate-linear to linear, 1–7(–13) mm wide, margins usually revolute, not notably ciliate ....5. *Cistus monspeliensis*

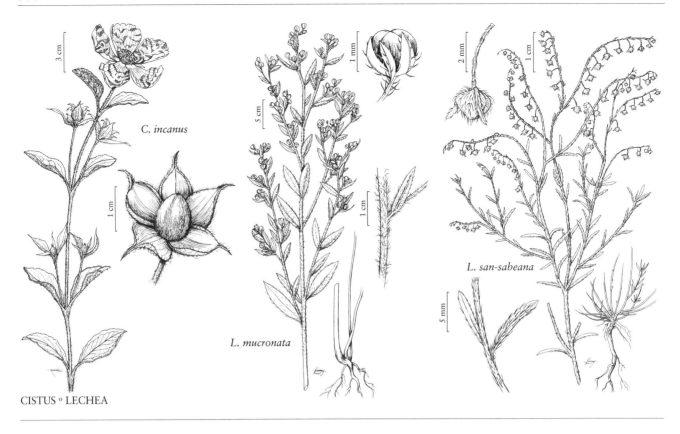

C. incanus

L. mucronata

L. san-sabeana

CISTUS ° LECHEA

1. Cistus incanus Linnaeus, Sp. Pl. 1: 524. 1753 (as incana) F I

Cistus creticus Linnaeus; *C. incanus* subsp. *corsicus* (Loiseleur-Deslongchamps) Heywood; *C. incanus* subsp. *creticus* (Linnaeus) Heywood

Shrubs to 10+ dm. **Leaves** petiolate; petiole 3–15 mm; blade 1-veined from base, narrowly elliptic, obovate, or ovate, (10–)25–50(–75) × (5–)8–30 mm, margins ± undulate, not revolute. **Inflorescences** solitary flowers or cymes. **Flowers:** sepals 5, apices acuminate; petals pink, purple, or reddish, sometimes yellowish near bases, 20–30 mm; styles 1–3+ mm; stigmas borne at level of anthers. **Capsules** 5-locular. $2n$ = 18 (Europe).

Flowering (Jan–)Mar–Aug. Disturbed sites, chaparral borders, clearings, oak woodlands, roadsides, abandoned plantings; 100–1500 m; introduced; Calif.; Europe.

If the types of the names *Cistus incanus* and *C. creticus* Linnaeus are not conspecific, plants in the flora area called *C. incanus* should be known as *C. creticus*. *Cistus villosus* Linnaeus has been misapplied to plants belonging to *C. incanus* (and/or *C. creticus*).

Cistus incanus is known from the north and south Central Coast Ranges, northern South Coast Ranges, San Gabriel Mountains, Western Transverse Ranges (Liebre Mountains), and the Peninsular Ranges.

2. Cistus salviifolius Linnaeus, Sp. Pl. 1: 524. 1753 (as salvifolia) I

Shrubs to 10 dm. **Leaves** petiolate; petiole 2–10 mm; blade 1-veined from base, narrowly elliptic, oblong-ovate, or ovate, 10–40 × 6–20 mm. **Inflorescences** usually solitary flowers, rarely cymes; pedicel 10–100 mm, longer in fruit. **Flowers:** sepals 5; petals white, sometimes yellow near bases, 15–25 mm; styles 0–0.5 mm. **Capsules** 5-locular. $2n$ = 18 (Europe).

Flowering Apr–May. Disturbed sites, chaparral, coastal sage scrub, abandoned plantings; 60–1000 m; introduced; Calif.; Europe.

Cistus salviifolius is known from the southern Central Coast Ranges, southern South Coast Ranges, and the San Gabriel Mountains.

3. Cistus ladanifer Linnaeus, Sp. Pl. 1: 523. 1753 (as ladanifera) [I]

Shrubs to 25 dm, stems usually viscid. **Leaves** sessile; blade 3-veined from base, elliptic, lanceolate-elliptic, lanceolate-linear, or oblong, 40–80(–120) × 6–25 mm. **Inflorescences** solitary flowers. **Flowers:** sepals 3, orbiculate; petals white, sometimes with red or yellow spot near bases, 30–50 mm; styles 0–0.5 mm. **Capsules** 6–12-locular. **2n** = 18 (Europe).

Flowering Feb–Aug. Disturbed sites, chaparral, roadsides, trails, abandoned plantings; 20–1600 m; introduced; Calif.; Europe.

Cistus ladanifer is known from the southern Central Coast Ranges, southern South Coast Ranges, Western Transverse Ranges, and the San Gabriel Mountains.

4. Cistus psilosepalus Sweet, Cistineae, plate 33. 1826 [I]

Shrubs to 10 dm. **Leaves** sessile; blade 3-veined from base, oblong-ovate to oblong-linear, (15–)30–60+ × (5–)10–20+ mm, margins not notably revolute, usually ciliate. **Inflorescences** usually corymbiform cymes, sometimes solitary flowers; pedicel 5–15+ mm. **Flowers:** sepals 5; petals white, 15–25 mm; styles 0–0.5 mm. **Capsules** 5-locular. **2n** = 18 (Europe).

Flowering May–Oct. Disturbed sites, abandoned plantings; 10–50 m; introduced; Calif.; s Europe; n Africa.

Plants of *Cistus psilosepalus* have been mistakenly identified as *C. monspeliensis*.

Cistus psilosepalus is known from the north coast area in California.

5. Cistus monspeliensis Linnaeus, Sp. Pl. 1: 524. 1753 [I]

Shrubs to 10 dm. **Leaves** sessile; blade 3-veined from base, mostly lanceolate-linear to linear, 15–45 (–70) × 1–7(–13) mm, margins usually revolute, not notably ciliate. **Inflorescences** usually scorpioid cymes, 3–10-flowered, sometimes solitary flowers; pedicel 5–15 mm. **Flowers:** sepals 5; petals white, 10–20 mm; styles 0–0.5 mm. **Capsules** 5-locular. **2n** = 18 (Europe).

Flowering May–Oct. Disturbed sites, abandoned plantings; 10–100 m; introduced; Calif.; s Europe; n Africa.

Cistus monspeliensis is known from the north and central coast areas in California.

2. LECHEA Linnaeus, Sp. Pl. 1: 90. 1753; Gen. Pl. ed. 5, 40. 1754 • Pinweed [For Johan Leche, 1704–1764, Swedish botanist]

David E. Lemke

Lechidium Spach

Herbs, perennial or biennial, **or subshrubs,** (0.7–)1.5–9 dm, sericeous, villous, or rarely glabrous. **Stems** dimorphic: basal stems produced late in growing season, prostrate to procumbent or ascending, overwintering; flowering stems produced in spring, erect to spreading-ascending. **Leaves** alternate, subopposite, opposite, or whorled, petiolate or sessile; blade 1-veined from base, linear to elliptic or ovate, oblanceolate, or orbiculate, margins revolute. **Inflorescences** paniculate or racemose, loose or congested, axils 1–3-flowered, bracteolate. **Pedicels** present. **Flowers:** chasmogamous; sepals 5, outer 2 linear to linear-lanceolate, inner 3 ovate to obovate, sometimes indurate in fruit; petals 3, maroon or green, usually shorter than sepals; stamens (3–)5–15(–25); carpels 3; style 0–0.5 mm; stigmas 3, dark red, fimbriate-plumose, sometimes persistent in fruit. **Capsules** 3-valved, exserted beyond calyx or not. **Seeds** 1–6 per capsule, ovoid, often compressed laterally, with or without easily separating membranous coat. *x* = 9 or unknown.

Species 18 (16 in the flora): North America, Mexico, West Indies (Cuba), Central America (Belize).

The flowers of *Lechea* species are at anthesis only early in the day and most collections comprise fruiting individuals. In the key here, specimens lacking fruit and seeds can be difficult to determine with certainty. Regional treatments (for example, K. Barringer 2004; D. D. Spaulding 2013) can be very useful for identification.

SELECTED REFERENCES Barringer, K. 2004. New Jersey pinweeds (*Lechea*, Cistaceae). J. Torrey Bot. Soc. 131: 261–276. Britton, N. L. 1894. A revision of the genus *Lechea*. Bull. Torrey Bot. Club 21: 244–253. Hodgdon, A. R. 1938. A taxonomic study of *Lechea*. Rhodora 40: 29–69, 87–131. Spaulding, D. D. 2013. Key to the pinweeds (*Lechea*, Cistacee) of Alabama and adjacent states. Phytoneuron 2013-99: 1–15. Wilbur, R. L. 1966. Notes on Rafinesque's species of *Lechea*. Rhodora 68: 192–208. Wilbur, R. L. and H. S. Daoud. 1961. The genus *Lechea* (Cistaceae) in the southeastern United States. Rhodora 63: 103–118.

1. Flowering stems spreading-villous.
   2. Basal stems not produced; leaves of flowering stems: blade 5–8 mm; capsules ellipsoid, distinctly longer than calyx . . . . .3. *Lechea divaricata*
   2. Basal stems produced; leaves of flowering stems: blade 10–30 mm; capsules subglobose, equaling or slightly longer than calyx. .9. *Lechea mucronata*
1. Flowering stems with appressed hairs or glabrous.
   3. Pedicels 3–5 mm; seeds (4–)6 . . . . 12. *Lechea san-sabeana*
   3. Pedicels 0.5–3 mm; seeds 1–3(–6).
      4. Outer sepals longer than or equaling inner sepals.
         5. Pedicels 1–2 mm; calyx indurate, shiny, yellow-brown basally in fruit . . . . . .
            . . . . .11. *Lechea racemulosa* (in part)
         5. Pedicels 0.5–2 mm; calyx not indurate, shiny, yellow-brown basally in fruit.
            6. Capsules ellipsoid or ellipsoid-ovoid to ovoid or obovoid, ± equaling or longer than calyx.
               7. Leaves of flowering stems whorled or opposite, blade 4–7 mm wide . . . . 8. *Lechea minor*
               7. Leaves of flowering stems alternate or subopposite, blade 0.3–1.5 mm wide . . . . .13. *Lechea sessiliflora* (in part)
            6. Capsules ovoid to broadly ovoid, ± equaling calyx.
               8. Seeds 1(2); capsules 1.2–1.3 mm diam. . . . . 7. *Lechea mensalis*
               8. Seeds 2 or 3(–5); capsules 1.3–1.5 mm diam. . . . . 15. *Lechea tenuifolia*
      4. Outer sepals shorter than inner sepals.
         9. Stems and leaves glabrous . . . . 5. *Lechea lakelae*
         9. Stems and leaves hairy, at least in part.
            10. Leaf blades densely hairy on both surfaces; pedicels 2 or 3 per axil . . . . 1. *Lechea cernua*
            10. Leaf blades glabrous or sparsely hairy adaxially, hairy abaxially at least on midvein and margins; pedicels 1 per axil.
               11. Capsules ellipsoid to ovoid or obovoid, length mostly 2+ times diam.
                  12. Stigmas not persistent on capsule; calyx indurate, shiny, yellow-brown basally in fruit . . . . .11. *Lechea racemulosa* (in part)
                  12. Stigmas persistent on capsule; calyx not indurate, shiny, yellow-brown basally in fruit . . . . .13. *Lechea sessiliflora* (in part)
               11. Capsules ovoid to subglobose, length mostly to 1.5 times diam.
                  13. Subshrubs; seeds 1(2) . . . . .2. *Lechea deckertii*
                  13. Herbs; seeds 2–6.
                     14. Capsules longer than calyx. . . . . .10. *Lechea pulchella*
                     14. Capsules shorter than or ± equaling calyx.
                        15. Seeds with membranous gray coat . . . . 4. *Lechea intermedia*
                        15. Seeds without membranous gray coat.

[16. Shifted to left margin.—Ed.]

16. Capsules mostly 2+ mm diam.; c Canada, midwestern United States . . . . .14. *Lechea stricta*
16. Capsules mostly to 2 mm diam.; coastal plain.
    17. Leaves of flowering stems: blade narrowly elliptic to oblanceolate, 1.2–4 mm wide;
        capsules 1.5–1.9 mm diam.; n coastal plain south to ne North Carolina . . . . 6. *Lechea maritima*
    17. Leaves of flowering stems: blade linear to narrowly elliptic, 1–1.5 mm diam.; s coastal
        plain north to se North Carolina . . . . 16. *Lechea torreyi*

1. Lechea cernua Small, Bull. Torrey Bot. Club 51: 384. 1924 • Nodding pinweed [E]

Subshrubs. Stems: basal not produced; flowering spreading-ascending, (10–)20–30 cm, densely sericeous. **Leaves** of flowering stems alternate; blade elliptic-ovate to orbiculate, 5–12 × 4–10 mm, apex acute, often mucronate, abaxial surface pilose to tomentose, adaxial pilose. **Pedicels** 2 or 3 per axil, 1.5–2.5 mm. **Flowers:** calyx 1.8–2 mm, outer sepals shorter than inner. **Capsules** ellipsoid to obovoid, 1.8–2 × 1.2–1.5 mm, ± equaling calyx. **Seeds** 1 or 2.

Flowering spring; fruiting summer–fall. Dry, open sand-scrub and flatwoods margins; 0–50 m; Fla.

Lechea cernua is endemic to much of peninsular Florida but is absent from the Florida Keys and panhandle.

2. Lechea deckertii Small, Torreya 27: 103. 1928 • Deckert's pinweed [E]

Lechea myriophylla Small

Subshrubs. Stems: basal not produced; flowering erect, 7–30 cm, sparsely sericeous. **Leaves** of flowering stems alternate to subopposite; blade linear, 3–7 × 0.5–1 mm, apex acute or obtuse, abaxial surface sparsely pilose on midvein and margins, adaxial glabrous. **Pedicels** 1 per axil, 1–1.5 mm. **Flowers:** calyx 1–1.2 mm, outer sepals shorter than inner. **Capsules** subglobose, 1.2–1.5 × 1.1–1.4 mm, longer than calyx, with persistent stigma. **Seeds** 1(2).

Flowering spring; fruiting summer–fall. Dry, white sand-scrub and oak woodland margins; 0–80 m; Fla., Ga.

Lechea deckertii is widespread in xeric, sandy habitats throughout much of peninsular Florida and also occurs in the eastern Florida panhandle and adjacent southern Georgia.

3. Lechea divaricata Shuttleworth ex Britton, Bull. Torrey Bot. Club 21: 249. 1894 • Dry-sand or spreading pinweed [C][E]

Lechea major Linnaeus var. *divaricata* (Shuttleworth ex Britton) A. Gray

Herbs, perennial. **Stems:** basal not produced; flowering erect, 25–60 cm, spreading-villous. **Leaves** of flowering stems alternate; blade lanceolate to elliptic-lanceolate, 5–8 × 1–3 mm, apex acute, abaxial surface villous on midvein and margins, adaxial glabrous. **Pedicels** 1 per axil, 1–1.5 mm. **Flowers:** calyx 1.3–1.5 mm, usually lightly colored basally, outer sepals shorter than inner. **Capsules** ellipsoid, 1.8–2 × 1.6–1.7 mm, longer than calyx. **Seeds** 1(–3), membranous coat white.

Flowering spring; fruiting summer–fall. Dry, open sand-scrub and flatwoods; of conservation concern; 0–100 m; Fla.

Lechea divaricata is endemic to peninsular Florida but apparently absent from the Florida Keys. A report of *L. divaricata* from Alabama was based on a misidentified specimen of *L. mucronata* (D. D. Spaulding 2013).

4. Lechea intermedia Leggett ex Britton, Bull. Torrey Bot. Club 21: 252. 1894 • Large-pod pinweed [E]

Herbs, perennial. **Stems:** basal produced; flowering erect to decumbent or arcuate-ascending, (3–)5–60 cm, sericeous. **Leaves** of flowering stems alternate or subopposite; blade elliptic lanceolate or lanceolate to oblanceolate, 10–25 × 1.5–5 mm, apex acute, abaxial surface hairy on midvein and margins, adaxial glabrous. **Pedicels** 1 per axil, 1.5–3 mm. **Flowers:** calyx 1.9–2.3 mm, outer sepals shorter than inner. **Capsules** broadly ovoid, depressed-globose, or subglobose, 1.8–2 × 1.7–2.1 mm, shorter to ± equaling calyx. **Seeds** 4–6, membranous coat gray.

Varieties 4 (4 in the flora): North America.

Lechea intermedia can be easily recognized by the membranous layer surrounding the seeds, a feature shared only with *L. divaricata*.

1. Capsules depressed-globose, ± equaling calyx; inner sepals broadly ovate, apex obtuse . . . . . .
. . . . 4a. *Lechea intermedia* var. *intermedia*
1. Capsules subglobose to broadly ovoid, shorter than calyx; inner sepals elliptic-ovate, apex subacute.
    2. Flowering stems decumbent to arcuate-ascending, 3–12 cm . . . .
    . . . . 4b. *Lechea intermedia* var. *depauperata*
    2. Flowering stems erect, 25–60 cm.
        3. Leaves of basal stems mostly 4-ranked; blades elliptic-lanceolate . . . .
        . . . . 4c. *Lechea intermedia* var. *juniperina*
        3. Leaves of basal stems mostly 2-ranked; blades lanceolate to oblanceolate . . . . . .
        . . . . 4d. *Lechea intermedia* var. *laurentiana*

4a. Lechea intermedia Leggett ex Britton var. **intermedia** • Léchéa intermédiaire [E]

Flowering stems erect, (15–)25–60 cm. **Leaves** of basal stems mostly 2-ranked; blade lanceolate. **Inner sepals** broadly ovate, apex obtuse. **Capsules** depressed-globose, ± equaling calyx.

Flowering summer; fruiting late summer–fall. Dry open woodlands and sand plains, burned areas and rock outcrops; 10–1500 m; Man., N.B., N.S., Ont., P.E.I., Que.; Conn., Ill., Iowa, Maine, Mass., Minn., Mont., N.H., N.J., N.Y., N.Dak., Ohio, Pa., R.I., S.Dak., Va., Wis., Wyo.

4b. Lechea intermedia Leggett ex Britton var. **depauperata** Hodgdon, Rhodora 40: 127, plate 491, figs. 12, 13. 1938 • Impoverished pinweed [C][E]

Flowering stems decumbent to arcuate-ascending, (3–)5–12 cm. **Leaves** of basal stems mostly 2-ranked; blade lanceolate to oblanceolate. **Inner sepals** elliptic-ovate, apex subacute. **Capsules** broadly ellipsoid, shorter than calyx.

Flowering summer; fruiting late summer–fall. Sandy lakeshores, open pine woodlands; of conservation concern; 500–600 m; Alta., Sask.

Variety *depauperata* is endemic to the Lake Athabasca sand dunes.

4c. Lechea intermedia Leggett ex Britton var. **juniperina** (E. P. Bicknell) B. L. Robinson, Rhodora 10: 34. 1908 [E]

Lechea juniperina E. P. Bicknell, Bull. Torrey Bot. Club 24: 88. 1897

Flowering stems erect, 25–50 cm. **Leaves** of basal stems mostly 4-ranked; blade elliptic-lanceolate. **Inner sepals** elliptic-ovate, apex subacute. **Capsules** subglobose, shorter than calyx.

Flowering summer; fruiting late summer–fall. Sandy soil of open woodlands and fields, usually near coast; 0–50 m; N.B., N.S., P.E.I.; Maine, Mass., N.H.

4d. Lechea intermedia Leggett ex Britton var. **laurentiana** Hodgdon, Rhodora 40: 127, plate 491, figs. 9–11. 1938 [E]

Flowering stems erect, 25–60 cm. **Leaves** of basal stems mostly 2-ranked; blade lanceolate to oblanceolate. **Inner sepals** elliptic-ovate, apex subacute. **Capsules** subglobose, shorter than calyx.

Flowering summer; fruiting late summer–fall. Dry sandy soil of open woodlands, streambanks, lake margins; 10–200 m; Ont., Que.; N.Y.

Variety *laurentiana* is found primarily along the Saint Lawrence and Ottawa rivers and their tributaries.

5. Lechea lakelae Wilbur, Rhodora 76: 481. 1974 • Lakela's pinweed [C][E]

Herbs, perennial. **Stems:** basal not produced; flowering erect, (20–)30–40 cm, glabrous. **Leaves** of flowering stems alternate; blade linear to narrowly elliptic, 7–15 × 0.5–1.5 mm, apex rounded to acute, surfaces glabrous. **Pedicels** 1 per axil, 0.7–2 mm. **Flowers:** calyx 1.5–1.9 mm, slightly indurate, discolored basally, outer sepals shorter than inner. **Capsules** ovoid to globose, 1.2–1.6 × 1–1.4 mm, ± equaling calyx. **Seeds** 2 or 3.

Flowering spring–summer; fruiting late summer–fall. Dry, sandy, open sites, coastal woodlands; of conservation concern; 0–10 m; Fla.

Lechea lakelae is known only from coastal scrub woodlands in Collier County; it has not been collected since 1987 and may have been extirpated.

6. **Lechea maritima** Leggett ex Britton, Catal. Fl. New Jersey, 13. 1881 • Beach or hoary or seaside pinweed E

Lechea minor Linnaeus var. *maritima* (Leggett ex Britton) A. Gray

Herbs, perennial. **Stems:** basal produced; flowering erect or procumbent to ascending, 15–35 cm, densely sericeous. **Leaves** of flowering stems whorled; blade narrowly elliptic to oblanceolate, 7–25 × 1.2–4 mm, apex acute, abaxial surface pilose, adaxial glabrous or sparsely pilose. **Pedicels** 1 per axil, 1–1.5 mm. **Flowers:** calyx 1.6–2.1 mm, outer sepals shorter than to ± equaling inner. **Capsules** subglobose to broadly ovoid, 1.3–2.1 × 1.5–1.9 mm, shorter than calyx. **Seeds** 2–5(6).

Varieties 3 (3 in the flora): e North America.

The varieties of *Lechea maritima* show appreciable overlap in many vegetative features but can be readily distinguished by the characters used here.

1. Flowering stems mostly branching distal to middle; outer sepals nearly equaling inner. . . . . .6b. *Lechea maritima* var. *subcylindrica*
1. Flowering stems mostly branching proximal to middle; outer sepals shorter than inner.
  2. Seeds 3 or 4(5), weakly 3-sided or 2-sided and convex; stems erect to procumbent, 1–2.5 mm diam.; sepals usually strongly maroon tinged, sometimes brown. . . . . 6a. *Lechea maritima* var. *maritima*
  2. Seeds 2(3), 2-sided and flat or convex; stems procumbent to ascending, 2–4 mm diam.; sepals usually brown, sometimes maroon tinged. . . . . .6c. *Lechea maritima* var. *virginica*

6a. **Lechea maritima** Leggett ex Britton var. **maritima** E

Flowering stems erect to procumbent, 15–35 cm × 1–2.5 mm, mostly branching proximal to middle. **Sepals** usually strongly maroon tinged, sometimes brown, outer sepals shorter than inner. **Seeds** 3 or 4(5), weakly 3-sided or 2-sided and convex.

Flowering summer; fruiting late summer–fall. Coastal sands, pine barrens; 0–20 m; Conn., Del., Maine, Md., Mass., N.H., N.J., N.Y., R.I.

Disjunct populations from eastern Canada are treated here as var. *subcylindrica*.

6b. **Lechea maritima** Leggett ex Britton var. **subcylindrica** Hodgdon, Rhodora 40: 109. 1938 • Gulf of St. Lawrence pinweed E

Flowering stems procumbent to ascending, 20–35 cm × 2–4 mm, mostly branching distal to middle. **Sepals** brown, outer sepals nearly equaling inner. **Seeds** 4 or 5(6), weakly 3-sided and convex or keeled.

Flowering summer; fruiting fall. Coastal sands, open pine woodlands; 0–30 m; N.B., P.E.I.

Variety *subcylindrica* has a limited distribution on stabilized barrier dunes along the eastern coast of New Brunswick and the northern shore of Prince Edward Island, frequently growing in association with *Hudsonia tomentosa*. The variety is also known to occur on old dunes in open pine woodlands on Fox and Portage islands in Miramichi Bay, New Brunswick.

6c. **Lechea maritima** Leggett ex Britton var. **virginica** Hodgdon, Rhodora 40: 109, plate 490, fig. 9. 1938 • Virginia pinweed E

Flowering stems procumbent to ascending, 20–30 cm × 2–4 mm, mostly branching proximal to middle. **Sepals** usually brown, occasionally maroon tinged, outer sepals shorter than inner. **Seeds** 2(3), 2-sided and flat or convex.

Flowering summer; fruiting fall. Coastal dunes, disturbed sandy sites near coast; 0–10 m; Del., Md., N.C., Va.

The morphologic variation and geographic distribution of var. *virginica* were documented by B. A. Sorrie and A. S. Weakley (2007).

7. **Lechea mensalis** Hodgdon, Rhodora 40: 92, plate 489, figs. 1–4. 1938 • Mountain pinweed C

Herbs, perennial. **Stems:** basal produced; flowering erect, 15–25 cm, sericeous. **Leaves** of flowering stems alternate; blade linear, 8–15 × 0.7–1.2 mm, apex acute to rounded, abaxial surface sparsely pilose on midvein and margins, adaxial glabrous. **Pedicels** 1 per axil, 1.5–2 mm. **Flowers:** calyx 1.9–2 mm, outer sepals longer than inner. **Capsules** ovoid, 1.6–1.8 × 1.2–1.3 mm, ± equaling calyx. **Seeds** 1(2).

Flowering late summer; fruiting fall. Oak-juniper woodlands; of conservation concern; 2000–2300 m; N.Mex., Tex.; Mexico (Coahuila).

Lechea mensalis is known from open woodlands dominated by species of oak and juniper at only two locations in the United States: the Guadalupe Mountains of southeastern New Mexico and adjacent Texas and the Chisos Mountains of west Texas; it also occurs in adjacent Coahuila, Mexico.

8. **Lechea minor** Linnaeus, Sp. Pl. 1: 90. 1753
   • Thymeleaved pinweed E

Lechea thymifolia Michaux

Herbs, biennial or perennial. **Stems:** basal produced; flowering erect, 20–50 cm, sparsely sericeous. **Leaves** of flowering stems whorled or opposite; blade elliptic to lanceolate, 8–15 × 4–7 mm, apex acute, abaxial surface pilose on midvein and margins, adaxial sparsely pilose or glabrous. **Pedicels** 1 per axil, 0.5–2 mm. **Flowers:** calyx 1.6–2 mm, outer sepals longer than inner. **Capsules** ellipsoid to ellipsoid-ovoid, 1.5–2 × 1–1.5 mm, ± equaling calyx. **Seeds** 2 or 3.

Flowering summer; fruiting late summer–fall. Dry sandy or gravelly soil of pine-oak woodlands, savannas, sandhills, disturbed sites; 10–600 m; Ont.; Ala., Conn., Del., Fla., Ga., Ill., Ind., Ky., La., Md., Mass., Mich., Miss., N.H., N.J., N.Y., N.C., Ohio, Pa., R.I., S.C., Tenn., Tex., Vt., Va., Wis.

Lechea minor may have been extirpated from the Canadian portion of its range.

9. **Lechea mucronata** Rafinesque, Précis Découv. Somiol., 37. 1814 • Hairy pinweed E F

Lechea minor Linnaeus var. *villosa* (Elliott) B. Boivin; *L. villosa* Elliott; *L. villosa* var. *macrotheca* Hodgdon; *L. villosa* var. *schaffneri* Hodgdon

Herbs, biennial or perennial. **Stems:** basal produced; flowering erect, (15–)30–90 cm, densely spreading-villous. **Leaves** of flowering stems opposite or whorled; blade elliptic to ovate, 10–30 × 3–4 mm, apex acute to obtuse, mucronate, abaxial surface villous, adaxial glabrous. **Pedicels** 1 per axil, 0.8–1.5 mm. **Flowers:** calyx 1.4–2 mm, outer sepals shorter than inner. **Capsules** subglobose, 1.4–1.7 × 1.3–1.6 mm, ± equaling calyx. **Seeds** 2–5.

Flowering late spring–summer; fruiting summer–fall. Dry, sandy or rocky open sites, sandy prairies, margins of oak-hickory and pine-oak woodlands; 10–1300 m; Ont.; Ala., Ark., Conn., Del., Fla., Ga., Ill., Ind., Iowa, Kans., Ky., La., Md., Mass., Mich., Miss.,

Mo., Nebr., N.H., N.J., N.Mex., N.Y., N.C., Ohio, Okla., Pa., R.I., S.C., Tenn., Tex., Vt., Va., W.Va., Wis.

A recent collection of *Lechea mucronata* from sandhills on the southern high plains of eastern New Mexico represents a significant range extension.

10. **Lechea pulchella** Rafinesque, New Fl. 1: 91. 1836 E

Herbs, perennial. **Stems:** basal produced; flowering erect, 25–80 cm, sparsely sericeous, glabrescent. **Leaves** of flowering stems opposite or subopposite; blade linear to narrowly elliptic, 15–30 × 1.5–3 mm, apex acute, indurate, abaxial surface hairy on midvein and, sometimes, margins, adaxial glabrous. **Pedicels** 1 per axil, 1–2.3 mm. **Flowers:** calyx 1.4–2 mm, outer sepals shorter than inner. **Capsules** subglobose, 1.5–2.2 × 1.3–1.6 mm, longer than calyx. **Seeds** 2 or 3(4).

Varieties 3 (3 in the flora): e North America.

Intergradation among the varieties of *Lechea pulchella* in areas of overlap is common (B. A. Sorrie and A. S. Weakley 2007b). Some authors (K. Barringer 2004; R. L. Wilbur 1966) have not recognized infraspecific taxa.

1. Capsules crowded-racemose to tightly clustered at branch tips; pedicels shorter than calyx. . . . . .
   . . . . 10a. *Lechea pulchella* var. *pulchella*
1. Capsules loosely racemose; pedicels ± equaling or longer than calyx.
   2. Seeds (2)3(4); calyces reddish purple at maturity, 1.7–2 mm . . . .
      . . . . 10b. *Lechea pulchella* var. *moniliformis*
   2. Seeds 2(3); calyces greenish purple at maturity, 1.6–1.8 mm . . . .
      . . . . 10c. *Lechea pulchella* var. *ramosissima*

10a. **Lechea pulchella** Rafinesque var. **pulchella**
   • Leggett's pinweed E

Lechea leggettii Britton & Hollick

Pedicels 1–1.5 mm, shorter than calyx. **Calyces** greenish purple at maturity, 1.6–1.8 mm. **Capsules** crowded-racemose to tightly clustered at branch tips. **Seeds** 3(4).

Flowering summer; fruiting late summer–fall. Dry oak-hickory and pine-oak woodlands, savannas, barrens, usually on sandy soil; 0–300 m; Conn., Del., Ind., Md., Mass., N.J., N.Y., N.C., Ohio, Pa., R.I., Va., W.Va.

10b. Lechea pulchella Rafinesque var. **moniliformis** (E. P. Bicknell ex Britton) Mohlenbrock, Ill. Fl. Illinois, Fl. Pl. Hollies–Loasas, 264. 1978 • Bead pinweed [E]

Lechea moniliformis E. P. Bicknell ex Britton, Man. Fl. N. States, 632. 1901; *L. leggettii* Britton & Hollick var. *moniliformis* (E. P. Bicknell ex Britton) Hodgdon

Pedicels 1.8–2.3 mm, ± equaling or longer than calyx. **Calyces** reddish purple at maturity, 1.7–2 mm. **Capsules** loosely racemose. **Seeds** (2)3(4).

Flowering summer; fruiting late summer–fall. Dry, sandy sites, old shorelines; 0–200 m; Ont.; Ill., Ind., Md., Mass., Mich., N.J., N.Y., Ohio, R.I.

Variety *moniliformis* has a disjunct distribution along the northeastern Atlantic coast and on old shorelines near Lake Erie and Lake Michigan.

10c. Lechea pulchella Rafinesque var. **ramosissima** (Hodgdon) Sorrie & Weakley, J. Bot. Res. Inst. Texas 1: 370. 2007 [E]

Lechea leggettii Britton & Hollick var. *ramosissima* Hodgdon, Rhodora 40, 119, plate 491, fig. 3. 1938

Pedicels 1.6–2.2 mm, ± equaling or longer than calyx. **Calyces** greenish purple at maturity, 1.6–1.8 mm. **Capsules** loosely racemose. **Seeds** 2(3).

Flowering summer; fruiting late summer–fall. Sandy soil of pine barrens, pine-oak woodlands, sandhills, savannas; 0–300 m; Ala., Fla., Ga., La., Miss., N.C., S.C., Tenn., Tex., Va.

11. Lechea racemulosa Michaux, Fl. Bor.-Amer. 1: 77. 1803 • Illinois pinweed [E]

Herbs, perennial. **Stems**: basal produced; flowering erect, 25–40 cm, sparsely sericeous. **Leaves** of flowering stems opposite to subopposite; blade linear-lanceolate to narrowly oblanceolate, 10–20 × 1–1.5 mm, apex acute, usually indurate, abaxial surface sparsely pilose on midvein and margins, adaxial glabrous. **Pedicels** 1 per axil, 1–2 mm. **Flowers**: calyx 1.8–2 mm, indurate, shiny, yellow-brown basally in fruit, outer sepals shorter than or equaling inner.

Capsules ellipsoid to ovoid, 1.5–2 × 1–1.2 mm, ± equaling to slightly longer than calyx. **Seeds** (1)2(3).

Flowering summer; fruiting late summer–fall. Old fields, woodland margins, other dry, sandy, open sites; 10–400 m; Ala., Conn., Del., D.C., Ga., Ind., Iowa, Ky., La., Md., Mo., N.J., N.Y., N.C., Ohio, Pa., S.C., Tenn., Va., W.Va.

12. Lechea san-sabeana (Buckley) Hodgdon, Rhodora 40: 49. 1938 • San Saba pinweed [E] [F]

Linum san-sabeanum Buckley, Proc. Acad. Nat. Sci. Philadelphia 13: 450. 1862; *Lechidium drummondii* Spach 1837, not *Lechea drummondii* Spach 1837

Herbs, perennial. **Stems**: basal produced; flowering erect, 15–35 cm, densely to sparsely sericeous. **Leaves** of flowering stems alternate to subopposite; blade linear, 6–18 × 0.6–1.3 mm, apex acute, abaxial surface appressed-pilose mostly on midvein and margins, adaxial glabrous. **Pedicels** 1 per axil, 3–5 mm. **Flowers**: calyx 1.8–2.2 mm, outer sepals longer than inner. **Capsules** secund, depressed-globose, (1.2–)1.8–2.2 × 2.2–2.4 mm, ± equaling to longer than calyx. **Seeds** (4–)6.

Flowering spring; fruiting summer–fall. Dry, sandy or gravelly, open sites, woodland margins; 10–600 m; Tex.

Lechea san-sabeana is known from a variety of sandy habitats in the eastern half of Texas.

13. Lechea sessiliflora Rafinesque, New Fl. 1: 97. 1836 • Pineland pinweed [E]

Lechea exserta Small; *L. patula* Leggett; *L. prismatica* Small

Herbs, perennial. **Stems**: basal produced; flowering erect, 15–70 cm, sericeous. **Leaves** of flowering stems alternate or subopposite; blade linear to narrowly elliptic, 4–15 × 0.3–1.5 mm, apex acute, abaxial surface pilose on midvein and margins, adaxial glabrous. **Pedicels** 1 per axil, 1–1.8 mm. **Flowers**: calyx 0.8–1.1 mm, outer sepals usually ± equaling or longer than, occasionally shorter than, inner. **Capsules** ellipsoid to obovoid, 1.6–1.9 × 0.9–1.1 mm, longer than calyx, with persistent stigma. **Seeds** 1(2).

Flowering summer; fruiting fall. Sandhills, pine-oak woodlands, flatwoods, dry, open sites; 10–100 m; Ala., Fla., Ga., Miss., N.C., S.C.

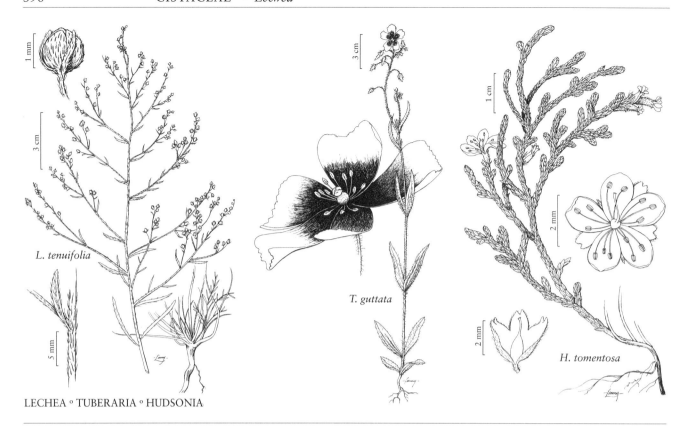

L. tenuifolia

T. guttata

H. tomentosa

LECHEA ∘ TUBERARIA ∘ HUDSONIA

14. **Lechea stricta** Leggett ex Britton, Bull. Torrey Bot. Club 21: 251. 1894 • Prairie pinweed E

Herbs, perennial. **Stems:** basal produced; flowering erect, 25–45 cm, densely sericeous. **Leaves** of flowering stems alternate or subopposite; blade narrowly oblanceolate, 13–20 × 1.5–3 mm, apex acute, often mucronate, abaxial surface pilose, adaxial glabrous. **Pedicels** 1 per axil, 1.6–2 mm. **Flowers:** calyx 1.6–1.8 mm, outer sepals shorter than inner. **Capsules** subglobose to broadly ovoid, (1.8–)2–2.5 × 2–2.5 mm, ± equaling calyx. **Seeds** 3 or 4.

Flowering summer–fall; fruiting fall. Sandy, open fields, grasslands, lakeshores, woodland margins; 100–800 m; Ont.; Ill., Iowa, Mich., Minn., Nebr., N.Dak., Wis.

Some individuals of *Lechea stricta* can be difficult to determine with certainty; the extremes of variation of *L. stricta* often grade into that of *L. intermedia* and *L. pulchella.*

15. **Lechea tenuifolia** Michaux, Fl. Bor.-Amer. 1: 77. 1803 • Narrowleaf pinweed E F

Lechea tenuifolia var. *occidentalis* Hodgdon

Herbs, biennial or perennial. **Stems:** basal produced; flowering erect, 12–40 cm, sparsely sericeous. **Leaves** of flowering stems opposite or whorled; blade linear to narrowly oblanceolate, 7–20 × 0.5–1.5 mm, apex rounded, abaxial surface sparsely pilose on midvein and margins, adaxial glabrous. **Pedicels** 1 per axil, 0.5–1.5 mm. **Flowers:** calyx 1.6–1.9 mm, outer sepals equaling or longer than inner. **Capsules** secund or not, broadly ovoid, 1.4–1.7 × 1.3–1.5 mm, shorter than or ± equaling calyx. **Seeds** 2 or 3(–5).

Flowering summer–fall; fruiting fall. Dry, sandy or gravelly soil in openings or along margins of oak woodlands and oak-pine forests; 100–300 m; Ala., Ark., Conn., Del., Ga., Ill., Ind., Iowa, Kans., Ky., La., Maine, Md., Mass., Minn., Miss., Mo., Nebr., N.H., N.J., N.Y., N.C., Ohio, Okla., Pa., R.I., S.C., Tenn., Tex., Vt., Va., W.Va., Wis.

16. Lechea torreyi Leggett ex Britton, Bull. Torrey Bot. Club 21: 251. 1894 • Piedmont pinweed

Herbs, perennial. **Stems:** basal produced; flowering erect, 30–50 cm, densely sericeous. **Leaves** of flowering stems alternate to subopposite; blade linear to narrowly elliptic, 10–20 × 1–1.5 mm, apex acute, abaxial surface pilose, adaxial glabrous. **Pedicels** 1 per axil, 0.5–1.5 mm. **Flowers:** calyx 1.8–2.1 mm, outer sepals shorter than inner. **Capsules** ovoid, 1.5–1.7 × 1.4–1.5 mm, shorter than or ± equaling calyx, with persistent stigma. **Seeds** 3–6.

Varieties 2 (2 in the flora): se United States, Central America (Belize).

1. Panicles loosely flowered; inner sepals brown in fruit, sometimes with some rust coloration; seeds 4–6 . . . . 16a. *Lechea torreyi* var. *torreyi*
1. Panicles congested; inner sepals rust colored in fruit; seeds 3 . . . . 16b. *Lechea torreyi* var. *congesta*

16a. Lechea torreyi Leggett ex Britton var. **torreyi** E

Panicles loosely flowered. **Inner sepals** brown in fruit, sometimes with some rust coloration. **Seeds** 4–6.

Flowering late spring–summer; fruiting summer–fall. Sandhills, pine barrens; 0–30 m; Fla.

Variety *torreyi* has the more restricted distribution of the two varieties, occurring only from the Florida panhandle to the vicinity of Lake Okeechobee.

16b. Lechea torreyi Leggett ex Britton var. **congesta** Hodgdon ex D. E. Lemke, Phytoneuron 2014-33: 1. 2014

Panicles congested. **Inner sepals** rust-colored in fruit. **Seeds** 3.

Flowering late spring–summer; fruiting summer–fall. Sandhills, pine barrens; 0–30 m; Ala., Fla., Ga., Miss., N.C., S.C.; Central America (Belize).

Variety *congesta* has an unusual disjunct distribution. It is widespread across the coastal plain from North Carolina to Mississippi at elevations below 30 m and also occurs in sandy pine-oak woodlands at approximately 500 m in the Cayo District of Belize.

3. TUBERARIA (Dunal) Spach, Ann. Sci. Nat., Bot., sér. 2, 6: 364. 1836, name conserved
  • [Latin *tuber*, swelling, and *-aria*, possession, alluding to swellings on roots] I

John L. Strother

Helianthemum Miller sect. *Tuberaria* Dunal in A. P. de Candolle and A. L. P. P. de Candolle, Prodr. 1: 270. 1824

Herbs annual [perennial], 0.3–2(–3)[–8] dm. **Leaves** mostly opposite, basal soon withering, sometimes in rosettes, distal cauline sometimes alternate, stipulate or estipulate, petiolate or sessile; blade usually 3[–5]-veined from base, margins sometimes revolute, surfaces hairy [glabrous], hairs sometimes clustered (stellate). **Inflorescences** racemiform [helicoid, scorpioid] cymes. **Pedicels** present; bracts [present or] absent. **Flowers** chasmogamous and cleistogamous, nodding or pendulous in bud. **Chasmogamous flowers:** sepals persistent, 5, outer smaller than [equaling] inner; petals 5, yellow, sometimes purple to brown at or near bases; stamens 10–15+; filaments distinct, outer stamens often sterile; carpels 3; styles 0; stigmas 1, ± sessile, hemispheric [obconic]. **Cleistogamous flowers** similar; petals 0; stamens 5–12. **Capsules** 3-valved. **Seeds** 6–50+ per capsule. $x = 9$.

Species 8–12 (1 in the flora): introduced, California; c, w Europe, n Africa.

Tuberaria differs from *Crocanthemum* and *Helianthemum* primarily in habit, mostly opposite leaves, and sessile or subsessile stigmas.

1. **Tuberaria guttata** (Linnaeus) Fourreau, Ann. Soc. Linn. Lyon, n.s. 16: 340. 1868 [F] [I]

Cistus guttatus Linnaeus, Sp. Pl. 1: 526. 1753; *Helianthemum guttatum* (Linnaeus) Miller

Leaf blades elliptic to obovate (proximal) or lanceolate-linear to linear, 10–50(–75) × (2–)5–10 (–18) mm, margins of distals sometimes revolute, surfaces hairy, hairs mixed: white and relatively long, white and in tufts, and red and relatively short. **Pedicels** 5–15 mm. **Flowers:** sepals 1–4 mm, villous; petals creamy yellow to yellow, usually brown to purple proximally, mostly obovate to rounded-cuneate, 4–10+ mm, apices truncate, erose to fimbrillate. **Capsules** 4–6 mm, villous. *2n* = 36 (Europe).

Flowering Apr–May. Disturbed sites, chaparral borders, flats, oak woodlands, along streams, vernal pools; 80–200 m; introduced; Calif.; Europe; n Africa.

Nomenclatural conservation of *Tuberaria* against *Xolantha* Rafinesque rendered the name *X. guttata* (Linnaeus) Rafinesque incorrect for *T. guttata*.

Tuberaria guttata is known from the northern and central Sierra Nevada Foothills and eastern Sacramento Valley.

4. HUDSONIA Linnaeus, Syst. Nat. ed. 12, 2: 323, 327. 1767; Mant. Pl. 1: 74. 1767

• [For William Hudson, 1730–1793, English botanist] [E]

John L. Strother

Shrubs, evergreen, sometimes forming clumps wider than high, 0.5–2(–4) dm. **Leaves** alternate, sessile; blade 1-veined from base, acerose to subulate or scalelike, margins sometimes ± revolute, surfaces glabrescent or hairy, hairs usually simple, not stellate. **Inflorescences** solitary flowers. **Pedicels** present or absent; bracts present or absent. **Flowers** chasmogamous; sepals persistent or tardily falling, 5; petals 5, usually yellow, sometimes white; stamens 8–30+; filaments distinct or bases weakly connate; carpels 3; styles 1; stigmas 1, minutely 3-toothed. **Capsules** 3-valved. **Seeds** 3–6 per capsule. *x* = 10.

Species 3 (3 in the flora): North America.

Species of *Hudsonia* are wiry shrubs or shrublets with crowded, acerose to subulate, or scalelike, leaves and the aspect of diminutive gymnosperms or overgrown mosses; they sometimes form relatively extensive stands.

SELECTED REFERENCES Morse, L. E. 1979. Systematics and Ecological Biogeography of the Genus *Hudsonia* (Cistaceae), the Sand Heaths. Ph.D. dissertation. Harvard University. Skog, J. E. and N. H. Nickerson. 1972. Variation and speciation in the genus *Hudsonia*. Ann. Missouri Bot. Gard. 59: 454–464.

1. Leaf blades mostly lanceolate-ovate, (scalelike), 1–2(–3+) mm; leaves usually appressed (to stems); pedicels 0–1(–5) mm; ovaries glabrous or glabrescent. . . . . 1. *Hudsonia tomentosa*
1. Leaf blades acerose to subulate, 2–7 mm; leaves weakly spreading; pedicels mostly 4–10 (–16) mm; ovaries usually hairy.
   2. Sepal apices acute to acuminate; ovaries proximally glabrous or glabrescent, distally hairy . . . . .2. *Hudsonia ericoides*
   2. Sepal apices acuminate to attenuate; ovaries hairy ± throughout. . . . . .3. *Hudsonia montana*

1. **Hudsonia tomentosa** Nuttall, Gen. N. Amer. Pl. 2: 5. 1818 [E] [F]

Hudsonia ericoides Linnaeus subsp. *tomentosa* (Nuttall) N. H. Nickerson & J. E. Skog

Plants to 40 cm. **Leaves** usually appressed (to stems); blade mostly lanceolate-ovate, scalelike, 1–2(–3+) mm, surfaces densely sericeous to tomentose. **Pedicels** 0–1(–5) mm. **Flowers:** sepal apices rounded, sometimes mucronate, retuse, or 2-fid; petals yellow; ovaries glabrous or glabrescent. $2n = 20$.

Flowering May–Jul. Mostly open, sandy sites, beaches, dunes, swales, pine barrens, pine-oak woods; 0–1400 m; Alta., Man., N.B., Nfld. and Labr. (Labr.), N.W.T., N.S., Ont., P.E.I., Que., Sask.; Conn., Del., Ill., Ind., Iowa, Maine, Md., Mass., Mich., Minn., N.H., N.J., N.Y., N.C., N.Dak., Ohio, R.I., S.C., Vt., Va., W.Va., Wis.

Some plants included here in *Hudsonia tomentosa* are intermediate to plants typical of *H. ericoides* for some traits. They may be hybrids or backcrosses.

2. **Hudsonia ericoides** Linnaeus, Syst. Nat. ed. 12, 2: 327. 1767; Mant. Pl. 1: 74. 1767 [E]

Hudsonia ericoides subsp. *andersonii* N. H. Nickerson & J. E. Skog; *H. ericoides* subsp. *intermedia* (Peck) N. H. Nickerson & J. E. Skog; *H. tomentosa* Nuttall var. *intermedia* Peck

Plants to 30 cm. **Leaves** weakly spreading; blade acerose to subulate, 2–7 mm, surfaces sericeous, glabrescent. **Pedicels** mostly 4–10(–16) mm. **Flowers:** sepal apices acute to acuminate; petals usually yellow, sometimes white; ovaries proximally glabrous or glabrescent, distally hairy. $2n = 20$.

Flowering May–Jun(–Jul). Open, sandy sites, beaches, pine and pine-oak woods, granite outcrops; 0–300 m; St. Pierre and Miquelon; Nfld. and Labr. (Nfld.), N.S., P.E.I., Que.; Conn., Del., Maine, Md., Mass., N.H., N.J., N.Y., R.I., S.C., Vt., Va.

3. **Hudsonia montana** Nuttall, Gen. N. Amer. Pl. 2: 5. 1818 [C] [E]

Hudsonia ericoides Linnaeus subsp. *montana* (Nuttall) N. H. Nickerson & J. E. Skog

Plants to 20 cm. **Leaves** ± spreading; blade acerose to subulate, 5–6 mm, surfaces sparsely sericeous, glabrescent. **Pedicels** 4–7 mm. **Flowers:** sepal apices acuminate to attenuate; petals yellow; ovaries hairy ± throughout. $2n = 20$.

Flowering Jun–Jul. Openings among shrubs, barrens, rocks at cliff edges; of conservation concern; 900–1200 m; N.C.

Hudsonia montana is known only from the vicinity of Table Mountain in western North Carolina.

5. CROCANTHEMUM Spach, Ann. Sci. Nat., Bot., sér. 2, 6: 370. 1836 • Frostweed, rushrose, rockrose [Greek *krokos*, saffron, and *anthemon*, flower, alluding to petal color]

Bruce A. Sorrie

Helianthemum Miller subg. *Lecheoides* (Dunal) Reiche

Herbs, perennial, **or subshrubs,** 0.4–8 dm. **Stems** not dimorphic, hairy, usually some hairs clustered (stellate); flowering stems divaricate to erect. **Leaves** cauline and, sometimes, basal, basal rosette- or mat-forming, alternate, petiolate or sessile; blade 1-veined from base, margins usually revolute. **Inflorescences** usually corymbs, cymes, panicles, racemes, thyrses, or umbels (cleistogamous inflorescences may be reduced to glomerules), seldom solitary flowers. **Pedicels** present; bracts present or absent. **Flowers** chasmogamous and, sometimes, cleistogamous. **Chasmogamous flowers:** sepals 5, outer 2 linear to lanceolate, inner 3 ovate to ovate-elliptic; petals 5, yellow, longer than sepals; stamens 10–50; carpels 2(3); styles 1; stigmas 1, capitate. **Cleistogamous flowers:** sepals 5, outer 2 linear to lanceolate or rudimentary; petals 0; stamens (3)4–6(–8); filaments glabrous; staminode 0; carpels 2 or 3; styles 1; stigmas 1, capitate. **Capsules** 2- or 3-valved. **Seeds** 1–46(–135) per capsule (chasmogamous), 1–20 per capsule (cleistogamous). *x* = 10.

Species 21 (15 in the flora): North America, Mexico, West Indies (Dominican Republic), Central America, South America.

Crocanthemum was created in 1836 to accommodate the New World species of *Helianthemum*, but most authors have not recognized the distinction. Studies by J. Arrington (2004) and B. Guzmán and P. Vargas (2009) showed that traditional *Helianthemum* is not monophyletic. *Helianthemum* is now a genus of some 80 species restricted to the Old World, and *Crocanthemum* is a New World genus most closely related to *Hudsonia*. Differences in leaf arrangement, pollen type, style architecture, and funicle and embryo shape all serve to distinguish the two genera. In particular, the style base in New World species is broad (broader than the apex) and confluent with the outline of the ovary, whereas in Old World species the style is tapered from apex to the narrower base. The style in New World species is short and straight, but long and curved in Old World species. In New World species, leaves are alternate and estipulate; in Old World species leaves are mostly opposite and stipulate. Regarding the apparent close relationship between *Crocanthemum* and *Hudsonia*, Guzmán and Vargas noted that these two genera have unique pollen types, and that both differ from Old World *Helianthemum*. More complete sampling of New World taxa may be necessary to resolve the issue of whether *Crocanthemum* is distinct from *Hudsonia*.

A prominent feature of *Crocanthemum* is production of dimorphic flowers in all but the Californian species. Showy, chasmogamous flowers are produced early in the growing season, followed one to three months later by apetalous, cleistogamous flowers (simultaneous in *C. georgianum* and *C. glomeratum*). Flowering dates cited are for chasmogamous flowers.

All species of *Crocanthemum* are fire-tolerant, readily resprouting from the woody caudex.

SELECTED REFERENCE Daoud, H. S. and R. L. Wilbur. 1965. A revision of the North American species of *Helianthemum* (Cistaceae). Rhodora 67: 63–82; 201–216; 255–312.

1. Capsules (chasmogamous flowers) stellate-pubescent distally.
   2. Chasmogamous capsules 3-valved; inflorescences umbellate; stems 5–15(–20) cm . . . . .2. *Crocanthemum arenicola*
   2. Chasmogamous capsules 2-valved; inflorescences thyrsiform; stems 20–35(–41) cm . . . . 11. *Crocanthemum nashii*

1. Capsules (chasmogamous flowers) glabrous.
    3. Leaf blades with lateral veins obscure abaxially.
        4. Cleistogamous flowers 2–10 per glomerule; w Texas eastward.
            5. Subshrubs; petioles 0–2 mm; leaf margins nonrevolute; inner sepals of cleistogamous flowers 2.4–5 mm . . . . .9. *Crocanthemum glomeratum* (in part)
            5. Herbs; petioles 1–4 mm; leaf margins revolute; inner sepals of cleistogamous flowers 1.5–1.8 mm . . . . 13. *Crocanthemum rosmarinifolium*
        4. Cleistogamous flowers 0; California.
            6. Pedicels stellate-pubescent and densely glandular-hairy; outer sepals lanceolate; calyx hairs 1–1.7 mm, simple, strigose (sometimes also short-stellate-hairy) . . . . .10. *Crocanthemum greenei*
            6. Pedicels with sparse or no glandular hairs; outer sepals linear; calyx hairs to 1 mm, stellate-pubescent.
                7. Petals 3–6 mm; stems 10–45 cm, spreading to erect-fastigiate; capsules 2.8–3.8 mm . . . . .14. *Crocanthemum scoparium*
                7. Petals 6–12 mm; stems 30–80 cm, erect; capsules 3.5–4.3 mm.
                    8. Stems glabrate to sparsely stellate-pubescent; leaves and inflorescences stellate-pannose; inflorescences broad, with spreading branches . . . . 1. *Crocanthemum aldersonii*
                    8. Stems, leaves, and inflorescences usually stellate-pannose, (sometimes sparsely stellate-pubescent); inflorescences narrow . . . . 15. *Crocanthemum suffrutescens*
    3. Leaf blades with lateral veins raised abaxially.
        9. Basal leaves rosette- or mat-forming.
            10. Basal leaf blades spatulate to obovate or elliptic, 5–18(–28) mm wide, sparsely stellate-pubescent adaxially; cleistogamous flowers rarely produced. . . . . .5. *Crocanthemum carolinianum*
            10. Basal leaf blades spatulate to oblanceolate, 5–11 mm wide, stellate-pannose adaxially; cleistogamous flowers 2–7 per cyme. . . . . 8. *Crocanthemum georgianum* (in part)
        9. Basal leaves absent.
            11. Cleistogamous flowers in terminal corymbiform cyme . . . . 6. *Crocanthemum corymbosum*
            11. Cleistogamous flowers in glomerules on lateral branches or terminal glomerules (*C. glomeratum*) or racemose cymes (*C. georgianum*).
                12. Cauline leaf blades stellate-pubescent, also with simple hairs; chasmogamous flowers 1–3; stems usually red tinged.
                    13. Stems ascending to erect; adaxial leaf surface ± lustrous under hairs; cleistogamous flowers with outer sepals 0.2–0.5 mm. . . . . . 4. *Crocanthemum canadense*
                    13. Stems ascending to divergent; adaxial leaf surface dull under hairs; cleistogamous flowers with outer sepals 0.4–1 mm . . . . 7. *Crocanthemum dumosum*
                12. Cauline leaf blades stellate-tomentose or stellate-pubescent, without simple hairs; chasmogamous flowers (1)2–18; stems not red tinged.
                    14. Subshrubs; petioles 0–2 mm . . . . .9. *Crocanthemum glomeratum* (in part)
                    14. Herbs (or subshrubs in *C. georgianum*); petioles 1–5 mm.
                        15. Chasmogamous and cleistogamous flowers produced together on same inflorescence; basal leaves present or absent. . . . . .8. *Crocanthemum georgianum* (in part)
                        15. Chasmogamous and cleistogamous flowers produced on separate inflorescences, cleistogamous flowering occurs 1–3 months later; basal leaves absent.
                            16. Stems (12–)20–50(–67) cm, clustered on vertical caudex; outer sepals of cleistogamous flowers linear, 0.5–1.5 mm . . . . .3. *Crocanthemum bicknellii*
                            16. Stems 10–27(–35) cm, scattered on horizontal rootstock; outer sepals of cleistogamous flowers rudimentary, 0.2–0.5 mm . . . . 12. *Crocanthemum propinquum*

1. Crocanthemum aldersonii (Greene) Janchen in H. G. A. Engler et al., Nat. Pflanzenfam. ed. 2, 21: 305. 1925 • Alderson rushrose

Helianthemum aldersonii Greene, Erythea 1: 259. 1893; *H. scoparium* Nuttall var. *aldersonii* (Greene) Munz

Subshrubs. Stems erect, 30–65 cm, glabrate to sparsely stellate-pubescent. **Leaves** cauline; petiole 0.5–1.5 mm; blade linear, 10–21(–26) × 0.9–2 mm, surfaces stellate-pannose, lateral veins obscure abaxially. **Inflorescences** terminal, broad panicles, branches spreading, usually curved, stellate-pannose; chasmogamous flowers 1–3 per branch; cleistogamous 0. **Pedicels** 5–13 mm, stellate-pubescent with no or sparse glandular hairs; bracts 1–2.5 × 0.3–0.5 mm. **Chasmogamous flowers:** outer sepals linear, 1.3–2 (–2.8) × 0.3 mm, inner sepals 4.7–7(–8) × 2–3 mm, apex acuminate or acute, calyx stellate-pubescent, hairs to 1 mm; petals obovate, 8–12 × 6–8 mm; capsules 3.5–4.3 × 2.3–2.7 mm, glabrous.

Flowering Mar–Jun. Chaparral on slopes and in canyons; 100–1200 m; Calif.; Mexico (Baja California).

Crocanthemum aldersonii occurs from Orange and San Bernardino counties south into Baja California. Although traditionally placed within a broadly circumscribed *C. scoparium*, the tall habit, paniculate inflorescence, and acuminate to acute inner sepals of *C. aldersonii* are more similar to those of *C. suffrutescens*. The calyx of *C. aldersonii* occasionally has stellate hairs with thickened bases, a characteristic always present in *C. suffrutescens*. The tall inflorescence, curved inflorescence branches, and relatively large corollas are unique among Pacific slope species of *Crocanthemum*.

2. Crocanthemum arenicola (Chapman) Barnhart in J. K. Small, Man. S.E. Fl., 879. 1933 • Gulf Coast frostweed [E]

Helianthemum arenicola Chapman, Fl. South. U.S., 35. 1860

Herbs. Stems ascending or spreading to erect, 5–15(–20) cm, stellate-pubescent. **Leaves** cauline; petiole 0.5–2 mm; blade oblanceolate to lanceolate, 8–20(–28) × 2–4(–10) mm, surfaces stellate-tomentose abaxially, stellate-pubescent adaxially, lateral veins obscure abaxially. **Inflorescences** usually terminal, sometimes lateral, umbellate; chasmogamous flowers 2–10 per umbel; cleistogamous 2–10 per umbel. **Pedicels** (2.5–)5–10(–15) mm; bracts 1–3.5 × 0.3–0.6 mm. **Chasmogamous flowers:** outer sepals linear, 2–4 × 0.5 mm, inner sepals 4–8 × 3.5–4.8 mm, apex acute; petals obovate, 8–9.5 × 7.5–9 mm; capsules 3-valved, 3.6–5.4 × 2.4–3.6 mm, stellate-pubescent distally. **Cleistogamous flowers:** outer sepals linear, 1–2 × 0.4 mm, inner sepals ovate, 3.6–5.6 × 2.2–2.8 mm, apex acute; capsules 3–4.6 × 2.4–3.4 mm, stellate-pubescent distally.

Flowering Mar–Apr. Maritime sand dunes and interdunes among pine-oak scrub; 0–10 m; Ala., Fla., Miss.

Crocanthemum arenicola occurs in 11 contiguous counties, from Franklin in Florida to Harrison in Mississippi. Only *C. arenicola* and *C. nashii* possess stellate-pubescent ovaries and capsules; they are allopatric, occupy different habitats, and present no identification difficulties.

3. Crocanthemum bicknellii (Fernald) Janchen in H. G. A. Engler et al., Nat. Pflanzenfam. ed. 2, 21: 307. 1925 • Hoary or plains frostweed [E] [F]

Helianthemum bicknellii Fernald, Rhodora 21: 36. 1919; *H. canadense* (Linnaeus) Michaux var. *walkerae* W. H. Evans

Herbs. Stems erect, clustered on vertical caudex, (12–) 20–50(–67) cm, simple or branched, stellate-pubescent to stellate-tomentose. **Leaves** cauline; petiole 1–4 mm; blade narrowly elliptic to oblanceolate, (10–)18–32(–40) × 4–7(–10) mm, surfaces stellate-tomentose abaxially, stellate-pubescent adaxially, without simple hairs, lateral veins raised abaxially. **Inflorescences** terminal, cymes; chasmogamous flowers 6–10(–18) per cyme; cleistogamous in glomerules, 1–10 flowers per glomerule, on lateral leafy branches 2.5–6 cm, flowering 1–3 months later than chasmogamous. **Pedicels** (1.5–)3–8(–12) mm; bracts 3–7 × 0.5–1 mm. **Chasmogamous flowers:** outer sepals linear, 3.5–4.5(–8) × 0.4–1 mm, inner sepals ovate-elliptic, 5–8 × 2.4–4 mm, apex acute; petals obovate, 8–12 × 5–10 mm; capsules 3–5.5 × 2.5–4.5 mm, glabrous. **Cleistogamous flowers:** outer sepals linear, 0.5–1.5 × 0.3 mm, inner sepals ovate-elliptic, 1.7–2.8 × 1.5–2.3 mm, apex acute; capsules 1.5–2.2 × 1.2–2.2 mm, glabrous.

Flowering Jun–Jul. Sandy or rocky barrens, glades, sandhills, prairies, fields, pine-oak and oak-hickory woodlands, montane outcrops and balds; 0–1500 m; Man., Ont.; Ark., Colo., Conn., Del., Ga., Ill., Ind., Iowa, Kans., Ky., Maine, Md., Mass., Mich., Minn., Mo., Nebr., N.H., N.J., N.Y., N.C., Ohio, Pa., R.I., S.C., S.Dak., Tenn., Vt., Va., W.Va., Wis., Wyo.

The dense, often crowded, cleistogamous flowers on branches of *Crocanthemum bicknellii* late in the season

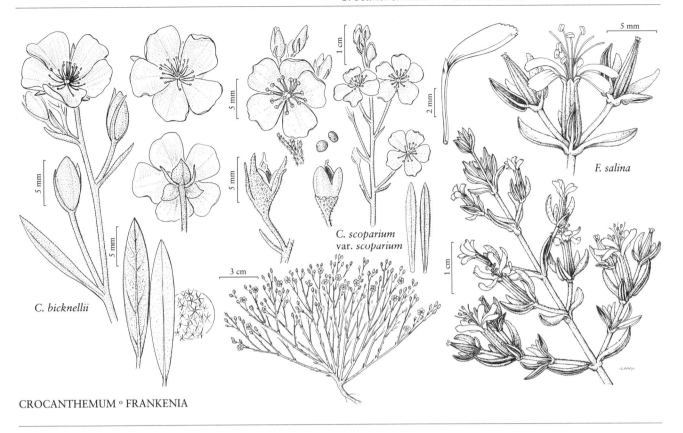

F. salina

C. scoparium
var. scoparium

C. bicknellii

CROCANTHEMUM ° FRANKENIA

contrast strongly with the relatively few, terminal, petaliferous flowers early in the season. Its closest relative is 12. *C. propinquum*; see that treatment for identification aids. The names *C. majus* (Linnaeus) Britton and *Helianthemum majus* (Linnaeus) E. P. Bicknell have been misapplied to *C. bicknellii*. They were based on *Lechea major* Linnaeus, which is a synonym of *C. canadense*, based on types.

4. Crocanthemum canadense (Linnaeus) Britton in
N. L. Britton and A. Brown, Ill. Fl. N. U.S. ed. 2, 2: 540. 1913 • Canada rockrose or frostweed, crocanthème du Canada E

Cistus canadensis Linnaeus, Sp. Pl. 1: 526. 1753; *Helianthemum canadense* (Linnaeus) Michaux; *H. canadense* var. *sabulonum* Fernald; *Lechea major* Linnaeus

Herbs. **Stems** ascending to erect, usually red tinged, 15–50 (–65) cm, stellate-pubescent to glabrate. **Leaves** cauline; petiole 1–3 mm; blade oblanceolate-elliptic to narrowly elliptic, tapered to base, 12–30(–38) × 4–7(–10) mm, apex acute, surfaces stellate-tomentose and with scattered simple hairs abaxially, ± lustrous, sparsely stellate-pubescent and with simple hairs adaxially, lateral veins raised abaxially. **Inflorescences** terminal or subterminal, cymes;

chasmogamous flowers 1–3 per cyme, cleistogamous 1–3 per glomerule, on lateral leafy branches 6–18 cm, flowering 1–3 months later than chasmogamous. **Pedicels** (1.5–)4–10(–17) mm, with stellate and simple hairs; bracts absent. **Chasmogamous flowers:** outer sepals narrowly lanceolate, 2–6 × 0.5–1 mm, inner sepals 5–9 × 3.5–5 mm, apex acute; petals obovate, 8–15 × 6–14 mm; capsules 5–8 × 4–7 mm, glabrous. **Cleistogamous flowers:** outer sepals rudimentary, 0.2–0.5 × 0.2–0.4 mm, inner sepals obovate, 2.5–4 × 2.5–3.5 mm, apex acute; capsules 2.3–3.5 × 2.5–3.5 mm, glabrous. **2n = 20.**

Flowering late Mar–Jul(–Aug). Sandy or rocky barrens, glades, sandhills, prairies, fields, roadsides, maritime grasslands and heathlands, interdunes, pine-oak and oak-hickory woodlands, rocky slopes; 0–700 m; N.S., Ont., Que.; Ala., Conn., Del., D.C., Ga., Ill., Ind., Iowa, Ky., Maine, Md., Mass., Mich., Minn., Mo., N.H., N.J., N.Y., N.C., Ohio, Pa., R.I., S.C., Tenn., Vt., Va., W.Va., Wis.

Crocanthemum canadense and *C. dumosum* are distinguished from sympatric species by simple hairs among the stellate ones on leaf surfaces and by reddish colored stems. Until the twentieth century, *C. canadense* was treated much more broadly, to include *C. bicknellii*, *C. dumosum*, and *C. propinquum*; it differs from *C. bicknellii* and *C. propinquum* by having simple hairs on foliage (versus stellate only) and larger cleistogamous capsules. Its closest relative is 7. *C. dumosum*; see that treatment for identification aids.

5. **Crocanthemum carolinianum** (Walter) Spach, Ann. Sci. Nat., Bot., sér. 2, 6: 370. 1836 • Carolina sunrose or frostweed [E]

Cistus carolinianus Walter, Fl. Carol., 152. 1788; *Helianthemum carolinianum* (Walter) Michaux

Herbs. Stems erect, (4–)10–30(–38) cm, stellate-pubescent, hairs 1.5–2.5 mm. **Leaves** basal, rosette- or mat-forming, and cauline, relatively few, similar to basal, margins not revolute; petiole 1–4 mm; blade spatulate to obovate or elliptic, 10–35(–60) × 5–18(–28) mm, tapered to base, apex obtuse, surfaces sparsely stellate-pubescent, lateral veins raised abaxially. **Inflorescences** terminal, scorpioid cymes; chasmogamous flowers 1–3 per cyme, cleistogamous rarely produced, flowering 1–3 months later than chasmogamous. **Pedicels** 4–15(–24) mm; bracts 2.5–6 × 0.5–1 mm. **Chasmogamous flowers:** outer sepals lanceolate, 3–6(–7.5) × 0.6–1.2 mm, inner sepals ovate, 7–12 × 4–6 mm, apex acuminate; petals broadly spatulate, 8–18 × 8–15 mm; capsules 6–9 × 4.5–9 mm, glabrous. **Cleistogamous flowers:** outer sepals linear, 1.6–2.8 × 0.4 mm, inner sepals ovate, 3–4.5 × 1.6 mm, apex acute; capsules not seen.

Flowering Mar–May. Dry to mesic pine savannas and flatwoods, sandy pine-oak woodlands, stable alluvial dunes; 0–200 m; Ala., Ark., Fla., Ga., La., Miss., N.C., S.C., Tex.

Crocanthemum carolinianum is one of the more distinctive members of *Crocanthemum* because of its short stature, basal rosettes, relatively large leaves, nonrevolute leaf margins, and long-stellate hairs on stems. Unlike other eastern species, it rarely produces cleistogamous flowers (fewer than 1% of specimens examined, according to H. S. Daoud and R. L. Wilbur 1965, all in Florida).

6. **Crocanthemum corymbosum** (Michaux) Britton in N. L. Britton and A. Brown, Ill. Fl. N. U.S. ed. 2, 2: 541. 1913 • Pine barren frostweed [E]

Helianthemum corymbosum Michaux, Fl. Bor.-Amer. 1: 307. 1803; *Cistus corymbosus* (Michaux) Poiret

Herbs or subshrubs. Stems ascending to erect, 10–30 (–50) cm, stellate-pubescent, glabrescent. **Leaves** cauline; petiole 1–5 mm; blade bicolor, obovate-elliptic to elliptic-lanceolate, 12–35(–47) × 3–10(–13) mm, base cuneate, apex obtuse, surfaces stellate-pubescent, lateral veins raised abaxially. **Inflorescences** terminal, compound dichasia in corymbiform cymes; chasmogamous flowers 1–6 per dichasium, overtopping cleistogamous, cleistogamous 10–45 per dichasium, produced simultaneously. **Pedicels** 6–15 mm, villous and stellate-pubescent; bracts 2–7 × 0.2–1.2 mm. **Chasmogamous flowers:** outer sepals spatulate-linear, 2.4–4.5 × 0.7–1.2 mm, apex obtuse, inner sepals 3–7 × 2.5–3.6 mm, apex acute to acuminate; petals obovate, 6–11 × 5–9.5 mm; capsules 3.6–5.4 × 3–4 mm, glabrous. **Cleistogamous flowers:** outer sepals linear, 1.8–3 × 0.3–0.9 mm, inner sepals ovate, 2.2–4.8 × 1.2–3 mm, apex acute; capsules 1.6–3.8 × 1.4–3 mm, glabrous.

Flowering late Feb–Apr. Stable maritime dunes, maritime forests, sandy pinelands, live-oak woodlands; 0–50 m; Ala., Fla., Ga., Miss., N.C., S.C.

Crocanthemum corymbosum is unique in *Crocanthemum* because of its corymbose inflorescence of long-pedicelled petaliferous flowers overtopping subsessile, apetalous flowers. Other species may produce long-pedicelled petaliferous flowers; the inflorescence shape is different. The two flower types develop synchronously, unlike in other species except *C. glomeratum*. The bicolored leaves, dark adaxially and pale abaxially, are a strong field and herbarium character.

7. **Crocanthemum dumosum** E. P. Bicknell, Bull. Torrey Bot. Club 40: 613. 1913 • Bushy rockrose or frostweed [E]

Helianthemum dumosum (E. P. Bicknell) Fernald

Herbs, cespitose, sometimes moundlike. **Stems** ascending to divergent, usually red tinged, 10–20(–30) cm, stellate-pubescent. **Leaves** cauline; petiole 1–2 mm; blade usually elliptic, rarely oblanceolate, 10–20(–26) × 3–7(–10) mm, tapered to base, apex acute, surfaces stellate-tomentose and with simple hairs abaxially, dull, stellate-pubescent and with simple and, sometimes, glandular hairs adaxially, lateral veins raised abaxially. **Inflorescences** terminal or subterminal, cymes; chasmogamous flowers 1–3 per cyme, cleistogamous 1(2) per glomerule, on lateral leafy branches 6–18 cm, flowering 1–3 months later than chasmogamous. **Pedicels** 2.4–6 mm, with stellate, simple, and glandular hairs; bracts absent. **Chasmogamous flowers:** outer sepals lanceolate, 2–6 × 0.8–3 mm, inner sepals 7–10 × 4–6.5 mm, apex acute; petals obovate, 8–15 × 5–13 mm; capsules 4–7 × 3–4 mm, glabrous. **Cleistogamous flowers:** outer sepals rudimentary, knoblike to triangular and acute, 0.4–1 × 0.3–0.6 mm, inner sepals ovate, 3–6 × 2.5–3.2 mm, apex acute; capsules 3.2–4.5 × 2.5–3.2 mm, glabrous.

Flowering late May–Jun. Sandplain grasslands, maritime heathlands; 0–50 m; Conn., Mass., N.Y., R.I.

Crocanthemum dumosum occurs in nine coastal counties. *Crocanthemum dumosum* and *C. canadense* are closely related; the former inhabits specialized maritime environments. They can be distinguished in the field by the ascending or divaricate, relatively numerous stems of *C. dumosum*, often giving it a bushy aspect (versus erect, single or relatively few stems, and nonbushy), and by the leaf surfaces dull under dense stellate hairs in *C. dumosum* (versus lustrous under sparser stellate hairs). Identification of occasional plants may be equivocal; those are a small minority.

Using AFLP markers, S. G. Obae et al. (2013) found that *Crocanthemum dumosum* nested within the same cluster as *C. canadense*. The authors recommended that it be considered as a subspecific variant of *C. canadense*, but did not make a new combination. Additional molecular work is warranted to solidify their findings.

8. **Crocanthemum georgianum** (Chapman) Barnhart in J. K. Small, Man. S.E. Fl., 879. 1933 • Georgia frostweed E

Helianthemum georgianum Chapman, Fl. South. U.S. ed. 3, 36. 1897

Herbs or subshrubs. Stems from caudices and elongate horizontal rootstocks, erect or spreading, (10–)15–30(–40) cm, stellate-tomentose. **Leaves** usually basal, mat-forming or sometimes absent, and cauline; petiole 1–2 mm; blade spatulate to oblanceolate, 10–28 × 5–11 mm, surfaces stellate-pannose, without simple hairs; cauline: petiole 1–3 mm; blade oblanceolate to narrowly elliptic, 20–40 × (2–)4.5–9(–12) mm, surfaces stellate-tomentose abaxially, stellate-pubescent adaxially, without simple hairs; lateral veins raised abaxially. **Inflorescences** terminal, racemose cymes; flowers (chasmogamous or cleistogamous) 2–7 per cyme, produced simultaneously. **Pedicels** 5–12 mm; bracts 1.2–5 × 0.3–1 mm. **Chasmogamous flowers:** outer sepals linear, 1.5–3.5 × 0.3–0.5 mm, inner sepals 3.6–6.6 × 3–4.5 mm, apex acute; petals obovate, 6–10(–12) × 4–8(–11) mm; capsules 3.8–5.7 × 3.2–4.5 mm, glabrous. **Cleistogamous flowers:** outer sepals linear, 1.4–2.2 × 0.3 mm, inner sepals ovate, 3–4.2 × 2.6–3.8 mm, apex acute; capsules 3–4.2 × 2.4–3.3 mm, glabrous.

Flowering Apr–Jun. Dry, sandy oak-pine woods, turkey oak sandhills, stable maritime dunes, maritime woodlands; 0–600 m; Ala., Ark., Fla., Ga., La., Miss., N.C., Okla., S.C., Tex.

Basal leaves occur often enough in *Crocanthemum georgianum* to be very useful for identification; in the eastern United States only *C. carolinianum* also regularly has basal leaves. Plants of maritime situations tend to be shorter than inland plants.

9. **Crocanthemum glomeratum** (Lagasca) Janchen in H. G. A. Engler et al., Nat. Pflanzenfam. ed. 2, 21: 305. 1925 • Clustered frostweed

Cistus glomeratus Lagasca, Gen. Sp. Pl., 16. 1816; *Helianthemum glomeratum* (Lagasca) Lagasca

Subshrubs. Stems from branched caudex and horizontal rootstocks, ascending to erect, 10–60(–80) cm, stellate-pubescent, glabrescent. **Leaves** cauline; petiole 0–2 mm; blade oblanceolate, 10–25(–35) × 2–8(–14) mm, base cuneate, margins nonrevolute, apex subacute, surfaces stellate-tomentose, without simple hairs, lateral veins obscure or raised abaxially. **Inflorescences** terminal and axillary, glomerules and panicles; chasmogamous flowers 1 or 2 per panicle, overtopping cleistogamous, cleistogamous 2–10 per glomerule, produced simultaneously with chasmogamous. **Pedicels** 10–20 mm; bracts 2–5 × 0.2–0.8 mm. **Chasmogamous flowers:** outer sepals linear, 1–4 × 0.3 mm, inner sepals 3–7 × 1.8–3.2 mm, apex acute to acuminate; petals obovate, 4–9 × 3.5–6 mm; capsules 3.5–4.5 × 2–3 mm, glabrous. **Cleistogamous flowers:** outer sepals linear, 1–3 × 0.2–0.4 mm, inner sepals ovate or ovate-lanceolate, 2.4–5 × 1.2–3.6 mm, apex acute or acuminate; capsules 1.6–3.6 × 1.2–2.5 mm, glabrous.

Flowering Jun–Aug. Dry montane oak woodlands; 2000–2400 m; Tex.; Mexico (Chihuahua, Durango, Nuevo León, San Luis Potosí, Sinaloa, Sonora, Zacatecas); Central America (Costa Rica, Guatemala, Honduras).

In the flora area, *Crocanthemum glomeratum* occurs only in the Chisos Mountains of western Texas. In Mexico and Central America, where it flowers throughout the year, it is variable in foliage, flowers, and fruits; H. S. Daoud and R. L. Wilbur (1965) were unable to discern any geographically meaningful patterns. Some plants produce cleistogamous flowers only, and some plants have prominent lateral leaf veins; it is not known if these variations occur within the flora area.

10. **Crocanthemum greenei** (B. L. Robinson) Sorrie, Phytologia 93: 270. 2011 • Island rushrose [C] [E]

Helianthemum greenei B. L. Robinson in A. Gray et al., Syn. Fl. N. Amer. 1(1,1): 191. 1895, based on *H. occidentale* Greene, Bull. Calif. Acad. Sci. 2: 144. 1886, not Nyman 1878

Subshrubs. Stems erect, 15–30 cm, sparsely to densely stellate-pubescent, distally with dense, dark, glandular hairs 0.4–0.8 mm. **Leaves** cauline; petiole 0–2.5 mm; blade oblanceolate to linear-lanceolate, 7–25(–35) × 2–4(–7) mm, margins nonrevolute, surfaces stellate-pubescent, lateral veins obscure abaxially. **Inflorescences** terminal, corymbose; chasmogamous flowers 3–25 per corymb, cleistogamous 0. **Pedicels** 1–4(–6) mm, stellate-pubescent and densely glandular-hairy; bracts 3–10 × 0.5–1.5 mm. **Chasmogamous flowers:** outer sepals lanceolate, 2.5–4 × 0.5–1 mm, inner sepals 4.5–8 × 3–4 mm, apex acuminate; calyx strigose (sometimes also short-stellate-hairy), hairs simple, 1–1.7 mm; petals obovate, 5–8 × 3–4 mm; capsules 4.3–6 × 3–3.5 mm, glabrous.

Flowering Apr–May.   Dry, rocky ridges; of conservation concern; 10–100 m; Calif.

Crocanthemum greenei has been documented from San Miguel, Santa Catalina, Santa Cruz, and Santa Rosa islands. According to the California Native Plant Society, it is extant on Santa Catalina, Santa Cruz, and Santa Rosa islands, totaling about 20 populations and listed federally as threatened. Threats are from non-native mammals and plants. The habit, dense, dark glandular hairs on stems and pedicels, and long, white hairs on the calyx distinguish *C. greenei*.

11. **Crocanthemum nashii** (Britton) Barnhart in J. K. Small, Man. S.E. Fl., 879. 1933 • Florida scrub frostweed or sunrose [E]

Helianthemum nashii Britton, Bull. Torrey Bot. Club 22: 147. 1895 (as nashi); *H. thyrsoideum* Barnhart

Herbs. Stems from caudices and horizontal rootstocks, erect to ascending, 20–35(–41) cm, stellate-pubescent. **Leaves** cauline; petiole 1–2(–3) mm; blade oblanceolate-elliptic to lanceolate-elliptic, 15–30 (–38) × 3–6(–8.4) mm, surfaces stellate-tomentose, without simple hairs, lateral veins slightly to moderately raised abaxially. **Inflorescences** terminal, thyrses; chasmogamous flowers 1–8 per thyrse, cleistogamous 8–40 per thyrse, flowering 1–3 months later than chasmogamous. **Pedicels** 3–8(–10) mm; bracts linear-lanceolate, 1–3.2 × 0.2–0.6 mm.   **Chasmogamous flowers:** outer sepals linear, 1–3 × 0.3 mm, inner sepals ovate-elliptic, 3.5–5 × 2.7–3.7 mm, apex acute; petals broadly cuneate, 6–9.5 × 3–6.5 mm; capsules 2-valved, 3–4.8 × 2.4–3.8 mm, stellate-pubescent distally. **Cleistogamous flowers:** outer sepals linear, 0.5–1.2 × 0.5 mm, inner sepals ovate-elliptic, 1–3.4 × 1–2.3 mm, apex acute; capsules 2-valved, 3–4 × 2.4–3.5 mm, stellate-pubescent distally.

Flowering Mar–Jun. Xeric sand-pine scrub and oak scrub, on stable maritime dunes, on inland sandhills; 0–100 m; Fla., N.C.

Crocanthemum nashii and *C. arenicola* are the only species in *Crocanthemum* with stellate-pubescent ovaries and capsules; *C. nashii* is unique in having two-valved capsules. It was discovered in New Hanover County, North Carolina, in 1997, disjunct some 540 km from the northernmost Florida populations. The uniformly gray-green foliage, thyrsoid inflorescence, and often patch-forming habit render it unmistakable within its range. Plants with simple hairs mixed with stellate hairs on the calyx were named *Helianthemum thyrsoideum*.

12. **Crocanthemum propinquum** (E. P. Bicknell) E. P. Bicknell, Bull. Torrey Bot. Club 40: 615. 1913 • Low or creeping frostweed [E]

Helianthemum propinquum E. P. Bicknell in N. L. Britton, Man. Fl. N. States ed. 2, 1069. 1905

Herbs.   Stems scattered on horizontal rootstocks, ascending to erect, 10–27(–35) cm, stellate-pubescent to stellate-tomentose. **Leaves** cauline; petiole 2–5 mm; blade narrowly elliptic to oblanceolate, gradually narrowed to base, 10–30 × 3–6(–8) mm, surfaces stellate-tomentose abaxially, stellate-pubescent adaxially, without simple hairs, lateral veins raised abaxially. **Inflorescences** terminal, cymes; chasmogamous flowers 2–6 per cyme, cleistogamous in glomerules, 1–6 flowers per glomerule, on lateral leafy branches 1–3 cm, flowering 1–3 months later than chasmogamous. **Pedicels** (2–)8–14(–22) mm; bracts 1.5–3.5 × 0.3 mm. **Chasmogamous flowers:** outer sepals linear, 1–3(–4) × 0.4–0.9 mm, inner sepals ovate-elliptic, 5–8 × 2.3–4.5 mm, apex acute; petals obovate, 8–10(–13) × 6–12 mm; capsules 3.7–5.3 × 3–4 mm, glabrous. **Cleistogamous flowers:** outer sepals rudimentary, 0.2–0.5 × 0.2 mm, inner sepals ovate, 2–2.5 × 1.5–2.2 mm, apex acute; capsules 1.5–2.2 × 1.3–2 mm, glabrous.

Flowering May–early Jul. Open woodlands, rock outcrops, sandplain grasslands, maritime heathlands, clearings, fields; 0–1500 m; Conn., Del., D.C., Ga., Md., Mass., N.H., N.J., N.Y., N.C., Pa., R.I., Tenn., Va.

The shorter stature, tiny outer sepals on cleistogamous flowers, and horizontal rootstocks distinguish *Crocanthemum propinquum* from *C. bicknellii*, which is often twice as tall, has elongate sepals, and has a caudex. *Crocanthemum propinquum* is disjunct from the coastal plain of Virginia to the southern Appalachian Mountains of Georgia, North Carolina, and Tennessee.

13. **Crocanthemum rosmarinifolium** (Pursh) Janchen in H. G. A. Engler et al., Nat. Pflanzenfam. ed. 2, 21: 307. 1925 • Rosemary frostweed

Helianthemum rosmarinifolium Pursh, Fl. Amer. Sept. 2: 364. 1813; *Crocanthemum domingense* (Urban) Janchen; *C. stenophyllum* (Urban) Janchen

Herbs. Stems ascending to erect, (13–)20–40(–51) cm, stellate-tomentose. **Leaves** cauline and, sometimes, basal, basal mat-forming; petiole 1–3 mm; blade obovate, 10–22 × 3–5 mm; cauline: petiole 1–4 mm; blade oblanceolate to narrowly lanceolate, 10–38(–48) × 2–5.5(–7.8) mm, margins revolute; surfaces stellate-tomentose abaxially, stellate-pubescent adaxially, without simple hairs; lateral veins obscure abaxially. **Inflorescences** axillary and terminal, solitary flowers or glomerules; chasmogamous flowers solitary, at tips of branches, overtopping cleistogamous, cleistogamous 2–7 per glomerule, flowering 1–3 months later than chasmogamous. **Pedicels** 10–22 mm; bracts 3–7 × 0.5–1 mm. **Chasmogamous flowers:** outer sepals linear, 1.2–2.5 × 0.2 mm, inner sepals 2.5–4.3 × 1.3–2 mm, apex acute; petals obovate, 4–6 × 3.5–5 mm; capsules 2–3 × 1.4–1.8 mm, glabrous. **Cleistogamous flowers:** outer sepals linear, 0.5–1 × 0.2 mm, inner sepals ovate, 1.5–1.8 × 1–1.5 mm, apex acute; capsules 1.3–1.7 × 1–1.3 mm, glabrous.

Flowering May–Jul. Dry, sandy roadsides, openings in pine and pine-oak woodlands, disturbed soil of clearings and fields; 0–200 m; Ala., Ark., Fla., Ga., La., Miss., N.C., Okla., S.C., Tex.; West Indies (Dominican Republic).

Crocanthemum rosmarinifolium is easily distinguished from sympatric species by its slender leaves, relatively small petaliferous flowers, occasional basal leaves, and overall pale color. As noted by R. M. Harper over a century ago, it seems not to occupy natural habitats and is found primarily on roadsides. Populations in the Dominican Republic occur up to at least 2200 m.

14. **Crocanthemum scoparium** (Nuttall) Millspaugh, Publ. Field Mus. Nat. Hist., Bot. Ser. 5: 175. 1923 • Broom or peak rushrose F

Helianthemum scoparium Nuttall in J. Torrey and A. Gray, Fl. N. Amer. 1: 152. 1838

Subshrubs. Stems spreading to erect-fastigiate, 10–45 cm, usually sparsely stellate-pubescent to glabrate, sometimes densely lanate. **Leaves** cauline, tending to be deciduous in summer; petiole 0–2 mm; blade linear, 5–11 × 0.5–2(–3.5) mm, surfaces stellate-pubescent to glabrate abaxially, sparsely stellate-pubescent to glabrate adaxially, lateral veins obscure abaxially. **Inflorescences** terminal, panicles or racemes; chasmogamous flowers 1–18 per panicle or raceme, cleistogamous 0. **Pedicels** 2–6 mm, sparsely or not glandular-hairy; bracts 2–4 × 0.3–0.5 mm. **Chasmogamous flowers:** outer sepals linear, 1.5–3.5 × 0.3 mm, inner sepals 3.5–5(–7.5) × 2–3 mm, apex acute to acuminate; calyx stellate-pubescent, hairs to 1 mm; petals obovate, 3–6 × 3–5 mm; capsules 2.8–3.8 × 2–2.5 mm, glabrous.

Varieties 2 (2 in the flora): California, nw Mexico.

Even with the recognition of *Crocanthemum aldersonii* and *C. suffrutescens* as separate species, morphological diversity within *C. scoparium* still remains considerable. The two varieties here recognized show differences in habit, average plant height, number of flowers, and distribution. Another variant occurs sporadically along the coast and on Santa Cruz Island, from Monterey to San Diego counties; vegetative parts (at least distal branches, pedicels, and sepals) are covered with white, lanate hairs. This variant has never been formally named. Another form from coastal Mendocino County was called "Helianthemum mendocinensis" by Alice Eastwood on a specimen (*H. E. Brown 785,* JEPS); the name was never published. These plants have densely stellate-pubescent stems and exceptionally elongate sepal tips.

1. Stems mostly 10–30 cm, divaricate and spreading or curved proximally and erect distally; mostly near-coastal. . . . .
. . . . .14a. *Crocanthemum scoparium* var. *scoparium*
1. Stems usually 30–45 cm, usually erect-fastigiate, sometimes curved proximally and erect distally; mostly inland . . . .
. . . . .14b. *Crocanthemum scoparium* var. *vulgare*

14a. Crocanthemum scoparium (Nuttall) Millspaugh var. **scoparium** [F]

Stems mostly 10–30 cm, divaricate and spreading with distal portions ascending, or curved proximally and erect distally. Flowers 1–10 per stem, outer sepals 1.5–2.5 mm.

Flowering Mar–Jul. Coastal sage scrub, chaparral, dry, sandy or rocky slopes and ridges; 0–500 m; Calif.; Mexico (Baja California).

Variety *scoparium* ranges from Mendocino County to Los Angeles County, with atypical plants south to San Diego County and northern Baja California.

14b. Crocanthemum scoparium (Nuttall) Millspaugh var. **vulgare** (Jepson) Sorrie, Phytologia 93: 271. 2011

Helianthemum scoparium Nuttall var. *vulgare* Jepson, Man. Fl. Pl. Calif., 641. 1925

Stems mostly 30–45 cm, usually erect-fastigiate, sometimes curved proximally and erect distally. Flowers 4–18 per stem, outer sepals 2–3.5 mm.

Flowering Mar–Jul. Mostly inland chaparral, dry, sandy or rocky slopes and ridges, sandy or stony mesas; 50–1200 m; Calif.; Mexico (Baja California).

Variety *vulgare* is the most abundant and widespread of the five western crocanthemums in the flora area. Variation ranges from plants that are relatively short and many-stemmed to those that are relatively tall and few-stemmed; all typically have erect stems and numerous flowers. It ranges from Lake and El Dorado counties to northern Baja California.

15. Crocanthemum suffrutescens (B. Schreiber) Sorrie, Phytologia 93: 271. 2011 • Ione rushrose [E]

Helianthemum suffrutescens B. Schreiber, Madroño 5: 81, fig. 1. 1939

Subshrubs. Stems erect, 35–80 cm, usually stellate-pannose, sometimes sparsely stellate-pubescent. Leaves cauline, tending to persist in summer; petiole 0–1 mm; blade linear to lanceolate, 10–30(–43) × 2–8 mm, surfaces stellate-pannose, lateral veins obscure abaxially. Inflorescences terminal, narrow panicles, usually stellate-pannose, sometimes sparsely stellate-pubescent; chasmogamous flowers 8–20 per panicle, cleistogamous 0. Pedicels 1–9 mm, sparsely or not glandular-hairy; bracts 5–25 × 0.3 mm. Chasmogamous flowers: outer sepals linear, 2 × 0.2 mm, inner sepals 5–7(–8) × 2.5–3.5 mm, apex acute to acuminate; calyx stellate-pubescent, hairs to 1 mm; petals obovate, 6–9 × 5–7 mm; capsules 3.5–4.3 × 2.5–2.7 mm, glabrous.

Flowering Apr–Jun. Xeric to dry chaparral in shallow soil of Ione Formation; 100–700 m; Calif.

Crocanthemum suffrutescens is known from Amador and Calaveras counties, where it is often associated with *Arctostaphylos myrtifolia*, grasses, and shrub oaks. The remarkable soils of the Ione Formation harbor other localized endemics, including *A. myrtifolia*, *Eriogonum apricum*, and *Horkelia parryi*.

Crocanthemum suffrutescens is similar to *C. aldersonii* in its tall habit, paniculate inflorescence, and long sepals; it differs in its denser stellate tomentum, shorter and straight inflorescence branches, smaller corollas, and thickened bases to the stellate hairs of the calyx. Leaves of *C. suffrutescens* tend to persist through summer; they tend to drop in *C. aldersonii* and *C. scoparium*; these are tendencies only and cannot be relied on for identification.

FRANKENIACEAE Desvaux

- Frankenia Family

Molly A. Whalen

Herbs, subshrubs, or shrubs, perennial or, rarely, annual. **Stems** rounded to noticeably angled. **Leaves** opposite, simple, petiolate or sessile, with salt glands, estipulate, paired leaves connate basally by membranous, sheathing margins with stiff, white hairs, flattened and appressed to stem; petiole narrow to broad, fleshy to flattened and chartaceous; blade margins entire. **Inflorescences** solitary flowers or terminal or axillary, simple or compound dichasia (frequently with some monochasial branching), bracteate; bracts [2]4, basally connate (rarely incompletely connate). **Flowers** bisexual [unisexual]; sepals 4–6[7], persistent, connate, ribbed; petals 4–6[7], distinct, clawed basally, limb spreading, scalelike appendage or ligule present [absent] on adaxial face of claw; stamens 3–8[–24]; filaments connate and thickened at base or distinct, tapering distally, somewhat flattened in proximal ½; anthers dehiscent longitudinally; ovary 1, superior, 1-locular, [1]2–4-carpellate; ovules 1–100; placentation parietal or basal; styles 1, filiform, [1]2–4-branched; stigmas terminal. **Fruits** capsular, included in persistent calyx, thin, papery, dehiscence loculicidal. **Seeds** 1–100, testa thin; embryo straight; endosperm present.

Genus 1, species ca. 70 (5 in the flora): United States, Mexico, South America, Eurasia, Africa, Atlantic Islands, Australia; introduced in West Indies.

K. Kubitzki (2003b) recognized two genera, *Frankenia* and *Hypericopsis* Boissier, in Frankeniaceae; molecular evidence supports inclusion of *Hypericopsis* within *Frankenia* (J. F. Gaskin et al. 2004). Molecular studies have placed Frankeniaceae and Tamaricaceae, which generally have been recognized as related families, in the Caryophyllales (for example, P. Cuénoud et al. 2002).

SELECTED REFERENCES Gaskin, J. F. et al. 2004. A systematic overview of Frankeniaceae and Tamaricaceae from nuclear rDNA and plastid sequence data. Ann. Missouri Bot. Gard. 91: 401–409. Kubitzki, K. 2003b. Frankeniaceae. In: K. Kubitzki et al., eds. 1990+. The Families and Genera of Vascular Plants. 10+ vols. Berlin etc. Vol. 5, pp. 209–212.

1. FRANKENIA Linnaeus, Sp. Pl. 1: 331. 1753; Gen. Pl. ed. 5, 154. 1754

- [For Johann Frankenius, 1590–1661, Swedish botanist]

Branches articulated, often swollen at nodes, usually differentiated into long and short shoots, hairs simple. **Leaves** petiolate or sessile, salt crystals usually present on leaf surfaces, sometimes forming solid crust, short-shoot leaves in axillary fascicles; petiole sometimes inconspicuous, narrow to broad distally, margins usually ciliate. **Inflorescences:** floral bracts leaflike, usually connate basally into verticels of 4, clasping calyx base (rarely adnate to calyx base). **Flowers:** calyx tube cylindric [campanulate], strongly rounded-ribbed, 4–6[7]-lobed; petals white, pink, violet, or blue-purple, oblong-oblanceolate to spatulate, claw linear, broadening slightly at apex, to narrowly cuneate, limb apex truncate or rounded, erose-denticulate to sinuous, glabrous; anthers yellow or red to yellowish pink to purplish red; ovary with [1]2–60[–100] ovules, funiculi ascending and bearing erect ovules or deflexed and bearing pendulous ovules, usually free, infrequently adnate to carpel wall for part of their lengths. **Seeds** ivory to golden brown, ovoid-conic or elongate-ovoid to oblong- or ovoid-ellipsoid or ellipsoid. *x* = 5.

Species ca. 70 (5 in the flora): United States, Mexico, South America, Eurasia, Africa, Atlantic Islands, Australia; introduced in West Indies.

Frankenia species vary in the degree of revoluteness of the leaf margins. This is most easily determined on fresh or softened material; loosely revolute margins will often tend to become somewhat more revolute on drying but can still be distinguished from tightly revolute margins that are not readily unrolled.

SELECTED REFERENCE   Whalen, M. A. 1987. Systematics of *Frankenia* (Frankeniaceae) in North and South America. Syst. Bot. Monogr. 17: 1–93.

1. Leaf blades subterete, margins tightly revolute, abaxial surface mostly concealed.
　　2. Styles (2)3(4)-branched; petioles sometimes absent, 0.1–1.5 mm, apex ± similar in
　　　　width to base of blade . . . . 1. *Frankenia jamesii*
　　2. Styles 2(3)-branched; petioles (0.2–)0.3–1 mm, apex abruptly and markedly wider
　　　　than base of blade . . . . 3. *Frankenia palmeri*
1. At least some leaf blades flat, margins slightly to loosely revolute, abaxial surface mostly exposed.
　　3. Herbs, annual; petals 2.5–5.2 mm; seeds 0.4–0.7 mm . . . . 4. *Frankenia pulverulenta*
　　3. Subshrubs or shrubs, perennial; petals 5.5–11.5 mm; seeds 0.8–3.1 mm.
　　　　4. Branches hairy, hairs appressed; petiole subcylindric and narrow distally, tapering
　　　　　　markedly toward blade; seeds 2–3.1 mm . . . . 2. *Frankenia johnstonii*
　　　　4. Branches glabrous or puberulous or short-pilose, hairs erect or suberect; petioles
　　　　　　not markedly tapering toward blade; seeds 0.8–1.5 mm . . . . 5. *Frankenia salina*

1. Frankenia jamesii Torrey ex A. Gray, Proc. Amer. Acad. Arts 8: 622. 1873   • James's sea-heath E

Shrubs, rounded, 1.5–5 dm; branches hairy, hairs retrorse-spreading to erect. **Leaves:** petiole sometimes absent, 0.1–1.5 mm, not or scarcely tapering toward blade, apex ± similar in width to base of blade; blade grayish yellow-green, narrowly ovate to linear, subterete, (1.5–) 2–7(–8.5) × 0.5–1 mm, margins tightly revolute, abaxial surface mostly concealed, adaxial surface glabrous or with scattered hairs. **Inflorescences** simple (compound) dichasia or solitary flowers. **Flowers:** calyx 4.5–7(–7.3) mm, lobes (4)5, (0.4–)0.5–1.2 mm; petals (4)5(6), white, spatulate, (5–)8.5–12 mm; stamens (3–)6(–8), exserted, 5.3–8.5 mm; anthers yellow; style exserted, (2)3(4)-branched; ovary (2)3(4)-carpellate; ovules (2)3(4),

attached to bases of sutures, pendulous on recurving funiculi. **Seeds** 1 per capsule, ovoid-conic, 2–3 mm.

Flowering May–Sep. Gypsiferous soil, sand dunes, sandy or silty soil or shale, near salt lakes and salt flats, shortgrass prairies; 1100–2000 m; Colo., N.Mex., Tex.

2. Frankenia johnstonii Correll, Rhodora 68: 424. 1966   • Johnston's frankenia C

Shrubs, ± sprawling, to 3 dm; branches hairy, hairs appressed. **Leaves:** petiole 0.7–2.2 mm, subcylindric and narrow distally, markedly tapering toward blade, apex narrower than base of blade; blade gray-green, narrowly obovate to oblong-elliptic, flat, (2.5–)3.5–9(–10.5) × (1–)1.3–3.2(–4) mm, margins slightly to loosely revolute, abaxial surface mostly exposed, adaxial surface sparsely hairy. **Inflorescences** usually solitary flowers, sometimes simple dichasia. **Flowers:** calyx 3.8–6.5 (–7.5) mm, lobes 5, (1–)1.3–2.5 mm; petals 5(6), white, sometimes pink-tinged, spatulate, 6–10 mm; stamens (5)6, longer 3 exserted, 3.5–6.8 mm; anthers yellow; style exserted, 3-branched; ovary 3-carpellate; ovules 3, attached basally, pendulous on recurved funiculi. **Seeds** 1 per capsule, elongate-ovoid, 2–3.1 mm.

Flowering year-round. Alkaline, saline and gypseous, clay and sandy clay to loam soil, hillsides and saline flats; of conservation concern; 50–200 m; Tex.; Mexico (Nuevo León, Tamaulipas).

Frankenia johnstonii occurs in Starr, Webb, and Zapata counties in southern Texas and in northern Mexico. In 1984, it was federally listed as endangered; it was thought to occur in relatively few localities. Extensive surveys conducted in southern Texas by Gena Janssen with the cooperation of local landowners revealed additional populations (G. K. Janssen 1999). In 2003, the U.S. Fish and Wildlife Service proposed to remove this species from the list of endangered plants; as of 2014, it had not yet been delisted.

3. **Frankenia palmeri** S. Watson, Proc. Amer. Acad. Arts 11: 124. 1876 • Palmer's frankenia

Shrubs, spreading, rounded, 1–3 dm; branches puberulous or scattered-hairy, hairs spreading to recurved. **Leaves:** petiole (0.2–)0.3–1 mm, not tapering toward blade, apex abruptly and markedly wider than base of blade; blade grayish yellow-green, oblong-ovate, oblong, or linear-oblong, subterete, 1.5–7 × 0.7–1 mm, margins tightly revolute, abaxial surface almost entirely concealed, adaxial surface scattered short-hairy to puberulous. **Inflorescences** usually compound or simple dichasia, sometimes solitary flowers. **Flowers:** calyx 3–5 mm, lobes (4)5, 0.5–1.1 mm; petals (4)5, white, sometimes pink-tinged proximally, spatulate, (4.5–)5–7 mm; stamens 4(5), exserted, (4–)5–8.5 mm; anthers red or yellowish pink; style strongly exserted from calyx tube, 2(3)-branched; ovary 2-carpellate; ovules 2, attached to bases of sutures, pendulous on recurving funiculi. **Seeds** 1 per capsule, ovoid-conic, 1.5–2 mm.

Flowering Mar–Jul. Coastal dunes, upper transitional zone of coastal salt marshes, sandy flats; 0–10 m; Calif.; Mexico (Baja California, Baja California Sur, Sonora).

In the flora area, *Frankenia palmeri* occurs in southwestern San Diego County.

4. **Frankenia pulverulenta** Linnaeus, Sp. Pl. 1: 332. 1753 • European sea-heath, wisp-weed [I]

Herbs, annual, often prostrate, to 1.5(–3) dm; branches glabrous or puberulous, hairs erect, usually curved, sometimes straight. **Leaves:** petiole 0.8–2 mm, markedly tapering toward blade, apex narrower than base of blade; blade gray-green, usually narrowly obovate or obovate to elliptic or oblong-elliptic, sometimes orbiculate, flat, 2–7 × 1–3 mm, margins slightly to loosely revolute, abaxial surface mostly exposed, adaxial surface usually glabrous, sometimes glabrate. **Inflorescences** usually compound, sometimes simple dichasia, sometimes solitary flowers. **Flowers:** calyx 2.5–4.5 mm, lobes 5, 0.5–1 mm; petals 5, pink to violet, oblong-oblanceolate to spatulate, 2.5–5.2 mm; stamens 6, included to ± exserted, 1.7–3.5 mm; anthers yellow; style included to ± exserted, 3-branched; ovary 3-carpellate; ovules 25–60, attached along sutures, funiculi erect. **Seeds** 20–60 per capsule, oblong-ellipsoid, 0.4–0.7 mm. $2n = 20$.

Flowering May–Sep. Ballast, moist, saline soil; 0–1300 m; introduced; Calif., Mass., N.J., Oreg., Utah; Eurasia; Africa; introduced also in South America, Australia.

In the United States, *Frankenia pulverulenta* is introduced and has been rarely collected on the east and west coasts and near Salt Lake City, Utah, where recent attempts to relocate it were unsuccessful (N. H. Holmgren 2005c).

5. **Frankenia salina** (Molina) I. M. Johnston, Contr. Gray Herb. 70: 92. 1924 • Alkali heath or seaheath, yerba reuma [F]

Ocimum salinum Molina, Sag. Stor. Nat. Chili, 139. 1782 (as Ocymum); *Frankenia grandifolia* Chamisso & Schlechtendal; *F. grandifolia* subsp. *campestris* (A. Gray) A. E. Murray; *F. grandifolia* var. *campestris* A. Gray

Subshrubs, sprawling, 1.2–6 dm; branches glabrous, or puberulous or short-pilose, hairs erect or suberect. **Leaves:** petiole (0.5–)0.7–2 (–2.5) mm, not markedly tapering toward blade, apex ± similar in width to or narrower than base of blade;

blade grayish yellow-green, usually obovate to narrowly oblanceolate, oblong-elliptic to linear, sometimes ovate or lanceolate, flat, 3–13.5 × 0.9–6(–7) mm, margins slightly to loosely revolute, abaxial surface mostly exposed, adaxial surface scattered-hairy to puberulous. **Inflorescences** simple or compound dichasia or solitary flowers. **Flowers:** calyx (4–)4.5–8.5(–9) mm, lobes (4)5(6), 0.5–1.5(–2) mm; petals (4)5(6), white to dark pink or blue-purple, spatulate to narrowly oblong-oblanceolate, 5.5–11(–11.5) mm; stamens (4–)6(–8), exserted, 5–11.5 mm; anthers yellowish to dark red or purple-red; style exserted, 3(4)-branched; ovary 3(4)-carpellate; ovules 8–26, attached along sutures, funiculi usually erect, sometimes curved. **Seeds** 1–20 per capsule, ellipsoid to ovoid-ellipsoid, 0.8–1.5 mm.

Flowering Apr–Nov. Saline and alkaline soil, salt marshes, alkali flats; 0–800 m; Calif., Nev.; Mexico (Baja California, Baja California Sur, Coahuila, San Luis Potosí, Sonora); South America (Chile); introduced in Pacific Islands (Hawaii).

Frankenia salina, as circumscribed here, occurs in the United States, Mexico, and Chile. The North American and South American populations, which previously had been recognized as *F. grandifolia* and *F. salina* respectively, are now considered conspecific (M. A. Whalen 1987). In North America, *F. salina* occurs in California, Nevada, and Mexico (Baja California and Sonora, with some highly disjunct populations in Coahuila and San Luis Potosí). In California, it occurs in both coastal and inland areas and is common in coastal salt marshes. The inland and coastal populations have sometimes been recognized as *F. grandifolia* var. *campestris* and var. *grandifolia*, respectively. Plants from inland regions tend to have proportionally narrower leaves and shorter, lighter-colored petals than those from coastal regions; there is considerable variation in these traits within regions (Whalen) and these infraspecific taxa are not recognized here. *Frankenia salina* was introduced on Tern Island in Hawaii, where it was collected in 1968 (*Herbst 1217*, BISH); W. L. Wagner et al. (1990) noted that this population did not survive a 1969 storm.

TAMARICACEAE Link

• Tamarisk Family

John F. Gaskin

Shrubs or trees [subshrubs], usually halophytes, rheophytes, or xerophytes. **Leaves** alternate, scalelike [subulate], small; stipules absent. **Inflorescences** simple or compound racemes [flowers solitary]. **Flowers** bisexual [rarely unisexual with staminate and pistillate on separate plants], small; perianth hypogynous; sepals persistent, 4 or 5, imbricate, distinct or, rarely, basally connate; petals sometimes persistent, 4 or 5, alternate with sepals; stamens 4–9[–14], usually equal or more in number to petals, distinct [connate basally or fasciculate], often attached to fleshy nectar disc; pistil (2)3 or 4(5)-carpellate; ovary 1-locular, sometimes almost plurilocular, ovules 2+ per placenta, anatropous, bitegmic; placentation parietal, basal, or parietal-basal; styles [2]3 or 4[5] [absent, stigmas sessile]. **Fruits** capsular, dehiscence loculicidal. **Seeds** comose at one end [hairy overall]; embryo straight; endosperm absent [scanty, starchy]; thin perisperm often present.

Genera 4, species ca. 80 (1 genus, 8 species in the flora): introduced; Europe, Asia, Africa, especially from Mediterranean to c Asia.

Tamaricaceae have traditionally been placed in the Violales, but recent analyses of molecular sequence data place the family within the Caryophyllales of the core eudicots. Frankeniaceae has retained its place as sister family to Tamaricaceae, sharing similarities in many characters, including secondary chemistry and salt gland structure.

SELECTED REFERENCES Cheng, Z. M., Pan H. X., and Yin L. K. 2000. Study on the phytochemistry taxonomy of *Tamarix* L. and *Myricaria* Desv. Acta Bot. Boreal.-Occid. Sin. 20: 275–282. Crins, W. J. 1989. The Tamaricaceae of the southeastern United States. J. Arnold Arbor. 70: 403–425. Gaskin, J. F. et al. 2004. A systematic overview of Frankeniaceae and Tamaricaceae from nuclear rDNA and plastid sequence data. Ann. Missouri Bot. Gard. 91: 401–409. Gupta, A. K. and Y. S. Murty. 1987. Floral anatomy in Tamaricaceae. J. Indian Bot. Soc. 66: 275–282. Niedenzu, F. 1925. Tamaricaceae. In: H. G. A. Engler et al., eds. 1924+. Die natürlichen Pflanzenfamilien, ed. 2. 26+ vols. Leipzig and Berlin. Vol. 21, pp. 282–289. Qaiser, M. 1987. Studies in the seed morphology of the family Tamaricaceae from Pakistan. Bot. J. Linn. Soc. 94: 469–484. Zhang, D. Y. et al. 2000. Systematic studies on some questions of Tamaricaceae based on ITS sequence. Acta Bot. Boreal.-Occident. Sin. 20: 421–431. Zhang, Y. M., Pan B. R., and Yin L. K. 1998. Seed morphology of Tamaricaceae in China arid areas and its systematic evolution. J. Pl. Resources Environm. 7(2): 22–27.

1. TAMARIX Linnaeus, Sp. Pl. 1: 270. 1753; Gen. Pl. ed. 5, 131. 1754 • Tamarisk [Arabic *tamr*, a tree with dark bark]　☐

Shrubs or trees. Leaves usually sessile, sessile or amplexicaul in *T. tetragyna*, or sheathing in *T. aphylla*. **Inflorescences:** flowers each subtended by 1 (rarely more) bract. **Flowers:** sepals ± connate basally; petals persistent or deciduous after anthesis, distinct, white to pink or reddish purple; stamens 4 or 5[–14], exserted, distinct, antisepalous, sometimes also antipetalous in *T. tetragyna*; ovary pear-shaped; styles 3(4), short. **Seeds** beaked. $x = 12$.

Species ca. 55 (8 in the flora): introduced; Eurasia, n, s Africa; introduced also in Mexico, West Indies, South America, Pacific Islands (Hawaii, New Zealand), Australia.

Tamarix has been introduced to many areas through horticultural specimens, which are still available in some areas of North America. The species commonly planted include all those treated here except *T. tetragyna*. The extensive use of *T. parviflora* in gardens throughout North America has led to an abundance of collections in herbaria, and thus the species has been reported as widespread in various works (see J. T. Kartesz and C. A. Meacham 1999). For this treatment, only naturalized plants were considered, which greatly reduced the reported distributions of some species. Additionally, many midwestern and eastern populations of *Tamarix* recorded from states such as Massachusetts and Missouri have not been found again for decades and are thought to no longer exist.

Identification of some *Tamarix* species is relatively simple based on leaf morphology (for example, leaves sheathing the stems in *T. aphylla* compared to sessile or amplexicaul leaves in all other species in the flora) or floral morphology (for example, four petals in *T. parviflora* compared to usually five in all other species in the flora). Sterile collections have led to many unidentified specimens, and the most common misidentifications are among the pentamerous species *T. canariensis*, *T. chinensis*, *T. gallica*, and *T. ramosissima*, which require careful study of the nectar disc and other minute characters. The following suggestions for dissection and clarifications of terminology may help. Soak dried flowers in hot water containing a drop of liquid soap. This will help keep parts pliable and complete during dissection. Under a dissecting scope, hold the pedicel firmly with tweezers and with other tweezers pull the pistil out of the flower. At this point, if the petals are not blocking the view, the nectar disc should be visible. Nectar disc lobes will be either confluent with the filaments (synlophic) or alternate with the filaments (hololophic) (see illustrations of *T. gallica* and *T. ramosissima* respectively). To inspect filament insertion (a character that distinguishes *T. chinensis* from *T. ramosissima*) it is necessary to remove floral parts from beneath the nectar disc and look at the lower surface of the disc. Filaments originate either from below the nectar disc, sometimes near the margin (hypodiscal insertion) or from the edge of the disc (peridiscal insertion).

Because the character states distinguishing *Tamarix chinensis* and *T. ramosissima* are at times difficult to determine and often are not satisfactorily unequivocal throughout a plant, various authors have, not unreasonably, synonymized these species in North American treatments (see S. L. Welsh et al. 1993; K. W. Allred 2002). The difficulty in distinguishing between these two species is probably due to their high rate of hybridization in North America (J. F. Gaskin and B. A. Schaal 2002). In Asia these two species were originally thought to come from overlapping regions, with *T. ramosissima* ranging from Turkey to Korea, and *T. chinensis* ranging from western China to Japan (B. R. Baum 1978). DNA sequences indicate that the two species are genetically distinct, and that there is no geographic overlap of the distinguishing genotypes, *T. ramosissima* being found west of central China, and *T. chinensis* to the east of central China.

There were no hybrid combinations of these species-specific genotypes found anywhere in Asia, thus the hybrids found in North America may be novel combinations produced since their introduction (Gaskin and Schaal).

Tamarix aralensis Bunge is most likely known in the flora area only from garden collections and is not naturalized in North America.

Various *Tamarix* species have been distributed horticulturally for their ornamental and erosion-control properties, especially in xeric habitats. Galls on certain species are a source of tannins, and some taxa are utilized in basket weaving and as firewood. *Tamarix mannifera* Ehrenberg ex Bunge (Egypt to Jordan) is a source of a so-called manna, a white, resinous exudate from twigs caused by insect damage that is still collected by Bedouins in the desert.

SELECTED REFERENCES Allred, K. W. 2002. Identification and taxonomy of *Tamarix* (Tamaricaceae) in New Mexico. Desert Pl. 18(2): 26–32. Baum, B. R. 1964. On the vernales-aestivales character in *Tamarix* and its diagnostic value. Israel J. Bot. 13: 30–35. Baum, B. R. 1967. Introduced and naturalized tamarisks in the United States and Canada. Baileya 15: 19–25. Baum, B. R. 1978. The Genus *Tamarix*. Jerusalem. Baum, B. R., I. J. Bassett, and C. W. Crompton. 1971. Pollen morphology of *Tamarix* species and its relationship to the taxonomy of the genus. Pollen & Spores 13: 495–521. Gaskin, J. F. 2002. Systematic and Population Studies of the Invasive Plant *Tamarix*. Ph.D. dissertation. Washington University. Gaskin, J. F. and B. A. Schaal. 2002. Hybrid *Tamarix* widespread in U.S. invasion and undetected in native Asian range. Proc. Natl. Acad. Sci. U.S.A. 99: 11256–11259. Gaskin, J. F. and B. A. Schaal. 2003. Molecular phylogenetic investigation of U.S. invasive *Tamarix*. Syst. Bot. 28: 86–95. Gaskin, J. F. and P. B. Shafroth. 2005. Hybridization of *Tamarix ramosissima* and *T. chinensis* (saltcedars) with *T. aphylla* (athel) (family Tamaricaceae) in the southwestern USA determined from DNA sequence data. Madroño 52: 1–10. Horton, J. S. 1964. Notes on the Introduction of Deciduous *Tamarix*. Fort Collins, Colo. [U.S.D.A. Forest Serv., Gen. Techn. Rep. RM-16.] Johri, B. M. and D. Kak. 1954. The embryology of *Tamarix* L. Phytomorphology 4: 230–247. Robinson, T. W. 1965. Introduction, Spread, and Aerial Extent of Saltcedar (*Tamarix*) in the Western States. [U.S. Geol. Surv., Profess. Pap. 491-A.] Sher, A. and M. F. Quigley, eds. 2013. *Tamarix*: A Case Study of Ecological Change in the American West. Oxford and New York. Zohary, M. and B. R. Baum. 1965. On the androecium of the *Tamarix* flower and its evolutionary trends. Israel J. Bot. 14: 101–111.

1. Leaves sheathing . . . . .2. *Tamarix aphylla*
1. Leaves sessile or amplexicaul.
    2. Antipetalous stamens 1–4 . . . . 8. *Tamarix tetragyna*
    2. Antipetalous stamens absent.
        3. Flowers 4-merous . . . . 6. *Tamarix parviflora*
        3. Flowers 5-merous.
            4. Nectar disc lobes alternate with filaments.
                5. Sepal margins entire; some or all filaments originating from below nectar disc . . . . .4. *Tamarix chinensis*
                5. Sepal margins denticulate; all filaments originating from edge of nectar disc . . . . 7. *Tamarix ramosissima*
            4. Nectar disc lobes confluent with filaments.
                6. Racemes 5–9 mm wide; petals 2–3 mm . . . . 1. *Tamarix africana*
                6. Racemes 4–5 mm wide; petals 1.2–2 mm.
                    7. Sepal margins denticulate; petals obovate, 1.2–1.5 mm . . . . 3. *Tamarix canariensis*
                    7. Sepal margins entire to subentire; petals elliptic to ovate, 1.5–2 mm . . .5. *Tamarix gallica*

1. Tamarix africana Poiret, Voy. Barbarie 2: 139. 1789

  • African tamarisk ☐

Shrubs or trees, to 5 m. Leaves: blade lanceolate, 1.5–2.5 mm. Inflorescences 3–7 cm × 5–9 mm; bract exceeding pedicel, rarely exceeding calyx tip. Flowers 5-merous; sepals 1.5 mm, margins subentire; petals ovate, 2–3 mm; antisepalous stamens 5, filaments confluent with nectar disc lobes, all originating from edge of disc. $2n = 24$.

Flowering spring–late summer. Ocean shorelines, riverways, waste areas, sandy soil; 0–300 m; introduced; La., S.C., Tex., Va.; s Europe; n Africa; introduced also in w Europe (England).

2. **Tamarix aphylla** (Linnaeus) H. Karsten, Deut. Fl.,
   641. 1882 • Athel tamarisk F I W

Thuja aphylla Linnaeus, Cent. Pl. I,
32. 1775

Trees, to 10+ m. **Leaves**
sheathing; blade abruptly
pointed, 2 mm. **Inflorescences**
3–6 cm × 4–5 mm; bract
exceeding pedicel, not reaching
calyx tip. **Flowers** 5-merous;
sepals 1–1.5 mm, margins
entire; petals oblong to elliptic, 2–2.5 mm; antisepalous
stamens 5, filaments alternate with nectar disc lobes, all
originating from edge of disc. $2n = 24$.

Flowering late summer–early winter. Lakeshores,
riverways, sandy soil; 0–600 m; introduced; Ariz., Calif.,
Nev., Tex., Utah; sw Asia; n Africa; introduced also in
Mexico (Baja California, Coahuila), Australia.

Tamarix aphylla forms hybrids (rarely) with
T. chinensis and *T. ramosissima*.

3. **Tamarix canariensis** Willdenow, Abh. Königl. Akad.
   Wiss. Berlin 1812–1813: 79. 1816 • Canary Island
   tamarisk I

Shrubs or trees, to 5 m. **Leaves:**
blade lanceolate, 1.5–2.5 mm.
Inflorescences 1.5–5 cm × 4–5
mm; bract reaching or exceeding
calyx tip. **Flowers** 5-merous;
sepals 1.5–1.8 mm, margins
denticulate; petals obovate,
1.2–1.5 mm; antisepalous
stamens 5, filaments confluent
with nectar disc lobes, all originating from edge of disc.

Flowering spring–summer. Ocean shorelines,
riverways, sandy soil; 0–300 m; introduced; Ga., La.,
N.C., S.C., Tex.; s Europe (Sardinia, Sicily); n Africa
(Tunisia); Atlantic Islands (Canary Islands).

Tamarix canariensis is morphologically very similar
to and may form hybrids with *T. gallica*.

4. **Tamarix chinensis** Loureiro, Fl. Cochinch. 1: 182.
   1790 • Saltcedar, five-stamen tamarisk I W

Tamarix juniperina Bunge

Shrubs or trees, to 8 m.
Leaves: blade lanceolate to
ovate-lanceolate, 1.5–3 mm.
Inflorescences 2–6 cm × 5–7
mm; bract reaching or exceeding
pedicel, not exceeding calyx tip.
Flowers 5-merous; sepals 0.5–1.5
mm, margins entire; petals
elliptic to ovate, 1.5–2 mm; antisepalous stamens 5,
filaments alternate with nectar disc lobes, some or all
originating from below disc. $2n = 24$.

Flowering early spring–fall. Riverways, lakeshores,
arroyos; 0–2500 m; introduced; Ariz., Calif., Colo.,
Idaho, Mont., Nev., N.Mex., Okla., Tex., Utah, Wyo.;
e Asia; introduced also in South America (Argentina).

Tamarix chinensis, morphologically very similar
to *T. ramosissima*, hybridizes with *T. ramosissima*
(commonly) and *T. aphylla* (rarely).

5. **Tamarix gallica** Linnaeus, Sp. Pl. 1 : 270. 1753
   • French tamarisk I W

Shrubs or trees, to 5 m. **Leaves:**
blade lanceolate, 1.5–2 mm.
Inflorescences 2–5 cm × 4–5
mm; bract exceeding pedicel,
not reaching calyx tip. **Flowers**
5-merous; sepals 0.5–1.5 mm,
margins entire or subentire;
petals elliptic to ovate, 1.5–2
mm; antisepalous stamens 5,
filaments confluent with nectar disc lobes, all originating
from edge of disc. $2n = 24$.

Flowering spring–early fall. Ocean shorelines,
riverways, sandy soil; 0–300 m; introduced; Ark., Calif.,
Ga., La., N.Mex., Okla., S.C., Tex.; s Europe; introduced
also in Mexico (Chihuahua, Durango, Sinaloa),
South America (Argentina).

Tamarix gallica is morphologically very similar to
and may form hybrids with *T. canariensis*.

6. **Tamarix parviflora** de Candolle in A. P. de Candolle
   and A. L. P. P. de Candolle, Prodr. 3: 97. 1828
   • Small-flower tamarisk F I W

Shrubs or trees, to 5 m. **Leaves:**
blade lanceolate, 2–2.5 mm.
Inflorescences 1.5–4 cm × 3–5
mm; bract exceeding pedicel,
not reaching calyx tip. **Flowers**
4-merous; sepals 1–1.5 mm,
margins entire or denticulate;
petals oblong to ovate, 2 mm;
antisepalous stamens 4, fila-
ments confluent with nectar disc lobes, all originating
from edge of disc. $2n = 24$.

Flowering early spring–early summer. Riverways,
lakeshores; 0–1500 m; introduced; Ariz., Calif., Colo.,
Kans., Miss., Nev., N.Mex., N.C., Okla., Oreg., Tex.,
Utah, Wash.; s Europe; n Africa; introduced also in
Mexico (Baja California), South America (Argentina),
Australia.

The name *Tamarix tetrandra* Pallas has been
misapplied to *T. parviflora*.

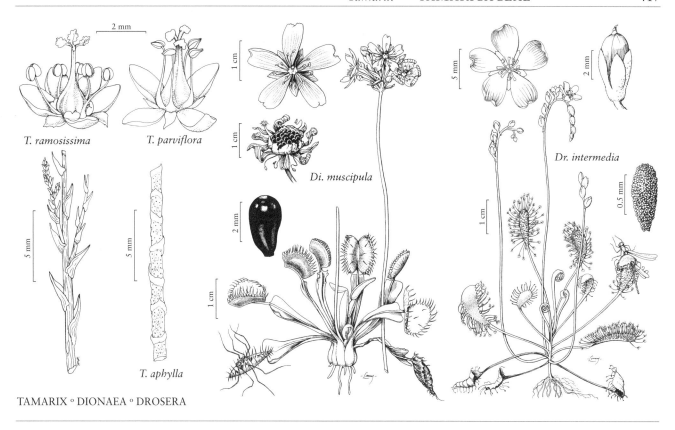

T. ramosissima

T. parviflora

T. aphylla

Di. muscipula

Dr. intermedia

TAMARIX ° DIONAEA ° DROSERA

7. Tamarix ramosissima Ledebour, Fl. Altaica 1: 424. 1829 [F] [I] [W]

Tamarix odessana Steven ex Bunge

Shrubs or trees, to 8 m. **Leaves:** blade lanceolate, 1.5–3.5 mm. **Inflorescences** 1.5–7 cm × 3–4 mm; bract exceeding pedicel, not reaching calyx tip. **Flowers** 5-merous; sepals 0.5–1.5 mm, margins denticulate; petals obovate to elliptic, 1.5–2 mm; antisepalous stamens 5, filaments alternate with nectar disc lobes, all originating from edge of disc. **2*n* = 24.**

Flowering early spring–late fall. Riverways, lakeshores, arroyos; 0–2500 m; introduced; Ariz., Ark., Calif., Colo., Idaho, Kans., La., Mont., Nebr., Nev., N.Mex., N.C., N.Dak., Okla., Oreg., S.Dak., Tex., Utah, Wash., Wyo.; Asia; introduced also in Mexico (Baja California, Baja California Sur, Chihuahua, Coahuila, Durango, Sinaloa, Sonora), South America (Argentina), Australia.

Morphologically very similar to *Tamarix chinensis*, *T. ramosissima* hybridizes with *T. chinensis* (commonly) and *T. aphylla* (rarely).

8. Tamarix tetragyna Ehrenberg, Linnaea 2: 258. 1827 [I]

Shrubs or trees, to 4 m. **Leaves** sessile or amplexicaul; blade lanceolate, 1.5–6 mm. **Inflorescences** 2–15 cm × 5–10 mm; bract exceeding pedicel, sometimes reaching calyx tip. **Flowers** 4- or 5-merous; sepals 2 mm, margins entire or denticulate; petals obovate to ovate, 2–5 mm; antisepalous stamens 4 or 5, filaments confluent with nectar disc lobes, antipetalous stamens 1–4, smaller, filaments alternate with nectar disc lobes, all originating from edge of disc.

Flowering spring(–summer). Coastal areas; 0–300 m; introduced; Ga.; sw Asia; ne Africa.

DROSERACEAE Salisbury

• Sundew Family

T. Lawrence Mellichamp

Herbs, annual or perennial, carnivorous, scapose. **Leaves** in basal rosettes (alternate-cauline in *Drosera intermedia*); stipulate or estipulate; petiolate; blade infolded or circinate in vernation, modified as hinged, jawlike trap or bearing mucilage-tipped, irritable, multicelled hairs. **Inflorescences** terminal, umbel-like cymes or lateral, circinate or scorpioid cymes, multiflowered (rarely 1-flowered). **Flowers:** perianth and androecium hypogynous; sepals 5, distinct or connate basally; petals 5, distinct; stamens usually 5 or (10–)15(–20), distinct or sometimes connate basally; pistils 1, compound, 3–5-carpellate; ovary superior, 1-locular; placentation basal or parietal; styles 1 and undivided or 3[5] and 2-fid; stigma plumose or capitate. **Fruits** capsular.

Genera 4, species ca. 175 (2 genera, 9 species in the flora): nearly worldwide.

Droseraceae comprise carnivorous plants with an unusual, worldwide distribution. They live mostly in sunny, low-nutrient, moist-to-wet acidic sands, clays, seeps, and peat bogs, often subjected to periodic fires. They catch rather small prey in jawlike traps (*Aldrovanda* Linnaeus and *Dionaea*) or on sticky glandular hairs (*Drosera* and *Drosophyllum* Link). All genera, except *Drosera*, are monospecific and sometimes have been placed in separate families for various reasons. F. Rivadavia et al. (2003) indicated that Droseraceae are monophyletic including these three genera but excluding *Drosophyllum*; K. M. Cameron et al. (2002) believed that the Old World *Aldrovanda* is more closely related to *Dionaea*, and both genera are relicts of a more widespread distribution based on fossil pollen records.

Some species of Droseraceae are grown worldwide as ornamental bog garden or terrarium specimens and as such have been given formal or informal cultivar names. Species of *Drosera* have been artificially hybridized for horticultural purposes; in the wild the rare hybrids that do occur are normally sterile.

SELECTED REFERENCES Williams, S. E., V. A. Albert, and M. W. Chase. 1994. Relations of the Droseraceae: A cladistic analysis of *rbc*L sequence and morphological data. Amer. J. Bot. 81: 1027–1037. Wood, C. E. Jr. 1960. The genera of Sarraceniaceae and Droseraceae in the southeastern United States. J. Arnold Arbor. 41: 152–163.

1. DIONAEA Solander ex J. Ellis, St. James's Chron. Brit. Eve. Post 1172: [4]. 1768

• Venus's-flytrap, flytrap, meadow-clam, tippitiwitchet [Greek *Dionaea*, daughter of Dione, referring to Aphrodite, goddess of beauty, whose Roman name was Venus, alluding to appearance of leaves and flowers]

Plants perennial, evergreen; scapes from bulblike rhizomes encased in fleshy petiole bases; leaves persisting, not forming overwintering buds (hibernaculae). **Roots** adventitious, white, unbranched. **Leaves:** stipules absent; petiole winged; blade greenish or bright red adaxially, of 2 subreniform, hinged lobes, margins with stout bristles (hinged along midrib, adaxial surfaces with 3 trigger hairs causing lobes to snap shut on prey when properly stimulated). **Inflorescences** umbel-like cymes. **Flowers:** petals marcescent, white; stamens (10–)15(–20), distinct; gynoecium 5-carpellate; style 1, not divided; stigma plumose. **Capsules** ovoid, opening irregularly. **Seeds** (20–)25(–30). $x = 16$.

Species 1: United States; introduced elsewhere.

Dionaea has been placed in its own family, Dionaeaceae Rafinesque. This is not without merit, as virtually all of its diagnostic features are different from those of *Drosera*. Molecular analysis places it in the Droseraceae in the narrow sense (K. M. Cameron et al. 2002; S. E. Williams et al. 1994; F. Rivadavia et al. 2003).

SELECTED REFERENCES Hodick, D. and A. Sievers. 1988. On the mechanism of trap closure of Venus flytrap (*Dionaea muscipula* Ellis). Planta 179: 32–42. Roberts, P. R. and H. J. Oosting. 1958. Responses of Venus fly trap (*Dionaea muscipula*) to factors involved in its endemism. Ecol. Monogr. 28: 193–218. Williams, S. E. and A. B. Bennett. 1982. Leaf closure in the Venus flytrap: An acid growth response. Science, ser. 2, 218: 1120–1122.

1. **Dionaea muscipula** J. Ellis, St. James's Chron. Brit. Eve. Post 1172: [4]. 1768 E F

Rosettes often clumping in older specimens; whole plant black upon drying. **Leaves:** petiole 2–7(–10) × 0.2–1(–2) cm, longer in shade-grown plants; blade lobes 1.2–2.5(–3) × 1–2 cm, marginal bristles 0.3–0.8 cm, rarely larger in all respects. **Inflorescences:** scape (15–)18–26 (–43) cm; bracts subtending pedicels 5 or 6, linear to lanceolate, 0.3–0.7 × 0.1–0.2 cm. **Flowers:** sepals green, lanceolate, 5–6(–10) × 2–3 mm; petals obovate to obspatulate, 10–12 × 5–6 mm, base cuneate; stigma papillate. **Capsules** 3–4 × 4–6 mm. **Seeds** shiny black, obovoid, 1–1.5 mm. $2n = 32$.

Flowering late May–Jun. Moist, sandy-peaty soil and sphagnum mats of roadsides, ditch banks, borrow-pits, longleaf pine savannas; 0–100 m; Ala., Calif., Fla., N.J., N.C., Pa., S.C., Va., Wash.

Dionaea muscipula has occurred historically in the outer Coastal Plain from Beaufort County, North Carolina, to Charleston County, South Carolina, and inland as far as Moore County, North Carolina. The present native range of the species is much reduced due to drainage, habitat conversion, fire exclusion, and development. It has been introduced into southern Alabama, northern California, eastern Pennsylvania, and Caroline County, Virginia, where it is doubtfully truly naturalized; it is reportedly naturalized in the Apalachicola region of Florida, southern New Jersey, and Skagit County, Washington.

Dionaea muscipula is one of the world's most famous plants. C. Darwin (1875) studied its actions extensively. The trigger hairs must be touched twice in close succession for the traps to snap shut (in less than a second), the bristles helping to keep perky prey from escaping. This double stimulation helps prevent inorganic debris and casual rain drops from triggering trap closure. When lightly closed, the two trap-lobes come together and appear convex, opening again the next day if no prey has been caught. If a protein meal is detected, the lobes will tighten completely, become slightly concave, squeeze the victim, and secrete digestive enzymes. After the prey is digested, the traps grow themselves open again, expose the indigestible carcass, and remain viable for some time. New leaf-traps are constantly formed (they curl open in a somewhat circinate fashion) as old ones die and turn black. Much research has been conducted on the mechanisms of stimulation and trap closure (B. E. Juniper et al. 1989).

There are many interesting accounts of encounters and explanations for flytraps in the older literature (F. E. Lloyd 1942) and the traps continue to be fascinating to old and young alike. Not the least of their mysterious features is their narrow distribution, which may be related to limited seed dispersal, unusual soil type (nutrient deficient), relatively high annual rainfall, and periodic fire.

In southeastern North Carolina and adjacent South Carolina *Dionaea muscipula* grows in association with no fewer than 16 other species of carnivorous plants, the highest diversity of carnivorous plants anywhere in the world. While usually locally abundant, the Venus's-flytrap is very sensitive to shading and drought and may go dormant during dry periods or if its habitat does not burn every five years or so. *Dionaea* requires sites with saturated but not inundated soils and strong sunlight, especially ecotones between pocosins and longleaf pine-dominated savannas or sandhills. Roadside populations of thousands of plants are managed for seed collection for nursery-grown specimens.

It is curious that the leaf petioles can vary from relatively long to short, apparently under genetic as well as environmental influences, and the adaxial surface of the traps may be bright red or greenish. Strong sunlight sometimes brings out the red color, but not always. Variants have been found in cultivation in which the entire plant is a rich red-burgundy color. There are also cultivars with variations in the lengths of the marginal bristles. The plants are easy to grow as long as fresh water low in solutes is available to keep the soil moist, and there is abundant bright light. They do not make good windowsill house plants.

2. DROSERA Linnaeus, Sp. Pl. 1: 281. 1753; Gen. Pl. ed. 5, 136. 1754 • Sundew, catch-fly, dew-threads, droséra, rossolis [Greek *droseros*, dewy, alluding to glistening glandular trichomes on leaves]

Plants annual or perennial [rarely subshrubs], deciduous, stems 1–2 cm (except also caulescent stems to 8(–20) cm in *D. intermedia*), usually forming over-wintering buds (hibernaculae). **Roots** adventitious, dark brown, fibrous. **Leaves:** stipules present (sometimes reduced to hairs) or absent; petiole present but, in some, not differentiated from blade; blade usually red on both surfaces in strong sunlight, greener in shade (except *D. tracyi*, which lacks red pigment even in full sun), unlobed, suborbiculate, orbiculate, spatulate, or obovate, or cuneate to linear or filiform, margins without stout bristles, surfaces with red, reddish purple, or pale green glandular trichomes that fold over prey when properly stimulated, in some species the blades fold as well. **Inflorescences** circinate or scorpioid cymes. **Flowers:** petals marcescent, white, pink, or rose to pinkish lavender; stamens 5, usually connate basally; gynoecium 3-carpellate; styles 3, deeply 2-fid; stigma capitate. **Capsules** obovoid, splitting between placentae. **Seeds** 20–70, minute. $x = 10$.

Species ca. 170 (8 in the flora): nearly worldwide.

Species of *Drosera* are concentrated in Latin America, South Africa, Madagascar, Australia, and New Zealand.

Droseras, like all carnivorous plants, have leaves that attract, catch, digest, and absorb nutrients from small, mostly arthropod prey. They are characterized by gland-tipped multicelled hairs that move in response to stimuli and that catch and appress prey to the leaf blade, where sessile glands secrete enzymes that dissolve the soft tissues. The released nutrients enhance growth by supplementing those available from the poor soils where they grow.

All species of *Drosera* are capable of moving their trichomes in response to contact with digestible prey. According to C. Darwin (1875), this movement can be induced by the mere touch of a part of a small insect with a single trichome. Besides having trichome movement, some species are able to curl their leaf blades to various degrees in order to maximize contact with prey.

Some species of *Drosera* may act as annuals, especially if the habitats dry out. The plants can be locally abundant. In most species, the flowers open only in the mornings on sunny days, or not at all on overcast days, and fruits may form from self-pollination. Some species, notably *D. intermedia*, may exhibit vegetative proliferation, portions of the flowers developing into leaves or plantlets. Some species form over-wintering buds called hibernaculae, requiring a cold period to break dormancy.

Some species of *Drosera* are reportedly utilized in herbal medicines to produce cough preparations and treat lung and skin ailments.

F. E. Wynne (1944) showed that seeds of North American *Drosera* are diagnostic for each species. The following key is adapted from various sources, and the species are presented in alphabetical order. Natural hybrids are rare in *Drosera*, and usually are sterile.

SELECTED REFERENCES Rivadavia, F. et al. 2003. Phylogeny of the sundews, *Drosera* (Droseraceae), based on chloroplast *rbc*L and nuclear 18S ribosomal DNA sequences. Amer. J. Bot. 90: 123–130. Shinners, L. H. 1962d. *Drosera* (Droseraceae) in the southeastern United States: An interim report. Sida 1: 53–59. Sorrie, B. A. 1998b. Distribution of *Drosera filiformis* and *D. tracyi* (Droseraceae): Phytogeographic implications. Rhodora 100: 239–260. Wynne, F. E. 1944. *Drosera* in eastern North America. Bull. Torrey Bot. Club 71: 166–174.

1. Leaf blades filiform, not differentiated from petioles; stem base bulbose-cormose (from expanded petiole base); petals 7–17(–20) mm.
    2. Petals 7–10(–12) mm; leaf blades 8–25(–30) cm × 1 mm, glandular trichomes red to reddish purple, drying dark brown; scapes 6–26 cm . . . . .4. *Drosera filiformis*
    2. Petals 12–17(–20) mm; leaf blades 30–50 cm × 1–2 mm, glandular trichomes pale green, drying pale greenish brown; scapes 25–60 cm. . . . . 8. *Drosera tracyi*
1. Leaf blades linear or suborbiculate to obovate, elongate-spatulate, or cuneate, usually differentiated from petioles; stem base not bulbose-cormose; petals 3–7(–8) mm.
    3. Scapes stipitate-glandular; stipules absent or reduced to minute hairs; seeds black, crateriform . . . . 2. *Drosera brevifolia*
    3. Scapes glabrous; stipules present; seeds reddish brown, light brown, brown, or black, striate, areolate, papillose, crateriform, or ridged.
        4. Leaf blades suborbiculate, (broader than long); seeds finely longitudinally striate . . . . 7. *Drosera rotundifolia*
        4. Leaf blades linear, orbiculate, or spatulate to obovate or elongate-spatulate (longer than broad); seeds areolate, crateriform, or papillose.
            5. Stipules adnate to petioles.
                6. Leaf blades obovate to elongate-spatulate; seeds fusiform, striate-areolate, 1–1.5 mm . . . . 1. *Drosera anglica*
                6. Leaf blades linear; seeds rhomboidal or oblong-obovoid, crateriform, 0.5–0.8 mm . . . . 6. *Drosera linearis*
            5. Stipules free from petioles or essentially so.
                7. Plants always rosulate; petioles flat, sparsely glandular-pilose; petals usually pink, 6–7 mm; seeds coarsely papillose-corrugated, 0.4–0.5 mm . . . . .3. *Drosera capillaris*
                7. Plants rosulate when young, developing leafy stems 1–8(–20) cm; petioles filiform, glabrous; petals white, 3–6 mm; seeds uniformly papillose, 0.7–1 mm . . . . 5. *Drosera intermedia*

1. Drosera anglica Hudson, Fl. Angl. ed. 2, 1: 135. 1778
 • Great or English sundew, droséra d'Angleterre

Plants forming winter hibernaculae, rosettes 2–6 cm diam.; stem base not bulbous-cormose. **Leaves** erect; stipules entirely adnate to petioles, 5 mm, margins fimbriate along distal ½; petiole differentiated from blade, 3–7 cm, glabrous or sparsely glandular-hairy; blade obovate to elongate-spatulate, 1.5–3.5(–5) cm × 3–7 mm. **Inflorescences** 1–12-flowered; scapes 3–25 cm, glabrous. **Flowers** 8–10 mm diam.; sepals connate basally, oblong, 5–6 × 4–5 mm, minutely glandular-denticulate; petals usually white, rarely pinkish, spatulate, 5–6 × 2–3.5 mm. **Capsules** 4–6 mm, minutely tuberculose. **Seeds** black, sigmoid-fusiform, 1–1.5 mm, length 1–2 times width, longitudinally striate-areolate. $2n = 40$.

Flowering Jun–Aug. Marly shores, fens, drainage tracks in peat bogs; 10–2600 m; Alta., B.C., Man., N.B., Nfld. and Labr., N.W.T., Ont., Que., Sask., Yukon; Alaska, Calif., Colo., Idaho, Maine, Mich., Minn., Mont., Oreg., Wash., Wis., Wyo.; Eurasia; Pacific Islands (Hawaii).

Drosera anglica is a boreal species that occurs on calcareous substrates. It often grows with *D. rotundifolia* in peat bogs, and with *D. linearis* and *D. rotundifolia* in marl fens, especially in the Great Lakes region. C. E. Wood Jr. (1955) presented a strong case for the hybrid origin of *D. anglica*, suggesting that it arose as a fertile amphiploid hybrid between *D. rotundifolia* and *D. linearis*. It is the only tetraploid North American species of *Drosera* with $2n = 40$. The sterile hybrid *D. rotundifolia* × *D. linearis* may be found whenever these species grow together. To avoid confusion, and because a formal name for the sterile hybrid has not been published, it should not be called *Drosera ×anglica* Hudson, as is commonly done. According to D. E. Schnell (2002), the fertile species may be distinguished from the sterile hybrid by its wider flowers (8–10 mm versus 6–7 mm) and wider scapes (1.5–2 mm versus 1–1.2 mm).

The hybrid between *Drosera rotundifolia* and *D. anglica* is a sterile triploid, and has been formally named *Drosera ×obovata* Mertens & W. D. J. Koch.

Because *Drosera longifolia* Linnaeus cannot be convincingly typified and has been so often used for plants of *D. anglica* and *D. intermedia* in the literature (F. E. Wynne 1944), the name *D. longifolia* has been rejected as ambiguous.

Drosera anglica is found in the Aalakai Swamp at 1500–2000 meters on the Hawaiian island of Kauai (perhaps brought by migrating birds from Alaska); it is otherwise found in cold northern climates.

D. E. Schnell (2002), who has grown the Kauai plants from seed, postulated that the high elevation provides a cooler temperature, and noted that plants from there do not form winter hibernaculae but only smaller winter rosettes.

SELECTED REFERENCE Wood, C. E. Jr. 1955. Evidence for the hybrid origin of *Drosera anglica*. Rhodora 57: 105–130.

2. Drosera brevifolia Pursh, Fl. Amer. Sept. 1: 211. 1813
 • Shortleaf or dwarf sundew

Drosera annua E. L. Reed;
D. leucantha Shinners

Plants not forming winter hibernaculae, rosettes to 2 (–3.5) cm diam.; stem base not bulbous-cormose. **Leaves** prostrate; stipules absent or reduced to 1 or 2 minute hairs; petiole often not differentiated from blade, dilated distally, 0.5–1 cm, glabrous; blade cuneate, 0.4–1 cm × 5–12 mm, usually longer than petiole. **Inflorescences** 1–8-flowered; scapes (1–)4–9 cm, stipitate-glandular. **Pedicels** stipitate-glandular. **Flowers** 15 mm diam.; sepals distinct, oblong-ovate, 2.5–3.5 × 1.5–2.5 mm, stipitate-glandular; petals white to rose-pink, obovate, 4–8 × 2–3 mm. **Capsules** 3 mm. **Seeds** black, obovoid or oblong, 0.3–0.5 mm, base caudate, crateriform, pits in 10–12 rows. $2n = 20$.

Flowering Apr–May(–Dec). Moist sandy-peaty pinelands and roadsides; 0–300 m; Ala., Ark., Fla., Ga., Kans., Ky., La., Miss., N.C., Okla., S.C., Tenn., Tex., Va.; Mexico; West Indies (Cuba); Central America; South America (Brazil).

Drosera brevifolia is the smallest and perhaps the most widespread species of the genus in the Southeast. It may be rare or local throughout its range, and may act as an annual, especially if the habitat dries out. The flowers are large for the size of the plant, and the stipitate-glandular scapes, pedicels, and sepals are quite distinctive. The species is easy to grow in cultivation.

3. Drosera capillaris Poiret in J. Lamarck et al., Encycl.
 6: 299. 1804 • Pink sundew

Plants not forming winter hibernaculae, rosettes (2–)3–4 (–12) cm diam.; stem base not bulbous-cormose. **Leaves** prostrate; stipules free from or adnate to petioles to 1 mm, then breaking into setaceous segments 3–5 mm; petiole differentiated from blade, 0.6–4 cm, sparsely glandular-pilose; blade broadly spatulate to orbiculate, 0.5–1 cm × 3–5 mm, usually at least slightly longer than broad, usually shorter than petiole. **Inflorescences** 2–20-

flowered; scapes 4–20(–35) cm, glabrous. **Flowers** 10 mm diam.; sepals connate basally, oblong-elliptic, 3–4 × 1–2 mm, apex obtuse, glabrous; petals usually pink, sometimes white, obovate, 6–7 × 2–3 mm. **Capsules** 4–5 mm, longer than sepals. **Seeds** brown, ellipsoid to oblong-ovoid, asymmetric, 0.4–0.5 mm, coarsely papillose-corrugated, 14–16-ridged. $2n = 20$.

Flowering May–Aug. Sandy soil of pine flatwoods and savannas, seepage slopes, peat-sedge bogs, pocosin borders, wet, sandy ditches; 0–300 m; Ala., Ark., Del., D.C., Fla., Ga., La., Md., Miss., N.C., S.C., Tenn., Tex., Va., W.Va.; Mexico; West Indies; Central America; South America.

Drosera capillaris is the most frequently encountered species of the genus in the South in moist habitats that can support carnivorous plants, especially in fire-maintained pinelands. Plants can be quite small, or form surprisingly large and robust rosettes (to 12 cm broad) in some places along the Gulf Coast. It is disjunct from the Coastal Plain to Arkansas and Tennessee, as are several other species from coastal wetland habitats. Since the leaf blades of *D. capillaris* can be somewhat orbiculate, it may be confused with the much more northern *D. rotundifolia*, which grows more typically in sphagnum (although in its northern range it often grows on moist sand substrates), has adnate stipules, white flowers, and forms hibernaculae.

Drosera capillaris is easy to grow and often behaves as an annual.

4. Drosera filiformis Rafinesque, Med. Repos., hexade 2, 5: 360. 1808 • Threadleaf sundew, dew-threads [E]

Plants forming winter hibernaculae, rosettes 1 cm diam.; stem base bulbous-cormose, composed of expanded petiole base, 1–2 cm, woolly. **Leaves** erect; stipules adnate to petioles for entire length, 10 mm, margins fimbriate, forming woolly appearance of cormose base; petiole not differentiated from blade; blade filiform, 8–25(–30) cm × 1 mm, surfaces glandular-hairy, trichomes red to reddish purple, drying dark brown. **Inflorescences** 4–12(–24)-flowered; scapes 6–26 cm, glabrous. **Pedicels** glandular-pilose. **Flowers** 10–20 mm diam.; sepals connate basally, oblong to elliptic, 4–7 × 2–2.5 mm, glandular-pilose; petals pink to purple-lavender, broadly ovate, 7–10(–12) × 5–15 mm, apical margins erose. **Capsules** 5–6 mm. **Seeds** black, ellipsoid, abruptly caudate at both ends, 0.5–0.8 mm, coarsely crateriform, pits in 16–20 lines. $2n = 20$.

Flowering late May–Sep. Exposed shores of freshwater ponds, streamside seepage bogs or fens, interdunal swales, coastal peat bogs, roadside depressions, moist borrow pits; 0–40 m; N.S.; Conn., Del., Fla., Md., Mass., N.J., N.Y., N.C., Pa., R.I., W.Va.

Drosera filiformis is disjunct in the outer Coastal Plain, from Nova Scotia and Connecticut to Maryland, southeastern North Carolina, and Bay and Washington counties, Florida. The species has been reported from South Carolina, but no specimens from there have been seen. It is introduced in Caroline County, Virginia, Preston County, West Virginia, and possibly other localities peripheral to its range; it is short-lived in some localities. B. A. Sorrie (1998b) gave a detailed account of the distribution and habitats of *D. filiformis* as well as convincing evidence that *D. tracyi* is a distinct species.

The species is very easy to cultivate, and the cormose-based clumps multiply and may be divided while dormant.

5. Drosera intermedia Hayne, J. Bot. (Schrader) 1800(1): 37. 1801 • Narrow- or oblong- or spatulate- or spoon-leaved or water sundew, droséra intermédiaire [F]

Drosera americana Willdenow

Plants forming winter hibernaculae, rosettes 3–6 cm diam. in young plants only; stem base not bulbous-cormose; stem 1–8(–20) cm (other species do not have distinct stem). **Leaves** erect to spreading, becoming widely separated as stem elongates; stipules adnate to petioles basally 1 mm, then dividing into several setaceous segments 2–5 mm; petiole differentiated from blade, 2–5 cm, filiform, not flat as in most other species, glabrous; blade spatulate to obovate, 0.8–2 cm × 4–5 mm. **Inflorescences** (1–)9–20-flowered, arising from stem and distinctly arching outward at base before growing upwards; scapes (2–)5–20(–25) cm, glabrous. **Flowers** 7–12 mm diam.; sepals connate basally, oblong, 5–6 × 1–1.5 mm, glabrous; petals white, sometimes pink tinged, obovate, 3–6 × 3–5 mm, apical margins erose. **Capsules** 4–5 mm. **Seeds** reddish brown, oblong-obovoid, 0.7–1 mm, densely, uniformly papillose. $2n = 20$.

Flowering Jun–Aug. Wet habitats, often in several cm of water, including peat bogs, beaver ponds, and wetland margins, often on wet logs (north), wet margins of streams, ponds, bays, and ditches (south); 0–500 m; St. Pierre and Miquelon; N.B., Nfld. and Labr., N.S., Nunavut, Ont., P.E.I., Que.; Ala., Ark., Conn., Del., D.C., Fla., Ga., Idaho, Ill., Ind., La., Maine, Md., Mass., Mich., Minn., Miss., Mo., N.H., N.J., N.Y., N.C., Ohio, Pa., R.I., S.C., Tenn., Tex., Vt., Va., W.Va., Wis.; West Indies; Central America; South America; Eurasia.

Drosera intermedia is the only species of the genus with a distinct stem, scattered cauline leaves, and an arching scape. Its habit of growing in the wettest places, often in standing water, is characteristic. The stems are longer in deeper water and may be up to 20 cm in the southern part of its range. The winter hibernaculae form at the tips of the elevated stems; they will fall to the ground and sprout the following spring. The plants may reproduce vegetatively by forming plantlets in place of flower parts. The plantlets can be rooted, and in general this species is easy to grow.

M. L. Fernald (1950) recognized forma *corymbosa* (de Candolle) Fernald with a corymb instead of a raceme, and forma *natans* Heuser with floating stems up to 20 cm long.

6. **Drosera linearis** Goldie, Edinburgh Philos. J. 6: 325. 1822 • Linear-leaved sundew, droséra à feuilles linéares [E]

Plants forming winter hibernaculae; rosettes 6–15 cm diam.; stem base not bulbous-cormose. **Leaves** erect; stipules adnate to petioles their entire length, 5 mm, margins fimbriate; petiole differentiated from blade, flattened, 3–7 cm, glabrous; blade linear, 1–6 cm × 1.5–3 mm. **Inflorescences** 1–8-flowered; scapes 2–15 cm, glabrous. **Flowers** 6–8 mm diam.; sepals connate basally, oblong-elliptic, 4–5 × 2 mm, minutely glandular-denticulate; petals white, obovate, 6 × 3–4 mm. **Capsules** 4–5 mm. **Seeds** black, rhomboidal or oblong-obovoid, 0.5–0.8 mm, densely, irregularly crateriform. $2n = 20$.

Flowering Jun–Aug. Marl fens, wet, limey shores, often in a centimeter of water; 10–400 m; Alta., B.C., Man., Nfld. and Labr. (Nfld.), N.W.T., Ont., Que., Sask.; Maine, Mich., Minn., Mont., Wis.

Drosera linearis is especially frequent on marly shores and fens of the western Great Lakes in Michigan. There, it is often found with *D. anglica* and *D. rotundifolia*. *Drosera linearis* is usually found on open marl, *D. rotundifolia* on sphagnum hummocks, and *D. anglica* more often around bases of hummocks and on fen margins (D. E. Schnell 2002). This is perhaps the most difficult species to maintain in cultivation, especially in warmer climates.

7. **Drosera rotundifolia** Linnaeus, Sp. Pl. 1: 281. 1753 • Roundleaf sundew, droséra à feuilles rondes

Plants forming winter hibernaculae, rosettes (2–)4–10(–15) cm diam.; stem base not bulbous-cormose. **Leaves** erect to prostrate; stipules adnate to petioles their entire length, 4–6 mm, margins fimbriate along distal ½; petiole differentiated from blade, 1.5–5 cm, glandular-pilose; blade suborbiculate, 0.4–1 cm × 5–12 mm, broader than long, much shorter than petiole. **Inflorescences** 2–15(–25)-flowered; scapes 5–35 cm, glabrous. **Flowers** 4–7 mm diam.; sepals connate basally, oblong, 4–5 × 1.5–2 mm, glabrous; petals white or pink, spatulate, 5–6 × 3 mm. **Capsules** 5 mm. **Seeds** light brown, fusiform, 1–1.5 mm, finely and regularly longitudinally striate, with metallic sheen. $2n = 20$.

Flowering Jun–Sep. Sphagnum bogs, fens, beaver ponds, swamps, peaty gravels, sandy soil, wet sand (for example, disturbed bottoms of old sand pits, emergent sandy shorelines) in the North, lake and stream margins, sphagnous streamheads, and seeps in the South; 0–3000 m; Greenland; St. Pierre and Miquelon; Alta., B.C., Man., N.B., Nfld. and Labr., N.W.T., N.S., Nunavut, Ont., P.E.I., Que., Sask., Yukon; Alaska, Calif., Colo., Conn., Del., D.C., Ga., Idaho, Ill., Ind., Iowa, Ky., Maine, Md., Mass., Mich., Minn., Miss., Mont., N.H., N.J., N.Y., N.C., N.Dak., Ohio, Oreg., Pa., R.I., S.C., Tenn., Vt., Va., Wash., W.Va., Wis.; Eurasia; Pacific Islands (New Guinea).

Drosera rotundifolia was the carnivorous plant most studied by C. Darwin (1875). The species is circumboreal and is widespread across North America, much more common northward and rarer in the South. It is difficult to grow in warmer climates.

M. L. Fernald (1950) recognized forma *breviscapa* (Regel) Domin, found in the Canadian Maritime Provinces, with scapes 1–4 cm long and one to three flowers, and var. *comosa* Fernald, found from Gaspé County, Quebec, to New England and northern New York, with the parts of the flowers modified to green gland-bearing leaves. Dwarf, few-leaved plants found in Alaska have been called var. *gracilis* Laested.

8. **Drosera tracyi** (Diels) Macfarlane in L. H. Bailey, Stand. Cycl. Hort. 2: 1077. 1914 • Tracy's or Gulf Coast threadleaf sundew [E]

Drosera filiformis Rafinesque var. *tracyi* Diels in H. G. A. Engler, Pflanzenr. 26[IV,112]: 92. 1906

Plants forming winter hibernaculae, rosettes 1 cm diam.; stem base bulbous-cormose, from expanded petiole base, 1–2 cm, woolly. **Leaves** erect; stipules adnate to petioles their entire length, 10 mm, margins fimbriate, forming woolly appearance of cormose base; petiole not differentiated from blade; blade filiform, 30–50 cm × 1–2 mm, glandular trichomes pale green, drying pale greenish brown. **Inflorescences** 4–12(–24)-flowered; scapes 25–60 cm, glabrous. **Pedicels** glandular-pilose. **Flowers** 15–30 mm diam.; sepals connate basally, oblong to elliptic, 4–7 × 2–3 mm, glandular-pilose; petals rose, (rarely white), broadly ovate, 12–17(–20) × 15 mm, apical margins erose. **Capsules** 5–6 mm. **Seeds** black, ellipsoid, abruptly caudate-truncate at both ends, 0.5–0.8 mm, coarsely crateriform, pits in 16–20 lines. $2n = 20$.

Flowering late May–Jun. Hillside seepage bogs and ecotones between pine savannas and bay-gum-cypress wetlands, wet roadside ditches and borrow pits, shores of sinkhole ponds; 0–70 m; Ala., Fla., Ga., La., Miss.

Drosera tracyi occurs on the Gulf Coastal Plain from southwestern Georgia to southeastern Louisiana. The species has been reported from South Carolina; no specimens from there have been seen. The leaves and flowers are larger than those of *D. filiformis*. An anthocyanin-free form with white flowers is known.

B. A. Sorrie (1998b) gave evidence from morphology and ecology that *Drosera tracyi* is distinct from *D. filiformis*. *Drosera filiformis* has been found growing with *D. tracyi* in Bay and Washington counties, Florida, without apparent hybrids (Sorrie). This seems to be a natural disjunction, but *D. filiformis* has been planted outside its range. An artificial hybrid, which is sterile, between these two species has resulted in a cultivar named 'California Sunset.'

Hybrids:

Many artificial hybrids have been made and given formal or informal cultivar names. The following sterile wild hybrids have been reported.

Drosera ×*beleziana* Camus (*D. intermedia* × *D. rotundifolia*): Nova Scotia, Michigan.

Drosera ×*hybrida* Macfarlane (*D. filiformis* × *D. intermedia*): New Jersey.

Drosera ×*linglica* Kusakabe ex Gauthier & Gervais (*D. linearis* × *D. anglica*): Quebec.

Drosera ×*obovata* Mertens & Koch (*D. anglica* × *D. rotundifolia*): British Columbia, southeastern Canada, Newfoundland, Quebec, Great Lakes region, northern California, Oregon, Washington, and New England.

Drosera ×*woodii* Gauthier & Gervais (*D. linearis* × *D. rotundifolia*): Quebec.

Drosera capillaris × *D. intermedia*, no hybrid name given: Pender County, North Carolina.

Literature Cited

Robert W. Kiger, Editor

This is a consolidated list of all works cited in volume 6, whether as selected references, in text, or in nomenclatural contexts. In citations of articles, both here and in the taxonomic treatments, and also in nomenclatural citations, the titles of serials are rendered in the forms recommended in G. D. R. Bridson and E. R. Smith (1991), Bridson (2004), and Bridson and D. W. Brown (http://fmhibd.library.cmu.edu/HIBD-DB/bpho/findrecords.php). When those forms are abbreviated, as most are, cross references to the corresponding full serial titles are interpolated here alphabetically by abbreviated form. In nomenclatural citations (only), book titles are rendered in the abbreviated forms recommended in F. A. Stafleu and R. S. Cowan (1976–1988) and Stafleu et al. (1992–2009). Here, those abbreviated forms are indicated parenthetically following the full citations of the corresponding works, and cross references to the full citations are interpolated in the list alphabetically by abbreviated form. Two or more works published in the same year by the same author or group of coauthors will be distinguished uniquely and consistently throughout all volumes of *Flora of North America* by lower-case letters (b, c, d, ...) suffixed to the date for the second and subsequent works in the set. The suffixes are assigned in order of editorial encounter and do not reflect chronological sequence of publication. The first work by any particular author or group from any given year carries the implicit date suffix "a"; thus, the sequence of explicit suffixes begins with "b". Works missing from any suffixed sequence here are ones cited elsewhere in the *Flora* that are not pertinent in this volume.

Abbiw, D. K. 1990. Useful Plants of Ghana: West African Uses of Wild and Cultivated Plants. London and Kew.

Abh. Königl. Akad. Wiss. Berlin = Abhandlungen der Königlichen Akademie der Wissenschaften in Berlin.

Abrams, L. and R. S. Ferris. 1923–1960. Illustrated Flora of the Pacific States: Washington, Oregon, and California. 4 vols. Stanford. (Ill. Fl. Pacific States)

Acta Bot. Boreal.-Occid. Sin. = Acta Botanica Boreali-occidentalia Sinica. [Xi-bei Zhiwu Xuebao.]

Acta Bot. Neerl. = Acta Botanica Neerlandica.

Acta Bot. Yunnan. = Acta Botanica Yunnanica. [Yunnan Zhiwu Yanjiu.]

Acta Hort. = Acta Horticulturae; Technical Communications of I S H S (International Society for Horticultural Science).

Adams, W. P. 1962. Studies in the Guttiferae. I. A synopsis of *Hypericum* section *Myriandra*. Contr. Gray Herb. 189: 1–51.

Adams, W. P. 1962b. Studies in the Guttiferae. II. Taxonomic and distributional observations in North American taxa. Rhodora 64: 231–242.

Adams, W. P. 1972. Studies in the Guttiferae. III. An evaluation of some putative spontaneous garden hybrids in *Hypericum* section *Myriandra*. Rhodora 74: 276–282.

Adams, W. P. 1973. Clusiaceae of the southeastern United States. J. Elisha Mitchell Sci. Soc. 89: 62–71.

Adansonia = Adansonia; Recueil Périodique d'Observations Botaniques.

Aiton, W. 1789. Hortus Kewensis; or, a Catalogue of the Plants Cultivated in the Royal Botanic Garden at Kew. 3 vols. London. (Hort. Kew.)

Alain (Hno.). 1985–1997. Descriptive Flora of Puerto Rico and Adjacent Islands: Spermatophyta. 5 vols. Río Piedras.

Alexander, S. N. et al. 2012. A subspecific revision of North American saltmarsh mallow *Kosteletzkya pentacarpos* (L.) Ledeb. (Malvaceae). Castanea 77: 105–122.

Ali, M. A., S. Karuppusamy, and F. M. A. Al-Hemaid. 2010. Molecular phylogenetic study of *Luffa tuberosa* Roxb. (Cucurbitaceae) based on internal transcribed spacer (ITS) sequences of nuclear ribosomal DNA and its systematic implication. Int. J. Bioinf. Res. 2: 42–60.

Allioni, C. 1773. Auctuarium ad Synopsim Methodicam Stirpium Horti Regii Taurinensis. Turin. [Preprinted from Mélanges Philos. Math. Soc. Roy. Turin 5: 53–96. 1774.] (Auct. Syn. Meth. Stirp. Taurin.)

Allioni, C. 1785. Flora Pedemontana sive Enumeratio Methodica Stirpium Indigenarum Pedemontii. 3 vols. Turin. (Fl. Pedem.)

Allison, J. R. 2011. Synopsis of the *Hypericum denticulatum* complex (Hypericaceae). Castanea 76: 99–115.

Allred, K. W. 2002. Identification and taxonomy of *Tamarix* (Tamaricaceae) in New Mexico. Desert Pl. 18(2): 26–32.

Allred, K. W. 2008. Flora Neomexicana. Las Cruces.

Alverson, W. S. et al. 1998. Circumscription of the Malvales and relationships to other Rosidae: Evidence from *rbc*L sequence data. Amer. J. Bot. 85: 876–887.

Alverson, W. S. et al. 1999. Phylogeny of the core Malvales: Evidence from *ndh*F sequence data. Amer. J. Bot. 86: 1474–1486.

Amer. Antiquity = American Antiquity.

Amer. J. Bot. = American Journal of Botany.

Amer. J. Sci. = American Journal of Science.

Amer. J. Sci. Arts = American Journal of Science, and Arts.

Amer. Midl. Naturalist = American Midland Naturalist; Devoted to Natural History, Primarily That of the Prairie States.

Amer. Mus. Novit. = American Museum Novitates.

Amer. Naturalist = American Naturalist....

Anais Congr. Soc. Bot. Brasil = Anais do [1+] Congresso da Sociedade de Botânica do Brasil.

Anales Inst. Biol. Univ. Nac. Autón. México, Bot. = Anales del Instituto de Biológia de la Universidad Nacional Autónoma de México. Série Botánica.

Andreasen, K. 2005. Implications of molecular systematic analyses of the conservation of rare and threatened taxa: Constrasting examples from Malvaceae. Conservation Genet. 6: 399–412.

Andreasen, K. and B. G. Baldwin. 2001. Unequal evolutionary rates between annual and perennial lineages of checker mallows (*Sidalcea*, Malvaceae): Evidence from 18S-26S rDNA internal and external transcribed spacers. Molec. Biol. Evol. 18: 936–944.

Andreasen, K. and B. G. Baldwin. 2003. Reexamination of relationships, habital evolution, and phylogeography of checker mallows (*Sidalcea*, Malvaceae) based on molecular phylogenetic data. Amer. J. Bot. 90: 436–444.

Andreasen, K. and B. G. Baldwin. 2003b. Nuclear ribosomal DNA sequence polymorphism and hybridization in checker mallows (*Sidalcea*, Malvaceae). Molec. Phylogen. Evol. 29: 563–581.

Andres, T. C. 1990. Biosystematics, theories on the origin, and breeding potential of *Cucurbita ficifolia*. In: D. M. Bates et al., eds. 1990. Biology and Utilization of the Cucurbitaceae. Ithaca, N.Y. Pp. 102–119.

Andres, T. C. and G. P. Nabhan. 1988. Taxonomic rank and rarity of *Cucurbita okeechobeensis*. Pl. Genet. Resources Newslett. 75: 21–22.

Andres, T. C. and H. B. Tukey. 1995. Complexities in the infraspecific nomenclature of the *Cucurbita pepo* complex. Acta Hort. 413: 65–91.

Angiosperm Phylogeny Group. 1998. An ordinal classification for the families of flowering plants. Ann. Missouri Bot. Gard. 85: 531–553.

Angiosperm Phylogeny Group. 2003. An update of the Angiosperm Phylogeny Group classification for the orders and families of flowering plants: APG II. Bot. J. Linn. Soc. 141: 399–436.

Angiosperm Phylogeny Group. 2009. An update of the Angiosperm Phylogeny Group classification for the orders and families of flowering plants: APG III. Bot. J. Linn. Soc. 161: 105–121.

Ann. Bot. Fenn. = Annales Botanici Fennici.

Ann. Bot. (Oxford) = Annals of Botany. (Oxford.)

Ann. Bot. Syst.—See: W. G. Walpers and K. Müller berol. 1848–1871

Ann. Jard. Bot. Buitenzorg = Annales du Jardin Botanique de Buitenzorg.

Ann. Lyceum Nat. Hist. New York = Annals of the Lyceum of Natural History of New York.

Ann. Missouri Bot. Gard. = Annals of the Missouri Botanical Garden.

Ann. Mus. Natl. Hist. Nat. = Annales du Muséum National d'Histoire Naturelle. ["National" dropped after vol. 5.]

Ann. Sci. Nat., Bot. = Annales des Sciences Naturelles. Botanique.

Ann. Soc. Linn. Lyon = Annales de la Société Linnéenne de Lyon.

Annuaire Conserv. Jard. Bot. Genève = Annuaire du Conservatoire et Jardin Botaniques de Genève.

Aquatic Bot. = Aquatic Botany; International Scientific Journal Dealing with Applied and Fundamental Research on Submerged, Floating and Emergent Plants in Marine and Freshwater Ecosystems.

Aquino Assis, J. G. de et al. 2000. Implications of the introgression between *Citrullus colocynthis* and *C. lanatus* characters in the taxonomy, evolutionary dynamics and breeding of watermelon. Pl. Genet. Resources Newslett. 121: 15–19.

Arbo, M. M. 1995. Turneraceae Parte I. *Piriqueta*. In: Organization for Flora Neotropica. 1968+. Flora Neotropica. 109+ nos. New York. No. 67.

Arbo, M. M. 2000. Estudios sistemáticos en *Turnera* (Turneraceae). II. Series *Annulares, Capitatae, Microphyllae* y *Papilliferae*. Bonplandia (Corrientes) 10: 1–82.

Arbo, M. M. 2005. Estudios sistemáticos en *Turnera* (Turneraceae). III. Series *Anomalae* y *Turnera*. Bonplandia (Corrientes) 14: 115–318.

Arbo, M. M. 2008. Estudios sistemáticos en *Turnera* (Turneraceae). IV. Series *Leiocarpae, Conciliatae* y *Sessilifoliae*. Bonplandia (Corrientes) 17: 107–334.

Arbo, M. M. and S. M. Espert. 2009. Morphology, phylogeny and biogeography of *Turnera* L. (Turneraceae). Taxon 58: 457–467.

Arbo, M. M. and Aveliano Fernández. 1987. Cruzamientos intra e interespecíficos en *Turnera*, serie *Canaligerae*. Bonplandia (Corrientes) 6: 23–38.

Arbo, M. M. and S. M. Mazza. 2011. The major biodiversity centre for neotropical Turneraceae. Syst. Biodivers. 9: 203–210.

Areces B., F. and P. A. Fryxell. 2007. Malvaceae. In: W. Greuter et al., eds. 1998+. Flora de la República de Cuba: Serie A, Plantas Vasculares. 14+ fascs. Koenigstein. Fasc. 13.

Arnott, G. A. W. 1832. Botany. Edinburgh. [Preprinted from M. Napier, ed. 1830–1842. Encyclopaedia Britannica..., ed. 7. 21 vols. Edinburgh. Vol. 5.] (Botany)

Arrington, J. M. 2004. Systematics of the Cistaceae. Ph.D. dissertation. Duke University.

Arrington, J. M. and K. Kubitzki. 2003. Cistaceae. In: K. Kubitzki et al., eds. 1990+. The Families and Genera of Vascular Plants. 10+ vols. Berlin etc. Vol. 5, pp. 62–70.

Arruda da Cámara, M. 1810. Dissertação Sobre as Plantas do Brazil.... Rio de Janeiro. (Diss. Pl. Brazil)

Asa Gray Bull. = The Asa Gray Bulletin....

Ashby, W. C. 1964. A note on basswood nomenclature. Castanea 29: 109–116.

Asiat. Res. = Asiatic Researches, or Transactions of the Society.

Aublet, J. B. 1775. Histoire des Plantes de la Guiane Françoise.... 4 vols. Paris. [Vols. 1 and 2: text, paged consecutively; vols. 3 and 4: plates.] (Hist. Pl. Guiane)

Auct. Syn. Meth. Stirp. Taurin.—See: C. Allioni 1773

Austral. J. Bot. = Australian Journal of Botany.

Austral. Syst. Bot. = Australian Systematic Botany.

Ayensu, E. S. and W. L. Stern. 1964. Systematic anatomy and ontogeny of the stem in Passifloraceae. Contr. U.S. Natl. Herb. 34: 45–73.

B. M. C. Evol. Biol. = B M C Evolutionary Biology.

Bahadur, B., B. Sailu, and N. R. Swamy. 1996. Pollen flow in heterostylous *Waltheria indica* L. J. Palynol. 32: 13–19.

Bailey, L. H. 1914–1917. The Standard Cyclopedia of Horticulture.... 6 vols. New York. [Vols. paged consecutively.] (Stand. Cycl. Hort.)

Bailey, L. H. 1943. Species of *Cucurbita*. Gentes Herbarum 6: 266–322.

Bailey, L. H., E. Z. Bailey, and Bailey Hortorium Staff. 1976. Hortus Third. A Concise Dictionary of Plants Cultivated in the United States and Canada. New York.

Baileya = Baileya; a Quarterly Journal of Horticultural Taxonomy.

Baillon, H. E. 1866–1895. Histoire des Plantes. 13 vols. Paris, London, and Leipzig. (Hist. Pl.)

Baillon, H. E. 1883–1884. Traité de Botanique Médicale Phanérogamique. 2 fascs. Paris. (Traité Bot. Méd. Phan.)

Baird, V. B. 1936. A natural violet hybrid. Madroño 3: 325–327.

Baird, V. B. 1942. Wild Violets of North America. Berkeley and Los Angeles.

Baker, M. S. 1935. Studies in western violets. I. Madroño 3: 51–57.

Baker, M. S. 1949. Studies in western violets. IV. Leafl. W. Bot. 5: 141–147.

Baker, M. S. 1949b. Studies in western violets. V. Leafl. W. Bot. 5: 173–177.

Baker, M. S. 1953. Studies in western violets. VII. Madroño 12: 8–18.

Baker, M. S. 1953b. A correction in the status of *Viola macloskeyi*. Madroño 12: 60.

Baker, M. S. 1957. Studies in western violets. VIII. The *Nuttallianae* continued. Brittonia 9: 217–230.

Baldwin, B. G. et al., eds. 2012. The Jepson Manual: Vascular Plants of California, ed. 2. Berkeley.

Ballard, H. E. 1985. *Viola epipsila* new to Michigan and the eastern United States. Michigan Bot. 24: 131–134.

Ballard, H. E. 1992. Systematics of *Viola* Section *Viola* in North America North of Mexico. M.S. thesis. Central Michigan University.

Ballard, H. E. 1993. Three new rostrate violet hybrids from Appalachia. Castanea 58: 1–9.

Ballard, H. E. 1994. Violets of Michigan. Michigan Bot. 33: 131–199.

Ballard, H. E. 2000. Violaceae. In: A. F. Rhoads and T. A. Block. 2000. The Plants of Pennsylvania: An Illustrated Manual. Philadelphia. Pp. 700–710.

Ballard, H. E. and S. C. Gawler. 1994. Distribution, habitat and conservation of *Viola novae-angliae*. Michigan Bot. 33: 35–52.

Ballard, H. E., K. J. Sytsma, and R. R. Kowal. 1998. Shrinking the violets: Phylogenetic relationships of infrageneric groups in *Viola* (Violaceae) based on internal transcribed spacer DNA sequences. Syst. Bot. 23: 439–458.

Ballard, H. E. and D. E. Wujek. 1994. Evidence for the recognition of *Viola appalachiensis*. Syst. Bot. 19: 523–538.

Barkman, T. J. et al. 2004. Mitochondrial DNA sequences reveal the photosynthetic relatives of *Rafflesia*, the world's largest flower. Proc. Natl. Acad. Sci. U.S.A. 101: 787–792.

Barringer, K. 2004. New Jersey pinweeds (*Lechea*, Cistaceae). J. Torrey Bot. Soc. 131: 261–276.

Barton, W. P. C. 1818. Compendium Florae Philadelphicae.... 2 vols. Philadelphia. (Comp. Fl. Philadelph.)

Bartonia = Bartonia; a Botanical Annual.

Bates, D. M. 1963. The Genus *Malacothamnus*. Ph.D. dissertation. University of California, Los Angeles.

Bates, D. M. 1965. Notes on the cultivated Malvaceae. 1. *Hibiscus*. Baileya 13: 56–130.

Bates, D. M. 1968. Notes on the cultivated Malvaceae. 2. *Abelmoschus*. Baileya 16: 99–112.

Bates, D. M. 1987. Chromosome numbers and evolution in *Anoda* and *Periptera* (Malvaceae). Aliso 11: 523–531.

Bates, D. M., L. J. Dorr, and O. J. Blanchard. 1989. Chromosome numbers in *Callirhoe* (Malvaceae). Brittonia 41: 143–151.

Bates, D. M., R. W. Robinson, and C. Jeffrey, eds. 1990. Biology and Utilization of the Cucurbitaceae. Ithaca, N.Y.

Baum, B. R. 1964. On the vernales-aestivales character in *Tamarix* and its diagnostic value. Israel J. Bot. 13: 30–35.

Baum, B. R. 1967. Introduced and naturalized tamarisks in the United States and Canada. Baileya 15: 19–25.

Baum, B. R. 1978. The Genus *Tamarix*. Jerusalem.

Baum, B. R., I. J. Bassett, and C. W. Crompton. 1971. Pollen morphology of *Tamarix* species and its relationship to the taxonomy of the genus. Pollen & Spores 13: 495–521.

Baum, D. A. et al. 2004. Phylogenetic relationships of Malvatheca (Bombacoideae and Malvoideae; Malvaceae sensu lato) as inferred from plastid DNA sequences. Amer. J. Bot. 91: 1863–1871.

Bayer, C. 1999. The bicolor unit—Homology and transformation of an inflorescence structure unique in core Malvales. Pl. Syst. Evol. 214: 187–198.

Bayer, C. et al. 1999. Support for an expanded family concept of Malvaceae with a recircumscribed order Malvales: A combined analysis of plastid *atp*B and *rbc*L DNA sequences. Bot. J. Linn. Soc. 129: 267–303.

Bayer, C., M. W. Chase, and M. F. Fay. 1998. Muntingiaceae, a new family of dicotyledons with malvalean affinities. Taxon 47: 37–42.

Bayer, C. and K. Kubitzki. 2003. Malvaceae. In: K. Kubitzki et al., eds. 1990+. The Families and Genera of Vascular Plants. 10+ vols. Berlin etc. Vol. 5, pp. 225–311.

Beattie, A. J. 1969. The floral biology of three species of *Viola*. New Phytol. 68: 1187–1201.

Beattie, A. J. 1969b. Studies in the pollination ecology of *Viola*. 1. The pollen content of stigmatic cavities. Watsonia 7: 142–156.

Beattie, A. J. 1971. Pollination mechanisms in *Viola*. New Phytol. 70: 343–360.

Beattie, A. J. 1972. The pollination ecology of *Viola*. 2. Pollen loads of insect visitors. Watsonia 9: 13–25.

Beattie, A. J. 1974. Floral evolution in *Viola*. Ann. Missouri Bot. Gard. 61: 781–793.

Beattie, A. J. 1976. Plant dispersion, pollination and gene flow in *Viola*. Oecologia 25: 291–300.

Beattie, A. J. 1978. Plant-animal interactions affecting gene flow in *Viola*. In: A. J. Richards, ed. 1978. The Pollination of Flowers by Insects. London. Pp. 151–164.

Beattie, A. J. and D. C. Culver. 1979. Neighborhood size in *Viola*. Evolution 33: 1226–1229.

Beattie, A. J. and D. C. Culver. 1981. The guild of myrmecochores in the herbaceous flora of West Virginia forests. Ecology 62: 107–115.

Beattie, A. J. and N. Lyons. 1975. Seed dispersal in *Viola* (Violaceae): Adaptations and strategies. Amer. J. Bot. 62: 714–722.

Becker, W. 1925. *Viola*. In: H. G. A. Engler et al., eds. 1924+. Die natürlichen Pflanzenfamilien..., ed. 2. 26+ vols. Leipzig and Berlin. Vol. 21, pp. 363–376.

Beih. Bot. Centralbl. = Beihefte zum Botanischen Centralblatt. Original Arbeiten.

Bell, C. R. 1965. In: Documented plant chromosome numbers 65: 3. Sida 2: 168–170.

Bellot, S. and S. S. Renner. 2014. The systematics of the worldwide endoparasite family Apodanthaceae (Cucurbitales), with a key, a map, and color photos of most species. PhytoKeys 36: 41–57.

Bemis, W. P. et al. 1978. The feral buffalo gourd, *Cucurbita foetidissima*. Econ. Bot. 32: 87–95.

Bemis, W. P. and T. W. Whitaker. 1965. Natural hybridization between *Cucurbita digitata* and *C. palmata*. Madroño 18: 39–47.

Bemis, W. P. and T. W. Whitaker. 1969. The xerophytic *Cucurbita* of northwestern Mexico and southwestern United States. Madroño 20: 33–41.

Benesh, D. L. 1999. Morphological variation in *Malacothamnus fasciculatus* (Torrey & A. Gray) E. Greene (Malvaceae) and related species. Madroño 46: 142–152.

Benson, W. W., K. S. Brown, and L. E. Gilbert. 1975. Coevolution of plants and herbivores: Passion flower butterflies. Evolution 29: 659–680.

Bentham, G. 1839[–1857]. Plantas Hartwegianas Imprimis Mexicanas.... London. [Issued by gatherings with consecutive signatures and pagination.] (Pl. Hartw.)

Bentham, G. 1844[–1846]. The Botany of the Voyage of H.M.S. Sulphur, under the Command of Captain Sir Edward Belcher...during the Years 1836–1842. 6 parts. London. [Parts paged consecutively.] (Bot. Voy. Sulphur)

Bentham, G. and J. D. Hooker. 1862–1883. Genera Plantarum ad Exemplaria Imprimis in Herbariis Kewensibus Servata Definita. 3 vols. London. (Gen. Pl.)

Berg, R. Y. 1975. Myrmecochorous plants in Australia and their dispersal by ants. Austral. J. Bot. 23: 475–508.

Bernhard, A. 1999. Flower structure, development, and systematics in Passifloraceae and in *Abatia* (Flacourtiaceae). Int. J. Pl. Sci. 160: 135–150.

Beutler, J. A. and A. DerMarderosian. 2004. Passion flower. Rev. Nat. Prod. 4.

Bhandari, M. M. and D. Singh. 1964. *Dactyliandra* (Hook. f.): A cucurbitaceous genus new to the Indian flora. Kew Bull. 19: 133–138.

Biblioth. Bot. = Bibliotheca Botanica; original Abhandlungen aus dem Gesammtgebiete der Botanik.

Biol. J. Linn. Soc. = Biological Journal of the Linnean Society.

Biosyst. Monogr. Cucumis—See: J. H. Kirkbride 1993

Blanchard, O. J. 1974. Chromosome numbers in *Kosteletzkya* Presl (Malvaceae). Rhodora 76: 64–66.

Blanchard, O. J. 1976. A Revision of Species Segregated from *Hibiscus* Sect. *Trionum* (Medicus) de Candolle Sensu Lato (Malvaceae). Ph.D. dissertation. Cornell University.

Blanchard, O. J. 2008. Innovations in *Hibiscus* and *Kosteletzkya* (Malvaceae, Hibisceae). Novon 18: 4–8.

Blanchard, O. J. 2012. Chromosome numbers, phytogeography, and evolution in *Kosteletzkya* (Malvaceae, Malvoideae, Hibisceae). Novon 18: 4–8.

Blarer, A., D. L. Nickrent, and P. K. Endress. 2004. Comparative floral structure and systematics in Apodanthaceae (Rafflesiales). Pl. Syst. Evol. 245: 119–142.

Blumea = Blumea; Tidjschrift voor die Systematiek en die Geografie der Planten (A Journal of Plant Taxonomy and Plant Geography).

Bodo Slotta, T. A. 2000. Phylogenetic Analysis of *Iliamna* (Malvaceae) Using the Internal Transcribed Spacer Region. M.S. thesis. Virginia Polytechnic Institute and State University.

Bodo Slotta, T. A. and D. M. Porter. 2006. Genetic variation within and between *Iliamna corei* and *I. remota* (Malvaceae): Implications for species delimitation. Bot. J. Linn. Soc. 151: 345–354.

Boissiera = Boissiera; Mémoires des Conservatoire et de l'Institut de Botanique Systématique de l'Université de Genève (later: Mémoires des Conservatoire et Jardin Botaniques de la Ville de Genève). Supplement de Candollea.

Boivin, B. 1967. Énumération des plantes du Canada. VII—Résumé statistique et régions adjacentes. Naturaliste Canad. 94: 625–655.

Bol. Soc. Bot. México = Boletín de la Sociedad Botánica de México.

Bonplandia (Corrientes) = Bonplandia. Corrientes.

Bonplandia (Hanover) = Bonplandia; Zeitschrift für die gesammte Botanik.

Borssum Waalkes, J. van. 1966. Malesian Malvaceae revised. Blumea 14: 1–251.

Bory de Saint-Vincent, J. B. et al. 1822–1831. Dictionnaire Classique d'Histoire Naturelle. 17 vols. Paris. (Dict. Class. Hist. Nat.)

Boston J. Nat. Hist. = Boston Journal of Natural History.

Bot. Beechey Voy.—See: W. J. Hooker and G. A. W. Arnott [1830–]1841

Bot. California—See: W. H. Brewer et al. 1876–1880

Bot. Explor. = The Botanical Explorer.

Bot. Gaz. = Botanical Gazette; Paper of Botanical Notes.

Bot. Handb.—See: C. Schkuhr [1787–]1791–1803

Bot. J. Linn. Soc. = Botanical Journal of the Linnean Society.

Bot. Jahrb. Syst. = Botanische Jahrbücher für Systematik, Pflanzengeschichte und Pflanzengeographie.

Bot. Mag. = Botanical Magazine; or, Flower-garden Displayed.... [Edited by Wm. Curtis.] [With vol. 15, 1801, title became Curtis's Botanical Magazine; or....]

Bot. Mag. (Kew Mag.) = The Kew Magazine; Incorporating Curtis's Botanical Magazine.

Bot. Misc. = Botanical Miscellany.

Bot. Repos. = Botanists' Repository, for New, and Rare Plants.

Bot. Rev. (Lancaster) = Botanical Review, Interpreting Botanical Progress.

Bot. Voy. Sulphur—See: G. Bentham 1844[–1846]

Bot. Zhurn. (Moscow & Leningrad) = Botanicheskii Zhurnal. (Moscow and Leningrad.)

Botany—See: G. A. W. Arnott 1832

Botany (Fortieth Parallel)—See: S. Watson 1871

Bothalia = Bothalia; a Record of Contributions from the National Herbarium, Union of South Africa.

Brainerd, E. 1908. *Viola chinensis* in the eastern United States. Rhodora 10: 38–40.

Brainerd, E. 1910. *Viola palmata* and its allies. Bull. Torrey Bot. Club 37: 181–190.

Brainerd, E. 1921. Violets of North America. Bull. Vermont Agric. Exp. Sta. 224.

Brainerd, E. 1924. Some natural violet hybrids of North America. Bull. Vermont Agric. Exp. Sta. 239.

Brewer, W. H et al. 1876–1880. Geological Survey of California.... Botany.... 2 vols. Cambridge, Mass. (Bot. California)

Brickell, C. D. and B. Mathew. 1976. *Daphne:* The Genus in the Wild and in Cultivation. Woking.

Bridson, G. D. R. 2004. BPH-2: Periodicals with Botanical Content. 2 vols. Pittsburgh.

Bridson, G. D. R. and E. R. Smith. 1991. B-P-H/S. Botanico-Periodicum-Huntianum/Supplementum. Pittsburgh.

Britton, N. L. 1881. A Preliminary Catalogue of the Flora of New Jersey.... New Brunswick. (Catal. Fl. New Jersey)

Britton, N. L. 1894. A revision of the genus *Lechea*. Bull. Torrey Bot. Club 21: 244–253.

Britton, N. L. 1901. Manual of the Flora of the Northern States and Canada. New York. (Man. Fl. N. States)

Britton, N. L. 1905b. Manual of the Flora of the Northern States and Canada, ed. 2. New York. (Man. Fl. N. States ed. 2)

Britton, N. L. and A. Brown 1913. An Illustrated Flora of the Northern United States, Canada and the British Possessions from Newfoundland to the Parallel of the Southern Boundary of Virginia, and from the Atlantic Ocean Westward to the 102d Meridian..., ed. 2. 3 vols. New York. (Ill. Fl. N. U.S. ed. 2)

Britton, N. L., E. E. Sterns, J. F. Poggenburg, et al. 1888. Preliminary Catalogue of Anthophyta and Pteridophyta Reported As Growing Spontaneously within One Hundred Miles of New York City. New York. [Authorship often attributed as B.S.P. in nomenclatural contexts.] (Prelim. Cat.)

Brittonia = Brittonia; a Journal of Systematic Botany....

Brizicky, G. K. 1961. The genera of Turneraceae and Passifloraceae in the southeastern United States. J. Arnold Arbor. 42: 204–218.

Brizicky, G. K. 1961b. The genera of Violaceae in the southeastern United States. J. Arnold Arbor. 42: 321–333.

Brizicky, G. K. 1966. The genera of Sterculiaceae in the southeastern United States. J. Arnold Arbor. 47: 60–74.

Brooks, R. E. and R. L. McGregor. 1986. Violaceae. In: Great Plains Flora Association. 1986. Flora of the Great Plains. Lawrence, Kans. Pp. 255–264.

Brown, J. P. 1941. *Napaea dioica* in New England. Rhodora 43: 94–95.

Brown, R. 1844. *Pterocymbium,* with Observations on Sterculieae, the Tribe to Which It Belongs. London. [Preprinted from J. J. Bennett et al. 1838–1852. Plantae Javanicae Rariores.... 4 parts. London. Pp. 219–238.] (Pterocymbium)

Brummitt, R. K. and C. E. Powell, eds. 1992. Authors of Plant Names. A List of Authors of Scientific Names of Plants, with Recommended Standard Forms of Their Names, Including Abbreviations. Kew.

Brunken, U. and A. N. Muellner. 2012. A new tribal classification of Grewioideae (Malvaceae) based on morphological and molecular phylogenetic evidence. Syst. Bot. 37: 699–711.

Bull. Bot. Res., Harbin = Bulletin of Botanical Research. [Zhiwi Yanjiu.]

Bull. Brit. Mus. (Nat. Hist.), Bot. = Bulletin of the British Museum (Natural History). Botany.

Bull. Calif. Acad. Sci. = Bulletin of the California Academy of Sciences.

Bull. Hawaii Agric. Exp. Sta. = Bulletin, Hawaii Agricultural Experiment Station.

Bull. Jard. Bot. État Bruxelles = Bulletin du Jardin Botanique de l'État à Bruxelles.

Bull. Misc. Inform. Kew = Bulletin of Miscellaneous Information, Royal Gardens, Kew.

Bull. Nat. Hist. Mus. London, Bot. = Bulletin of the Natural History Museum. Botany Series.

Bull. S. Calif. Acad. Sci. = Bulletin of the Southern California Academy of Sciences.

Bull. Soc. Bot. France = Bulletin de la Société Botanique de France.

Bull. Torrey Bot. Club = Bulletin of the Torrey Botanical Club.

Bull. Vermont Agric. Exp. Sta. = Bulletin of the Vermont Agricultural Experiment Station.

Bunge, A. A. [1833.] Enumeratio Plantarum, Quas in China Boreali Collegit.... St. Petersburg. [Preprinted from Mém. Acad. Imp. Sci. St.-Pétersbourg Divers Savans 2: 75–148. 1835.] (Enum. Pl. China Bor.)

Burch, D. G., D. B. Ward, and D. W. Hall. 1988. Checklist of the Woody Cultivated Plants of Florida. Gainesville.

Burger, Y. et al. 2010. Genetic diversity of *Cucumis melo*. Hort. Rev. 36: 165–198.

Burnett, G. T. 1835. Outlines of Botany.... 2 vols. London. [Volumes paged consecutively.] (Outlines Bot.)

Butterwick, M. L. 1980. *Cucurbita digitata* (Cucurbitaceae) in Texas. Sida 8: 315.

Calder, J. A. and R. L. Taylor. 1968. Flora of the Queen Charlotte Islands. 2 vols. Ottawa.

Callihan, R. H., S. L. Carson, and R. T. Dobbins. 1995. NAWEEDS, Computer-aided Weed Identification for North America. Illustrated User's Guide plus Computer Floppy Disk. Moscow, Idaho.

Cameron, K. M., K. Wurdack, and R. W. Johnson. 2002. Molecular evidence for the common origin of snap-traps among carnivorous plants. Amer. J. Bot. 89: 1503–1509.

Canad. Field-Naturalist = Canadian Field-Naturalist.

Canad. J. Bot. = Canadian Journal of Botany.

Candolle, A. L. P. P. de and C. de Candolle, eds. 1878–1896. Monographiae Phanerogamarum.... 9 vols. Paris. (Monogr. Phan.)

Candolle, A. P. de and A. L. P. P. de Candolle, eds. 1823–1873. Prodromus Systematis Naturalis Regni Vegetabilis.... 17 vols. Paris etc. [Vols. 1–7 edited by A. P. de Candolle, vols. 8–17 by A. L. P. P. de Candolle.] (Prodr.)

Canne, J. M. 1987. Determinations of chromosome numbers in *Viola* (Violaceae). Canad. J. Bot. 65: 653–655.

Canotia = Canotia; a New Journal of Arizona Botany.

Carter, G. F. 1954. The disharmony between Asiatic flower-birds and American bird-flowers. Amer. Antiquity 20: 176–177.

Caryologia = Caryologia; Giornale di Citologia, Citosistematica e Citogenetica.

Castanea = Castanea; Journal of the Southern Appalachian Botanical Club.

Castetter, E. F. 1925. Horticultural groups of cucurbits. Proc. Amer. Soc. Hort. Sci. 22: 338–340.

Cat. Afr. Pl.—See: W. P. Hiern 1896–1901

Cat. Pl. Cub.—See: A. H. R. Grisebach 1866

Cat. Pl. Upper Louisiana—See: T. Nuttall 1813

Catal. Fl. New Jersey—See: N. L. Britton 1881

Cavanilles, A. J. 1785–1790. Monadelphiae Classis Dissertationes Decem. 10 parts. Madrid. [Parts (dissertations) paged consecutively.] (Diss.)

Cent. Pl. I—See: C. Linnaeus 1755

Chapman, A. W. 1860. Flora of the Southern United States.... New York. (Fl. South. U.S.)

Chapman, A. W. [1892.] Flora of the Southern United States..., ed. 2 reprinted. New York, Cincinnati, and Chicago. [Second Supplement added.] (Fl. South. U.S. ed. 2 repr. 1892)

Chapman, A. W. 1897. Flora of the Southern United States..., ed. 3. Cambridge, Mass. (Fl. South. U.S. ed. 3)

Chase, M. W. et al. 1993. Phylogenetics of seed plants: An analysis of nucleotide sequences from the plastid gene *rbc*L. Ann. Missouri Bot. Gard. 80: 528–580.

Chase, M. W. et al. 2002. When in doubt, put it in Flacourtiaceae: A molecular phylogenetic analysis based on plastid *rbc*L DNA sequences. Kew Bull. 57: 141–181.

Chattaway, M. M. 1933. Tile-cells in the rays of the Malvales. New Phytol. 32: 261–273.

Cheek, M. 1989. Lectotypification and authorship of *Hibiscus schizopetalus* (Malvaceae). Taxon 38: 261–263.

Chen, J. F. et al. 1997. Successful interspecific hybridization between *Cucumis sativus* L. and *C. hystrix* Chakr. Euphytica 96: 413–419.

Chen, Y. S. et al. 2007. Violaceae. In: Wu Z. and P. H. Raven, eds. 1994+. Flora of China. 20+ vols. Beijing and St. Louis. Vol. 13, pp. 72–111.

Cheng, Z. M., Pan H. X., and Yin L. K. 2000. Study on the phytochemistry taxonomy of *Tamarix* L. and *Myricaria* Desv. Acta Bot. Boreal.-Occid. Sin. 20: 275–282.

Chevallier, A. 1996. The Encyclopedia of Medicinal Plants. New York.

Chung, S. M., J. E. Staub, and Chen J. F. 2006. Molecular phylogeny of *Cucumis* species as revealed by consensus chloroplast SSR marker length and sequence variation. Genome 49: 219–229.

Cistineae—See: R. Sweet 1825–1830

Clapham, A. R., T. G. Tutin, and D. M. Moore. 1987. Flora of the British Isles, ed. 3. Cambridge.

Clarke, A. C. et al. 2006. Reconstructing the origins and dispersal of the Polynesian bottle gourd (*Lagenaria siceraria*). Molec. Biol. Evol. 23: 893–900.

Clausen, J. 1929. Chromosome number and relationship of some North American species of *Viola*. Ann. Bot. (Oxford) 43: 741–764.

Clausen, J. 1964. Cytotaxonomy and distributional ecology of western North American violets. Madroño 17: 173–197.

Clausen, J. 1964b. New combinations in western North American violets. Madroño 17: 295.

Clausen, J., R. B. Channell, and U. Nur. 1964. *Viola rafinesquii,* the only *Melanium* violet native to North America. Rhodora 66: 32–46.

Clement, W. L. et al. 2004. Phylogenetic position and biogeography of *Hillebrandia sandwicensis* (Begoniaceae): A rare Hawaiian relict. Amer. J. Bot. 91: 905–917.

Clokey, I. W. 1945. Notes on the flora of the Charleston Mountains, Clark County, Nevada. Madroño 8: 56–61.

Cody, W. J. 2000. Flora of the Yukon Territory, ed. 2. Ottawa.

Cogniaux, C. A. 1876–1877. Diagnoses de Cucurbitacées Nouvelles.... 2 vols. Brussels. (Diagn. Cucurb. Nouv.)

Collectanea—See: N. J. Jacquin 1786[1787]–1796[1797]

Collevatti, R. G., L. A. O. Campos, and A. F. Da Silva. 1998. Pollination ecology of the tropical weed *Triumfetta semitriloba* Jacq. (Tiliaceae), in the south-eastern Brazil. Revista Brasil. Biol. 58: 383–392.

Comp. Fl. Philadelph.—See: W. P. C. Barton 1818

Compt.-Rend. Trav. Carlsberg Lab., Sér. Physiol. = Comptes-Rendus des Travaux du Carlsberg Laboratoriet; Série Physiologique.

Conservation Genet. = Conservation Genetics.

Consp. Regn. Veg.—See: H. G. L. Reichenbach 1828

Contr. Dudley Herb. = Contributions from the Dudley Herbarium of Stanford University.

Contr. Gray Herb. = Contributions from the Gray Herbarium of Harvard University. [Some numbers reprinted from (or in?) other periodicals, e.g. Rhodora.]

Contr. Herb. Franklin Marshall Coll. = Contributions from the Herbarium of Franklin and Marshall College.

Contr. U.S. Natl. Herb. = Contributions from the United States National Herbarium.

Contr. Univ. Michigan Herb. = Contributions from the University of Michigan Herbarium.

Contr. W. Bot. = Contributions to Western Botany.

Coombs, R. E. 2003. Violets: The History & Cultivation of Scented Violets, ed. 2. London.

Cooperrider, T. S. 1986. *Viola ×brauniae (Viola rostrata × V. striata)*. Michigan Bot. 25: 107–109.

Correll, D. S. and M. C. Johnston. 1970. Manual of the Vascular Plants of Texas. Renner, Tex.

Cortés-Palomec, A. C. and H. E. Ballard. 2006. Influence of annual fluctuations in environmental conditions on chasmogamous flower production in *Viola striata*. J. Torrey Bot. Soc. 133: 312–320.

Cosson, E. S.-C. and J. N. E. Germain de Saint-Pierre. 1859. Synopsis Analytique de la Flore des Environs de Paris..., ed. 2. Paris. (Syn. Anal. Fl. Paris ed. 2)

Coulter, J. M. 1886. Revision of North American Hypericaceae I. Bot. Gaz. 11: 78–88.

Coulter, J. M. 1890. Upon a collection of plants made by Mr. G. C. Nealley, in the region of the Rio Grande, in Texas, from Brazos Santiago to El Paso County. Contr. U.S. Natl. Herb. 1: 29–62.

Coulter, J. M. 1891. Botany of western Texas. A manual of the phanerogams and pteridophytes of western Texas. Contr. U.S. Natl. Herb. 2: 1–588.

Cowan, C. W. and B. D. Smith. 1993. New perspectives on a wild gourd in eastern North America. J. Ethnobiol. 13: 17–54.

Crantz, H. J. N. von. 1762–1767. Stirpium Austriarum Fasciculus I [–III]. 3 fascs. Vienna and Leipzig. (Stirp. Austr. Fasc.)

Crantz, H. J. N. von. 1766. Institutiones Rei Herbariae.... 2 vols. Vienna. (Inst. Rei Herb.)

Craven, L. A. et al. 2011. A taxonomic re-evaluation of *Hibiscus trionum* (Malvaceae) in Australasia. New Zealand J. Bot. 49: 27–40.

Crins, W. J. 1989. The Tamaricaceae of the southeastern United States. J. Arnold Arbor. 70: 403–425.

Cristóbal, C. L. 1960. Revisión del género *Ayenia* (Sterculiaceae). Opera Lilloana 4: 1–230.

Crockett, S. L. 2003. Phytochemical and Biosystematic Investigations of New and Old World *Hypericum* Species (Clusiaceae). Ph.D. dissertation. University of Mississippi.

Cronquist, A. 1981. An Integrated System of Classification of Flowering Plants. New York.

Cronquist, A. et al. 1972+. Intermountain Flora. Vascular Plants of the Intermountain West, U.S.A. 6+ vols. in 7+. New York and London.

Crow, G. E. and C. B. Hellquist. 2000. Aquatic and Wetland Plants of Northeastern North America.... 2 vols. Madison.

Cruzan, M. B. 2005. Patterns of introgression across an expanding hybrid zone: Analyzing historical patterns of gene flow using nonequilibrium approaches. New Phytol. 167: 267–278.

Cuénoud, P. et al. 2002. Molecular phylogenetics of Caryophyllales based on nuclear 18S rDNA and plastid *rbc*L, *atp*B, and *mat*K DNA sequences. Amer. J. Bot. 89: 132–144.

Culley, T. M. 2000. Inbreeding depression and floral type fitness differences in *Viola canadensis* (Violaceae), a species with chasmogamous and cleistogamous flowers. Canad. J. Bot. 78: 1420–1429.

Culley, T. M. 2002. Reproductive biology and delayed selfing in *Viola pubescens* (Violaceae), an understorey herb with chasmogamous and cleistogamous flowers. Int. J. Pl. Sci. 163: 113–122.

Culver, D. C. and A. J. Beattie. 1978. Myrmecochory in *Viola*: Dynamics of seed–ant interactions in some West Virginia species. J. Ecol. 66: 53–72.

Culver, D. C. and A. J. Beattie. 1980. The fate of *Viola* seeds dispersed by ants. Amer. J. Bot. 67: 710–714.

Culwell, D. E. 1970. A Taxonomic Study of the Section *Hypericum* in the Eastern United States. Ph.D. dissertation. University of North Carolina.

Cybele Columb. = Cybele Columbiana; a Series of Studies in Botany, Chiefly North American.

Cycl.—See: A. Rees [1802–]1819–1820

Cytologia = Cytologia; International Journal of Cytology.

Dahlgren, G. 1989b. The last Dahlgrenogram: System of classification of the dicotyledons. In: K. Tan, ed. 1989. The Davis & Hedge Festschrift...Plant Taxonomy, Phytogeography, and Related Subjects. Edinburgh. Pp. 249–260.

Dane, F. and P. Lang. 2004. Sequence variation at cpDNA regions of watermelon and related wild species: Implications for the evolution of *Citrullus* haplotypes. Amer. J. Bot. 91: 1922–1929.

Dane, F., P. Lang, and R. Bakhtiyarova. 2004. Comparative analysis of chloroplast DNA variability in wild and cultivated *Citrullus* species. Theor. Appl. Genet. 108: 958–966.

Dane, F. and J. Liu. 2007. Diversity and origin of cultivated and citron type watermelon *(Citrullus lanatus)*. Genet. Resources Crop Evol. 54: 1255–1265.

Dansereau, P. M. 1939. Monographie du genre *Cistus* L. Boissiera 4: 1–90.

Daoud, H. S. and R. L. Wilbur. 1965. A revision of the North American species of *Helianthemum* (Cistaceae). Rhodora 67: 63–82, 201–216, 255–312.

Darwin, C. 1875. Insectivorous Plants, ed. 2. London.

Davidse, G. 1976. A study of some intermountain violets (*Viola* sect. *Chamaemelanium*). Madroño 23: 274–283.

Davis, C. C. et al. 2007. Floral gigantism in Rafflesiaceae. Science, ser. 2, 315: 1812.

Davis, C. C. and M. W. Chase. 2004. Elatinaceae are sister to Malphigiaceae; Peridiscaceae belong to Saxifragales. Amer. J. Bot. 91: 262–273.

de Wilde, W. J. J. O. 1971. The systematic position of the Paropsieae, in particular the genus *Ancistrothyrsus*, and a key to the genera of Passsifloraceae. Blumea 14: 99–104.

de Wilde, W. J. J. O. 1974. The genera of tribe Passifloreae (Passifloraceae) with special reference to flower morphology. Blumea 22: 37–50.

de Wilde, W. J. J. O. and B. E. E. Duyfjes. 2010. *Cucumis sativus* L. forma *hardwickii* (Royle) W. J. de Wilde & Duyfjes and feral forma *sativus*. Thai Forest Bull., Bot. 38: 98–107.

Deam, C. C. 1940. Flora of Indiana. Indianapolis.

DeCandido, R. 1991. *Hibiscus militaris* L. (Malvaceae) new to New York state. Bull. Torrey Bot. Club 118: 329.

Decker, D. S. 1988. Origin(s), evolution and systematics of *Cucurbita pepo* (Cucurbitaceae). Econ. Bot. 42: 4–15.

Decker, D. S. 1990. Evidence for multiple domestications of *Cucurbita pepo*. In: D. M. Bates et al., eds. 1990. Biology and Utilization of the Cucurbitaceae. Ithaca, N.Y. Pp. 96–101.

Decker, D. S. et al. 1993. Isozymic characteriztion of wild populations of *Cucurbita pepo*. J. Ethnobiol. 13: 55–72.

Decker, D. S. et al. 2002. The origin and genetic affinities of wild populations of melon (*Cucumis melo*, Cucurbitaceae) in North America. Pl. Syst. Evol. 233: 183–197.

Decker, D. S. et al. 2002b. Diversity in free-living populations of *Cucurbita pepo* (Cucurbitaceae) as assessed by random amplified polymorphic DNA. Syst. Bot. 27: 19–28.

Decker, D. S. et al. 2004. Discovery and genetic assessment of wild bottle gourd [*Lagenaria siceraria* (Mol.) Standley; Cucurbitaceae] from Zimbabwe. Econ. Bot. 58: 501–508.

Decker, D. S., T. W. Walters, and U. Posluszny. 1990. Genealogy and gene flow among annual domesticated species of *Cucurbita*. Canad. J. Bot. 68: 782–789.

Decker, D. S. and H. D. Wilson. 1987. Allozyme variation in the *Cucurbita pepo* complex: *C. pepo* var. *ovifera* vs. *C. texana*. Syst. Bot. 12: 263–273.

DeLaney, K. R. 2010. *Sida planicaulis* (Malvaceae; Malveae) in Florida, a new addition to the flora of North America. Bot. Explor. 4: 127–134.

Delesalle, V. A. 1989. Year-to-year changes in phenotypic gender in a monoecious cucurbit, *Apodanthera undulata*. Amer. J. Bot. 76: 30–39.

Deleuze, J. P. F. 1804. Notice historique sur André Michaux. Ann. Mus. Natl. Hist. Nat. 3: 191–227.

Demonstr. Pl.—See: C. Linnaeus [1753]

Denkschr. Königl.-Baier. Bot. Ges. Regensburg = Denkschriften der Königlich-baierischen botanischen Gesellschaft in Regensburg.

Descr. Pl. Nouv.—See: É. P. Ventenat [1800–1803]

Desert Pl. = Desert Plants; a Quarterly Journal Devoted to Broadening Our Knowledge of Plants Indigenous or Adaptable to Arid and Sub-arid Regions.

Deut. Fl.—See: H. Karsten 1880–1883

DeVeaux, J. S. and E. B. Shultz. 1985. Development of buffalo gourd (*Cucurbita foetidissima*) as a semiaridland starch and oil crop. Econ. Bot. 39: 454–472.

Diagn. Cucurb. Nouv.—See: C. A. Cogniaux 1876–1877

Dict. Class. Hist. Nat.—See: J. B. Bory de Saint-Vincent et al. 1822–1831

Dippel, L. 1889–1893. Handbuch der Laubholzkunde. 3 vols. Berlin. (Handb. Laubholzk.)

Diss.—See: A. J. Cavanilles 1785–1790

Diss. Bot. Pl. Doming.—See: C. F. von Ledebour and J. P. Alderstam 1805

Diss. Pl. Brazil—See: M. Arruda da Câmara 1810

Domke, F. W. 1934. Untersuchungen über die systematische und geographische Gliederung der Thymelaeaceen. Biblioth. Bot. 27: 1–151.

Don, G. 1831–1838. A General History of the Dichlamydeous Plants.... 4 vols. London. (Gen. Hist.)

Dorr, L. J. 1990. A revision of the North American genus *Callirhoe* (Malvaceae). Mem. New York Bot. Gard. 56: 1–75.

Dorr, L. J. 1994. Plants in peril, 21. *Callirhoe scabriuscula*. Bot. Mag. (Kew Mag.) 11: 146–151.

Dorr, L. J. 1996. *Ayenia saligna* (Sterculiaceae), a new species from Colombia. Brittonia 48: 213–216.

Douglas, G. W. et al. 1998–2002. Illustrated Flora of British Columbia. 8 vols. Victoria.

Douglas, G. W. and M. Ryan. 1998. Status of the yellow montane violet, *Viola praemorsa* ssp. *praemorsa* (Violaceae) in Canada. Canad. Field-Naturalist 112: 491–495.

Douglas, G. W., G. B. Straley, and D. V. Meidinger. 1989–1994. The Vascular Plants of British Columbia. 4 vols. Victoria. [B.C. Minist. Forests, Special Rep. 1–4.]

Duchen, P. and S. S. Renner. 2010. The evolution of *Cayaponia* (Cucurbitaceae): Repeated shifts from bat to bee pollination and long-distance dispersal to Africa 2–5 million years ago. Amer. J. Bot. 97: 1129–1141.

Duchesne, A. N. 1786. Essai sur l'Histoire Naturel des Courges. Paris. [Preprinted from J. Lamarck et al. 1783–1817. Encyclopédie Méthodique. Botanique.... 13 vols. Paris and Liège. Vol. 2.] (Essai Hist. Nat. Courges)

Duncan, W. H. 1964. New *Elatine* populations in the southeastern United States. Rhodora 66: 47–53.

Duncan, W. H. and M. Mellinger. 1972. *Edgeworthia* (Thymelaeaceae) new to the Western Hemisphere. Rhodora 74: 436–439.

Durand, E. M. 1855. Plantae Prattenianae Californicae.... Philadelphia. [Preprinted from J. Acad. Nat. Sci. Philadelphia, n. s. 3: 79–104. 1855.] (Pl. Pratten. Calif.)

Durand, E. M. and T. C. Hilgard. 1854. Plantae Heermannianae.... Philadelphia. [Preprinted from J. Acad. Nat. Sci. Philadelphia, n. s. 3: 37–46. 1855.] (Pl. Heermann.)

Dutt, B. and R. P. Roy 1990. Cytogenetics of the Old World species of *Luffa*. In: D. M. Bates et al., eds. 1990. Biology and Utilization of the Cucurbitaceae. Ithaca, N.Y. Pp. 134–140.

Ecklon, C. F. and K. L. P. Zeyher. [1834–]1835–1836[–1837]. Enumeratio Plantarum Africae Australis Extratropicae.... 3 parts. Hamburg. [Parts paged consecutively.] (Enum. Pl. Afric. Austral.)

Ecol. Monogr. = Ecological Monographs.

Ecology = Ecology, a Quarterly Journal Devoted to All Phases of Ecological Biology.

Econ. Bot. = Economic Botany; Devoted to Applied Botany and Plant Utilization.

Edinburgh J. Nat. Geogr. Sci. = The Edinburgh Journal of Natural and Geographical Science.

Edinburgh Philos. J. = Edinburgh Philosophical Journal.

Edwards's Bot. Reg. = Edwards's Botanical Register....

Elliott, S. [1816–]1821–1824. A Sketch of the Botany of South-Carolina and Georgia. 2 vols. in 13 parts. Charleston. (Sketch Bot. S. Carolina)

Emory, W. H. 1857–1859. Report on the United States and Mexican Boundary Survey, Made under the Direction of the Secretary of the Interior. 2 vols. in parts. Washington. (Rep. U.S. Mex. Bound.)

Encycl.—See: J. Lamarck et al. 1783–1817

Endress, P. K. 1996. Diversity and Evolutionary Biology of Tropical Flowers, pbk. ed. 1 corr. Cambridge and New York.

Engl. Bot.—See: J. E. Smith et al. 1790–1866

Engler, H. G. A., ed. 1900–1953. Das Pflanzenreich.... 107 vols. Berlin. [Sequence of vol. (Heft) numbers (order of publication) is independent of the sequence of series and family (Roman and Arabic) numbers (taxonomic order).] (Pflanzenr.)

Engler, H. G. A. et al., eds. 1924+. Die natürlichen Pflanzenfamilien..., ed. 2. 26+ vols. Leipzig and Berlin. (Nat. Pflanzenfam. ed. 2)

Engler, H. G. A. and K. Prantl, eds. 1887–1915. Die natürlichen Pflanzenfamilien.... 254 fascs. Leipzig. [Sequence of fasc. (Lieferung) numbers (order of publication) is independent of the sequence of division (Teil) and subdivision (Abteilung) numbers (taxonomic order).] (Nat. Pflanzenfam.)

Enum.—See: P. C. Fabricius 1759

Enum. Hort. Berol. Alt.—See: J. H. F. Link 1821–1822

Enum. Pl.—See: C. L. Willdenow 1809–1813[–1814]

Enum. Pl. Afric. Austral.—See: C. F. Ecklon and K. L. P. Zeyher [1834–]1835–1836[–1837]

Enum. Pl. China Bor.—See: A. A. Bunge [1833]

Enum. Pl. Zeyl.—See: G. H. K. Thwaites [1858–]1864

Enum. Subst. Braz.—See: A. L. P. da Silva Manso 1836

Enum. Syst. Pl.—See: N. J. Jacquin 1760

Erikson, D. L. et al. 2005. An Asian origin for a 10,000-year-old domesticated plant in the Americas. Proc. Natl. Acad. Sci. U.S.A. 102: 18315–18320.

Erythea = Erythea; a Journal of Botany, West American and General.

Essai Hist. Nat. Courges—See: A. N. Duchesne 1786

Estill, J. C. and M. B. Cruzan. 2001. Phytogeography of rare plant species endemic to the southeastern United States. Castanea 66: 3–23.

Euphytica = Euphytica. Netherlands Journal of Plant Breeding.

Evolution = Evolution; International Journal of Organic Evolution.

Exot. Fl.—See: W. J. Hooker [1822–]1823–1827

Fabijan, D. M., J. G. Packer, and K. E. Denford. 1987. The taxonomy of the *Viola nuttallii* complex. Canad. J. Bot. 65: 2562–2580.

Fabricius, P. C. 1759. Enumeratio Methodica Plantarum Horti Medici Helmstadiensis.... Helmstedt. (Enum.)

Fassett, N. C. 1939. Notes from the herbarium of the University of Wisconsin. No. 17. *Elatine* and other aquatics. Rhodora 41: 367–377.

Fawcett, W. and A. B. Rendle. 1910–1936. Flora of Jamaica.... 5 vols. London. (Fl. Jamaica)

Fay, M. F. et al. 1998. Plastid *rbc*L sequence data indicate a close affinity between *Diegodendron* and *Bixa*. Taxon 47: 43–50.

Feddes Repert. = Feddes Repertorium.

Felger, R. S. 2000. Flora of the Gran Desierto and Río Colorado of Northwestern Mexico. Tucson.

Feng, M. 2005. Floral Morphogenesis and Molecular Systematics of the Family Violaceae. Ph.D. dissertation. Ohio University.

Feng, M. and H. E. Ballard. 2005. Molecular systematic, floral developmental and anatomical revelations on generic relationships and evolutionary patterns in the Violaceae. [Abstract.] In: International Botanical Congress. 2005. XVII International Botanical Congress, Vienna, Austria, Europe, Austria Center Vienna, 17–23 July 2005. Abstracts Vienna. P. 169.

Ferguson, I. K. and J. Muller, eds. 1976. The Evolutionary Significance of the Exine. London.

Fernald, M. L. 1917. The genus *Elatine* in eastern North America. Rhodora 19: 10–15.

Fernald, M. L. 1941b. *Elatine americana* and *E. triandra*. Rhodora 43: 208–211.

Fernald, M. L. 1942c. *Hibiscus moscheutos* and *H. palustris*. Rhodora 44: 266–278.

Fernald, M. L. 1943g. The fruit of *Dirca palustris*. Rhodora 45: 117–119.

Fernald, M. L. 1950. Gray's Manual of Botany, ed. 8. New York.

Fernández, Aveliano. 1997. Estudio citogenético en hibridos entre una especie octoploide, *Turnera aurelii* y dos diploides, *T. caerulea* y *T. joelii*. Bonplandia (Corrientes) 9: 281–286.

Fernández, Aveliano and M. M. Arbo. 2000. Cytogenetic relationships between *Turnera aurelii*, *T. cuneiformis* ($2n = 8x = 40$) and *T. orientalis* ($2n = 6x = 30$) (Turneraceae). Cytologia 65: 97–102.

Feuillet, C. and J. M. MacDougal. 2003. A new infrageneric classification of *Passiflora* L. (Passifloraceae). Passiflora 13: 34–35, 37–38.

Feuillet, C. and J. M. MacDougal. 2007. Passifloraceae. In: K. Kubitzki et al., eds. 1990+. The Families and Genera of Vascular Plants. 10+ vols. Berlin etc. Vol. 9, pp. 270–281.

Field & Lab. = Field & Laboratory.

Filipowicz, N. and S. S. Renner. 2010. The worldwide holoparasitic Apodanthaceae confidently placed in the Cucurbitales by nuclear and mitochondrial gene trees. B. M. C. Evol. Biol. 10: 219.

Fl. Altaica—See: C. F. von Ledebour 1829–1833

Fl. Amer. Sept.—See: F. Pursh [1813]1814

Fl. Angl. ed. 2—See: W. Hudson 1778

Fl. Austriac.—See: N. J. Jacquin 1773–1778

Fl. Bor.-Amer.—See: W. J. Hooker [1829–]1833–1840; A. Michaux 1803

Fl. Bras.—See: C. F. P. von Martius et al. 1840–1906

Fl. Bras. Merid.—See: A. St.-Hilaire et al. 1824[–1833]

Fl. Brit. W. I.—See: A. H. R. Grisebach [1859–]1864

Fl. Calif.—See: W. L. Jepson 1909–1943

Fl. Carol.—See: T. Walter 1788

Fl. Chil.—See: C. Gay 1845–1854

Fl. Cochinch.—See: J. de Loureiro 1790

Fl. Francisc.—See: E. L. Greene 1891–1897

Fl. Ind. ed. 1832—See: W. Roxburgh 1832

Fl. Jamaica—See: W. Fawcett and A. B. Rendle 1910–1936

Fl. Less. Antill.—See: R. A. Howard 1974–1989

Fl. Miami—See: J. K. Small 1913b

Fl. N. Amer.—See: J. Torrey and A. Gray 1838–1843

Fl. N.W. Amer.—See: T. J. Howell 1897–1903

Fl. Pedem.—See: C. Allioni 1785

Fl. Ross.—See: C. F. von Ledebour [1841]1842–1853

Fl. S.E. U.S.—See: J. K. Small 1903

Fl. Seneg. Tent.—See: J. B. A. Guillemin et al. 1831–1833

Fl. South. U.S.—See: A. W. Chapman 1860

Fl. South. U.S. ed. 2 repr. 1892—See: A. W. Chapman [1892]

Fl. South. U.S. ed. 3—See: A. W. Chapman 1897

Fl. Tellur.—See: C. S. Rafinesque 1836[1837–1838]

Fl. Trop. Afr.—See: D. Oliver et al. 1868–1937

Flexner, S. B. and L. C. Hauck, eds. 1987. The Random House Dictionary of the English Language, ed. 2 unabridged. New York.

Flora of Taiwan Editorial Committee, eds. 1975–1979. Flora of Taiwan. 6 vols. Taipei.

Forrest, L. L., M. Hughes, and P. M. Hollingsworth. 2005. A phylogeny of Begonia using nuclear ribosomal sequence data and morphological characters. Syst. Bot. 30: 671–682.

Fosberg, F. R. and M.-H. Sachet. 1972. Thespesia populnea (L.) Solander ex Correa and Thespesia populneoides (Roxburgh) Kosteletzky (Malvaceae). Smithsonian Contr. Bot. 7: 1–13.

Freitas, L. and M. Sazima. 2003. Floral biology and pollination mechanisms in two Viola species—From nectar to pollen flowers? Ann. Bot. (Oxford) 91: 311–317.

Fries, R. E. 1908. Entwurf einer Monographie der Gattungen Wissadula und Pseudabutilon. Kongl. Svenska Vetensk. Acad. Handl., n. s. 43(4): 1–114.

Fryxell, J. E. and P. A. Fryxell. 1983. A revision of Abutilon sect. Oligocarpae (Malvaceae), including a new species from Mexico. Madroño 30: 84–92.

Fryxell, P. A. 1969. The genus Cienfuegosia Cav. Ann. Missouri Bot. Gard. 56: 179–250.

Fryxell, P. A. 1978. Neotropical segregates from Sida L. (Malvaceae). Brittonia 30: 447–462.

Fryxell, P. A. 1979. The Natural History of the Cotton Tribe. College Station, Tex.

Fryxell, P. A. 1980. A Revision of the American Species of Hibiscus Section Bombicella (Malvaceae). Washington. [U.S.D.A. Techn. Bull. 1624.]

Fryxell, P. A. 1985. Sidus sidarum V. The North and Central American species of Sida. Sida 11: 62–91.

Fryxell, P. A. 1987. Revision of the genus Anoda (Malvaceae). Aliso 11: 485–522.

Fryxell, P. A. 1988. Malvaceae of Mexico. Syst. Bot. Monogr. 25.

Fryxell, P. A. 1992. A revised taxonomic interpretation of Gossypium L. (Malvaceae). Rheedea 2: 108–165.

Fryxell, P. A. 1997. The American genera of Malvaceae—II. Brittonia 49: 204–269.

Fryxell, P. A. 1997b. A revision and redefinition of Pseudabutilon (Malvaceae). Contr. Univ. Michigan Herb. 21: 175–195.

Fryxell, P. A. 1998. A synopsis of the neotropical species of Triumfetta (Tiliaceae). In: P. Mathew and M. Sivadasan, eds. 1998. Diversity and Taxonomy of Tropical Flowering Plants. Calicut. Pp. 167–192.

Fryxell, P. A. 1999. Pavonia. In: Organization for Flora Neotropica. 1968+. Flora Neotropica. 109+ nos. New York. No. 76.

Fryxell, P. A. 2002. An Abutilon nomenclator. Lundellia 5: 79–118.

Fryxell, P. A., A. Krapovickas, and D. W. Crewz. 1984. Sidus sidarum IV. A new record of Sida in North America, S. santaremensis. Sida 10: 319–320.

Fryxell, P. A. and J. Valdès R. 1991. Observations of Fryxellia pygmaea (Malvaceae). Sida 14: 399–404.

Fuchs, A. 1938. Beiträge zur Embryologie der Thymelaeaceae. Oesterr. Bot. Z. 87: 1–41.

Fursa, T. B. 1972. K sistematike roda Citrullus Schrad. (On the taxonomy of genus Citrullus Schrad.) Bot. Zhurn. (Moscow & Leningrad) 57: 31–41.

Fursa, T. B. 1981. Intraspecific classification of water-melon under cultivation. Kulturpflanze 29: 297–300.

Gard. Chron. = Gardener's Chronicle.

Gard. Dict. ed. 8—See: P. Miller 1768

Gard. Dict. Abr. ed. 4—See: P. Miller 1754

Gartenflora = Gartenflora; Monatsschrift für deutsche und schweizerische Garten- und Blumenkunde.

Gaskin, J. F. 2002. Systematic and Population Studies of the Invasive Plant Tamarix. Ph.D. dissertation. Washington University.

Gaskin, J. F. et al. 2004. A systematic overview of Frankeniaceae and Tamaricaceae from nuclear rDNA and plastid sequence data. Ann. Missouri Bot. Gard. 91: 401–409.

Gaskin, J. F. and B. A. Schaal. 2002. Hybrid Tamarix widespread in U.S. invasion and undetected in native Asian range. Proc. Natl. Acad. Sci. U.S.A. 99: 11256–11259.

Gaskin, J. F. and B. A. Schaal. 2003. Molecular phylogenetic investigation of U.S. invasive Tamarix. Syst. Bot. 28: 86–95.

Gaskin, J. F. and P. B. Shafroth. 2005. Hybridization of Tamarix ramosissima and T. chinensis (saltcedars) with T. aphylla (athel) (family Tamaricaceae) in the southwestern USA determined from DNA sequence data. Madroño 52: 1–10.

Gay, C. 1845–1854. Historia Física y Política de Chile.... Botánica [Flora Chilena]. 8 vols., atlas. Paris and Santiago. (Fl. Chil.)

Gen. Amer. Bor.—See: A. Gray 1848–1849

Gen. Hist.—See: G. Don 1831–1838

Gen. N. Amer. Pl.—See: T. Nuttall 1818

Gen. Pl.—See: G. Bentham and J. D. Hooker 1862–1883

Gen. Pl. ed. 5—See: C. Linnaeus 1754

Gen. Sp. Pl.—See: M. Lagasca y Segura 1816b

Genet. Molec. Biol. = Genetics and Molecular Biology.

Genet. Resources Crop Evol. = Genetic Resources and Crop Evolution; an International Journal.

Gentes Herbarum = Gentes Herbarum; Occasional Papers on the Kinds of Plants.

Gershoy, A. 1928. Studies in North American violets. I. General considerations. Bull. Vermont Agric. Exp. Sta. 279.

Gershoy, A. 1934. Studies in North American violets. III. Chromosome numbers and species characters. Bull. Vermont Agric. Exp. Sta. 367.

Gettys, L. A. 2012. Genetic control of white flower color in scarlet rosemallow (*Hibiscus coccineus* Walter). J. Heredity 103: 594–597.

Ghebretinsae, A. G., M. Thulin, and J. C. Barber. 2007. Relationships of cucumbers and melons unraveled: Molecular phylogenetics of *Cucumis* and related genera (Benincaseae, Cucurbitaceae). Amer. J. Bot. 94: 1256–1266.

Gil-Ad, N. L. 1975. Systematics and Evolution of *Viola* L. Subsection *Boreali-Americanae* (W. Becker) Brizicky. Ph.D. dissertation. University of Michigan.

Gil-Ad, N. L. 1997. Systematics of *Viola* L. subsection *Boreali-Americanae*. Boissiera 53: 1–130.

Gil-Ad, N. L. 1998. The micromorphologies of seed coats and petal trichomes of the taxa of *Viola* subsect. *Boreali-Americanae* (Violaceae) and their utility in discerning orthospecies from hybrids. Brittonia 50: 91–121.

Gilbert, L. E. 1982. The coevolution of a butterfly and a vine. Sci. Amer. 247: 110-121.

Gilg, E. 1894. Thymelaeaceae. In: H. G. A. Engler and K. Prantl, eds. 1887–1915. Die natürlichen Pflanzenfamilien.... 254 fascs. Leipzig. Fasc. 106[III,6a], pp. 217–245.

Gillett, J. M. and N. K. B. Robson. 1981. The St. John's-worts of Canada (Guttiferae). Publ. Bot. (Ottawa) 11: 1–40.

Gleason, H. A. 1947. Notes on some American plants: *Triadenum*. Phytologia 2: 288–291.

Gleason, H. A. and A. Cronquist. 1991. Manual of Vascular Plants of Northeastern United States and Adjacent Canada, ed. 2. Bronx.

Gmelin, J. F. 1791[–1792]. Caroli à Linné...Systema Naturae per Regna Tria Naturae.... Tomus II. Editio Decima Tertia, Aucta, Reformata. 2 parts. Leipzig. (Syst. Nat.)

Godfrey, R. K. and J. W. Wooten. 1981. Aquatic and Wetland Plants of Southeastern United States: Dicotyledons. Athens, Ga.

Goldberg, A. 1967. The genus *Melochia* L. (Sterculiaceae). Contr. U.S. Natl. Herb. 35: 191–363.

Goldman, D. H. 2003. Two species of *Passiflora* (Passifloraceae) in the Sonoran Desert and vicinity: A new taxonomic combination and an introduced species in Arizona. Madroño 50: 243–264.

Goldman, D. H. 2004. Lectotypification of *Passiflora affinis* (Passifloraceae) and discussion of its geographic range within the United States. Sida 21: 275–285.

Gray, A. 1848–1849. Genera Florae Americae Boreali-orientalis Illustrata. The Genera of the Plants of the United States.... 2 vols. Boston, New York, and London. (Gen. Amer. Bor.)

Gray, A. 1850. Plantae Lindheimerianae. Part 2. Boston J. Nat. Hist. 6: 141–240.

Gray, A. 1854. Plantae Novae Thurberianae.... Cambridge, Mass. [Preprinted from Mem. Amer. Acad. Arts, n. s. 5: 297–328. 1855.] (Pl. Nov. Thurb.)

Gray, A. 1867. A Manual of the Botany of the Northern United States..., ed. 5. New York and Chicago. [Pteridophytes by D. C. Eaton.] (Manual ed. 5)

Gray, A., S. Watson, B. L. Robinson, et al. 1878–1897. Synoptical Flora of North America. 2 vols. in parts and fascs. New York etc. [Vol. 1(1,1), 1895; vol. 1(1,2), 1897; vol. 1(2), 1884; vol. 2(1), 1878.] (Syn. Fl. N. Amer.)

Great Basin Naturalist Mem. = Great Basin Naturalist Memoirs.

Great Plains Flora Association. 1986. Flora of the Great Plains. Lawrence, Kans.

Grebenscikov, I. 1953. Die Entwicklung der Melonsystematik. Kulturpflanze 1: 121–138.

Greene, E. L. 1891–1897. Flora Franciscana. An Attempt to Classify and Describe the Vascular Plants of Middle California. 4 parts. San Francisco. [Parts paged consecutively.] (Fl. Francisc.)

Greene, E. L. 1898b. New or noteworthy violets. Pittonia 3: 313–320.

Gremillion, K. J. 1989. The development of a mutualistic relationship between humans and maypops (*Passiflora incarnata* L.) in the southeastern United States. J. Ethnobiol. 9: 135–155.

Greuter, W., A. T. Leiva Sánchez, and P. Pérez Alvarez, eds. 1998+. Flora de la República de Cuba: Serie A, Plantas Vasculares. 14+ fascs. Koenigstein.

Grisebach, A. H. R. [1859–]1864. Flora of the British West Indian Islands. 7 parts. London. [Parts paged consecutively.] (Fl. Brit. W. I.)

Grisebach, A. H. R. 1866. Catalogus Plantarum Cubensium Exhibens Collectionem Wrightianam Aliasque Minores ex Insula Cuba Missas. Leipzig. (Cat. Pl. Cub.)

Guillemin, J. B. A., G. S. Perrottet, and A. Richard. 1831–1833. Florae Senegambiae Tentamen.... Tomus Primus. 8 parts. Paris and London. [All published; parts paged consecutively.] (Fl. Seneg. Tent.)

Gupta, A. K. and Y. S. Murty. 1987. Floral anatomy in Tamaricaceae. J. Indian Bot. Soc. 66: 275–282.

Gustafsson, M. H. G., V. Bittrich, and P. F. Stevens. 2002. Phylogeny of Clusiaceae based on *rbc*L sequences. Int. J. Pl. Sci. 163: 1045–1054.

Guymer, G. P. 1988. A taxonomic revision of *Brachychiton* (Sterculiaceae). Austral. Syst. Bot. 1: 199–323.

Guzmán, B. and P. Vargas. 2005. Systematics, character evolution, and biogeography of *Cistus* L. (Cistaceae) based on ITS, *trn*L-*trn*F, and *mat*K sequences. Molec. Phylogen. Evol. 37: 644–660.

Guzmán, B. and P. Vargas. 2009. Historical biogeography and character evolution of Cistaceae (Malvales) based on analysis of plastid *rbc*L and *trn*L-*trn*F sequences. Organisms Diversity Evol. 9: 83–99.

Haines, A. 2001b. The Genus *Viola* of Maine: A Taxonomic and Ecological Reference. Bar Harbor.

Haines, A. 2011. New England Wildflower Society's Flora Novae Angliae.... New Haven.

Halda, J. J. 2001. The Genus *Daphne*. Dobré.

Halse, R. R., B. A. Rottink, and R. Mishaga. 1989. Studies in *Sidalcea* taxonomy. NorthW. Sci. 63: 154–161.

Hamilton, W. 1825. Prodromus Plantarum Indiae Occidentalis.... London, Paris, and Strasbourg. (Prodr. Pl. Ind. Occid.)

Hammel, B. E. 2009. A new species of *Cyclanthera* (Cucurbitaceae) from Alajuela Province, Costa Rica. Novon 19: 49–51.

Hammer, K., P. Hanelt, and P. Perrino. 1986. Carosello and the taxonomy of *Cucumis melo* L. especially of its vegetable races. Kulturpflanze 34: 249–259.

Hammer, R. L. 2002. Everglades Wildflowers.... Guilford, Conn.

Handb. Laubholzk.—See: L. Dippel 1889–1893

Hanelt, P., ed. 2001. Mansfeld's Encyclopedia of Agricultural and Horticultural Crops.... 6 vols. Berlin and New York.

Hansen, A. K. et al. 2006. Phylogenetic relationships and chromosome number evolution in *Passiflora*. Syst. Bot. 31: 138–150.

Hardin, J. W. 1990. Variation patterns and recognition of varieties of *Tilia americana* s.l. Syst. Bot. 15: 33–48.

Harrington, H. D. 1954. Manual of the Plants of Colorado. Denver. (Man. Pl. Colorado)

Haston, E. M. et al. 2007. A linear sequence of Angiosperm Phylogeny Group II families. Taxon 56: 7–12.

Hatch, S. L., K. N. Gandhi, and L. E. Brown. 1990. Checklist of the Vascular Plants of Texas. College Station, Tex.

Hayden, W. J. and J. Clough. 1990. Methyl salicilate in roots of *Viola arvensis* and *V. rafinesquii* (Violaceae). Castanea 55: 65–70.

Hayek, A. von. 1904–1912. Schedae Floram Stiriacum Exsiccatum.... 26 parts in 13 vols. Vienna. (Sched. Fl. Stiriac.)

Heiser, C. B. 1979. The Gourd Book. Norman.

Heiser, C. B. and E. E. Schilling. 1990. The genus *Luffa*: A problem in phytogeography. In: D. M. Bates et al., eds. 1990. Biology and Utilization of the Cucurbitaceae. Ithaca, N.Y. Pp. 120–133.

Hekking, W. H. A. 1988. Violaceae Part I—*Rinorea* and *Rinoreocarpus*. In: Organization for Flora Neotropica. 1968+. Flora Neotropica. 109+ nos. New York. No. 46.

Heredity = Heredity; an International Journal of Genetics.

Hernández, T. et al. 2003. Ethnobotany and antibacterial activity of some plants used in traditional medicine of Zapotitlán de las Salinas, Puebla (México). J. Ethnopharmacol. 88: 181–188.

Herrera, C. M. 1990. The adaptedness of the floral phenotype in a relict endemic, hawkmoth-pollinated violet. 1. Reproductive correlates of floral variation. Biol. J. Linn. Soc. 40: 263–274.

Heywood, V. H., ed. 1978. Flowering Plants of the World. Oxford.

Hickman, J. C., ed. 1993. The Jepson Manual. Higher Plants of California. Berkeley, Los Angeles, and London.

Hickok, L. G. and J. C. Anway. 1972. A morphological and chemical analysis of geographical variation in *Tilia* L. of eastern North America. Brittonia 21: 2–8.

Hiern, W. P. 1896–1901. Catalogue of the African Plants Collected by Dr. Friedrich Welwitsch in 1853–61.... 2 vols in parts. London. [Parts paged consecutively within volumes.] (Cat. Afr. Pl.)

Hill, S. R. 1981. Supplement to the Flora of the Texas Coastal Bend by F. B. Jones. Sida 9: 43–54.

Hill, S. R. 1982. A monograph of the genus *Malvastrum* A. Gray (Malvaceae: Malveae). Rhodora 84: 1–83, 159–264, 317–409.

Hill, S. R. 1993. *Sidalcea*. In: J. C. Hickman, ed. 1993. The Jepson Manual. Higher Plants of California. Berkeley, Los Angeles, and London. Pp. 755–760.

Hill, S. R. 2009. Notes on California Malvaceae including nomenclatural changes and additions to the flora. Madroño 56: 104–111.

Hill, S. R. 2012. *Hibiscus*. In: B. G. Baldwin et al., eds. 2012. The Jepson Manual: Vascular Plants of California, ed. 2. Berkeley. P. 881.

Hill, S. R. 2012b. *Sidalcea*. In: B. G. Baldwin et al., eds. 2012. The Jepson Manual: Vascular Plants of California, ed. 2. Berkeley. Pp. 887–896.

Hist. Nat. Îles Canaries—See: P. B. Webb and S. Berthelot [1835–]1836–1850

Hist. Nat. Vég.—See: É. Spach 1834–1848

Hist. Pl.—See: H. E. Baillon 1866–1895

Hist. Pl. Guiane—See: J. B. Aublet 1775

Hitchcock, C. L. 1957. A Study of the Perennial Species of *Sidalcea*. Seattle. [Univ. Wash. Publ. Biol. 18.] (Perenn. Sp. Sidalcea)

Hitchcock, C. L. et al. 1955–1969. Vascular Plants of the Pacific Northwest. 5 vols. Seattle. [Univ. Wash. Publ. Biol. 17.] (Vasc. Pl. Pacif. N.W.)

Hochreutiner, B. P. G. 1900. Révision du genre *Hibiscus*. Annuaire Conserv. Jard. Bot. Genève 4: 23–190.

Hodgdon, A. R. 1938. A taxonomic study of *Lechea*. Rhodora 40: 29–69, 87–131.

Hodges, S. A. et al. 1995. Generic relationships in the Violaceae: Data from morphology, anatomy, chromosome numbers and *rbc*L sequences. [Abstract.] Amer. J. Bot. 82(6, suppl.): 136.

Hodgson, W. C. 2001. Food Plants of the Sonoran Desert. Tucson.

Hodick, D. and A. Sievers. 1988. On the mechanism of trap closure of Venus flytrap (*Dionaea muscipula* Ellis). Planta 179: 32–42.

Holloway, J. K. 1957. Weed control by an insect. Sci. Amer. 54: 57–62.

Holm, T. 1921. *Dirca palustris* L. A morphological study. Amer. J. Sci., ser. 5, 2: 177–182.

Holmgren, N. H. 2005c. Frankeniaceae. In: A. Cronquist et al. 1972+. Intermountain Flora. Vascular Plants of the Intermountain West, U.S.A. 6+ vols. in 7+. New York and London. Vol. 2, part B, pp. 72–74.

Holmgren, N. H. 2005d. Violaceae. In: A. Cronquist et al. 1972+. Intermountain Flora. Vascular Plants of the Intermountain West, U.S.A. 6+ vols. in 7+. New York and London. Vol. 2, part B, pp. 53–68.

Hooker, W. J. [1822–]1823–1827. Exotic Flora, Containing Figures and Descriptions of New, Rare, or Otherwise Interesting Exotic Plants.... 3 vols. in 38 fascs. Edinburgh. (Exot. Fl.)

Hooker, W. J. [1829–]1833–1840. Flora Boreali-Americana; or, the Botany of the Northern Parts of British America.... 2 vols. in 12 parts. London, Paris, and Strasbourg. (Fl. Bor.-Amer.)

Hooker, W. J. and G. A. W. Arnott. [1830–]1841. The Botany of Captain Beechey's Voyage; Comprising an Account of the Plants Collected by Messrs Lay and Collie, and Other Officers of the Expedition, during the Voyage to the Pacific and Bering's Strait, Performed in His Majesty's Ship Blossom, under the Command of Captain F. W. Beechey...in the Years 1825, 26, 27, and 28. 10 parts. London. [Parts paged and plates numbered consecutively.] (Bot. Beechey Voy.)

Hooker's J. Bot. Kew Gard. Misc. = Hooker's Journal of Botany and Kew Garden Miscellany.

Hornschuch, C. F. [1822–]1824–1828. Sylloge Plantarum Novarum.... 2 vols. Regensburg. (Syll. Pl. Nov.)

Hort. Berol.—See: C. L. Willdenow 1803–1816

Hort. Brit.—See: R. Sweet 1826

Hort. Kew.—See: W. Aiton 1789

Hort. Rev. = Horticultural Reviews.

Hort. Suburb. Calcutt.—See: J. O. Voigt 1845

Horton, J. S. 1964. Notes on the Introduction of Deciduous *Tamarix*. Fort Collins, Colo. [U.S.D.A. Forest Serv., Gen. Techn. Rep. RM-16.]

HortTechnol. = HortTechnology.

Howard, R. A. 1974–1989. Flora of the Lesser Antilles: Leeward and Windward Islands. 6 vols. Jamaica Plain. (Fl. Less. Antill.)

Howell, T. J. 1897–1903. A Flora of Northwest America. 1 vol. in 8 fascs. Portland. [Fascs. 1–7 (text) paged consecutively, fasc. 8 (index) independently.] (Fl. N.W. Amer.)

Hrusa, F. et al. 2002. Catalogue of non-native vascular plants occurring spontaneously in California beyond those addressed in The Jepson Manual—Part I. Madroño 49: 61–98.

Hsieh, C. F. 1977. Violaceae. In: Flora of Taiwan Editorial Committee, eds. 1975–1979. Flora of Taiwan. 6 vols. Taipei. Vol. 3, pp. 769–784.

Hudson, W. 1778. Flora Anglica..., ed. 2. 2 vols. London. (Fl. Angl. ed. 2)

Hultén, E. 1941–1950. Flora of Alaska and Yukon. 10 vols. Lund and Leipzig. [Vols. paged consecutively and designated as simultaneous numbers of Lunds Univ. Årsskr. (= Acta Univ. Lund.) and Kungl. Fysiogr. Sällsk. Handl.]

Hultén, E. 1968. Flora of Alaska and Neighboring Territories: A Manual of the Vascular Plants. Stanford.

Hultén, E. and M. Fries. 1986. Atlas of North European Vascular Plants North of the Tropic of Cancer. 3 vols. Königstein.

Humboldt, A. von, A. J. Bonpland, and C. S. Kunth. 1815[1816]–1825. Nova Genera et Species Plantarum Quas in Peregrinatione Orbis Novi Collegerunt, Descripserunt.... 7 vols. in 36 parts. Paris. (Nov. Gen. Sp.)

Hunkins, S. 2012. New records for the flora of Arizona. Canotia 8: 66.

Iamonico, D. and L. Peruzzi. 2014. Typification of Linnaean names in Malvaceae for the Italian flora. Taxon 63: 161–166.

Icon. Pl. = Icones Plantarum....

Iconogr. Bot. Pl. Crit.—See: H. G. L. Reichenbach 1823–1832

Ill. Fl. Illinois, Fl. Pl. Hollies–Loasas—See: R. H. Mohlenbrock 1978

Ill. Fl. N. U.S. ed. 2—See: N. L. Britton and A. Brown 1913

Ill. Fl. Pacific States—See: L. Abrams and R. S. Ferris 1923–1960

Ill. Pl. Orient.—See: H. F. Jaubert and É. Spach 1842–1857

Illinois Biol. Monogr. = Illinois Biological Monographs.

Iltis, H. H. 1963. *Napaea dioica* (Malvaceae): Whence came the type? Amer. Midl. Naturalist 70: 90–109.

Iltis, H. H. and S. Kawano. 1964. Cytotaxonomy of *Napaea dioica* (Malvaceae). Amer. Midl. Naturalist 72: 76–81.

Index Kew.—See: B. D. Jackson et al. [1893–]1895+

Index Seminum (Dorpat) = Index Seminum Horti Academici Dorpatensis.

Index Seminum (Göttingen) = Index Seminum Horti Academici Goettingensis Anno...Collecta.

Index Seminum (Tokyo) = Catalogus Seminum et Sporarum in Horto Botanico Universitatis Imperialis Tokyoensis....

Inst. Rei Herb.—See: H. J. N. von Crantz 1766

Int. J. Bioinf. Res. = International Journal of Bioinformatics Research.

Int. J. Pl. Sci. = International Journal of Plant Sciences.

International Botanical Congress. 2005. XVII International Botanical Congress, Vienna, Austria, Europe, Austria Center Vienna, 17–23 July 2005. Abstracts. Vienna.

Israel J. Bot. = Israel Journal of Botany.

Ives, J. C. 1861. Report upon the Colorado River of the West, Explored in 1857 and 1858 by Lieutenant Joseph C. Ives.... 5 parts, appendices. Washington. (Rep. Colorado R.)

J. Acad. Nat. Sci. Philadelphia = Journal of the Academy of Natural Sciences of Philadelphia.

J. Arizona-Nevada Acad. Sci. = Journal of the Arizona-Nevada Academy of Science.

J. Arnold Arbor. = Journal of the Arnold Arboretum.

J. Biosci. = Journal of Biosciences.

J. Bot. = Journal of Botany, British and Foreign.

J. Bot. (Hooker) = Journal of Botany, (Being a Second Series of the Botanical Miscellany), Containing Figures and Descriptions....

J. Bot. Res. Inst. Texas = Journal of the Botanical Research Institute of Texas.

J. Bot. (Schrader) = Journal für die Botanik. [Edited by H. A. Schrader.] [Volumation indicated by nominal year date and vol. number for that year (1 or 2); e.g. 1800(2).]

J. Ecol. = Journal of Ecology.

J. Elisha Mitchell Sci. Soc. = Journal of the Elisha Mitchell Scientific Society.

J. Ethnobiol. = Journal of Ethnobiology.

J. Ethnopharmacol. = Journal of Ethnopharmacology; Interdisciplinary Journal Devoted to Bioscientific Research on Indigenous Drugs.

J. Heredity = Journal of Heredity.

J. Indian Bot. Soc. = Journal of the Indian Botanical Society.

J. New York Bot. Gard. = Journal of the New York Botanical Garden.

J. Palynol. = Journal of Palynology.

J. Pl. Res. = Journal of Plant Research. [Shokubutsu-gaku zasshi.]

J. Pl. Resources Environm. = Journal of Plant Resources and Environment.

J. Torrey Bot. Soc. = Journal of the Torrey Botanical Society.

J. Wash. Acad. Sci. = Journal of the Washington Academy of Sciences.

Jackson, B. D. et al., comps. [1893–]1895+. Index Kewensis Plantarum Phanerogamarum.... 2 vols. + 21+ suppls. Oxford. (Index Kew.)

Jacquin, N. J. 1760. Enumeratio Systematica Plantarum, Quas in Insulis Caribaeis Vicinaque Americes Continente Detexit Novas.... Leiden. (Enum. Syst. Pl.)

Jacquin, N. J. 1764–1771. Observationum Botanicarum.... 4 parts. Vienna. (Observ. Bot.)

Jacquin, N. J. 1773–1778. Florae Austriaceae.... 5 vols. Vienna. (Fl. Austriac.)

Jacquin, N. J. 1786[1787]–1796[1797]. Collectanea ad Botanicam, Chemiam, et Historiam Naturalem Spectantia.... 5 vols. Vienna. (Collectanea)

Jäger-Zürn, I. 2005. Morphology and morphogenesis of ensiform leaves, syndesmy of shoots and an understanding of the thalloid plant body in species of *Apinagia, Mourera* and *Marathrum* (Podostemaceae). Bot. J. Linn. Soc. 147: 47–71.

Jahrb. Königl. Bot. Gart. Berlin = Jahrbuch des Königlichen botanischen Gartens und des Botanischen Museums zu Berlin.

Jahresber. Naturwiss. Vereins Halle = Jahresbericht des Naturwissenschaftlichen Vereins in Halle.

Janssen, G. K. 1999. Site Characteristics and Management of Johnston's Frankenia *(Frankenia johnstonii)*. Austin.

Jarret, R. L. and M. Newman. 2000. Phylogenetic relationships among species of *Citrullus* and the placement of *C. rehmii* De Winter as determined by internal transcribed spacer (ITS) sequence heterogeneity. Genet. Resources Crop Evol. 47: 215–222.

Jaubert, H. F. and E. Spach. 1842–1857. Illustrationes Plantarum Orientalium.... 5 vols. in 50 parts. Paris. [Vols. paged independently, parts numbered consecutively.] (Ill. Pl. Orient.)

Jeffrey, C. 1962. Notes on Cucurbitaceae, including a proposed new classification of the family. Kew Bull. 15: 337–371.

Jeffrey, C. 1969. A review of the genus *Bryonia* L. (Cucurbitaceae). Kew Bull. 23: 441–461.

Jeffrey, C. 1971. Further notes on Cucurbitaceae: II. Kew Bull. 25: 191–236.

Jeffrey, C. 1975. Further notes on Cucurbitaceae: IV. Some New World taxa. Kew Bull. 33: 347–380.

Jeffrey, C. 1980b. A review of the Cucurbitaceae. Bot. J. Linn. Soc. 81: 233–247.

Jeffrey, C. 1980c. Further notes on Cucurbitaceae: V. The Cucurbitaceae of the Indian subcontinent. Kew Bull. 34: 789–809.

Jeffrey, C. 1990. An outline classification of the Cucurbitaceae. In: D. M. Bates et al., eds. 1990. Biology and Utilization of the Cucurbitaceae. Ithaca, N.Y. Pp. 449–463.

Jeffrey, C. 2001. Cucurbitaceae. In: P. Hanelt, ed. 2001. Mansfeld's Encyclopedia of Agricultural and Horticultural Crops.... 6 vols. Berlin and New York. Vol. 3, pp. 1510–1557.

Jeffrey, C. 2005. A new system of Cucurbitaceae. Bot. Zhurn. (Moscow & Leningrad) 90(3): 332–335.

Jeffrey, C. and W. J. J. O. de Wilde. 2006. A review of the subtribe Thladianthinae (Cucurbitaceae). Bot. Zhurn. (Moscow & Leningrad) 91(5): 766–776.

Jenny, M. 1985. Struktur, Funktion und systematische Bedeutung des Gynoeciums bei Sterculiaceen. Ph.D. thesis. Universität Zurich.

Jenny, M. 1988. Different gynoecium types in Sterculiaceae: Ontogeny and functional aspects. In: P. Leins et al., eds. 1988. Aspects of Floral Development: Proceedings of the Double Symposium "Floral Development: Evolutionary Aspects and Special Topics," Held at the XIV International Botanical Congress, Berlin (West), Germany, July 24 to August 1, 1987. Berlin. Pp. 225–236.

Jepson, W. L. 1909–1943. A Flora of California.... 3 vols. in 12 parts. San Francisco etc. [Pagination consecutive within each vol.; vol. 1 page sequence independent of part number sequence (chronological); part 8 of vol. 1 (pp. 1–32, 579–index) never published.] (Fl. Calif.)

Jepson, W. L. [1923–1925.] A Manual of the Flowering Plants of California.... Berkeley. (Man. Fl. Pl. Calif.)

Jepson, W. L. 1951. A Manual of the Flowering Plants of California. Berkeley.

Johnson-Fulton, S. B. 2014. Systematics, Biogeography, and Ethnobotany of the Pantropical Family Cochlospermaceae (Malvales). Ph.D. dissertation. Miami University.

Johri, B. M. and D. Kak. 1954. The embryology of *Tamarix* L. Phytomorphology 4: 230–247.

Jones, C. E. 1969. A Revision of the Genus *Cyclanthera* (Cucurbitaceae). Ph.D. dissertation. Indiana University.

Jones, G. N. 1968. Taxonomy of American species of linden *(Tilia)*. Illinois Biol. Monogr. 39: 1–156.

Jonsell, B. and T. Karlsson, eds. 2000+. Flora Nordica. 3+ vols. Stockholm.

Judd, W. S. and S. R. Manchester. 1997. Circumscription of Malvaceae (Malvales) as determined by a preliminary cladistic analysis of morphological, anatomical, palynological, and chemical characters. Brittonia 49: 384–405.

Juniper, B. E., R. J. Robins, and D. M. Joel. 1989. The Carnivorous Plants. London and San Diego.

Kapoor, B. M. 1972. In: IOPB chromosome number reports XXXV. Taxon 21: 161–166.

Karsten, H. 1880–1883. Deutsche Flora. Pharmaceutisch-medicinische Botanik.... 13 Lieferungen. Berlin. (Deut. Fl.)

Kartesz, J. T. 1994. A Synonymized Checklist of the Vascular Flora of the United States, Canada, and Greenland, ed. 2. 2 vols. Portland.

Kartesz, J. T. and C. A. Meacham. 1999. Synthesis of the North American Flora, ver. 1.0. Chapel Hill. [CD-ROM.]

Kearney, T. H. 1935. The North American species of *Sphaeralcea* subgenus *Eusphaeralcea*. Univ. Calif. Publ. Bot. 19: 1–128.

Kearney, T. H. 1951. The genus *Malacothamnus* Greene (Malvaceae). Leafl. W. Bot. 6: 113–140.

Kearney, T. H. 1954. A tentative key to the North American species of *Sida* L. Leafl. W. Bot. 7: 138–150.

Kearney, T. H. 1955. A tentative key to the North American species of *Abutilon* Miller. Leafl. W. Bot. 7: 241–254.

Kearney, T. H. 1955b. A tentative key to the North American species of *Hibiscus* L. Leafl. W. Bot. 7: 274–284.

Kearns, D. M. 1994. The genus *Ibervillea* (Cucurbitaceae). An enumeration of the species and two new combinations. Madroño 41: 13–22.

Kearns, D. M. 1994b. A revision of *Tumamoca* (Cucurbitaceae). Madroño 41: 23–29.

Kearns, D. M. and E. C. Jones. 1992. A re-evaluation of the genus *Cremastopus* (Cucurbitaceae). Madroño 39: 301–303.

Keating, R. C. 1968. Comparative morphology of Cochlospermaceae. I. Synopsis of the family and wood anatomy. Phytomorphology 18: 379–392.

Keating, R. C. 1970. Comparative morphology of the Cochlospermaceae. II. Anatomy of the young vegetative shoot. Amer. J. Bot. 57: 889–898.

Keating, R. C. 1972. The comparative morphology of the Cochlospermaceae. III. The flower and pollen. Ann. Missouri Bot. Gard. 59: 282–296.

Kelman, W. M. 1991. A revision of *Fremontodendron* (Sterculiaceae). Syst. Bot. 16: 3–20.

Kelman, W. M. et al. 2006. Genetic relationships among *Fremontodendron* (Sterculiaceae) populations of the central Sierra Nevada foothills of California. Madroño 53: 380–387.

Kew Bull. = Kew Bulletin.

Kiger, R. W. and D. M. Porter. 2001. Categorical Glossary for the Flora of North America Project. Pittsburgh.

Killip, E. P. 1938. The American species of Passifloraceae. Publ. Field Mus. Nat. Hist., Bot. Ser. 19.

Kirkbride, J. H. 1993. Biosystematic Monograph of the Genus *Cucumis* (Cucurbitaceae): Botanical Identification of Cucumbers and Melons. Boone, N.C. (Biosyst. Monogr. Cucumis)

Klaber, D. 1976. Violets of the United States. South Brunswick, N.J.

Klips, R. A. 1995. Genetic affinity of the rare eastern Texas endemic *Hibiscus dasycalyx* (Malvaceae). Amer. J. Bot. 82: 1463–1472.

Knerr, L. D. and J. E. Staub. 1991. A multivariate re-evaluation of biochemical genetic diversity in *Cucumis sativus* L. Rep. Cucurbit Genet. Coop. 14: 25–28.

Kocyan, A. et al. 2007. A multi-locus chloroplast phylogeny for the Cucurbitaceae and its implications for character evolution and classification. Molec. Phylogen. Evol. 44: 553–577.

Köhler, E. 1971. Zur Pollenmorphologie der Gattung *Waltheria* L. (Sterculiaceae). Feddes Repert. 82: 125–153.

Köhler, E. 1976. Pollen dimorphism and heterostyly in the genus *Waltheria* L. (Sterculiaceae). In: I. K. Ferguson and J. Muller, eds. 1976. The Evolutionary Significance of the Exine. London. Pp. 147–161.

Kongl. Svenska Vetensk. Acad. Handl. = Kongl[iga]. Svenska Vetenskaps Academiens Handlingar.

Korotkova, N. et al. 2009. Phylogeny of the eudicot order Malpighiales: Analysis of a recalcitrant clade with sequences of the *pet*D group II intron. Pl. Syst. Evol. 282: 201–228.

Kostermans, A. J. G. H. 1954. A note on some African Sterculiaceae. Bull. Jard. Bot. État Bruxelles 24: 335–338.

Kostermans, A. J. G. H. 1956. The genus *Firmiana* (Sterculiaceae). Pengum. Balai Besar Penjel. Kehut. Indonesia 54: 3–33.

Kostermans, A. J. G. H. 1957. The genus *Firmiana* (Sterculiaceae). Reinwardtia 4: 281–310.

Kostermans, A. J. G. H. 1989. Notes on *Firmiana* Marsili (Sterculiaceae). Blumea 34: 117–118.

Krapovickas, A. 1996. La identidad de *Wissadula amplissima* (Malvaceae). Bonplandia (Corrientes) 9: 89–94.

Krapovickas, A. 2003. Las especies Austroamericanas del género *Cienfuegosia* Cav. Bonplandia (Corrientes) 12: 5–47.

Krapovickas, A. 2003b. *Sida* sección *Distichifolia* (Monteiro) Krapov. comb. nov., stat. nov. (Malvaceae–Malveae). Bonplandia (Corrientes) 12: 83–121.

Krapovickas, A. 2007. Las especies de *Sida* secc. *Malachroideae* del Cono Sur de Sudamérica. Bonplandia (Corrientes) 16: 209–225.

Krosnick, S. E. et al. 2013. New insights into the evolution of *Passiflora* subgenus *Decaloba* (Passifloraceae): Phylogenetic relationships and morphological synapomorphies. Syst. Bot. 38: 692–713.

Krosnick, S. E. and J. V. Freudenstein. 2005. Monophyly and floral character homology of Old World *Passiflora* (subgenus *Decaloba*: supersection *Disemma*). Syst. Bot. 30: 139–152.

Krutzsch, W. 1989. Paleogeography and historical phytogeography (paleochorology) in the Neophyticum. Pl. Syst. Evol. 162: 5–61.

Kubitzki, K. 2003b. Frankeniaceae. In: K. Kubitzki et al., eds. 1990+. The Families and Genera of Vascular Plants. 10+ vols. Berlin etc. Vol. 5, pp. 209–212.

Kubitzki, K. et al., eds. 1990+. The Families and Genera of Vascular Plants. 10+ vols. Berlin etc.

Kugler, E. E. and L. A. King. 2004. A brief history of the passionflower. In: T. Ulmer and J. M. MacDougal. 2004. *Passiflora*: Passionflowers of the World. Portland. Pp. 15–26.

Kulturpflanze = Kulturpflanze. Berichte und Mitteilungen aus dem Institut für Kulturpflanzenforschung der Deutschen Akademie der Wissenschaften zu Berlin in Gatersleben Krs. Aschersleben.

Kurz, H. and R. K. Godfrey. 1962. Trees of Northern Florida. Gainesville.

L'Héritier de Brutelle, C.-L. 1784[1785–1805]. Stirpes Novae aut Minus Cognitae.... 2 vols. in 9 fascs. Paris. [Fascicles paged and plates numbered consecutively.] (Stirp. Nov.)

La Duke, J. C. and D. K. Northington. 1978. The systematics of *Sphaeralcea coccinea* (Nutt.) Rydberg. SouthW. Naturalist 23: 651–660.

La Rosa, A. M. 1984. The Biology and Ecology of *Passiflora mollissima* in Hawaii. Honolulu.

Lagasca y Segura, M. 1816b. Genera et Species Plantarum.... Madrid. (Gen. Sp. Pl.)

Lamarck, J. et al. 1783–1817. Encyclopédie Méthodique. Botanique.... 13 vols. Paris and Liège. [Vols. 1–8, suppls. 1–5.] (Encycl.)

Lay, K. K. 1950. The American species of *Triumfetta* L. Ann. Missouri Bot. Gard. 37: 315–395.

Leafl. Bot. Observ. Crit. = Leaflets of Botanical Observation and Criticism.

Leafl. W. Bot. = Leaflets of Western Botany.

Ledebour, C. F. von. 1829–1833. Flora Altaica. 4 vols. Berlin. (Fl. Altaica)

Ledebour, C. F. von. [1841]1842–1853. Flora Rossica sive Enumeratio Plantarum in Totius Imperii Rossici Provinciis Europaeis, Asiaticis, et Americanis Hucusque Observatarum.... 4 vols. Stuttgart. (Fl. Ross.)

Ledebour, C. F. von and J. P. Alderstam. 1805. Dissertatio Botanica Sistens Plantarum Domingensium Decadem.... Greifswald. (Diss. Bot. Pl. Doming.)

Leinfellner, W. 1960. Zur Entwicklungsgeschichte der Kronblatter der Sterculiaceae-Buettnerieae. Oesterr. Bot. Z. 107: 153–176.

Leins, P. et al., eds. 1988. Aspects of Floral Development: Proceedings of the Double Symposium "Floral Development: Evolutionary Aspects and Special Topics," Held at the XIV International Botanical Congress, Berlin (West), Germany, July 24 to August 1, 1987. Berlin.

Leitão, C. A. E. et al. 2005. Anatomy of the floral, bract, and foliar nectaries of *Triumfetta semitriloba* (Tiliaceae). Canad. J. Bot. 83: 279–286.

Lengyel, S. et al. 2010. Convergent evolution of seed dispersal by ants, and phylogeny and biogeography in flowering plants: A global survey. Perspect. Pl. Ecol. Evol. Syst. 12: 43–55.

Les, D. H. et al. 1997c. The phylogenetic placement of river-weeds (Podostemaceae): Insights from *rbc*L sequence data. Aquatic Bot. 57: 5–27.

Lévesque, F. L. and P. M. Dansereau. 1966. Études sur les violettes jaunes caulescentes de l'est de l'Amérique du Nord. 1. Taxonomie, nomenclature, synonymie et bibliographie. Naturaliste Canad. 93: 489–569.

Levi, A. et al. 2010. DNA markers and pollen morphology reveal that *Praecitrullus fistulosus* is more closely related to *Benincasa hispida* than to *Citrullus* spp. Genet. Resources Crop Evol. 57: 1191–1205.

Levi, A. and C. E. Thomas. 2005. Polymorphisms among chloroplast and mitochondrial genomes of *Citrullus* species and subspecies. Genet. Resources Crop Evol. 52: 609–617.

Liang, G. X. and Xing F. W. 2010. Infrageneric phylogeny of the genus *Viola* based on *trn*L-*trn*F, *psb*A-*trn*H, *rp*L16, ITS sequences, cytological and morphological data. Acta Bot. Yunnan. 32: 477–488.

Link, J. H. F. 1821–1822. Enumeratio Plantarum Horti Regii Berolinensis Altera.... 2 parts. Berlin. (Enum. Hort. Berol. Alt.)

Linnaea = Linnaea; ein Journal für die Botanik in ihrem ganzen Umfange.

Linnaeus, C. [1753.] Demonstrationes Plantarum in Horto Upsaliensi.... Uppsala. (Demonstr. Pl.)

Linnaeus, C. 1753. Species Plantarum.... 2 vols. Stockholm. (Sp. Pl.)

Linnaeus, C. 1754. Genera Plantarum, ed. 5. Stockholm. (Gen. Pl. ed. 5)

Linnaeus, C. 1755. Centuria I. Plantarum.... Uppsala. (Cent. Pl. I)

Linnaeus, C. 1758[–1759]. Systema Naturae per Regna Tria Naturae..., ed. 10. 2 vols. Stockholm. (Syst. Nat. ed. 10)

Linnaeus, C. 1762–1763. Species Plantarum..., ed. 2. 2 vols. Stockholm. (Sp. Pl. ed. 2)

Linnaeus, C. 1766–1768. Systema Naturae per Regna Tria Naturae..., ed. 12. 3 vols. Stockholm. (Syst. Nat. ed. 12)

Linnaeus, C. 1767[–1771]. Mantissa Plantarum. 2 parts. Stockholm. [Mantissa [1] and Mantissa [2] Altera paged consecutively.] (Mant. Pl.)

Linnaeus, C. f. 1781[1782]. Supplementum Plantarum Systematis Vegetabilium Editionis Decimae Tertiae, Generum Plantarum Editionis Sextae, et Specierum Plantarum Editionis Secundae. Braunschweig. (Suppl. Pl.)

Lira, R. 2001. Cucurbitaceae. In: J. Rzedowski and G. C. de Rzedowski, eds. 1991+. Flora del Bajío y de Regiones Adyacentes. 99+ fascs. Pátzcuaro. Fasc. 92.

Lira, R., G. Téllez, and P. D. Dávila. 2009. The effects of climate change on the geographic distribution of Mexican wild relatives of domesticated Cucurbitaceae. Genet. Resources Crop Evol. 56: 691–703.

Lira, R., J. L. Villaseñor, and P. D. Dávila. 1997. A cladistic analysis of the subtribe Sicyinae (Cucurbitaceae). Syst. Bot. 22: 415–425.

Little, R. J. 1993. Violaceae. In: J. C. Hickman, ed. 1993. The Jepson Manual. Higher Plants of California. Berkeley, Los Angeles, and London. Pp. 1089–1092.

Little, R. J. 2001. Vascular plants of Arizona: Violaceae. J. Arizona-Nevada Acad. Sci. 33: 73–82.

Little, R. J. 2010. *Viola howellii* A. Gray (Violaceae). Noteworthy collection, California. Madroño 57: 209, 211.

Little, R. J. and G. Leiper. 2012. Capsule dehiscence in *Viola betonicifolia* Sm. (Violaceae). Austrobaileya 8: 624–633.

Lloyd, F. E. 1942. The Carnivorous Plants. Waltham, Mass.

London J. Bot. = London Journal of Botany.

Long, R. W. and O. Lakela. 1971. A Flora of Tropical Florida: A Manual of the Seed Plants and Ferns of Southern Peninsular Florida. Coral Gables. [Reprinted 1976, Miami.]

Loureiro, J. de. 1790. Flora Cochinchinensis.... 2 vols. Lisbon. [Vols. paged consecutively.] (Fl. Cochinch.)

Löve, Á. and D. Löve. 1982b. In: IOPB chromosome number reports LXXV. Taxon 31: 342–368.

Lu, A. M. and Zhang Zhi Y. 1982. A revision of genus *Thladiantha* Bunge (Cucurbitaceae). Bull. Bot. Res., Harbin 1: 61–96.

Lundellia = Lundellia; Journal of the Plant Resources Center of the University of Texas at Austin.

Mabberley, D. J. 2008. The Plant-book: A Portable Dictionary of the Vascular Plants..., ed. 3. Cambridge.

MacDougal, J. M. 1983. Revision of *Passiflora* L. Sect. *Pseudodysosmia* (Harms) Killip Emend. J. MacDougal, the Hooked Trichome Group. Ph.D. dissertation. Duke University.

MacDougal, J. M. 1994. Revision of *Passiflora* subgenus *Decaloba* section *Pseudodysosmia* (Passifloraceae). Syst. Bot. Monogr. 41: 1–146.

MacDougal, J. M. and R. McVaugh. 2001. Passifloraceae. In: R. McVaugh and W. R. Anderson, eds. 1974+. Flora Novo-Galiciana: A Descriptive Account of the Vascular Plants of Western Mexico. 8+ vols. Ann Arbor. Vol. 3, pp. 406–459.

Madroño = Madroño; Journal of the California Botanical Society [from vol. 3: a West American Journal of Botany].

Maggs-Kolling, G., S. Madsen, and J. L. Christiansen. 2000. A phenetic analysis of morphological variation in *Citrullus lanatus* in Namibia. Genet. Resources Crop Evol. 47: 385–393.

Maguire, B. 1976. Apomixis in the genus *Clusia* (Clusiaceae)—A preliminary report. Taxon 25: 241–244.

Malvenfam.—See: F. K. Medikus 1787

Man. Fl. N. States—See: N. L. Britton 1901

Man. Fl. N. States ed. 2—See: N. L. Britton 1905b

Man. Fl. Pl. Calif.—See: W. L. Jepson [1923–1925]

Man. Pl. Colorado—See: H. D. Harrington 1954

Man. Pl. Oregon—See: M. E. Peck 1941

Man. S.E. Fl.—See: J. K. Small 1933

Manchester, S. R. 1994. Inflorescence bracts of fossil and extant *Tilia* in North America, Europe, and Asia: Patterns of morphologic divergence and biogeographic history. Amer. J. Bot. 81: 1176–1185.

Mant. Pl.—See: C. Linnaeus 1767[–1771]

Manual ed. 5—See: A. Gray 1867

Marcussen, T. et al. 2010. Evolution of plant RNA polymerase IV/V genes: Evidence of subneofunctionalization of duplicated NRPD2/NRPE2-like paralogs in *Viola* (Violaceae). B. M. C. Evol. Biol. 10: 45.

Marcussen, T. et al. 2011. Establishing the phylogenetic origin, history, and age of the narrow endemic *Viola guadalupensis* (Violaceae). Amer. J. Bot. 98: 1978–1988.

Marcussen, T. et al. 2012. Inferring species networks from gene trees in high-polyploid North American and Hawaiian violets (*Viola*, Violaceae). Syst. Biol. 61: 107–126.

Marcussen, T. and T. Karlsson. 2010. Violaceae. In: B. Jonsell and T. Karlsson, eds. 2000+. Flora Nordica. 3+ vols. Stockholm. Vol. 6, pp. 12–52.

Marr, K. L., X. Y. Mei, and N. Bhattarai. 2004. Allozyme, morphological, and nutritional analysis bearing on the domestication of *Momordica charantia* L. (Cucurbitaceae). Econ. Bot. 58: 435–455.

Martin, W. C. and C. R. Hutchins. 1980. A Flora of New Mexico. 2 vols. Vaduz.

Martius, C. F. P. von, A. W. Eichler, and I. Urban, eds. 1840–1906. Flora Brasiliensis. 15 vols. in 40 parts, 130 fascs. Munich, Vienna, and Leipzig. [Vols. and parts numbered in systematic sequence, fascs. numbered independently in chronological sequence.] (Fl. Bras.)

Maskas, S. D. and M. B. Cruzan. 2000. Patterns of intraspecific diversification in the *Piriqueta caroliniana* complex in southeastern North America and the Bahamas. Evolution 54: 815–827.

Mathew, P. and M. Sivadasan, eds. 1998. Diversity and Taxonomy of Tropical Flowering Plants. Calicut.

Matlack, G. 1994. Plant species migration in a mixed-history forest landscape in eastern North America. Ecology 75: 1491–1502.

Mayers, A. and E. Lord. 1983. Comparative flower development in the cleistogamous species *Viola odorata*. I. A growth rate study. Amer. J. Bot. 70: 1548–1555.

Mayers, A. and E. Lord. 1983b. Comparative flower development in the cleistogamous species *Viola odorata*. II. An organographic study. Amer. J. Bot. 70: 1556–1563.

McBreen, K. and M. B. Cruzan. 2004. A case of recent long distance dispersal in the *Piriqueta caroliniana* complex. Heredity 95: 356–361.

McCarthy, D. 2012. Systematics and Phylogeography of the Genus *Tilia* in North America. Ph.D. dissertation. University of Illinois, Chicago.

McCreary, C. S. 2005. Genetic Relationships, Morphological Divergence and Ecological Correlates in Three Species of the *Viola canadensis* Complex in Western North America. Ph.D. dissertation. Ohio University.

McGuire, C. M. 1999. *Passiflora incarnata* (Passifloraceae): A new fruit crop. Econ. Bot. 53: 161–176.

McKinney, L. E. 1986. The taxonomic status of *Viola appalachiensis* Henry. Bartonia 52: 42–43.

McKinney, L. E. 1992. A Taxonomic Revision of the Acaulescent Blue Violets (*Viola*) of North America. Fort Worth. [Sida Bot. Misc. 7.]

McKinney, L. E. and N. H. Russell. 2002. Violaceae of the southeastern United States. Castanea 4: 369–379.

McMinn, H. and B. Forderhase. 1935. Notes on western leatherwood, *Dirca occidentalis* Gray. Madroño 3: 117–120.

McPherson, G. D. and J. G. Packer. 1974. A contribution to the taxonomy of *Viola adunca*. Canad. J. Bot. 52: 895–902.

McVaugh, R. 1941. The fruit of the eastern leatherwood. Castanea 6: 83–86.

McVaugh, R. 2001. Bixaceae. In: R. McVaugh and W. R. Anderson, eds. 1974+. Flora Novo-Galiciana: A Descriptive Account of the Vascular Plants of Western Mexico. 8+ vols. Ann Arbor. Vol. 3, pp. 335–349.

McVaugh, R. 2001b. Cucurbitaceae. In: R. McVaugh and W. R. Anderson, eds. 1974+. Flora Novo-Galiciana: A Descriptive Account of the Vascular Plants of Western Mexico. 8+ vols. Ann Arbor. Vol. 3, pp. 483–652.

McVaugh, R. and W. R. Anderson, eds. 1974+. Flora Novo-Galiciana: A Descriptive Account of the Vascular Plants of Western Mexico. 8+ vols. Ann Arbor.

Med. Repos. = Medical Repository.

Meded. Bot. Mus. Herb. Rijks Univ. Utrecht = Mededeelingen van het Botanisch Museum en Herbarium van de Rijks Universiteit te Utrecht.

Medikus, F. K. 1787. Ueber einige künstliche Geschlechter aus der Malven-Familie.... Mannheim. (Malvenfam.)

Meerman, J. C. 1993. Provisional annotated checklist of the flora of the Shipstern Nature Reserve. Occas. Pap. Belize Nat. Hist. Soc. 2: 8–36.

Meglic, V., F. Serquen, and J. E. Staub. 1996. Genetic diversity in cucumber (*Cucumis sativus* L.): I. A reevaluation of the U.S. germplasm collection. Genet. Resources Crop Evol. 43: 533–546.

Meijer, W. 1993. Rafflesiaceae. In: K. Kubitzki et al., eds. 1990+. The Families and Genera of Vascular Plants. 10+ vols. Berlin etc. Vol. 2, pp. 557–563.

Melet. Bot.—See: H. W. Schott and S. L. Endlicher 1832

Mem. Amer. Acad. Arts = Memoirs of the American Academy of Arts and Science.

Mem. New York Bot. Gard. = Memoirs of the New York Botanical Garden.

Mém. Soc. Phys. Genève = Mémoires de la Société de Physique et d'Histoire Naturelle de Genève.

Mem. Torrey Bot. Club = Memoirs of the Torrey Botanical Club.

Mentzelia = Mentzelia; Journal of the Northern Nevada Native Plant Society.

Menzel, M. Y. 1986. Genetic relationships among the relatives of *Hibiscus cannabinus* and *H. sabdariffa*. In: K. A. Siddiqui and A. M. Faruqui, eds. 1986. New Genetical Approaches to Crop Improvement. Karachi. Pp. 445–456.

Menzel, M. Y., P. A. Fryxell, and F. D. Wilson. 1983. Relationships among New World species of *Hibiscus* section *Furcaria* (Malvaceae). Brittonia 35: 204–221.

Merriam-Webster. 1988. Webster's New Geographical Dictionary. Springfield, Mass.

Methodus—See: C. Moench 1794

Meyer, H. W. and S. R. Manchester. 1997. The Oligocene Bridge Creek flora of the John Day Formation, Oregon. Univ. Calif. Publ. Geol. Sci. 111: 1 195.

Michaux, A. 1803. Flora Boreali-Americana.... 2 vols. Paris and Strasbourg. (Fl. Bor.-Amer.)

Michener, C. D., R. J. McGinley, and B. N. Danforth. 1994. The Bee Genera of North and Central America (Hymenoptera: Apoidea). Washington.

Michigan Bot. = Michigan Botanist.

Middendorff, A. T. von. 1847–1867. Reise in den äussersten Norden und Osten Siberiens während der Jahre 1843 und 1844.... 4 vols. in parts and fascs. St. Petersburg. (Reise Siber.)

Miller, P. 1754. The Gardeners Dictionary.... Abridged..., ed. 4. 3 vols. London. (Gard. Dict. Abr. ed. 4)

Miller, P. 1768. The Gardeners Dictionary..., ed. 8. London. (Gard. Dict. ed. 8)

Moench, C. 1794. Methodus Plantas Horti Botanici et Agri Marburgensis.... Marburg. (Methodus)

Moerman, D. E. 1986. Medicinal Plants of Native America. 2 vols. Ann Arbor. [Univ. Michigan, Mus. Anthropol., Techn. Rep. 19.]

Moerman, D. E. 1998. Native American Ethnobotany. Portland.

Mohlenbrock, R. H. 1978. The Illustrated Flora of Illinois. Flowering Plants: Hollies to Loasas. Carbondale. (Ill. Fl. Illinois, Fl. Pl. Hollies–Loasas)

Molec. Biol. Evol. = Molecular Biology and Evolution.

Molec. Phylogen. Evol. = Molecular Phylogenetics and Evolution.

Molina, G. I. 1782. Saggio sulla Storia Naturale del Chili.... Bologna. (Sag. Stor. Nat. Chili)

Monogr. Malv. Bras.—See: H. C. Monteiro 1936

Monogr. Phan.—See: A. L. P. P. de Candolle and C. de Candolle 1878–1896

Monro, A. K. and P. J. Stafford. 1998. A synopsis of the genus *Echinopepon* (Cucurbitaceae: Sicyeae), including three new taxa. Ann. Missouri Bot. Gard. 85: 257–272.

Monteiro, H. C. 1936. Monographia das Malvaceas Brasileiras (Monographia Malvacearum Brasiliensis). 1 fasc. only. Rio de Janeiro. (Monogr. Malv. Bras.)

Moore, G. 1989. A checklist of the vascular plants of Cumberland County, New Jersey. Bartonia 55: 25–39.

Morse, L. E. 1979. Systematics and Ecological Biogeography of the Genus *Hudsonia* (Cistaceae), the Sand Heathers. Ph.D. dissertation. Harvard University.

Morton, C. V. 1944. The genus *Hybanthus* in continental North America. Contr. U.S. Natl. Herb. 29: 74–82.

Morton, C. V. 1971. Some types and range extensions in *Hybanthus* (Violaceae). Phytologia 21: 56–62.

Morton, J. F. 1981. Atlas of Medicinal Plants of Middle America: Bahamas to Yucatan. Springfield, Ill.

Moscosoa = Moscosoa; Contribuciones Científicas del Jardin Botánico Nacional "Dr. Raphael M. Moscosa".

Muenscher, W. C. 1980. Weeds, ed. 2, [new] foreword and appendixes by Peter A. Hyypio. Ithaca, N.Y., and London.

Muhlenbergia = Muhlenbergia; a Journal of Botany.

Munger, H. M. and R. W. Robinson. 1991. Nomenclature of *Cucumis melo* L. Rep. Cucurbit Genet. Coop. 14: 43–44.

Munz, P. A. 1959. A California Flora. Berkeley and Los Angeles.

Munz, P. A. 1974. A Flora of Southern California. Berkeley.

Murray, B. G., L. A. Craven, and P. J. de Lange. 2008. New observations on chromosome number variation in *Hibiscus trionum* s.l. (Malvaceae) and their implications for systematics and conservation. New Zealand J. Bot. 46: 315–319.

Murray, J. A. 1770. Prodromus Designationis Stirpium Gottingensium.... Göttingen. (Prodr. Stirp. Gott.)

Muschner, V. C. et al. 2003. A first molecular phylogenetic analysis of *Passiflora* (Passifloraceae). Amer. J. Bot. 90: 1229–1238.

Myers, O. 1963. Studies of Reproduction and Hybridization in Five Species of *Hypericum* Native to Eastern North America. Ph.D. dissertation. Cornell University.

Nandi, O. I., M. W. Chase, and P. K. Endress. 1998. A combined cladistic analysis of angiosperms using *rbc*L and non-molecular data sets. Ann. Missouri Bot. Gard. 85: 137–212.

Nat. Pflanzenfam.—See: H. G. A. Engler and K. Prantl 1887–1915

Nat. Pflanzenfam. ed. 2—See: H. G. A. Engler et al. 1924+

Native Pl. J. = Native Plants Journal.

Natl. Hort. Mag. = National Horticultural Magazine.

Naturaliste Canad. = Naturaliste Canadien. Bulletin de Recherches, Observations et Découvertes se Rapportant à l'Histoire Naturelle du Canada.

Naudin, C. V. 1859. Essais d'une monographie des espèces et des variétés du genre *Cucumis*. Ann. Sci. Nat., Bot., sér. 4, 11: 5–87.

Nayar, N. M. and T. A. More, eds. 1998. Cucurbits. Enfield, N.H.

Nee, M. 1990. The domestication of *Cucurbita* (Cucurbitaceae). Econ. Bot. 44(3,suppl.): 56–68.

Neff, J. L. 2003. The passionflower bee: *Anthemurgus passiflorae*. Passiflora 13: 7–9.

Neff, J. L. and J. G. Rozen. 1995. Foraging and nesting biology of the bee *Anthemurgus passiflorae* (Hymenoptera: Apoidea), descriptions of its immature stages and observations on its floral host (Passifloraceae). Amer. Mus. Novit. 3138: 1–19.

Nesom, G. L. 2011. New state records for *Citrullus*, *Cucumis*, and *Cucurbita* (Cucurbitaceae) outside of cultivation in the USA. Phytoneuron 2011-1: 1–7.

Nesom, G. L. 2011b. Taxonomy of *Sicyos* (Cucurbitaceae) in the USA. Phytoneuron 2011-15: 1–11.

Nesom, G. L. 2011c. Toward consistency of taxonomic rank in wild/domesticated Cucurbitaceae. Phytoneuron 2011-13: 1–33.

Nesom, G. L. 2012f. Terms for surface vestiture and relief of Cucurbitaceae fruits. Phytoneuron 2012-108: 1–4.

Nesom, G. L. 2014. Taxonomy of *Cyclanthera* (Cucurbitaceae) in the USA. Phytoneuron 2014-11: 1–17.

Nevling, L. I. Jr. 1962. Thymelaeaceae in the southeastern United States. J. Arnold Arbor. 43: 428–434.

New Fl.—See: C. S. Rafinesque 1836[–1838]

New Phytol. = New Phytologist; a British Botanical Journal.

New Zealand J. Bot. = New Zealand Journal of Botany.

Newsom, L. A., S. D. Webb, and J. S. Dunbar. 1993. History and geographic distribution of *Cucurbita pepo* gourds in Florida. J. Ethnobiol. 13: 75–97.

Niedenzu, F. 1925. Tamaricaceae. In: H. G. A. Engler et al., eds. 1924+. Die natürlichen Pflanzenfamilien..., ed. 2. 26+ vols. Leipzig and Berlin. Vol. 21, pp. 282–289.

Nomencl. Bot. ed. 2—See: E. G. Steudel 1840–1841

NorthW. Sci. = Northwest Science.

Notes Roy. Bot. Gard. Edinburgh = Notes from the Royal Botanic Garden, Edinburgh.

Notizbl. Bot. Gart. Berlin-Dahlem = Notizblatt des Botanischen Gartens und Museums zu Berlin-Dahlem.

Nov. Gen. Sp.—See: A. von Humboldt et al. 1815[1816]–1825

Nov. Pl. Descr. Dec.—See: C. G. Ortega 1797–1800

Novak, S. J. and R. N. Mack. 1995. Allozyme diversity in the apomictic vine *Bryonia alba* (Cucurbitaceae): Potential consequences of multiple introductions. Amer. J. Bot. 82: 1153–1162.

Novak, S. J. and R. N. Mack. 2000. Clonal diversity within and among introduced populations of the apomictic vine *Bryonia alba* (Cucurbitaceae). Canad. J. Bot. 78: 1469–1481.

Novon = Novon; a Journal for Botanical Nomenclature.

Nuttall, T. 1813. A Catalogue of New and Interesting Plants Collected in Upper Louisiana.... London. (Cat. Pl. Upper Louisiana)

Nuttall, T. 1818. The Genera of North American Plants, and Catalogue of the Species, to the Year 1817.... 2 vols. Philadelphia. (Gen. N. Amer. Pl.)

Nyffeler, R. et al. 2005. Phylogenetic analysis of the Malvadendrina clade (Malvaceae s.l.) based on plastid DNA sequences. Organisms Diversity Evol. 5: 109–123.

Obae, S. G., B. A. Connolly, and M. H. Brand. 2013. Genetic relationship among four *Crocanthemum* species (Cistaceae) revealed by amplified fragment length polymorphism markers. J. Torrey Bot. Soc. 140: 170–180.

Observ. Bot.—See: N. J. Jacquin 1764–1771

Occas. Pap. Belize Nat. Hist. Soc. = Occasional Papers of the Belize Natural History Society.

Occas. Pap. Rancho Santa Ana Bot. Gard. = Occasional Papers of the Rancho Santa Ana Botanical Garden.

Oesterr. Bot. Z. = Oesterreichische botanische Zeitschrift. Gemeinütziges Organ für Botanik.

Ohio J. Sci. = Ohio Journal of Science.

Oliver, D. et al., eds. 1868–1937. Flora of Tropical Africa.... 10 vols. London. (Fl. Trop. Afr.)

Organisms Diversity Evol. = Organisms, Diversity and Evolution; Journal of the Gesellschaft für Biologische Systematik.

Organization for Flora Neotropica. 1968+. Flora Neotropica. 109+ nos. New York.

Ornduff, R. 1970c. Relationships in the *Piriqueta caroliniana–P. cistoides* complex (Turneraceae). J. Arnold Arbor. 51: 492–498.

Ortega, C. G. 1797–1800. Novarum, aut Rariorum Plantarum Horti Reg. Botan. Matrit. Descriptionum Decades.... 10 decades in 4 parts. Madrid. [Parts paged consecutively.] (Nov. Pl. Descr. Dec.)

Outlines Bot.—See: G. T. Burnett 1835

Pacif. Railr. Rep.—See: War Department 1855–1860

Pacific (San Francisco) = The Pacific.

Paris, H. S. 1986. A proposed subspecific classification for *Cucurbita pepo*. Phytologia 61: 133–138.

Paris, H. S. 1989. Historical records, origins, and development of the edible cultivar groups of *Cucurbita pepo* (Cucurbitaceae). Econ. Bot. 43: 423–443.

Paris, H. S. 1996. Summer squash: History, diversity, and distribution. HortTechnol. 6: 6–13.

Paris, H. S. 2001. History of the cultivar-groups of *Cucurbita pepo*. Hort. Rev. 25: 71–170.

Paris, H. S. et al. 2006. First known image of *Cucurbita* in Europe, 1503–1508. Ann. Bot. (Oxford), n. s. 98: 41–47.

Patterson, D. T. et al. 1989. Composite List of Weeds. Champaign.

Peck, M. E. 1941. A Manual of the Higher Plants of Oregon. Portland. (Man. Pl. Oregon)

Peguero, B. and F. Jiménez. 2005. *Cucurbita okeechobeensis* (Small) L. Bailey (Cucurbitaceae): Nuevo reporte para La Isla Española. Moscosoa 14: 56–64.

Pengum. Balai Besar Penjel. Kehut. Indonesia = Pengumuman Balai Besar Penjelidikan Kehutanan Indonesia.

Perenn. Sp. Sidalcea—See: C. L. Hitchcock 1957

Perspect. Pl. Ecol. Evol. Syst. = Perspectives in Plant Ecology, Evolution and Systematics.

Peterson, B. J., W. R. Graves, and J. Sharma. 2009. Color of pubescence on bud scales conflicts with keys for identifying species of *Dirca* (Thymelaeaceae). Rhodora 111: 126–130.

Pfeil, B. E. et al. 2002. Phylogeny of *Hibiscus* and the tribe Hibisceae (Malvaceae) using chloroplast DNA sequences of *ndh*F and the *rpl*16 intron. Syst. Bot. 27: 333–350.

Pfeil, B. E. and M. D. Crisp. 2005. What to do with *Hibiscus*? A proposed nomenclatural resolution for a large and well known genus of Malvaceae and comments on paraphyly. Austral. Syst. Bot. 18: 49–60.

Pflanzenr.—See: H. G. A. Engler 1900–1953

Philbrick, C. T. 1984. Aspects of floral biology, breeding system, and seed and seedling biology in *Podostemum ceratophyllum* Michx. (Podostemaceae). Syst. Bot. 9: 166–174.

Philbrick, C. T. and G. E. Crow. 1983. Distribution of *Podostemum ceratophyllum* Michx. (Podostemaceae). Rhodora 85: 325–341.

Philbrick, C. T. and G. E. Crow. 1992. Isozyme variation and population structure in *Podostemum ceratophyllum* Michx. (Podostemaceae): Implications for colonization of glaciated North America. Aquatic Bot. 43: 311–325.

Philbrick, C. T. and A. Novelo R. 1995. New World Podostemaceae: Ecological and evolutionary enigmas. Brittonia 47: 210–222.

Philbrick, C. T. and A. Novelo R. 2004. Monograph of *Podostemum*. Syst. Bot. Monogr. 70: 1–106.

Philbrick, R. N. 1980. Distribution and evolution of endemic plants of the California islands. In: D. M. Power, ed. 1980. The California Islands: Proceedings of a Multidisciplinary Symposium. Santa Barbara. Pp. 173–187.

Phytologia = Phytologia; Designed to Expedite Botanical Publication.

Phytomorphology = Phytomorphology; an International Journal of Plant Morphology.

Phyton (Horn) = Phyton; Annales Rei Botanica.

Picotte, J. J., J. M. Rhode, and M. B. Cruzan. 2009. Leaf morphological responses to variation in water availability for plants in the *Piriqueta caroliniana* complex. Pl. Ecol. 200: 267–275.

Pigott, D. 2012. Lime-trees and Basswoods: A Biological Monograph of the Genus *Tilia*. Cambridge and New York.

Pijl, L. van der. 1937. Disharmony between Asiatic flower-birds and American bird-flowers. Ann. Jard. Bot. Buitenzorg 48: 2–26.

Pilger, R. K. F. 1925. Cochlospermaceae. In: H. G. A. Engler et al., eds. 1924+. Die natürlichen Pflanzenfamilien..., ed. 2. 26+ vols. Leipzig and Berlin. Vol. 21, pp. 316–320.

Pitrat, M., ed. 2008. Cucurbitaceae 2008: Proceedings of the IXth EUCARPIA Meeting on Genetics and Breeding of Cucurbitaceae, Avignon (France), May 21–24th, 2008. Avignon.

Pitrat, M. et al. 2000. Some comments on infraspecific classification of cultivars of melon. Acta Hort. 510: 29–36.

Pl. Biosyst. = Plant Biosystems; Giornale Botanico Italiano.

Pl. Ecol. = Plant Ecology.

Pl. Genet. Resources Newslett. = Plant Genetic Resources Newsletter.

Pl. Hartw.—See: G. Bentham 1839[–1857]

Pl. Heermann.—See: E. M. Durand and T. C. Hilgard 1854

Pl. Nov. Thurb.—See: A. Gray 1854

Pl. Pratten. Calif.—See: E. M. Durand 1855

Pl. Select.—See: C. J. Trew 1750–1773

Pl. Spec. Biol. = Plant Species Biology; an International Journal.

Pl. Syst. Evol. = Plant Systematics and Evolution.

Planchon, J. É. 1847. Sur la nouvelle famille des Cochlospermaceae. London J. Bot. 6: 294–311.

Planta = Planta. Archiv für wissenschaftliche Botanik. (Zeitschrift für wissenschaftliche Biologie. Abt. E.)

Pohl, R. W. 1955. *Thymelaea passerina*, new weed in the United States. Proc. Iowa Acad. Sci. 62: 152–154.

Poiret, J. 1789. Voyage en Barbarie.... 2 vols. Paris. (Voy. Barbarie)

Pollard, C. L. 1898. Further observations on the eastern acaulescent blue violets. Bot. Gaz. 26: 325–342.

Pollen & Spores = Pollen et Spores.

Poppendieck, H. H. 1980. A monograph of the Cochlospermaceae. Bot. Jahrb. Syst. 101: 191–265.

Poppendieck, H. H. 1981. Cochlospermaceae. In: Organization for Flora Neotropica. 1968+. Flora Neotropica. 109+ nos. New York. No. 27.

Poppendieck, H. H. 2003. Bixaceae. In: K. Kubitzki et al., eds. 1990+. The Families and Genera of Vascular Plants. 10+ vols. Berlin etc. Vol. 5, pp. 33–35.

Poppendieck, H. H. 2003b. Cochlospermaceae. In: K. Kubitzki et al., eds. 1990+. The Families and Genera of Vascular Plants. 10+ vols. Berlin etc. Vol. 5, pp. 71–74.

Porter-Utley, K. E. 2003. Revision of *Passiflora* Subgenus *Decaloba* Supersection *Cieca* (Passifloraceae). Ph.D. dissertation. University of Florida.

Possley, J. et al. 2007. A common passion. Multiple agencies and volunteers unite to reintroduce goatsfoot passionflower to rockland hammocks of Miami, Florida. Native Pl. J. 8: 252–258.

Power, D. M., ed. 1980. The California Islands: Proceedings of a Multidisciplinary Symposium. Santa Barbara.

Précis Découv. Somiol.—See: C. S. Rafinesque 1814

Prelim. Cat.—See: N. L. Britton et al. 1888

Prendergast, H. D. V. 1982. Pollination of *Hibiscus rosa sinensis*. Biotropica 14: 287.

Presl, C. B. 1825–1835. Reliquiae Haenkeanae seu Descriptiones et Icones Plantarum, Quas in America Meridionali et Boreali, in Insulis Philippinis et Marianis Collegit Thaddeus Haenke.... 2 vols. in 7 parts. Prague. (Reliq. Haenk.)

Pritchard, P. C. H., ed. 1978–1982. Rare and Endangered Biota of Florida. Gainesville.

Proc. Acad. Nat. Sci. Philadelphia = Proceedings of the Academy of Natural Sciences of Philadelphia.

Proc. Amer. Acad. Arts = Proceedings of the American Academy of Arts and Sciences.

Proc. Amer. Assoc. Advancem. Sci. = Proceedings of the American Association for the Advancement of Science.

Proc. Amer. Soc. Hort. Sci. = Proceedings of the American Society for Horticultural Science.

Proc. Biol. Soc. Wash. = Proceedings of the Biological Society of Washington.

Proc. Calif. Acad. Sci. = Proceedings of the California Academy of Sciences.

Proc. Iowa Acad. Sci. = Proceedings of the Iowa Academy of Science.

Proc. Natl. Acad. Sci. U.S.A. = Proceedings of the National Academy of Sciences of the United States of America.

Proc. Roy. Soc. Biol. Sci. Ser. B = Proceedings of the Royal Society. Biological Sciences Series B.

Prodr.—See: A. P. de Candolle and A. L. P. P. de Candolle 1823–1873

Prodr. Fl. Ind. Orient.—See: R. Wight and G. A. W. Arnott 1834

Prodr. Pl. Cap.—See: C. P. Thunberg 1794–1800

Prodr. Pl. Ind. Occid.—See: W. Hamilton 1825

Prodr. Stirp. Gott.—See: J. A. Murray 1770

Pterocymbium—See: R. Brown 1844

Publ. Bot. (Ottawa) = Publications in Botany, National Museum of Natural Sciences, Canada.

Publ. Field Columbian Mus., Bot. Ser. = Publications of the Field Columbian Museum. Botanical Series.

Publ. Field Mus. Nat. Hist., Bot. Ser. = Publications of the Field Museum of Natural History. Botanical Series.

Puri, V. 1948. Studies in floral anatomy. V. On the structure and nature of the corona in certain species of Passifloraceae. J. Indian Bot. Soc. 27: 130–149.

Pursh, F. [1813]1814. Flora Americae Septentrionalis; or, a Systematic Arrangement and Description of the Plants of North America. 2 vols. London. (Fl. Amer. Sept.)

Qaiser, M. 1987. Studies in the seed morphology of the family Tamaricaceae from Pakistan. Bot. J. Linn. Soc. 94: 469–484.

Raamsdonk, L. W. D. van, A. P. M. Nijs, and M. C. Jongerius. 1989. Meiotic analysis of Cucumis hybrids and an evolutionary evaluation of the genus Cucumis. Pl. Syst. Evol. 163: 133–146.

Radford, A. E., H. E. Ahles, and C. R. Bell. 1968. Manual of the Vascular Flora of the Carolinas. Chapel Hill.

Radhamani, T. R., L. Sudarshana, and R. Krishnan. 1995. Defense and carnivory: Dual role of bracts in Passiflora foetida. J. Biosci. 20: 657–664.

Rafinesque, C. S. 1814. Précis des Découvertes et Travaux Somiologiques.... Palermo. (Précis Découv. Somiol.)

Rafinesque, C. S. 1836[–1838]. New Flora and Botany of North America.... 4 parts. Philadelphia. [Parts paged independently.] (New Fl.)

Rafinesque, C. S. 1836[1837–1838]. Flora Telluriana.... 4 vols. Philadelphia. (Fl. Tellur.)

Ray, M. F. 1995. Systematics of Lavatera and Malva (Malvaceae)—A new perspective. Pl. Syst. Evol. 198: 29–53.

Ray, M. F. 1998. New combinations in Malva (Malvaceae: Malveae). Novon 8: 288–295.

Rees, A. [1802–]1819–1820. The Cyclopaedia; or, Universal Dictionary of Arts, Sciences, and Literature.... 39 vols. in 79 parts. London. [Pages unnumbered.] (Cycl.)

Reichenbach, H. G. L. 1823–1832. Iconographia Botanica seu Plantae Criticae. 10 vols. Leipzig. [Vols. 6 and 7 each published in two half-centuries paged independently.] (Iconogr. Bot. Pl. Crit.)

Reichenbach, H. G. L. 1828. Conspectus Regni Vegetabilis.... Leipzig. (Consp. Regn. Veg.)

Reise Siber.—See: A. T. von Middendorff 1847–1867

Relaz. Cult. Coton.—See: A. Todaro 1877–1878

Reliq. Haenk.—See: C. B. Presl 1825–1835

Renner, S. S. and H. Schaefer. 2008. Phylogenetics of Cucumis (Cucurbitaceae) as understood in 2008. In: M. Pitrat, ed. 2008. Cucurbitaceae 2008: Proceedings of the IXth EUCARPIA Meeting on Genetics and Breeding of Cucurbitaceae, Avignon (France), May 21–24th, 2008. Avignon. Pp. 53–58.

Renner, S. S., H. Schaefer, and A. Kocyan. 2007. Phylogenetics of Cucumis (Cucurbitaceae): Cucumber (C. sativus) belongs in an Asian/Australian clade far from melon (C. melo). B. M. C. Evol. Biol. 7: 58.

Rep. Colorado R.—See: J. C. Ives 1861

Rep. Cucurbit Genet. Coop. = Report, Cucurbit Genetics Cooperative.

Rep. U.S. Mex. Bound.—See: W. H. Emory 1857–1859

Repert. Bot. Syst.—See: W. G. Walpers 1842–1847

Repert. Spec. Nov. Regni Veg. = Repertorium Specierum Novarum Regni Vegetabilis.

Rev. Nat. Prod. = The Review of Natural Products.

Revista Brasil. Biol. = Revista Brasileira de Biologia.

Rheede, H. A. van. 1678–1693. Hortus Indicus Malabaricus.... 12 vols. Amsterdam.

Rheedea = Rheedea; Official Journal of Indian Association for Angiosperm Taxonomy.

Rhoads, A. F. and T. A. Block. 2000. The Plants of Pennsylvania: An Illustrated Manual. Philadelphia.

Rhodes, A. M. et al. 1968. Numerical taxonomic study of Cucurbita. Brittonia 20: 251–266.

Rhodora = Rhodora; Journal of the New England Botanical Club.

Richards, A. J., ed. 1978. The Pollination of Flowers by Insects. London.

Rickett, H. W. [1966–1973.] Wild Flowers of the United States. 6 vols. in 14 parts. New York.

Ridley, H. N. 1930. The Dispersal of Plants throughout the World. Ashford.

Ridley, H. N. 1934. Firmiana and Erythropsis. Bull. Misc. Inform. Kew 1934: 214–217.

Rivadavia, F. et al. 2003. Phylogeny of the sundews, Drosera (Droseraceae), based on chloroplast rbcL and nuclear 18S ribosomal DNA sequences. Amer. J. Bot. 90: 123–130.

Roberts, M. L. and R. L. Stuckey. 1992. Distribution patterns of selected aquatic and wetland plants in relation to the Ohio canal system. Bartonia 47: 50–74.

Roberts, P. R. and H. J. Oosting. 1958. Responses of Venus fly trap (Dionaea muscipula) to factors involved in its endemism. Ecol. Monogr. 28: 193–218.

Robinson, R. W. and D. S. Decker. 1997. Cucurbits. New York.

Robinson, R. W. and J. T. Puchalski. 1980. Synonymy of Cucurbita martinezii and C. okeechobeensis. Rep. Cucurbit Genet. Coop. 3: 45–46.

Robinson, T. W. 1965. Introduction, Spread, and Aerial Extent of Saltcedar (Tamarix) in the Western States. Washington. [U.S. Geol. Surv., Profess. Pap. 491-A.]

Robson, N. K. B. 1977. Studies in the genus Hypericum L. (Guttiferae) 1. Infrageneric classification. Bull. Brit. Mus. (Nat. Hist.), Bot. 5: 291–355.

Robson, N. K. B. 1980. The Linnaean species of Ascyrum (Guttiferae). Taxon 29: 267–274.

Robson, N. K. B. 1981. Studies in the genus Hypericum L. (Guttiferae) 2. Characters of the genus. Bull. Brit. Mus. (Nat. Hist.), Bot. 8: 55–226.

Robson, N. K. B. 1990. Studies in the genus Hypericum L. (Guttiferae) 8. Sections 29. Brathys (part 2) and 30. Trigynobrathys. Bull. Brit. Mus. (Nat. Hist.), Bot. 20: 1–151.

Robson, N. K. B. 1994. Studies in the genus Hypericum L. (Guttiferae) 6. Sections 20. Myriandra to 28. Elodes. Bull. Nat. Hist. Mus. London, Bot. 26: 75–217.

Robson, N. K. B. 2001. Studies in the genus Hypericum L. (Guttiferae) 4(1). Sections 7. Roscyna to 9. Hypericum sensu lato (part 1). Bull. Nat. Hist. Mus. London, Bot. 31: 37–88.

Robson, N. K. B. 2002. Studies in the genus Hypericum L. (Clusiaceae) 4(2). Section 9. Hypericum sensu lato (part 2): subsection 1. Hypericum series 1. Hypericum. Bull. Nat. Hist. Mus. London, Bot. 32: 61–123.

Robson, N. K. B. 2006. Studies in the genus Hypericum L. (Clusiaceae) 4(3). Section 9. Hypericum sensu lato (part 3): subsection 1. Hypericum series 2. Senanensis, subsection 2. Erecta and section 9b. Graveolentia. Syst. Biodivers. 4: 19–98.

Robson, N. K. B. 2012. Studies in the genus *Hypericum* L. (Hypericaceae) 9. Addenda, corrigenda, keys, lists and general discussion. Phytotaxa 72: 1–111.

Robson, N. K. B. and W. P. Adams. 1968. Chromosome numbers in *Hypericum* and related genera. Brittonia 20: 95–106.

Rodríguez, J. C. 1995. Distribución geográfica del género *Echinopepon*. Anales Inst. Biol. Univ. Nac. Autón. México, Bot. 66: 171–181.

Roemer, J. J., J. A. Schultes, and J. H. Schultes. 1817[–1830]. Caroli a Linné...Systema Vegetabilium...Editione XV.... 7 vols. Stuttgart. (Syst. Veg.)

Rose, J. N. 1897. A synopsis of American species of *Hermannia*. Contr. U.S. Natl. Herb. 5: 130–131.

Rose, J. N. 1897b. Studies of Mexican and Central American plants—No. 1. Contr. U.S. Natl. Herb. 5: 109–122.

Roubik, D. W. et al. 1982. Stratum, tree, and flower selection by tropical bees: Implications for the reproductive biology of outcrossing *Cochlospermum vitifolium* in Panama. Ecology 63: 712–720.

Roush, E. M. F. 1931. A monograph of the genus *Sidalcea*. Ann. Missouri Bot. Gard. 18: 117–244.

Roxburgh, W. 1832. Flora Indica; or, Descriptions of Indian Plants. 3 vols. Serampore. (Fl. Ind. ed. 1832)

Ruhfel, B. R. et al. 2011. Phylogeny of the clusioid clade (Malpighiales): Evidence from the plastid and mitochondrial genomes. Amer. J. Bot. 98: 306–325.

Russell, N. H. 1955. The taxonomy of the North American acaulescent white violets. Amer. Midl. Naturalist 54: 481–494.

Russell, N. H. 1955b. Morphological variation in *Viola rotundifolia* Michx. Castanea 20: 144–153.

Russell, N. H. 1965. Violets *(Viola)* of the central and eastern United States: An introductory survey. Sida 2: 1–113.

Russell, N. H. and M. Cooperrider. 1955. Prediction of an introgressant in *Viola*. Amer. Midl. Naturalist 54: 42–51.

Russell, N. H. and A. C. Risser. 1960. The hybrid nature of *Viola emarginata* (Nuttall) Leconte. Brittonia 12: 298–305.

Rutishauser, R. 1997. Structural and developmental diversity in Podostemaceae (river-weeds). Aquatic Bot. 57: 29–70.

Rutishauser, R. et al. 2003. Developmental morphology of roots and shoots of *Podostemum ceratophyllum* (Podostemaceae–Podostemoideae). Rhodora 105: 337–353.

Rzedowski, J. and G. C. de Rzedowski, eds. 1991+. Flora del Bajío y de Regiones Adyacentes. 99+ fascs. Pátzcuaro.

Sag. Stor. Nat. Chili—See: G. I. Molina 1782

Saggi Sci. Lett. Accad. Padova = Saggi Scientifici e Letterarj dell' Accademia di Scienze, Lettere ed Arti in Padova.

Sanjur, O. I. et al. 2002. Phylogenetic relationships among domesticated and wild species of *Cucurbita* (Cucurbitaceae) inferred from a mitochondrial gene: Implications for crop plant evolution and areas of origin. Proc. Natl. Acad. Sci. U.S.A. 99: 535–540.

Saunders, J. G. 1995. Evolution and Systematics of *Waltheria* (Sterculiaceae). Ph.D. dissertation. University of Texas.

Savolainen, V. et al. 2000b. Phylogeny of the eudicots: A nearly complete familial analysis based on *rbc*L gene sequences. Kew Bull. 55: 257–309.

Schaefer, H. 2007. *Cucumis* (Cucurbitaceae) must include *Cucumella, Dicoelospermum, Mukia, Myrmecosicyos,* and *Oreosyce*: A recircumscription based on nuclear and plastid DNA data. Blumea 52: 165–177.

Schaefer, H., C. Heibl, and S. S. Renner. 2008. Gourds afloat: A dated phylogeny reveals an Asian origin of the gourd family (Cucurbitaceae) and numerous oversea dispersal events. Proc. Roy. Soc. Biol. Sci. Ser. B 276: 843–851.

Schaefer, H. and S. S. Renner. 2010. A three-genome phylogeny of *Momordica* (Cucurbitaceae) suggests seven returns from dioecy to monoecy and recent long-distance dispersal to Asia. Molec. Phylogen. Evol. 54: 553–560.

Schaefer, H. and S. S. Renner. 2011. Cucurbitaceae. In: K. Kubitzki et al., eds. 1990+. The Families and Genera of Vascular Plants. 10+ vols. Berlin etc. Vol. 10, pp. 112–174.

Schaefer, H. and S. S. Renner. 2011b. Phylogenetic relationships in the order Cucurbitales and a new classification of the gourd family (Cucurbitaceae). Taxon 60: 122–138.

Sched. Fl. Stiriac.—See: A. von Hayek 1904–1912

Schkuhr, C. [1787–]1791–1803. Botanisches Handbuch.... 3 vols. Wittenberg. (Bot. Handb.)

Schlising, R. A. 1969. Seedling morphology in *Marah* (Cucurbitaceae) related to the California mediterranean climate. Amer. J. Bot. 56: 552–561.

Schlising, R. A. 1993. Cucurbitaceae. In: J. C. Hickman, ed. 1993. The Jepson Manual. Higher Plants of California. Berkeley, Los Angeles, and London. Pp. 535–538.

Schnell, D. E. 2002. Carnivorous Plants of the United States and Canada, ed. 2. Portland.

Schott, H. W. and S. L. Endlicher. 1832. Meletemata Botanica. Vienna. (Melet. Bot.)

Schrader, J. A. and W. R. Graves. 2004. Systematics of *Dirca* (Thymelaeaceae) based on ITS sequences and ISSR polymorphisms. Sida 21: 511–524.

Schrader, J. A. and W. R. Graves. 2005. Seed germination of *Dirca* (leatherwood); pretreatments and interspecific comparisons. HortScience 40: 1838–1842.

Schubert, B. G. 1954. *Begonia cucullata* and its varieties. Natl. Hort. Mag. 1954(Oct.): 244–248.

Sci. Amer. = Scientific American.

Sci. Hort. = Scientia Horticulturae; International Journal Sponsored by the International Society for Horticultural Science.

Science = Science; an Illustrated Journal [later: a Weekly Journal Devoted to the Advancement of Science]. American Association for the Advancement of Science.]

Scoggan, H. J. 1978–1979. The Flora of Canada. 4 parts. Ottawa. [Natl. Mus. Nat. Sci. Publ. Bot. 7.]

Scott, J. A. 1986. The Butterflies of North America: A Natural History and Field Guide. Stanford.

Sebastian, P. et al. 2010. Cucumber *(Cucumis sativus)* and melon *(C. melo)* have numerous wild relatives in Asia and Australia, and the sister species of melon is from Australia. Proc. Natl. Acad. Sci. U.S.A. 107: 14269–14273.

Seo, M. N., A. M. Sanso, and C. C. Xifreda. 2010. Chromosome numbers and meiotic behaviour in South American species of *Hybanthus* Jacq. and *Anchietea* A. St. Hil. (Violaceae). Pl. Biosyst. 144: 340–347.

Seo, M. N., A. M. Sanso, and C. C. Xifreda. 2011. Foliar micromorphology in the classification of South American *Hybanthus* species (Violaceae). Ann. Bot. Fenn. 48: 247–255.

Sher, A. and M. F. Quigley. 2013. *Tamarix*: A Case Study of Ecological Change in the American West. Oxford and New York.

Shinners, L. H. 1962d. *Drosera* (Droseraceae) in the southeastern United States: An interim report. Sida 1: 53–59.

Shore, J. S. and S. C. H. Barrett. 1985. The genetics of distyly and homostyly in *Turnera ulmifolia* L. (Turneraceae). Heredity 55: 167–174.

Shultze, J. H. 1946. The introduction of *Viola lanceolata* into the Pacific Northwest. Madroño 8: 191–193.

Sida = Sida; Contributions to Botany.

Siddiqui, K. A. and A. M. Faruqui, eds. 1986. New Genetical Approaches to Crop Improvement. Karachi.

Siedo, S. J. 1999. A taxonomic treatment of *Sida* sect. *Ellipticifolia* (Malvaceae). Lundellia 2: 100–127.

Silberstein, L. et al. 1999. Molecular variation in melon (*Cucumis melo* L.) as revealed by RFLP and RAPD markers. Sci. Hort. 79: 101–111.

Silva Manso, A. L. P. da. 1836. Enumeração das Substancias Brazileiras.... Rio de Janeiro. (Enum. Subst. Braz.)

Singh, D. and A. S. R. Dathan. 1998. Morphology and embryology. In: N. M. Nayar and T. A. More, eds. 1998. Cucurbits. Enfield, N.H. Pp. 67–84.

Singh, F. and T. N. Khoshoo. 1970. Chromosomal polymorphism within the *Hibiscus rosa-sinensis* complex. Caryologia 23: 19–27.

Sketch Bot. S. Carolina—See: S. Elliott [1816–]1821–1824

Skog, J. E. and N. H. Nickerson. 1972. Variation and speciation in the genus *Hudsonia*. Ann. Missouri Bot. Gard. 59: 454–464.

Skovsted, A. 1941. Chromosome numbers in the Malvaceae II. Compt.-Rend. Trav. Carlsberg Lab., Sér. Physiol. 23: 195–242.

Small, J. K. 1903. Flora of the Southeastern United States.... New York. (Fl. S.E. U.S.)

Small, J. K. 1913b. Flora of Miami.... New York. (Fl. Miami)

Small, J. K. 1913c. Flora of the Florida Keys.... New York.

Small, J. K. 1922. Wild pumpkins. J. New York Bot. Gard. 23: 19–23.

Small, J. K. 1933. Manual of the Southeastern Flora, Being Descriptions of the Seed Plants Growing Naturally in Florida, Alabama, Mississippi, Eastern Louisiana, Tennessee, North Carolina, South Carolina and Georgia. New York. (Man. S.E. Fl.)

Small, R. L. 2004. Phylogeny of *Hibiscus* sect. *Muenchhusia* (Malvaceae) based on chloroplast *rpL*16 and *ndh*F, and nuclear ITS and GBSSI sequences. Syst. Bot. 29: 385–392.

Smith, A. C. 1979–1996. Flora Vitiensis Nova: A New Flora of Fiji (Spermatophytes Only). 5 vols. + index. Lawai.

Smith, B. D. 1997. The initial domestication of *Cucurbita pepo* in the Americas 10,000 years ago. Science, ser. 2, 276: 932–934.

Smith, C. F. 1976. A Flora of the Santa Barbara Region, California. Santa Barbara.

Smith, J. E. et al. 1790–1866. English Botany; or, Coloured Figures of British Plants.... 36 vols. + 5 suppls. London. (Engl. Bot.)

Smith, L. B. et al. 1986. Begoniaceae. Part I: Illustrated key. Part II: Annotated species list. Smithsonian Contr. Bot. 60.

Smith, N. P. et al., eds. 2004. Flowering Plants of the Neotropics. Princeton.

Smithsonian Contr. Bot. = Smithsonian Contributions to Botany.

Smithsonian Contr. Knowl. = Smithsonian Contributions to Knowledge.

Smithsonian Misc. Collect. = Smithsonian Miscellaneous Collections.

Snow, A. A. and D. W. Roubik. 1987. Pollen deposition and removal by bees visiting two tree species in Panama. Biotropica 19: 57–63.

Snow, N. and J. M. MacDougal. 1993. New chromosome reports in *Passiflora* (Passifloraceae). Syst. Bot. 18: 261–273.

Soerjani, M. 1987. An introduction to the weeds of rice in Indonesia. In: M. Soerjani et al., eds. 1987. Weeds of Rice in Indonesia. Jakarta. Pp. 2–4.

Soerjani, M., A. J. G. H. Kostermans, and G. Tjitrosoepomo, eds. 1987. Weeds of Rice in Indonesia. Jakarta.

Solís Neffa, V. G. and Aveliano Fernández. 2000. Chromosome studies in *Turnera* (Turneraceae). Genet. Molec. Biol. 23: 925–930.

Soltis, D. E. et al. 2000. Angiosperm phylogeny inferred from 18S rDNA, *rbc*L, and *atp*B sequences. Bot. J. Linn. Soc. 133: 381–461.

Sorrie, B. A. 1998b. Distribution of *Drosera filiformis* and *D. tracyi* (Droseraceae): Phytogeographic implications. Rhodora 100: 239–260.

Sorrie, B. A. and A. S. Weakley. 2007. Notes on *Lechea maritima* var. *virginica* (Cistaceae). J. Bot. Res. Inst. Texas 1: 367–368.

Sorrie, B. A. and A. S. Weakley. 2007b. Recognition of *Lechea pulchella* var. *ramosissima* (Cistaceae). J. Bot. Res. Inst. Texas 1: 369–371.

Sorsa, M. 1968. Cytological and evolutionary studies on *palustres* violets. Madroño 19: 165–179.

SouthW. Naturalist = Southwestern Naturalist.

Sp. Pl.—See: C. Linnaeus 1753; C. L. Willdenow et al. 1797–1830

Sp. Pl. ed. 2—See: C. Linnaeus 1762–1763

Spach, É. 1834–1848. Histoire Naturelle des Végétaux. Phanérogames.... 14 vols., atlas. Paris. (Hist. Nat. Vég.)

Spaulding, D. D. 2013. Key to the pinweeds (*Lechea*, Cistaceae) of Alabama and adjacent states. Phytoneuron 2013-99: 1–15.

Spencer, N. R. 1984. Velvetleaf, *Abutilon theophrasti* (Malvaceae), history and economic impact in the United States. Econ. Bot. 38: 407–416.

Spooner, D. M. et al. 1985. Observations on the distribution and ecology of *Sida hermaphrodita* (L.) Rusby (Malvaceae). Sida 11: 215–225.

Sprague, T. A. 1922. A revision of *Amoreuxia*. Bull. Misc. Inform. Kew 1922: 96–105.

Sprengel, K. [1824–]1825–1828. Caroli Linnaei...Systema Vegetabilium. Editio Decima Sexta.... 5 vols. Göttingen. [Vol. 4 in 2 parts paged independently; vol. 5 by A. Sprengel.] (Syst. Veg.)

St.-Hilaire, A., J. Cambessèdes, and A. H. L. de Jussieu. 1824[–1833]. Flora Brasiliae Meridionalis.... 3 vols. in 24 parts. Paris. [Vols. paged independently, parts and plates numbered consecutively.] (Fl. Bras. Merid.)

St. James's Chron. Brit. Eve. Post = The St. James's Chronicle; or the British Evening Post.

Stafleu, F. A. et al. 1992–2009. Taxonomic Literature: A Selective Guide to Botanical Publications and Collections with Dates, Commentaries and Types. Supplement. 8 vols. Königstein.

Stafleu, F. A. and R. S. Cowan. 1976–1988. Taxonomic Literature: A Selective Guide to Botanical Publications and Collections with Dates, Commentaries and Types, ed. 2. 7 vols. Utrecht etc.

Stand. Cycl. Hort.—See: L. H. Bailey 1914–1917

Stebbins, G. L. et al. 1963. Identification of the ancestry of an amphiploid Viola with the aid of paper chromatography. Amer. J. Bot. 50: 830–839.

Stepansky, A., I. Kovalski, and R. Perl-Treves. 1999. Intraspecific classification of melons (Cucumis melo L.) in view of their phenotypic and molecular variation. Pl. Syst. Evol. 217: 313–332.

Steudel, E. G. 1840–1841. Nomenclator Botanicus Enumerans Ordine Alphabetico Nomina atque Synonyma tum Generica tum Specifica..., ed. 2. 2 vols. Stuttgart and Tübingen. (Nomencl. Bot. ed. 2)

Stevens, P. F. 1980. A revision of the Old World species of Calophyllum (Guttiferae). J. Arnold Arbor. 61: 117–699.

Stevens, P. F. 2007. Clusiaceae-Guttiferae. In: K. Kubitzki et al., eds. 1990+. The Families and Genera of Vascular Plants. 10+ vols. Berlin etc. Vol. 9, pp. 48–66.

Stevens, P. F. 2007b. Hypericaceae. In: K. Kubitzki et al., eds. 1990+. The Families and Genera of Vascular Plants. 10+ vols. Berlin etc. Vol. 9, pp. 194–210.

Steyermark, J. A. 1963. Flora of Missouri. Ames.

Stirp. Austr. Fasc.—See: H. J. N. von Crantz 1762–1767

Stirp. Nov.—See: C.-L. L'Héritier de Brutelle 1784[1785–1805]

Stocking, K. M. 1955. Some considerations of the genera Echinocystis and Echinopepon in the United States and northern Mexico. Madroño 13: 84–100.

Stocking, K. M. 1955b. Some taxonomic and ecological considerations of the genus Marah. Madroño 13: 113–137.

Stuckey, R. L. 1968b. Aquatic flowering plants new to the Erie Islands. Ohio J. Sci. 68: 180–187.

Suppl. Pl.—See: C. Linnaeus f. 1781[1782]

Sweet, R. 1825–1830. Cistineae. The Natural Order of Cistus, or Rock-rose.... London. (Cistineae)

Sweet, R. 1826. Hortus Britannicus.... 2 parts. London. [Parts paged consecutively.] (Hort. Brit.)

Syll. Pl. Nov.—See: C. F. Hornschuch [1822–]1824–1828

Symb. Antill.—See: I. Urban 1898–1928

Syn. Anal. Fl. Paris ed. 2—See: E. S.-C. Cosson and J. N. E. Germain de Saint-Pierre 1859

Syn. Fl. N. Amer.—See: A. Gray et al. 1878–1897

Syst. Biodivers. = Systematics and Biodiversity.

Syst. Biol. = Systematic Biology.

Syst. Bot. = Systematic Botany; Quarterly Journal of the American Society of Plant Taxonomists.

Syst. Bot. Monogr. = Systematic Botany Monographs; Monographic Series of the American Society of Plant Taxonomists.

Syst. Nat.—See: J. F. Gmelin 1791[–1792]

Syst. Nat. ed. 10—See: C. Linnaeus 1758[–1759]

Syst. Nat. ed. 12—See: C. Linnaeus 1766[–1768]

Syst. Veg.—See: J. J. Roemer et al. 1817[–1830]; K. Sprengel [1824–]1825–1828

Takhtajan, A. L. 1980. Outline of the classification of flowering plants (Magnoliophyta). Bot. Rev. (Lancaster) 46: 225–359.

Takhtajan, A. L. 1987. Systema Magnoliophytorum. [Sistema Magnoliofitov.] Leningrad.

Takhtajan, A. L. 1997. Diversity and Classification of Flowering Plants. New York.

Tan, K. 1980. Studies in the Thymelaeaceae: 2. A revision of the genus Thymelaea. Notes Roy. Bot. Gard. Edinburgh 38: 189–246.

Tan, K., ed. 1989. The Davis & Hedge Festschrift...Plant Taxonomy, Phytogeography, and Related Subjects. Edinburgh.

Taxon = Taxon; Journal of the International Association for Plant Taxonomy.

Taylor, N. 1915. Flora of the vicinity of New York. Mem. New York Bot. Gard. 5.

Teppner, H. 2000. Cucurbita pepo (Cucurbitaceae)—History, seed coat types, thin coated seeds and their genetics. Phyton (Horn) 40: 1–42.

Teppner, H. 2004. Notes on Lagenaria and Cucurbita (Cucurbitaceae)—Review and new contributions. Phyton (Horn) 44: 245–308.

Thai Forest Bull., Bot. = Thai Forest Bulletin. Botany.

Theor. Appl. Genet. = Theoretical and Applied Genetics; International Journal of Breeding Research and Cell Genetics.

Thomas, R. D. and C. M. Allen. 1993–1998. Atlas of the Vascular Flora of Louisiana. 3 vols. Baton Rouge.

Thorne, R. F. 1992. An updated phylogenetic classification of the flowering plants. Aliso 13: 365–389.

Thunberg, C. P. 1794–1800. Prodromus Plantarum Capensium, Quas in Promontorio Bonae Spei Africes, Annis 1772–1775, Collegit.... 2 parts. Uppsala. [Parts paged consecutively.] (Prodr. Pl. Cap.)

Thwaites, G. H. K. [1858–]1864. Enumeratio Plantarum Zeylaniae: An Enumeration of Ceylon Plants.... 5 parts. London. [Parts paged consecutively.] (Enum. Pl. Zeyl.)

Tirel, C., J. Jérémie, and D. Lobreau-Callen. 1996. Corchorus neocaladonicus (Tiliaceae), véritable identité de l'enigmatique Oceanopapaver. Adansonia 18: 35–43.

Todaro, A. 1877–1878. Relazione sulla Cultura dei Cotoni in Italia.... 1 vol. text + 1 vol. ills. Rome and Palermo. (Relaz. Cult. Coton.)

Tokuoka, T. 2008. Molecular phylogenetic analysis of Violaceae (Malpighiales) based on plastid and nuclear DNA sequences. J. Pl. Res. 121: 253–260.

Tornadore, N., R. Marcucchi, and S. Marchiori. 2000. Karyology, pollen and seed morphology, and distribution of eight endangered species in the Veneto region (northern Italy). Pl. Biosyst. 134: 71–82.

Torrey, J. and A. Gray. 1838–1843. A Flora of North America.... 2 vols. in 7 parts. New York, London, and Paris. (Fl. N. Amer.)

Torreya = Torreya; a Monthly Journal of Botanical Notes and News.

Traité Bot. Méd. Phan.—See: H. E. Baillon 1883–1884

Trans. Linn. Soc. = Transactions of the Linnean Society.

Trans. Linn. Soc. London = Transactions of the Linnean Society of London.

Trans. Wisconsin Acad. Sci. = Transactions of the Wisconsin Academy of Sciences, Arts and Letters.

Trew, C. J. 1750–1773. Plantae Selectae.... 10 vols. Nuremberg. (Pl. Select.)

Truyens, S., M. M. Arbo, and J. S. Shore. 2005. Phylogenetic relationships, chromosome and breeding system evolution in Turnera (Turneraceae): Inferences from ITS sequence data. Amer. J. Bot. 92: 1749–1758.

Tucker, G. C. 1986. The genera of Elatinaceae in the southeastern United States. J. Arnold Arbor. 67: 471–483.

Tuckey, J. H. 1818. Narrative of an Expedition to Explore the River Zaire.... London.

Türk Bot. Derg. = Türk Botanik Dergisi.

Turner, B. L. 2008. Infraspecific categories of Hibiscus moscheutos (Malvaceae) in Texas. Phytologia 90: 378–381.

Turquie Agric.—See: P. M. Zhukovsky 1933

U.S. Expl. Exped.—See: C. Wilkes et al. 1854–1876

U.S.D.A. Bur. Pl. Industr. Bull. = U S Department of Agriculture. Bureau of Plant Industry. Bulletin.

Ulmer, T. and J. M. MacDougal. 2004. Passiflora: Passionflowers of the World. Portland.

Univ. Calif. Publ. Bot. = University of California Publications in Botany.

Univ. Calif. Publ. Geol. Sci. = University of California Publications in Geological Sciences.

University of Chicago Press. 1993. The Chicago Manual of Style, ed. 14. Chicago.

Urban, I. 1883. Monographie der Familie der Turneraceen. Jahrb. Königl. Bot. Gart. Berlin 2: 81–148.

Urban, I., ed. 1898–1928. Symbolae Antillanae seu Fundamenta Florae Indiae Occidentalis.... 9 vols. Berlin etc. (Symb. Antill.)

Utech, F. H. 1970. Preliminary reports on the flora of Wisconsin no. 60. Tiliaceae and Malvaceae—basswood and mallow families. Trans. Wisconsin Acad. Sci. 58: 301–323.

Utech, F. H. and H. H. Iltis. 1970. Preliminary reports on the flora of Wisconsin no. 61. Hypericaceae—St. John's wort family. Trans. Wisconsin Acad. Sci. 58: 325–351.

Uzunhisarckili, M. E. and M. Vural. 2012. The taxonomic revision of Alcea and Althaea (Malvaceae) in Turkey. Türk Bot. Derg. 36: 603–636.

van Royen, P. 1951. The Podostemaceae of the New World. Part I. Meded. Bot. Mus. Herb. Rijks Univ. Utrecht 107: 1–151.

van Royen, P. 1953. The Podostemaceae of the New World. Part II. Acta Bot. Neerl. 2: 1–20.

van Royen, P. 1954. The Podostemaceae of the New World. Part III. Acta Bot. Neerl. 3: 215–263.

Vanderplank, J. 2000. Passion Flowers, ed. 3. Cambridge, Mass.

Vasc. Pl. Pacif. N.W.—See: C. L. Hitchcock et al. 1955–1969

Veg. Hist. Archaeobot. = Vegetation History and Archaeobotany.

Ventenat, É. P. [1800–1803.] Description des Plantes Nouvelles et Peu Connues Cultivés dans le Jardin de J. M. Cels.... 10 parts. Paris. [Plates numbered consecutively.] (Descr. Pl. Nouv.)

Verdoorn, I. C. 1980. Revision of Hermannia subg. Hermannia in southern Africa. Bothalia 13: 1–63.

Verh. Vereins Beförd. Gartenbaues Königl. Preuss. Staaten = Verhandlungen des Vereins zur Beförderung des Gartenbaues in den Königlich preussischen Staaten.

Vogel, S. 1981. Die Klebstoffhaare an den Antheren von Cyclanthera pedata (Cucurbitaceae). Pl. Syst. Evol. 137: 291–316.

Vogel, S. 2000. The floral nectaries of Malvaceae s.l.—A conspectus. Kurtziana 28: 155–171.

Vogelman, H. A. 1953. Comparison of Dirca palustris and Dirca occidentalis (Thymelaeaceae). Asa Gray Bull. 2: 77–82.

Voigt, J. O. 1845. Hortus Suburbanus Calcuttensis. Calcutta. (Hort. Suburb. Calcutt.)

Volz, S. M. and S. S. Renner. 2009. Phylogeography of the ancient Eurasian medicinal plant genus Bryonia (Cucurbitaceae) inferred from nuclear and chloroplast sequences. Taxon 58: 550–560.

Vorles. Churpfälz. Phys.-Ökon. Ges. = Vorlesungen der Churpfälzischen Physicalisch-Ökonomischen Gesellschaft.

Voy. Barbarie—See: J. Poiret 1789

Wagner, W. L., D. R. Herbst, and S. H. Sohmer. 1990. Manual of the Flowering Plants of Hawai'i. 2 vols. [Honolulu.] [Vols. paged consecutively.]

Wahlert, G. A. et al. 2014. Phylogeny of the Violaceae (Malpighiales) inferred from plastid DNA sequences: Implications for generic diversity and intrafamilial classification. Syst. Bot. 39: 239–252.

Walpers, W. G. 1842–1847. Repertorium Botanices Systematicae.... 6 vols. Leipzig. (Repert. Bot. Syst.)

Walpers, W. G. and K. Müller berol. 1848–1871. Annales Botanices Systematicae.... 7 vols. Leipzig. (Ann. Bot. Syst.)

Walter, T. 1788. Flora Caroliniana, Secundum Systema Vegetabilium Perillustris Linnaei Digesta.... London. (Fl. Carol.)

Walters, T. W. and D. S. Decker. 1993. Systematics of the endangered Okeechobee gourd (Cucurbita okeechobeensis: Cucurbitaceae). Syst. Bot. 18: 175–187.

Wang, Jiongqian and M. B. Cruzan. 1998. Interspecific mating in the Piriqueta caroliniana (Turneraceae) complex: Effects of pollen load size and composition. Amer. J. Bot. 85: 1172–1179.

War Department [U.S.]. 1855–1860. Reports of Explorations and Surveys, to Ascertain the Most Practicable and Economical Route for a Railroad from the Mississippi River to the Pacific Ocean. Made under the Direction of the Secretary of War, in 1853[–1856].... 12 vols. in 13. Washington. (Pacif. Railr. Rep.)

Ward, D. B. 1980. Plants. In: P. C. H. Pritchard, ed. 1978–1982. Rare and Endangered Biota of Florida. Gainesville. Vol. 5.

Ward, D. B. 2006. Keys to the flora of Florida—14, *Viola* (Violaceae). Phytologia 88: 242–252.

Warrier, P. K. et al., eds. 1993+. Indian Medicinal Plants: A Compendium of 500 Species. 5+ vols. Madras.

Wasylikowa, K. and M. Van der Veen. 2004. An archaeobotanical contribution to the history of watermelon, *Citrullus lanatus* (Thunb.) Matsum. Veg. Hist. Archaeobot. 13: 213–217.

Waterhouse, B. M. and A. A. Mitchell. 1998. Northern Australia Quarantine Strategy: Weeds Target List, ed. 2. Canberra.

Watson, S. 1871. United States Geological Expolration [sic] of the Fortieth Parallel. Clarence King, Geologist-in-charge. [Vol. 5] Botany. By Sereno Watson.... Washington. [Botanical portion of larger work by C. King.] [Botany (Fortieth Parallel)]

Watsonia = Watsonia; Journal of the Botanical Society of the British Isles.

Webb, D. H. 1980. A Biosystematic Study of *Hypericum* Section *Spachium* in Eastern North America. Ph.D. dissertation. University of Tennessee.

Webb, P. B. and S. Berthelot. [1835–]1836–1850. Histoire Naturelle des Îles Canaries.... 3 vols. in 9. Paris. [Tome troisième, Botanique: Première partie, 1 vol; deuxième partie, 4 vols.] (Hist. Nat. Îles Canaries)

Weckesser, W. 2011. A new hybrid of *Hibiscus* (Malvaceae) from Texas. J. Bot. Res. Inst. Texas 5: 41–44.

Welsh, S. L. et al., eds. 1987. A Utah flora. Great Basin Naturalist Mem. 9.

Welsh, S. L. et al., eds. 1993. A Utah Flora, ed. 2. Provo.

Welsh, S. L. et al., eds. 2003. A Utah Flora, ed. 3. Provo.

West, H. 1793. Bidrag til Beskrivelse over Ste Croix.... Copenhagen.

Whalen, M. A. 1987. Systematics of *Frankenia* (Frankeniaceae) in North and South America. Syst. Bot. Monogr. 17: 1–93.

Whetstone, R. D. 1983. The Sterculiaceae in the flora of the southeastern United States. Sida 10: 15–23.

Whitaker, T. W. and W. P. Bemis. 1975. Origin and evolution of the cultivated *Cucurbita*. Bull. Torrey Bot. Club 102: 362–368.

Whitlock, B. A., C. Bayer, and D. A. Baum. 2001. Phylogenetic relationships and floral evolution of the Byttnerioideae ("Sterculiaceae" or Malvaceae s.l.) based on sequences of the chloroplast gene, *ndh*F. Syst. Bot. 26: 420–437.

Whitlock, B. A. and A. M. Hale. 2011. The phylogeny of *Ayenia*, *Byttneria*, and *Rayleya* (Malvaceae s.l.) and its implications for the evolution of growth forms. Syst. Bot. 36: 129–136.

Whitlock, B. A., K. G. Karol, and W. S. Alverson. 2003. Chloroplast DNA sequences confirm the placement of the enigmatic *Oceanopapaver* within *Corchorus* (Grewioideae: Malvaceae sensu lato, formerly Tiliaceae). Int. J. Pl. Sci. 164: 35–41.

Wiggins, I. L. 1936. A resurrection and revision of the genus *Iliamna* Greene. Contr. Dudley Herb. 1: 213–229.

Wight, R. and G. A. W. Arnott. 1834. Prodromus Florae Peninsulae Indiae Orientalis.... 1 vol. only. London. (Prodr. Fl. Ind. Orient.)

Wilbur, R. L. 1966. Notes on Rafinesque's species of *Lechea*. Rhodora 68: 192–208.

Wilbur, R. L. and H. S. Daoud. 1961. The genus *Lechea* (Cistaceae) in the southeastern United States. Rhodora 63: 103–118.

Wilcox, E. V. and V. S. Holt. 1913. Ornamental *Hibiscus* in Hawaii. Bull. Hawaii Agric. Exp. Sta. 29: 1–60.

Wilder, G. J. and M. R. McCombs. 2002. New records of vascular plants for Ohio and Cuyahoga County, Ohio. Rhodora 104: 350–372.

Wilkes, C. et al. 1854–1876. United States Exploring Expedition. During the years 1838, 1839, 1840, 1841, 1842. Under the Command of Charles Wilkes, U.S.N..... 18 vols. (1–17, 19). Philadelphia. [Vol. 15: Botany, Phanerogamia (A. Gray), 1854; Atlas, 1856. Vol. 16: Botany, Cryptogamia, Filices (W. D. Brackenridge), 1854; Atlas, 1855. Vol. 17: incl. Phanerogamia of Pacific North America (J. Torrey), 1874. Vol. 19 (2 parts): Geographical Distribution of Animals and Plants (C. Pickering), 1854. Vol. 18: Botany, Phanerogamia, part 2 (A. Gray) not published.] (U.S. Expl. Exped.)

Wilkie, P. et al. 2006. Phylogenetic relationships within the subfamily Sterculioideae (Malvaceae/Sterculiaceae-Sterculieae) using the chloroplast gene *ndh*F. Syst. Bot. 31: 160–170.

Willdenow, C. L. 1803–1816. Hortus Berolinensis.... 2 vols. in 10 fascs. Berlin. [Fascs. and plates numbered consecutively.] (Hort. Berol.)

Willdenow, C. L. 1809–1813[–1814]. Enumeratio Plantarum Horti Regii Botanici Berolinensis.... 2 parts + suppl. Berlin. [Parts paged consecutively.] (Enum. Pl.)

Willdenow, C. L., C. F. Schwägrichen, and J. H. F. Link. 1797–1830. Caroli a Linné Species Plantarum.... Editio Quarta.... 6 vols. Berlin. [Vols. 1–5(1), 1797–1810, by Willdenow; vol. 5(2), 1830, by Schwägrichen; vol. 6, 1824–1825, by Link.] (Sp. Pl.)

Williams, C. E. 2004. Mating system and pollination biology of the spring-flowering shrub, *Dirca palustris*. Pl. Spec. Biol. 19: 101–106.

Williams, K. S. and L. E. Gilbert. 1981. Insects as selective agents on plant vegetative morphology: Egg mimicry reduces egg laying by butterflies. Science, ser. 2, 212: 467–469.

Williams, S. E., V. A. Albert, and M. W. Chase. 1994. Relations of the Droseraceae: A cladistic analysis of *rbc*L sequence and morphological data. Amer. J. Bot. 81: 1027–1037.

Williams, S. E. and A. B. Bennett. 1982. Leaf closure in the Venus flytrap: An acid growth response. Science, ser. 2, 218: 1120–1122.

Wilson, F. D. 1994. The genome biogeography of *Hibiscus* L. section *Furcaria* DC. Genet. Resources Crop Evol. 41: 13–25.

Wilson, F. D. and M. Y. Menzel. 1964. Kenaf *(Hibiscus cannabinus)*, roselle *(Hibiscus sabdariffa)*. Econ. Bot. 18: 80–91.

Wilson, H. D. 1990. Gene flow in squash species. BioScience 40: 449–455.

Wilson, H. D., J. Doebley, and M. Duvall. 1992. Chloroplast DNA diversity among wild and cultivated members of *Cucurbita* (Cucurbitaceae). Theor. Appl. Genet. 84: 859–865.

Winters, H. F. 1970. Our hardy *Hibiscus* species as ornamentals. Econ. Bot. 24: 155–164.

Wofford, B. E. et al. 2004. The rediscovery of the South American *Hybanthus parviflorus* (Violaceae) in North America. Sida 21: 1209–1214.

Wood, C. E. Jr. 1955. Evidence for the hybrid origin of *Drosera anglica*. Rhodora 57: 105–130.

Wood, C. E. Jr. 1960. The genera of Sarraceniaceae and Droseraceae in the southeastern United States. J. Arnold Arbor. 41: 152–163.

Wooton, E. O. and P. C. Standley. 1915. Flora of New Mexico. Contr. U.S. Natl. Herb. 19.

Wrightia = Wrightia; a Botanical Journal.

Wu, Z. and P. H. Raven, eds. 1994+. Flora of China. 20+ vols. Beijing and St. Louis.

Wunderlin, R. P. and B. F. Hansen. 2003. Guide to the Vascular Plants of Florida, ed. 2. Gainesville.

Wurdack, K. and C. C. Davis. 2009. Malpighiales phylogenetics: Gaining ground on one of the most recalcitrant clades in the angiosperm tree of life. Amer. J. Bot. 96: 1551–1570.

Wynne, F. E. 1944. *Drosera* in eastern North America. Bull. Torrey Bot. Club 71: 166–174.

Yetman, D. and T. R. Van Devender. 2002. Mayo Ethnobotany: Land, History, and Traditional Knowledge in Northwest Mexico. Berkeley.

Yockteng, R. and S. Nadot. 2004. Phylogenetic relationships among *Passiflora* species based on the glutamine synthetase nuclear gene expressed in chloroplast (ncpGS). Molec. Phylogen. Evol. 31: 379–396.

Z. Naturwiss. = Zeitschrift für Naturwissenschaften.

Zhang, D. Y. et al. 2000. Systematic studies on some questions of Tamaricaceae based on ITS sequence. Acta Bot. Boreal.-Occid. Sin. 20: 421–431.

Zhang, Y. M., Pan B. R., and Yin L. K. 1998. Seed morphology of Tamaricaceae in China arid areas and its systematic evolution. J. Pl. Resources Environm. 7: 22–27.

Zhukovsky, P. M. 1933. La Turquie Agricole. Moscow and Leningrad. (Turquie Agric.)

Zohary, M. and B. R. Baum, B. R. 1965. On the androecium of the *Tamarix* flower and its evolutionary trends. Israel J. Bot. 14: 101–111.

Index

Names in *italics* are synonyms, casually mentioned hybrids, or plants not established in the flora. Page numbers in **boldface** indicate the primary entry for a taxon. Page numbers in *italics* indicate an illustration. Roman type is used for all other entries, including author names, vernacular names, and accepted scientific names for plants treated as established members of the flora.

Flora of North America — Index to families/volumes of vascular plants, current as of June 2015. Bolding denotes published volume: page number.